ANALYSIS OF
SURFACTANTS

SURFACTANT SCIENCE SERIES

ANALYSIS OF SURFACTANTS

Second Edition
Revised and Expanded

Thomas M. Schmitt

BASF Corporation
Wyandotte, Michigan

CRC Press
Taylor & Francis Group
Boca Raton London New York

CRC Press is an imprint of the
Taylor & Francis Group, an **informa** business

CRC Press
Taylor & Francis Group
6000 Broken Sound Parkway NW, Suite 300
Boca Raton, FL 33487-2742

First issued in paperback 2019

© 2001 by Taylor & Francis Group, LLC
CRC Press is an imprint of Taylor & Francis Group, an Informa business

ISBN-13: 978-0-8247-0449-0 (hbk)
ISBN-13: 978-0-367-39762-3 (pbk)

Visit the Taylor & Francis Web site at
http://www.taylorandfrancis.com

and the CRC Press Web site at
http://www.crcpress.com

Preface to the Second Edition

It gives me deep pleasure to offer the second edition of *Analysis of Surfactants*. The nine years since the appearance of the first edition have seen a great expansion in the literature on the analytical chemistry of surface active agents. Incorporating this literature into an updated text has been my only hobby during this time.

The overall organization of the volume is the same. As in the previous edition, the chapters on individual instrumental analytical methods do not describe the theory and practice of the techniques. Rather, they tell generalists what results they can expect from employing the particular method, and allow specialists to immediately discover which approaches their colleagues have found useful. The sections on molecular spectroscopy have been expanded and a chapter on capillary electrophoresis has been added. The chapter on titrations has been reworked to discuss the exciting developments made in the past decade. Some additional surfactants have been added because of their commercial importance—namely, ether carboxylates and ester quats.

Titles of journal articles are included in the references, but they have been truncated, dropping words like *analysis, determination, quantitative, trace, new, improved,* and so on. Whenever possible, hard-to-get references are omitted in favor of those more recent or more easily obtained. Since about half the copies of the first edition were sold outside the United States, I have taken steps to make the book more useful worldwide by including foreign-language references and translating units of measure as necessary. ISO standards are referenced when available.

Many procedures are described in enough detail to permit an experienced analyst to understand the principles behind the procedures and even to use them for exploratory work. It is expected that the original reference will be obtained if the analyst intends to use a method routinely. It was necessary to omit many of the details required to obtain precise results, particularly in the case of ISO and ASTM standards.

In order to keep the size of the volume within practical limits, I have allowed myself to be more critical in the second edition than the first. Published methodology is not taken at face value, but is put into the context of previous work and evaluated in light of my own thirty years of experience in industrial chemistry. Techniques that are only of academic interest are covered in much less detail than those that have immediate application. I have benefited from reader comments on the first edition, as well as from addressing the hundreds of technical questions asked by my colleagues at BASF and our customers over the years. The purpose remains to give the chemist a ready reference for deciding how best to approach the analytical chemistry challenges that present themselves in the world of surfactants.

Thomas M. Schmitt

Preface to the First Edition

This volume is intended to serve as a handbook to the chemist involved in surfactant analysis. Literature is reviewed through 1990, with applications evaluated in terms of their utility in providing accurate analytical results at a reasonable cost. In order to illustrate the various techniques applied to surfactants, some procedures are given in the text. These examples are most often taken from literature that is not readily available because it is somewhat dated or perhaps not written in English. While accurately transcribed, the examples are necessarily abbreviated, and the reader is urged to consult the original source for more information before beginning laboratory work.

I had two objectives: to give the chemist new to a particular area the perspective to begin an analysis, and to provide the more experienced analyst an up-to-date reference book, made accessible by a thorough index. In recognition of usage in modern analytical chemistry laboratories, instrumental techniques are described in preference to "wet chemistry" for both qualitative and quantitative analysis. Chromatographic procedures are emphasized. For the literature prior to 1972, I recommend study of the comprehensive volume of Rosen and Goldsmith.

In addition to covering the determination of surfactants, I have included an extensive section on the characterization of commercial surfactants, including the measurement of parameters affecting purity and suitability for use. This section should help chemists addressing the very important question of quality of the products their employers buy and sell. Knowledge of the critical parameters of a commercial product will also help the analyst perform a more complete analysis of a formulation containing a surfactant, obtaining enough information to permit prediction and perhaps duplication of its properties.

Thomas M. Schmitt

Contents

Contents   ix

1
Characterization of Anionic Surfactants

Anionic surfactants account for about 50% of surfactant use in Europe and about 60% in the United States. Most are high-foaming but sensitive to hard water and therefore require the addition of substances to complex calcium and magnesium ions (i.e., detergent builders). They are more effective than other surfactants in particulate soil removal, especially from natural fabrics. As a rule, they are easily spray-dried and thus are favored for detergent powders.

I. DESCRIPTION AND TYPICAL SPECIFICATIONS

A. Alkylbenzenesulfonates

1. Description

Sodium tetrapropylenebenzensulfonate (ABS)

Sodium 5-dodecylbenzenesulfonate (LAS)

The term alkylbenzenesulfonate (ABS) is generally applied to the branched-chain products, which are only biodegradable with difficulty. ABS is not used in developed countries except for specialty applications where it will not reach natural waters (e.g., as an emulsifier for agricultural chemicals). As opposed to ABS, linear alkylbenzenesulfonate (LAS) is rapidly biodegradable under aerobic conditions. The feedstocks are synthesized by Friedel–Crafts alkylation of benzene. Depending on the alkylation catalyst, phenyl substitution may be uniform on the secondary carbons, or higher in the 2 position. (The 1-phenyl isomer is not found.) Subsequent sulfonation with oleum or sulfur trioxide produces mainly the *para* isomers. LAS has the lowest cost of any surfactant and is used throughout the world. It is sensitive to water hardness.

2. Typical Specifications

Parameter	Test method
Assay	Titration with cationic surfactant
Alkyl chain length (average molecular wt)	C/S ratio by elemental analysis; desulfonation/gas chromatography; titration with cationic surfactant
Disulfonated alkylbenzene	Extraction
Alkyl chain distribution	Gas chromatography; HPLC
Alcohol-insoluble matter	Gravimetry
Neutral oil	Petroleum ether extraction; gas chromatography; HPLC
Moisture	Azeotropic distillation; Karl Fischer titration
Sodium sulfate	Potentiometric titration
Sodium chloride	Potentiometric titration
pH	0.1% aqueous

B. Alkyl Sulfates

1. Description

Sodium *n*-dodecyl sulfate

Alkyl sulfates, also called alcohol sulfates, are formed by making the sulfuric acid esters of linear alcohols. Alkyl chain lengths range from C_{10} to C_{18}. The properties of the alkyl sulfates vary with the alkyl chain length distribution. The alcohol source can be either natural (linear) or synthetic oxo (some branching) alcohols. About 15% of the detergent range alcohols in the United States come from saponification of natural oil and fat. Products are called, for example, tallow alcohol sulfate or coconut alcohol sulfate. They tend to be sensitive to water hardness but are very widely used in cosmetics and detergents. Lauryl alcohol sulfate is a major hair shampoo ingredient.

2. Typical Specifications

Parameter	Test Method
Active agent	Titration with cationic surfactant
Identity of alkyl components	GC or HPLC
Combined alcohols	Hydrolysis and extraction
Ester SO_3	Hydrolysis followed by determination of sulfate by titration or gravimetry
Unsulfated material	Extraction
Alcohol-soluble matter	Gravimetry
Sodium sulfate	Gravimetric analysis of alcohol-insolubles
Other tests	Chloride; water; pH, 0.1% aqueous; alkalinity; viscosity

C. Ether Sulfates

1. Description

$$CH_3(CH_2)_{10}CH_2O(CH_2CH_2O)_4SO_3^- \quad Na^+$$

Sodium *n*-dodecyltetraethoxysulfate (sodium laureth sulfate)

Sodium nonylphenoltetraethoxy sulfate

Alkyl ether sulfates, also called alcohol ethoxysulfates, are prepared by addition of one to four oxyethylene groups to an alcohol which is then sulfated. Oxyethylation enhances water solubility and foaming over the analogous alcohol sulfate, giving a product useful in shampoos and in liquid and powdered detergents. The raw material for these products can be either natural fatty alcohols or primary or secondary synthetic alcohols, usually of C_{12}–C_{18} chain length. The analogous alkylphenol ether sulfates are found in industrial applications. Ether sulfates are not as sensitive to water hardness as are other anionic surfactants.

2. Typical Specifications

Parameter	Test method
Active agent	Titration with cationic surfactant; gravimetry
Nature of ethoxylated alcohol or alkylphenol	Acid hydrolysis followed by HPLC or GC
Alkyl/ethoxy ratio	NMR
Average molecular weight	Titration
Molecular weight distribution	GC or HPLC
Unethoxylated alkyl sulfate	Acid hydrolysis followed by esterification and column chromatography
Unsulfated material	Ion exchange/gravimetry; HPLC
Polyethylene glycol	Extraction/gravimetry

Parameter	Test method
Polyethylene glycol sulfate and disulfate	Extraction or ion exchange
Cloud point	5% aqueous
Unsaturation	Iodine number
Coupling agent (solubilizer)	Gas chromatography
Other tests	Inorganic salts; water; pH, 10% aqueous; Gardner color; 1,4-dioxane

D. α-Olefin Sulfonates

1. Description

Sodium hydroxytetradecanesulfonate

Sodium *n*-tetradecenesulfonate

These are produced by reacting linear α-olefins (usually in the C_{11}–C_{20} range) with SO_3 and air to give mainly 2,3-alkenesulfonic acid and 1,3-sultone. Subsequent alkaline hydrolysis converts the sultones to alkene sulfonates and hydroxyalkane sulfonate. A typical product ratio is 7:3:1 alkene sulfonate/hydroxyalkane sulfonate/disulfonate. Low levels of vinylidene sulfonates are also present. Compounds with an alkyl chain length in the range below C_{14} are used in liquid detergents, while those above C_{14} are used in spray-dried powders. The hydroxy group gives the product mixture enhanced water solubility, but the double bond makes the mixture unstable to bleach. α-Olefin sulfonates are less sensitive to water hardness than most anionics and are readily biodegradable. α-Olefin sulfonates are widely used in the United States and the Pacific rim countries but are less common in Europe.

2. Typical Specifications

Parameter	Test method
Assay	Titration with cationic surfactant
Sultones	Infrared spectroscopy; HPLC
Hydroxyalkanesulfonate content	Hydroxyl number (acetylation method); HPLC
Alkenesulfonate content	Hydrogenation; HPLC
Alkyl chain distribution	GC
Other tests	Unsulfonated oil; alkalinity; pH, 1% aqueous; water; sodium sulfate; sodium chloride; color, platinum-cobalt scale

E. Alkanesulfonates

1. Description

Sodium 2-hexadecanesulfonate

These are commonly called paraffin sulfonates or secondary alkane sulfonates and are used extensively, especially in Europe. Their commercial synthesis requires a relatively high capital investment, so they do not find worldwide application in household detergents. They have high biodegradability, low toxicity, and, depending on chain length and degree of sulfonation, reasonably good water solubility. Paraffin sulfonates are made by sulfoxidation or sulfochlorination of n-paraffins in the C_{14}–C_{17} range. About 90% of the product of sulfoxidation consists of monosulfonates, with the balance being di- and polysulfonates. Eighty to ninety percent of sulfonation occurs at the internal carbon atoms, rather than at the ends of the chain, with the sulfonate group substituted almost uniformly on secondary carbons throughout the chain.

2. Typical Specifications

Parameter	Test method
Total active agent	Two-phase titration; gravimetric analysis
Unsulfonated material	Pentane extraction from neutral or weakly basic 50:50 ethanol/water
Monosulfonated material	Titration with added sodium sulfate; petroleum ether extraction from acidic ethanol/water
Di- and more highly sulfonated material	Calculated from total and monosulfonate; column chromatography separation
Molecular weight	Titration/gravimetric analysis
Alkyl chain distribution	HPLC
Other tests	As for other sulfonates

F. Petroleum Sulfonates

1. Description

A petroleum sulfonate component

Petroleum sulfonates are prepared by sulfonation of hydrocarbons; most commercial products are actually byproducts of petroleum refining. In the case of "natural" petroleum sulfonates, they are produced by sulfonation of crude oil fractions. The purpose of the sulfonation is removal of a portion of the aromatics from the oil and the resulting byproduct/coproduct stream is a complex mixture of sulfonated aliphatic and (mainly) aromatic hydrocarbons. To complicate matters, the term petroleum sulfonates is also applied to materials produced by sulfonation of synthetic materials, such as alkylbenzenes (detergent alkylate). In this case, a higher boiling alkylbenzene feedstock is used than for the production of surfactants for household detergents. The products therefore have only limited water solubility.

The natural petroleum sulfonates are the more complex because the original stream has fused-ring aromatics which permit multiple sulfonation to occur. They are mixtures of alkylarylsulfonates, where the length of the alkyl chains and their points of attachment to the ring vary. Individual molecules may be mono-, di-, or polysulfonated.

Petroleum sulfonates are oil-soluble compounds that find use as emulsifiers in lubricating oil and other functional fluids and in secondary oil recovery. They are usually supplied as sodium salts in mixtures with oil. They may contain unsulfonated oil or be diluted in a light petroleum distillate. "Overbased" products contain high alkalinity additives, such as colloidal calcium carbonate.

Alkylnaphthalene sulfonates, though a distinct class, have many of the properties of petroleum sulfonates and are analyzed by the same general procedures. Their application is mainly in textiles.

2. Typical Specifications

Parameter	Test method
Active agent	Column chromatography; two-phase titration (if eq. wt. is known)
Equivalent weight	Gravimetry
Molecular weight	Vapor pressure osmometry of derivatives or SEC
% Mono-, di-, and polysulfonated material	HPLC
Unsulfonated oil	Column chromatography
Other tests	Specific gravity; alkalinity; water content; inorganic salts

G. Lignin Sulfonate

1. Description

A lignin sulfonate fragment

Lignin is a complex polymer found in wood which is separated from cellulose during manufacture of paper and other products by various means, including sulfonation. Lignin sulfonates are obtained as byproducts of the sulfite process. These are widely disperse, water-soluble lignosulfonic acids, with an average molecular weight greater than 100,000. They find a variety of uses as low cost emulsifiers in areas such as ore processing and oil-field chemicals. Impurities include sugars, such as glucose, mannose, xylose, and arabinose. Both dried powder and aqueous solutions are sold, usually without stringent specifications. Some purified products find their way into higher value applications, such as pesticide formulations.

2. Typical Specifications

Parameter	Test method
Molecular weight	Gel permeation chromatography
Other tests	Ash; total sulfur; % solids

H. Ester Sulfonates

1. Description

$$\underset{\displaystyle C_{16}H_{33}\overset{\textstyle |}{C}HCOOCH_3}{SO_3^- \ Na^+}$$

Sodium α-sulfooctadecanoic acid, methyl ester

$$\underset{\displaystyle C_{16}H_{33}\overset{\textstyle |}{C}HCOO^- \ Na^+}{SO_3^- \ Na^+}$$

Disodium α-sulfooctadecanoate

Also called α-sulfo fatty acid esters, these products currently have significant market share only in Japan. They possess good biodegradability and excellent hard-water detergency. Products of interest for laundry detergents are derived from C_{14}–C_{18} fatty acids [1,2]. Two sulfonated species are normally present: monosodium-α-sulfomethyl fatty acid salt and the unesterified disodium α-sulfo fatty acid salt. The disodium salt is a less effective surfactant than the ester.

2. Typical Specifications

Parameter	Test method
Total anionic active agent	Two-phase titration
Sodium α-sulfo fatty acid, methyl ester	TLC; extraction and titration; ion chromatography
Disodium α-sulfo fatty acid salt	TLC; extraction and titration; ion chromatography
Alkyl chain length distribution	Pyrolysis GC; ion chromatography
Sodium methyl sulfate	Ion chromatography; saponification and titration
Unsulfonated fatty acid salt	Extraction and titration
Unsulfonated methyl ester	Extraction; calculation by difference
Other tests	Sodium sulfate; sodium chloride; methanol; water

I. Phosphate Esters

1. Description

$$CH_3(CH_2)_{10}CH_2O(CH_2CH_2O)_4\overset{\displaystyle O}{\underset{\displaystyle OH}{\overset{\|}{P}}}O^-\ \ Na^+$$

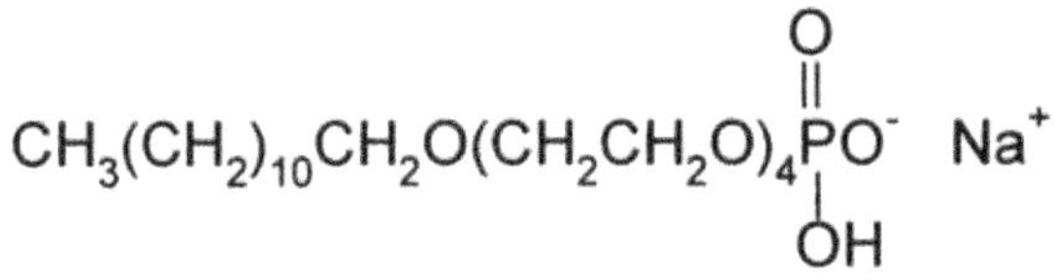

Sodium dodecyltetraethoxy phosphate

These are usually prepared by reaction of fatty alcohols or alkoxylated fatty alcohols with polyphosphoric acid or phosphoric anhydride. They resist hydrolysis in alkaline solutions, have good biodegradability, and are more compatible with hypochlorite bleach than the corresponding ethoxylated nonionics. Many members of this class have excellent skin compatibility, and are used in cosmetics as emulsifiers. Commercial processes usually give a mixture of mono- and diesters, to which the name sesquiphosphate is sometimes given. The monoesters tend to have better detergency and are better foamers, while the diesters are better emulsifiers. Specialty products based on alkylphenolethoxylates are also available for industrial applications (3).

2. Typical Specifications

Parameter	Test method
% Mono- and diester	Potentiometric titration; HPLC; ^{31}P NMR
Free phosphate content	Ion chromatography; titration
Free alcohol content	Petroleum ether extraction
Alkyl chain distribution	Hydrolysis and GC
Other tests	pH, 1% aqueous; pour point; specific gravity; foam height; viscosity; cloud point, 1% aqueous; surface tension, 0.1% aqueous; water

J. Sulfosuccinate Esters

1. Description

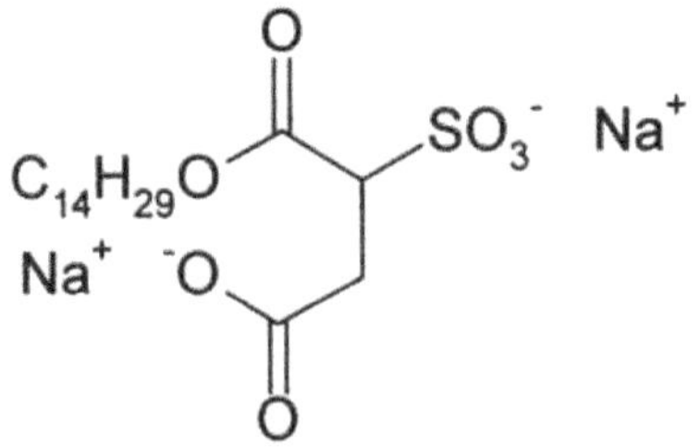

Disodium tetradecyl sulfosuccinate

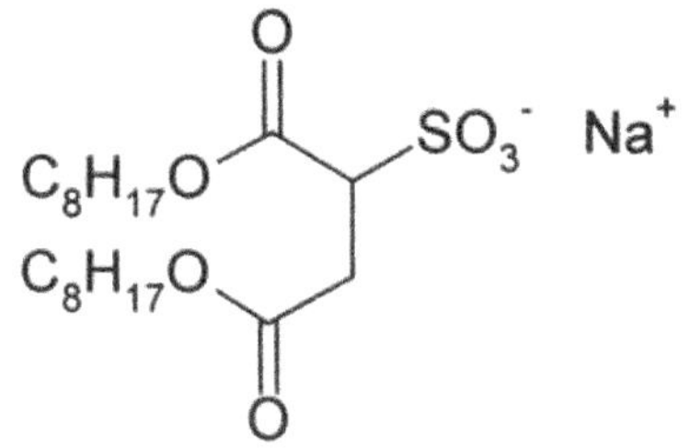

$C_8H_{17}O$ ... SO_3^- Na^+

$C_8H_{17}O$

Sodium *bis*(2-ethylhexyl)sulfosuccinate

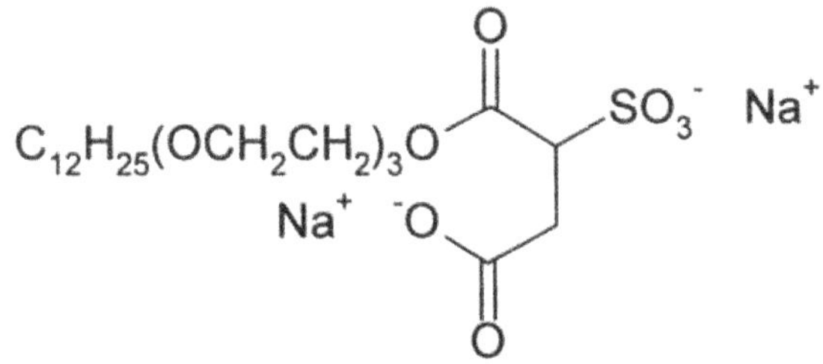

$C_{12}H_{25}(OCH_2CH_2)_3O$ ... SO_3^- Na^+

Na^+ ^-O

Disodium salt of sulfosuccinate monoester of dodecanol 3-mole ethoxylate

The most common examples of this class are diesters of 2-ethylhexanol. However, there are commercial products which are monoesters, diesters, or mixtures of the two. The alcohol portion of the ester may be fatty alcohol, fatty acid alkanolamide, or ethoxylated fatty amine. They are typically synthesized by forming the ester or amide with maleic anhydride, followed by reaction of the alkyl fumarate with sodium bisulfite. The monoesters are often used in cosmetics and shampoos.

2. Typical Specifications

Parameter	Test method
Active agent	Gravimetric (ethanol solubles) or titration method, depending on the individual product
Average composition	NMR
Confirmation of identity	TLC
Diester content	Two-phase titration; gravimetric determination with *p*-toluidine ion
Monoester	Potentiometric titration at controlled pH
Nature of hydrophobe	Hydrolysis with acid or base and extraction, followed by GC analysis of resulting products
Nonionic material	Ion exchange analysis
Maleic/fumaric acid	HPLC
Other tests	Sodium sulfate; sodium sulfite; water content; ester number

K. Isethionate Esters

1. Description

O
‖
~~~~~~~~~OCH$_2$CH$_2$SO$_3^-$ Na$^+$

Sodium dodecyl isethionate

This class of specialty surfactants consists of the fatty acid esters of isethionic acid (2-hydroxyethane sulfonic acid). They have good foaming and dispersing properties and are suitable for synthetic bar "soap" and cosmetic applications. The ammonium salts are very water soluble (~30%), whereas the sodium salts are not (~0.1%). Commercial products are generally made with coco fatty acids and contain residual free fatty acid in the 10% range.

2. Typical Specifications

Parameter	Test method
Assay	Potentiometric titration
Alkyl chain length	Acid hydrolysis, followed by extraction and GC analysis of fatty acid
Sodium isethionate	HPLC
Free fatty acid	Extraction
Other tests	As for other sulfonate surfactants

L. Acyl Taurates

1. Description

O
‖
~~~~~~~~~~~~NCH$_2$
|
CH$_3$

Sodium N-methyl-N-oleoyltaurate

These are named as derivatives of taurine (2-aminoethanesulfonic acid) and are also called taurides. The typical synthetic route involves reaction of sodium bisulfite and ethylene oxide to form sodium isethionate, reaction of the isethionate with methylamine to form sodium N-methyltaurate, then subsequent reaction with the acyl chloride of a fatty acid to give the N-acyl-N-methyltaurate. Taurates have good tolerance to water hardness and find specialty applications in textiles and in detergent bars and shampoos. Taurates are sometimes classified as amphoteric surfactants.

2. Typical Specifications

Parameter	Test method
Alkyl chain length	Acid hydrolysis, followed by extraction and GC analysis of fatty acid
Free fatty acid	Extraction
Other tests	Ethylene glycol; chloride, sulfate, and sulfite; methylamine

M. N-Acylated Amino Acids

1. Description

$$\text{CH}_3(\text{CH}_2)_{10}\text{C(=O)NCH}_2\text{COO}^- \text{ Na}^+$$

Sodium lauryl sarcosinate

These are condensation products of saturated or unsaturated fatty acids with amino acids or N-alkylamino acids. The most frequently encountered examples are sarcosine compounds (N-methylglycine derivatives). These specialty surfactants have been used in toothpaste, shampoo, and hand cleaners as lather boosters. Unlike most anionics, they are compatible with cationic surfactants. Products are available as the free acids or as the sodium or ammonium salts. These also are sometimes classified as amphoteric surfactants.

2. Typical Specifications

Parameter	Test method
Assay	Acid-base titration
Free fatty acids	Ion exchange/titration
Free amino acids	Ion exchange/titration
Acid number	Titration
Amide functionality	Titration
Alkyl chain distribution	HPLC
NaCl	Ion chromatography
Other tests	Color; pH, 10% aqueous

N. Ether Carboxylates

1. Description

$$-\text{O(CH}_2\text{CH}_2\text{O)}_4\text{CH}_2\text{COO}^- \text{ Na}^+$$

Sodium nonylphenolpentaethoxy carboxylate

This class consists of carboxylates of alcohol ethoxylates and alkylphenolethoxylates. Another name given to the compounds is carboxymethylated ethoxylates. Compared to other anionics, they are insensitive to water hardness. They are good foamers and mild, suitable for use in cosmetics. Biodegradability is similar to that of the corresponding ethoxylates. They are produced from the ethoxylates, either by direct oxidation or by reaction with chloroacetic acid.

2. Typical Specifications

Parameter	Test method
Assay	Two-phase titration
Residual AE or APE	HPLC or TLC; OH number
Acid number	Titration
Chloride, chloroacetate, acetate ions	Ion chromatography
PEG, dioxane, other tests	As for the parent ethoxylate compounds

O. Soap

1. Description

Sodium stearate

Historically, the first synthetic surfactants were the salts of fatty acids. These are no longer generally used in domestic laundry and dishwashing applications because they form in- soluble curds with calcium and magnesium ions. However, they are found in many other applications. Potassium soaps have better solubility in nonaqueous media than do sodium soaps.

2. Typical Specifications

Parameter	Test method
Alkyl chain length distribution	HPLC or GC
Melting point	Classical technique
Glycerin	Gas chromatography; liquid chromatography; periodic acid reaction and titration
Unsaponified matter	Extraction
Iodine number	Titration
Free acid or alkali	Titration of ethanol solubles
Na and K content	Atomic absorption or emission spectroscopy
Inorganic salts	Ethanol insolubles
Other tests	Water; chloride; sulfate

II. GENERAL TEST METHODS

Methods in this section are applicable to most anionic surfactants. Methods applicable only to specific compounds are presented in Section III.

A. Assay of Anionic Surfactants

Active agent content of anionics is usually determined by titration with a cationic surfac- tant such as benzethonium chloride. Two-phase titration is suitable for assay of alkylaryl sulfonates; alkyl sulfates and hydroxysulfates, alkanesulfonates, fatty alcohol ethoxysul- fates and alkylphenolethoxysulfates; and dialkylsulfosuccinates (4,5). The results on materials containing more than one sulfate or sulfonate group per molecule must be determined by experiment. One-phase titration to a potentiometric end point using a surfactant-selective electrode is also generally applicable to anionic surfactants and in fact is preferred. Both procedures are the subjects of national and international standards and are described in Chapter 16.

B. Equivalent Weight

Equivalent weight of sulfated and sulfonated anionics can be determined directly by titration with base or by titration with a cationic surfactant. This requires that the active agent be isolated in relatively pure condition. A more general procedure is the *para*-toluidine hydrochloride method, which includes purification of the anionic by extraction to eliminate interferences commonly found in detergents. These procedures are described in Chapter 16.

Liquid chromatographic methods are generally applicable to determination of the molecular weight of anionics. Gas chromatography may also be used, although conversion to the original hydrocarbon or alcohol is generally required prior to analysis. These procedures are described in Chapters 7 and 8. The chromatographic procedures are not as suitable for routine quality control as are titration methods because the former give values for the concentration of each homolog. Careful calibration and calculation are required to determine the composite molecular weight. The titration methods give the number average equivalent weight directly.

C. Analysis of Hydrophobe

While spectroscopic techniques will give a qualitative identification of the surfactant, detailed characterization of the hydrophobic portion of the molecule is almost always performed by gas chromatography. Because most anionic surfactants lack volatility, the compounds are desulfonated or derivatized before gas chromatographic analysis. Liquid chromatography is also suitable for the analysis. Although HPLC provides less information about isomers than does GC, it is usually possible to analyze the sample directly, without desulfonation or derivatization. These techniques are discussed in Chapters 7 and 8.

D. Unsulfonated or Unsulfated Material

This parameter is also called "neutral oil," or, in the case of higher molecular weight materials, "wax." The determination is most often performed by extraction from water with a nonpolar solvent, such as petroleum ether. The nonpolar hydrocarbon will be extracted, while the surfactant and salts remain in the aqueous phase. Sulfones, if present, are included with the neutral oil. Some surfactants are not amenable to analysis by extraction because of emulsion formation. These are analyzed by ion exchange or adsorption chromatography procedures. The chromatography methods are more amenable to automation.

1. Liquid-Liquid Extraction

There are a number of standard methods for this analysis, for example, ASTM D1568 for alkylaryl sulfonates, ASTM D1570 for analysis of alkyl sulfates, ISO 1104 for alkanesulfonates and certain non-LAS alkylarylsulfonates, and ASTM D3673 for α-olefin sulfonates (6–9). As with any extraction, it is advisable to check the isolated oil by IR spectroscopy to determine if any anionic material has also been extracted under the conditions used. Recovery of oil is good for sulfonates, but several extractions are necessary for alkyl sulfates, while recovery from alkyl ether sulfates is quite low (10).

Procedure (7)

Dissolve a 10–20 g sample of the alcohol-soluble fraction of the product in about 50 mL hot denatured 95% ethanol in a 250-mL extraction cylinder. Bring the volume up to 80 mL with ethanol and add 50% KOH solution until the phenolphthalein end point is reached. Add water to the 150-mL mark and cool to avoid boiling the solvent in the next step. Extract five times with 50-mL portions of petroleum ether, adding a few grams NaH_2PO_4, if necessary, to clear emulsions. Combine the extracts in a separatory funnel and wash with 30 mL water. Carefully evaporate the combined extracts to dryness on a steam bath, avoiding excessive heating. Cool and weigh.

2.   HPLC

Low-resolution HPLC procedures may be used to determine unsulfonated oil:

(a)   *Ion exchange chromatography.*   HPLC on a mixed-bed ion exchange column with refractive index detection is suitable for determination of unsulfonated, nonionic material in sulfates, sulfonates, and esters. It is not suitable for olefin sulfonates. The column must be regenerated regularly (11).

(b)   *Conventional reversed-phase HPLC.*   This is opposite to ion exchange chromatography in that the nonionic material is last to elute. Gradient elution is usually required, so that this approach is most useful when the UV or evaporative light scattering detector can be used (12,13).

(c)   *Back-flush reversed-phase HPLC.*   HPLC on a reversed-phase column is suitable for determination of unsulfonated material in many anionics. The nonionic material is detected in the back-flush mode, since it is strongly retained at the head of the column. Refractive index or evaporative light scattering detection may be used, although very low molecular weight oil is not detected with the latter (10,13–15).

3. Column Chromatography

(a)   *Adsorption.*   Liquid-solid extraction has an advantage over liquid-liquid extraction in that sulfonic acids need not be neutralized before analysis.

Procedure (16)

One gram of sample is weighed into a beaker, dissolved in 2 mL 50:50 ethanol/water, and mixed with 4 g Johns Manville Celite 545 diatomaceous earth. The mixture is packed firmly, but not too tightly, into a 19 × 400 mm glass column and eluted with 20 mL 93:7 petroleum ether/CH_2Cl_2. The eluate is evaporated under reduced pressure at 60°C and the residue weighed as neutral oil. For LAS, use pure petroleum ether as eluent. Alkylethoxysulfates may require a higher volume of eluent.

This procedure may give slightly higher results for LAS because of extraction of traces of sulfonate. The following procedure is recommended for LAS:

Procedure (17)

One gram of sample is mixed with 6 g Tolso (Madrid) #38017 sepiolite and packed into a glass column to give a bed height of about 5 cm. Oil is eluted with 60–100 mL 85:15 petroleum ether/CH_2Cl_2 at 2–3 mL/min. The eluate is evaporated at 80°C and the residue weighed as neutral oil.

(b) Ion exchange. A methanol/water solution of the anionic can be passed through a mixed-bed ion exchange column and eluted with methanol. The eluate contains the non-ionic material, which may be determined gravimetrically. This procedure is quantitative for oil in alkyl sulfates and alkylether sulfates, but gives low recovery with LAS and almost no recovery when applied to paraffin sulfonates and α-olefin sulfonates (10).

E. Inorganic Salts

Impurities in anionic surfactants include sulfate salts from the sulfonation reaction and perhaps residual catalyst from previous steps, such as the KOH catalyst that may have been used to prepare the ethoxylated alcohol which was subsequently sulfated. Individual ions can be determined by dilution with water and application of the usual wet chemical tests or ion chromatography. Direct ion chromatographic analysis of a surfactant may require the addition of a low molecular weight alcohol to the aqueous buffer.

1. Total Inorganic Salts

An accurate value for total inorganic salts can be obtained by weighing that portion of the sample insoluble in alcohol. The following procedure specifies use of isopropanol. ASTM methods D1568 and D1570 specify ethanol (6,7).

Procedure: Gravimetric Method for Determination of Total Inorganic Salts in Anionic Surfactants (Alcohol Insoluble Matter)

Dissolve an accurately weighed quantity of sample of about 10 g in 200 mL anhydrous 2-PrOH in a 400-mL beaker. Boil for 5 min with stirring. Cool and filter through a tared Gooch crucible containing a fiberglass filter pad. Add another 200 mL 2-PrOH to the beaker containg the residual salt and again boil for 5 min and filter. Transfer any remaining salt to the filter crucible, rinsing the beaker with hot 2-PrOH. Wash the precipitate with about 100 mL hot 2-PrOH, dry at 105°C for 1 hr, and weigh.

2. Sodium Sulfate

Sodium sulfate is the main inorganic salt expected in sulfonated surfactants. It may also be deliberately added to formulations as a filler or processing aid. Classical methods for sulfate determination are gravimetric or based on $BaCl_2$ titration. It has been found that non-aqueous acid-base titration permits sulfate to be differentiated from sulfonates (18). Sulfate ion has been titrated with lead chloride solution using dithizone as indicator and acetone/water as solvent (19). Alternatively, it may be titrated with barium perchlorate with Sulfonazo III as indicator (9).

(a) Aqueous potentiometric titration of sulfate. This determination is based on the titration of sulfate with lead nitrate, giving a precipitate. When an excess of lead nitrate is added, it is detected at a platinum electrode by its effect on ferriferrocyanide indicator (20,21). The method is effective with linear sodium alkanesulfonates, alcohol ethoxysulfates, LAS, and α-olefin sulfonates, although in the latter case interference from the surfactant makes it preferable to spike a known excess of sulfate into the sample to obtain a better end point. Depending on the surfactant, modifications in the solvent system may be required. For example, acetic acid/water/acetone is often used (22).

Procedure (20)

Weigh out sufficient sample to give a 10-mL titration and dissolve in 50 mL water and 50 mL EtOH. Add 0.1 mL 0.005 M $K_4Fe(CN)_6$, 1.0 mL 0.1 M $K_3Fe(CN)_6$, and 3.0 mL 0.1 M

HCl. Titrate potentiometrically with 0.05 M Pb(NO$_3$)$_2$ solution, using platinum and saturated calomel electrodes to monitor the titration. Standardize with sodium sulfate.

(b)   Nonaqueous potentiometric titration of sulfate.   By use of a nonleveling solvent, the neutralization of the relatively weakly acidic second proton of sulfuric acid can be differentiated from the neutralization of the strongly acidic first proton and the sulfonic acid. This has been demonstrated for analysis of α-olefin sulfonate (21), alkylbenzene sulfonate (23–25), sulfated alcohols, and sulfated ethoxylated alcohols (18). Tetrabutylammonium hydroxide titrant was used in the earlier work, with acetone or acetone/water solvent. Later, cylcohexylamine in methanol or ethanol (23,24) and triethylamine in 50:50 acetonitrile/ethylene glycol (26) were recommended as more stable titrants. (The sulfonic and sulfuric acids do not require a strongly basic titrant.) This method requires that the sulfonates and sulfates be present as the free acids. Salts may be put into the acid form by passage through a strongly acidic cation exchange column in the acid form (21).

Procedure: Titration of Sulfuric Acid in Sulfonic Acids with Cyclohexylamine (23,24)

Prepare cyclohexylamine titrant in MeOH or EtOH and standardize by potentiometric titration against sulfamic acid. Weigh about 200 mg sulfonic acid sample into a 100-mL beaker and dissolve in 50 mL methanol or ethanol. Titrate potentiometrically with 0.1 M cyclohexylamine, monitoring the titration with a glass combination pH electrode. The first potentiometric break (see Fig. 1) corresponds to neutralization of the sulfonic acid and the first proton of sulfuric acid. The second break corresponds to neutralization of the second proton of sulfuric acid.

(c)   Conductometric titration

Procedure: Conductometric Determination of Sulfate in Sulfonates (27)

An automatic titrator is used, connected to a conductivity meter with millivolt output. A 5-mL buret glassware unit is suitable. Weigh a sample of about 4 g into a 100-mL volumetric flask and dissolve and dilute to volume with 1:1 THF/water. Depending on the expected SO$_4^{2-}$ content, transfer an aliquot of 1 to 10 mL into the conductometric titration cell. Add 5 mL acetate buffer solution (pH 4.62) and 5 mL THF as well as 1 or 2 drops of an aqueous suspension of barium sulfate (freshly shaken). If the liquid level of the cell is too low, add 1:1 THF/water as necessary to properly immerse the electrodes. Thermostat the cell at 25°C (or other convenient temperature) and titrate with 0.1 M barium acetate solution (dissolved in 1:1 THF/water, in which 20% of the water is the pH 4.62 acetate buffer solution) with stirring.

Notes: The total titrant volume should be limited to 3 mL. For sodium sulfate concentrations less than 1%, the titration should be repeated with the addition of 1.00 mL 0.1 M Na$_2$SO$_4$ solution. The limit of detection is about 0.2% Na$_2$SO$_4$. NaOH and Na$_2$CO$_3$ will, of course, interfere with the titration. The interference can be eliminated by dissolving the original sample in water, rather than the THF/water mixture, acidifying with HCl to pH 1 and heating to boiling. The solution is then adjusted to pH 7 with dilute NaOH solution and diluted with sufficient THF to obtain a 50% THF solution as specified in the normal procedure.

(d)   Gravimetric determination.   The standard barium precipitation of sulfate can be performed if the bulk of the sulfonate surfactant is first removed by extraction. If the surfactant is not removed, its lowered water solubility in the presence of the added salt will

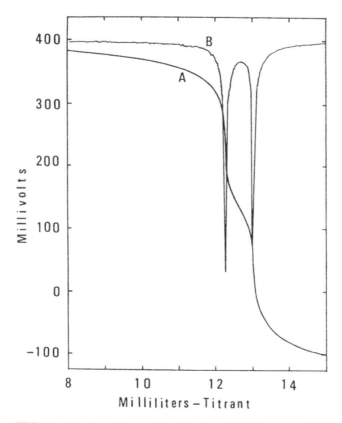

FIG. 1 Titration with cyclohexylamine of alkylbenzenesulfonic acid containing sulfuric acid. (Reprinted with permission from Ref. 24. Copyright 1978 by the American Oil Chemists' Society.)

interfere by increasing the mass of the precipitate. Washing the precipitate free of the surfactant is tedious, even if its solubility is acceptable in the medium.

(e)   Liquid chromatography (ion chromatography).   Sulfate is most often determined in surfactants by ion chromatography, with conductometric or inverse photometric detection. Manufacturers of IC equipment provide literature illustrating this analysis. Sulfite ion will interfere with the determination by being oxidized to sulfate, unless stabilized by, for example, formation of the sulfite-aldehyde complex.

(f)   Turbidimetry.   A turbidimetric determination of sulfate ion can be performed without a preliminary separation of the surfactant. A ratio of alcohol/water must be chosen that maintains the solubility of the surfactant without causing the precipitation of the barium chloride reagent. Sulfite ion concentrations above about 0.1% will cause positive interference due to the low solubility of barium sulfite.

Procedure (28)

Dissolve 1 g sample in a total volume of 20 mL ethanol. Add 1.0 mL 4 M HCl, then 9.0 mL water, and mix. Add 2.0 mL 0.1 M BaCl$_2$ solution, mix, let stand 15 min, and measure the absorbance at 650 nm against a sample blank, prepared similarly but with 2 mL water substituted for the $BaCl_2$ solution. The calibration graph is prepared from surfactant purified of salts by precipitation from ethanol at –40°C, with sodium sulfate added to give concentrations in the 0.1–0.6% range. It may be necessary to modify the ratio of ethanol/water, depending on the particular surfactant analyzed. This procedure was optimized for analysis of a sulfosuccinate. For higher sensitivity, turbidity measurements may be made at a lower wavelength, down to about 400 nm.

3. Sodium Sulfite

Some anionic surfactants are prepared using sulfite compounds. Such products are analyzed for residual sulfite ion. The standard method for sulfite determination is iodometric titration. Determination of sulfite in α-olefin sulfonates presents special problems, covered in detail by Ferm and Griffin (29).

(a)   Titration

Procedure (28)

Dissolve 10 g sample in 50 mL $CHCl_3$ and add 10 mL Wijs reagent (0.2 N, Riedel de Haen, no. 35071—this is a solution of iodine monochloride in glacial acetic acid). Let stand 30 min, then add 20 mL 15% KI solution and 100 mL water. Titrate with 0.1 N sodium thiosulfate solution to the starch end point, with vigorous stirring. Determine the reagent blank in the same manner. Calculation:

$$\% \text{ sodium sulfite } = \frac{(B - V)(N)6.3}{W}$$

where B and V are milliliters of titrant required for the blank and sample titrations, N is the concentration of the titrant (normality), and W is the sample weight in grams.

> *Note*: The sample is added to excess reagent to minimize interference from oxygen, which can also oxidize sulfite. The similarities between this procedure and that for unsaturation determination are obvious, and neither parameter can be determined in the presence of the other by this procedure. A more gentle reaction is sometimes conducted with 20 mL 0.05 M I_2 solution, 2 mL conc. HCl, 10 g sample, and 50 mL water. After 10 min the excess iodine is titrated as above (30).

(b)   Ion chromatography.   Sulfite ion is often determined by liquid chromatography with conductimetric detection. Since sulfite is readily oxidized to sulfate by air which permeates through the plastic tubing of conventional instruments, formaldehyde is generally added to the sample to form the stable bisulfite/formaldehyde adduct. An ion chromatography text may be consulted for details.

4. Sodium Chloride

Chloride ion may be an impurity in anionic surfactants, depending on the process used for sulfonation and the purity of the alkali used to neutralize the product after sulfonation. Chloride ion is determined by titration or ion chromatography. Ion chromatography conditions must be worked out for each type of surfactant because of differences in polarity and water solubility. Titration is more universally applicable. Potentiometric titration has replaced titration to a color change end point in most laboratories.

Procedure (7)

Weigh out a sample of no more than 10 g, dissolve in 250 mL hot water, and acidify to the methyl orange end point by addition of 3 M HNO_3. Add 50 mL acetone. Titrate potentiometrically with 0.2 M $AgNO_3$ solution using an automatic titrator equipped with a silver/silver chloride combination electrode. Use a span of 250 mV, full scale. The end point is the center of the potentiometric break. Run a reagent blank in the same manner. Calculation:

$$\% \text{ NaCl} = \frac{(A - B)(M)5.85}{W}$$

where A and B are the volumes of titrant required for the sample and blank, respectively, M is the molar concentration of the silver nitrate titrant, and W is the sample weight in grams.

F. Solubilizer Alcohols

Some surfactants are supplied as blends with alcohol to provide a liquid product. All suppliers of gas chromatography equipment provide recommended procedures for the determination of low molecular weight alcohols. An example: J&W DB-PS1 column, 0.53 mm × 30 m, 0.3 μm film thickness. An initial hold at 40°C for 5 min, followed by a temperature program to 260°C at 10°C/min will give baseline separation of alcohols from methanol to 1-decanol in 23 min. Flame ionization detection is most appropriate.

G. Unsaturation (Iodine Number)

Unsaturation is described in Chapter 2, under characterization of nonionic surfactants. The Wijs procedure is most often specified for anionic surfactants.

H. Water Content

Water content is almost always determined by titration by the Karl Fischer method (31). Weight loss methods are suitable for certain products, but the presence of volatile solvents or unsulfonated oil will lead to errors. Azeotropic distillation is used only in special cases.

1. Karl Fischer Method

Special reagents are available which make Karl Fischer titration suitable for the analysis of all common matrices (e.g., Hydranal® from Riedel de Haen). One or the other of these may be specified by an individual procedure, such as ASTM D1568 for analysis of alkylaryl sulfonates (6).

Procedure: Karl Fischer Titration with Hydranal Reagent
Set up a commercial Karl Fischer titrator according to the manufacturer's instructions. Add about 60 mL MeOH to the titration vessel and allow the moisture in the solvent to be titrated. Standardize the titrant by titrating 2–3 drops of water weighed by difference from a dropping bottle, calculating the calibration factor from three determinations. Add the sample to the titration vessel from a weighing bottle, dropper bottle, or disposable syringe, weighing by difference in each case. For an expected water content of 10%, a sample size of 0.3 g is appropriate. Problem samples, such as very viscous materials, may be dissolved in 10 mL THF before introduction. In this case, the water content of the THF must be measured separately, and the result corrected for the contribution of the solvent.

2. Distillation

Water, especially at high levels, is often still determined by azeotropic distillation (6,32,33). However, with the development of Karl Fischer reagents that are not susceptible to interference, this technique is fading from use and is not described here.

I. Apparent pH

Anionic surfactants are normally adjusted to neutral pH after sulfonation. A low product pH shows that hydrolysis may have occurred, while high pH may indicate improper neutralization. pH is usually measured on an aqueous solution of sample concentration 1% or less (34,35).

J. Acidity or Alkalinity

Product specifications often include titration to neutrality. This test is simple to perform and gives an indication of consistency and impurity level. There are, however, some pitfalls which become apparent as instrumentation is applied. First of all, the phenolphthalein end point is about pH 8.1, rather than 7.0. Thus if an end point titrator is used, it cannot be set for the theoretical neutrality point of 7.0 and yield the same value as the manual titration to the usual indicator end point. Second, if an alcohol/water medium is used for titration, the neutrality point will not be at apparent pH 7.0, but at some higher value. Complicating the situation is the use by most modern laboratories of recording titrators. Many of these laboratories adopt the method of titrating to a potentiometric "break," rather than to a specific pH. The center of the break will probably not be at pH 7.0. This gives rise to the apparent contradiction of a buffered sample yielding positive values for both acid number and alkalinity number. There are ISO methods for determination of acidity and alkalinity (36,37).

1. Acid Number

By convention, acidity is reported as "acid number," the number of milligrams of KOH required to neutralize one gram of sample.

Procedure (38)
For water-soluble materials, add 50 mL water to a 250-mL conical flask. For materials with poor water solubility, some or all of the water may be replaced with ethanol or isopropanol. Add 0.5 mL phenolphthalein indicator solution (1% in ethanol) and titrate with 0.05 M NaOH solution to the pink end point. Re-zero the buret. Add 10 g sample, mix until dissolved, and again titrate to the end point. Calculation:

$$\text{Acidity, mg KOH/g sample} = \frac{(V)(N)56.1}{W}$$

where V is the titrant volume, N is the titrant molarity, and W is the sample weight in grams.

2. Alkalinity

Procedures for alkalinity are best arrived at by contract with suppliers and customers. ASTM D1570 and ASTM D3673 call for use of phenolphthalein and methyl orange indicator solutions, with results reported in terms of percent sodium hydroxide, sodium carbonate, sodium bicarbonate, or sodium bisulfate (7,9). Obviously, agreement over conventions is required here. A common approach is to follow the same procedure as for acid number, titrating with HCl rather than NaOH. Use the same calculation, reporting apparent milligrams KOH per gram sample.

K. Ash

Ash determination gives a measure of gross cation level. In the case of sulfates and sulfonates, the ash will consist mainly of sodium sulfate, both from sulfate originally present and sulfate formed as a combustion product of the organosulfur compounds. The ash may be subjected to qualitative analysis by X-ray fluorescence or emission spectroscopy to determine whether other ions are present. A general procedure for ash determination is given in ASTM D482 (39).

Knowledge of the ash content tells little about the purity of anionic surfactants. A more useful parameter is the content of alcohol-insoluble inorganic salts, discussed previously.

L. Iron

Specifications for iron generally refer to ASTM D1068, which includes procedures for determining dissolved and total iron by a variety of spectrophotometric and atomic absorption spectroscopic methods (40). ASTM E394, a 1,10-phenanthroline spectrophotometric procedure, may also be referenced (41). As with other trace analyses, it is necessary to validate the procedure with the particular family of products in question, since it is possible for both ashing and for direct dilution techniques to give low results in individual cases.

M. Color

Depending on the raw materials used and the process conditions, sulfonated products have colors ranging from light yellow to black. As a coarse test of product uniformity, a color test may be specified. There are many color scales in use. For sulfonates, the Gardner (42,43) or iodine (44,45) color are usually specified, although many other color scales are in use. A widely applied scale for slightly colored substances is the APHA, or platinum-cobalt scale (46–48). In general, color values can be converted from one scale to another mathematically or graphically, but the comparison is not exact since the shade, as well as the intensity, of the color standard varies from one scale to the other.

Some laboratories measure the absorbance at a specified wavelength using a spectrophotometer and convert the absorbance reading to the iodine or platinum-cobalt color scale using a linear calibration plot. These calibrations are, strictly speaking, only valid when the absorbance spectrum of the colored impurity of interest is exactly the same as that of iodine/iodide or of platinum/cobalt solution.

Modern spectrophotometers are capable of specifying color with much more precision on a three-dimensional basis (49). Such methods are not yet widely used for industrial chemicals, where color is used as a gross estimate of purity rather than a critical specification.

III. ANALYSIS OF INDIVIDUAL SURFACTANTS

Besides the above procedures that are generally applicable to anionic surfactants, specialized methods have been developed to characterize individual surfactants.

A. Alkylbenzene Sulfonates

There is an ASTM standard, D1568, for analysis of LAS (6). It recommends determination of water by azeotropic distillation, unsulfonated material by extraction, and chloride by titration. pH is determined on a 0.1% aqueous solution and a procedure is given for determining ethanol-insoluble matter. ISO 1104 covers methods for specification analysis of alkylarylsulfonates other than LAS, including sulfonated mono-, di-, and trialkyl derivatives of toluene and naphthalene (8). There is not currently an ISO standard for specification analysis of LAS. Additional background information on alkylarylsulfonates can be found in other volumes of the Surfactant Science Series (50,51).

1. Alkyl Chain Length Distribution

As the alkyl chain length of LAS increases, surface activity, sensitivity to water hardness, and solution viscosity increase while water solubility decreases. For most applications, the C_{11}–C_{13} range provides the best mix of properties.

The average alkyl chain length must be known for proper calibration of the titration method for determination of LAS. Knowledge of the actual chain length distribution is sometimes required, as in determination of biodegradability. Gas chromatography after desulfonation is most often used to determine the alkyl distribution. HPLC, while giving less resolution than GC, has the advantage that desulfonation or derivatization is unnecessary. GC is superior if it is necessary to determine the distribution of isomers of each chain length. For example, it might be important to know what portion of the LAS is the 2-phenyl isomer. The 2-phenyl isomer comprises about 27–30% of alkylbenzenes produced by $AlCl_3$-catalyzed alkylation, but only 15–24% of alkylbenzenes from HF-catalyzed alkylations. Sulfonation does not change the distribution, and the high 2-phenyl isomer content LAS is more water soluble, other things being equal.

Analytical methods are available for very thorough characterization of the impurities in the alkylbenzene raw material used to produce LAS. These include trace levels of tetralins and of di- and polyaromatic compounds. Analysis is via preparative HPLC separation into the major classes, followed by capillary GC/MS identification and quantification of individual components (52). In principle, this approach could also be applied to LAS, after desulfonation.

(a) Elemental analysis. Elemental analysis for the atomic ratio of carbon and sulfur will give the average chain length by simple calculation. This value must be used with caution, since if the surfactant contains appreciable quantities of impurities, such as sodium sulfate or disulfonated alkylbenzene, these must be quantitatively determined and suitable corrections made.

(b) Desulfonation and gas chromatography. As discussed further in Chapter 8, GC analysis of alkylarylsulfonates is possible if the methyl ester derivatives are made or if the compounds are desulfonated.

Procedure: Determination of Alkyl Chain Length Distribution by Sealed-Tube Desulfonation and Gas Chromatography (53)

A sample of 50–100 mg is weighed into a combustion tube, 12×120 mm. Two milliliters 85% phosphoric acid is added, and the tube is sealed in a flame such that the wall thickness is maintained in the area of the seal. For safety, the tube is contained within a screw-cap metal tube. The nested tubes are placed in a silicone-oil bath for 15 min at 150°C, then removed and cooled. The contents of the combustion tube are transferred to a test tube of similar dimensions which is equipped with a ground-glass stopper. Two mL of petroleum ether is used to rinse the combustion tube, and the rinsings are added to the test tube. The test tube is shaken, and the petroleum ether layer is pipetted into a second test tube. The contents of the first test tube are extracted with another 2-mL aliquot of petroleum ether, and this extract is added to the first. The combined extracts are washed four times with 1-mL portions of 4 M NaOH solution, transferring the petroleum ether layer to a clean test tube after each wash. The petroleum ether phase is dried by addition of 1 g sodium sulfate, transferred to yet another test tube, and evaporated to a small volume in a water bath at 60–70°C. This solution is then diluted 1:1 with very pure acetone and analyzed by gas chromatography. GC conditions: silica capillary column, OV-101 or equivalent, 0.25 mm × 20–50 m; isothermal, 170–200°C, or programmed from 140–170°C to 180–210°C at 0.5–2.0°C/min, with FID detection. Peaks are identified by use of standard alkylbenzenes of known composition, with quantification by area percent.

(c) Liquid chromatography determination of alkyl chain distribution. Alkylbenzene-sulfonates can be separated according to the length and degree of branching of the alkyl chain by paired-ion HPLC. Changes can be made in the pH, ionic strength, and solvent ratio of the mobile phase to control the separation.

Procedure: LAS Characterization by Paired-Ion HPLC (12)

A sample of 50–100 mL is deposited on a 3-mL silica SPE column (J. T. Baker) conditioned in $CHCl_3$. The unsulfonated oil is removed by washing the column four times with 0.5-mL aliquots of $CHCl_3$. The sulfonate is then recovered with four 0.5-mL methanol washes. The methanol solution is evaporated to dryness to expel $CHCl_3$, and the residue is redissolved in water or methanol.

The HPLC system should be capable of ternary gradient operation and have a column compartment thermostatted at 35°C. Detection is by UV absorbance at 225 nm. A guard column of Alltech RSIL prep C_{18}, 4.6 × 32 mm, is used with an analytical column of DuPont Zorbax C_{18}, 4.6 × 250 mm. The mobile phase is a mixture of water, acetonitrile, and 0.1 M aqueous tetrabutylammonium bisulfate solution, adjusted to pH 5.0 with NaOH. Flow is 2.0 mL/min. A linear gradient is used, beginning with 10:50:40 TBAHSO$_4$/water/acetonitrile. During the first 30 min, H_2O is lowered to 10%, increasing CH_3CN to 80%. During the following 10 min, TBAHSO$_4$ is dropped to 0%, resulting in a H_2O/CH_3CN isocratic system, 20:80, which is held for 15 min. Identification is by comparison with known materials or by mass spectrometry analysis.

2. Sulfones and Unsulfonated Alkylbenzenes

Free oil in LAS consists not only of unsulfonated alkylbenzene but also of sulfones. Anhydrides are also formed in the sulfonation reaction but are completely eliminated from the commercial product by a hydrolysis step during LAS production (54):

A sulfone byproduct

An anhydride byproduct

Usually, the extraction/gravimetric determination of unsulfonated material, as described in Section II, is sufficient. For more precise characterization of the the nature of the oil, GC, LC, and MS may be used. HPLC is capable of simultaneous determination of unsulfonated alkylbenzenes and of sulfones (54–56). HPLC usually gives higher values for unsulfonated material than does the gravimetric method, presumably because volatile material is lost when performing the latter (13).

Procedure: Determination of Unsulfonated Alkylbenzenes by Extraction and GC (57)

A 100-mg solid sample is extracted for 1 hr in an ultrasonic bath with 10 mL *n*-hexane. The slurry is washed in a separatory funnel with 30 mL 3 M HCl. The hexane is separated, and the aqueous phase re-extracted with four more aliquots of hexane. The combined extracts are reduced in volume to 0.3 mL, purified by chromatography on silica gel (0.5 × 18 cm; discard first 5 mL of hexane eluent, save next 13 mL for analysis), evaporated to dryness at a low temperature, and dissolved in ethanol. The alkylbenzenes are analyzed by gas chromatography on a 25 m SE-54 column, using flame ionization detection. The column temperature is raised from 50°C to 120°C at 30°C/min, then to 220°C at 3°C/min.

Procedure: Determination of Unsulfonated Alkylbenzene and Sulfone Content by HPLC (13)

A methanol solution of the surfactant, about 10% concentration, is injected on an HPLC system consisting of a C_{18} reversed-phase column, 4.6 × 150 mm, a mobile phase of 94:6 MeOH/H_2O at 1 mL/min, and both refractive index and UV detectors. The RI detector is thermostatted at 35°C, while the UV detector is used at 220 nm. In the normal mode of operation, LAS is eluted immediately, followed by unsulfonated material in order of chain length, followed by sulfones over the range 10–35 min. For rapid quantitative analysis, the system is back-flushed after 4.5 min, after elution of the LAS. The total nonpolar material is then eluted through the RI detector, at a retention time of 9 min after injection.

3.   Disulfonated Alkylbenzenes

Müller and Noffz report that the monosulfonate may be quantitatively extracted from hydrochloric acid solution with ethyl ether, with a subsequent extraction of disulfonate from the same aqueous solution with isoamyl alcohol (27).

4.   Indane and Tetralin Sulfonates

Dialkyltetralin sulfonate

Dialkylindanesulfonate

Linear alkylbenzene produced from chloroparaffins contains 6–8% dialkyltetralin and dialkylindane. Alkylbenzene produced from olefins contains about 1% of these compounds. Thus, LAS may contain about 1%, or about 7% of the sulfonated indanes and tetralins, with the indanes being at very low concentration compared to the tetralins. These byproducts are surfactants, and their presence is beneficial in liquid formulations since they act as hydrotropes and viscosity improvers. They can be determined by GC or GC/MS by the same methods as are used for determining the alkyl chain distribution. The compounds are made volatile by desulfonation or by derivatization as discussed in Chapter 8 (58,59).

B. Alkyl Sulfates

Sampling and analysis of alkyl sulfates is the subject of ASTM standard D1570 (7). It recommends determination of water by azeotropic distillation, pH on a 0.1% aqueous solution, alkalinity by sequential titration, ethanol-insolubles, unsulfated material by liquid-liquid extraction, combined alcohol by saponification and extraction, and ester SO_3 by saponification and titration or gravimetric determination of sulfate. Methods for free chloride and sulfate are also specified.

There are separate ISO standards for primary and secondary alkyl sulfates (i.e., for products made from primary or secondary alcohols). The ISO standard covering primary alkyl sulfates specifies neutral oil by extraction from ethanol/water solution, total combined alcohols by ether extraction after acid hydrolysis, pH of a 10% aqueous solution, water by titration if below 10% or by azeotropic distillation if above 5%, and chloride by titration (60). The standard covering secondary alkyl sulfates is similar, but combined alcohols are not determined. Assay is by determination of total solids after extraction of the neutral oil, corrected for the presence of other impurities (61).

Determination of alkyl sulfates is usually performed by HPLC as discussed in Chapter 7. GC may also be used after cleavage of the sulfate group, as discussed in Chapter 8.

1. Assay

Assay is conventionally performed by two-phase or one-phase titration with benzethonium chloride, as described in Chapter 16. Assay may also be performed by determination of ester SO$_3$ as described subsequently. For example, ISO 2271 specifies purity determination of sodium lauryl sulfate (used as a reagent) by a variation of the ester SO$_3$ method (5,30).

2. Alkyl Chain Length Distribution

The alkyl chain length distribution is readily determined by HPLC or GC. These methods are described in Chapters 7 and 8.

3. Total and Combined Alcohols

The alcohol sulfates are decomposed by acid hydrolysis to yield the free alcohol and sulfate ion. The alcohols are isolated and weighed. The ASTM procedure is summarized below. The ISO procedure differs in that the starting sample has already been extracted free of neutral oil, the solvent for extraction after hydrolysis is ethyl ether, and 1 M HCl rather than ethanol/water is used to wash the extract (60).

Procedure (7)

Weigh a sample of 1–5 g into a suitable flask and dissolve in about 50 mL water. Adjust to the methyl orange end point with HCl or KOH solution, then add 50 mL 1 M HCl solution. Reflux overnight on a steam bath (to minimize foaming), then boil at least 0.5 hr to complete hydrolysis. Cool to about 50°C and rinse the condenser into the flask with water and petroleum ether. This step is vital because the alcohols are steam-distilled from the flask and trapped in the condenser (62). Transfer the flask contents to a separatory funnel and extract several times with 40-or 50-mL aliquots of petroleum ether. Make the first extraction gently to minimize formation of emulsions. Wash the combined extracts with 30-mL aliquots of 10:90 EtOH/H$_2$O until the washings are neutral to methyl orange, shaking gently to avoid forming emulsions. Salt may be added to break emulsions. Carefully evaporate down the washed extract and weigh. This is the "total fatty alcohol." "Combined fatty alcohol" is obtained by subtracting from this value the value for unsulfated material.

4. Free Alcohol

Besides the extraction procedure described in Section II, unsulfated alcohol can be determined by TLC and by HPLC, as described below. In the special case of a mixture of alkyl sulfate and alkylarylsulfonate, unsulfated alcohol can be separated from unsulfonated alkylbenzenes by forming the urea adduct of the alcohol and separating by centrifugation (63).

(a)   TLC

Procedure: Determination of Free Alcohol by TLC (64)

Silica gel G plates are used, with development by 80:20 ethyl ether/benzene. The sample is spotted as 30 or 40 µL of a 5% MeOH solution. Modified Dragendorff reagent is suitable for visualization. Residual alcohol appears as a light spot on a yellow background, with an Rf value of about 0.7. The sulfated material remains at the origin. The detection limit is about 0.1%.

(b)   HPLC.   Free alcohol may be selectively determined in alkyl sulfates by HPLC of urethane derivatives. Alcohols are separated according to their alkyl chain length. Naphthylisocyanate is preferred to the more frequently used phenylisocyanate because fewer byproducts are formed and because of the higher molar absorptivity of the derivatives. The following method is suitable for determining low levels of free alcohol, down to

about 0.001 mole percent (65). Room temperature derivatization is specified because yield-reducing side reactions are minimized (66).

Procedure (67)

About 100 mg sample is dissolved in 1 mL N,N-dimethylformamide and mixed with 25 µL 1-naphthylisocyanate in a glass-stoppered test tube. The mixture is let stand for 30 min, then reacted with 25 µL MeOH to destroy excess naphthylisocyanate. A column of LiChrosorb RP-8, 4.6 × 250 mm, is used with a mobile phase of 78:22 CH_3CN/H_2O. Detection is by UV absorbance at 290 or, for more sensitivity, 222 nm.

5. Ester SO_3

Acid-catalyzed hydrolysis of, for example, sodium lauryl sulfate, yields lauryl alcohol and sodium hydrogen sulfate. The sodium bisulfate is easily titrated. Depending on the expected impurities, this determination may be made on either the original sample or on the alcohol-soluble portion. If on the latter, the determination should be combined with the total fatty alcohol method, above, by titrating the aqueous phases. The gravimetric method is superior to the titration method if interferences are present in the product.

(a) Titration method

Procedure (7,30)

Accurately weigh about 5 g sample or about 1 g alcohol-soluble material into a flask, dissolve in about 50 mL water, and adjust to the methyl orange end point with dilute HCl. Add exactly 35 mL 1 M HCl and boil with reflux for 2 hr after foaming has stopped (or remains constant). Cool and titrate to methyl orange with 1 M NaOH. Calculation:

$$\% \text{ Ester SO}_3 = \frac{(V_1 B - V_2 A)8.0}{W}$$

where V_1 and V_2 are the volumes of base and acid used, respectively, and B and A are their molar concentrations. W is the sample weight in grams.

(b) Gravimetric method

Procedure (7)

Accurately weigh about 1 g of the alcohol-soluble fraction into a flask, add 50–100 mL H_2O and 5–10 mL 1 M HCl. Hydrolyze as above. Rinse the contents of the flask into a 250-mL volumetric flask, cool, and dilute to volume with water. Mix thoroughly and let settle. The fatty alcohol layer should be above the mark; discard it. Transfer 100 mL of the solution to a beaker and neutralize to methyl orange. Add 0.5 mL 1 M HCl and dilute to 175–200 mL. Boil, adding 20 mL 10% $BaCl_2 \cdot 2H_2O$ solution. Continue boiling a few minutes, then digest at 70°C for 1 hr. Filter with filter paper, wash with hot water until the washings test negative for chloride ion using silver nitrate solution, ash, and weigh. Calculation:

$$\% \text{ Ester SO}_3 = \frac{(\text{wt BaSO}_4)34.3}{\text{sample wt}}$$

C. Ether Sulfates

Due to the kinetics of ethoxylation, low-mole ethoxylates such as these normally contain high levels of unethoxylated material, in this case alcohol sulfate or alkylphenol sulfate.

Other impurities may be present: sulfated PEG, unsulfated alcohol ethoxylate or alkyl-phenol ethoxylate, sodium sulfate, sodium chloride, PEG.

Noninstrumental analysis of ether sulfates is complex because of the various types of organic material which must be distinguished. Puschmann and Wickbold have described comprehensive procedures for quantitative determination of the components in alkyl ether sulfates by wet chemical methods (68,69). Nowadays, these tests are usually performed by HPLC. Cloud point determination, a measure of EO content, is described in Chapter 2 with characterization of nonionic surfactants.

1.   Total Active Matter

ISO 6842 defines total active matter as ethanol solubles, which includes not only the ether sulfate but also alkyl or alkylphenol sulfate, PEG sulfate, and unsulfated nonionic material including PEG (70). ISO 6843 describes wet chemistry methods for separately determining PEG, sulfonated PEG, and inorganic salts in order to arrive at a value for the ether sulfate itself, but this value will still include alkylsulfate or alkylphenolsulfate (71).

Procedure: Gravimetric Determination of Active Matter (70)

Weigh a sample containing about 1 g active matter into a 250-mL conical flask, add 100 mL 99% ethanol and 0.1 g Na_2SO_4, fit a reflux condenser, and boil for 30 min. While still hot, filter with fast filter paper, rinsing the apparatus with hot ethanol. Evaporate the filtrate to dryness, twice adding 10-mL portions of CH_2Cl_2 and again evaporating. Cool and weigh. Determine the amount of sodium chloride impurity by dissolving the residue in acetone and titrating with $AgNO_3$ with K_2CrO_4 indicator. Active matter is equal to the ethanol-soluble fraction less chloride, calculated as sodium chloride.

2.   Ether Sulfate Content

The ether sulfate can be determined by two-phase titration. However, interference from sulfated impurities (sulfated alcohol/alkylphenol and sulfated PEG) in the titration requires that a preliminary separation step be performed first for high accuracy. This analysis may also be performed by HPLC, as described in Chapter 7.

ISO 6843 specifies a variation of the extraction method for determination of PEG in nonionic surfactants. In this case, an aqueous salt solution of the surfactant is extracted with 9:1 ethyl acetate/1-butanol. Almost all of the ether sulfate goes into the organic phase, leaving behind PEG, PEG sulfate, and sulfate ion, as well as a small amount of the ether sulfate. Alkyl- or alkylphenol ethoxylate and alkyl- or alkylphenol sulfate are expected to partition primarily in the organic phase. The amount of anionic surfactant remaining in the aqueous phase is determined by two-phase titration, as is the equivalent weight of the extracted material (71).

3.   Alkyl Chain Distribution

The determination of alkyl chain distribution of alcohol ether sulfates can be most easily performed by gas chromatography after cleavage of the sulfated ethoxy moiety with hydriodic acid. This procedure is similar to that described under the analysis of ethoxylated nonionic surfactants. The alkyl iodides from cleavage may also be determined by HPLC (72). An alternative approach requires the decomposition of the sulfate and ethoxy groups with the mixed anhydrides of paratoluenesulfonic and acetic acid, followed by GC determination of the alcohol acetates (73).

High temperature gas chromatography may be performed on the trimethylsilyl derivatives of the AE resulting from acid hydrolysis of AES. This procedure simultaneously gives the alkyl and ethoxy chain length, although the chromatogram requires careful interpretation. This technique is described in Chapter 8.

Procedure: Determination of the Chain Length Distribution of Alkyl Ether Sulfates by Gas Chromatography (74,75)

A sample containing about 25 mg sulfated fatty alcohol ethoxylate is placed in a glass stoppered test tube to which is added 1.5 mL hydriodic acid. The head space is purged with nitrogen, then the tube is stoppered and placed within a pressure tube in a heating block. After heating at 185°C for 10 min, the tube is cooled and 3 mL water is added. The resulting alkyl iodides are extracted with three 1-mL portions of petroleum ether. The combined extracts are washed free of iodine with two 1-mL portions of 20% aqueous sodium thiosulfate solution, then evaporated to dryness on a water bath. Analysis is by gas chromatography on a 5-ft column of 10% polyethylene glycol diadipate on Celite. Calibration is effected by carrying known alcohols through the procedure.

4. Alkyl Sulfate or Alkylphenol Sulfate Content

This analysis is normally performed by HPLC (see Chapter 7) or, after hydrolysis of the sulfate group, by GC. Wet chemical methodology may also be applied: if the sulfate groups are hydrolyzed, the determination of unethoxylated material can be performed in the same way as for alcohol ethoxylates. Hydrolysis conditions must be mild to avoid rupture of ether linkages.

The method below is based on hydrolysis of sulfate groups, followed by column chromatographic separation of derivatives of the resulting alcohols and alcohol ethoxylates. Only the unethoxylated material elutes from a silica gel column.

Procedure: Gravimetric Determination of Alkyl Sulfate in Alcohol Ethoxysulfates (76)

Dissolve about 10 g sample, accurately weighed, in 250 mL 1 M H_2SO_4 in a flask with ground glass neck. Fit a condenser and heat at reflux for 90 min. Dilute with an equal volume of EtOH in a 1-L separatory funnel and extract 3 times with 150-mL aliquots of pentane, combining the extracts in a flask with a ground glass neck. Add 5 g 3,5-dinitrobenzoyl chloride, fit a reflux condenser, and heat to reflux for 90 min. Let cool, remove the esters by filtration, and dilute to 50 mL with toluene. Fill a chromatography column with silica gel of 0.2–0.5 mm particle size which has been activated at 150°C for 3 hr. Apply 10 mL of the toluene solution, eluting with 750 mL toluene. Strip the toluene from the eluate, transfer to a flask with ground glass neck, add ethanolic KOH, and heat to reflux for 30 min. Extract the free alcohols from the saponification mixture with pentane or petroleum ether, evaporate the solvent, and weigh the residue.

5. Unsulfated AE or APE

This determination is best made by HPLC or CE, as described in Chapters 7 and 11. Under the proper conditions, a single peak is obtained for nonionic impurities. The standard extraction method for unsulfated material in alcohol sulfates (ASTM D 1570) is not appropriate for ethoxysulfates because the solubility characteristics of ethoxylated fatty alcohols differ from those of fatty alcohols. The products with higher degrees of ethoxylation will not be completely extracted and emulsion formation may altogether prevent extraction.

(a) Ion exchange/gravimetric method. An ion exchange procedure is used to remove the sulfated surfactant, as well as inorganic salts. The remaining unsulfated material is dried and weighed.

Procedure (69,77)

Dissolve 3 g sample, accurately weighed, in 50 mL methanol and filter through course filter paper into a glass column, 25 × 200 mm, packed with a mixture of 25 mL of a cation exchange resin such as Bio-Rad AG 50W-X4, acid form, and 25 mL of an anion exchange resin such as Bio-Rad AG 1-X4, hydroxide form. (If determination of active agent and sulfated PEG is later to be performed on the same ion exchange fractions, use two separate columns attached together, with the cation exchanger first. Suitable dimensions are 20 × 200 mm for the cation exchange column and 30 × 170 mm for the anion exchange column.) Allow the solution to pass through the column(s) at about 3 mL/min, collecting the eluent in a 500-mL beaker. Wash the apparatus several times with a total volume of 450 mL methanol. Evaporate the eluate using a rotary evaporator on a steam bath until a constant weight is obtained. The residue is the unsulfated material.

   Note: If unfamiliar with the particular product, check the residue for sulfated material by two-phase titration for anionic surfactant. The value for alcohol ethoxylate may be corrected for any polyethylene glycol present by analyzing the residue by TLC or one of the other methods given in Section II of Chapter 2.

(b) HPLC. Free alcohol ethoxylate can be determined in the sulfated alcohol ethoxylates by HPLC. Under ordinary reversed-phase conditions, the unsulfated material is highly retained, making quantification of low levels difficult. This can be avoided by using a reversed-phase column with ion exchange functionality, so that the nonionic material elutes first (78). The procedure described below avoids the need of quantifying multiple peaks for AE by use of a back-flush technique. A reversed-phase method is used which causes ionic materials to be eluted first. The direction of flow of the column is then reversed, with the refractive index detector still monitoring the effluent as the AE then comes off in a single peak. If other nonionic materials are present (for example, free alcohol), a very short reversed-phase column is added to the system so that two peaks are registered for nonionic material, one for AE and the other for nonethoxylated materials (14). A similar procedure has been described using an evaporative light scattering detector rather than a refractive index detector (15).

Procedure: Back-Flush HPLC Method for Determination of Ethoxylated Alcohol in Alcohol Ethoxysulfates (14)

The sample is dissolved in MeOH and the AE separated from ionic materials by a back-flush technique using the C_{18} Unisil QC-18 column (Gasukuro, Japan), 4.6 × 150 mm, for separation from ionic substances, and Merck LiChrosorb RP-2, 4.6 × 50 mm, for separation of AE from unethoxylated nonionics. The mobile phase is 85:15 MeOH/H_2O and refractive index detection is adequate. Valving is used so that the flow is either forward or backward through the QC-18 column, then through the RP-2 column (one direction only) to the detector. If a flow of 1 mL/min is used, the valve position is switched from the forward position to the back-flush position at 5.5 min.

6. Sulfated PEG

The ISO 6843 standard specifies separation of the surface-active material by liquid-liquid extraction from salt water into 9:1 ethyl acetate/1-butanol. The material remaining in the aqueous phase is purified from excess NaCl by evaporation and re-solution in methanol, then subjected to ion exchange to separate PEG from salts and PEG sulfate. PEG sulfate is determined gravimetrically, by difference, after calculating PEG, Na_2SO_4, NaCl, and titratible anionic surfactant not removed in the original extraction (71).

7. 1,4-Dioxane Content

Ether sulfates may contain low levels of 1,4-dioxane, both because of impurities in the starting ethoxylate and because of formation of dioxane during the sulfation reaction (79). This analysis is discussed with the characterization of nonionic surfactants (Chapter 2).

D. α-Olefin Sulfonates

Olefin sulfonates are analyzed in the same general manner as other anionics. For certain purposes it may be necessary to determine the relative proportions of the main components: the alkenesulfonates and hydroxyalkanesulfonates. The concentration of residual sultones, especially unsaturated sultones caused by interaction of alkenesulfonate with bleach, is carefully controlled. Other low level impurities are disulfonated and dialkyl compounds. Beranger and Holt have thoroughly discussed the characterization of middle and heavy α-olefin sulfonates used in enhanced oil recovery (80). This included use of the methods below, as well as others not suitable for routine quality control.

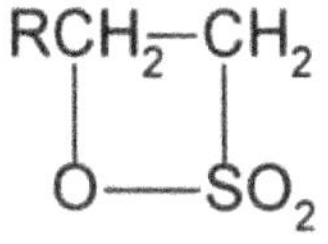

$$RCH_2-CH_2$$
$$\quad | \qquad |$$
$$\quad O-SO_2$$

A 1,3-sultone

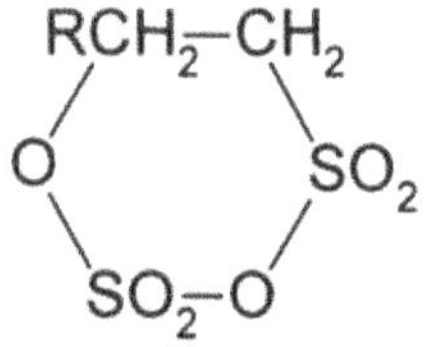

$$RCH_2-CH_2$$

O ⟍ SO₂

SO₂-O

A 2-sulfatosulfonate

ASTM standard D3673 specifies methods for gross characterization of α-olefin sulfonates: assay by two-phase titration, water by azeotropic distillation, unsulfonated material by extraction, sulfate and chloride by titration, alkalinity by sequential titrations. pH is determined on a 1% aqueous solution and color is according to the platinum-cobalt scale (9). Internal olefin sulfonates may be analyzed by the same methods used for α-olefin sulfonates (81).

1. Assay

Assay is most often performed by titration with a cationic surfactant, as described in Chapter 16. Potentiometric one-phase titration gives higher assay values by about 3% compared to two-phase titration. This is attributed to the presence of disulfonates, which are only measured by potentiometric titration (82).

The acid-base titration described below may also be used. This procedure can be more accurate than benzethonium chloride titration, especially for determining compounds of relatively short (C_8-C_{10}) chain length which have greater water solubility.

Procedure: Determination of Active Agent and Sulfate in α-Olefin Sulfonates by
Nonaqueous Potentiometric Titration (83)

In this determination, 5 mL of a 0.1 g/L aqueous solution of the surfactant is passed
through a cation exchange column (1 × 15 cm, containing 15 mL Dowex 50W-X8, acid
form) to convert the salts to the corresponding acids. The acids are eluted with two 3-mL
portions of water and four 6-mL portions of water, then titrated with tetrabutylammonium
hydroxide, 0.1 M in 70:30 benzene/methanol. The first inflection corresponds to the total
sulfonate, plus the first proton of sulfuric acid. The second inflection corresponds to neutralization of the second proton from sulfuric acid. A third inflection may be observed due
to solvent impurities. If necessary because of poor water solubility of the surfactant, the
initial sample dissolution and ion exchange is conducted at 55°C.

2.   Hydroxyalkanesulfonate Content

This determination is readily performed by HPLC (Chapter 7), as well as by the wet chemical methods described below.

(a)   Hydroxyl number.   Conventional methods for determination of hydroxyl functionality can be used to estimate the concentration of hydroxyalkanesulfonates. These are based
upon infrared or near infrared spectroscopy or wet chemical methodology and are described with methods for characterization of nonionic surfactants.

(b)   Titrimetric determination.   The 3- and 4-hydroxyalkanesulfonates are converted to
the corresponding sultones by acid treatment and the petroleum ether extractible material
is weighed and titrated. The sultones are not titrated as acids, while the alkenesulfonates,
which survive the acid treatment and extraction unchanged, are titrated. The difference in
titration before and after conversion to sultones gives the level of hydroxyalkanesulfonates
originally present. 2-Hydroxyalkanesulfonates will not dehydrate to the sultones under
these conditions and are not measured.

Procedure (21)

A sample containing 1–2 g active ingredient is dissolved in 50 mL H_2O in a 250-mL separatory funnel and the solution is extracted twice with 50-mL portions of petroleum ether to
remove oil and sultones. To the aqueous layer are added 25 mL 6 M HCl and 50 mL
ethyl ether. The separatory funnel is shaken thoroughly but cautiously, the layers are allowed to separate, and the aqueous layer is discarded. The ether phase is washed 3 times
with 15-mL portions of 3 M HCl, then transferred to a tared extraction flask and evaporated to dryness. Drying is continued in an oven at 120°C to insure removal of HCl, then
the residue is cooled, weighed, and titrated with NaOH solution.

3.   Alkenesulfonate Content

Alkenesulfonate content is determined by HPLC, although GC and TLC may also be applied (Chapters 7–9). The determination may also be made by performing an unsaturation
analysis, the result of which must be corrected for the presence of any unsulfonated olefin
starting material. It has been noted that the conventional determination of unsaturation by
bromination or Wijs reagent gives erratic results for the alkenesulfonates, while even hydrogenation can give low values for certain compounds (84). The unusual technique of
ozone titration yields more quantitative results, but the apparatus is not commercially
available (85).

On occasion, it may be necessary to determine the position of the double bond in an alkenesulfonate. This can be accomplished by oxidation with permanganate/periodate reagent to cleave the compound into two carboxylic acids (86,87) which are determined by gas or liquid chromatography. For an alkenesulfonate of a single chain length, NMR spectroscopy is suitable for determining the position of the double bond.

4. Sultones

Sultones can be hydrolyzed to the corresponding sulfonates, which are easily determined by titration or spectrophotomety. A preliminary separation of the sultones from the sulfonate surfactant is necessary. Alternatively, the sultones may be determined directly by HPLC. Other specialized techniques for characterization of sultones themselves have been developed (88).

(a) Extraction. This method is suitable for relatively high levels of sultones, greater than 500 ppm. The procedure given here uses ethanol/water as the polar phase. For high molecular weight α-olefin sulfonates, 2:1 isopropanol/water is preferred (80).

Procedure (89)

A sample of about 100 g is weighed, dissolved in 850 mL 50:50 EtOH/H$_2$O and transferred to a continuous liquid-liquid extraction device. This solution is extracted for 48 hr using 200 mL petroleum ether. The petroleum ether extract is evaporated to dryness and the residue dissolved in EtOH and diluted to 50 mL. A 10-mL aliquot is hydrolyzed by boiling for 1 hr with 0.1 g KOH. The mixture is evaporated and analyzed by two-phase titration in the usual manner.

(b) Column chromatography. This procedure is suitable for levels of sultone in the 10–500 ppm range.

Procedure (89)

Two grams of an aqueous 40% sample solution is mixed with 20 mL of 2:1 acetone/water and applied to the top of a silica gel column, 50–150 mesh, 20 × 250 mm. The column is eluted with 500 mL petroleum ether. The eluate is collected, split into two portions, and evaporated to dryness, then an aliquot representing 50% of the eluate is hydrolyzed by boiling for 1 hr with 0.1 g KOH and again evaporated down. The residue is analyzed colorimetrically: The residue is dissolved in 85 mL water and transferred to a separatory funnel. Fifteen milliliters 0.1 M H$_2$SO$_4$, 5 mL methylene blue solution (0.35 g methylene blue, 6.5 mL conc. H$_2$SO$_4$ to 1 L water), and 15 mL CHCl$_3$ are added. As a blank, the remainder of the unhydrolyzed column eluate is analyzed in another separatory funnel. The 15-mL 0.1 M H$_2$SO$_4$ solution is omitted with the blank. The extraction is performed, then repeated with a second 15-mL portion of CHCl$_3$. The extracts are combined into 50-mL volumetric flasks, diluted to volume, and the absorbance is measured at 642 nm, calibrating with solutions of sodium hexadecanesulfonate analyzed in the same way.

(c) Liquid chromatography. All published HPLC methods for determination of sultones require a preliminary separation to eliminate ionic material. The final normal phase HPLC analysis is performed with a silica column and refractive index detection.

Macmillan and Wright describe a method suitable for trace analysis, down to levels of 0.2 ppm (90). Ion exchange is used to remove ionic material. (Three columns, 2.2 × 28.5 cm, in series; one of Bio-Rad AG 50W-X8, H$^+$ form, two of a strongly basic anion exchange resin. Elution of 8 g sample is performed with 85:10:5 isopropanol/water/methylene chloride until 250 mL eluent is collected.) TLC (silica gel 60, 250 mm on Al plates: development with 50:50:2 pentane/ethyl ether/methanol; 25 mg residue per plate; sulfuric acid charring of standard, placed on a removable strip on the plate edge, to identify sultone zone) is used to further concentrate the sultones, which are finally removed with ethyl ether and determined by HPLC with a mobile phase of 96:4 isooctane/ethanol.

Slagt and coworkers recommend pre-concentration of sultones by either ion exchange or extraction (91). (Ion exchange: 2×40 cm, either the mixed resin Amberlite MB-3 or the anion resin Bio Rad AG 1-X2, Cl^- form; elution of 5 g sample is performed with 400 mL 80:20 methanol/methylene chloride. Extraction: hexane/50:50 ethanol/water.) HPLC is performed on the residue with a mobile phase of 20:80 ethyl acetate/hexane.

Roberts and coworkers recommend a solid-liquid extraction for preliminary separation of the sultones prior to HPLC analysis:

Procedure (92)

A 5-g sample of α-olefin sulfonate is mixed with 50 g of basic alumina, activity grade I, which has been activated at 450°C for 3 hr. The mixture is then extracted four times with 50-mL aliquots of ethyl ether. The extracts are combined and evaporated to dryness, and the residue is dissolved in the HPLC mobile phase. The analysis is performed on a CPS-Hypersil normal phase HPLC column, 4.6×20 cm, with a mobile phase of 90:10 hexane/ethyl acetate and differential RI detection. Peak identification and quantification is made versus authentic sultone standards. The method was demonstrated for determination of hexadecane-1,4-sultone, hexadecane-1,3-sultone, hexadec-1-ene-1,3-sultone, and 2-alkylalkane-1,4-sultone, as well as 2-hexadecylhexadec-1-ene-1-sulfonate.

(d)   Gas chromatography.   Gas chromatographic analysis of sultones concentrated from α-olefin sulfonates by ion exchange or extraction is feasible (91). This has been demonstrated using a packed column and flame ionization detection. However, the chromatograms of commercial products are quite complex, and the investigators recommended that HPLC be used.

5. Sulfatosulfonate Content

The content of sulfatosulfonates can be estimated by measuring the active agent content by titration with benzethonium chloride before and after acid hydrolysis. Hydrolysis converts the sulfatosulfonates to hydroxyalkane sulfonates, which add to the assay value (81).

6. Separation Methods

HPLC methodology has made these separation procedures unnecessary for routine analysis. They are still valuable for a thorough characterization of a new product.

(a)   Column chromatography

(1)   Salting out chromatography.   Hydroxyalkanesulfonates and alkenesulfonates may be separated from each other by a variant of ion exchange chromatography called salting-out chromatography (93).

Procedure (93)

The separation is performed on a column of Amberlite CG-50, 25×400 mm, mixed Na^+ and H^+ form, thermostatted at 35°C. An eluent of 30:70 2-PrOH/0.5 M NaCl solution is used, with the column effluent checked batchwise by the methylene blue method to detect the anionic surfactants. The hydroxyalkanesulfonates elute first, followed by alkene-sulfonates.

(2) *Partition chromatography.* Using silanized silica gel, the product can be separated into two fractions, one corresponding to combined disulfonates, sulfatosulfonates, and sodium sulfate, and the second containing monosulfonates. Hydroxysulfonates are not separated from other sulfonate and disulfonates. Sulfatosulfonates are detected by hydrolyzing the disulfonate fraction with 3 M HCl and testing by TLC for the presence of the hydroxy-alkanesulfonate hydrolysis products.

Procedure (94)

A glass chromatography column, 20 × 350 mm, with a glass frit, is packed with 40 g silanized silica gel (Merck no. 7719) suspended in 80 mL 30:70 2-propanol/water. After draining off the excess solvent, 1.5 g acid-washed sea sand is placed over the packing. A solution is prepared of 1.25 g sample in 25 mL 2-PrOH/H_2O. A 2.0-mL aliquot is transferred to the column, followed by 250 mL 2-PrOH/H_2O at a rate of 4 mL/min, using suction to increase the flow. The first fraction consists of the first 70 mL eluate and contains the disulfonates and Na_2SO_4. The second and third fractions are each of 5 mL volume. The fourth fraction, containing the remaining eluate, has the monosulfonates. All fractions are evaporated to dryness and freed of silica gel by resuspending in 60:40:17 n-BuOH/acetone/H_2O and filtering. The fractions are again evaporated to dryness and weighed. The identity of the fractions is checked by TLC as described in Chapter 9. The second and third fractions belong to either the disulfonate or the monosulfonate fractions, as determined by TLC. To check for sulfatosulfonates, the disulfonate fraction is boiled with 3 M HCl and tested by TLC for the presence of the resulting hydroxyalkanesulfonates. Na_2SO_4 is determined by titration of the disulfonate fraction with $PbCl_2$ solution in acetone/H_2O to the dithizone end point.

(3) *Adsorption chromatography (95).* After removal of disufonates, alkene monosulfonates may be separated from hydroxyalkanemonosulfonates by silica gel chromatography of the methyl esters. About a gram of the monosulfonates is reacted with diazomethane, taken up in chloroform (stabilized with amylene, not ethanol), and applied to a column of Merck silica gel (0.05–0.2 mm particle size), 20 × 350 mm. The column is eluted with 500 mL chloroform, monitoring the effluent with a flow-through IR detector at 2900 cm^{-1}. The first fraction to elute contains unsulfonated olefin. The second contains alkenesulfonate, methyl ester. Elution is continued with 500 mL 98:2 chloroform/ethanol, which removes the hydroxyalkane sulfonate, methyl ester.

(b) Liquid-liquid extraction. Extraction of monosulfonates from disulfonates by extraction is not completely straightforward. The literature disagrees as to what the best conditions are, presumably at least in part because of differences in chain length and degree of sulfonation of the substances tested.

Milwidsky reports that olefin monosulfonates may be extracted into ethyl ether from 3 M HCl solution, while disulfonates remain in the aqueous phase. Repeated extractions or a continuous extractor is necessary. The acidic aqueous phase may then be extracted with n-butanol or isobutanol to remove the disulfonates. The mass and molecular weight of each extract is determined by volatilizing HCl, then neutralizing with standard alkali and weighing the dried residue (96,97).

Kupfer and Künzler found that a continuous extractor was necessary for this separation of monosulfonates from disulfonates (98). A 2.5-g sample is dissolved in 20 mL

40:60 ethanol/water, 20 mL concentrated HCl is added, and the solution is extracted for 4 hr with 50:50 disopropylether/*n*-hexane. In some cases, 6 hr is required for complete extraction. The average equivalent weight of the monosulfonates in the diisopropyl ether/hexane extract may be determined by quantitatively neutralizing a portion to phenolphthalein with 0.5 M NaOH solution, evaporating the solvent and any unsulfonated oil, and weighing the sodium salts.

The acidic ethanol/water phase, besides disulfonates, will also contain any sulfatosulfonates which were present. These may be determined by boiling the residue with HCl, then again subjecting to continuous extraction. Any sulfatosulfonates will be converted to hydroxyalkanesulfonates and extracted into the isopropylether/hexane phase, which can be titrated and weighed as above. The ethanol/water phase contains the disulfonates, as well as sulfate ion present in the original sample and sulfate produced from hydrolysis of sulfatosulfonates. Quantitative evaluation of this fraction can be made by steaming off excess HCl, subjecting to cation exchange to remove sodium, boiling further to remove chloride, then neutralizing to phenolphthalein and weighing.

E. Alkanesulfonates

Commercial alkanesulfonates consist predominantly of monosulfonates and also contain di- and polysulfonates (typically 10% of alkanesulfonates made by sulfoxidation) and sodium sulfate (3–5% of the total for the sulfoxidation process; less for sulfochlorination products).

ISO standard 893 covers analysis of alkanesulfonates, specifying assay by gravimetry, alkane monosulfonates by titration or extraction (Sections 2(a) and 2(c), below), neutral oil by petroleum ether extraction of an isopropanol/water solution, pH of a 5% aqueous solution, water by Karl Fischer titration or azeotropic distillation, sulfate by titration with lead nitrate, sulfite by iodometric titration, and chloride by potentiometric titration (99). The value for neutral oil can be checked for losses due to evaporation by confirming that the weight of the sulfonates and other salts in the 2-propanol/water phase plus the water content and weight of neutral oil is equal to 100%.

1. Total Alkanesulfonates

(a) Gravimetric method.   A straightforward analysis of the commercial product can be performed gravimetrically, presuming that the only compounds present are the alkanesulfonates, sodium sulfate, sodium chloride, and unsulfonated paraffin. The product is taken up in a solvent in which sodium sulfate is insoluble (additional sodium sulfate is added to aid in the formation of easily-filtered crystals). Paraffin is simply driven off by heating, and sodium chloride is determined by titration.

Procedure (100,101)

The sodium alkanesulfonate mixture (0.5–1.0 g) is slurried in 15 mL of warm aqueous Na_2SO_4 solution (200 g/L). 150 mL 60:40 1-butanol/acetone is then added, slowly and without heating. The solution will first become clear, then cloudy as sodium sulfate decahydrate begins to crystallize. After cooling and filtering off the sodium sulfate, the solution is evaporated to dryness at steam bath temperature, treated three times with acetone, evaporating the acetone off each time, then finally heated in a vacuum oven at 120°C to remove the last traces of moisture and paraffin. The final weight is corrected for the sodium chloride content.

(b) *Titration.* Two-phase titration using methylene blue indicator, as described in Chapter 16, will determine total alkanesulfonate (102). Only one sulfonate group is titrated with di- and polysulfonates. Unlike the standard method for two-phase titration, use of methylene blue requires carefully matching the colors of the aqueous and organic phases for end point detection, leading to relatively high standard deviation. Alternatively, monosulfonate can be determined according to the titration procedure described below, under "Alkanemonosulfonate Content," and a factor can be used to calculate the total sulfonate. The factor is expected to remain constant for the product of a given production facility and is typically 1.11 for sulfoxidation products (i.e., 90:10 ratio of mono- to higher sulfonates) (102).

Potentiometric one-phase titration under most conditions will give higher values than two-phase titration since each sulfonate group of disulfonates and polysulfonates is titrated. Titration in 0.5 M H_2SO_4 will give two breaks, the first of which corresponds to the monosulfonate. However, the value obtained is not exactly the same as that obtained from two-phase titration, so an agreement between buyer and seller is needed as to which analytical method should be used (82).

2. Alkanemonosulfonate Content

(a) *Titration.* If the usual two-phase titration is applied to this product, the monosulfonates are titrated quantitatively, but the di- and polysulfonates are incompletely titrated. This problem is avoided if sodium sulfate is added to the titration vessel. Under these conditions, only the monosulfonate is titrated (100,102). The ISO method specifies addition of 10 mL of a 200 g/L sodium sulfate solution to the titration vessel containing 20 mL sample and 10 mL indicator solution (103).

(b) *Ion exchange method*

Procedure (104)

The anion exchange column from Section 3(a), below, is eluted with 500 mL 0.3 N NH_4HCO_3 in 60:40 PrOH/H_2O. The eluate is evaporated on a water bath until free of NH_4HCO_3, then dissolved in 50 mL 50:50 EtOH/H_2O and rinsed through the cation exchange column with 75 mL EtOH. The eluate from the cation exchanger is evaporated down to 50 mL and titrated just to the bromcresol purple end point with 0.1 M KOH. The residue is dried and weighed to give the amount and average equivalent weight of the monosulfonate.

(c) *Liquid-liquid extraction.* Liquid-liquid extraction may be used to separate the monosulfonates from di- and polysulfonates, as well as sulfate salts. The monosulfonates are extracted into petroleum ether, while the more polar compounds remain in the ethanol/water phase. A continuous extraction device is required because of unfavorable distribution coefficients (105,106).

Procedure (106)

The surfactant should have been first made free of unsulfonated material by extraction with pentane, although a trace of paraffin is permissible. The surfactant is evaporated free of solvent, then a 0.7 g sample is dissolved in 50 mL 50:50 EtOH/H_2O and 20 mL concentrated HCl. A continuous extraction device is used to extract the solution with petroleum ether for 5 hr. The petroleum ether extract is evaporated to dryness, redissolved in EtOH, passed through a cation exchange column in the acid form to remove traces of alkali, titrated to the bromophenol blue end point with standard base, then evaporated again to dryness and weighed as the sodium salt. The amount and the equivalent weight of the monosulfonate are calculated from these values.

(d)   Column chromatography

Procedure (107)

A glass column, 1.2 × 25 cm, is filled to a height of 25 cm with Amberlite XAD-2 resin and washed with H_2O and MeOH. A sample is weighed out and applied to the top of the column, which is eluted with water to remove H_2SO_4 and alkanedisulfonic acid. (The column eluate may be monitored by acid-base titration.) The monosulfonic acid is then eluted with MeOH.

3.   Di- and Polysulfonates

Paraffin sulfonate of higher degree of sulfonation is usually determined by difference, based on values of total and monosulfonate determined by titration or other methods described above. Following are other procedures which may be used for confirmation.

(a)   Ion exchange

Procedure (104)

The EtOH-soluble portion of a sample containing 2 meq sulfonate, dissolved in 50–100 mL EtOH, is applied to an EtOH-washed cation and anion exchange column connected in series. The cation exchanger is the strongly acidic Bio-Rad AG 50W-X8, H^+ form, 17 × 100 mm, and the anion exchanger is the weakly basic DEAE Sephadex A25, OH^- form. The sample is followed by 90 mL EtOH, then the columns are separated. The anion exchange column is rinsed with 50 mL H_2O, then eluted with 1 N NH_4HCO_3 solution in H_2O or $MeOH/H_2O$. (For dodecanesulfonate, 100 mL H_2O; for hexadecanesulfonate, 500 mL H_2O; for octadecanesulfonate, 500 mL 70:30 $H_2O/MeOH$; for mixed paraffin sulfonates, 350 mL H_2O.) The eluate is evaporated until free of NH_4HCO_3, dissolved in 50 mL 1:1 $EtOH/H_2O$, and passed through the cation exchange column. The column is washed with 75 mL EtOH, and the combined eluate is evaporated until free of HCl. The residue is dissolved in 50 mL EtOH and titrated with alkali to the bromcresol purple end point. The solution is then evaporated to dryness and weighed to give the amount and average equivalent weight of the combined di- and polysulfonates.

(b)   Liquid-liquid extraction.   After the extraction of monosulfonates described in Section 2(c), the disulfonate fraction is put into the acid form, then quantitatively titrated with alkali.

Procedure (105)

The $EtOH/H_2O$ mixture from the monosulfonate determination (Section 2.c), which has been extracted free of unsulfonated and monosulfonated material, is freed of excess HCl by successive evaporations and additions of H_2O. The residue is dissolved in 50 mL H_2O and passed through a cation exchange column (in the H^+ form). The column is rinsed with an additional 50 mL H_2O, and the eluate evaporated to dryness. The evaporation is repeated three times, with 15 mL H_2O added each time, to ensure the removal of HCl. The residue is dissolved in H_2O and titrated to the phenolphthalein end point with standard NaOH solution. This solution is evaporated to dryness and weighed. This titration and weighing measures the sum of Na_2SO_4 and of disulfonated paraffin. The values are corrected for the amount of Na_2SO_4 determined by titration, then calculations are performed to obtain the percent and equivalent weight of disulfonate.

(c) TLC. Thin layer chromatography, described in more detail in Chapter 9, can give the relative amounts of mono-, di-, and polysulfonates (105). A suitable system consists of plates of Merck Silica Gel GF_{254}, developed with 70:30 *n*-propanol/ammonium hydroxide for approximately 2 hr to a height of 10 cm. Visualization is with 4,5-dichlorofluorescein, 2.5% in ethanol, and visible or UV light. Rf values are 0.32 for polysulfonates, 0.36 for disulfonates, and 0.55 for monosulfonates.

4. Alkyl Chain Distribution

(a) HPLC. The alkyl chain distribution can be determined by gradient elution HPLC using indirect photometric detection. Use of an RP-8 column will give sharp peaks, each of which represents the sum of all isomers of a specific alkyl chain length. A C_{18} column will give multiple peaks for each chain length, corresponding to isomers with varying locations of the sulfonate group. Disulfonates elute with the dead volume.

Procedure (108,109)

Procedure A: Separation by alkyl chain length. A column of LiChrosorb RP-8, 4.6×200 mm is used, thermostatted at 40°C. A mobile phase gradient is used, where solvent A is 30:70 $MeOH/H_2O$ and solvent B is 80:20 $MeOH/H_2O$, each 0.25 mmolar in *N*-methylpyridinium chloride. The gradient is linear, from 25 to 95% B in 20 min. Differential UV detection at 260 nm is suitable.

Procedure B: Separation by alkyl chain length and by isomer composition. Two columns in series of Shandon Hypersil ODS are used, 4.6×200 mm, thermostatted at 40°C. Solvent A is 30:80 $MeOH/H_2O$; solvent B is 80:20 $MeOH/H_2O$, each 0.25 mmolar in *N*-methylpyridinium chloride. Linear gradient, 40 to 100% B in 40 min. Differential UV detection at 260 nm.

(b) GC. Esterification with diazomethane prepares derivatives of sufficient volatility for analysis. In general, only the monosulfonate can be separated well according to chain length. Paraffin sulfonates may also be desulfonated, with the resulting olefins analyzed directly or, for greater precision, after hydrogenation to the original paraffins. More information on GC analysis is given in Chapter 8.

F. Petroleum Sulfonates

Because their solubility properties differ from those of other anionic surfactants, petroleum sulfonates are analyzed by specialized methods. ASTM standard D3712 applies to characterization of petroleum sulfonates (110). Alkylnaphthalene sulfonates are generally characterized according to the same procedures given below for petroleum sulfonates.

ISO standard 1104 applies to the specification analysis of alkylnaphthalene sulfonates (8). This and most other standard methods referenced in this section specify assay by gravimetric analysis after removal of unsulfonated material and correction for other impurities.

1. Qualitative Characterization

Unlike many other sulfonates, alkali and alkaline earth metal salts of natural and synthetic petroleum sulfonates are soluble in chloroform. The IR spectra of the petroleum sulfonates are very similar to those of LAS and alkylnaphthalenesulfonates. GC or GC-MS of the unsulfonated oil or of methyl esters of the sulfonates is usually conclusive, since petroleum sulfonates have molecular weights at least 100 units higher than LAS. Some products are formulated with oil, so caution should be used in assuming that the unsulfonated hydrocarbon was the feedstock for the sulfonated portion of the sample (95).

2.   Unsulfonated Oil

Oil determination is performed either by HPLC (see Section II of this Chapter, as well as chapter 7) or by a physical separation and gravimetry, as described below. HPLC is most suitable for routine analysis. Under the proper low resolution conditions, sharp peaks can be obtained for inorganic salts, sulfonates, and oil, respectively (15).

3.   Separation of Sulfonates from Unsulfonated Oil

Petroleum sulfonates contain much more unsulfonated material than other surfactants, often 25–50%. The liquid-liquid extraction methods suitable for removal of oil from other anionics give less than quantitative results for petroleum sulfonates because of emulsion formation. While extraction is suitable for special cases, column chromatography separation of the oil is universally applicable. It is prudent to check the effectiveness of the separation by inspecting the IR spectrum of the oil fraction for the absence of sulfonate absorption at 1050 and 1200 cm^{-1} (95). There is no similar method to quickly check the purity of the sulfonate fraction, although an HPLC procedure could presumably be developed.

(a)   Column chromatography.   The separation of sulfonated from unsulfonated material can be performed by either ion exchange or by adsorption on silica, with equivalent results (95). The adsorption procedure is more widely used.

(1)   *Adsorption chromatography.*   The ASTM procedure given below is widely used. Methanol is sometimes substituted for ethanol to elute the sulfonate (111). It has been reported that methylene chloride is more universally applicable for elution of oil than is chloroform and that a 50:50 mixture of methylene chloride/ethanol is sometimes required for elution of the anionic (95).

Although sodium sulfate may be eluted by a final wash with water, this fraction will be contaminated with silica gel adsorbant. It is preferred to determine sulfate on the original sample by one of the procedures in Section II of this chapter.

Procedure (110)

Conversion of salts to sodium salts.   If the product is a calcium, barium, magnesium, or ammonium petroleum sulfonate, it must first be converted to the sodium salt: Dissolve a weighed quantity of sample, about 10 g, in 50 mL ethyl ether. Add 100 mL 6 M HCl. If BaCl$_2$ precipitates from a barium petroleum sulfonate solution, add enough water to redissolve. Mix and let separate in a 500-mL separatory funnel, then transfer the aqueous phase to a 250-mL separatory funnel. Wash the aqueous phase with three 50-mL portions of ether, combining the extracts in the 500-mL separatory funnel. Wash the ether extract with 50 mL 3 M HCl, combining the aqueous phases and again washing them with 50 mL ethyl ether. Wash the combined ether extracts in the separatory funnel with 50-mL portions of 1.7 M Na$_2$SO$_4$ solution until the washings are neutral to methyl orange. Dry the ether by shaking carefully with about 10 g Na$_2$SO$_4$ and filter. Evaporate to about 10 mL volume, then add about 50 mL isopropanol (containing less than 1% H$_2$O). Neutralize to phenolphthalein by titration with 0.1 M NaOH solution and evaporate to dryness. Dissolve the residue of sodium sulfonate in chloroform and dilute to volume in a 100-mL volumetric flask.

Chromatography. Petroleum sulfonate is separated from unsulfonated oil by chromatography on 15 g silica gel, Davison grade 62, packed into a glass column, 22×300 mm. Ten to 25 mL of a $CHCl_3$ solution containing about 2 g of the sodium salt and oil is placed on the column and the oil is eluted with 210 mL $CHCl_3$, followed by 10 mL denatured EtOH. The sulfonate itself is eluted with 250 mL denatured EtOH. For petroleum sulfonates of molecular weight above 500, hot pyridine is used instead of EtOH (112). The amount of sodium petroleum sulfonate and oil is determined by evaporating the eluates to dryness and weighing. The oil fraction should receive its final evaporation in a vacuum desiccator to avoid loss of light fractions. If the product was diluted with a light distillate, it will not be possible to obtain a constant weight for the unsulfonated oil by this test.

(2) *Ion-exchange chromatography.* The sample is prepared as a 2% solution in ethanol, adding a little water if necessary to eliminate turbidity. Cation and anion exchange columns of 26 × 260 mm are used, in the H^+ and OH^- form, respectively. The 4-g sample is passed through the cation exchanger to put the surfactant into the acid form. The eluate is passed through the anion exchanger, where the surfactant is retained. The mineral oil is recovered from the eluate. The anionic and any sulfate ion should be completely recoverable by elution with ethanolic HCl, although this was not pursued (95).

(b) *Liquid-liquid extraction.* The applicability of extraction is a function of the solubility and emusifying property of the product to be characterized. The ISO standard for alkylnaphthalene sulfonates calls for analysis by petroleum ether extraction from 10:4 isopropanol/water (8).

Procedure (113)

A 50% solution of the surfactant is prepared in 2-PrOH/H_2O, then extracted several times with equal volumes of pentane until no more material is removed by the pentane. The 2-PrOH/H_2O is removed by evaporation, and the residue is taken up in hot absolute EtOH, leaving an insoluble residue of salts. The EtOH extract is evaporated and any remaining oil is removed by extracting the residue with boiling pentane.

4. Equivalent and Molecular Weight

(a) *Equivalent weight*

(1) *Titration method.* Average equivalent weight is determined by simple acid-base titration of the purified sulfonate, after conversion to the free acid by passage through an ion exchange column (95,113).

Procedure (113)

About 0.5 g of deoiled and desalted sulfonate from above is dissolved in 50:50 2-PrOH/H_2O and eluted through a column of 50 g cation exchange resin in the acid form (Fisher Rexyn 101 strong acid sulfonated polystyrene copolymer resin). The column is rinsed with three bed volumes of 2-PrOH/H_2O. The combined column effluent is titrated with 0.1 M NaOH to a pH of 8.5. The average equivalent weight of the petroleum sulfonate is equal to the number of milligrams sample divided by the milliequivalents titrant required.

(2) *Ashing method*

Procedure (110)

Burn the purified sodium petroleum sulfonate fraction in a platinum dish. Add a few drops concentrated H_2SO_4 and heat again until fuming ceases, then heat to 800°C in an oven, cool, and weigh the Na_2SO_4 residue. The equivalent weight is equal to 71 times the ratio of the sodium petroleum sulfonate weight to the sulfated ash weight.

(b) Molecular weight. Determination of average molecular weight is performed by vapor pressure osmometry. The sulfonates are converted to their methyl esters before analysis to prevent micelle formation. Analysis is then performed by conventional osmometry methods (114). Molecular weight determination has also been reported using size exclusion chromatography, using THF as solvent (115).

5. Fractionation According to Equivalent Weight

Procedure: Liquid-Liquid Extraction (113)

Fifty grams deoiled, desalted surfactant is dissolved in 750 mL 50:50 2-PrOH/H_2O and extracted serially with 300, 100, 100, and 200 mL benzene and 100, 100, and 200 mL $CHCl_3$. The 2-PrOH/H_2O phase is then evaporated to dryness, and the residue is redissolved in H_2O-saturated 2-BuOH (about 250 mL H_2O and 500 mL 2-BuOH). This solution is extracted with 100 mL H_2O, then with smaller volumes of H_2O, until no more residue is found in the evaporated extracts. Solvent is stripped from all fractions, and their mass and equivalent weight determined.

Alternatively, fractionation of petroleum sulfonates according to equivalent weight may be made in a 10-stage Craig countercurrent extraction apparatus (a series of 100-mL centrifuge tubes may also be used). The solvents are prepared from equal volumes of water, isopropanol, and chloroform. Sixty milliliters of the chloroform phase is added to each stage or tube. Twenty-five milliliters aqueous phase solvent is added to the first stage, equilibrated, and transferred to the second stage, with fresh aqueous phase then added to stage 1. The process is continued until all stages contain two solvent phases. All 20 fractions are worked up individually.

6. Mono-, Di-, and Polysulfonated Components

This analysis is performed by HPLC of the deoiled product as described in Chapter 7. Anion exchange chromatography is almost always the method of choice.

7. Alkalinity

Alkaline additives are used with petroleum sulfonates in lubricating oils. These can be determined by nonaqueous potentiometric titration with perchloric acid (116). Additives such as colloidal calcium carbonate may also be determined by infrared spectroscopy, based on the carbonate bands (117).

G. Lignin Sulfonate

Lignosulfonates are polymers containing 4-hydroxypropyl-2-methoxyphenol and 4-hydroxypropyl-2,6-dimethoxyphenol repeating units, approximately half of which have undergone alkyl chain sulfonation. The repeating unit is taken as about 223 daltons.

1. Molecular Weight Distribution

The maximum molecular weight of a commercial lignosulfonate was reported to be 144,000, with M_w and M_n values of 16,700 and 2,700, respectively. Dissolved lignosulfonates found in paper mill effluents were lower in molecular weight, with maximum, M_w, and M_n values of about 20,000, 8,400, and 3,000, respectively (118). Because these compounds are so disperse in molecular weight, it is difficult to reproducibly calculate average values. It is usually more valuable to simply compare the SEC curves visually and qualitatively (119).

SEC of lignin sulfonates is discussed in Chapter 7. Aqueous SEC is usually applied, but is fraught with problems. SEC with organic eluents is more straightforward and can be applied if the ion pair of the lignin sulfonate with a quaternary amine is extracted from the aqueous solvent by ethyl acetate, evaporated down, and then redissolved in THF (119).

2. Methoxyphenol/Dimethoxyphenol Ratio

Pyrolysis GC-MS may be used to determine this ratio, which is related to the type of wood from which the compounds originated. All compounds judged to be pyrolysis products of 2-methoxyphenol and 2,6-dimethoxyphenol structural units are summed and the ratio determined (118).

H. Ester Sulfonates

The main components of these products are the nominal active agent, α-sulfo fatty acid methyl ester, and the chief impurity, unesterified α-sulfo fatty acid. The concentration of the latter, generally present as the disodium salt, is increased by hydrolysis over the lifetime of the product. The two components are most easily differentiated by chromatography, although a combination of extraction and titration may also be used.

1. Total Sulfonated Material

The total of α-sulfo fatty acid methyl ester and α-sulfo fatty acid can be determined by the standard two-phase titration method (Chapter 16) with benzethonium chloride titrant and either methylene blue or mixed indicator. The titration is carried out at acid pH so that the carboxylic acid groups of unesterified material and free fatty acid are not measured (120,121).

2. α-Sulfo Fatty Acid and α-Sulfo Fatty Acid Ester

(a)   Titration methods

Titration at alkaline pH.   The two-phase titration is performed twice at pH 9, using bromcresol green and then phenol red to indicate the titration end point. With bromcresol green, only the α-sulfo fatty acid methyl ester is determined. With phenol red, the total of sulfonate and carboxylate groups is determined. The difference between these values and that for total sulfonate (determined by titration to the methylene blue end point) yields the carboxylate content. If soap is present, its concentration must be subtracted to determine α-sulfo fatty acid (120,121).

Selective extraction and titration.   The surfactant mixture is freeze-dried, mixed with sand, and subjected to Soxhlet extraction with isopropanol. The ester is soluble in isopropanol, while the acid salt is not. The surfactant concentration of each fraction is determined by two-phase titration (122).

(b) TLC analysis.   α-Sulfo fatty acid ester and α-sulfo fatty acid may be determined by thin layer chromatography on silica gel 60. Plates are developed with 2:11:27 0.05 M H_2SO_4/methanol/chloroform and the spots visualized with a spray of 1% aqueous pinacryptol yellow. The fluorescence is measured at 254 and 366 nm (122).

(c) Ion chromatography analysis.   Use of "mobile phase ion chromatography" with an acetonitrile solvent gradient allows differentiation of the alkyl chain length distribution simultaneously with the determination of the α-sulfo fatty acid ester and α-sulfo fatty acid components. A poly(styrene/divinylbenzene) column is used with a mobile phase gradient formed from acetonitrile and 0.005 M aqueous ammonium hydroxide. Conductivity detection is suitable, with a membrane suppressor regenerated with 0.005 M H_2SO_4 (122).

3. Alkyl Chain Distribution

Pyrolysis GC at 400°C in the presence of P_2O_5 results in desulfonation, yielding free fatty acids and methyl esters. The resulting chromatogram shows peaks for acid and ester for each alkyl chain length. A 30 m DB 5 silica capillary is used, with FID detection (122).

4. Unsulfonated Fatty Acid and Fatty Acid Ester

An ethanol/water solution of the sample is extracted with petroleum ether at pH 3 and the residue weighed. This fraction corresponds to the unsulfonated fatty acid and fatty acid ester. The extract is titrated with alkali to determine the fatty acid content, and the ester is determined by difference (120).

5. Sodium Methyl Sulfate

Methyl sulfate ion is most easily determined by ion chromatography. Titration is applicable but complicated: The sample is saponified in ethylene glycol solvent, converting sodium methyl sulfate to methanol and sodium sulfate. Sodium methyl sulfate concentration is calculated from the alkali consumption. Since some of the active agent will also be saponified under these conditions, the saponification number must be corrected for the amount of the active agent saponified. This is determined by performing the two-phase titration with benzethonium chloride at pH 9 before and after saponification. This will indicate the increase in carboxylate concentration due to saponification of the active agent. This value is subtracted from the saponification number to give the amount of methyl sulfate (120).

I. Phosphate Esters

These products are characterized by using the procedures (appropriately modified) for characterization of the parent compound. For example, phosphate esters of ethoxylated alcohols can be analyzed for EO content, alkyl chain length, and free alcohol by the methods described in Chapter 2 for analysis of alcohol ethoxylates.

In general, phosphate esters will respond poorly if at all to the paired-ion procedures used to determine sulfate and sulfonate anionic surfactants by titration or spectrophotometry. This is presumably due to their hydrophilicity. Some success has been reported with individual products of higher lipophile character, such as didodecyl phosphate (123).

1. Mono- and Diester Content

The mono- and diester content is most accurately obtained by ^{31}P NMR or by HPLC, as discussed in Chapters 14 and 7, respectively. For quality control purposes, an approximate value for mono- and diester, as well as excess phosphate ion, can be obtained by potentiometric titration in a suitable solvent.

Procedure: Analysis of a Phosphate Ester by Titration (124)

Dissolve 1–2 g accurately weighed sample in 65:35 EtOH/H$_2$O and titrate potentiometrically, using a glass combination pH electrode, with 0.5 M KOH solution. There will be two inflection points in the titration curve. Another aliquot of sample (preferably of identical weight) is dissolved and titrated past the first inflection point. An excess of silver nitrate solution is added to precipitate free phosphate ion as Ag$_3$PO$_4$. The titration is then continued to the inflection point of the released HNO$_3$. This "third" inflection point is at a higher titrant volume than was the second end point of the previous titration.

The monoester contributes to the first and the second potentiometric breaks. The diester contributes only to the first. The phosphate ion contributes to all three. By appropriate calculation, the percentage of each may be determined. The values are usually only approximate because of other equilibria. In particular, if the free acidity is neutralized during manufacture, there will be a profound effect on the above titration.

2.   Characterization of Alcohol

The average composition of the hydrophobe is generally determined by NMR spectroscopy. Common phosphate esters cannot be analyzed directly or as derivatives by gas chromatography because they are not volatile below their decomposition temperature. While HPLC seems appropriate for direct analysis of phosphates, very few applications have been published, and these relate only to determination of relative content of monoester and diester (see Chapter 7). Supercritical fluid chromatography of the methyl ester derivatives has been demonstrated to show the homolog distribution of phosphate esters of ethoxylated alcohols (125).

Gas chromatographic characterization of phosphates is possible. The procedure of Tsuji and Konishi consists of decomposition of the compound with acetic and paratoluenesulfonic anhydrides, resulting in formation of the volatile acetate esters of the starting alcohols (73). These are readily characterized by gas chromatography. In another method, the ester is hydrolyzed by digestion with methanol, under neutral conditions, for 8 hr at 200°C in a steel pressure vessel. The reaction products consist both of the starting alcohol (or ethoxylated alcohol) and the methyl ethers (126). The chain length distribution may then be determined by gas chromatography, as described in Chapter 8.

3.   Free Alcohol

A sample is dissolved in ethanol or ethanol/water, neutralized with aqueous triethanolamine, and extracted with petroleum ether. The petroleum ether solution is evaporated to dryness at low temperature and the residue of free alcohol is weighed (124).

J.   Sulfosuccinate Esters

Sulfosuccinates are available as either monoesters or diesters. In order to have the desirable surfactant properties, monoesters tend to be made with alcohols of longer alkyl chain than the diesters. It is difficult to generalize about the analytical behavior of monoesters because these depend on the nature of the alkyl chain.

Because their surfactant properties are destroyed by saponification, these materials can be determined in mixtures by titration with a cationic surfactant before and after alkaline hydrolysis (127).

1. Assay

(a) Diesters. In the case of the most common alkyl sulfosuccinate, *bis*(2-ethyl-hexyl)sulfosuccinate, the diester content is determined by the standard two-phase titration method for anionic surfactants under either acid or alkaline conditions. The relatively hydrophilic monoester, if present as an impurity, is not titrated at pH 1 (30). Potentiometric titration with a surfactant-selective electrode is commonly used for assay of *bis*(2-ethyl-hexyl)sulfosuccinate. Monester will also be titrated, but can be differentiated by performing the titration at acid and alkaline pH, as discussed in the next paragraph. Gravimetric analysis using *p*-toluidine is also suitable for assay of the diester, although this approach is rarely used nowadays (128). Monoesters are too hydrophilic to form a precipitate with *p*-toluidine hydrochloride.

(b) Monoesters. Most monosuccinates are not hydrophobic enough to be quantitatively determined by two-phase titration with indicators. Even in the case of those that can be determined, end points are not sharp and the titration is therefore difficult. However, the one-phase potentiometric version of titration with a cationic surfactant is successful with monoesters, including the monoesters of alcohol ethoxylates. At pH 1–3, the sulfonate is titrated. At pH 7–10, the titration value is twice as high, corresponding to titration of both the sulfonate and carboxylate groups. At pH 5, no titration is observed. pH adjustment should be made immediately prior to titration to minimize hydrolysis of the ester group (82,129).

2. Nonionic Impurities

Sulfosuccinate monoesters may be produced from an ethoxylated alcohol or ethoxylated alkanolamide. In such cases, possible low level impurities in the final product are free alcohol, alkanolamide, ethoxylated alcohol, ethoxylated alkanolamide, and PEG.

(a) Ion exchange. The surfactant sample is dried, dissolved in 95% ethanol (0.3 g/100 mL) and passed through a Dowex 50W-X4 cation exchange column, acid form, 2 × 10 cm, and a Dowex 1-X2 anion exchange column (in the OH⁻ form). The columns are rinsed with an additional 100 mL ethanol. The eluate is evaporated free of solvent and the residue weighed and identified by infrared spectrometry (128).

(b) Extraction. Nonionic impurities can be quantitatively extracted from the dried solids with carbon tetrachloride and identified by infrared spectrometry (128). For sound ecological reasons very few laboratories permit use of carbon tetrachloride; presumably another nonpolar solvent may be substituted.

3. Identification of Alcohol

NMR spectroscopy is in most cases suitable for determination of the composition of sulfosuccinates of a single alcohol. The alcohol portion of the ester may also be identified by the classical technique of saponifying the ester and separating the alcohol (128). The sample is refluxed with 2 N KOH solution in ethanol. The resulting fatty alcohol is extracted with ethyl ether from ethanol/water solution and determined by gas chromatography. In the case of esters where the alcohol is an alkanolamide, the remaining solution may be further treated by evaporating the ethanol and boiling with hydrochloric acid. The amide bond is cleaved, allowing the fatty acids to be extracted into ethyl ether and analyzed by gas chromatography, either directly or as the methyl esters.

4. Ester Number

A measure of the purity of either the mono- or diester can be made by saponifying the ester by reflux with excess alkali in a suitable solvent and determining the alkali consumed by the reaction. By convention, ester number is equivalent to the saponification number corrected for the acid number. The conventional units are milligrams KOH required to saponify 1 g sample.

K. Isethionate Esters

1. Assay

Assay by two-phase or potentiometric titration is made difficult by the tendency of the surfactant to hydrolyze and by the presence of free fatty acid in the 10% range. Best results are obtained by dissolving the sample by warming in a minimum quantity of methanol/water (5 mL each), then adding 80 mL cold water and 2 mL 0.1 M HCl so that fatty acids precipitate out. Potentiometric titration then proceeds successfully, although with a rather flat curve. A more concentrated titrant should be used, 0.01–0.05 M rather than the usual 0.004 M (82). Since most commercial products are based on coconut oil, TEGOtrant® gives more quantitative results than Hyamine 1622 for the determination because of the 7% or so of the product with the more soluble C_8 alkyl chain length (82). It is likely that potentiometric titration at alkaline pH will give the sum of the isethionate and fatty acid.

2. Sodium Isethionate Content

The presence of unreacted sodium isethionate can be detected from the NMR spectrum (130). Quantitative analysis is usually performed by ion chromatography. Nonsuppressed ion chromatography is suitable for analysis of isethionate esters, using a Vydac 302 IC column, 4.6 × 250 mm, using a mobile phase prepared from 150 mL methanol and 250 mL 0.020 M phthalic acid, diluted to 1 L with water (131). Conductivity detection is suitable. Separation of the isethionate from chloride and surfactant is excellent. Extended contact of the sample with the mobile phase is to be avoided, as this can cause hydrolysis of the ester to produce additional isethionate ion.

For samples other than pure esters, which may contain high levels of chloride ion, better results are obtained with suppressed ion chromatography using conventional Dionex columns and membrane suppressors (132).

3. Identification of Acyl Chain

The CTFA method for determining the acyl substituents calls for hydrolyzing the product with constant-boiling hydrochloric acid. Warming the acid solution overnight prior to boiling prevents excessive foaming. The fatty acids are extracted from acidic hydrolysis mixture (made 50:50 ethanol/water) with petroleum ether and analyzed by gas chromatography (133).

L. Acyl Taurates

Determination of the alkyl chain length distribution is readily performed by reversed-phase HPLC, as described in Chapter 7 (134). Taurates behave as strongly anionic surfactants and can be separated according to the ion exchange methods given in Chapter 6 (135). Taurates can be separated from other surfactants by TLC, as covered in Chapter 9 (136).

M. N-Acylated Amino Acids

Most published information deals with the analysis of sarcosine derivatives. Sarcosine is often made by reaction of chloroacetic acid with methylamine. N-acylated sarcosine is made by reaction of the acyl chloride with sarcosine in the presence of alkali. Possible impurities in the final product thus include glycolic acid and N-methylalkylamide. Fatty acid and chloride salts are expected as byproducts of the acylation reaction, and ethyl esters of fatty acids and acylsarcosine may be formed during workup of the synthesis product with ethanol (137).

1. Assay

These compounds can be determined by either acid-base titration as carboxylic acids or by ion-pair titration. In the case of ion-pair titration, sarcosinates, like other carboxylic acids, can be titrated by two-phase titration at high pH with a cationic titrant (137). In addition, their titration has been demonstrated at low pH either by two-phase titration with an anionic surfactant (138) or one-phase titration with tetraphenylborate using a surfactant-selective electrode (139). These procedures are described in Chapter 16. Both acid-base titration and titration with a cationic surfactant will measure free fatty acid along with the acylsarcosine content. While titration at low pH with anionic titrants is presumably not subject to interference by free fatty acid, this has not been confirmed in the literature.

2. Other Titration Procedures

Total anion content.   The product is dissolved in isopropanol and passed through a column of Dowex 50W-X8 cation exchange resin in the acid form. The eluate is titrated with tetrabutylammonium hydroxide solution to determine the sum of the active agent, free fatty acid, and chloride (140).

Total amide content.   Another aliquot of the eluate from the ion exchange column is titrated with perchloric acid in acetic anhydride solvent to determine the total amide content (140).

Free amine and amino acid.   An aliquot of the original sample is titrated with perchloric acid in acetic acid solvent, with addition of mercuric acetate. This titration determines the sum of free amine, free amino acid, and chloride concentrations (140). Chloride is easily determined separately by ion chromatography or by silver nitrate titration.

3. Characterization

Qualitative information is obtained by hydrolyzing the amides at 160–180°C for 4 hr with 6 M HCl (140). The resulting fatty acids and amino acids are separated by liquid-liquid extraction and characterized by GC, LC, and TLC techniques. Direct HPLC analysis without hydrolysis can be used for determination of the alkyl chain distribution (141).

4. GC Analysis

GC analysis of the methyl esters gives the alkyl chain length distribution of the acyl sarcosines and of free fatty acids. The sample is first acidified and extracted with n-hexane or petroleum ether to isolate the acylsarcosinate and soap in the form of the free carboxylic acids. About 1 g of the acids are reacted with 25 mL methanol and 2 drops of concentrated sulfuric hours for 2 hr under reflux to form the methyl esters. These are isolated by adding 75 mL saturated aqueous NaCl and extracting with 25 mL n-hexane. The hexane phase is washed with saturated NaCl until neutral and evaporated down. (If an emulsion forms, a few milliliters ethanol should clear it up.) GC analysis of the methyl esters on a 2 m Chromosorb W-AW-DMCS 10% SE-30 column at 300 or 320°C results in the elution according to increased alkyl chain length of first the fatty acids, then the acylsarcosine compounds (137).

5. Nonionic Impurities

Nonionic impurities include ethyl esters of the fatty acids and acylsarcosine resulting from workup of the reaction mass with ethanol. They also include the acylation products of any methylamine contaminating the sarcosine raw material.

    The total of all nonionic impurities can be determined by a batch ion exchange method. About 0.6 g sample is dissolved in 70 mL ethanol and stirred with 30 g anion exchange resin (OH⁻ form) and 20 g cation exchange resin (H⁺ form) for an hour. The filtrate from this treatment is evaporated down and weighed (137).

N. Ether Carboxylates

Ethoxy chain length distribution can be determined by HPLC or TLC, as described in Chapters 7 and 9.

1. Assay

Carboxylates in the acid form can be titrated potentiometrically with alkali in 97% methanol. Observed pK_a values range from 5.6 for a carboxylated nonylphenol 3-mole ethoxylate to 4.1 for a 12-mole ethoxylate (142). Thus, "acid number" is a measure of the purity of the carboxylate.

    Two-phase titration is applicable for determination of the carboxylates of alcohol ethoxylates and alkylphenol ethoxylates. The usual indicators do not give sharp end points; Safranine T is used instead with a borax pH 8.6 buffer. Poor results are obtained using electrochemical detection of the end point (142,143). Other anionics interfere (144).

2. Nonionic Material

Either HPLC or TLC analysis is suitable for determining noncarboxylated ethoxylate impurities, as described in Chapters 7 and 9. These impurities can also be determined on a preparative scale by passing an ethanolic solution through an anion exchange column; the nonionics pass through unretained (142). A simple hydroxyl number determination, as described for the analysis of nonionics in Chapter 2, is used for quality control determination of residual nonionic surfactant and PEG (145).

O. Soap

Knowledge of the source and content of the fatty acids is important, as well as content of glycerin and other impurities and additives (146). Very detailed test methods for soap and soap products are given in ASTM D460 (147) and AOCS Official Methods, Section D (148). There is a full suite of ISO methods under the control of Technical Committee 91. Additionally, analysis of soap-based detergents is covered in ASTM D820 (149).

    Perfluorinated alkylcarboxylic acids are analyzed in the same way as their hydrocarbon analogs. In general, the perfluorinated acids are more hydrophobic than the hydrocarbon acids of the same chain length, so the analytical methods are generally applicable to lower chain lengths of perfluorinated acids (150).

Soap can be determined by potentiometric titration with TEGOtrant®, provided the fatty acids are at least C_{12} in chain length. This approach can sometimes be used in formulation analysis (82).

1. Glycerin

In soap made by the traditional route of alkali saponification of triglycerides, glycerin is an impurity. In soaps made directly from fatty acids, glycerin may be deliberately added. Glycerin affects the texture and other properties of solid soap, and its concentration is closely controlled. Levels of 0.5–3.0% are typical. The spectrophotometric method is most suitable for low levels of glycerin, less than 0.5% (151).

(a) *Gas chromatography.* Ten grams soap is mixed in a blender with 200 mL *N,N*-dimethylformamide for 5 min. A portion of the mixture is filtered, and 0.5 mL filtrate is mixed with 0.25 mL *bis*(trimethylsilyl)trifluoroacetamide in a septum-sealed vial. One microliter injections are appropriate. The column is a 0.2 mm × 12 m methyl silicone coated silica capillary, held at 100°C, isothermal, except to bake out impurities. FID detection is suitable (152).

(b) *Liquid chromatography.* Five grams soap is mixed in a blender with 50 mL mobile phase for 10 min, then filtered—first with filter paper, then with a 0.45 µ membrane filter. The solution is mixed 50:50 with a solution of 1,2,4-butanetriol internal standard, 6.2 g/L, and injected in the HPLC. A Waters Carbohydrate Analysis column, 3.9 × 300 mm, is used with a mobile phase of 92.5:7.5 acetonitrile/water. Differential refractive index allows determination of glycerin as well as other alcohols. Calibration is made with solutions of glycerin and the internal standard (153).

(c) *Spectrophotometry.* Two to three grams soap is treated with H_2SO_4 and warmed on a water bath until a clear layer of fatty acids is seen. Water is added and the fatty acids are extracted away with petroleum ether. The aqueous phase is filtered to remove additives, if necessary, then diluted to 100 mL. An aliquot of 2 mL or less is transferred to another 100-mL flask to which is added 1 mL 0.03 M sodium periodate in 25 g/L H_2SO_4. The mixture is let stand for 15 min to convert glycerin to formaldehyde and formic acid. One milliliter of freshly prepared tin (II) chloride solution (3 g of the dihydrate and 3 mL concentrated HCl in 100 mL water) and 10 mL chromotropic acid (0.23 g 1,8-dihydroxy-naphthalene-3,6-disulfonate, disodium salt, dissolved in 10 mL water and diluted to 250 mL with 15 M H_2SO_4) are added. After mixing, the solution is heated on a water bath for 30 min, cooled, diluted to volume with 10 M H_2SO_4, and its absorbance is measured at 571 nm at constant temperature versus water. A calibration curve including a reagent blank is made corresponding to 10–50 µg glycerin in the final aliquot (151).

2. Alkyl Chain Length Distribution

Fatty acids are easily determined by gas chromatography, either high temperature GC of the free acids or lower temperature analysis of ester derivatives. All vendors of GC columns will provide application information. The alkyl chain length may also be determined by reversed-phase HPLC, as described in Chapter 7.

3. Chloride Ion

Potentiometric titration of chloride ion in soap is awkward because of the necessity to perform the analysis in acidic conditions, causing the precipitation of fatty acids. Molever recommends formation of the methyl esters of the fatty acids in a simple reaction performed in the titration vessel (154).

Procedure: Titration of Chloride in Bar Soap (154)

Three grams chopped soap are weighed into a titration beaker and treated with 40 mL 0.6 M methanolic H_2SO_4. The mixture is stirred for 10 min, then 40 mL water is added and the solution is titrated potentiometrically with 0.1 M $AgNO_3$ solution using a silver combination electrode. Calculation:

$$\% \text{ NaCl} = \frac{(V)(M)5.85}{W}$$

where V is the titrant volume, M is the molar concentration of the silver nitrate titrant, and W is the sample weight in grams.

REFERENCES

1. Schwuger, M. J., H. Lewandowski, α-Sulfomonocarboxylic esters, in H. W. Stache, ed., *Anionic Surfactants. Organic Chemistry*, Marcel Dekker, New York, 1996.
2. Schwuger, M. J., H. Lewandowski, α-Sulfomonocarboxylic esters, in K. Holmberg, ed., *Novel Surfactants*, Marcel Dekker, New York, 1998.
3. Behler, A., K. Hill, A. Kusch, S. Podubrin, H. Raths, G. Uphues, Nonionics as intermediates for ionic surfactants, in N. M. van Os, ed., *Nonionic Surfactants: Organic Chemistry*, Marcel Dekker, 1998.
4. American Society for Testing and Materials, Synthetic anionic ingredient by cationic titration, D3049-89. West Conshohocken, PA 19428.
5. International Organization for Standardization, Surface active agents—detergents—determination of anionic-active matter by manual or mechanical direct two-phase titration procedure, ISO 2271:1989. Geneva, Switzerland.
6. American Society for Testing and Materials, Sampling and chemical analysis of alkylbenzene sulfonates, D1568-97. West Conshohocken, PA 19428.
7. American Society for Testing and Materials, Sampling and chemical analysis of fatty alkyl sulfates, D1570-95. West Conshohocken, PA 19428.
8. International Organization for Standardization, Surface active agents—technical sodium alkylarylsulfonates (excluding benzene derivatives)—methods of analysis, ISO 1104:1977. Geneva, Switzerland.
9. American Society for Testing and Materials, Chemical analysis of alpha olefin sulfonates, D3673-89. West Conshohocken, PA 19428.
10. Kok, R., R. Riemersma, A. K. van Helden, Unreacted material in anionic surfactants by HPLC, *Proc. 3rd CESIO Intern. Surf. Cong.*, 1992, *Sect. E, F, & LCA*, 256–265.
11. Van Dijk, J. H., R. R. Lamping, C. Slagt, A. W. J. Raaijmakers, Nonionics in surfactants by high speed ion exchange chromatography, *Tenside*, 1975, *12*, 261–263.
12. Bear, G. R., Surfactant characterization by reversed-phase ion pair chromatography, *J. Chromatogr.*, 1986, *371*, 387–402.
13. Senden, W. A. A., R. Riemersma, Alkylaryl sulfonates by HPLC, *Tenside, Surfactants, Deterg.*, 1990, *27*, 46–51.
14. Yoshimura, H., T. Sugiyama, T. Nagai, HPLC analysis of polyethoxylated nonionics, *J. Am. Oil Chem. Soc.*, 1987, *64*, 550–555.
15. Bear, G. R., Universal detection and quantitation of surfactants by HPLC by means of the evaporative light-scattering detector, *J. Chromatogr.*, 1988, *459*, 91–107.
16. Cross, C. K., Neutral oil in detergent products by solid phase extraction, *J. Am. Oil Chem. Soc.*, 1990, *67*, 142–143.

17. Cohen, L., R. Vergara, A. Moreno, J. L. Berna, Free oil components in LAS, *Tenside, Surfactants, Deterg.*, 1995, *32*, 238–239.

18. Carasik, W., M. Mausner, G. Spiegelman, Detergent intermediates analysis, *Soap Chem. Spec.*, 1967, *(5)*, 106–112, 277.

19. International Organization for Standardization, Surface active agents—determination of mineral sulfate content—titrimetric method, ISO 6844:1983. Geneva, Switzerland.

20. Crabb, N. T., H. E. Persinger, Potentiometric titration of sulfate in some typical anionic sulfonate and sulfate surfactants, *J. Am. Oil Chem. Soc.*, 1967, *44*, 229–231.

21. Ranky, W. O., G. T. Battaglini, Alpha olefin sulfonates, *Soap Chem. Spec.*, 1968, *43(4)*, 36–39,78–86.

22. Hughes, J. P., Trace components in surfactants, in M. R. Porter, ed., *Recent Developments in the Analysis of Surfactants*, Elsevier, London, 1991.

23. Yamaguchi, S., Sodium sulfate in anionic surfactants by nonaqueous titration, *J. Am. Oil Chem. Soc.*, 1978, *55*, 673–674.

24. Yamaguchi, S., S. Nukui, M. Kubo, K. Konishi, Nonaqueous titrimetric analysis of sulfuric and alkylbenzene sulfonic acids in detergent intermediates, *J. Am. Oil Chem. Soc.*, 1978, *55*, 359–362.

25. American Society for Testing and Materials, Sulfonic and sulfuric acids in alkylbenzene sulfonic acids, D4711-89. West Conshohocken, PA 19428.

26. Sak-Bosnar, M., L. Zelenka, M. Budimir, Sulfuric and alkylbenzenesulfonic acids, *Tenside, Surfactants, Deterg.*, 1992, *29*, 289–291.

27. Müller, K., D. Noffz, Length and the degree of branching of the alkyl chains of surface-active alkylbenzene derivatives (in German), *Tenside*, 1965, *2*, 68–75.

28. Tavernier, S. M. F., R. Gijbels, Sulfate content of anionic surfactants: sodium di-2-ethylhexyl sulfosuccinate, *Talanta*, 1981, *28*, 221–224.

29. Ferm, D. J., T. S. Griffin, Stability of Kathon CG/ICP microbicide in α-olefin sulfonate based systems, *J. Am. Oil Chem. Soc.*, 1990, *67*, 116–122.

30. Arens, M., G. Krusche, G. Schneeweis, Organic surface active compounds XII (in German), *Fette, Seifen, Anstrichm.*, 1986, *88*, 9–13.

31. International Organization for Standardization, Surface active agents and detergents—determination of water content—Karl Fischer method, ISO 4317:1991. Geneva, Switzerland.

32. American Oil Chemists' Society, Moisture: Distillation method, Official Method Da 2b-42. 2211 West Bradley Ave., Champaign, IL 61826.

33. International Organization for Standardization, Surface active agents and soaps—determination of water content—azeotropic distillation method, ISO 4318:1989. Geneva, Switzerland.

34. American Society for Testing and Materials, pH of aqueous solutions of soaps and detergents, D1172-95. West Conshohocken, PA 19428.

35. International Organization for Standardization, Surface active agents—determination of pH of aqueous solutions—potentiometric method, ISO 4316:1977. Geneva, Switzerland.

36. International Organization for Standardization, Surface active agents—determination of free alkalinity or free acidity—titrimetric method, ISO 4314:1977. Geneva, Switzerland.

37. International Organization for Standardization, Surface active agents—determination of alkalinity—titrimetric method, ISO 4315:1977. Geneva, Switzerland.

38. American Society for Testing and Materials, Acidity in volatile solvents and chemical intermediates used in paint, varnish, lacquer, and related products, D1613-96. West Conshohocken, PA 19428.

39. American Society for Testing and Materials, Ash from petroleum products, D482-95. West Conshohocken, PA 19428.

40. American Society for Testing and Materials, Iron in water, D1068-96. West Conshohocken, PA 19428.

41. American Society for Testing and Materials, Iron in trace quantities using the 1,10-phenanthroline method, E394-94. West Conshohocken, PA 19428.

42. International Organization for Standardization, Binders for paints and varnishes—estimation of color of clear liquids by the Gardner color scale, ISO 4630:1997. Geneva, Switzerland.
43. American Society for Testing and Materials, Color of transparent liquids (Gardner color scale), D1544-80. West Conshohocken, PA 19428.
44. Deutsches Institut für Normung, Testing of fabric softeners: measurement of light transmission with the iodine color scale (in German), DIN 53403. Deutsches Informationszentrum für technische Regeln. Berlin: Beuth.
45. Deutsches Institut für Normung, Determination of iodine color number (in German), DIN 6162. Deutsches Informationszentrum für technische Regeln. Berlin: Beuth.
46. International Organization for Standardization, Clear liquids—estimation of color by the platinum-cobalt scale, ISO 6271:1997. Geneva, Switzerland.
47. Deutsches Institut für Normung, Testing of plasticizers and solvents. Determination of Hazen color (platinum cobalt color, APHA method) (in German), DIN 53409. Deutsches Informationszentrum für technische Regeln. Berlin: Beuth.
48. American Society for Testing and Materials, Color of clear liquids (platinum-cobalt scale), D1209-93. West Conshohocken, PA 19428.
49. Gohlke, F. J., Colorimetric characterization of optically clear colored liquids on X, Y, Z basis, *Tenside, Surfactants, Deterg.*, 1995, *32*, 340–346.
50. Hons, G., Alkylarylsulfonates: history, manufacture, analysis, and environmental properties, in H. W. Stache, ed., *Anionic Surfactants: Organic Chemistry*, Marcel Dekker, New York, 1996.
51. Matheson, K. L., Physical properties and behavior of LAS in mixtures with other surfactants in household detergent products, in H. W. Stache, ed., *Anionic Surfactants: Organic Chemistry*, Marcel Dekker, New York, 1996.
52. Cavalli, L., A. Landone, T. Pellizzon, Linear alkylbenzenes for detergency: characterization of the secondary components, *Comun. Jorn. Com. Esp. Deterg.*, 1988, *19*, 41–52.
53. International Organization for Standardization, Surface active agents—technical straight-chain sodium alkylbenzenesulfonates—determination of mean relative molecular mass by GC, ISO 6841:1988. Geneva, Switzerland.
54. Cohen, L., A, Moreno, J. L. Berna, Minor products in linear alkylbenzene sulfonation, *Tenside, Surfactants, Deterg.*, 1996, *33*, 441–446.
55. Moreno, A., J. Bravo, J. L. Berna, Influence of unsulfonated material and its sulfone content on the physical properties of LAS, *J. Am. Oil Chem. Soc.*, 1988, *65*, 1000–1006.
56. Bravo, J., A. Moreno, C. Bengoeches, Sulfones in free oil of LAS by HPLC, *Comun. Jorn. Com. Esp. Deterg.* (in Spanish), 1988, *19*, 359–367.
57. Takada, H., R. Ishiwatari, Linear alkylbenzenes in urban riverine environments in Tokyo: distribution, source, and behavior, *Environ. Sci. Technol.*, 1987, *21*, 875–883.
58. Trehy, M. L., W. E. Gledhill, R. G. Orth, LAS and dialkyltetralinsulfonates in water and sediment by GC/MS, *Anal. Chem.*, 1990, *62*, 2581–2586.
59. Field, J. A., J. A. Leenheer, K. A. Thorn, L. B. Barber, II, C. Rostad, D. L. Macalady, S. R. Daniel, Persistent anionic surfactant-derived chemicals in sewage effluent and groundwater, *J. Contam. Hydrol.*, 1992, *9*, 55–78.
60. International Organization for Standardization, Surface active agents—technical sodium primary alkylsulfates—methods of analysis, ISO 894:1977. Geneva, Switzerland.
61. International Organization for Standardization, Surface active agents—technical sodium secondary alkylsulfates—methods of analysis, ISO 895:1977. Geneva, Switzerland.
62. Meissner, C., H. Engelhardt, Surfactants derived from fatty alcohols—hydrolysis and enrichment, *Chromatographia*, 1999, *49*, 12–16.
63. Kahovcova, J., M. Ranny, Unbranched fatty alcohols in the presence of alkylbenzenes in the unsulfonated fraction of anionic surfactants by use of urea inclusion compounds (in German), *Tenside, Surfactants, Deterg.*, 1968, *5*, 83–86.

64. Czichocki, G., D. Vollhardt, H. Seibt, Preparation and characterization of interfacially-chemically pure sodium dodecylsulfate (in German), *Tenside, Surfactants, Deterg.*, 1981, *18*, 320–327.

65. Czichocki, G., P. Müller, D. Vollhardt, M. Krüger, Alcohols in sodium alkyl sulphate mixtures using HPLC and surface tension measurements, *J. Chromatogr.*, 1992, *604*, 213–218.

66. Vollhardt, D., G. Czichocki, Traces of alcohols in sodium alkylsulfates, *Tenside, Surfactants, Deterg.*, 1993, *30*, 349–355.

67. Czichocki, G., H. Much, D. Vollhardt, Dodecanol in sodium dodecyl sulphate in the nanogram range by HPLC with UV detection, *J. Chromatogr.*, 1983, *280*, 109–118.

68. Puschmann, H., Ether sulfates using silanized silica gel (in German), *Chem., Phys. Chem. Anwendungstech. Grenzflächenaktiven Stoffe, Ber. Int. Kongr., 6th*, 1972, *Vol. 1*, 397–406. Carl Hanser Verlag. Munich.

69. Wickbold, R., Sulfated alcohol ethoxylates (in German), *Tenside, Surfactants, Deterg.*, 1976, *13*, 181–184.

70. International Organization for Standardization, Surface active agents—sulfated ethoxylated alcohols and alkylphenols—determination of total active matter content, ISO 6842:1989. Geneva, Switzerland.

71. International Organization for Standardization, Surface active agents—sulfated ethoxylated alcohols and alkylphenols—estimation of the mean relative molecular mass, ISO 6843:1988. Geneva, Switzerland.

72. Benning, M., H. Locke, R. Ianniello, Identification of hydrophobes in normal alcohol ethoxylates by hydriodic acid cleavage and reversed-phase HPLC analysis with UV detection. *J. Liq. Chromatogr.*, 1989, *12*, 757–770.

73. Tsuji, K., K. Konishi, GC analysis of ester-type surfactants by using mixed anhydride reagent, *J. Am. Oil Chem. Soc.*, 1975, *52*, 106–109.

74. Lee, S., N. A. Puttnam, GC determination of chain length distribution in fatty acid ethanolamides, *J. Am. Oil Chem. Soc.*, 1965, *42*, 744.

75. Lee, S., N. A. Puttnam, GLC determination of chain length distribution in AE and sulfated derivatives, *J. Am. Oil Chem. Soc.*, 1966, *43*, 690.

76. Krusche, G., Unethoxylated alcohol in ethoxylates (in German), *Tenside, Surfactants, Deterg.*, 1975, *12*, 35–38.

77. International Organization for Standardization, Surface active agents—sulfated ethoxylated alcohols and alkylphenols—determination of content of unsulfated matter, ISO 8799:1988. Geneva, Switzerland.

78. Austad, T., I. Fjelde, Chromatographic analysis of commercial products of ethoxylated sulfonates. *Anal. Lett.*, 1992, *25*, 957–971.

79. Matheson, K. L., P. A. Schwab, Effect of ethylene oxide adduct distribution on the formation of 1,4-dioxane during sulfation of alcohol ethoxylates, *Proc. 3rd CESIO Intern. Surf. Cong.*, 1992, *Sect. E, F, & LCA*, 211–220.

80. Beranger, A., T. Holt, Middle and heavy α-olefin sulfonates, *Tenside, Surfactants, Deterg.*, 1986, *23*, 247–254.

81. Radici, P., L. Cavalli, C. Maraschin, Internal *n*-olefin sulfonates, *Comun. Jorn. Com. Esp. Deterg.*, 1992, *23*, 205–218.

82. Schulz, R., *Titration of Surfactants and Pharmaceuticals* (in German), Verlag für Chemische Industrie, Augsburg, 1996.

83. Griffin, E. H., E. W. Albaugh, α-Olefin-derived sodium sulfonates by ion exchange and potentiometric titrimetry, *Anal. Chem.*, 1966, *38*, 921–923.

84. Kuemmel, D. F., S. J. Liggett, Level and position of unsaturation in alpha olefin sulfonates, *J. Am. Oil Chem. Soc.*, 1972, *49*, 656–659.

85. McClure, J. D., Alkene sulfonates in olefin sulfonates by ozone titration, *J. Am. Oil Chem. Soc.*, 1978, *55*, 905–908.

86. Von Rudloff, E., Periodate-permanganate oxidations. IV: Determination of the position of double bonds in unsaturated fatty acids and esters, *J. Am. Oil Chem. Soc.*, 1956, *33*, 126–128.

87. Boyer, J. L., J. P. Canselier, V. Castro, Analysis of SO_3-sulfonation products of 1-alkenes by spectrometric methods, *J. Am. Oil Chem. Soc.*, 1982, *59*, 458–464.

88. Roberts, D. W., D. L. Williams, Formation of sultones in olefin sulfonation, *J. Am. Oil Chem. Soc.*, 1990, *67*, 1020–1027.

89. Martinsson, E., K. Nilsson, Sultones in α-olefin sulfonates, *Tenside, Surfactants, Deterg.*, 1974, *11*, 249–251.

90. Macmillan, W. D., H. V. Wright, Sultones in anionic surfactants, *J. Am. Oil Chem. Soc.*, 1977, *54*, 163–166.

91. Slagt, C., W. G. B. Huysmans, A. W. J. Raaijmakers, Sultones in α-olefin sulfonates by GC and HPLC, *Tenside, Surfactants, Deterg.*, 1976, *13*, 185–187.

92. Roberts, D. W., J. G. Lawrence, I. A. Fairweather, C. J. Clemett, C. D. Saul, Possibility of formation of alk-1-ene-1,3-sultones in α-olefin sulfonates, *Comun. Jorn. Com. Esp. Deterg.*, 1987, *18*, 101–117.

93. Fudano, S., and K. Konishi, α-Olefin sulfonates by salting-out chromatography, *J. Chromatogr.*, 1971, *62*, 467–470.

94. Puschmann, H., Olefin sulfonates (in German), *Fette, Seifen, Anstrichm.*, 1973, *75*, 434–437.

95. Kupfer, W., K. Künzler, Petroleum sulfonates (in German), *Fette, Seifen, Anstrichm.*, 1972, *74*, 287–291.

96. Holtzman, S., B. M. Milwidsky, New olefin sulfonation for detergents, *Soap Chem. Spec.*, 1967, *43(11)*, 64,66,68,112–115.

97. Milwidsky, B. M., Continuous liquid/liquid extraction, *Soap Chem. Spec.*, 1969, *45(12)*, 79–80,84,86,88,117–118.

98. Kupfer, W., K. Künzler, Olefin sulfonates (in German), *Chem., Phys. Chem. Anwendungstech. Grenzflächenaktiven Stoffe, Ber. Int. Kongr., 6th 1972*, pub. 1973, Vol. *1*, 381–395.

99. International Organization for Standardization, Surface active agents—technical alkane sulfonates—methods of analysis, ISO 893:1989. Geneva, Switzerland.

100. Wickbold, R., Characterization of alkanesulfonates (in German), *Tenside, Surfactants, Deterg.*, 1971, *8*, 130–134.

101. International Organization for Standardization, Surface active agents—technical alkane sulfonates—determination of total alkane sulfonates content, ISO 6122:1978. Geneva, Switzerland.

102. Klotz, H., R. Spilker, Active agent in alkanesulfonates, *Tenside, Surfactants, Deterg.*, 1992, *29*, 13–15.

103. International Organization for Standardization, Surface active agents—technical alkane sulfonates—determination of alkane monosulfonates content by direct two-phase titration, ISO 6121:1988. Geneva, Switzerland.

104. Mutter, M., Alkylsulfonates using ion exchange resins (in German), *Tenside, Surfactants, Deterg.*, 1968, *5*, 138–140.

105. Kupfer, W., J. Jainz, H. Kelker, Alkyl sulfonates (in German), *Tenside, Surfactants, Deterg.*, 1969, *6*, 15–21.

106. International Organization for Standardization, Surface active agents—technical alkane sulfonates—determination of the mean relative molecular mass of the alkane monosulfonates and the alkane monosulfonate content, ISO 6845:1989. Geneva, Switzerland.

107. Scoggins, M. W., J. W. Miller, Separation technique for mono- and disulfonic acids, *Anal. Chem.*, 1968, *40*, 1155–1157.

108. Eppert, G., G. Liebscher, Indirect photometric detection with gradient elution: characterization of commercial alkylsulfonates (in German), *J. Chromatogr.*, 1986, *356*, 372–378.

109. Liebscher, G., G. Eppert, H. Oberender, H. Berthold, H. G. Hauthal, HPLC of alkanemonosulfonates (in German), *Tenside, Surfactants, Deterg.*, 1989, *26*, 195–197.

110. American Society for Testing and Materials, Analysis of oil-soluble petroleum sulfonates by LC, D3712-91. West Conshohocken, PA 19428.

111. Desbène, P. L., C. Rony, B. Desmazières, J. C. Jacquier, Analysis of alkylaromatic sulfonates by HPCE, *J. Chromatogr.*, 1992, *608*, 375–383.

112. Brewer, P. L., Oil-soluble sulfonates by two-phase titration, *J. Inst. Petrol., London*, 1972, *58*, 41–46.

113. Sandvik, E. I., W. W. Gale, M. O. Denekas, Petroleum sulfonates, *Soc. Pet. Eng. J.*, 1977, *17*, 184–192.

114. Metzger, G., Molecular weight of sulfonate soaps, *J. Am. Oil Chem. Soc.*, 1964, *41*, 495–496.

115. Márquez, N., S. Gonzalez, N. Subero, B. Bravo, G. Chavez, R. Bauza, F. Ysambertt, Isolation and characterization of petroleum sulfonates, *Analyst*, 1998, *123*, 2329–2332.

116. American Society for Testing and Materials, Base number of petroleum products by potentiometric perchloric acid titration, D2896-96. West Conshohocken, PA 19428.

117. Said, E. Z., I. H. Al-Wahaib, H. H. Nima, IR determination of the alkalinity of overbased petroleum sulfonates, *Analyst*, 1987, *112*, 499–500.

118. Van der Hage, E. R. E., W. M. G. M. van Loon, J. J. Boon, H. Lingeman, U. A. Th. Brinkman, HPSEC and pyrolysis-GC-MS study of lignosulfonates in pulp mill effluents, *J. Chromatogr.*, 1993, *634*, 263–271.

119. Majcherczyk, A., A. Hüttermann, SEC of lignin as ion-pair complex, *J. Chromatogr. A*, 1997, *764*, 183–191.

120. Battaglini, G. T., J. L. Larsen-Zobus, T. G. Baker, α-Sulfo methyl tallowate, *J. Am. Oil Chem. Soc.*, 1986, *63*, 1073–1077.

121. Marcinkiewicz-Salmonowicz, J., A. Górska, W. Zwierzykowski, Production of α-sulfonated fatty acid methyl esters, *Tenside, Surfactants, Deterg.*, 1995, *32*, 240–242.

122. Köhler, M., E. Keck, G. Jaumann, Sulfonated esters (in German), *Fett. Wiss. Technol.*, 1988, *90*, 241–243.

123. Drozd, A. V., B. G. Klimov, Extraction-spectrophotometric determination of anionic surfactants with Rhodamine 6G, *J. Anal. Chem.*, 1998, *53*, 711–713.

124. Kurosaki, T., J. Wakatsuki, T. Imamura, A. Matsunaga, H. Furugaki, Y. Sassa, Monoalkylphosphate: phosphorylation, analysis and properties, *Comun. Jorn. Com. Esp. Deterg.*, 1988, *19*, 191–205.

125. Sandra, P., F. David, Microcolumn chromatography for the analysis of detergents and lubricants. Part 1: High temperature capillary GC and capillary supercritical fluid chromatography, *J. High Resolut. Chromatogr.*, 1990, *13*, 414–417.

126. Vonk, H. J., A. J. van Wely, L. G. J. van der Ven, A. J. J. de Breet, F. P. B. van der Maeden, M. E. F. Biemond, A. Venema, W. G. B. Huysmans, Ethoxylated surfactants, *Tr.-Mezhdunar. Kongr. Poverkhn.-Akt. Veshchestvam, 7th*, 1976, *1*, 435–449.

127. International Organization for Standardization, Surface active agents—detergents—anionic-active matter hydrolyzable under alkaline conditions—determination of hydrolyzable and non-hydrolyzable anionic-active matter, ISO 2869:1973. Geneva, Switzerland.

128. König, H., Sulfosuccinate half-esters (in German), *Fresenius' Z. Anal. Chem.*, 1971, *254*, 198–209.

129. Schulz, R., R. Gerhards, Titrant for the potentiometric titration of anionic surfactants, *Tenside Surfactants, Deterg.*, 1995, *32*, 6–11.

130. Van Gorkom, L. C. M., A. Jensen, NMR Spectroscopy, in J. Cross, ed., *Anionic Surfactants: Analytical Chemistry*, 2nd ed., Marcel Dekker, New York, 1998.

131. Ianniello, R. M., Ion chromatographic determination of sodium isethionate, *J. Liq. Chromatogr.*, 1988, *11*, 2305–2314.

132. Rasmussen, H. T., N. Omeiczenko, B. P. McPherson, Sodium isethionate in soap and lye process streams by suppressed ion chromatography, *J. Am. Oil Chem. Soc.*, 1993, *70*, 733–734.

133. Cosmetics, Toiletry and Fragrance Association, Separation of fatty acids from acyl isethionates, CTFA Method A 5-1. 1110 Vermont Ave. NW, Washington, DC 20005.

134. König, H., W. Strobel, Surfactants in toothpastes by HPLC (in English), *Fresenius' Z. Anal. Chem.*, 1988, *331*, 435–438.
135. Voogt, P., Ion-exchangers in detergent analysis. Part II, *Rec. Trav. Chim.*, 1959, *78*, 899–912.
136. Bey, K., TLC analysis of surfactants (in German), *Fette, Seifen, Anstrichm.*, 1965, *67*, 217–221.
137. Daradics, L., J. Pálinkás, Synthesis and analysis of fatty acid sarcosides, *Tenside, Surfactants, Deterg.*, 1994, *31*, 308–313.
138. Buschmann, N., Zwitterionic surfactants by electrochemically indicated amphimetry, *Tenside, Surfactants, Deterg.*, 1991, *28*, 329–332.
139. Buschmann, N., R. Schulz, Cationic and zwitterionic surfactants using ion selective electrodes, *Tenside, Surfactants, Deterg.*, 1992, *29*, 128–130.
140. Arens, M., G. Krusche, Organic surface active compounds IX (in German), *Fette, Seifen, Anstrichm.*, 1981, *83*, 449–451.
141. Noguchi, H., S. Matsutani, S. Tanaka, Y. Horiguchi, T. Hobo, Alkyl homolog distribution of ionic surfactants by two-dimensional HPLC (in Japanese), *Bunseki Kagaku*, 1998, *47*, 473–479.
142. Gerhardt, W., G. Czichocki, H.-R. Holzbauer, C. Martens, B. Weiland, Synthesis and analysis of ether carboxylic acids. Characterization of ether carboxylic acids during the synthesis process (in German), *Tenside, Surfactants, Deterg.*, 1992, *29*, 285–288.
143. König, H., W. Strobel, Alkyl- and alkylphenol ether carboxylates by HPLC. *Fresenius. J. Anal. Chem.*, 1990, *338*, 728–731.
144. Spilker, R., Hüls AG, Marl, Germany. Personal communication, 1993.
145. Gerhardt, W., G. Czichocki, H.-R. Holzbauer, C. Martens, Synthesis and analysis of ether carboxylic acids. Preparation of ether carboxylic acids (in German), *Tenside, Surfactants, Deterg.*, 1992, *29*, 169–174.
146. Wood, T. E., Analytical methods, evaluation techniques, and regulatory requirements, in L. Spitz, ed., *Soap Technology for the 1990's*, American Oil Chemists' Society, Champaign, IL, 1990.
147. American Society for Testing and Materials, Sampling and chemical analysis of soaps and soap products, D460-91. West Conshohocken, PA 19428.
148. American Oil Chemists' Society, Official and tentative methods. Section D: Sampling and analysis of soap and soap products, fatty alkyl sulfates, alkyl benzene sulfonates. 2211 West Bradley Ave., Champaign, IL 61826.
149. American Society for Testing and Materials, Chemical analysis of soaps containing synthetic detergents, D820-93. West Conshohocken, PA 19428.
150. Sharma, R., R. Pyter, P. Mukerjee, Spectrophotometric determination of perfluorocarboxylic acids (heptanoic to decanoic) and sodium perfluorooctanoate and decyl sulfate in mixtures by dye-extraction, *Anal. Lett.*, 1989, *22*, 999–1007.
151. International Organization for Standardization, Surface active agents—soaps—determination of low contents of free glycerol by molecular absorption spectrometry, ISO 2272:1989. Geneva, Switzerland.
152. Molever, K., Glycerin in soap by capillary gas chromatography, *J. Am. Oil Chem. Soc.*, 1987, *64*, 1356–1357.
153. George, E. D., J. A. Acquaro, HPLC for the analysis of glycerol and other glycols in soap, *J. Liq. Chromatogr.*, 1982, *5*, 927–938.
154. Molever, K., Sample preparation for chloride analysis in bar soaps, *J. Am. Oil Chem. Soc.*, 1995, *72*, 161–162.

2
Characterization of Nonionic Surfactants

I. DESCRIPTION AND TYPICAL SPECIFICATIONS

Nonionics account for roughly 40% of worldwide surfactant use. Nonionics are generally more tolerant than anionics of water hardness, which makes the requirements for "builders" in laundry detergents less demanding. They tend to be more effective than other surfactants for removal of oily soil from synthetic fabrics. Most nonionics are considered low-foaming products, have good cold water solubility, and have a low critical micelle concentration, making them effective at low concentration. Their compatibility with cationic fabric softeners makes them preferable to anionics in certain formulations. They are more common in industrial applications than are anionics.

A. Ethoxylated Alcohols

1. Description

$O(CH_2CH_2O)_8CH_2CH_2OH$

Dodecanol 9-mole ethoxylate

The largest use of these compounds is in laundry detergents, where the EO chain length is usually in the range 7–9, although specialty products may contain 30 or more moles. The primary alcohol ethoxylates are the common item of commerce, but secondary alcohol ethoxylates are also available.

The materials are generally made by solventless addition of ethylene oxide to fatty alcohols using KOH catalysis. Most often, the alcohol starting material consists of a range of alkyl chain lengths, almost always linear. The ethylene oxide adds so as to give a Poisson distribution of EO chain length. Unless the product is made with special precautions, unethoxylated alcohol can remain after the ethoxylation reaction. If there is any source of water, for example from the catalyst or from leakage of the reactor heat exchanger, polyethylene glycol (PEG) will also be present. Since neither free alcohol nor PEG have surfactant properties, their presence is undesirable. Catalyst may be present if not deliberately removed by filtration or by extraction. It may be present as traces of the free base or as a salt. As with all polymers, care must be taken that unreacted monomer is not present. Thus free ethylene oxide is determined, along with combinations and rearrangements thereof such as 1,4-dioxane and acetaldehyde.

The most important surfactant properties are controlled by the average percent EO and the average chain length of the starting alcohol. However, since there is a lesser effect on properties from the presence of the lower or higher MW members of the oligomer distribution, it is often necessary to perform a complete oligomer analysis, rather than simply report the bulk properties (1).

2. Typical Specifications

Parameter	Test method
Oxyethylene content	HI cleavage, followed by titration; infrared spectroscopy; NMR spectrometry
Identification of starting alcohol	GC; HPLC; MS
EO/alcohol ratio	NMR; calculation based on OH number and known alcohol starting material
Ethoxylate distribution	HPLC, GC, or SFC
Unethoxylated alcohol	Gas chromatography; column chromatography
Ethylene oxide, acetaldehyde, 1,4-dioxane	Headspace GC
Branching of hydrophobe chain	NMR spectrometry
pH	5% aqueous
Other tests	Appearance; color; water; number average molecular weight; cloud point; polyethylene glycol; ash; iron; unsaturation; acidity or basicity; refractive index; iodine number; viscosity; surface tension, 0.01% aqueous

B. Ethoxylated Alkylphenols

1. Description

Nonylphenol 9-mole ethoxylate

Most commercial products contain branched, rather than linear, alkyl groups. Since these compounds only slowly reach complete biodegradation upon release to the environment,

APE use is limited to commercial and industrial applications which are not considered likely to make a large contribution to municipal sewage treatment plants, such as in textile chemicals. The alkyl chain is usually C_8 (diisobutylene), C_9 (propylene trimer), or C_{12} (propylene tetramer). The nonylphenol ethoxylates (NPE) are used in much greater volume than the others, perhaps 80% of the market. Octylphenol ethoxylates (OPE) have about 15% of the market and dinonylphenol- and dodecylphenolethoxylates are at about 1% each. All are excellent for the removal of oily soils and can be produced at lower cost than the more abundant alcohol ethoxylates.

Because the alkyl chain is typically monodisperse, it is not necessary to characterize the hydrophobe moiety beyond a simple confirmation of identity. It should be noted that even though monodisperse in respect to molecular weight, the alkyl chain may be a mixture of isomers, especially in the case of the triisopropylene product. The surfactant is produced by alkali-catalyzed ethoxylation of the alkylphenol. Normally, a Poisson distribution of ethoxylated oligomers is produced, with some unethoxylated alkylphenol remaining, and perhaps with some polyethylene glycol formed. Reaction residues, such as ethylene oxide, 1,4-dioxane, and acetaldehyde, are removed by vacuum stripping (2).

2. Typical Specifications

Parameter	Test method
Oxyethylene content	NMR; IR spectroscopy; HI cleavage and titration; GC if PO is also present
Molecular weight distribution	Normal phase HPLC, GC, or SFC; GPC for degree of ethoxylation >40
Residual alkylphenol and phenol	HPLC
pH	1% aqueous
Acidity or basicity	Titration in 3:1 isopropanol/water
Ethylene oxide, acetaldehyde, 1,4-dioxane	Head space GC
Other tests	Appearance; color; water; number average molecular weight; cloud point; poly(ethylene glycol); ash; iron; unsaturation; iodine number

C. Ethoxylated Acids (PEG Esters)

1. Description

Dodecanoic acid, nonaethylene glycol ester

These can be considered to be esters of fatty acids and polyethylene glycol, although they are generally made by direct addition of EO to the acid. They contain, besides the main monoester ingredient, diester, free acid, and free PEG. This class includes tall oil ethoxylates. Tall oil fatty acids are linear, long-chain, and unsaturated, being obtained by distillation of tall oil, a water-insoluble, ether-soluble fraction of pulp mill waste (3).

2. Typical Specifications

Parameter	Test method
Assay	Gas chromatography; tetraphenylborate titration
Mono- and diester content	Calculation, after removal of PEG and determination of hydroxyl number and saponification number; HPLC; proton NMR, after derivatization
Free acid	Titration, directly or after first removing KOH catalyst
Other tests	PEG; ethylene oxide, acetaldehyde, 1,4-dioxane

D. Fatty Acid Alkanolamides

1. Description

Dodecanoic acid diethanolamide

The most common examples of this class are the mono- and diethanolamides of linear alkyl acids. They are effective for increasing the viscosity of liquid formulations. Even normally water-insoluble products, like the monoethanolamides and isopropanolamides, can be used as viscosity improvers in the presence of large amounts of other surfactants. The monoethanolamides are widely used to stabilize the foam formed by other surfactants. Alkanolamides are also effective surfactants in their own right, although very sensitive to water hardness in the absence of other surfactants, as well as being subject to attack by acid and base. They are commonly used in shampoos and liquid "soaps."

Commercial products are mixtures of homologs made by reaction of an alkanola-mine with either a fatty acid or a fatty acid ester, usually triglycerides or methyl esters. An example is the reaction of diethanolamine with coconut oil, which results chiefly in the formation of diethanolamides along with glycerin. Some esters may also be formed, with or without free amine functionality. Frequently encountered ratios of alkanolamine to acid are 2:1 (Kritchevsky type) and 1:1 (4).

2. Typical Specifications

Parameter	Test method
Assay	Titration in acetic anhydride; gas chromatography
Free amine and amino esters	Gas chromatography; titration in acetic acid
Free ethanolamine or diethanolamine	Titration; gas chromatography
Free fatty acid	Titration in isopropanol
Ester content	Gas chromatography; TLC
Amine/acid ratio	Hydrolysis, extraction, and titration
Alkyl chain length distribution	GC, HPLC, or MS
N,N'-bis(2-hydroxyethyl)piperazine	Gas chromatography; TLC
NaCl	AgNO$_3$ titration for Cl$^-$; flame photometry for Na$^+$
N-Nitrosodiethanolamine	HPLC
Other tests	Unsaturation

E. Ethoxylated Alkanolamides

1. Description

Ethoxylated dodecanoic acid diethanolamide

These are produced by addition of ethylene oxide to the hydroxyl group of the alkanola-mides described above. Ethanolamine and fatty acid impurities, if present in the starting alkanolamide, may also be ethoxylated to give ethoxylated ethanolamine and ethoxylated fatty acid. Other impurities may include polyethylene glycol as well as fatty acid esters of all hydroxyl-containing impurities (4).

2. Typical Specifications

Parameter	Test method
Assay	Titration
Free acid	Ion exchange and titration
Ethoxylated amine	Extraction and ion exchange
Ethoxylated ester	Calculated from other values
Alkyl/EO ratio	NMR
Characterization of acid	Gas chromatography after hydrolysis
Other tests	Ethylene oxide, acetaldehyde, 1,4-dioxane, PEG

F. Ethoxylated Amines

1. Description

Ethoxylated dodecylamine

These are made by ethoxylation of amines prepared from naturally occuring fatty acids (5).

2. Typical Specifications

Parameter	Test method
Alkyl chain distribution	Reversed-phase HPLC
Ethoxy chain distribution	Normal phase HPLC
Amine nitrogen content	Potentiometric titration
Apparent pH	5% in 50:50 methanol/water
Distribution of primary, secondary, and tertiary amine	Titration; NMR
Oxyethylene content	HI cleavage, corrected
Ash	Gravimetry (sulfated)

Parameter	Test method
Free PEG	Ion exchange/gravimetry
Unsaturation	Bromide/bromate titration
Other tests	Ethylene oxide; acetaldehyde; 1,4-dioxane; hydroxyl number; water content

G. Esters of Polyhydroxy Compounds

1. Description

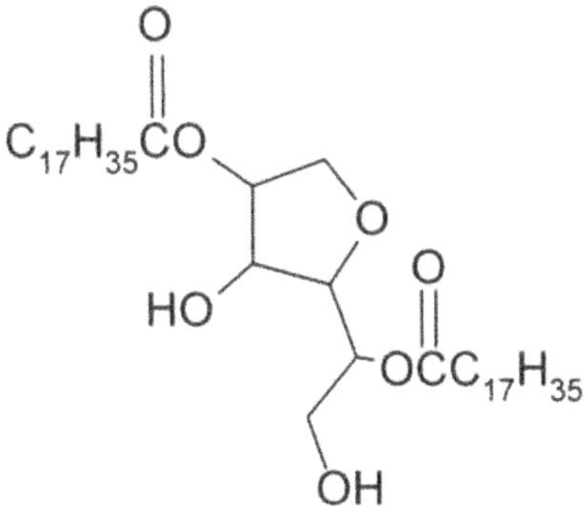

A sorbitan distearate

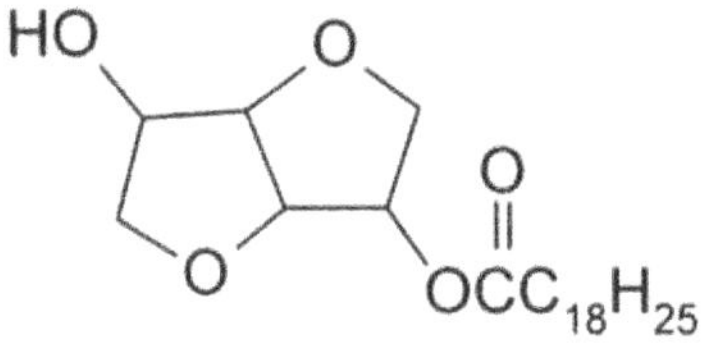

A disorbitan monostearate

A broad class of food emulsifiers consists of fatty acid esters of glycerin, sorbitol, and similar polyfunctional alcohols. These may be made from sorbitan (dehydrated sorbitol) which has been partially esterified with fatty acids, from reaction of sucrose with methyl esters of fatty acids, or, in the case of the sucroglycerides, from reaction of sucrose with triglycerides in a transesterification reaction which produces mainly mixed mono- and diesters of sucrose with a lesser concentration of mono- and diesters of glycerin. Most ester products are mixtures, containing free acids; mono-, di-, and triesters; and perhaps oligomers of the base alcohol. Multiple isomers may be seen, for example, with sucrose, where each molecule has eight hydroxyl groups suitable for esterification with the three primary hydroxyl groups being the most reactive. Sucrose esters made with three moles or less of fatty acid per sugar molecule have surfactant properties. Higher esters are used as fat substitutes and are not treated in this volume. If N,N-dimethylformamide is used as catalyst and solubilizer, residual nitrogen compounds can make the product unsuitable for food use (6). The esters most commonly encountered are those of sorbitol, with sucrose esters a distant second.

2. Typical Specifications

Parameter	Test method
Total sorbitol content (of sorbitol esters)	Spectroscopy; determination of formaldehyde after oxidation
Free sucrose content (of sucrose esters)	Extraction and spectrophotometry or HPLC
Total sucrose content (of sucrose esters)	Optical rotation
Average degree of substitution	Saponification number
Free fatty acids and methyl esters	HPLC; titration
Monoester, diester	HPLC; GC of silyl derivatives
Higher esters	TLC
Composition of alkyl constituents	Saponification, followed by GC of acids or their methyl esters; HPLC of underivatized sample
Isomer content	TLC; HPLC
Nitrogen content	Elemental analysis

H. Ethoxylated Esters

1. Description

Ethoxylated sorbitan distearate

These are esters, as described above, where ethylene oxide has been added to the unesterified hydroxyl functional groups. They are very widely used as emulsifiers in foods and cosmetics and in various specialty products (6).

2. Typical Specifications

Parameter	Test method
Ester content	Saponification number
EO content	HI cleavage and titration; NMR analysis
Low MW impurities	Gas chromatography
Other tests	Color; viscosity; cloud point; hydroxyl number; acid number; PEG; ash

I. Alkyl Polyglycosides

1. Description

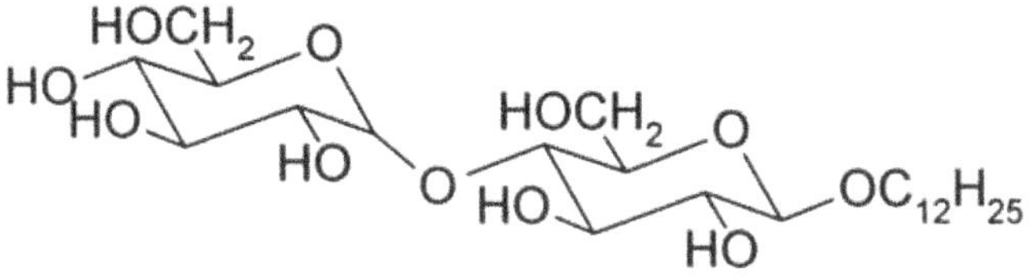

Lauryl diglucoside

Commercial glycosides are based on glucose and thus are also called glucosides. These compounds are made by acid-catalyzed condensation of the alcohol with glucose, either directly or after forming an intermediate condensate of glucose and a lower-boiling alcohol. Alkyl polyglycosides typically have single alkyl chains with lengths in the C_8–C_{16} range, with an average degree of polymerization of the glycoside moiety of only between 1 and 2. Thus, in spite of the name, the commercial product is mainly a monoglucoside rather than a polyglycoside. They are marketed as made from "all natural" raw materials and are mainly encountered in applications where mildness is important, as in cosmetics. Alkyl polyglycosides are thoroughly described in recent publications by Henkel chemists (7,8).

2. Typical Specifications

Parameter	Test method
Alkyl chain length distribution	GC or HPLC
Glucose/alcohol ratio	NMR
Degree of polymerization	NMR or GC
Residual free alcohol content	Gas chromatography after solid phase extraction
Other tests	Cloud point; color; viscosity; ash

J. Ethylene Oxide/Propylene Oxide Block Copolymers

1. Description

$$HO(CH_2CH_2O)_y(\overset{\overset{\displaystyle CH_3}{|}}{C}HCH_2O)_x(CH_2CH_2O)_zH$$

Copolymer of the poloxamer type

These compounds are generally made by KOH-catalyzed polymerization of propylene oxide to form a hydrophobic base of poly(propylene glycol). Ethylene oxide is then added to give a block copolymer with terminal poly(ethylene glycol) groups. The process can be reversed to give products with terminal hydrophobe groups on a hydrophilic backbone. Besides the desired product, other materials which may be present are the catalyst, polyethylene or polypropylene glycols, monofunctional material, residual EO or PO, and low molecular weight impurities from side reactions of the oxides. Some or all of the catalyst may remain in the product, although free alkalinity is neutralized. Neutralization is via a buffering acid, such as phosphoric, citric, or acetic acid.

Analogous products are made substituting butylene oxide for propylene oxide, although these are much less common than the EO/PO copolymers.

2. Typical Specifications

Parameter	Test method
Average molecular weight	Hydroxyl number
Molecular weight distribution	Gel permeation chromatography
Monofunctional material (unsaturation)	Gel permeation chromatography; titration
Antioxidant	Ultraviolet spectrophotometry; liquid chromatography
EO/PO ratio	NMR spectroscopy
Carbonyl content	HPLC determination of aldehydes
Peroxide content	Iodometric titration
Other tests	Color; cloud point; viscosity; melting point; appearance; water, Karl Fischer method; pH, 5% aqueous; ethylene oxide; acetaldehyde; 1,4-dioxane

K. Amine Oxides

1. Description

Lauryl dimethylamine oxide

These are made by oxidation of tertiary amines, usually with hydrogen peroxide. At acid pH, these products have cationic properties and are sometimes classified as cationic or amphoteric surfactants. They are useful as foam stabilizers and are used extensively in cosmetics as well as household products.

2. Typical Specifications

Parameter	Test method
Assay	Potentiometric titration
Unreacted amine	Potentiometric titration; TLC
Chain length distribution	Gas chromatography after reduction
N-Nitrosamines	Thermal energy analysis
Other tests	Water; total nitrogen; color

II. GENERAL TEST METHODS

A. Oxyethylene or Oxypropylene Content

In the research laboratory, percent EO or PO composition is most often determined by NMR. Infrared spectrometry may also be used for determination of degree of ethoxylation, but the need for calibration makes this approach inherently less accurate than NMR. These techniques are discussed in chapters 13 and 14. For routine quality control of compounds with the same hydrophobe, the correlation between cloud point or refractive index and EO content is quite precise (9). Similarly, it has been demonstrated that the heat of fusion, easily measured by differential scanning calorimetry, is linearly related to the degree of ethoxylation of at least some surfactants (10).

Chromatographic procedures give the distribution of the individual oligomers of the alkoxylate. This is more information than is needed for routine analysis, but invaluable knowledge when characterizing a material from a new process or made with a new catalyst. GC with derivatization is preferred for measuring lower MW oligomers, perhaps up to the 6-ethoxylate, while HPLC is used for higher degrees of ethoxylation. High temperature GC or supercritical fluid chromatography have been demonstrated for determination of degree of ethoxylation up to about 12. Details can be found in later chapters. Gas chromatography of cleavage products after chemical reaction or pyrolysis, rarely used nowadays except for qualitative analysis, is discussed in Chapter 8.

The titration procedure described below is more time-consuming than the instrumental methods, but is more suitable for use in capital-poor laboratories.

1. Titration

This method is optimized to determine the length of the poly(ethylene oxide) chain in, for example, an ethoxylated alcohol surfactant (11,12). It is also applicable to determination of oxypropylene content and to analysis of other products, such as EO/PO block copolymers (13,14). Quantitative results are not obtained from analysis of alkoxylated amines due to incomplete cleavage of the ethoxy group adjacent to the nitrogen.

The poly(ethylene oxide) and poly(propylene oxide) chains are cleaved by reaction with hydriodic acid.

$$—CH_2CH_2O— + 2HI \rightarrow ICH_2CH_2I + H_2O$$

$$—CH(CH_3)CH_2O— + 2HI \rightarrow ICH(CH_3)CH_2I + H_2O$$

The diiodoalkanes are unstable and subsequently decompose:

$$ICH_2CH_2I \rightarrow CH_2CH_2 + I_2 \quad \text{about 30\% yield}$$

$$ICH_2CH_2I + HI \rightarrow CH_3CH_2I + I_2 \quad \text{about 70\% yield}$$

(1,2-Diiodopropane reacts similarly to give isopropyliodide and byproducts) One mole of titratable iodine is produced for each alkylene oxide unit. The liberated iodine is titrated with sodium thiosulfate solution. Alternatively, the cleavage products may be determined by gas chromatography, permitting distinguishing between oxyethylene and oxypropylene content. A number of studies of this reaction have appeared. They indicate that only about 80% yield is obtained in the reaction of PO units with HI (15). For practical reasons, the unstable HI reactant is usually replaced with KI and phosphoric acid (12).

Procedure: Determination of Ethylene Oxide Content of Ethoxylated Surfactants by HI Cleavage (11)

Assemble an apparatus consisting of two 100-mL round bottom flasks fitted with reflux condensers and a continuous CO_2 purge in the headspace of the flasks. For a substance containing about 60% ethylene oxide, weigh out a sample aliquot of 0.15 g into one of the flasks. Adjust the sample size for higher or lower EO contents. Pipet in exactly 5.00 mL unstabilized hydriodic acid from a freshly opened bottle. Pipet another 5.00 mL into the second flask to serve as the reagent blank. Add a glass bead to each. Connect the reflux condensers, using Teflon sleeves or a small amount of stopcock grease to insure a good seal. Start the flow of cooling water and the CO_2 purge. After a 10-min purge, apply heat.

Reflux for 90 min, maintaining the CO_2 flow. When flasks cool, wash down the condensers into the flasks with about 15 mL each 20% KI solution, then rinse with about 20 mL water. Rinse the contents of the flasks into 250-mL conical flasks with water, rinsing also the ends of the condensers. If a tarry residue remains, transfer it with a small amount of methanol. Titrate with sodium thiosulfate solution to the starch end point. Calculation:

$$\% \text{ EO} = \frac{(A-B)(N)2.20}{W} \quad \text{or} \quad \% \text{ PO} = \frac{(A-B)(N)2.90}{W}$$

where A and B are the mL titrant required by the sample and blank, respectively, N is the normality of the titrant, and W is sample weight in grams.

2. Elemental Analysis

For ordinary ethoxylated surfactants, where the ethylene oxide moiety is the only source of oxygen, elemental analysis (where percent oxygen is usually determined by difference) gives an accurate value for the length of the EO chain (16). The advantage of this technique is that modern instruments for elemental analysis provide the results in a few minutes. The disadvantage—that contamination from water, alcohol, or other materials will cause inaccurate results—is the same as with the spectroscopic methods.

B. Molecular Weight

For routine control analysis, molecular weight is determined by end group analysis. This can be performed by spectroscopic means, notably by infrared and NMR spectrometry. However, wet chemical methods are somewhat more precise and do not require the availability of standards of known molecular weight. Wet chemical methods measure equivalent weight rather than molecular weight. This can be related to number average MW. Such values do not usually correspond exactly with direct measurement of molecular weight by freezing point depression (number average), vapor pressure osmometry (number average), viscosity (closest to weight average), or other direct techniques because of the presence of impurities having more or less of the theoretical number of functional groups per molecule. With surfactants, one must be aware of the formation of micelles, which can make the apparent molecular weight by colligative property measurement too high by several times (17). In some cases, accurate molecular weight determinations of surfactants may be made by mass spectroscopy. This is discussed in Chapter 15.

1. Hydroxyl Number

There are various wet chemical procedures for determination of hydroxyl end groups, with more than a dozen having official status with ASTM, AOCS, or similar bodies. Investigations into hydroxyl number methods are dominated by the plastics industry, since precise knowledge of hydroxyl content is essential to polyurethane reaction stoichiometry. Surfactant applications rarely require such exactitude. Esterification methods, especially the methods based on phthalic anhydride esterification, are used most often for surfactants (18–20). Some compounds (such as those initiated with polyhydroxy compounds like sucrose, or those with phenolic hydroxyl functionality) do not give accurate results with phthalation, and other reagents such as acetic anhydride are used (21–24). Most wet chemical methods for hydroxyl number determination rely on titration of unused reagent and are thus inaccurate for very low hydroxyl numbers, such as would be encountered with high molecular weight products or if the analysis were conducted to determine a minor component or impurity. In such cases, a method based on direct titration, such as the *p*-toluenesulfonyl isocyanate method, is preferred.

(a) Phthalic anhydride method. In this classical wet chemical procedure, the terminal hydroxyl groups of the surfactant are esterified by reaction with phthalic anhydride. The excess phthalic anhydride is hydrolyzed to phthalic acid by subsequent addition of water and the acid titrated with NaOH solution. The hydroxyl content is then calculated by difference. Phthalic anhydride esterification is appropriate for primary and secondary hydroxyl groups. It is not suitable for determination of tertiary or phenolic hydroxyl groups or some other sterically hindered groups.

There are many variations of the phthalic anhydride method accepted in industry. Since surfactant properties are not very sensitive to hydroxyl number, the small bias between results obtained by different methods is rarely significant.

Procedure (19)

Reagent grade pyridine is adequate only if freshly distilled. Burdick and Jackson distilled-in-glass grade is suitable without further purification. Phthalic anhydride solution is prepared from 98 g reagent-grade phthalic anhydride per 700 mL pyridine, allowed to stand overnight before use. It may be stored one week in a brown bottle. Phenolphthalein indicator solution is 1% in pyridine. Since pyridine has an obnoxious odor, all operations, including cleaning of glassware, should be conducted in a fume hood.

Accurately weigh into a clean, dry 250-mL conical flask with ground glass joint a sample size sufficient to give a titration in the proper range. This equation may be used: grams of sample = molecular weight/($100 \times$ functionality). Do not use more than 10 g sample. Add 25.00 mL phthalic anhydride reagent to the sample flask and a blank flask. Swirl to dissolve, add a few glass beads, attach a dry reflux condenser, and boil for 1 hr on a hotplate. Raise the flasks from the hotplate and allow to cool. Wash down the condenser into the flask with 25 mL pyridine, then with 25 mL water. Remove the flasks from the condenser and titrate to the phenolphthalein end point with 1 M NaOH solution. The difference between the sample and blank titration should be about 10 mL. A correction must be made for the acidity or alkalinity of the sample. Calculation:

$$\text{OH number} = \frac{(B - S)(N)56.1}{W}$$

where B and S represent milliliters titrant required by the blank and sample, respectively, N is the titrant concentration, and W is the sample weight in grams.

$$\text{Acidity or alkalinity correction} = \frac{(A)(N)56.1}{W}$$

where A represents milliliters titrant required to neutralize the sample to phenolphthalein. The acid correction is added to the OH number, and the alkalinity correction is subtracted.

$$\text{Molecular weight} = \frac{56100 \ (\text{OH groups per molecule})}{\text{hydroxyl number}}$$

(b) Near IR spectrophotometry. Most industrial quality control laboratories have adopted methods based on NIR analysis for determination of hydroxyl end groups as well as other parameters like water content. The main advantage of NIR is the speed of analysis. Such methods are always calibrated and validated using titration methods. These ap-

plications are usually developed by the manufacturers of the instrument. NIR analysis is discussed in Chapter 13.

2. Chromatographic Methods

Determination of molecular weight distribution by GPC, HPLC, GC, and SFC is discussed in Chapters 7–10. Only GPC is suitable for routine analysis. For most applications, the HPLC and GC procedures require the formation of derivatives to increase detectability or volatility, respectively. SFC instrumentation is rarely found in the quality control laboratory.

Although tedious, the chromatographic methods are irreplaceable for showing the actual oligomer and homolog distribution of ethoxylated polymers. HPLC is able to give both the alkyl chain distribution (by a reversed-phase method) and the ethoxy chain distribution (usually by a normal phase method) in two separate, easily quantified chromatograms. For simple surfactants, such as an alcohol ethoxylate of moderate molecular weight based on a simple mixture of C_{12} and C_{14} linear alcohols, GC or SFC can show the alkyl and ethoxy distribution in a single chromatogram.

C. Polyethylene Glycol

Polyethylene glycol (PEG) is an impurity in ethoxylated surfactants, typically found in the range 1–10%. It does not have surfactant properties. It is formed by homopolymerization of ethylene oxide in competition to the desired reaction of addition of ethylene oxide to the hydrophobic initiator. Formation of PEG is initiated by traces of water in the reaction medium. The oligomer distribution of PEG is usually similar to, but broader than, the oligomer distribution of the surfactant (25).

The established method for determination of PEG in fatty alcohol and alkylphenol ethoxylates is the gravimetric procedure (often called the Weibull method). The gravimetric procedure gives low recoveries with PEG of low molecular weight, since low oligomers cannot be easily extracted from brine into chloroform. Failure of extraction at low molecular weight is more marked if methylene chloride is substituted for chloroform (26). The Weibull method is therefore not suitable for the analysis of products with a degree of ethoxylation of less than 6 units, while the HPLC and TLC procedures are suitable over both low and high MW ranges.

The Weibull method is not suitable for determination of polypropylene glycol in propoxylates. HPLC is suitable for the analysis, using almost the same normal phase or reversed-phase conditions as for PEG determination (27).

1. Gravimetric Methods

(a) Extraction procedure, for typical nonionics.   This wet chemical method is based on the different solubilities of a surfactant and PEG. The most widely used method is that of Weibull, reproduced below (28). Methods based on extraction with other solvents are also in common use, such as partition with 2-butanone/water (29) or *n*-butanol/aqueous sodium bicarbonate solution. The accuracy of the determination is dependant upon the matrix. It is known that the method gives low results for determination of PEG in ethoxylated amines, so these materials are normally analyzed by an ion exchange procedure (30). It is thought that the extraction method also gives low results for other products (31). For AE or APE, it has been shown that the PEG fraction isolated by the extraction method is contaminated with low levels of the active surfactant, so that large errors can occur when only low levels of PEG are present (32).

The ISO procedure for PEG determination specifies extraction at the inconvenient temperature of 35°C (33). A clever stationary glass apparatus which takes up no more space than a conventional separatory funnel has been developed to perform this extraction, using a stream of nitrogen for agitation. Results for PEG determination are more precise with this equipment, which optimizes temperature control and minimizes operator exposure to organic vapors (34).

Procedure: Determination of Polyethylene Glycol in Ethoxylated Products by Extraction (19)

Accurately weigh about 10 g sample into a 250-mL separatory funnel. Add 50 mL ethyl acetate and 50 mL saturated NaCl solution (293 g/L H_2O). Shake 2 min, then let stand until the phases have separated. Drain the lower, aqueous phase into a second separatory funnel. Extract the ethyl acetate in the original separatory funnel with an additional 50 mL NaCl solution. Add the aqueous extract to the second funnel, and wash the combined extracts with 50 mL ethyl acetate. Transfer the washed aqueous layer to a third separatory funnel, and extract with 50 mL $CHCl_3$. Drain the lower, $CHCl_3$ layer through a small wad of glass wool into a tared evaporating dish. Repeat the extraction with a second 50-mL aliquot of $CHCl_3$, draining the extract into the same evaporating dish. Evaporate the $CHCl_3$ on a steam bath. Purify the extract by dissolving in acetone and filtering out NaCl crystals, then evaporating the acetone. Dry at 50°C for 1 hr, cool, and weigh to determine the mass of the PEG.

(b)   Ion exchange procedure for ethoxylated amines.   The sample is passed through an ion exchange column where the amine ethoxylate is retained, while the PEG passes through. The ion exchange separation of PEG from amines gives a fraction which may be contaminated with other products, such as the substituted morpholine: *N*-alkyl-2,3,5,6-tetra-hydro-1,4-oxazine (35).

Procedure (30)

The sample is dissolved in 40 mL water or methanol and passed through a cation exchange column (strongly acidic, 2% crosslinking agent, H^+ form, 2×25 cm) at about 1 mL/min. Wash with 500 mL water. The eluate is evaporated to dryness and weighed to obtain the mass of PEG. The column is regenerated with HCl and washed with water.

(c)   Column chromatography.   PEG can be quantitatively separated from common ethoxylates by chromatography on either silica or on C_{18} reversed-phase material. Since PEG is only eluted from silica with great difficulty, the reversed-phase method is more practical. Experimental details are given in Chapter 6.

2.   HPLC Method

This method has been demonstrated for analysis of ethoxylates of alcohols, alkylphenols, fatty acids, sorbitan esters, glycerides, alkanolamides, and other materials. The HPLC method typically gives higher results than the extraction method. This procedure is readily adaptable to ordinary reversed-phase columns with methanol/water or acetonitrile/water mobile phases, provided that ordinary optimization procedures are followed. If desired, the molecular weight distribution of the PEG can be determined by HPLC analysis of the original sample on a LiChrosorb Diol (Merck) column, with isocratic mobile phase of 105:95:10:1 *n*-hexane/2-propanol/water/acetic acid and refractive index detection (25). An occasional back-flush of the column with the same mobile phase is required.

A source of error in the HPLC method is the very short retention time of PEG in reversed-phase systems. This makes quantification by differential refractive index subject to interference by peaks from water, solvent impurities, and other unretained substances. This source of error may be eliminated by coupling the HPLC column to a GPC column, so that PEG is resolved from both the surfactant and from other impurities (36). Even with this modification, if phenol impurity were present in the initial alkylphenol, the resulting ethoxylated phenol could interfere with PEG determination by this method. Its presence is detected by using a UV detector in series with the RI detector. Another source of interference is the anion, such as acetate or lactate, added during the neutralization of the catalyst. For careful work, this should be removed by ion exchange prior to HPLC analysis (36).

Procedure: HPLC Determination of PEG in Ethoxylates (25,31)

Samples are prepared as 20% solutions in the eluent, 95:5 methanol/water. The column packing is C_{18}-based, such as Merck LiChrosorb RP-18, 4.6 × 250 mm. An injection volume of 0.5–2.5 µL is used. Quantification is via a refractive index detector, using pure PEG of appropriate molecular weight as external standard.

3. GC Method

Lower molecular weight ethoxylates are readily analyzed by gas chromatography, provided that they are first derivatized. Ethoxylates of a single alcohol or alkylphenol generally give a series of peaks with the PEG oligomers interspersed among the surfactant oligomers. In commercial products, where the hydrophobe portion of the molecule is usually a mixture, the chromatograms are too complex for accurate quantification of low level PEG. Thus, the determination is performed by first extracting the PEG, with subsequent GC analysis of the extract. Szewczyk and Szymanowski showed that this technique gave PEG values lower, by about 1.3% absolute, than did extraction alone (32). The difference is due to co-extracted surfactant, as well as unidentified impurities.

Procedure: Determination of PEG by Gas Chromatography (32)

This procedure is used after preliminary separation of the PEG from the surfactant by the extraction or ion exchange methods described above. Compounds are analyzed as either the trimethylsilyl or the acetate derivatives. Standard techniques for derivatization are used, as available in the literature of vendors such as Supelco. An ordinary packed-column GC apparatus is used, with flame ionization detection. The column is stainless steel, 2.7 × 600 mm, packed with 2% OV-17 on Chromosorb G-AW-DMCS. Helium carrier flow is at 50 cc/min. Injector and detector are at 370°C; the column is programmed from 130°C to 300°C at 4°C/min.

4. TLC Method

PEG can be determined either by normal phase TLC, where it has a lower Rf value than ethoxylates, or by reversed-phase TLC, where it has a higher Rf than ethoxylates. In either case, modified Dragendorff reagent is most commonly used for detection.

Procedure: Determination of PEG in AE by Normal Phase TLC (37)

Modified Dragendorff reagent may be prepared by dissolving 170 mg $BiONO_3 \cdot H_2O$ in 22 mL HOAc and mixing with 4 g KI dissolved in 10 mL water. When dissolved, dilute to 100 mL and protect from light. Just before use, mix 10 mL of this solution with 1 mL 85% H_3PO_4, 10 mL ethanol, and 5 mL 20% $BaCl_2 \cdot 2H_2O$. The reagent is commercially available, but without the necessary barium chloride (Sigma).

The sample is diluted in absolute ethanol in the proportion 50 mg surfactant to 10 mL ethanol. A 5-μL aliquot is spotted on a ready-made silica gel plate. In the meantime, the developing chamber should have been filled with 100 mL of the developing mixture (40:30:30 ethyl acetate/H_2O/HOAc). Saturation of the vapor space is encouraged by putting a sheet of filter paper in the liquid to act as a wick. After allowing 30 min for equilibration, the TLC plate is placed in the chamber and allowed to develop until the solvent front has advanced 10 cm from the spotting point. The plate is dried in an oven at 100°C until the solvent is evaporated, sprayed with modified Dragendorff reagent until the orange spots are just visible, then let stand a few minutes until color development is complete. PEG has an Rf value of about 0.4, while ethoxylated alcohols, for example, have an Rf value from 0.74 to 1.0.

D. Refractive Index

The refractive index of a nonionic surfactant is dependent upon the starting material and the degree of ethoxylation. It is used in quality control to serve as a confirmation of identity. For example, within a limited range, the refractive index of AE and NPE is a linear function of the degree of ethoxylation (38–40). Over a much larger range, the refractive index is linearly related to the recriprocal of molecular weight. For example, Meszlényi et al. report the following relationship for ethoxylates of *tris*(*t*-butyl)phenol (41):

$$M = \frac{5.45}{\eta_D^{20} - 1.4740}$$

where M is the molecular weight. The relationship between refractive index and degree of alkoxylation is generally true of other alkoxylates, although the constants must be determined for each product family. A general test method for determination of refractive index is given in ASTM D1218 (42).

E. Cloud Point

Ethoxylated materials exhibit water solubility which is lower, rather than higher, at higher temperature (43). This is the basis of the cloud point test, an identity check for nonionic surfactants which reflects the extent of ethoxylation (44). A solution of the surfactant is warmed until it becomes cloudy, and the temperature recorded. For a family of materials, cloud point is directly proportional to extent of ethoxylation. Usually, the cloud point is not much affected by concentration in the 1–10% range. Highly ethoxylated materials (EO > 75%) often do not have a cloud point below the boiling point of water, although there may be a measurable cloud point in 1 M sodium chloride solution.

The presence of impurities, particularly salts and PEG, can have a significant impact on the cloud point. Contamination by ionic surfactants raises the cloud point of nonionics, while most salts have a lowering effect (45). The effect of salts is negligible until a critical concentration is reached (dependent on the salt, order of magnitude 0.1 molar), after which the cloud point is linearly related to salt concentration. The ionic surfactants have a gradual, concentration-dependent effect until the CMC is reached, beyond which the cloud point is very strongly affected (46).

This test is often called the compatibility index when solvents other than water are used. Isopropanol/water, 50:50, is generally used for less water-soluble materials. Mixtures of water and ethylene glycol monobutyl ether or diethylene glycol monobutyl ether are also applicable (47). For the CTFA cloud point, a 25 wt% solution of the surfactant is prepared in denatured 95% ethanol. Water is added until turbidity develops at 25°C. This volume of water is reported as the cloud point (48).

Procedure (49)

Weigh 1.0 g surfactant into a beaker and add 100 mL deionized water. Stir until dissolved. Transfer 50 mL to a test tube, 25 mm i.d., and warm in a hot water bath until cloudy, stirring occasionally with a thermometer. Remove from heat, place in a stand, and continue stirring occasionally until the solution is again clear. Often, the transition from cloudy to clear is most apparent when looking through the solution to a printed sheet of white paper. Record the temperature to the nearest 0.1°C.

F. Salts

Ethoxylation reactions are normally initiated with sodium or potassium hydroxide as catalyst. When the reaction is complete, the catalyst may be removed by filtration or liquid-liquid extraction, or may simply be neutralized and left in the product. The presence of neutralized catalyst may cause problems in certain applications, or may sometimes be detectable as turbidity or even crystals in the product. Normally, if the catalyst is removed, inorganic residues are reduced to the low parts per million level. If the catalyst is simply neutralized, the resulting salt is present in the 0.5–1.0% range.

Most often, catalyst is neutralized with phosphoric or acetic acid, because the buffering effect of these ions makes it easy to adjust the product to near-neutral pH. In general, organic anions will be less likely to form crystals which are insoluble in the product. Low levels of formic and acetic acid may also be formed by oxidation of polyethers (50).

Individual inorganic and organic ions can be determined by many methods. Nowadays, the surfactant is most often dissolved in water or a water/alcohol mixture and injected directly into an ion chromatograph for both qualitative and quantitative analysis. As an alternative, sulfated ash content may be determined as a measure of total salt content (see below).

Procedure: Determination of Organic Anions in an Ethoxylated Surfactant (51)

In this procedure, the anions are first isolated from the surfactant matrix on anion exchange SPE columns before determination by ion exclusion chromatography. This step avoids the column degradation which can occur from exposure to large quantities of organic material. The method was developed specifically for determination of acetate and lactate ion in APE, but should be universally applicable. For low concentrations, a narrower diameter column may be used, such as 2.6 × 100 mm. While this procedure is based on ion exclusion, conventional ion chromatography with conductivity detection will work as well.

Anion exchange SPE cartridges (Alltech) are put in the OH⁻ form by rinsing with 50 mL each of water, 1 M NaOH solution, and water. The sample is passed through as a 5% solution in water, adjusted to pH 9 with dilute NaOH. Typically, 100 mL of sample solution is used. (The capacity of the SPE cartridge is 0.5 meq.) The cartridge is rinsed with 150 mL of water. More water may be used if the eluent still foams. The cartridge is dried by vacuum, then the anions are eluted with 6 mL of 4.5 M H_2SO_4. A 1-mL aliquot of the eluate is diluted to 5 mL with water and analyzed by ion exclusion chromatography on a Polypore H (Brownlee/ABI) anion exclusion column, 7 × 250 mm, with Polypore H guard column, 3×50 mm. A mobile phase of 0.005 M H_2SO_4 is used, with detection by UV absorbance at 235 nm.

G. Low Molecular Weight Impurities: Ethylene Oxide, Propylene Oxide, 1,4-Dioxane, Acetaldehyde, Ethylene Glycol, Diethylene Glycol

All polymers must be checked for the presence of residual monomers, which by their nature are reactive, toxic materials. During manufacture, other low molecular weight materials may be unintentionally synthesized from EO and PO, such as acetaldehyde and 1,4-dioxane (from EO), and propionaldehyde, allyl alcohol, allyloxy-2-propanol, and substituted dioxolanes (from PO). Good manufacturing practice requires that these low molecular weight compounds be vacuum stripped from the surfactant at the end of the alkoxylation reaction. Generally, the presence of a high level of one of these is associated with the presence of others, so that it is not necessary to determine each individually for routine quality control. Dioxanes and dioxolanes from surfactant synthesis seem not to be the cause of odor complaints, unlike the similiar family of compounds produced as a byproduct of polyester resin manufacture (52,53).

Although low levels of volatile impurities may be determined by direct injection GC or GC-MS, this approach has disadvantages: (1) much nonvolatile material remains in the injection port or precolumn and must be routinely removed and (2) pyrolysis products and high-boiling impurities will elute at high retention volumes, requiring long elapsed times before another sample can be injected. Thus, some sample pretreatment is almost always performed. One approach has been a column-switching GC method, in which the first column serves the function of a distillation column, allowing only the highly volatile materials to be swept to a second column for analysis (54). While quite satisfactory, this procedure has been largely displaced by the more convenient headspace method.

The most popular cleanup method now is headspace GC or headspace GC-MS, which permits automated analysis of dozens of samples and standards in a single campaign. Vacuum distillation is at least as sensitive and accurate as headspace GC, but only a few samples and standards can be analyzed in a day. A number of liquid-liquid and liquid-solid extraction methods have been proposed to isolate 1,4-dioxane from surfactants prior to GC determination, but recovery varies with each matrix (55). The purge-and-trap methods used in environmental analysis are rarely applied to analysis of surfactants, since the impurities of interest are not readily adsorbed from the purge gas stream. In order to gain the high sensitivity of electron capture detection, ethylene oxide may be stripped and converted to ethylene iodohydrin before GC analysis (59). Detection limits in the low parts per billion range are claimed for this technique.

HPLC with UV detection at 200 nm has been proposed for determination of dioxane in ether sulfates and in cosmetics, with preliminary separation by solid phase extraction (56,57). While the HPLC procedure is too prone to give false positives to be very useful for trace analysis, it may be that solid phase extraction could be applied to GC analysis, making an expensive headspace apparatus or time-consuming distillation unnecessary. HPLC determination of EO and PO can be performed after formation of the dithiocarbamoyl esters, with detection limits in the 0.5 ppm range (58).

The analyst should be aware that compendial methods exist for the determination of some volatile impurities in ethoxylates. These are found in the publications of the Cosmetic, Toiletries, and Fragrances Association, the Association of Official Analytical

Chemists, and the *United States Pharmocopeia/National Formulary*, as well as the publications of other countries.

There is not often cause to determine ethylene glycol and diethylene glycol in surfactants. If the analysis becomes necessary, two procedures for determining EG and DEG in PEG found in the *National Formulary* are suitable (60). The first is a colorimetric method, while the second is a gas chromatography procedure. The GC procedure specifies use of a 5% phenyl/95% methyl polysiloxane capillary column, 0.32 mm × 50 m, 5 µm film thickness. Determination of EG by headspace GC requires higher temperatures and achieves higher detection limits than is the case for EO and 1,4-dioxane (61–63).

1. Headspace Gas Chromatography

As this technique is practiced, the surfactant is placed in a small vial and equilibrated at elevated temperature. The volatile materials partition themselves between the liquid phase and the vapor phase in the headspace of the vial. A portion of the headspace gas is transferred to a gas chromatograph for analysis. It is important that standards be prepared with a similar surfactant to that being analyzed because different materials have different affinities for the substances of interest, which, especially at low temperature, can result in different ratios of the concentrations in the headspace and the liquid phase (64). The problem can be circumvented by using a standard addition procedure so that a surfactant standard is not required (65,66).

Often, ordinary packed columns do not resolve ethylene oxide from acetaldehyde. Mass spectrometric detection does not help here, since the two compounds have the same weight. High resolution MS detection will at least differentiate the two compounds from CO_2. A convincing confirmation technique is determination of the compounds using two different GC columns.

As a rule, the lower the temperature used for equilibration, the fewer artifacts from decomposition of the surfactant clutter the chromatogram. On the other hand, the lower the equilibration temperature, the higher the detection limit for the analyte. Thus the optimal headspace GC conditions are somewhat dependent on the value chosen for minimum quantifiable concentration. Artifacts produced from decomposition of the surfactant include aldehydes and ethylene oxide itself. Some investigators add stabilizers like butylated hydroxytoluene to inhibit decomposition.

For extreme confidence in determination of 1,4-dioxane, headspace GC-MS may be used with an isotope dilution procedure, making use of the commercially available deuterated analog of dioxane (67). The precision of results becomes poor with decreasing analyte concentration. A coefficient of variation of about 30% is reported for determination of EO at the 0.2–30 ppm level in an alkyl ether sulfate, using the standard addition method of calibration (a single standard addition per sample) (68).

2. Vacuum Distillation–GC Determination of 1,4-Dioxane

Headspace GC is the method of choice for 1,4-dioxane (69). If headspace GC is not available, dioxane and other volatile materials may be concentrated by distillation prior to GC analysis (70).

3. HPLC Determination of Aldehydes

Aldehydes can be determined with great sensitivity in ethoxylated surfactants by HPLC of the 2,4-dinitrophenylhydrazine derivatives. This procedure allows individual determination of formaldehyde, acetaldehyde and propionaldehyde. (Formaldehyde is not a byproduct of alkoxylation and is normally absent from pure surfactants. However, it is an oxidation product and can form in aged ethoxylates.) The level of acetaldehyde reported by this HPLC method is often much higher than that found by headspace GC. This is thought to be an artifact of the analysis due to acid-catalyzed hydrolysis of acetal linkages which occur randomly in the polyether chain. The headspace GC method does not provide conditions for hydrolysis, so only free acetaldehyde is determined. The HPLC method, because of the derivatization reaction, exposes the polymer to harsher conditions, resulting in release of acetaldehyde.

Procedure (71)

The derivatizing reagent is prepared by dissolving 124 g 2,4-dinitrophenylhydrazine in 100 mL CH_3CN and adding 2 mL 2 M HCl. DNPH derivatives of formaldehyde, acetaldehyde, and propionaldehyde are used as standards. They are prepared by reaction of the aldehydes with 2M HCl saturated with 2,4-dinitrophenylhydrazine, separated from the reaction mixture by filtration, washed with water and 2 M HCl, air-dried, and stored in a desiccator.

About 1 g of surfactant is weighed out and dissolved in 4.0 mL of derivatizing reagent. After 5 min reaction, the solution is diluted to 10.0 mL with CH_3CN, then filtered and diluted 1:10 with HPLC mobile phase. The resulting solution is analyzed promptly by HPLC. An Ultrasphere ODS (Beckman) column is used, 4.6 × 250 mm. The mobile phase consists of 60:40 acetonitrile/water, 0.1 M in HCl, pumped at 1 mL/min. Detection is by UV absorbance at 365 nm. The order of elution is formaldehyde, acetaldehyde, acetone, propionaldehyde. The samples should be analyzed within 2 hr of derivatization as there are side reactions which form formaldehyde from the reagents. Detection at 254 nm is not suitable because impurities contribute to the baseline, causing poor sensitivity.

H. Unsaturation (Iodine Number)

During the synthesis of ethoxylated polymers, chain termination can occur by a mechanism which results in the formation of a $RC{=}CH_2$ group, rather than the desired terminal hydroxyl group. These compounds typically have a lower average molecular weight than the main component. Unsaturation determination is also required when characterizing other compounds, especially those made from natural fatty acids. In this case, the double bonds are usually not terminal but internal.

Several different techniques may be used for determination of unsaturation. The techniques give different results for various compounds. The referee method, which gives the highest results, is based on reaction with hydrogen in the presence of platinum catalyst and is not suitable for routine use. The Wijs procedure, based on reaction with iodine monochloride, is most often used in industry, but does not give total unsaturation in systems having conjugated unsaturation. Bromine number (72) is suitable for determination of total unsaturation in many compounds, although it gives high values in the presence of hindered phenol stabilizers. The mercuric acetate method, which determines specifically terminal (vinyl) unsaturation and certain *cis* unsaturation, as in cyclohexene, is most often used for ethoxylates.

Unsaturation may also be determined by spectroscopic techniques like IR, NIR, and NMR. NIR in particular may be more convenient if other parameters are determined simultaneously. Instrumental techniques are tailored to specific products and are usually verified by comparison to the wet chemical methods.

1. Mercuric Acetate Method

Procedure (73)

Prepare methanolic mercuric acetate reagent, 40 g/L, by dissolving 40 g reagent-grade mercuric acetate in methanol containing a few drops glacial HOAc (the quantity of HOAc should be sufficient to give a blank titration of 1–10 mL 0.1 M KOH per 50 mL reagent), diluting to 1 L. Keep no longer than one week, and filter before using.

Determine the acidity of the sample by titrating 30 g in 50 mL MeOH potentiometrically or to the phenolphthalein end point using 0.1 M methanolic KOH titrant. In a separate flask, determine unsaturation by dissolving 30 g sample in 50 mL mercuric acetate reagent, letting stand 30 min, then adding 8–10 g NaBr and titrating with 0.1 M methanolic KOH potentiometrically or to the phenolphthalein end point. Determine the reagent blank by titrating 50 mL mercuric acetate reagent. Calculation:

$$\text{Unsaturation, meq/g} = \frac{(V-B)(N)}{W} - \text{acidity, meq/g}$$

where V and B are the titrant volumes required by the sample and blank, respectively, N is the titrant concentration, and W is the sample weight in grams.

Note: The mercuric acetate method has been problematic for two generations of analytical chemists because of the toxicity of the reagents and the dependence of the results on chemical structure. ASTM subcommittee D202.22 has suspended its efforts to arrive at a value for acceptable reproducibility between laboratories.

2. Iodine Number

Procedure: Determination of Iodine Number by the Wijs Method (74)

This method relies on reaction of unsaturated compounds with iodine/chlorine reagent, and back-titration of excess reagent. Wijs reagent, 0.2 N, may be obtained from Riedel de Haen, no. 35071. It is a solution of iodine monochloride in glacial acetic acid.

The sample must be absolutely dry. Heat to melt, if necessary, and filter to remove solids. Weigh out enough sample to give 2.0–2.5 meq unsaturation into a dry 500-mL flask containing 20 mL $CHCl_3$. Pipet 25 mL Wijs solution into the flask. Carry out at least two reagent blank determinations at the same time. Mix and let stand in the dark for 1 hr at 25 ± 5°C. Add 20 mL 15% KI solution and 100 mL water. Titrate with 0.1 N sodium thiosulfate solution to the starch end point, with vigorous stirring. Calculation:

$$\text{Iodine value, g } I_2/100 \text{ g sample} = \frac{(B-V)(N)12.69}{W}$$

where B and V are milliliters titrant required for the blank and sample titrations, respectively, N is the concentration of the titrant (normality), and W is the sample weight in grams.

I. N-Nitrosamines

N-Nitrosamines are potent animal carcinogens which may be present at trace levels in many products. Since the development of the thermal energy analyzer made their detection a relatively simple matter, it has become routine to specify that commercial products be free of these compounds. This is particularly true in cosmetic raw materials, most importantly for the dialkanolamides [see Section III.D.7]. Amine oxides may also contain *N*-nitrosamines.

The European Community has been the most aggressive in nitrosamine testing. Typically, products are screened by a nonspecific method in which the samples are reacted with HBr to decompose *N*-nitroso compounds, forming NO gas. NO is swept into the thermal energy analyzer, where it reacts with ozone to give chemiluminescence (75,76). [The same detection technique may be applied by using a commercial NO detector, rather than the proprietary thermal energy analyzer (77).] This screening test may give false positives, but will not give false negatives. Thus, a product which "passes" this test is not subjected to further testing.

In the United States, it is more common to use the product synthesis chemistry to predict which nitrosamines might logically be present, then determine these by specific methods. The methods normally used are GC with MS or thermal energy analyzer detection or HPLC with thermal energy analyzer detection. The pitfall in this approach is that nitrosamines can be present from unexpected sources such as contaminants or trace additives. Such nitrosamines will probably only be detected by the nonspecific methodology.

J. Peroxide Content

Low levels of hydroperoxide groups are formed in polyether compounds upon exposure to oxygen and elevated temperature. Peroxide formation is also characteristic of photodegradation of nonionic surfactants (78). As such, the concentration of peroxide groups is an index of the degradation which has occurred to the polymer structure. This index is helpful only to a degree, since the peroxide group is an intermediate in the degradation, not the final product. Further decomposition gives carbonyl compounds, such as aldehydes and acids (79,80). The total peroxide content reaches a maximum value, generally in the parts per million region, early on. After this point, further polymer degradation does not cause the peroxide content to rise, since a steady state is maintained.

Some products contain stabilizers against oxidative degradation. Many of these stabilizers interfere with the conventional peroxide tests, so that a negative value is returned for peroxide content until the stabilizer is consumed. There are a variety of methods used for the determination of peroxides. Peroxides differ in reactivity, some being much more easily reduced than others. The iodometric procedures are most generally used for detection of peroxide impurities in polyethers (81,82). Iodine may be determined either by titration or spectrophotometry, with the titration methods sometimes giving low values due to complex formation between polyethers and iodine (83). Data on peroxide content should always indicate which method was used for the determination.

Although not accepted as a standard method, IR or near IR spectrophotometry is often used for determination of peroxide in well-characterized products. Comparison is made to a standard material, such as *tert*-butyl hydroperoxide.

1. Titration

Procedure (84)

Add a weighed sample of appropriate size to 20 mL acetic acid in a 250-mL conical flask, purged with CO_2. Add 10 mL CH_2Cl_2, stopper, and mix well. Add 5 mL freshly prepared saturated NaI solution, stopper, mix well, and let stand in the dark for 15 min. Add 50 mL water and titrate to the starch end point with 0.1 N $Na_2S_2O_3$ solution. Titrate a reagent blank prepared in the same manner. Calculation:

$$\text{Peroxide oxygen, ppm} = \frac{(A - B)(N)8000}{W}$$

where A and B represent milliliters titrant required for the sample and blank, respectively, N is the concentration of the sodium thiosulfate titrant, and W is the sample weight in grams.

Note: By using deaerated water for all solutions, by maintaining a CO_2 purge, and by generally using careful technique to exclude atmospheric oxygen, the value for the reagent blank can be maintained at less than 0.05 mL of 0.1 N $Na_2S_2O_3$.

2. Spectrophotometry

Procedure (85)

Dissolve no more than 15 g of sample in 2:1 HOAc/CHCl$_3$ in a 25-mL volumetric flask, and dilute to volume. Sparge the solution with nitrogen for 1–2 min and add 1 mL freshly prepared KI solution (20 g KI/20 mL deaerated water). Stopper, mix, and let stand in the dark for 1 hr. Measure the absorbance at 470 nm versus water, using a 1-cm cuvette. Measure a reagent blank prepared in the same manner. Compare the absorbance to a calibration curve prepared by adding 1–5 mL of a solution of 1.27 mg I_2 per mL HOAc/CHCl$_3$ to 25-mL volumetric flasks in the same manner as the samples. 1.27 mg I_2 is equivalent to 0.08 mg peroxide oxygen.

Note: While the ASTM procedure is widely used, it is not optimized for nonionic surfactants. The wavelength of maximum absorbance of iodine is shifted in the presence of ethoxylated materials, so that 360 nm is usually more appropriate than 470 nm, especially for low concentrations of peroxide. The 1-hr reaction time permits fading of the iodine color, so that the results of the ASTM spectrophotometric procedure are low. This can be compensated by taking readings every 5 min for the first 20 min after reagent addition and extrapolating the concentration, on a semi-log plot, to zero time (83).

K. Carbonyl Content

Carbonyl content is normally determined for polyethoxylated surfactants, such as EO/PO copolymers, that do not have a carbonyl group as part of their nominal structure. The presence of carbonyl compounds in polyethers indicates that some degradation of the product has taken place, or that lower molecular weight aldehydes were not stripped properly. Oxidation of ethoxylates results in the formation of formaldehyde, formic acid, acetaldehyde, and acetic acid, with the carbonyl functionality of primary interest being the aldehyde group (79,80). There is some interest in formate esters of the type R(OCH$_2$CH$_2$)$_n$OCHO, since they can be formed in photoinduced degradation (86,156).

For a thorough study of a specific surfactant, standards can be synthesized. For example, standards of the type C$_{12}$H$_{25}$(OCH$_2$CH$_2$)$_n$OCH$_2$CHO were prepared to study aldehyde formation in a dodecyl alcohol 5-mole ethoxylate, where n ranged from 0 to 4 (87). Determination of aldehydes produced by oxidation of the surfactant was then performed by GC-MS.

1. Titration

Carbonyl compounds will oxidize hydroxylamine hydrochloride, releasing acid:

$$R_1R_2CO + 2NH_2OH \cdot HCl \rightarrow R_1R_2CHOH + N_2 + 2H_2O + 2HCl$$

Procedure (88)

Add 50 mL of 10 g/L methanolic hydroxylamine hydrochloride solution to a 250-mL conical flask. Add 50 mL MeOH to a second flask and add thymol blue indicator solution (0.04 g/100 mL MeOH) to each. Use 0.05 M methanolic KOH or HCl to adjust the color of each solution to yellow with a slight orange tint. Add 50 g sample to each flask, mix well, then stopper and let stand for 15 min. Titrate each solution back to the original color, which should be stable for 5 min, with alcoholic KOH or, in the case of the blank, with HCl, as necessary. Calculation:

$$\text{Aldehyde content, as ppm acetaldehyde } = \frac{(V)(M)44050}{W} + \text{correction}$$

where V is the volume, in milliliters, of KOH titrant, M is its molar concentration, and W is the sample weight in grams. The inherent acidity of the sample, as shown by the blank titration, is subtracted from the result, while any inherent alkalinity is added to the result.

2. Visible Spectrophotometry

Aldehydes and ketones form derivatives with 2,4-dinitrophenylhydrazine. These derivatives develop a red color upon treatment with alkali. Although the intensity of the color is somewhat dependent upon the type of aldehyde compound, reaction temperature, and time allowed for color development, the method is generally suitable for assessing degradation in alkoxylates (89,90).

Procedure (91,92)

Weigh about 2 g sample, containing 0.5–50 µg carbonyl (as CO) into a 25-mL volumetric flask. Add 2 mL MeOH to a second flask for the reagent blank. To each flask, add 2 mL 2,4-DNPH solution (0.1 g 2,4-DNPH in 50 mL MeOH containing 4 mL conc. HCl; dilute to 100 mL and discard after 2 weeks), stopper, and let stand 30±2 min. Dilute to the mark with KOH solution (100 g KOH, 200 mL water, and methanol to give a final volume of 1 L), stopper, and mix well. Read the absorbance at 480 nm of sample and blank versus water. Compare to a calibration curve prepared with dilute solutions of a suitable carbonyl compound (usually acetaldehyde in the 10–50 µg range).

Notes: To measure carbonyl compounds bound as acetals, react with 2,4-DNPH reagent at 60°C rather than at room temperature, cooling before addition of KOH solution. Carbonyl compounds containing conjugated unsaturation will absorb at a different wavelength than that given here. Not all carbonyl compounds will react quantitatively under these conditions. Because of the sensitivity of the test, carbonyl compounds must be excluded from the room in which the test is performed. This particularly applies to that universal laboratory solvent, acetone. For determination of low levels of carbonyl compounds, the methanol may be purified by distilling from 2,4-dinitrophenylhydrazine hydrochloride. Care must be taken not to distill to dryness, since 2,4-DNPH is capable of detonating at high temperature.

3. Infrared Spectroscopy

A rapid method of detecting polyether degradation is direct examination by infrared spectroscopy. The sample is examined in a fixed pathlength cell, either neat or in solution, recording the absorbance in the 1700 cm^{-1} region. Calibration is with amyl acetate or a similar stable carbonyl compound. Sensitivity is generally low, in the 0.1% range.

Thus, the infrared method can detect gross degradation of the product, but not trace contamination.

4. HPLC Determination of Aldehydes and Acids

The procedure for aldehyde determination is described above in the section on EO, PO, dioxane, and acetaldehyde. Generally, HPLC is preferred for determination of all alde-hydes but acetaldehyde. Formic and acetic acid can be determined by a number of proce-dures, including gas chromatography and ion chromatography. A method for HPLC determination of these acids as their 2-nitrophenylhydrazone derivatives has been worked out specifically for ethoxylated esters of sorbitan (93).

L. Water Content

Water determination is generally performed by Karl Fischer titration, described in Chapter 1.

M. Apparent pH

Apparent pH is usually measured of an aqueous solution of 5 or 10% concentration. Ap-parent pH should be about neutral, although there will be variations due to the difficulty in measuring potentials in the relatively nonconducting medium of nonionic surfactant dis-solved in distilled water. The main value of pH determination is indication of gross devia-tions from product identity, such as an oxidized product or the presence of unneutralized or overneutralized catalyst.

N. Acidity or Alkalinity

By convention, acidity is defined as milligrams KOH required to bring 1 g of sample to the pH of the phenolphthalein end point. Alkalinity is equivalent milligrams KOH neutralized by addition of acid to reach the end point. This gives a value which can be used directly to correct the hydroxyl number determination (section B.1). This measurement is easily per-formed and gives an indication of consistency and impurity level for quality control pur-poses. The pitfalls of acidity and alkalinity determination are discussed in Chapter 1 with the analysis of anionic surfactants. The same procedures are used as for the analysis of anionics.

O. Ash

Ash determination gives a measure of gross impurity level and remains a quite useful tech-nique for quality control and troubleshooting. In the case of ethoxylated materials, the ash value simply indicates whether the alkaline catalyst has been removed or only neutralized. If the ash content is high, it is normally submitted to qualitative analysis, most often by X-ray fluorescence or emission spectroscopy. (Anion chromatography analysis is futile, since the ashing temperatures convert the original salts to oxides, carbonates, or, if H_2SO_4 is used, sulfates.) A general procedure for ash determination is given in ASTM D482 (94).

Most often, sulfuric acid is added during the ashing procedure (95). This drives off most anions as their acids. In such cases, sodium or potassium ions are weighed as sodium sulfate or potassium sulfate, regardless of which anions were originally used to neutralize the NaOH or KOH catalyst. Typical ash contents for ethoxylates are 1% if the catalyst was neutralized or 0.01% if it was removed.

III. ANALYSIS OF INDIVIDUAL SURFACTANTS

A. Alcohol Ethoxylates

Methods for analysis of AE are well established. ASTM standard D4252 specifies determination of water by Karl Fischer titration, pH of a 1% aqueous solution, hydroxyl number by phthalation, and PEG by extraction, as well as other tests (19). NMR determination of ethoxy content is the subject of CTFA Method J 3-1 (96).

1. Assay (Determination of Active Agent)

It is unusual to perform an assay on AE, since the commercial products are reasonably pure. Assay is not a regular part of the certificate-of-analysis quality check. Simple determination of AE in aqueous solution may be performed by potentiometric titration with tetraphenylborate ion as described in Chapter 16. This titration is less effective for AE with less than 10 moles of ethylene oxide. In this case, the product may be titrated after conversion to its sulfated derivative, also described in Chapter 16, although this is rarely done. Titration by the HI cleavage method (Section II of this chapter) may also be used for assay. AE can be determined by all of the chromatographic techniques described in later chapters. These are generally unsuitable for assay because of the 1–5% uncertainty associated with chromatographic determinations.

2. Free Alcohol Content

GC, LC, and SFC are all appropriate for determination of free alcohol, with SFC being best. In the case of oxo alcohols, there are so many isomeric forms of the various chain length alcohols that the GC and SFC retention times of some alcohols overlap with those of the monoethylene oxide adducts of other alcohols. Various derivatization techniques can be used to improve the resolution of the compounds. The acetate derivatives can usually be analyzed without difficulty by GC. These techniques are discussed in Chapters 7–10.

3. Alkyl Distribution

The chain length of the alcohol initiator may be determined by either gas, liquid, or supercritical fluid chromatography, as discussed in later chapters. Reversed-phase HPLC can be applied to the complete molecule, after derivatization, to aid detection. However, at the present time the resolution is not adequate to separate isomers. Highest resolution in GC requires the fission of the molecule to increase the volatility. For example, reaction with hydriodic acid will produce iodide derivatives of the starting alcohol which can be resolved into isomers of the various alkyl chain lengths by conventional GC. As discussed in Chapter 8, derivatized AE may also be analyzed without cleavage with somewhat less resolution of isomers. Supercritical fluid chromatography has been demonstrated for analysis of alcohol ethoxylates. Less resolution is obtained than with GC, but SFC is more useful than direct GC for analysis of higher molecular weight adducts.

B. Alkoxylated Alkylphenols

ASTM standard D4252 covers analysis of APE, specifying test methods generally identical to those recommended for AE (19). Most commercial APE is produced from 4-nonylphenol. This raw material contains perhaps 10% 2-nonylphenol and low levels of octyl- and decylphenol, and this same composition carries through to the APE. Fine structure of the alkylphenol portion of APE is best determined by GC or GC-MS after cleavage of the ethoxylate chain.

1. Assay

APE may be titrated in aqueous solution with tetraphenylborate ion, as described in Chapter 16. As with other ethoxylates, this titration is incomplete for products with less than about 10 moles of ethylene oxide. Chapter 16 describes how APE of lower degree of ethoxylation may be converted to the corresponding sulfate and titrated as an anionic surfactant. APE is sometimes assayed simply by UV spectrophotometry. APE is easily determined by HPLC and indeed by all of the chromatographic methods described in this volume.

2. General Characterization

High-field proton or ^{13}C NMR spectroscopy is suitable for determination of the average composition of the surfactant, as discussed in Chapter 14. If this information is not available, then cleavage is performed to permit GC or GC-MS characterization of the hydrophobe. Reversed-phase HPLC analysis of the intact surfactant is sufficient to show whether it is an octyl-, nonyl-, or dodecylphenol ethoxylate. Normal phase HPLC gives the ethoxy distribution, as described in Chapter 7. The ethoxy distribution of lower ethoxylates may also be determined by GC of the trimethylsilyl derivatives, while higher ethoxylates may be analyzed by SFC. Ultraviolet absorption spectra and refractive index have been suggested as rapid quality control identity tests for APE, although IR is more generally applicable (38).

C. Ethoxylated Acids

Anyone seeking to characterize ethoxylated acids should study the specifications and descriptions of the fatty acid starting materials, for which excellent brochures are available from major suppliers. For example, the oleic acid of commerce is a mixture of carboxylic acids but also contains other materials, which can total several percent. The purchaser has a choice of several grades of this so-called oleic acid, with the pharmaceutical grades not necessarily being the purest. Consider the fate of the impurities during the ethoxylation reaction to predict what might be likely impurities in the final product.

Ethoxylated acids can be differentiated from AE and APE (other than by spectroscopy) because the former are esters and thus are hydrolyzed in the presence of acid or (more often) base to form the starting acid and PEG. This operation may be performed in methanol to directly give methyl esters of the fatty acids for further analysis.

1. Assay

There are no truly satisfactory methods for assay of these products. They can be determined by most chromatographic methods, as discussed in later chapters. An approximation of active agent content may be made by determining the main impurities, namely free fatty acid, PEG, and water, and subtracting from 100%. Concentration of acid ethoxylates in solution can be determined by potentiometric titration with tetraphenylborate ion as described in Chapter 16. Free PEG is also determined by this titration.

2. Free Fatty Acid

Free fatty acid cannot usually be determined by direct acid-base titration because of the presence of the alkaline catalyst from the ethoxylation or its neutralization products. The two-phase titration usually applied to the determination of anionic surfactants does not suffer from interference from inorganic ions, and may be used. Titration in an alkaline medium with bromcresol green indicator permits titration of the fatty acid, which is not determined by titration at low pH.

Procedure: Two-Phase Titration of Free Fatty Acid (97)

Weigh 0.2–0.7 g acid ethoxylate sample into a stoppered 100-mL graduated cylinder and dissolve in 2.0 mL n-propanol and 7.0 mL water. Add 24 mL phosphate buffer (300 mL 0.065 M aqueous Na_2HPO_4, 100 mL 0.065 M Na_3PO_4, 80 mL n-propanol) and 20 mL $CHCl_3$, and adjust to pH 11.6 with 0.1 M NaOH or 0.1 M H_3PO_4. Add 0.8 mL bromcresol green (0.06 g/L water), and titrate with 0.004 M benzethonium chloride solution in 20:80 n-propanol/water. After each addition of titrant, shake and let the layers separate. The end point is taken as the time when the blue color leaves the aqueous layer and is only observed in the $CHCl_3$ layer. A blank must also be determined. This is in the range 0.4–0.8 mL. Calculation:

$$\% \text{ Free fatty acid} = \frac{(A - B)(N)(MW)}{(W)10}$$

where A and B are the mL titrant required by the sample and blank, respectively, N is the normality of the titrant, MW is the average molecular weight of the fatty acid, and W is the sample weight in grams. Presumably, this titration can also be performed using a surfactant-selective electrode for end point detection.

3. Mono- and Diester

(a) Calculation of mono- and diester content. PEG is determined by extraction (Section II of this chapter), and free acid by titration, as discussed above. The saponification number and hydroxyl number are obtained on recovered sample after extraction of PEG, as well as on the original sample. Mono- and diester content can then be calculated by simple arithmetic (98).

(b) HPLC determination. HPLC methods for analysis of acid ethoxylates are discussed in Chapter 7. HPLC often gives too much detail for routine characterization of these products, since resolution is by length of the PEG chain as well as by length of the alkyl chain, and by whether the compounds are mono- or diesters. Size exclusion chromatography is most useful for routine determination of the ratio of monoester to diester.

(c) NMR determination. Mono- and diester content can be determined by high-field proton NMR. Free polyethylene glycol must first be removed by extraction. The esters are first analyzed directly, then after addition of trichloroacetylisocyanate derivatizing agent. This reagent converts the hydroxyl groups of the monoester to the trichloroacetylcarbamate. Quantification is by integration of the signals from the methylene protons adjacent to the carbamate group and those from the protons adjacent to the ester linkages. Derivatization is necessary since the signal of the hydroxyl proton is rarely observed directly (99).

D. Fatty Acid Alkanolamides

The analyst must be aware that the esters and amides in these products exist in a temperature-dependent equilibrium, which can cause confusion when interpreting the analytical data. The product should not be melted before sampling.

Gas chromatography is the most useful technique for examination of the total product, while HPLC is appropriate if only the alkyl chain length distribution is of interest. Gas chromatography of the trimethylsilyl derivatives permits determination of the main components and impurities in a single analysis (see Chapter 8). If improper synthesis conditions were used, it is possible that some N,N'-bis(2-hydroxyethyl)piperazine would be present from condensation of two moles of diethanolamine. Conditions leading to formation of the diethanolpiperazine also favor formation of amine ester and amide ester impurities, so the level of the piperazine is an indicator of general product quality.

TLC allows determination of the main components and impurities simultaneously (Chapter 9). Reversed-phase HPLC analysis is most useful for the rapid determination of chain length distribution of the fatty acid portion of the alkanolamide. The materials are injected without derivatization, and detected by UV absorbance at 210 nm (Chapter 7). Although some impurities appear in the chromatogram, the analysis is not as comprehensive as that obtained with GC.

1. Determination of Fatty Alkanolamide

(a) Column chromatography. As an approximation of the active agent content, the product may simply be passed through a mixed-bed ion exchange column to remove ionic impurities, with quantification by weighing the residue (100). For more precise work, the more complex procedures of this section are followed.

(b) Nonaqueous titration. The sample can be titrated directly by dissolving in acetic anhydride and titrating with perchloric acid (see the procedure for ethoxylated amides, Section III.E). This will suffer from positive interference from amines and other basic compounds.

(c) Titration after derivatization. Alkanolamides may be derivitized to convert hydroxyl groups to sulfate groups, then titrated potentiometrically with a cationic surfactant as described in Chapter 16 (101). Diethanolamides can be titrated potentiometrically in this manner under either acidic or basic conditions.

2. Free Alkanolamine Content

The content of free diethanolamine in particular is controlled in these products because secondary amines open the possibility of N-nitrosamine formation if they should come in contact with nitrites. The methods most frequently used nowadays for alkanolamine analysis are gas chromatography, ion chromatography, and capillary electrophoresis. The titration methods have become less useful over the last decade as the concentration of the free alkanolamines in commercial products continues to decline.

(a) Determination of diethanolamine in diethanolamides by gas chromatography

Procedure (102)

The analysis is performed with wide-bore capillary columns, using two different columns to increase confidence at low concentrations. System 1: Restek Rtx-1 dimethylsilicone, 0.53 mm × 30 m, 5 µm film. Column temperature from 120°C to 170°C at 6°C/min, then to 250°C at 30°C/min; 30 min hold; splitless injection, FID. System 2: Supelco SPB-5 (5% diphenyl/95% dimethyl silicone), 0.53 mm × 30 m, 1.5 µm film. Column temperature from 120°C to 170°C at 6°C/min, then to 280°C at 30°C/min; 30 min hold; splitless injection, FID.

Samples of about 10 mg are dissolved in 10 mL MeOH; 2-µL injections are made. Calibration is with solutions of diethanolamine in MeOH. The limit of detection was calculated at 0.05 ppm, although the lowest result reported in the publication is 1%. Peak identity may be confirmed by GC-MS.

(b) *Determination of free diethanolamine in diethanolamides by titration.*   The free diethanolamine content may be calculated by determining total amine by titration in acetic acid solvent and subtracting the value obtained for amino ester content. The titration is typically performed with 0.1 M perchloric acid to either the methyl violet or the potentiometric end point (103,104).

3. Free Fatty Acid

Fatty acid can be determined by direct titration. The sample is dissolved in isopropanol or ethanol and titrated with alkali (104,105).

4. Amido Ester

Amido esters can be formed by thermal degradation of alkanolamides. These impurities are thought to inhibit foam stabilization performance of the product.

Procedure: Determination of Amido Ester Content by IR Spectrometry (106)

A conventional IR spectrometer is used with a sealed liquid absorption cell of 0.1 mm pathlength. Calibration is performed by adding a standard ester, such as N-2-lauroylethyl-N-2-hydroxyethyllauramide, to an ester-free alkanolamide such as N-bis(2-hydroxyethyl)lauramide. The five calibration solutions each contain 1 g of amide + amido ester, with ester contents of 0, 2, 5, 10, and 20%. The mixtures are dissolved in $CHCl_3$ and diluted to exactly 10 mL. The IR absorbance spectrum from 1600–2000 cm^{-1} is recorded for each standard. A baseline is drawn from 1700–1780 cm^{-1} and used to determine the net absorbance at 1740 cm^{-1}, and a calibration curve is drawn. The sample is analyzed in the same way, drying the $CHCl_3$ solution with Na_2SO_4 if it appears cloudy. If exactly 1.0 g sample and standard is not used throughout, the specific absorbance [net absorbance/(path length)(sample wt)(10)] should be used for calibration.

Methyl palmitate may be used for calibration rather than the lauroyl amido ester. The calibration curve is still plotted in terms of the lauroyl compound, with the actual methyl palmitate concentration multiplied by 1.93 to give the corresponding percentage of lauroyl ester.

5. Total Amine Content

Total amine content of alkanolamides can be determined by potentiometric titration of the sample in acetic acid solvent with a titrant of perchloric acid or trifluoromethanesulfonic acid. In either case the titrant is diluted in acetic acid (107,108).

6. Determination of Amino Ester

The sample is dissolved in ethyl ether and washed free of ionic material with 10% sodium chloride solution. The ether phase is evaporated, and the residue dissolved in dry acetic acid and titrated with perchloric acid (103).

7. *N*-Nitrosodiethanolamine

N-Nitrosamines are formed by reaction of a secondary amine (or an amide containing a single N–H moiety) with a nitrosating agent, usually nitrite ion. Although such compounds would not normally be expected in alkanolamides, trace quantities can be formed if they or the amine raw materials come in contact with nitrosating agents during production or storage. Analysis may be required for those products destined to become cosmetic raw materials. Although there are general tests which can indicate the presence of any member of the general class of N-nitrosoamines, these are only considered suitable for screening. For quantitative analysis, it is necessary to determine specific N-nitroso compounds by methods which can discriminate between them. If the only secondary amine present in the product is diethanolamine, then the only nitrosamine which can be formed is N-nitrosodiethanolamine (NDElA).

GC-MS analysis, while normally considered definitive for confirmation of nitrosamines, is not suitable for determination of the nonvolatile NDElA unless silylation is first performed. Although procedures have been published for determination of NDElA by polarography (109), by a nonspecific "total nitrosamine" analysis train (77), and by spectrophotometry (110), the only widely accepted method for determination of NDElA in amide surfactants is HPLC with detection by the thermal energy analyzer (TEA) (111,112). Ultraviolet detection is also used, but the results must be confirmed with the TEA detector (113,114). Most recently, a non-TEA HPLC method has been proposed incorporating extensive preliminary sample workup: liquid-liquid extraction of NDElA from the matrix, alkaline decomposition to produce nitrite ion, and derivatization of the nitrite with 4-methyl-7-aminocoumarin, followed by hydrolysis, to give the fluorescent 7-hydroxycoumarin (115). The eventual acceptance of this method will depend on how well its results track those obtained from TEA detection.

Chou recommends that 4-hexyloxyaniline be added to alkyldiethanolamide samples before HPLC-TEA analysis as an antioxidant to prevent artifact formation of NDElA (102).

8. Characterization of Alkanolamide by Hydrolysis

Procedure (116,117)

About 500 mg of the isolated material is refluxed for 8 hr with 20 mL 6 M aqueous HCl, then the liberated fatty acids are extracted into petroleum ether (once with a 100-mL, then twice with 50-mL portions). The extract is washed with 10-mL aliquots of water until free of acid, dried with sodium sulfate, filtered, and reduced in volume to about 15 mL. Fifteen milliliters 96% ethanol is added, and the acids are titrated to the phenolphthalein end point with 0.1 M ethanolic KOH under a CO_2 blanket. The solution is evaporated to dryness and weighed. The equivalent weight is equal to the final weight divided by the milliequivalents titrant added, the whole less 38 to correct for the weight of the K^+ added.

The aqueous phase from the petroleum ether extraction contains the amine from hydrolysis of the amide. This may be identified by neutralizing the aqueous phase with alkali and analyzing by GC.

9. Column Chromatography

A comprehensive procedure has been worked out for characterization of ethanolamides (116,118): A sample is dissolved in ethanol and passed through ganged cation and anion exchange columns, rinsing with ethanol. The eluate contains the nonionic components, namely the alkanolamide and the alkanolamide ester. This fraction is evaporated to dryness, dried with absolute EtOH and acetone, and weighed. A portion is fractionated on a silica gel column to separate the alkanolamide and the ester. Then, the ion exchange columns are separated and the free ethanolamine is eluted from the cation exchanger and quantified by titration. Free fatty acid is eluted from the anion exchanger and also determined by titration. A simpler procedure is given below, where dialkanolamides are separated into their components by chromatography on silica gel for subsequent characterization by spectroscopic methods.

Procedure (119)

A 1–2 g sample is dissolved in 20 mL $CHCl_3$ and added to a column, 1 × 20 inches, containing 75 g silica gel, Davison no. 992. Sequential elution is performed with 70 mL $CHCl_3$, 100 mL 99:1 $CHCl_3$/ethyl ether, 70 mL 50:50 $CHCl_3$/ethyl ether, 80 mL 50:50 $CHCl_3$/acetone, 100 mL 95:5 $CHCl_3$/MeOH, 70 mL 90:10 $CHCl_3$/MeOH, 75 mL 2:1 $CHCl_3$/MeOH, and 100 mL MeOH. If desired, free amines are removed in a final rinse of aqueous 2% HCl. Ten-milliliter fractions are collected throughout the elution, weighed, and identified by infrared spectroscopy and other means.

E. Ethoxylated Amides

Two complete schemes have been worked out for characterization of these products. That of Mutter et al. relies on column chromatography separations and gravimetry, with TLC to confirm the efficiency of separations (118). Krusche recommends similar techniques, along with nonaqueous titration for much of the quantification (120).

1. Assay of Ethoxylated Ethanolamide

The determination of active agent is complicated by the presence of various impurities in the commercial product: ethoxylated ethanolamine, ethoxylated ester, polyethylene glycol, fatty acid. The nonaqueous titration is rapid, but requires knowledge of the equivalent weight of the surfactant. This makes the method unsuitable for the analysis of unknown materials unless further characterization is performed.

(a) Titration with acid (120). Nonaqueous potentiometric titration in acetic anhydride will give the sum of the ethoxylated amide, ethoxylated amine, and the ethoxylated ester. Titration in glacial acetic acid will give the amines, but not the more weakly basic amide. The active agent can thus be determined by difference.

Procedure: Assay by Titration (120,121)

Perchloric acid titrant: Add about 250 mL glacial HOAc to a 1-L volumetric flask on a magnetic stirrer. While mixing, add 8.6 mL (14.4 g) 70% perchloric acid and 22 mL (24 g) acetic anhydride. [If 60% perchloric acid is used, add 10.9 mL (16.8 g) acid and 35 mL (38 g) acetic anhydride.] Fill to the mark with acetic acid. Mix well and allow to reach room temperature before standardizing against 4-aminopyridine.

Dissolve a sample containing about 0.5 meq total base in 50 mL acetic anhydride. Titrate potentiometrically, using a glass combination electrode, with 0.1 M $HClO_4$ in acetic acid. Titrate to the second, weak break. This value corresponds to the sum of the ethoxylated amide, ethoxylated amine, and ethoxylated ester. Dissolve a second portion of the sample, about five times as much as titrated above, in glacial acetic acid. Titrate potentiometrically with 0.1 M $HClO_4$ in acetic acid. This titration gives the sum of the ethoxylated amine and ethoxylated ester. Calculation:

$$\% \text{ Ethoxylated amide} = \left(\frac{V_1 N}{W_1} - \frac{V_2 N}{W_2} \right) \frac{MW}{10}$$

where V_1 and V_2 are milliliters titrant required in the first and second titrations, W_1 and W_2 are the sample weights used in the titrations, and MW is the molecular weight of the ethoxylated ethanolamide.

(b)   *Titration with cationic surfactant.*   Potentiometric titration of an ethoxylated mono-ethanolamide was demonstrated after formation of the sulfate derivative (101). The titration was effective only under basic conditions, pH 10.

(c)   *Gravimetric determination of active agent*

Procedure (120)

Dissolve a portion of the sample in water/methanol and let pass through, in series, a cation exchange column (in the H^+ form) and an anion exchange column (in the OH^- form). Wash the columns with methanol/water, collecting the eluate. Evaporate the combined eluate to dryness and weigh. This nonionic fraction corresponds to the ethoxylated alkanolamide and PEG.

Dissolve the residue in aqueous saturated NaCl solution and extract three times with ethyl acetate. Wash the combined ethyl acetate layers with fresh NaCl solution, evaporate the ethyl acetate solution to dryness, and weigh. This is the ethoxylated alkanolamide. Extract the brine solution three times with chloroform. Evaporate the extract to dryness. If salt crystals are visible, redissolve in acetone, filter, and again evaporate. Weigh the residue. This is PEG.

(d)   *Characterization of the isolated ethoxylated alkanolamide.*   The separated amide may be hydrolyzed with HCl at 160°C, and the fatty acid portion extracted into petroleum ether. The acids are then derivatized and analyzed by gas chromatography of the methyl esters. Once the alkyl chain length has been determined in this way, the ratio of ethoxy groups to alkyl groups, obtained by NMR, allows calculation of the molecular weight of the compound (120).

2.   Ethoxylated Alkanolamine

(a)   *Determination of ethoxylated amine by extraction and column chromatography*

Procedure (120)

A portion of the original sample is dissolved in concentrated aqueous NaCl solution and extracted three times with ethyl acetate. The combined ethyl acetate layers are back-washed once with fresh NaCl solution. The combined NaCl solutions are then extracted three times with $CHCl_3$. The $CHCl_3$ layers are combined and evaporated to dryness. If salt crystals are evident, the extract is redissolved in acetone, filtered, and again evaporated and weighed. This is the total of the polyethylene glycol and the ethoxylated amine.

The residue is dissolved in $MeOH/H_2O$ and passed through a cation exchange column (hydrogen form), rinsing the column with $MeOH/H_2O$. The combined eluate is evaporated to dryness and weighed. This is the polyethylene glycol. The weight of the ethoxylated amine is calculated from the difference in weights of the total extract, above, and the polyethylene glycol. If the equivalent weight of the ethoxylated amine is required for further calculations, the amine may be eluted from the cation exchange column with methanolic HCl. It is separated from the HCl by extraction and titrated with $HClO_4$ in HOAc solvent.

(b)   Determination of ethoxylated amine by extraction and titration.   If the equivalent weight of the ethoxylated amine is already known, the cation exchange procedure described above may be replaced by a simple titration with $HClO_4$ in acetic acid solvent (120).

3.   Free Fatty Acid

The original sample is titrated potentiometrically with tetrabutylammonium hydroxide solution in methanol. If the equivalent weight of the fatty acid is known, this titration suffices for the acid determination (120).

4.   Polyethylene Glycol

The PEG value is obtained during the assay of the ethoxylated alkanolamide by gravimetry or during the determination of ethoxylated amine by ion exchange and extraction, as described above. Both procedures consist of extraction and ion exchange separation; only the order is different (120).

5.   Ethoxylated Ester

The ethoxylated ester concentration is obtained by difference. During the titration for assay, a value, in milliequivalents per gram, was obtained for total amine ethoxylate and ester ethoxylate. This value is corrected for the ethoxylated amine content and equivalent weight, as determined above (120).

F.   Ethoxylated Amines

An ISO standard, 6384-1981, specifies methods of analysis for ethoxylated fatty amines: Apparent pH is determined at a 5% concentration in 50:50 methanol/water, water by Karl Fischer titration, amine types by sequential titrations, EO content by HI cleavage, corrected, and PEG by ion exchange chromatography (30). Most of these procedures are discussed in Section II of this chapter. A note on apparent pH: the method of calibration is with buffers prepared in methanol/water. This differs from the practice in most chemical laboratories.

Unsaturation is a typical specification, since these are often based on natural products like tallow or coconut oil. The common test for iodine value by the Wijs method gives high results for ethoxylated amines. Results better related to the true degree of unsaturation are obtained by methods which titrate directly with the halogenation reagent, rather than react with an excess as in the Wijs method. Accurate results are reported using bromide/bromate reagent and electrometric end point detection, as in ASTM D1159 (122). The 1,1,1-trichloroethane specified in the solvent mixture may be substituted by octamethyl cyclotetrasiloxane (123).

1.   Assay

Amines of low degree of ethoxylation can sometimes be titrated at low pH according to the methods for cationic surfactant titration described in Chapter 16. Alternatively, amine ethoxylates with alkoxy chain length less than about 10 can be converted to the corresponding ether sulfates and titrated as anionic surfactants, also discussed in Chapter 16 (101). Two-phase titration is required, since these compounds respond poorly to surfactant-selective electrodes. Higher ethoxylates are true nonionics and can be titrated as such with tetraphenylborate solution, also discussed in Chapter 16 (124).

2. Homolog Distribution

Both the alkyl chain length distribution and the ethoxy chain distribution can be character-ized by HPLC, using two different column and eluent systems. HPLC is covered in more detail in Chapter 7. Great selectivity for ethoxylated amines in the presence of other non-ionic materials is attained by using a selective detection system based on extracting the ion pair of the eluting compound and measuring its fluorescence online (125). The ethoxy dis-tribution may also be determined by supercritical fluid chromatography (126) or mass spectrometry (157).

3. Primary, Secondary, and Tertiary Amines

The classical titration methods for determination of amines may be applied to ethoxylates. Primary amines are reacted with salicylaldehyde, allowing titration of secondary and ter-tiary amines, as the first break of the curve, then of the primary amines, to the second potentiometric break. In a second titration, primary and secondary amines are reacted with acetic anhydride, allowing selective titration of tertiary amines.

Procedure: Determination of Primary, Secondary, and Tertiary Amines by Titration (30)

(a) *Primary and total amine.*   Weigh a sample of about 1 g into a beaker and dissolve in 20 mL solvent. (The ISO method specifies 1,4-dioxane, a suspected carcinogen. This is usually replaced with other solvents, such as 2-PrOH or $CHCl_3$.) Add 5 mL glacial acetic acid and 5 mL salicylaldehyde, mix, and let stand 30 min. Add 50 mL 1,4-dioxane and 20 mL nitroethane and titrate potentiometrically with 0.1 M $HClO_4$ in 1,4-dioxane, using a glass combination pH electrode. The first potentiometric break of the titration curve is due to the total of the secondary and tertiary amines. The second break is the end point for the primary amine.

(b) *Tertiary amine.*   Weigh a sample of about 1 g into a 50-mL conical flask, add 10 mL acetic anhydride, fit with a condenser, and heat to reflux for 1 hr. Cool and transfer the contents to a 250-mL beaker with 100 mL glacial HOAc. Titrate potentiometrically with 0.1 M $HClO_4$ in 1,4-dioxane. This titration corresponds to the tertiary amine.

4. Oxyethylene Content

The procedure given in Section II of this chapter may be used. However, the calculation must be adjusted because the HI cleavage method does not remove the EO units attached to N atoms (30). Calculation:

$$\% \text{ EO} = \text{EO}_x + 4.4(2X_3 + X_2)$$

where EO_x is the percent EO determined by the HI cleavage method and X_2 and X_3 are the secondary alkalinity and tertiary alkalinity, respectively (in meq/g). This calculation will give erroneous results if the sample contains free, unethoxylated amine.

G. Esters of Polyhydroxy Compounds

Condensation of sugars with fatty acids by commercial processes yields a product hetero-geneous as to number of acyl groups per molecule as well as heterogeneous in isomer con-tent. HPLC or GC may be used to determine free fatty acids and fatty acid methyl esters. Simple determination of acid number is usually sufficient for quantifying total unreacted fatty acid. These esters can be saponified to yield the original starting materials. This property sometimes permits them to be distinguished from other surfactants in mixtures.

1. Titration

Potentiometric titration of these compounds has been demonstrated after formation of the sulfate esters (101). They are titrated with a cationic surfactant, as described in Chapter 16. Since there are multiple hydroxyl groups available for the derivatization reaction, response factors must be developed for each commercial product.

2. Total Acid and Total Alcohol Content

Characterization of esters generally requires that they be saponified for subsequent analysis of the acid and alcohol. Determination of the degree of esterification is generally carried out by GC or HPLC, as described in Chapters 7 and 8.

(a) Determination of total sucrose and fatty acid content of sucrose esters by saponification and gas chromatography

Procedure (127)

Mix 0.050 g sample and 50 mL 4% NaOH in EtOH in a 100-mL volumetric flask and let stand overnight at room temperature. After this saponification, neutralize to the bromthymol blue end point with 1 M HCl and dilute to volume with EtOH.

   Sucrose determination: Transfer a 1-mL aliquot of the saponified solution to a 5-mL test tube and evaporate to dryness at 90°C. Add 1 mL pyridine and 0.2 mL each of trimethylchlorosilane and N-trimethylsilylimidazole. Shake well and let stand 30 min. Add 1 mL of 100 ppm n-octacosane ($C_{28}H_{58}$) as internal standard, dilute to 3 mL with pyridine, and analyze 5 µL aliquots by gas chromatography, comparing to a calibration curve made by treating pyridine solutions of sucrose in the same way. GC conditions: 2% OV-17 on Chromosorb W (AW-DMCS), 0.3 × 225 cm, 230°C, FID.

   Fatty acid determination: Transfer all remaining saponified solution to a separatory funnel, add 100 mL water and 5 mL HCl, and extract twice with 50-mL portions of ethyl ether. Dilute the combined ether extracts to volume in a 100-mL volumetric flask. Transfer an aliquot to a 10-mL test tube, evaporate to dryness at 40°C, and add BF_3-methanol reagent. React at 65°C for 5 min, then transfer to a 50-mL separatory funnel, add 15 mL saturated NaCl solution, and extract with 10 mL ethyl ether. Add 1 mL of 250 ppm n-eicosane internal standard solution, dilute to 10 mL with ether, and analyze by gas chromatography, comparing to ethyl ether solutions of fatty acids (100–500 ppm) derivatized in the same way. GC Conditions: 2% DEGS + 0.5% H_3PO_4 on Chromosorb W (AW-DMCS), 0.3 × 225 cm, 165°C, FID.

(b) Determination of the sorbitol content of sorbitan esters.   The sorbitol portion of the molecule can be quantified by saponification to yield the polyhydroxy alcohol. The sorbitol can be further decomposed with periodic acid to yield formaldehyde, which is determined by the chromotropic acid procedure (128). This test is not suitable for application to unknown mixtures since ethoxylated materials will also be oxidized to formaldehyde. Although few recent publications have addressed this area, it seems reasonable that a modern approach to this determination would concentrate on the more specific determination of sorbitol in the saponified material by infrared or liquid chromatographic analysis. Most manufacturers of HPLC apparatus supply columns designed for the analysis of monosaccharides such as sorbitol.

3. Unesterified Alcohol

Unreacted sucrose may be separated by dissolving the mixture in n-butanol and extracting the sucrose with water at a pH of 5–6 (129,130). This extraction also removes alkali metal ions as their phosphate salts.

Procedure: Determination of Free Sucrose in a Sucroglyceride (130)

Approximately 1 g sample is weighed, dissolved in 25 mL water-saturated n-butanol, transferred to a separatory funnel and carefully adjusted to a pH of 5–6 (in the aqueous layer) by dropwise addition of 20% H_3PO_4. The butanol phase is washed three times with 5-mL aliquots of water, swirling gently to avoid emulsion formation. The combined aqueous extracts are transferred to a 100-mL volumetric flask and diluted to volume with water. Quantification of sucrose is by the anthrone method recounted in Snell and Snell (131): Place 2.0 mL sample solution, containing less than 160 mg sucrose, in a beaker and add 0.5 mL 2% anthrone in ethyl acetate. Carefully layer 5 mL concentrated sulfuric acid below. Swirl until the anthrone dissolves, indicating hydrolysis of the ethyl acetate. Swirl vigorously to mix and measure absorbance at 620 nm after 10 min versus a reagent blank. Prepare a calibration curve versus pure sucrose oven-dried at 110°C for 2 hr.

4. Monoester Content of Glycerides

Monoglyceride esters are almost entirely 1-glycerides. Thus, the monoglyceride content can be determined by the periodic acid cleavage method, which is specific for determination of adjacent hydroxyl groups. The standard methods of the AOCS and AOAC International specify dissolving the sample in chloroform and reacting with periodic acid in a two-phase system, determining the periodic acid consumed using iodometric titration (132,133). Free glycerol or other compounds with vicinal hydroxyl groups cause interference. A correction is made for free glycerol content by performing the titration on an aqueous extraction of the sample solution, which will extract glycerol but not glycerides (133). GC analysis is a more generally informative method for determining the composition of monoglycerides, as described in Chapter 8.

5. Mono-, Di-, and Triester Distribution

Glycerides may be separated into the classes of mono-, di-, and triglycerides by column chromatography on silica [(133); see also Chapter 6]. Reversed-phase HPLC analysis is capable of separating homologs according to total alkyl content, i.e., according to number of acyl groups per molecules and according to the chain length of the acyl groups. Normal phase HPLC or TLC is capable of differentiating isomers. Details are in Chapters 7 and 9. The most interesting new work on esters in the surfactant range is being performed by high temperature GC (Chapter 8).

Procedure: Determination of Ester Distribution of Sucroglycerides by TLC (129)

TLC plates of Merck Silica Gel 60 are used. For determining glycerides and fatty acids, plates are developed with 90:30:1 petroleum ether/diethyl ether/acetic acid and visualized with copper acetate/H_3PO_4, followed by charring at 175°C for 20 min. The developing reagent for sucrose esters is 2:1:1 toluene/ethyl acetate/95% ethanol, with visualization by spraying with urea/H_3PO_4/n-butanol and heating at 110°C for 30 min.

6. Preparative Separation of Sucrose Esters

The crude product is dissolved in hot methyl ethyl ketone. Soap (emulsifier for the process) and sucrose are insoluble, and are filtered out. The solution is acidified to convert remaining soap to the soluble fatty acids, then cooled to precipitate salt and the sucrose ester (134).

H. Ethoxylated Esters

Ethoxylated esters of fatty acids and sorbitol, sucrose, and glycerin are widely used in food and pharmaceuticals, and their analysis is performed according to compendial methods. The specified methods are generally those listed in Section II of this chapter, as well as modifications of the above procedures applied to unethoxylated esters. Like other polyethoxy compounds, these can be determined in aqueous solution by titration with tetraphenylborate ion as described in Chapter 16 (124).

I. Alkylglycosides

Alkylpolyglycosides (APG) are complex mixtures in that there is a distribution of mono-, di-, and higher glycosides as well as an alkyl chain length distribution. Commercial APG consists of about 50% monoglycosides, with decreasing quantities of higher oligomers

Glucosides can form either 5- or 6-membered rings, so that alkylmonoglucoside has four possible isomers (135):

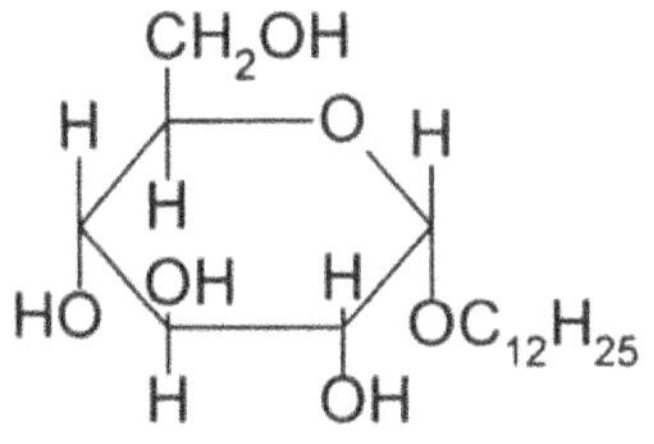

Lauryl α-glucopyranoside

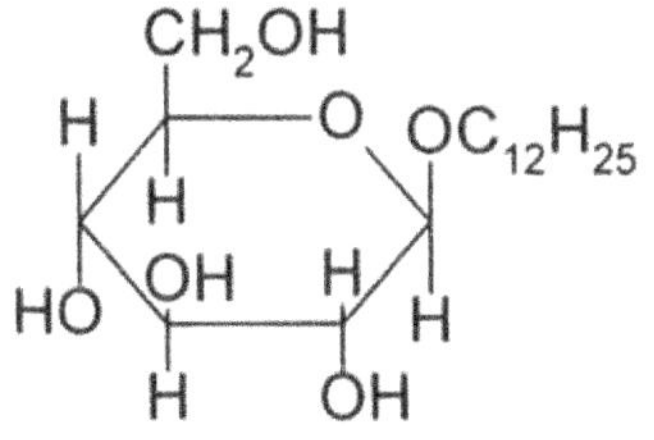

Lauryl β-glucopyranoside

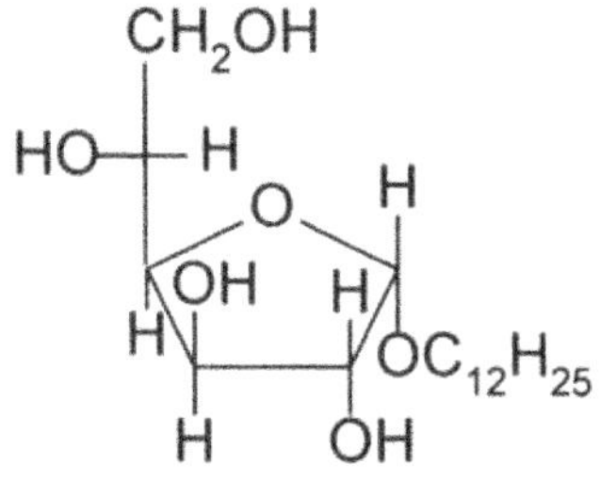

Lauryl α-glucofuranoside

Lauryl β-glucofuranoside

A recent book chapter discusses analysis of alkyl polyglycosides (136). APG is usually determined in formulated products by HPLC or TLC.

1. Assay

Alkylpolyglucosides can be converted to sulfates by reaction with DMF · SO$_3$. These anionic derivatives can then be titrated potentiometrically with a cationic surfactant (137). The calibration factors differ depending on the alkylpolyglucoside products; other surfactants interfere.

Alkylpolyglucosides can be determined in formulations by a colorimetric procedure. Acid hydrolysis converts polyglucosides to glucose, which forms a green complex with anthrone. Cellulose derivatives and propylene glycol interfere, but not glycerol and sorbitol (138). A similar procedure consists of enzymatic conversion of APG to glucose, with determination of glucose with an enzymatic electrode. Unfortunately, an enzyme mixture has not yet been found which will cleave all APG linkages to glucose, so this determination remains somewhat emperical (139). LC and TLC are also used for detection of APG in formulations (see Chapters 7 and 9).

2. Alkyl Chain Length

Alkyl polyglucosides are separated according to alkyl chain length by reversed-phase HPLC, reversed-phase TLC, and high-temperature GC, as described in the respective chapters later in this volume. Alternatively, alcohols can be liberated by acid hydrolysis of the APG and determined by HPLC (140).

3. Degree of Polymerization

Degree of polymerization for polyglucosides is taken to be the ratio of the number of moles of glucose bound in the alkylpolyglycoside to the number of moles of alcohol bound in the alkylpolyglycoside (135). This can be determined by proton NMR or, more accurately, by ^{13}C NMR (135). In either case, unreacted alcohol and glucose raw materials must be removed prior to the analysis. Degree of polymerization can also be determined by high-temperature GC, as described in Chapter 8, as part of a complete characterization of the product.

4. Isomer Composition

Gas chromatography is capable of resolving alkylmonoglucosides into their α- and β-pyranoside and furanoside isomers (see Chapter 8).

5. Free Alcohol

Unreacted fatty alcohol may be separated from APG by applying 20 mL of a 1% aqueous suspension of APG to a silica solid phase extraction column, eluting with 100 mL n-hexane, and concentrating by evaporation under vacuum. Analysis is carried out by gas chromatography on a Perkin-Elmer SPB5 30-m capillary column at 200°C (141). Analysis can also be carried out by direct GC after derivatization (136). These conditions are given in Chapter 8. HPLC is also applicable to determination of free alcohol in most types of APG, as described in Chapter 7.

6. Free Glucose

High-temperature GC analysis of APG gives early-eluting peaks for free α- and β-anomers of glucose. Glucose can also be determined enzymatically using test kits available from Sigma-Aldrich, Boehringer, and other suppliers (136).

J. Ethylene Oxide/Propylene Oxide Block Copolymers

An excellent summary of the characterization of these materials appeared in 1996 (142).

1. Composition

Qualitative infrared analysis is sufficient to show that the product is an EO/PO copolymer. However, this is not adequate as an identity check because of the large number of products in this category. The differentiation between products consists in differences in the molecular weight and in the EO/PO ratio. Quantitative IR spectroscopy, with standards, can answer these two questions by giving the concentration of terminal hydroxyl groups (and thus the equivalent weight) and the ratio of methylene to methyl groups (and thus the EO/PO ratio).

It should be noted that NMR is more reliable for the determination of EO/PO ratio than IR. IR requires calibration with standards, and the calibration curve is not linear over all EO/PO ratios (143). GC determination of the reaction products after cleavage with acetyl chloride may be used for determination of EO/PO ratio, but a calibration curve is also required for this method, because of nonlinear response above an EO content of 30% (144).

Another test occasionally run on these compounds is percent secondary hydroxyl content, as determined by ^{13}C NMR spectroscopy (145). This is a measure of polypropylene oxide base material with one end uncapped by polyethylene oxide groups. This value is generally very low for EO/PO copolymers in the surfactant range, but can be more significant for analogous polymers made with butylene oxide rather than propylene oxide (146).

2. Antioxidant

These surfactants often contain a hindered phenol antioxidant to stabilize the polymer against degradation. Usually, a single antioxidant is present, and it may be determined by ultraviolet spectroscopy. The determination is also conventionally performed by reversed-phase HPLC or by gas chromatography, with or without prior derivatization. Positive identification of the stabilizer can usually be made by direct mass spectrometry.

In some cases, the stabilizer can be separated by extraction with basic aqueous solution from an organic solution of the surfactant. However, the solubility and emulsifying properties of most of these copolymers prevent such enrichment. There is some evidence that the stabilizer can be removed by passage of an aqueous solution of the surfactant through a silica gel column (147).

3. Impurities

The free polyalkalene glycol found in these copolymers is normally PEG, since ethylene oxide is generally added last during synthesis. Its formation is initiated by water impurities during the addition of EO to the polypropylene oxide base. The molecular weight of PEG is usually significantly less than that of the main product.

Especially for higher molecular weight materials, chain termination reactions occur which leave one end of the polymer with terminal unsaturation, rather than a hydroxyl group. This leads to a bimodal molecular weight distribution, with the unsaturated component having a broad MW distribution lower than that of the main component. The amount of this component can be determined semi-quantitatively by size exclusion chromatography (Chapter 7).

In addition to the small molecules expected in ethoxylates (ethylene oxide, 1,4-dioxane), various three-carbon oxygenated compounds (allyl alcohol, acrolein, dioxolanes) may also be present. They are determined by the procedures described in Section II. The presence of carbonyl impurities such as develop after exposure to air is indicated by the strengthening of the UV absorbance band in the 210-nm region.

K. Amine Oxides

1. Determination of Amine Oxide and Unreacted Amines

High-field NMR may be used to obtain the approximate percentage of free amine in an amine oxide. For some products, the analysis can also be performed by GC, using any of a number of columns recommended by suppliers for amine analysis. However, most amine oxides are heat sensitive and decompose to the free amine in the GC injection port. The methods described below are generally selected for routine quality control.

(a) Titration.   A number of titration procedures may be used for rapid quality control of amine oxide containing unreacted tertiary amine.

Titration in acetic acid, with and without chemical reaction (148): The total amine oxide and tertiary amine content of the product may be determined by direct potentiometric titration with perchloric acid in acetic acid solvent. Another portion of the sample is reacted for 10 min with acetic anhydride, converting tertiary amine oxide to an amide and an aldehyde. The reaction product is titrated with perchloric acid in 2:1 acetic acid/acetic anhydride. Under these conditions, only tertiary amine is titrated.

Note: Primary and secondary amines, if present, will be measured as amine oxide by this procedure.

Stepwise titration in acetonitrile or MEK: If the titration of an amine oxide is followed potentiometrically, two inflections are observed in certain solvents. The first corresponds to the titration of one-half of the equivalents of amine oxide; the second is the sum of the remainder of the amine oxide plus any unreacted free amine. This procedure is limited to amine oxides containing an *N*-methyl group, since only these will give the double inflection (149).

Procedure (149)

The titration is conducted on an automatic recording titrator equipped with a glass indicator electrode and platinum reference. A sample containing about 1.5 meq amine oxide is dissolved in 60 mL acetonitrile or methyl ethyl ketone and titrated with 0.1 M $HClO_4$ in acetonitrile. The amine oxide is calculated as corresponding to twice the milliequivalents determined at the first inflection; the free amine corresponds to the difference between the first and second inflection, less the milliliters to reach the first inflection.

Titration in isopropanol, with and without chemical reaction: In this solvent, the two compounds give a single inflection point (150,151). The procedure consists of using isopropanol as solvent and alcoholic HCl as titrant. In one determination, methyl iodide is added to the mixture to react with the amine; only the amine oxide is titrated. By making a second titration without added methyl iodide, the sum of amine plus amine oxide can be determined.

Stepwise titration in isopropanol/water: In 50:50 isopropanol/water, two inflections are seen in the potentiometric titration with HCl, corresponding to the titration of the free amine and of the amine oxide, respectively. Free hydroxide, if present, will appear as an inflection prior to the others.

Procedure (152)

Spike Solution: Mix 250 mL 2-propanol, 250 mL water, 400 mg NaOH, and 2 g tri-*n*-butylamine. Protect from atmospheric CO_2. Weigh out an aliquot of sample containing about 1 meq amine oxide and dissolve in 100 mL 50:50 2-propanol/water (carbonate-free). Add 10.00 mL spike solution. Titrate potentiometrically with 0.1 M HCl, dissolved in water or 2-propanol. There will be three inflection points. Titrate a 10-mL aliquot of the spike solution separately. There will be two inflection points. Calculation:

Tertiary amine, meq $= [(V_2 - V_1) - (B_2 - B_1)](M)$

Amine oxide, meq $= [(V_3 - V_2) - (B_1 - V_1)](M)$

where V_1, V_2, and V_3 and B_1 and B_2 are the cumulative volumes of titrant required to reach the inflection points in the titration of the sample and blank, respectively, and M represents the titrant concentration.

Discussion: Weak acids may be present in the sample which partially neutralize the amine. This interference is avoided by spiking with NaOH to convert the amine salt to the free amine. The anion from the acid will interfere with the third break of the titration. This interference is equivalent to the OH^- consumed by the acid, and thus can be corrected by calculation. Commercial amine oxide samples contain only low levels of free amine, giving an inflection which is difficult to measure quantitatively. To produce a better titration curve, the titration vessel is spiked with a known quantity of amine.

Titration with an anionic surfactant: At low pH, long-chain amine oxides and the free amines can be determined by titration with an anionic surfactant or tetraphenylborate, using either the two-phase indicator method or the single-phase ion selective electrode method, as described in Chapter 16. The free amine can be separated from the aqueous sample solution by extraction at pH 10.5–12.0 with petroleum ether. The amine content is then determined by stripping off the petroleum ether, dissolving the residue in water, and titrating at pH 2 with an anionic. The amine oxide content is determined by difference (153).

(b) TLC

Procedure (154)

Standards are prepared by extracting an alkyldimethylamine oxide solution (pH 8, aqueous) with petroleum ether three times to remove amines, then spiking with known amounts of alkyldimethylamine. Standards of 0.1%, 0.3%, and 0.5% are usually sufficient. In the case of a 30% active product, it is applied as a 50% solution in methanol to a silica gel 60 (no fluorescing agent) on aluminum plate, interspersed with the standard solutions. After the spots have dried, the plate is developed with 90:10:1 chloroform/ethanol/NH$_3$ in a tank lined with chromatography paper. The plate is removed when the solvent front has advanced about 10 cm, then oven-dried at 110°C for 2 min. Initial visualization is with 1% bromcresol green in ethanol, followed by oven-drying for 5 min. The amine and amine oxide components show as blue spots. The amine spots are marked, then a second visualizing agent, 15% molybdophosphoric acid in ethanol, is applied. The plate is cured at 200°C for 2 min and the amine is quantified by comparison to the standards.

2.   GC Determination of Alkyl Chain Distribution

Alkyldimethylamine oxides cannot be analyzed directly by gas chromatography because they are thermally unstable above 100°C. However, they may be reduced by triphenylphosphine to form the alkyldimethylamines, which are readily determined by GC.

Procedure (155)

A sample containing about 1 g amine oxide is mixed with 1 to 1.5 g triphenylphosphine and 30 mL glacial acetic acid and heated to reflux for 1 to 1.5 hr. Concentrated NaOH solution is added until the solution is basic, then it is extracted with ethyl ether. The alkyldimethylamines in the ether extract are determined by gas chromatography on a 3-ft Carbowax/KOH column, using a temperature program of 170–220°C at 4°C/min.

If the original sample is an alkylamidopropylamine oxide, an additional reaction step is added: After the initial reduction with triphenylphosphine and acetic acid, the solution is evaporated to dryness, the residue is redissolved in 25 mL methanol and 3.6 g concentrated H$_2$SO$_4$, and the solution is heated with reflux for 18 hr to break the amide linkage. The methanol is stripped, 5 mL H$_2$O is added, and the solution is extracted twice with 10-mL aliquots of chloroform. The chloroform phase, containing the methyl esters of the fatty acids, is analyzed by GC on an OV-17 column, using a temperature program of 180–250°C at 8°C/min. The aqueous phase, containing the salt of *N,N*-dimethylaminopropylamine, is made alkaline by addition of NaOH solution and analyzed by GC with the Carbowax/KOH column.

REFERENCES

1.   Edwards, C. L., Polyoxyethylene alcohols, in N. M. van Os, ed., *Nonionic Surfactants: Organic Chemistry*, Marcel Dekker, 1998.
2.   Weinheimer, R. M., P. T. Varineau, Polyoxyethylene alkylphenols, in N. M. van Os, ed., *Nonionic Surfactants: Organic Chemistry*, Marcel Dekker, 1998.
3.   Kosswig, K., Polyoxyethylene esters of fatty acids, in N. M. van Os, ed., *Nonionic Surfactants: Organic Chemistry*, Marcel Dekker, 1998.

4. Lif, A., M. Hellsten, Nonionic surfactants containing an amide group, in N. M. van Os, ed., *Nonionic Surfactants: Organic Chemistry*, Marcel Dekker, 1998.

5. Hoey, M. D., J. F. Gadberry, Polyoxyethylene alkylamines, in N. M. van Os, ed., *Nonionic Surfactants: Organic Chemistry*, Marcel Dekker, 1998.

6. Lewis, J. J., Polyol ester surfactants, in N. M. van Os, ed., *Nonionic Surfactants: Organic Chemistry*, Marcel Dekker, 1998.

7. Hill, K., W. von Rybinski, G. Stoll, ed., *Alkyl Polyglycosides: Technology, Properties and Applications*, VCH, Weinheim, 1997.

8. Von Rybinski, W., K. Hill, Alkyl polyglycosides, in K. Holmberg, ed., *Novel Surfactants*, Marcel Dekker, New York, 1998.

9. Ivanova, M. A., A. D. Nikolov, I. B. Ivanov, L. Petrova, Refraction method for determining the degree of ethoxylation of nonionic surfactants, *J. Dispersion Sci. Technol.*, 1995, *16*, 495–510.

10. Rao, V. M., S. Uma, C. G. Gajelli, R. Y. Kelkar, Low-temperature DSC for determination of ethylene oxide content in some nonionic surfactants, *J. Surfactants Deterg.*, 1998, *1*, 515–517.

11. American Society for Testing and Materials, Ethylene oxide content of polyethoxylated nonionic surfactants, D2959-95. West Conshohocken, PA 19428.

12. International Organization for Standardization, Nonionic surface active agents—polyethoxylated derivatives—iodometric determination of oxyethylene groups, ISO 2270:1989. Geneva, Switzerland.

13. Siggia, S., A. C. Starke, Jr., J. J. Garis, Jr., C. R. Stahl, Oxalkylene groups in glycols and glycol and polyglycol ethers and esters, *Anal. Chem.*, 1958, *30*, 115–116.

14. Siggia, S., J. G. Hanna, *Quantitative Organic Analysis via Functional Groups*, 4th ed., Robert E. Krieger, Malabar, FL, 1988.

15. Cross, J., ed., *Nonionic Surfactants: Chemical Analysis*, Marcel Dekker, New York, 1987.

16. Diez, R., A. Morra, Degree of ethoxylation of nonionic surfactants by elemental analysis, *J. Am. Oil Chem. Soc.*, 1988, *65*, 1202–1203.

17. Cowie, J. M. G., A. F. Sirianni, Solution properties of some polyoxyethylene-polyoxypropylene surfactants in nonaqueous solvents, *J. Am. Oil Chem. Soc.*, 1966, *43*, 572–575.

18. American Society for Testing and Materials, Hydroxyl groups by phthalic anhydride esterification, E326-96. West Conshohocken, PA 19428.

19. American Society for Testing and Materials, Chemical analysis of alcohol ethoxylates and alkylphenol ethoxylates, D4252-89 (1995). West Conshohocken, PA 19428.

20. International Organization for Standardization, Nonionic surface active agents—polyalkoxylated derivatives—determination of hydroxyl value—phthalic anhydride method, ISO 4327:1979. Geneva, Switzerland.

21. American Oil Chemists Society, Hydroxyl value, Official Method Cd 13-60. 2211 West Bradley Ave., Champaign, IL 61826

22. American Society for Testing and Materials, Hydroxyl groups by acetic anhydride acetylation, E222-94. West Conshohocken, PA 19428.

23. International Organization for Standardization, Nonionic surface active agents—polyethoxylated derivatives—determination of hydroxyl value—acetic anhydride method, ISO 4326:1980. Geneva, Switzerland.

24. Arens, M., R. Spilker, B. Menzebach, Hydroxyl number—German standard methods for analysis of fats, fatty products, surfactants, and related products, contribution 126: analysis of surface active materials XXIX (in German), *Fett/Lipid*, 1996, *98*, 221–223.

25. Zeman, I., M. Paulovic, Comparison of HPLC determination of free PEG in ethoxylated surfactants with results of extraction procedure, *Proc. 2nd World Surfactants Congress*, Paris, 1988, 384–398.

26. Szymanski, A., Z. Lukaszewski, PEG in environmental samples by the indirect tensammetric method, *Analyst*, 1996, *121*, 1897–1901.

27. Zeman, I., M. Paulovic, HPLC analysis of propylene oxide derivatives of nonylphenol, *Petrochemia*, 1990, *30*, 141–149.

28. Weibull, B., PEG in surface-active ethylene oxide condensates, *Vorträge Originalfassung Intern. Kongr. Grenzflächenaktive Stoffe, 3.*, Cologne, 1960, *3*, 121–124.

29. Bürger, K., Separation of surface-active ethylene oxide adducts of fatty acids, fatty alcohols, fatty amines, fatty acid amides, and alkylphenols from PEG and determination by liquid-liquid extraction (in German), *Fresenius' Z. Anal. Chem.*, 1963, *196*, 22–26.

30. International Organization for Standardization, Surface active agents—technical ethoxylated fatty amines—methods of analysis, ISO 6384:1981. Geneva, Switzerland.

31. Coupkova, M., K. Janes, J. Sanitrak, J. Coupek, PEG in ethoxylated derivatives by liquid chromatography, *J. Chromatogr.*, 1978, *160*, 73–80.

32. Szewczyk, H., J. Szymanowski, PEG in oxyethylation products by GC, *Tenside*, 1982, *19*, 357–359.

33. International Organization for Standardization, Nonionic surface active agents—polyethoxylated derivatives—determination of polyethylene glycol, ISO 2268:1972. Geneva, Switzerland.

34. Hreczuch, W., Z. Krasnodebski, J. Szymanowski, PEG in ethoxylates by a modified Weibull method, *Comun. Jorn. Com. Esp. Deterg.*, 1992, *23*, 587–594.

35. Szymanowski, J., H. Szewczyk, W. Jerzykiewicz, Products obtained in the first stages of the ethoxylation of alkylamines, *Tenside Deterg.*, 1981, *18*, 130–136.

36. Winkle, W., Quantitative analysis in the V_0 zone. Chromatographic approach by coupling HPLC with GPC, *Chromatographia*, 1990, *29*, 530–536.

37. Wickbold, R., Sulfated AE (in German), *Tenside*, 1976, *13*, 181–184.

38. Drugarin, C., P. Getia, I. Jianu, Ethoxylation degree of nonylphenols, *Tenside*, 1981, *18*, 308–309.

39. Chiu, Y. C., L. J. Chen, Refractive index of nonionic surfactant solutions containing polyoxyethylene mono-*n*-alkylether, *Colloids Surf.*, 1989, *41*, 239–244.

40. Hreczuch, W., Refractive index—hydrophilic lipophilic balance relationship for alcohol ethoxylates, *Ind. Eng. Chem. Res.*, 1995, *34*, 410–412.

41. Meszlényi, G., É. Juhász, M. Lelkes, 2,4,6-Tri-tert-butylphenol polyethylene glycol ethers, *Tenside, Surfactants, Deterg.*, 1994, *31*, 83–85.

42. American Society for Testing and Materials, Refractive index and refractive dispersion of hydrocarbon liquids, D1218-92. West Conshohocken, PA 19428.

43. Karlstroem, G., B. Lindman, Phase behavior of nonionic polymers and surfactants of the oxyethylene type in water and in other polar solvents, in S. E. Friberg and B. Lindman, eds., *Organized Solutions*, Marcel Dekker, New York, 1992.

44. International Organization for Standardization, Nonionic surface active agents—determination of cloud point index—volumetric method, ISO 4320:1977. Geneva, Switzerland.

45. Lad, K., A. Bahadur, P. Bahadur, Clouding behavior of an ethylene oxide-propylene oxide block copolymer, *Tenside, Surfactants, Deterg.*, 1997, *34*, 37–42.

46. Goel, S. K., Critical phenomena in the clouding behavior of nonionic surfactants induced by additives, *J. Colloid Interface Sci.*, 1999, *212*, 604–606.

47. International Organization for Standardization, Nonionic surface active agents obtained from ethylene oxide and mixed nonionic surface active agents—determination of cloud point, ISO 1065:1991. Geneva, Switzerland.

48. Cosmetic, Toiletry and Fragrance Association, Cloud point of nonionic surfactants (water titration method), CTFA Method C16-1. 1110 Vermont Ave. NW, Washington, DC 20005.

49. American Society for Testing and Materials, Cloud point of nonionic surfactants, D2024-65 (1997). West Conshohocken, PA 19428.

50. Santacesaria, E., D. Gelosa, M. Di Serio, R. Tesser, Thermal stability of nonionic polyoxyalkylene surfactants, *J. Appl. Polym. Sci.*, 1991, *42*, 2053–2061.

51. Ianniello, R. M., Organic acids in APE by anion exclusion HPLC, *Anal. Lett.*, 1988, *21*, 87–99.
52. Preti, G., T. S. Gittelman, P. B. Staudte, P. Luitweiler, Letting the nose lead the way. Malodorous components in drinking water, *Anal. Chem.*, 1993, *65*, 699A–702A.
53. Romero, J., F. Ventura, J. Caixach, J. Rivera, L. X. Godé, J. M. Niñerola, 1,3-Dioxanes and 1,3-dioxolanes as malodorous compounds in river water, groundwater, and tap water, *Environ. Sci. Technol.*, 1998, *32*, 206–216.
54. Stafford, M. L., K. F. Guin, G. A. Johnson, L. A. Sanders, S. L. Rockey, 1,4-Dioxane in ethoxylated surfactants, *J. Soc. Cosmet. Chem.*, 1980, *31*, 281–287.
55. Black, D. B., R. C. Lawrence, E. G. Lovering, J. R. Watson, GC method for 1,4-dioxane in cosmetics, *J. Assoc. Off. Anal. Chem.*, 1983, *66*, 180–183.
56. Scalia, S., Reversed-phase HPLC method for the assay of 1,4-dioxane in sulfated AE surfactants, *J. Pharm. Biomed. Anal.*, 1990, *8*, 867–870.
57. Scalia, S., M. Guarneri, E. Menegatti, 1,4-Dioxane in cosmetic products by HPLC, *Analyst*, 1990, *115*, 929–931.
58. Van Damme, F., A. C. Oomens, Free epoxide in polyether polyols by derivatization with diethylammonium *N,N*-diethyldithiocarbamate and HPLC, *J. Chromatogr. A*, 1995, *696*, 41–47.
59. Sasaki, K., K. Kijima, M. Takeda, S. Kojima, Ethylene oxide and ethylene chlorohydrin in cosmetics and polyoxyethylated surfactants by gas chromatography with electron capture detection, *J. AOAC Int.*, 1993, *76*, 292–296.
60. United States Pharmacopeial Convention, Inc., *United States Pharmacopeia*, 24th Revision/ *National Formulary*, 19th ed. 12601 Twinbrook Parkway, Rockville, MD 20852, 1999.
61. Wala-Jerzykiewicz, A., J. Szymanowski, Headspace GC of toxic contaminants in ethoxylated alcohols and alkylamines, *Chromatographia*, 1998, *48*, 299–304.
62. Wala-Jerzykiewicz, A., W. Hreczuch, J. Szymanowski, Toxic contaminants in AE, *Tenside, Surfactants, Deterg.*, 1999, *36*, 122–126.
63. Wala-Jerzykiewicz, A., W. Jerzykiewicz, A. Sobczynska, J. Szymanowski, Toxic contaminants in polyoxyethylene alkylamines, *Tenside, Surfactants, Deterg.*, 1999, *36*, 173–177.
64. Gaia, G., E. Moretti, A. Penati, GC determination of 1,4-dioxane in ethoxylated products using the head-space technique, *Riv. Ital. Sostanze Grasse*, 1989, *66*, 575–580.
65. Rümenapp, J., J. Hild, Dioxane in cosmetic materials, *Lebensmittelchem. Gerichtl. Chem.*, 1987, *41*, 59–61.
66. TEGEWA, Determination of monomeric ethylene oxide in pharmaceutical raw materials, *Fresenius' Z. Anal. Chem.*, 1989, *333*, 26–28.
67. Rastogi, S. C., Headspace analysis of 1,4-dioxane in products containing polyethoxylated surfactants by GC-MS, *Chromatographia*, 1990, *29*, 441–445.
68. Leskovšek, H., A. Grm, J. Marsel, Head space GC determination of residual ethylene oxide in cosmetic products, *Fresenius' Z. Anal. Chem.*, 1991, *341*, 720–722.
69. Wala-Jerzykiewicz, A., J. Szymanowski, Free oxirane and 1,4-dioxane content of ethoxylated surface-active compounds by GC with headspace sample injection, *Chem. Anal.* (Warsaw), 1996, *41*, 253–261.
70. Birkel, T. J., C. R. Warner, T. Fazio, GC determination of 1,4-dioxane in polysorbate 60 and polysorbate 80, *J. Assoc. Off. Anal. Chem.*, 1979, *62*, 931–936.
71. Dahlgran, J. R., M. N. Jameson, Formaldehyde and other aldehydes in industrial surfactants by HPLC separation of their respective 2,4-dinitrophenylhydrazone derivatives, *J. Assoc. Off. Anal. Chem.*, 1988, *71*, 560–563.
72. American Society for Testing and Materials, Total bromine number of unsaturated aliphatic chemicals, E234-90. West Conshohocken, PA 19428.
73. American Society for Testing and Materials, Polyurethane raw materials: unsaturation of polyols, D4671-93. West Conshohocken, PA 19428.
74. American Society for Testing and Materials, Iodine value of drying oils and fatty acids, D1959-97. West Conshohocken, PA 19428.

75. Downes, M. J., M. W. Edwards, T. S. Elsey, C. L. Walters, Non-volatile nitrosamine by using denitrosation and a chemiluminescence analyser, *Analyst*, 1976, *101*, 742–748.
76. Fletcher, H., J. Collins, *N*-nitrosamines in surfactants, *Proc. 3rd CESIO International Surfactants Cong. Exhib.*, London, 1992, *Sect. E,F,LCA*, 239–246.
77. Brennan, S., C. W. Frank, *N*-Diethanolnitrosamine in cosmetics, *J. Soc. Cosmet. Chem.*, 1983, *34*, 41–46.
78. Hidaka, H., J. Zhao, S. Suenaga, N. Serpone, E. Pelizzetti, Photodegradation of surfactants. VII: Peroxide and aldehyde formation in the photocatalyzed oxidation of nonionic surfactants, *Yukagaku*, 1990, *39*, 963–966.
79. Donbrow, M., Stability of the polyoxyethylene chain, in M. J. Schick, ed., *Nonionic Surfactants: Physical Chemistry*, Marcel Dekker, New York, 1987.
80. Evetts, S., C. Kovalski, M. Levin, M. Stafford, High-temperature stability of alcohol ethoxylates, *J. Am. Oil Chem. Soc.*, 1995, *72*, 811–816.
81. Mair, R. D., R. T. Hall, Organic peroxides, in I. M. Kolthoff and P. J. Elving, eds., *Treatise on Analytical Chemistry*, Vol. 14. Wiley-Interscience, New York, 1971.
82. Johnson, R. M., I. W. Siddiqi, *Determination of Organic Peroxides*, Pergamon Press, New York, 1970.
83. Azaz, E., M. Donbrow, R. Hamburger, Hydroperoxide and iodine in aqueous nonionic surfactant solutions by a spectrophotometric method, *Analyst*, 1973, *98*, 663–672.
84. American Society for Testing and Materials, Assay of organic peroxides, E298-91. West Conshohocken, PA 19428.
85. American Society for Testing and Materials, Trace amounts of peroxides in organic solvents, E299-97. West Conshohocken, PA 19428.
86. Brand, N., G. Mailhot, M. Bolte, Degradation photoinduced by Fe(III): APE removal in water, *Environ. Sci. Technol.*, 1998, *32*, 2715–2720.
87. Bergh, M., L. P. Shao, G. Hagelthorn, E. Gäfvert, J. L. G. Nilsson, A. Karlberg, Atmospheric oxidation of polyoxyethylene alcohols, formation of ethoxylated aldehydes, and their allergenic activity, *J. Pharm. Sci.*, 1998, *87*, 276–282.
88. American Society for Testing and Materials, Determination of carbonyls in C_4 hydrocarbons, D4423-91 (1996). West Conshohocken, PA 19428.
89. Felippone, F., E. Moretti, P. Fanelli, A. Penati, Carbonyl group determination in ethoxylated products: stability study (in Italian), *Riv. Ital. Sostanze Grasse*, 1990, *67*, 621–624.
90. Felippone, F., A. Penati, Carbonyl compounds in oxyalkylated products, *Tenside, Surfactants, Deterg.*, 1994, *31*, 236–238.
91. American Society for Testing and Materials, Trace quantities of carbonyl compounds with 2,4-dinitrophenylhydrazine, E411-92. West Conshohocken, PA 19428.
92. Lappin, G. R., L. C. Clark, Colorimetric method for carbonyl compounds, *Anal. Chem.*, 1951, *23*, 541–542.
93. Miwa, H., M. Yamamoto, Volatile acids induced by autoxidation of nonionic surfactants by LC with direct derivatization, *J. AOAC Int.*, 1996, *79*, 418–422.
94. American Society for Testing and Materials, Ash from petroleum products, D482-95. West Conshohocken, PA 19428.
95. International Organization for Standardization, Nonionic surface active agents—determination of sulfated ash—gravimetric method, ISO 4322:1977. Geneva, Switzerland.
96. Cosmetic, Toiletry and Fragrance Association, Mole ratio of ethylene oxide in PEG alkyl ethers, CTFA Method J 3-1. 1110 Vermont Ave. NW, Washington, DC 20005.
97. Bareš, M., J. Zajíc, Fatty acids in nonionic surfactants, *Tenside, Surfactants, Deterg.*, 1974, *11*, 251–254.
98. Malkemus, J. D., J. D. Swan, PEG esters, *J. Am. Oil Chem. Soc.*, 1957, *34*, 342–344.
99. Stefanova, R., D. Rankoff, S. Panayotova, S. L. Spassov, Proton NMR determination of linoleic acid mono- and diesters of polyethyleneglycols via reaction with trichloroacetyl isocyanate, *J. Am. Oil Chem. Soc.*, 1988, *65*, 1516–1518.

100. Cosmetic, Toiletry and Fragrance Association, Total amide in alkylolamides, CTFA Method D 22-1. 1110 Vermont Ave. NW, Washington, DC 20005.

101. Buschmann, N., F. Hülskötter, Titration of low ethoxylated nonionic surfactants, *Tenside, Surfactants, Deterg.*, 1997, *34*, 8–11.

102. Chou, H. J., Diethanolamine and *N*-nitrosodiethanolamine in fatty acid diethanolamides, *J. AOAC Int.*, 1998, *81*, 943–947.

103. Kroll, H., H. Nadeau, Chemistry of lauric acid-diethanolamine condensation products, *J. Am. Oil Chem. Soc.*, 1957, *34*, 323–326.

104. Arens, M., H. König, M. Teupel, German Standard Methods for analysis of fats, fatty products, surfactants, and related materials: analysis of organic surface active compounds X (in German), *Fette, Seifen, Anstrichm.*, 1982, *84*, 105–111.

105. Cosmetic, Toiletry and Fragrance Association, Free fatty acids in alkylolamides, CTFA Method D 23-1. 1110 Vermont Ave. NW, Washington, DC 20005.

106. Cosmetic, Toiletry and Fragrance Association, Total esters in alkylolamides, CTFA Method G 7-1, 1110 Vermont Ave. NW, Washington, DC 20005.

107. Cosmetic, Toiletry and Fragrance Association, Total titratable amines in alkylolamides, CTFA Method D 24-1. 1110 Vermont Ave. NW, Washington, DC 20005.

108. Arens, M., M. Hirschen, H. Klotz, Potentiometric titration of total basic nitrogen content of surfactants—German standard methods for analysis of fats, fatty products, surfactants, and related materials: analysis of surface active materials XXXI (in German), *Fett/Lipid*, 1997, *99*, 369–371.

109. Chou, H. J., R. L. Yates, R. J. Gajan, H. M. Davis, Differential pulse polarographic determination of *N*-nitrosodiethanolamine in cosmetic products, *J. Assoc. Off. Anal. Chem.*, 1982, *65*, 850–854.

110. Rosenberg, I. E., J. Gross, T. Spears, U. Caterbone, Nitrite and nitrosamines in cosmetic raw materials and finished products, *J. Soc. Cosmet. Chem.*, 1979, *30*, 127–135.

111. Erickson, M. D., D. B. Lakings, A. D. Drinkwine, J. L. Spigarelli, *N*-nitrosodiethanolamine by HPLC-thermal energy analyzer detection, *J. Soc. Cosmet. Chem.*, 1985, *36*, 213–222.

112. Erickson, M. D., D. B. Lakings, A. D. Drinkwine, J. L. Spigarelli, *N*-nitrosodiethanolamine in cosmetic ingredients, *J. Soc. Cosmet. Chem.*, 1985, *36*, 223–230.

113. Rosenberg, I. E., J. Gross, T. Spears, *N*-nitrosodiethanolamine in linoleamide DEA by HPLC and UV detection, *J. Soc. Cosmet. Chem.*, 1980, *31*, 323–327.

114. Rosenberg, I. E., J. Gross, T. Spears, P. Rahn, Nitrosamines in cosmetic raw materials and finished product by HPLC, *J. Soc. Cosmet. Chem.*, 1980, *31*, 237–252.

115. Diallo, S., J. Y. Zhou, C. Dauphin, P. Prognon, M. Hamon, Determination of *N*-nitrosodiethanolamine as nitrite in ethanolamine derivative raw materials by HPLC with fluorescence detection after alkaline denitrosation, *J. Chromatogr. A*, 1996, *721*, 75–81.

116. Mutter, M., G. W. van Galen, P. W. Hendrikse, Fatty acid monoethanolamide foam stabilizer (in German), *Tenside, Surfactants, Deterg.*, 1968, *5*, 33–36.

117. Cosmetic, Toiletry and Fragrance Association, Total fatty acids in alkylolamides, CTFA Method D 25-1. 1110 Vermont Ave. NW, Washington DC 20005.

118. Mutter, M., G. W. van Galen, P. W. Hendrikse, Fatty acid monoethanolamide-ethylene oxide condensate nonionic surfactant (in German), *Tenside, Surfactants, Deterg.*, 1968, *5*, 36–39.

119. Hejna, J. J., D. Daly, Fatty alkanolamides, *J. Soc. Cosmetic Chem.*, 1970, *21*, 107–118.

120. Krusche, G., Amides in surfactants (in German), *Tenside, Surfactants, Deterg.*, 1973, *10*, 182–185.

121. Fritz, J. S., *Acid-Base Titrations in Nonaqueous Solvents*, Allyn and Bacon, Boston, 1973.

122. American Society for Testing and Materials, Bromine numbers of petroleum distillates and commercial aliphatic olefins by electrometric titration, D1159-93. West Conshohocken, PA 19428.

123. Datta, S., J. M. Plantinga, Bromine addition method for the determination of the iodine value in fatty amines and their derivatives, *World Surfactants Congr., 4th*, 1996, *1*, 418–431.

124. Schulz, R., Potentiometric titration of nonionic surfactants with the surfactant-selective electrode (in German), *SÖFW-Journal*, 1996, *122*, 1022,1024–1028.

125. Schreuder, R. H., A. Martijn, H. Poppe, J. C. Kraak, Ethoxylated alkylamines in pesticide formulations by HPLC using ion-pair extraction detection, *J. Chromatogr.*, 1986, *368*, 339–350.

126. Sandra, P., F. David, Microcolumn chromatography for the analysis of detergents and lubricants. Part 1: High temperature capillary GC and capillary SFC, *J. High Resolut. Chromatogr.*, 1990, *13*, 414–417.

127. Tsuda, T., H. Nakanishi, GC determination of sucrose fatty acid esters, *J. Assoc. Off. Anal. Chem.*, 1983, 66, 1050–1052.

128. Gatewood, L., H. D. Graham, Adaptation of the chromotropic acid method to the assay of SPANs, *J. Amer. Pharm. Assoc.*, 1960, *49*, 678–680.

129. Zeringue, H. J., R. O. Feuge, Purification of sucrose esters by selective adsorption, *J. Am. Oil Chem. Soc.*, 1976, *53*, 567–571.

130. De la Nuez Fonseca, L. R., P. L. Gutiérrez Moreno, Sucrose in a commercial sucroglyceride used as a surfactant in the sugar industry (in Spanish), *Cent. Azucar*, 1985, *10*, 133–139.

131. Snell, F. D., C. T. Snell, *Colorimetric Methods of Analysis*, 3rd ed., Vol. III. D. Van Nostrand, New York, 1953.

132. American Oil Chemists Society, α-Monoglycerides, Official Method Cd 11-57. 2211 West Bradley Ave., Champaign, IL 61826

133. AOAC International, *Official Methods of Analysis of AOAC International*, 16th ed., Gaithersburg, MD 20877, 1995.

134. Osipow, L. I., W. Rosenblatt, Micro-emulsion process for the preparation of sucrose esters, *J. Am. Oil Chem. Soc.*, 1967, *44*, 307–309.

135. Spilker, R., B. Menzebach, U. Schneider, I. Venn, Alkylpolyglucosides (in German), *Tenside, Surfactants, Deterg.*, 1996, *33*, 21–25.

136. Waldhoff, H., J. Scherler, M. Schmitt, J. R. Varvil, Alkyl polyglycosides and determination in consumer products and environmental matrices, in K. Hill, W. von Rybinski, and G. Stoll, eds., *Alkyl polyglycosides. Technology, Properties and Applications*, VCH, Weinheim, 1997.

137. Buschmann, N., F. Hülskötter, A. Kruse, S. Wodarczak, Alkylpolyglucosides, *Fett/Lipid*, 1996, *98*, 399–402.

138. Buschmann, N., S. Wodarczak, Alkylpolyglucosides. Part I: Colorimetry, *Tenside, Surfactants, Deterg.*, 1995, *32*, 336–339.

139. Kroh, L. W., T. Neubert, E. Raabe, H. Waldhoff, Enzymatic analysis of alkyl polyglycosides, *Tenside, Surfactants, Deterg.*, 1999, *36*, 19–21.

140. Meissner, C., H. Engelhardt, Surfactants derived from fatty alcohols—hydrolysis and enrichment, *Chromatographia*, 1999, *49*, 12–16.

141. Facino, R. M., M. Carini, G. Depta, P. Bernardi, B. Casetta, Atmospheric pressure ionization MS analysis of new anionic surfactants: alkylpolyglucoside esters, *J. Am. Oil Chem. Soc.*, 1995, *72*, 1–9.

142. Kalinoski, H. T., Polyoxyalkylene block copolymers, in V. M. Nace, ed., *Nonionic Surfactants: Polyoxyalkylene Block Copolymers*, Marcel Dekker, New York, 1996.

143. Zgoda, M. M., Surfactants in the group of EO-PO copolymers. II: Application of IR and proton NMR spectroscopy to the determination of the hydrophilic-lipophilic balance in polyetherdiols (in Polish), *Acta Pol. Pharm.*, 1988, *45*, 63–70.

144. Kusz, P., J. Szymanowski, K. Pyzalski, E. Dziwinski, Degradation and analysis of EO-PO block copolymers in the presence of acetyl chloride, *LC-GC*, 1990, *8*, 48,50.

145. American Society for Testing and Materials, Primary hydroxyl contents of polyether polyols, D4273-94. West Conshohocken, PA 19428.

146. Nace, V. M., R. H. Whitmarsh, M. W. Edens, Effects and measurements of polyoxyethylene block length in polyoxyethylene-polyoxybutylene copolymers, *J. Am. Oil Chem. Soc.*, 1994, *71*, 777–781.

147. Lowe, K. C., B. A. Furmidge, S. Thomas, Hemolytic properties of Pluronic® surfactants and effects of purification, *Artif. Cells, Blood Substitutes, Immobilization Biotechnol.*, 1995, *23*, 135–139.

148. Comité Européen des Agents de Surface et leurs Intermédiaires Organiques, Fatty alkyl dimethyl amine oxides—determination of the amine oxide content, CESIO/AIS Analytical Method 6-89, available from European Chemical Industry Council, Brussels.

149. Wang, C. N., L. D. Metcalfe, Unreacted amines in long chain amine oxides by potentiometric titration, *J. Am. Oil Chem. Soc.*, 1985, *62*, 558–560.

150. Metcalfe, L. D., Potentiometric titration of long chain amine oxides using alkyl halide to remove tertiary amine interference, *Anal. Chem.*, 1962, *34*, 1849.

151. Toney, C. J., F. E. Friedli, P. J. Frank, Kinetics and preparation of amine oxides, *J. Am. Oil Chem. Soc.*, 1994, *71*, 793–794.

152. Donkerbroek, J. J., C. N. Wang, Cationic surfactant analysis, *Proc. 2nd World Surfactants Cong.*, Paris, 1988, *III*, 134–152.

153. Walton, B., Potentiometric titration of long chain tertiary amine oxides and tertiary amines using a poly(vinyl chloride) membrane electrode, *Analyst*, 1994, *119*, 2202–2203.

154. Comité Européen des Agents de Surface et leurs Intermédiaires Organiques, Determination of free amine in amine oxides by TLC, CESIO/AIS Analytical Method 7-91, available from European Chemical Industry Council, Brussels.

155. Langley, N. A., D. Suddaby, K. Coupland, Carbon chain length composition of amine oxides, *Int. J. Cosmet. Sci.*, 1988, *10*, 257–261.

156. Bergh, M., L. P. Shao, K. Magnusson, E. Gäfvert, J. L. G. Nilsson, A. Karlberg, Atmospheric oxidation of AE. Ethoxylated formates as oxidation products and study of their contact allergenic activity, *J. Pharm. Sci.*, 1999, *88*, 483–488.

157. Lang, R. F., D. Parra-Diaz, D. Jacobs, Ethoxylated fatty amines. Comparison of methods for the determination of molecular weight, *J. Surfactants Deterg.*, 1999, *2*, 503–513.

3

Characterization of
Cationic Surfactants

I. DESCRIPTION AND TYPICAL SPECIFICATIONS

Cationic surfactants are useful as fabric softeners, corrosion inhibitors, and antimicrobial agents. They are not used in general purpose detergents because they do not provide effective cleaning at neutral pH. They are adsorbed rapidly to textiles so that their solution concentration drops very quickly to low levels, making them unsuitable for industrial processing baths, although they see some specialty use in connection with anionic dyes. Except for applications where biological activity is critical, as in pesticides and pharmaceuticals, only nitrogen-based compounds are used as cationic surfactants. More information about these compounds can be obtained from other volumes of the Surfactant Science Series (1).

A. Alkyl Quaternary Ammonium Salts

1. Description

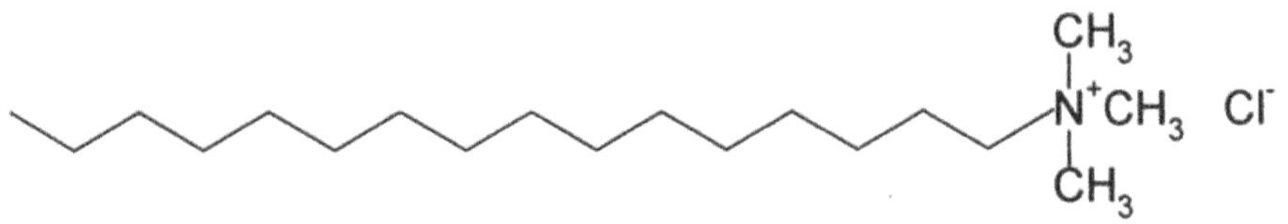

$$CH_3CH_2(...)N^+CH_3 \quad Cl^-$$

N-Hexadecyltrimethylammononium chloride

Commercial alkyl quaternaries are made from naturally occurring fatty acids. Distearyldimethylammonium salts are the high volume compounds in the cationic surfactant group because of their use as fabric softeners in North America.

2. Typical Specifications

Parameter	Test method
Assay	Two-phase titration
Average molecular weight	Titration
Distribution of mono-, di-, and tri- alkylammonium compounds	HPLC
Alkyl chain length distribution	HPLC
Acid value	Titration
Amine value	Titration
pH	5% in isopropanol, ASTM D2081 (2); 5% in water, AOCS Tu 1a-64 (3)
Solids	105°C, reduced pressure, ASTM D2079 (4)
Alcohols	Gas chromatography
Anions	Ion chromatography
Other tests	Ash, ASTM D2077 (5); Karl Fischer water; Gardner or platinum-cobalt scale color; unsaturation

B. Benzylalkyldimethylammonium Salts

1. Description

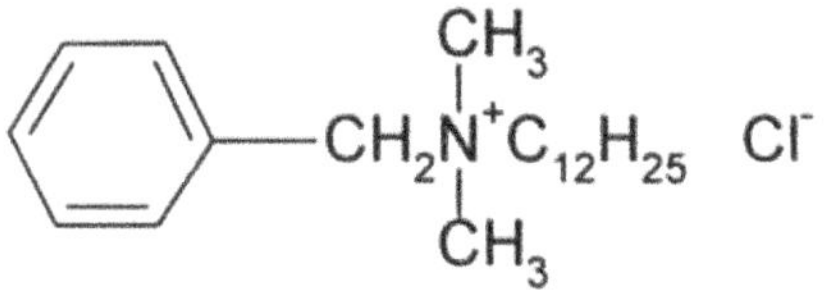

Benzyldodecyldimethylammonium chloride

Called benzalkonium salts, these are widely used, especially in pharmaceuticals. Both benzyltrialkyl and alkylbenzyltrialkyl compounds are common. Typical products contain a range of alkyl chain lengths.

2. Typical Specifications

Parameter	Test method
Assay	Iodometric titration
Homolog distribution	Hydrogenation and GC; paired-ion HPLC with UV detection
Other tests	Insoluble material; ash content; odor; free amines; anions; water; alcohols

C. Amidoamine Quaternaries

1. Description

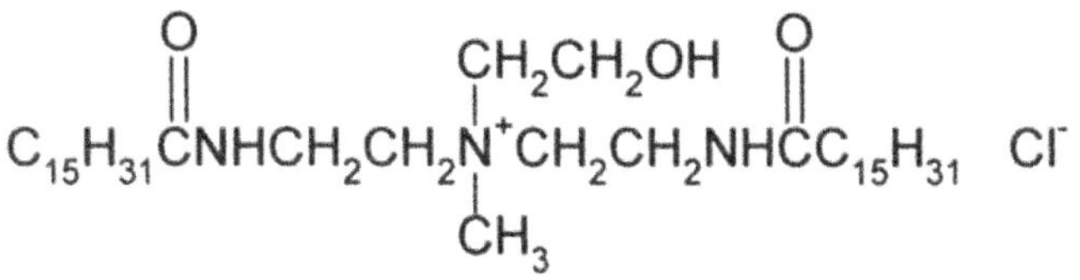

Methyl-*bis*(hexadecylamidoethyl)-2-hydroxyethyl ammonium chloride

The most common products are prepared from tallow fatty acids and diethylenetriamine. These are widely used as fabric softeners and antistatic agents.

2. Typical Specifications

Parameter	Test method
Alkyl chain distribution	GC or HPLC
Residual solvents	Gas chromatography
Other tests	Color; odor; apparent pH; unsaturation

D. Quaternary Imidazolium Compounds

1. Description

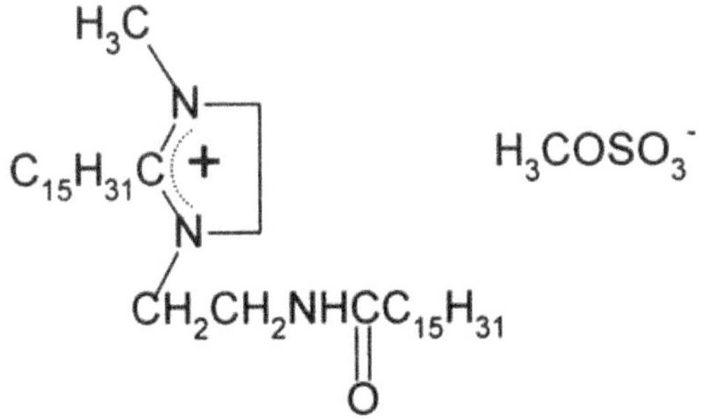

A di-C_{16} quaternary imidazolium methyl sulfate

These are prepared by quaternization of an imidazoline compound, the latter usually made by reaction of diethylenetriamine with fatty acids, followed by formation of the imidazoline ring. Imidazolium compounds are frequently simply called imidazolines. Quaternization greatly stabilizes the imidazoline structure. These cationic surfactants are used as fabric softeners and antistatic agents.

2. Typical Specifications

Parameter	Test method
Alkyl chain distribution	GC or HPLC
Residual solvents	Gas chromatography
Other tests	Color; odor; apparent pH; unsaturation

E. Ester Quats

1. Description

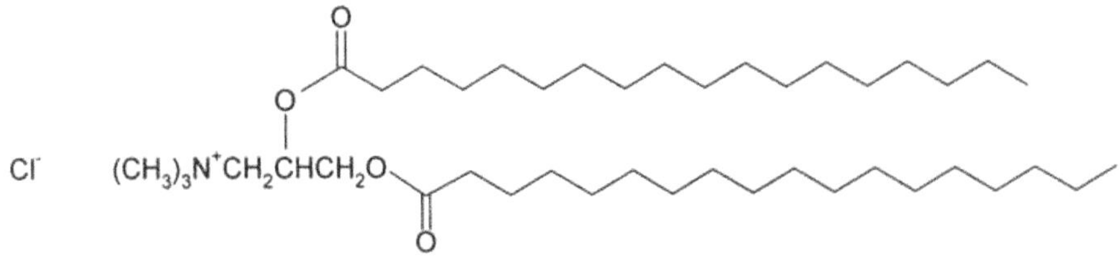

Ditallow ester of 2,3-dihydroxypropanetrimethylammonium chloride

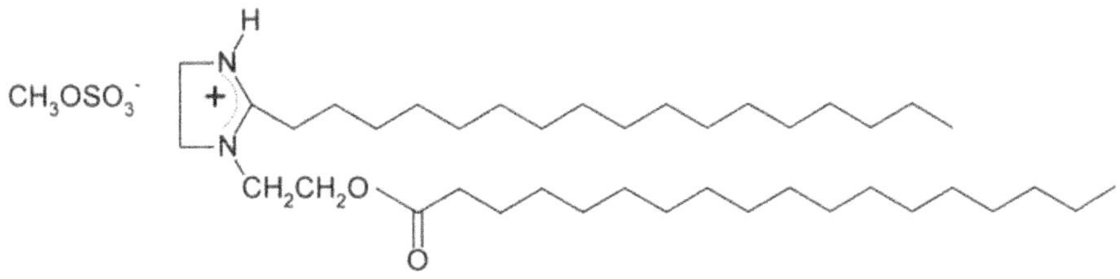

Ditallow imidazolium ester

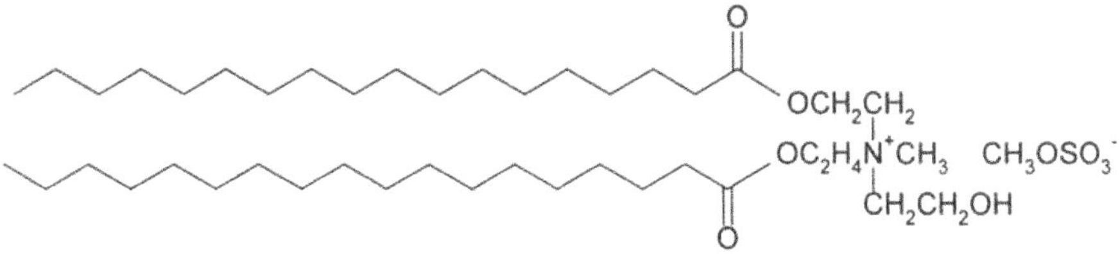

N-methyl-*bis*(tallowacyloxyethyl)-2-hydroxyethylammonium methylsulfate

These are analogs of previously described quaternaries that contain ester linkages to speed biodegradation (6). The most common are fatty acid esters of triethanolamine, quaternized with dimethyl sulfate or methyl chloride. They are mixtures of quaternized mono-, di-, and triesters.

2. Typical Specifications

Parameter	Test method
Alkyl chain distribution	GC or HPLC
Residual solvents	Gas chromatography
Other tests	Color; odor; apparent pH; unsaturation

II. GENERAL TEST METHODS

A. Assay

Quaternary amines may be assayed by titration with an anionic surfactant by the potentiometric or two-phase procedures described in Chapter 16. The ISO procedure specifies titration with sodium dodecyl sulfate, which is available in high purity (7). Sodium tetraphenylborate may also be used if potassium and ammonium ions are absent. Other approaches to assay are based on acid-base titration or quantification of the anion associated with the quaternary.

B. Acid-Base Titration

1. General Considerations

Quaternary amine salts may be titrated with acid by conventional nonaqueous procedures. To be precise, the anion associated with the quaternary is titrated. In the case of halides, mercuric acetate is added to replace free halide ion with easily titratable acetate ion [see Section III] (8,9). Such methods are suitable for analysis of pure samples, which do not contain other salts. A more specific assay of cationic surfactants is the two-phase titration with an anionic surfactant, described in Chapter 16.

2. Acid Value

The procedure given here is ASTM D2076 (10). The AOCS procedure, Te 3a-64, is almost identical (11). Both are written specifically for analysis of quaternary amines.

Procedure (10)

Dissolve 5–20 g of well-mixed sample in 100 mL neutralized isopropanol. Add phenol-phthalein indicator and titrate with 0.1 M NaOH solution. If the sample is already alkaline, titrate instead with 0.2 M HCl solution. Calculations:

$$\text{Acid value, mg KOH/g} = \frac{(V)(M)56.1}{W}$$

$$\% \text{ Free NaOH} = \frac{(V)(M)4.0}{W}$$

where V is the titrant volume, M is its concentration, and W is the sample weight in grams.

3. Amine Value

The significance of amine value varies with the product. The procedure below, based on simple titration to a visual end point, is useful for quality control of a well-characterized material. For quaternaries, a high amine value may indicate that the compound is not com-pletely neutralized, or it may indicate that primary, secondary, or tertiary amine is present. For new products, it is preferable to conduct the analysis with a recording potentiometric titrator. In the case of a quaternary amine hydroxide, two inflections may be observed. The first is due to the strongly basic quaternary, the second to more weakly basic amines.

Procedure (10,11)

Weigh a sample of 5–20 g into a flask and dissolve in 100 mL isopropanol. Titrate to the yellow end point of bromphenol blue using 0.2 M HCl. Calculation:

$$\text{Amine value, mg KOH/g} = \frac{(V)(M)56.1}{(W)}$$

where V is the titrant volume, M is its concentration, and W is the sample weight in grams.

C. Iodine Value

The usual procedure for determining unsaturation in quaternaries is the Wijs method, de-scribed in Chapter 2. The ASTM test optimized for determination of unsaturation of the alkyl group in fatty quaternary ammonium chlorides is D2078 (12). The corresponding AOCS method is Tg 3a-64 (11). For other compounds, the recommended AOCS method is Cd 1-25 (13).

D. Anions and Salts

Thorough characterization of commercial products requires determination of the anion as-sociated with the cationic surfactant as well as determination of other salts present. Deter-mination of chloride and sulfate ions is described in Chapter 1. These and other anions are usually determined by ion chromatography, although wet chemical methods may also be applied after qualitative analysis has shown which ions are present.

Chloroacetate ion may be determined by conventional HPLC rather than ion chromatography (14). Chloroacetate may also be determined by gas chromatography after formation of the ethyl ester (15–17). Common cations present in formulations, such as sodium, potassium, ammonium, and simple organic amines, are determined by ion chromatography.

Procedure: Determination of Halide, Sulfate, *p*-Toluenesulfonate and Methyl Sulfate Ions in Quaternary Compounds by Ion Chromatography (18)

The determination is performed by single-column ion chromatography with inverse photometric detection. A Vydac 302IC4.6 anion exchange column is used at 35°C with a mobile phase of aqueous 0.003 M potassium hydrogen phthalate solution. Ultraviolet detection is at 290 nm; conductivity detection may also be used. Iodide ion may be used as an internal standard if absent from the formulation tested. Samples and standards are diluted in water and injected directly, with quantification by peak area.

Procedure: Determination of Mono- and Dichloroacetic acid in Betaines and Formulations by Gas Chromatography (15,16)

A sample solution of 0.5 mL, representing 0.5 g shampoo or 0.1 g of surfactant, is transferred to a 25-mL stoppered test tube. To this is added 2 mL ethanol containing 25 ppm monobromoacetic acid internal standard. After addition of 0.25 mL concentrated H_2SO_4, the test tube is capped, agitated, and heated at 50°C for 15 min. After cooling, 5 mL 5% NaCl solution and 5 mL cyclohexane are added and the mixture is shaken or stirred magnetically for 10 min. After phase separation, the upper cyclohexane phase is analyzed by capillary GC in split mode on a Restek RTx-35 (35% diphenyl, 65% dimethyl polysiloxane) column, 0.32 mm × 60 m, with 15 min initial hold at 80°C, followed by a programmed rise to 240°C at 10°C/min. A collaborative study of this method gave a coefficient of variation of about 10% for concentrations in the 10 ppm range and about 35% for the 1 ppm concentration range.

E. Alkyl Chain Length Distribution

The homolog distribution of the fatty alkyl portion of surfactants such as the alkyltrimethylammonium chlorides and the benzylalkyldimethylammonium chlorides is determined by HPLC or, after appropriate decomposition of the sample, by GC. Details of these determinations are found in Chapters 7 and 8.

F. N-Nitrosamines

Although cationic surfactants are generally free of *N*-nitrosamines, those used in cosmetics may occasionally be analyzed by the method for total nitrosamines described under characterization of nonionic surfactants (Chapter 2, Section II)

G. Other Tests

A number of other tests are routinely applied to cationic surfactants. Nitrogen content is determined by standard elemental analysis procedures or by acid-base titration in acetic acid solvent. Total dry matter is generally determined by oven weight loss or by individual determination of water and solvent content. Apparent pH is determined as with anionic and nonionic surfactants, at a specified concentration in a specified solvent.

III. ANALYSIS OF INDIVIDUAL SURFACTANTS

A. Fatty Quaternary Ammonium Salts

A large body of standard test methods exists for quaternary compounds, issued by ASTM, AOCS, and ISO. The current standards should be consulted prior to choosing methods of analysis. The analyst must be aware that these products are often mixtures. For example, a nominal dialkyldimethylammonium chloride will usually contain monoalkyltrimethyl- and trialkylmonomethylammonium salts. The results from potentiometric titration with an anionic surfactant vary according to the alkyl chain length. For high molecular weight products like the distearyldimethylammonium salts, a single potentiometric end point is seen for the sum of the mono-, di-, and trialkyl compounds. For lower molecular weight products such as didecyldimethylammonium chloride, an ititial break is seen for the didecyldimethyl- and tridecylmethylammonium salts, with a second break for the monodecyltrimethylammonium chloride, depending on the alcohol content of the titration medium (19).

1. Average Molecular Weight

The ASTM and AOCS standards specify determination of molecular weight by titration after addition of mercuric acetate. Mercuric ion complexes halides, resulting in the re-placement of halide ions in solution with acetate ion. Acetate is titrated potentiometrically with perchloric acid in acetic acid solvent. The method is valid if a correction is made for the concentration of any halides not associated with the quaternary amine, for example, sodium chloride.

Procedure (8,20)

Weigh out 1.0–1.5 g sample and dissolve in 100 mL glacial acetic acid. Add 15 mL of a solution of freshly prepared mercuric acetate in glacial acetic acid (6 g in 100 mL). Stir for 5 min, then titrate potentiometrically with 0.1 M perchloric acid in acetic acid, using a glass combination pH electrode and an automatic titrator.

2. Unreacted Primary and Secondary Amines

Amines may be reacted with carbon disulfide to form the dithiocarbamic acids, which are titrated potentiometrically with base (21).

Procedure (22)

A sample of about 20 g is weighed and dissolved in 100 mL 90:10 isopropanol/water. The sample solution should be held at 45°C during the analysis. The solution is put on the auto-matic recording titrator, stirring is begun, and phenolphthalein indicator is added. If the sample is not acidic, add a little HCl until it is acidic. Titrate the sample with 0.1 M NaOH solution to the phenolphthalein end point, noting the millivolt reading on the recorder at the indicator end point. Five milliliters of carbon disulfide is added, and the titration is continued, taking as end point the same millivolt reading as noted above. The second titra-tion corresponds to the total of primary and secondary amines.

3. Mono-, Di-, and Trialkylammonium Compounds

HPLC methods are the most convenient for determining, for example, monoalkyltri-methyl- and trialkylmethylammonium chloride impurities in dialkyldimethylammonium chloride. This analysis is described in Chapter 7.

B. Benzylalkyldimethylammonium Salts

The compendial method for assay of benzalkonium chloride is based upon reversal of the quaternization reaction. The compound is reacted with NaI, the benzyl iodide and tertiary amine are removed by extraction, and the iodide consumed is determined by titration with KIO_3 (9). The alkyl chain length distribution is determined by HPLC. Other tests for characterization of benzalkonium salts are similar to those performed on alkyl quaternaries. These salts are also readily determined by two-phase or potentiometric titration with an anionic surfactant.

Quats made from reaction of benzyl chloride with tertiary amines potentially contain residual benzyl chloride. However, it has been demonstrated by HPLC analysis that benzyl chloride is unstable in the reaction product and is undetectable in a matter of hours (23). Benzyl alcohol and benzaldehyde may be determined in this same HPLC analysis, but the preferred method for these substances is gas chromatography, following the recommendations of GC column manufacturers.

C. Quaternary Imidazolinium Compounds

These compounds are readily titrated with anionic surfactants, either according to the two-phase method or to the potentiometric end point, as discussed in Chapter 16. Alcohol is added as necessary for solubility (19). Determination of the alkyl chain length distribution may be performed directly, by HPLC, or after liberation of the fatty acids, by GC. More information is given in Chapters 7 and 8.

D. Ester Quats

Fabric softeners based on quaternized triethanolamine esters of long-chain fatty acids can be analyzed by HPLC, as described in Chapter 7. Separate peaks are resolved for the mono-, di-, and triesters as well as the main impurities (24). Fatty acid composition may be determined more precisely by gas chromatography after transesterification to form methyl esters (24).

These compounds can be titrated with anionic surfactants, but special care is necessary to minimize hydrolysis during the analysis. The best approach is to adjust the sample to pH 3 and then immediately perform a potentiometric titration. Any unquaternized amine impurity is also determined at this pH (19). Some investigators prefer to make the determination at pH 2 (25).

NMR analysis can give the gross characterization of the material, including the average composition in terms of alkyl chain length, unsaturation, ester number, and degree of quaternization.

Ester quats may be separated using a silica solid phase extraction cartridge into fractions consisting of unquaternized triesteramine, mixed triesterquat/diesteramine/diesterquat, diester quat, and monoester quat. The conditions are given in Chapter 6.

REFERENCES

1. Richmond, J. M., ed., *Cationic Surfactants: Organic Chemistry*, Marcel Dekker, New York, 1990.
2. American Society for Testing and Materials, pH of fatty quaternary ammonium chlorides, D2081-64 (1987). West Conshohocken, PA 19428.

3. American Oil Chemists' Society, pH of fatty quaternary ammonium chlorides, Official Method Tu 1a-64. 2211 West Bradley Ave., Champaign, IL 61826.

4. American Society for Testing and Materials, Nonvolatile matter (solids) in fatty quaternary ammonium chlorides, D2079-92. West Conshohocken, PA 19428.

5. American Society for Testing and Materials, Ash in fatty quaternary ammonium chlorides, D2077-64 (1987). West Conshohocken, PA 19428.

6. Kruger, G., D. Boltersdorf, K. Overkempe, Esterquats, in K. Holmberg, ed., *Novel Surfactants*, Marcel Dekker, New York, 1998.

7. International Organization for Standardization, Surface active agents—detergents—determination of cationic-active matter content—Part 1: High-molecular-mass cationic-active matter, ISO 2871-1:1988. Geneva, Switzerland.

8. American Society for Testing and Materials, Average molecular weight of fatty quaternary ammonium chlorides, D2080-92. West Conshohocken, PA 19428.

9. United States Pharmacopeial Convention, Inc., *United States Pharmacopeia*, 24th Revision/ *National Formulary*, 19th ed. 12601 Twinbrook Parkway, Rockville, MD 20852, 1999.

10. American Society for Testing and Materials, Acid value and amine value of fatty quaternary ammonium chlorides, D2076-92. West Conshohocken, PA 19428.

11. American Oil Chemists' Society, Acid value and free amine value of fatty quaternary ammonium chlorides, Official Method Te 3a-64. 41 East University Ave., Champaign, IL 61820.

12. American Society for Testing and Materials, Iodine value of fatty quaternary ammonium chlorides, D2078-86 (1995). West Conshohocken, PA 19428.

13. American Oil Chemists' Society, Iodine value of fats and oils, Official Method Cd 1-25. 2211 West Bradley Ave., Champaign, IL 61826.

14. Husain, S., R. Narsimha, S. N. Alvi, R. N. Rao, Chloroacetic acids by HPLC, *J. High Resolut. Chromatogr.*, 1993, *16*, 381–383.

15. Cetinkaya, M., GC determination of monochloroacetic acid in surfactants and surfactant-containing body cleaning products (in German), *Parfüm. Kosmet.*, 1991, *72*, 816–818.

16. Arens, M., R. Spilker, Fatty acid amidopropylbetaines—Concentration of mono- and dichloroacetic acid—Collaboration of the DGF, Communication 151: German standard methods for study of fats, fatty products, surfactants and related materials. Communication 117: Analysis of surface active materials XXVI (in German), *Fett/Lipid*, 1996, *98*, 37–40.

17. Humbert, M., A. Hernandez, Monochloroacetate, dichloroacetate, and glycolate in betaines, *World Surfactants Congr., 4th*, 1996, *1*, 379–388.

18. Wigman, L. S., M. L. Thomson, R. S. Wayne, Single column IC determination of anions in agriculturally useful quaternary compounds using indirect UV detection, *J. Liq. Chromatogr.*, 1989, *12*, 3219–3229.

19. Schulz, R., *Titration of Surfactants and Pharmaceuticals* (in German), Verlag für chemische Industrie, H. Ziolkowsky GmbH, Augsburg, 1996.

20. American Oil Chemists' Society, Average molecular weight of fatty quaternary ammonium chlorides, Official Method Tv 1a-64. 2211 West Bradley Ave., Champaign, IL 61826.

21. Arens, M., M. Hirschen, H. Klotz, Potentiometric titration of primary, secondary, and tertiary amine nitrogen content of surfactants—German Standard Methods for analysis of fats, fatty products, surfactants, and related materials: Analysis of surface active materials XXXII (in German), *Fett/Lipid*, 1997, *99*, 372–376.

22. Meijer, J. A., Primary and secondary amines in ditallowdimethylammonium chloride (in German), *Tenside Surfactants, Deterg.*, 1977, *14*, 198–199.

23. Prieto-Blanco, M. C., P. López-Mahía, D. Prada-Rodríguez, Residual products in benzalkonium chloride by HPLC, *J. Chromatogr. Sci.*, 1999, *37*, 295–299.

24. Wilkes, A. J., C. Jacobs, G. Walraven, J. M. Talbot, Quaternized triethanolamine esters (esterquats) by HPLC, HRCGC, and NMR, *World Surfactants Congr., 4th*, 1996, *1*, 389–412.

25. Gerlache, M., Z. Sentürk, J. C. Viré, J. M. Kauffmann, Potentiometric analysis of ionic surfactants by an ion-selective electrode, *Anal. Chim. Acta*, 1997, *349*, 59–65.

4

Characterization of Amphoteric Surfactants

I. DESCRIPTION AND TYPICAL SPECIFICATIONS

Amphoteric surfactants can function either as anionic or cationic surfactants, depending on the pH of the system. They contain both anionic and cationic functions in the same molecule. More costly to produce than ionic surfactants, amphoteric surfactants represent only about 3% of surfactant volume in Europe and less than 1% in the United States. They are less irritating than other materials and are largely used in personal care products. A distinction can be made between amphoteric and zwitterionic surfactants. This distinction does not affect their analysis. For the analyst, a more important distinction is between amphoterics with a secondary or tertiary amine group and those containing a quaternary amine function. The former only have cationic properties when protonated at low pH, while the quaternary amines have cationic properties even under alkaline conditions.

A very thorough discussion of these compounds comprises a volume of the Surfactant Science Series (1). We adopt the organization and terminology of Lomax (1) and the names used by the CTFA (2).

A. Alkylamino Acids

1. Description

N-Dodecylaminoacetic acid, sodium salt

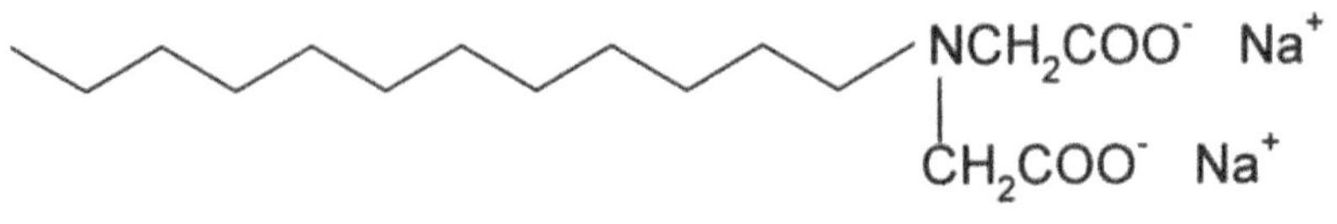

$$\text{NCH}_2\text{COO}^- \text{ Na}^+$$
$$\text{CH}_2\text{COO}^- \text{ Na}^+$$

N-Dodecyliminodiacetic acid, disodium salt

These amphoteric surfactants may be prepared by reaction of a fatty primary amine with chloroacetic acid. They are also produced by reaction of formaldehyde and HCN with a fatty amine. *N*-alkyl-β-aminopropionates and *N*-alkyl-β-iminodipropionates are usually made by reaction of the alkylamine with methyl acrylate, followed by hydrolysis of the methyl esters (3).

2. Typical Specifications

Parameter	Test method
Active agent	Alcohol-soluble portion
Free amines	Gas chromatography
Di- or tri-acids	Gas chromatography of methyl esters
NaCl and Na_2SO_4	Titration or ion chromatography
Methanol	Gas chromatography

B. Alkylbetaines
1. Description

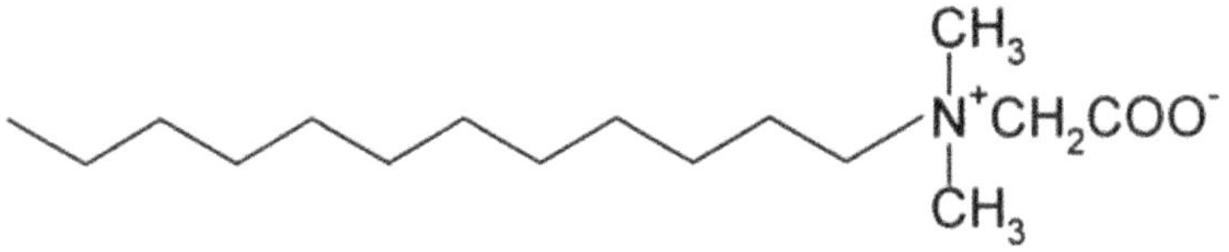

$$\begin{array}{c} \text{CH}_3 \\ | \\ \text{N}^+\text{CH}_2\text{COO}^- \\ | \\ \text{CH}_3 \end{array}$$

Dodecyldimethylammononiomethane carboxylate

Although sulfur- and phosphorus-containing betaine derivatives are effective as amphoteric surfactants, only the carboxybetaines are widely used. They are internal salts. [Betaine is trimethylammonium acetate, $(CH_3)_3N^+CH_2COO^-$]. The most frequently encountered example is (alkyldimethylammonio)methane carboxylate (DAMC). These are most often made by quaternization of a tertiary amine with chloroacetic acid. Commercial products are generally mixtures of homologs with alkyl chain lengths in the C_8–C_{18} range (4).

2. Typical Specifications

Parameter	Test method
Assay	Titration
Sodium chloride; sodium glycolate; sodium chloroacetate	Ion chromatography
Free tertiary amine	Titration

C. Alkylamidobetaines

1. Description

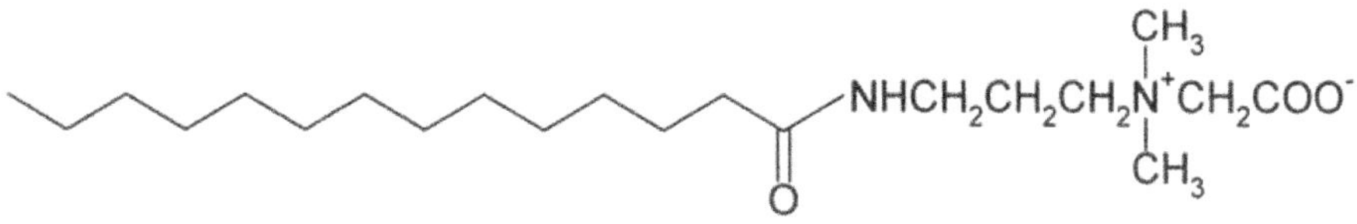

$$CH_3$$
$$\text{--NHCH}_2\text{CH}_2\text{CH}_2\overset{+}{\text{N}}\text{CH}_2\text{COO}^-$$
$$CH_3$$

A cocoamidopropylbetaine component

Amidobetaines are the most commercially important betaines. They tend to be high foam-ing, have good detergent properties, and are outstanding in their lack of irritation to skin and eyes. They are produced in a way analogous to alkylbetaines, first by reaction of a primary/tertiary diamine, such as dimethylaminopropylamine, with a fatty acid (either free fatty acid or a fatty acid methyl ester or a triglyceride) to form an alkylamidodialkylamine, then by quaternization with chloroacetic acid.

2. Typical Specifications

Parameter	Test method
Assay	Titration
Identity	IR
Solids content	Oven weight loss
Total N	Kjeldahl nitrogen
Glycerin	GC or HPLC
Free fatty acid content	GC
Sodium chloride; sodium glycolate; sodium chloroacetate; sodium dichloroacetate	Ion chromatography
Other tests	Saponification number; acid number; pH, 5% aqueous; color

D. Imidazoline-Derived Amphoterics

1. Description

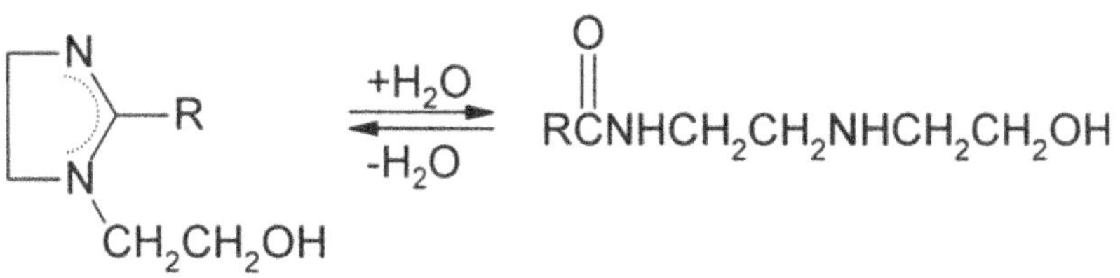

1-Hydroxyethyl-2-alkyl-imidazoline   An alkyl amidoamine

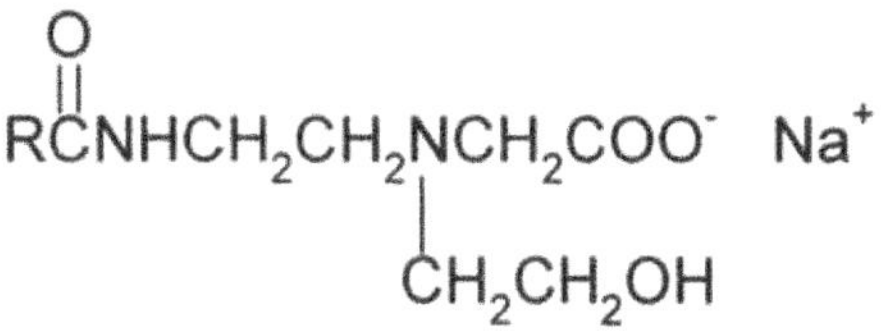

$$O$$
$$\text{R}\overset{\parallel}{\text{C}}\text{NHCH}_2\text{CH}_2\overset{}{\text{N}}\text{CH}_2\text{COO}^- \ \text{Na}^+$$
$$CH_2CH_2OH$$

Monocarboxylated imidazoline derivative (CTFA cocoamphoacetate)

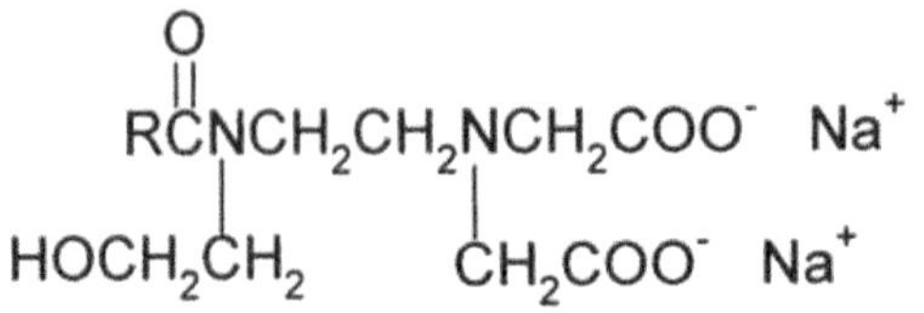

$$O$$
$$\|$$
$$RCNCH_2CH_2NCH_2COO^- \ Na^+$$

$$HOCH_2CH_2 \qquad CH_2COO^- \ Na^+$$

Dicarboxylated imidazoline derivative (CTFA cocoamphocarboxyglycinate)

In former times it was thought that imidazole-derived amphoterics had betaine structure, but they are now classified as alkylamidoamino acids. Imidazolines are formed by condensation of a fatty acid (introduced as the free acid or as its ester with glycerin or methanol) with aminoethylethanolamine $(NH_2C_2H_4NHC_2H_4OH)$, or aminopropylethanolamine. Other amines may be used, such as ethylenediamine or diethylenetriamine. Heating of the amide intermediate under vacuum results in formation of the imidazoline ring structure. The imidazoline is reacted further to add functional groups and it is during this functionalization reaction that the imidazoline structure is lost again by hydrolysis. Usually this reaction is with sodium chloroacetate. The hydrolysis product of 1-hydroxyethyl-2-alkylimidazoline, shown upper left, is the parent amidoamine, above right. A typical commercial product is the carboxylated product pictured above, with the dicarboxylate present as a trace impurity. This is true whether the product is described as being monocarboxylated or dicarboxylated (1,5,6).

2. Typical Specifications

Parameter	Test method
Assay	Titration
Identity	IR
Solids content	Oven weight loss
Total N	Kjeldahl nitrogen
Glycerin	GC or HPLC
Free fatty acid content	GC
Piperazine	GC
Sodium chloride; sodium glycolate; sodium chloroacetate; sodium dichloroacetate	Ion chromatography
Other tests	Saponification number; acid number; pH, 5% aqueous; color

E. Sulfur-Containing Amphoterics

1. Description

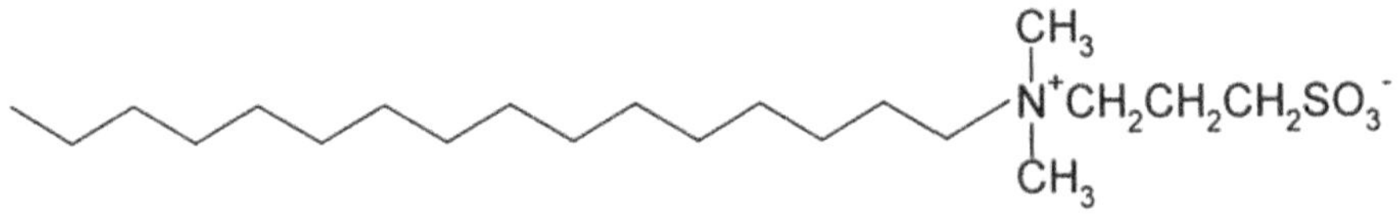

$$CH_3$$
$$|$$
$$N^+CH_2CH_2CH_2SO_3^-$$
$$|$$
$$CH_3$$

N-Cetyl-N,N-dimethylammoniumpropanesultaine; 3-(hexadecyldimethylammonio)-1-propanesulfonate (a component of CTFA coco-sultaine)

Sulfobetaines are the most common in this category, but even they are not widely used. Sulfobetaines are insensitive to water hardness and have excellent solubility in electrolyte solutions. They are good foamers and also function as lime soap dispersing agents, thus synergistically increasing the detergency obtained with soap formulations. They also are effective antistats. For laundry applications, a lower wash temperature can be used with sulfobetaine surfactants than is generally effective with other surfactants.

2. Typical Specifications

Parameter	Test method
Alkyl chain length	HPLC
Free tertiary amine	HPLC
Unsulfonated quat intermediate	HPLC
Other tests	Sulfate; bisulfite; chloride; apparent pH

F. Lecithin

1. Description

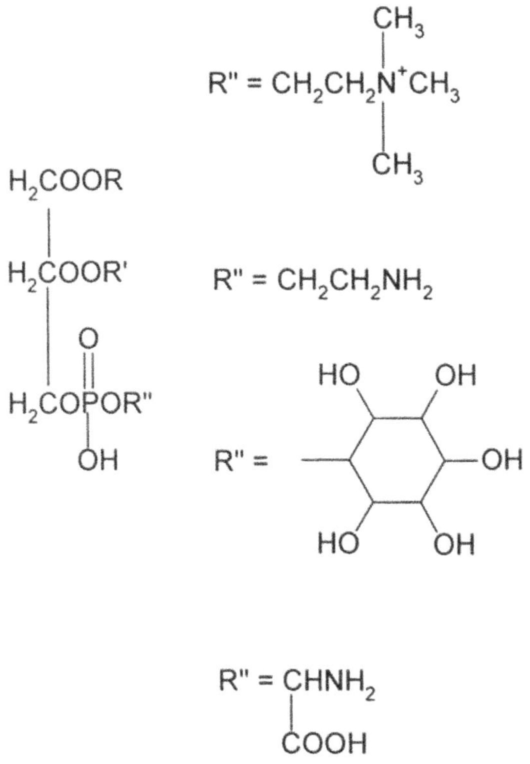

$$R'' = CH_2CH_2\overset{+}{N}(CH_3)_3 \qquad \text{Phosphatidylcholine}$$

$$R'' = CH_2CH_2NH_2 \qquad \text{Phosphatidylethanolamine}$$

$$R'' = \qquad \text{Phosphatidylinositol}$$

$$R'' = CHNH_2(COOH) \qquad \text{Phosphatidylserine}$$

Lecithin is a phosphoglyceride (phosphatide) natural product isolated commercially from soybean oil and, in much lower quantity, from egg yolks. The terminology is somewhat confused for historical reasons (7). While the term lecithin is sometimes used synonymously with α-phosphatidyl choline, it more correctly denotes a crude mixture also containing β-phosphatidyl choline and phosphatidyl esters of other compounds, chiefly ethanolamine, inositol, and serine. (While phosphatidyl choline and phosphatidyl ethanolamine are amphoteric, phosphatidyl inositol and phosphatidyl serine are anionic.) For soybean lecithin, R and R' represent C_{16} saturated fatty acid and C_{18} saturated and unsaturated fatty acids. Egg lecithin also contains C_{20} unsaturated fatty acid. Soybean lecithin usually contains a substantial amount of soybean oil, including free fatty acids and carbohydrates.

Lecithin is widely used as an emulsifier in foods and pharmaceuticals, as well as in coatings, printing inks, and other products. Pharmaceutical compendia in various countries give specifications for purified lecithin, generally not in terms of chemical composition but in terms of gross parameters like ash, volatile matter, nitrogen content, and the like. There is a wide range of specially treated products described as lecithin. For example, if deoiled lecithin is treated with alcohol, one fraction will be greatly enriched in alcohol-soluble phosphatidyl choline, while the other will be enriched in alcohol-insoluble phosphatidyl inositol (8).

2. Typical Specifications

Parameter	Test method
Acetone-insoluble matter	AOCS Ja 4-46
Petroleum ether insolubles	AOCS Ja 3-55
Water	Karl Fischer titration; azeotropic distillation (AOCS Ja 2-46)
Total phosphorus	Alkaline fusion, precipitation of molybdophosphoric acid, titration (AOCS Ja 5-55)
Determination of individual phosphatides	TLC (AOCS Ja 7-86); HPLC
Free fatty acid value	Titration of acetone solubles
Other tests	Apparent pH, 1% in 30:70 ethanol/water; viscosity, Brookfield, 80°C; Gardner color; peroxide value, AOCS Cd 8-53; iodine number, AOCS Cd 1-25; acid number, AOCS Ja 6-55

II. GENERAL TEST METHODS

A. Assay

The standard two-phase titration usually applied to cationics can also be used with amphoteric surfactants if the pH is adjusted. This is discussed in chapter 16. The ethanol concentration of the aqueous phase must be controlled within narrow limits for accurate results (9).

B. Acid-Base Titration

1. General Considerations

Potentiometric titration with HCl gives a characteristic value proportional to the content of amphoteric surfactant (1,10,11). In some cases, solvent systems have been optimized so that acid-base titration is suitable for assay of the product, as described below for characterization of alkylbetaines (12).

2. Amine Value

The significance of amine value varies with the product. The procedure given in Chapter 3 for analysis of cationic surfactants is based on simple titration to a visual end point and is useful for quality control of a well-characterized material. For new products, it is preferable to conduct the analysis with a recording potentiometric titrator.

C. Iodine Value

The usual procedure for determining unsaturation is the Wijs method, described in Chapter 2. The recommended ASTM and AOCS methods are D2075 and Cd 1-25, respectively (13,14).

D. Anions and Salts

The same remarks apply as for cationic surfactants in Chapter 3 (15).

E. Alkyl Chain Length Distribution

The homolog distribution of the fatty alkyl portion of these surfactants is determined by HPLC or, after appropriate decomposition of the sample, by GC. Details of these determinations are found in Chapters 7 and 8.

F. *N*-Nitrosamines

Although amphoteric surfactants are generally free of *N*-nitrosamines, those used in cosmetics may occasionally be analyzed by the method for total nitrosamines described under characterization of nonionic surfactants (Chapter 2, Section II).

III. ANALYSIS OF INDIVIDUAL SURFACTANTS

A. Alkylbetaines

The precise characterization of commercial betaines is difficult because of the variety of similar compounds which may be present in the product (1,11). Free amine concentration is a critical parameter because of the odor it gives to the product.

Alkyldimethylbetaine and free amine can be determined by potentiometric acid-base titration (12). For titration of the betaine, the sample is dissolved in a mixture of 10:1 methyl isobutyl ketone/isopropanol to which a little HCl has been added. Three breaks are observed on titration with ethanolic KOH, corresponding respectively to excess HCl, the betaine, and combined impurities: amine, glycolic acid, and monochloracetic acid (see Fig. 1). Carbon dioxide must be excluded during the titration. Sodium chloride is insoluble in the solvent system and precipitates. The titration is very sensitive to water concentration (16)

For titration of the amine, the sample is dissolved in 50:50 isopropanol/water to which excess NaOH and a small, precisely measured, amount of tri-*n*-butylamine are added. Again, CO_2 must be excluded. Titration with aqueous HCl gives two breaks, the first corresponding to excess NaOH and the second to the free amine (see Fig. 2). The amine titration is corrected for the spiked amount; spiking allows a readily visible end point even when the sample has a low free amine content. The amine determination was subjected to a collaborative study which showed that at the 0.2–0.5% level repeatability by the same laboratory was in the range of 10%, and was 20% between laboratories (17).

B. Alkylamidobetaines

Alkyl chain length is most readily determined by HPLC, as discussed in Chapter 7. NMR analysis has been proposed for assay of betaines, using the signal at 3.3 ppm versus trimethylsilylpropionate. Trimethylsilylpropionate can be used as the internal standard for

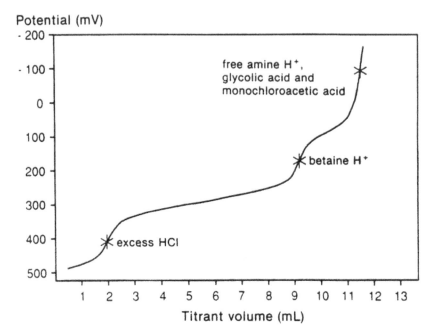

Potential (mV)

FIG. 1 Nonaqueous titration of alkyldimethylbetaine with potassium hydroxide solution. (Reprinted with permission from Ref. 12. Copyright 1993 by the American Oil Chemists' Society.)

quantification (16). Amidobetaines can be hydrolyzed with hydrochloric acid to yield the starting fatty acids.

1. Assay

These products can be determined by titration. Although they are cationic at low pH, their ion pairs with anionic surfactants are not sufficiently insoluble that the regular anionic/cationic titration method can be used. This is especially true of the lower chain length components of, for example, a cocoamidoalkylbetaine. A better system is to use tetraphenylborate ion as titrant with the electrode system typically chosen for the titration of nonionics (18).

Certain alkylamidobetaines may contain organic acids as hydrotropes (perhaps trimethylglycine or citric acid) to allow a highly concentrated product. In such cases, the active agent should be isolated by solid phase extraction prior to titration (19).

The alkylamidobetaine can be titrated in one of three ways (16,18,19):

1. Excess acid is added, and the product is titrated potentiometrically in a nonaqueous medium with alkali. The first break is due to the excess acid, the second is due to the active agent. A third break may be observed due to impurities. Acetone/2-propanol, 4:1, is a suitable solvent for use with methanolic KOH titrant. The titration is sensitive to the presence of water and CO_2.
2. Excess base is added with a little sodium acetate, and the product is titrated potentiometrically with acid in a nonaqueous system. The first break is due to excess base, the second to impurities, and the third to the active agent (Fig. 3).

Potential (mV)

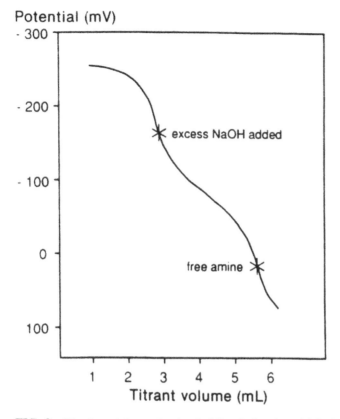

FIG. 2 Titration of free amine in alkyldimethylbetaine with hydrochloric acid. (Reprinted with permission from Ref. 12. Copyright 1993 by the American Oil Chemists' Society.)

Ethylene glycol monomethyl ether/methanol, 3:1, is a suitable solvent with perchloric acid titrant in dioxane. Other solvent systems may be used, but then citric acid (sometimes added as a hydrotrope) will interfere. Glycine, dimethylglycine, and trimethylglycine interfere regardless.

3. The product is acidified and titrated with sodium tetraphenylborate in aqueous solution using a surfactant-sensitive electrode or other electrode for end point detection, as described in Chapter 16. The addition of gum arabic to the titration vessel smooths the titration curve by preventing the deposition of the cationic/tetraphenylborate precipitate on the electrode.

In a comparison of the titrations, Buschmann and Wille report that the titration with alkali (after SPE purification) gives lower results than the titration with acid (again after purification). Titration with tetraphenylborate gives significantly lower results with poor repeatability (19).

Procedure: Separation of Cocoamidopropylbetaine from Hydrophilic Organic Acids and Alkylamidoamine by Solid Phase Extraction (19)

Use a reversed-phase cartridge such as Macherey-Nagel C_8, 6 g in a 15-mL cartridge. Dissolve 800 mg product in 3 mL 2:1 MeOH/H_2O and add to cartridge. Elute organic acid

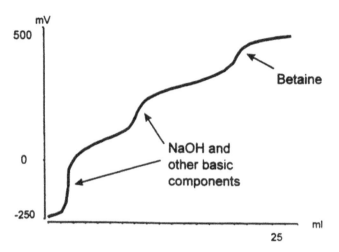

FIG. 3 Nonaqueous titration of cocoamidopropylbetaine with perchloric acid. (Reprinted with permission from Ref. 16. Copyright 1996 by Karl Hanser Verlag.)

hydrotropes with 12 mL 2:1 MeOH/H_2O, then elute the cocoamidopropylbetaine with 18 mL MeOH. Any amidoamine remains in the cartridge.

2. Determination of Unquaternized Amine

Alkylamidobetaines may contain residual unquaternized alkylamidoamine. A procedure developed specifically for determination of C_{12}–C_{18} alkylamidopropylamine in the corresponding alkylamidopropylbetaine consists of extracting the amine with hexane from a 25:75 isopropanol/0.5 M aqueous NaOH solution of the betaine. The amine is isolated from the hexane by evaporation, then determined by two-phase titration with dodecylsulfate with dimidium bromide/disulfine blue mixed indicator under acid conditions according to the method described in Chapter 16 (20). The method gives good precision for amine concentrations in the 0.3–1.0% range.

C. Imidazoline-Derived Amphoterics

Products described commercially as being monocarboxylates and dicarboxylates are produced by reaction of the imidazoline intermediate with monochloroacetic acid, 1:1 or 1:2, respectively. It has been found that the product of this reaction, the active agent, is in each case the monocarboxylate, with at most a trace of the dicarboxylate being present. The difference between the two products is that those made with an excess of monochloroacetic acid contain more of the active agent and less of the unreacted amidoamine intermediate (6).

Amphoglycinates can be determined by potentiometric titration with perchloric acid using a pH electrode or by titration with tetraphenylborate using a nonionic surfactant-selective electrode (18).

Characterization of an imidazoline-derived amphoteric requires considering the reaction pathway used to prepare the compound. For example, an amidobetaine may be made by sodium chloroacetate treatment of the imidizoline/linear amide mixture obtained by reacting a fatty acid or ester with N-hydroxyethylethylenediamine. Possible components will then be free fatty acid, fatty acid esters of as many kinds as there are alcohols, free amine, free amides and diamides, amphoacetate (main component), amphodiacetate, sodium glycolate, residues from catalysts, etc. Excess sodium chloroacetate is destroyed at the end of the synthesis by hydrolysis with alkali, yielding sodium glycolate and sodium chloride. The aminoethyl ethanolamine used to prepare the intermediate amide can cyclize to form piperazine.

Schwarz and coworkers used a number of titration procedures to characterize the intermediates of the synthesis. They differentiated the content of primary, secondary, and tertiary amine as well as amide functionality (21). Hydroxyethyl groups are determined by titration with periodate ion, with a preliminary extraction step serving to separate amino-ethylethanolamine from the alkylamidoethylethanolamine.

1. Chromatographic Analysis

Imidazoline derivatives can be hydrolyzed with sulfuric acid into their starting materials: diamines and fatty acids. These can then be analyzed by gas chromatography, as described in Chapter 8. Some impurities, as well as the main components, can be detected (22,23). As described in Chapter 9, TLC may be used for semiquantitative determination of various compounds: main component, imidazoline intermediate, secondary and tertiary amide inter-mediates, and *N*-hydroxyethylethylenediamine starting material (5). HPLC methods used to analyze imidazoline derivatives are summarized in Chapter 7. Separation is typically accord-ing to the length of the alkyl chain, with further differentiation of impurities and intermediates within each alkyl chain group (24,25). HPLC has been demonstrated for the determination of free dimethylaminopropylamine after derivatization with salicylaldehyde (26).

2. Determination of Imidazoline Structure

Amidobetaines were once thought of as containing the imidazoline structure. However, it is now known that the imidazoline ring is only formed as an intermediate structure which is hydrolyzed under the alkaline conditions of the final carboxymethylation reaction to form an amphoteric surfactant. Amidobetaines contain no more than a trace of the imida-zoline ring (5,6). The imidazoline ring gives a relatively strong UV absorption at 235–244 nm. It is necessary to have standard compounds to perform a quantitative determination based on this phenomenon. The amide functionality also gives an absorbance near this region, making it necessary to compensate with an imidazoline-free amide solution of the same concentration (21,27).

D. Lecithin

Lecithin is analyzed according to the AOCS Ja method series (28). For food uses, the *Food Chemicals Codex* should also be consulted. These procedures are mainly based on gross physical properties, such as acetone solubility. An exception is the TLC method, which gives the composition in terms of individual phosphatides (29). Because of their impor-tance in biology, phosphatides are the focus of intense method development. Mass spec-trometry is the most powerful tool to characterize mixtures of phophatides (see Chapter 15), but there is usually no need to perform a complete MS analysis of lecithin. Commer-cial lecithin is a complex mixture, the properties of which are profoundly influenced by the considerable amount of nonphosphatide materials present. Further information on lec-ithin can be found in specialized publications (30).

1. Total Phospholipids

Total phospholipid content of soybean lecithin can be obtained by solid phase extraction on silica. Lipids are eluted with 20:80 hexane/ethyl ether, then phospholipids are eluted with methanol. The amount of each fraction is determined gravimetrically after solvent removal (31).

2. Phospholipid Profile

HPLC is the most common technique applied to the determination of the chemical composition of lecithin. Normal phase HPLC is convenient for the determination of the major constituents (i.e., phosphatidylcholine, phosphatidylethanolamine, etc), as described in Chapter 7. ^{31}P NMR is also suitable for this analysis, as discussed in Chapter 14. The biochemical literature contains many enzymatic methods, mainly for specific determination of phosphatidylcholine and its hydrolysis product, choline (32). For instance, phosphatidylcholine can be hydrolyzed by phospholipase C to a diacylglycerol and the phosphate ester of choline, which itself can be hydrolyzed by alkaline phosphatase to form choline and phosphate ion. Alternatively, action of phospholipase D on phosphatidylcholine yields phosphatidic acid and choline. These methods are not applied to analysis of the commercial lecithin used as a surfactant.

3. Fatty Acid Composition of the Phospholipids

In soybean lecithin, the acyl groups consist of oleic, linoleic, linolenic, stearic, and palmitic acids. Hydrogenated lecithin contains only stearic and palmitic acid groups. Lecithin can be characterized in terms of its total fatty acid composition. Total fatty acid composition is determined by saponification of the ester and esterification of the isolated acids with BF_3 in methanol, followed by gas chromatography of the fatty acid methyl esters (33). There is evidence that methanolic sodium methoxide gives the best yields in transesterification (34).

Reversed-phase HPLC is capable of separating the individual phosphatides according to the acyl substitution. Details of the HPLC analysis are found in Chapter 7. Enzymatic techniques allow selective hydrolysis of either the phosphorus-containing component or the fatty acids in the beta position, so that the resulting fragments can be analyzed by GC, GC-MS, and other methods (35–37).

REFERENCES

1. Lomax, E. G., ed., *Amphoteric Surfactants*, 2nd ed., Marcel Dekker, New York, 1996.
2. Cosmetics, Toiletries and Fragrances Association, *International Cosmetic Ingredient Dictionary and Handbook*, 7th ed., Washington, DC, 1997.
3. Rosenblatt, W., Amino acid amphoterics, in E. G. Lomax, ed., *Amphoteric Surfactants*, 2nd ed., Marcel Dekker, New York, 1996.
4. Domingo, X., Betaines, in Lomax, E. G., ed., *Amphoteric Surfactants*, 2nd ed., Marcel Dekker, New York, 1996.
5. Li, Z., Z. Zhang, Amphoteric imidazoline surfactants, *Tenside, Surfactants, Deterg.*, 1994, *31*, 128–132.
6. Derian, P. J., J. M. Ricca, F. Marcenac, R. Vukov, D. Tracy, M. Dahanayake, Imidazoline-derived amphoteric surfactants, *SÖFW Journal*, 1995, *121*, 399–410.
7. Parnham, M. J., Phospholipid terminology, *Inform*, 1996, *7*, 1168–1175.
8. Krawczyk, T., Lecithin: consider the possibilities, *Inform*, 1996, *7*, 1158–1159, 1162–1167.

9. Rosen, M. J., F. Zhao, D. S. Murphy, Two-phase mixed indicator method for the determination of zwitterionic surfactants, *J. Am. Oil Chem. Soc.*, 1987, *64*, 439–441.

10. Holzman, S., N. Avram, Amphoteric surfactants, *Tenside, Surfactants, Deterg.*, 1986, *23*, 309–313.

11. Lomax, E., Amphoteric surfactants, in M. R. Porter, ed., *Recent Developments in the Analysis of Surfactants*, Elsevier, London, 1991.

12. Plantinga, J. M., J. J. Donkerbroek, R. J. Mulder, Betaine and free amine in alkyldimethyl betaine by potentiometric titrations, *J. Am. Oil Chem. Soc.*, 1993, *70*, 97–99.

13. American Society for Testing and Materials, Iodine value of fatty amines, amidoamines, and diamines, D2075-66 (1981). West Conshohocken, PA 19428.

14. American Oil Chemists' Society, Iodine value of fats and oils, Official Method Cd 1-25, 2211 West Bradley Ave., Champaign, IL 61826.

15. Humbert, M., A. Hernandez, Monochloroacetate, dichloroacetate, and glycolate in betaines, *World Surfactants Congr.*, 4th, 1996, *1*, 379–388.

16. Gerhards, R., I. Jussofie, D. Käseborn, S. Keune, R. Schulz, Cocoamidopropyl betaines, *Tenside, Surfactants, Deterg.*, 1996, *33*, 8–14.

17. Arens, M., H. P. Wingen, Alkyldimethylbetaines—Determination of free amine—collaboration of the DGF. Communication 149: German standard methods for study of fats, fatty products, surfactants and related materials. Communication 115: Analysis of surface active materials XXIV (in German), *Fett Wiss. Technol.*, 1995, *97*, 430–431.

18. Schulz, R., *Titration of Surfactants and Pharmaceuticals* (in German), Verlag für chemische Industrie, H. Ziolkowsky GmbH, Augsburg, 1996.

19. Buschmann, N., H. Wille, Titration of amphoteric surfactants—comparison of methods, *Comun. Jorn. Com. Esp. Deterg.*, 1997, *27*, 65–70.

20. Arens, M., R. Spilker, Fatty acid amidopropylbetaines—determination of fatty acid amidopropylamine—collaboration of the DGF. Communication 150: German standard methods for study of fats, fatty products, surfactants and related materials. Communication 116: Analysis of surface active materials XXV (in German), *Fett Wiss. Technol.*, 1995, *97*, 468–470.

21. Schwarz, G., P. Leenders, U. Ploog, Condensation products of fatty acids or their methyl esters with aminoethylethanolamine (in German), *Fette, Seifen, Anstrichm.*, 1979, *81*, 154–158.

22. Takano, S., K. Tsuji, Structural analysis of the amphoteric surfactants obtained by the reaction of 1-(2-hydroxyethyl)-2-alkyl-2-imidazoline with ethyl acrylate, *J. Am. Oil Chem. Soc.*, 1983, *60*, 1798–1806.

23. Takano, S., K. Tsuji, Structural analysis of the amphoteric surfactants obtained by the reaction of 1-(2-hydroxyethyl)-2-alkyl-2-imidazoline with sodium monochloroacetate, *J. Am. Oil Chem. Soc.*, 1983, *60*, 1807–1815.

24. Kawase, J., K. Tsuji, Y. Yasuda, K. Yashima, Amphoteric surfactants by LC with post-column detection. II: Imidazoline-type amphoteric surfactants derived from sodium chloroacetate, *J. Chromatogr.*, 1983, *267*, 133–148.

25. Kawase, J., K. Tsuji, Y. Yasuda, Amphoteric surfactants by LC with post-column detection. III: Salt-free-type imidazoline amphoteric surfactants, *J. Chromatogr.*, 1983, *267*, 149–166.

26. Prieto-Blanco, M. C., P. López-Mahía, D. Prada Rodríguez, Dimethylaminopropylamine in alkylaminoamides by HPLC, *J. Chromatogr. Sci.*, 1997, *35*, 265–269.

27. Bailles, F. Ch. Paquot, Synthesis of 2-alkyl-1-(hydroxyethyl)imidazolines and appropriate analytical methods (in French), *Chim. Phys. Appl. Prat. Ag. Surface, C. R. Congr. Int. Deterg.*, 5th, Sept 9–13, 1968, *1*, 267–273.

28. American Oil Chemists' Society, Official and tentative methods. Section J: Analysis of lecithin. 2211 West Bradley Ave., Champaign, IL 61826.

29. Erdahl, W. L., A. Stolyhwo, O. S. Privett, Soybean lecithin by TLC and HPLC, *J. Am. Oil Chem. Soc.*, 1973, *50*, 513–515.

30. Szuhaj, B. F., ed., *Lecithins: Sources, Manufacture and Uses*, American Oil Chemists' Society, Champaign, IL, 1989.

31. Merton, S. L., Soybean lecithins and beef phospholipids by HPLC with an evaporative light scattering detector, *J. Am. Oil Chem. Soc.*, 1992, *69*, 784–788.
32. Kotsira, V. P., Y. D. Clonis, Colorimetric assay for lecithin using two co-immobilized enzymes and an indicator dye conjugate, *J. Agric. Food Chem.*, 1996, *46*, 3389–3394.
33. Dornbos, D. L., R. E. Mullen, E. G. Hammond, Phospholipids of environmentally stressed soybean seeds, *J. Am. Oil Chem. Soc.*, 1989, *66*, 1371–1373.
34. Eder, K., A. M. Reichlmayr-Lais, M. Kirchgessner, Methanolysis of small amounts of purified phospholipids for GC analysis of fatty acid methyl esters, *J. Chromatogr.*, 1992, *607*, 55–67.
35. Blank, M. L., L. J. Nutter, O. S. Privett, Structure of lecithins, *Lipids*, 1966, *1*, 132–135.
36. Rezanka, T., M. Podojil, Separation of algal polar lipids and of individual molecular species by HPLC and their identification by GC-MS, *J. Chromatogr.*, 1989, *463*, 397–408.
37. Ohshima, T., C. Koizumi, Selected ion monitoring GC-MS of 1,2-diacylglycerol *tert*-butyldimethylsilyl ethers derived from glycerophospholipids, *Lipids*, 1991, *26*, 940–947.

5
Qualitative Analysis

I. INTRODUCTION

It is often desirable to perform a test to determine whether a surfactant is present and, if so, whether it is anionic, cationic, or nonionic. There are a number of booklength works of varying age presenting wet chemical methods for qualitative identification of surfactants and other organic compounds (1–3). However, these are not much used because the most convenient tests for qualitative analysis are those which require equipment already in daily use in the laboratory. Recently, this means sophisticated apparatus designed for quantitative analysis. Thus, most of the qualitative chemical tests have been supplanted by molecular spectroscopic analyses, especially IR and MS analyses.

Visual examination of the sample may be sufficient to show the presence of a surfactant. If the sample foams on being shaken, or if an aqueous fluid wets the sides of its container without droplet formation, or if an emulsion forms upon addition of water and hydrophobic solvent, it can be presumed that a surfactant is present. More evidence is obtained by measuring the surface tension of the solution, either directly or electrochemically. An often-overlooked method for detecting additives is elemental analysis. Frequently, an inexpensive determination of total sulfur or nitrogen content is sufficient to provide confirmation of the presence of an anionic or cationic surfactant.

If formulated products are examined, it is generally necessary to perform an initial separation of the surfactant before any of the qualitative tests can be used. The exceptions are the color indicator tests, which often give useful information even in the presence of inorganic salts and other compounds. Other tests require that the surfactant, if not pure, at least be present as a major component of the mixture. Most often, surfactants are isolated by a simple extraction of a solid formulation with 95% or 100% ethanol or by partition of a solution between methylene chloride and water. The resulting fraction, after evaporation of solvent, is usually suitable for direct qualitative analysis. Reliance upon these simple techniques can sometimes lead to errors. Some surface-active materials can be lost by evaporation during the solvent-stripping step, while others are so hydrophilic that they can only be extracted from an absolutely dry sample matrix.

Nowadays, the most common approach to qualitative analysis is examination of the isolated surfactant by molecular spectroscopy: IR, NMR, and MS analyses. The result of this examination is normally sufficient to tell whether the surfactant fraction is pure and what classes of compounds are present. Further analytical strategy is developed based upon this information.

II. TECHNIQUES

A. Chromatography

1. Thin-Layer Chromatography

If a mixture of surfactants is present, then TLC separation is a powerful tool in the hands of experienced analysts. Separation, identification, and estimation of concentration may be performed in a single analysis. TLC methods are described in detail in Chapter 9. Here, we will mention schemes designed for qualitative analysis (see also Table 1).

Henrich developed a comprehensive TLC method for identification of surfactants in formulations (4). She specified two reversed-phase and four normal phase systems, with detection by fluorescence quenching, pinacryptol yellow and rhodamine B, and iodine. Prior to visualization, one plate was scanned with a densitometer at 254 nm, and UV reflectance spectra were recorded for each spot detected. Tables were prepared showing the Rf values of 150 standard surfactants in each of the six systems, along with the reflectance spectra and response to the visualizers. This system allows for systematic identification of compounds of a number of surfactant types (LAS, alcohol sulfates and ether sulfates, alkane sulfonates, sufosuccinate esters, phosphate compounds, AE, APE, ethoxylated sorbitan esters, mono- and dialkanolamides, EO/PO copolymers, amine oxides, quaternary amines, amphoterics and miscellaneous compounds). Supplementary analysis by normal phase HPLC aided in exactly characterizing ethoxylated compounds. For confirmation, the separated spots may be scraped from one of the silica gel plates and the surfactant extracted from the silica with methanol and identified by IR spectroscopy.

Matissek proposed a comparable scheme, using plates of silica, alumina, and 90:10 silica/ammonium sulfate (5). Six different development systems were used individually, as well as a two-dimensional system. Detection was by pinacryptol yellow, Dragendorff reagent, ninhydrin, iodine, leucomalachite green, molybdophosphoric acid, cobalt thiocyanate, and supression of fluorescence of the F-254 plates. Compounds distinguished were a range of sulfates, sulfonates, carboxylates, quaternary amines, betaines, amine oxides, and fatty acid alkanolamides.

Dieffenbacher and Bracco reported a general method for detection of ester and ethoxylated ester emulsifiers in food product (6). The extract containing the lipids from the sample was analyzed on two silica gel plates using different developing solutions. Again, several different visualizing reagents were used to gain selectivity.

These schemes are not for amateurs. To use this technology effectively it is necessary to have laboratory space dedicated to the procedure, as well as chemists experienced in the characterization of formulations. A simpler TLC procedure, suitable for distinguishing between a smaller number of possible surfactants, is based on use of alumina plates (7). The plates are simply developed in isopropanol, and sprayed with pinacryptol yellow for visualization. Other visualizers, such as iodine and cobalt thiocyanate, may also be used in conjunction with pinacryptol yellow. Identification is made on the basis of Rf value and color.

Block distinguished between classes of surfactants using two TLC systems (8). The first used Silica Gel G F-254 with 90:10 ethanol/acetic acid. For visualization, the upper part of the plate was sprayed with pinacryptol yellow to detect anionic surfactants and soap, while the lower portion was sprayed with Dragendorff reagent to detect nonionics and cationics. For further separation of nonionics and cationics, a Silica Gel G plate was used with 80:19:1 methanol/chloroform/0.05 M sulfuric acid. Detection was by spraying with palatine blue solution. For exact characterization, the spots were removed, eluted with solvent, and the surfactants identified by IR spectroscopy.

Some comprehensive methods for surfactant analysis use two-dimensional TLC, where the unknown mixture is first separated into the classes nonionic, cationic, and anionic by reversed-phase or normal phase TLC. After this first development, the spots are developed in a second direction to separate individual surfactants (9–13).

Bare and Read describe a system by which the unknown surfactant mixture is first separated by TLC into the classes of anionics, cationics, and nonionics (14). The spots are then transferred to the FAB probe of a mass spectrometer for identification.

Surfactants may be classified by multiple development of a single silica plate. Successive development with 3:1 MeOH/2 M NH$_3$, 9:1 acetone/THF, 9:1 CHCl$_3$/MeOH, and 80:19:1 CHCl$_3$/MeOH/0.05 M H$_2$SO$_4$ separates surfactants into zones of, in order of increasing Rf values, cationics, amphoterics, anionics, and nonionics (15).

Use of silica gel plates impregnated with oxalic acid permits narrowing down the identity of a nonionic surfactant to one of a number of classes. More exact identification can be performed by IR spectroscopy of spots separated from the TLC medium (16,17). Alternatively, once other compounds have been removed, the nonionic fraction can be characterized by paper chromatography (18). By using two solvent systems and tables of Rf values compiled previously, it is possible to identify the class of nonionic surfactant present.

2. Liquid Chromatography

Except in the case of LC-MS, HPLC is most useful as a quantitative technique, separating mixtures which contain few unidentified compounds. Most HPLC methods use a combination of column and mobile phase which separate a small range of similiar compounds from each other, but which relegate the vast majority of other substances to an initial peak of materials that elute with the void volume and another, invisible group of materials that do not elute from the column during the analysis. Compounds are identified by retention time, hardly a unique characteristic.

Despite these drawbacks, some so-called universal LC methods have been developed to screen unknowns. These give chromatograms which, while not of textbook quality, will resolve many surfactants from each other with sufficient reproducibility to permit tentative assignment of identity. These are useful because they can quickly tell the chemist which of a limited number of possible compounds may be present in a particular sample.

TABLE 1 TLC Methods for Qualitative Analysis of Surfactants

Compounds identified	Stationary phase	Development	Visualization	Comments	Ref.
Anionics (C_{12} AS, C_{12} LAS, dioctylsulfosuccinate, C_{12} soap), nonionics (AE_{10}, OPE_{10}, NPE_5), and cationics (cetyltrimethylammonium bromide, cetylpyridinium chloride, cetyltrimethylammonium chloride); demonstration of 2-dimensional TLC with hybrid NP/RP plates	Whatman hybrid Multi-K CS5 (strip of C_{18} adsorbent on a silica gel plate), activated 2 hr at 115°C	First development: 75:25 EtOH/H_2O; second development: for anionics: 8:1 CH_2Cl_2/MeOH; for cationics: 16:2:1 CH_2Cl_2/MeOH/HOAc	Iodine vapor	Mixture is spotted on C_{18} strip and developed along this strip for class separation of anionics, nonionics, and cationics. The regions containing the anionics and cationics are then cut apart and developed individually in the direction of the silica portion of the plate to separate individual surfactants.	11
LAS, alkylsulfate, ether sulfate, soap, AE, PEG, alkyl polyglycoside, betaine; 2-dimensional TLC for identification of residues extracted from detergents, floor polish, and textiles	Merck silica gel 60	First development: 27:11:1 $CHCl_3$/MeOH/H_2SO_4; second development: 2-butanone saturated with H_2O	Primulin, 0.005% in EtOH, 1-s contact, or modified Dragendorff reagent, 1-s contact	Class separation of anionics and nonionics occurs in the first development. In the second development, soaps are separated from LAS and AE is separated according to degree of ethoxylation and separated from PEG.	12
ABS, ether sulfate, dioctyl sulfosuccinate, soap, toluene sulfonate, AE, APE, alkanolamide, EO/PO copolymer; 2-dimensional TLC for identification in detergents	Merck silica gel G	First development: 10:10:5:2 PrOH/CHCl$_3$/MeOH/10 M NH$_3$; second development: 90:10:5 ethyl acetate/NH$_3$/MeOH	Pinacryptol yellow and modified Dragendorff reagent	Class separation in first development. Homolog separation in second development.	9,10

LAS, soap, an ethoxylate, benzethonium chloride; two TLC systems for identification in toilet bowl products	Silica Gel G F-254	System 1: 90:10 EtOH/HOAc; system 2: 80:19:1 MeOH/CHCl$_3$/0.05 M H$_2$SO$_4$	System 1: pinacryptol yellow for anionics; modified Dragendorff reagent for nonionics and cationics; system 2: palatine blue	Colors and Rf values differ between systems, allowing identification of a number of surfactants.	8
ABS, soap, ether sulfate, amine oxide, alkanolamide, ethoxylates, hydrotropes; general method	Alumina G, 0.25 mm, activated at 75°C for 4 hr	2-PrOH	Pinacryptol yellow: iodine and cobalt thiocyanate used for confirmation	Distinguishes between a number of possible surfactants; identification made by Rf value and color.	7
Esters and ethoxylated esters; phosphorus compounds; detection in food product	Merck silica gel 60 F-254	60:40:1 petroleum ether/ethyl ether/HOAc or 65:25:4 CHCl$_3$/MeOH/H$_2$O	2,7'-Dichlorofluorescein; anisidine/KIO$_4$; modified Dragendorff reagent; acidic naphthoresorcinol; molybdic acid	Identification by Rf values with the two systems and by color.	6
LAS, alcohol sulfates, ether sulfates, alkane sulfonates, sulfosuccinate esters, phosphate compounds, AE, APE, ethoxylated sorbitan esters, mono- and dialkanolamides, EO/PO copolymers, amine oxides, quaternary amines, amphoterics, others; identification in formulations	Merck silica gel 60 F-254 and C$_{18}$	Class separation: 75:25 EtOH/H$_2$O (C$_{18}$): anionics 1: 8:1 CH$_2$Cl$_2$/MeOH (SiO$_2$): anionics 2: 8:2:1 acetone/THF/MeOH (SiO$_2$): nonionics: 8:2 EtOH/2% borax (C$_{18}$): cationics: 8:1:0.75 CH$_2$Cl$_2$/MeOH/HOAc (SiO$_2$): sulfonates and ethoxylate oligomers: 14:6:1 MIBK/PrOH/0.1 M HOAc (SiO$_2$)	Direct densitometry (254 nm), fluorescence quenching, pinacryptol yellow, rhodamine B, and iodine	Identification by Rf values in the 6 systems, along with reflectance spectra and response to visualizers.	4

(Continued)

TABLE 1 (*Continued*)

Compounds identified	Stationary phase	Development	Visualization	Comments	Ref.
AE, NPE, acid ethoxylates, amine ethoxylates, esters, ethoxylated esters, alkanolamides, ethoxylated alkanolamides, EO/PO copolymers	Silica gel impregnated with oxalic acid	9:1 $CHCl_3$/MeOH	Modified Dragendorff reagent or $BaCl_2/I_2$	Identification by Rf value and color.	16
LAS, alkyl sulfates, ether sulfates, isethionates, sulfosuccinate half ester, carboxylates, sarcosides, taurides, fatty acid alkanolamides, amine oxides, quaternary amines, betaines, imidazoline derivatives; detection in personal care formulations	Silica gel 60, silica gel 60 F_{254}, aluminum oxide 60 F_{254} (type E), silica gel G, 90:10 silica/ammonium sulfate	For silica gel 60: 10:10:5:2 n-PrOH/$CHCl_3$/MeOH/10 M NH_3 or 45:5:2.5 ethyl acetate/MeOH/10 M NH_3 or 9:1 EtOH/HOAc; for silica gel 60 at 30–50% relative moisture: 9:1 acetone/THF; for silica gel G with 10% $(NH_4)_2SO_4$: 80:19:1 $CHCl_3$/MeOH/0.05 M H_2SO_4; for alumina: 2-PrOH; two-dimensional system on silica gel 60 at 30–50% relative moisture: step 1: 9:1 acetone/THF: step 2: 10:10:5:2 n-PrOH/$CHCl_3$/MeOH/10 M NH_3	Pinacryptol yellow, modified Dragendorff reagent, ninhydrin, iodine, leucomalachite green, molybdophosphoric acid, cobalt thiocyanate, and supression of fluorescence of the F-254 plates	Identification by Rf value and color.	5

The results are then confirmed by further analysis. For example, a general method to identify the surfactants most often used in shampoos is based on use of two connected reversed-phase columns with 25:75 water/methanol mobile phase, 0.25 M in sodium perchlorate and adjusted to pH 2.5. Refractive index detection allowed the detection of four nonionics, seven anionics, and four amphoteric surfactants (19).

A clever way to make a universal LC method more reliable for identification of surfactants in shampoos and hair conditioners was described by Kadano et al. (20). Two detectors were used, refractive index and UV. The ratio of the two responses was recorded versus retention time, giving a more specific signal than the output of either detector alone. Of course, calibration with the proper standards is all-important. Modern photodiode array UV detectors permit using a similar approach: detection at two or more wavelengths, with automatic calculation of the ratio of the absorbance at different wavelengths.

An HPLC method optimized for scanning mixtures of unknown nonionic surfactants can be used for qualitative and semi-quantitative analyses (21). Use of a C_{18} column, a mobile phase of 90:10 methanol/water, and a refractive index detector allows detection of all major classes of nonionics except the EO/PO copolymers. These last can be determined using a C_8 column and 100% methanol.

3. Gas Chromatography

Ordinary gas chromatography is of little use in the qualitative analysis of surfactants. The low volatility of most surfactants combined with the low selectivity provided by ordinary GC detectors make this technique most suitable for quantitative analysis. However, pyrolysis gas chromatography, in the hands of an operator experienced in surfactant analysis, can be used to identify classes of surfactants (22–24). In addition to considerable experience, access to a library of chromatograms of surfactants of known identity is necessary. Pyrolysis GC is most useful when coupled to mass spectroscopy detection.

GC-MS analysis after partial cleavage with boron tribromide and formation of trimethylsilane derivatives is suggested for qualitative analysis of ethoxylates. Some differentiation of compounds is possible (25).

B. Spectroscopy

1. Ultraviolet/Visible Spectrophotometry and Indicator Tests

Information provided by UV spectroscopy is usually limited to showing the presence or absence of an aryl group in the structure, although an experienced spectroscopist can often also distinguish phenyl from naphthyl character. Once other tests have indicated that an unknown surfactant is anionic, the UV spectrum will indicate if it is an alkylbenzenesulfonate or alkylnaphthalene sulfonate. If the surfactant is cationic, the UV spectrum will indicate whether or not one of the substituents is benzyl or pyridyl. If the surfactant is nonionic, the UV absorbance spectrum indicates if a phenol or a naphthol group is present. With experience, it is possible to identify certain other functionalities of pure surfactant compounds (26).

Colored indicators provide a quick method of checking for the presence of classes of ionic surfactants, and are widely used in QC applications. They are less useful in the analytical laboratory because of the limited information provided. Many procedures have been published for detection of various surfactants by observing a color change in a single- or multi-phase system (27).

Valea Pérez and González Arce propose a series of test solutions to classify common surfactants: ceric ammonium nitrate in HNO_3, $AlCl_3$ in $CHCl_3$, bromine water, $KMnO_4$ solution, KOH/ethanol, acetic anhydride in H_2SO_4. Each reagent gives a positive test for a particular functional group or groups. When an isolated surfactant is tested with this battery of reagents, its identity can usually be narrowed down to a class (28).

In general, any of the visible spectrophotometric methods described in Chapter 12 can be converted to a qualitative test by replacing the instrumental measurement of color intensity with a visual inspection. A typical procedure is use of a mixture of methylene blue (a blue, cationic dye) and pyrocatechol violet (a yellow, anionic dye) to provide a general test for surfactants (29). The aqueous sample (pH 5–6), mixed indicator, and petroleum ether are mixed. Test results are as follows:

No surfactant	Aqueous phase remains green; organic phase is colorless; interface is sharp.
Anionic present	Aqueous phase yellow; interface is blue.
Cationic present	Aqueous phase blue; interface yellow.
Nonionic present	Milky interface.

A sensitive confirmatory test for the presence of cationic surfactants is based upon a modification of the above ion-pair extraction with pyrocatechol violet (29). At pH 5–6, the cationic is extracted into the interface of a water/petroleum ether mixture. The aqueous phase is discarded, and after washing, the petroleum ether phase is evaporated to dryness and the residue reacted with diphenyl tin dichloride to give the intense blue color of the extracted dye.

Another scheme calls for adjusting two surfactant aliquots to pH 1 and pH 11, respectively. The solutions are mixed with chloroform and treated with the mixture of two indicators used in two-phase titration (Chapter 16), dimidium bromide and disulfine blue. If only soap is present, the chloroform phase will be colorless at low pH and red at high pH; other anionics will show the pink color in the chloroform phase at each pH. Quaternary amines will color the chloroform phase blue at either pH, while other amines will only yield the blue color at low pH; nonionics show no color. Obviously, mixtures of surfactants will give confused results (27,30).

Methyl green, purified to remove crystal violet impurity, can be used as a reagent to determine trace contamination of water with anionic surfactant. If a 200-mL sample, acidified and containing methyl green, is extracted with 5 mL chloroform, any green color in the extract signifies the presence of anionic surfactant (31).

In a mixture of anionic surfactant and starch, aqueous iodine exhibits a color range from pure blue to violet, depending on the characteristics of the surfactant (32,33). With proper standards, it is possible to distinguish between short and long alkyl chains, and between linear and branched chains.

The cobaltithiocyanate test for nonionic surfactants, described in Chapter 12, can be used as a qualitative test. Since it requires a two-phase system, it is less convenient than traditional indicator tests.

2. IR and NMR Spectroscopy

IR spectroscopy is the most generally used method for qualitative analysis of organic compounds. Its low cost and rapid turnaround time make it suitable for initial examination of a mixture, as well as more detailed examination of purified fractions prepared by extraction or ion exchange chromatography. The small sample size requirement makes it possible to identify compounds collected from the eluent of a liquid chromatograph or removed from a TLC plate. This is accomplished using diffuse reflectance Fourier transform IR, or, more commonly, IR microscopy (34). Hummel has collected the spectra of all commercial surfactants (35), as has Nyquist in the Sadtler spectral library (36). In addition, ASTM D2357 gives a table of absorbance bands of common surfactants (37), and Nettles has described the IR spectroscopy of common surfactants and their raw materials (38).

More than other techniques, the organic spectroscopy methods can mislead the inexperienced into making false identifications. Chemists new to a product area should approach the identification of unknown compounds in mixtures with caution. A frequent error of the novice is to claim the presence of a compound based on observing one or two absorbance bands, even though some others of its bands are absent. If minor bands are present but one or more of the major bands are absent, the identification is incorrect. The assignment of identity can be made only if every major band is observed when making comparisons to standard spectra.

NMR spectroscopy is very useful for identifying organic compounds, provided that they can be obtained in a reasonably pure state. König has published a table of chemical shifts of functional groups found in common surfactants (39). This allows use of proton magnetic resonance to identify components of commercial products, where the range of possible structures is limited. Carminati and coworkers recommend the use of ^{13}C NMR for the identification of unknown surfactants, both alone and in formulated products. With experience, not only the surfactants, but other components of products can be identified (40).

3. Mass Spectrometry

In the hands of an experienced operator, a modern mass spectrometer is a very powerful and extremely rapid tool for identifying surfactants in formulations. At the time this is written, ion-spray ionization is the most generally useful technique for MS of surfactants (41). Ideally, the instrument is capable of operation in the negative ion mode for detection of anionics and in positive ion mode to detect nonionic and cationic surfactants. MS-MS is helpful for positive identification of the ions. MS is covered in detail in Chapter 15.

C. Other Methods

Polarography at the dropping mercury electrode can be used to estimate the concentration of total surfactants of natural and synthetic origin (42). Measurement is based on the suppression of the streaming maximum of the reduction wave of Hg(II) in aqueous solution. A similar indication of the presence of surfactants is the suppression of the oxygen overpotential. By use of ion exchange or other separation methods, this technique can be made selective for anionic surfactants (43). In general, electrochemical methods give little qualitative information, but simply indicate that a surfactant of a particular class (i.e., anionic, cationic) is present. Identification of the surfactant requires other methods.

A test developed specifically to detect lignin sulfonates consists of treating the solution with aniline. A precipitate that is insoluble in hydrochloric acid indicates the presence of the sulfonate (44).

Many of the tests developed to determine proteins suffer from interference from nonionic surfactants. The Bio-Rad protein assay is an example. Such tests are usually not amenable to detecting traces of surfactant, but are suitable for higher concentrations.

Modified Dragendorff reagent forms a precipitate with even small amounts of poly-alkoxylates (45). The following procedure is applicable to amounts of nonionic down to about 0.1 mg:

Procedure: Qualitative Test for Nonionic Surfactants Using Modified Dragendorff Reagent (45).

Reagents   Solution A: Dissolve 1.7 g basic bismuth nitrate or bismuth nitrate pen-tahydrate in 20 mL glacial acetic acid. Make up to 100 mL with water. Solution B: Dissolve 40 g KI in 100 mL H_2O. Test Solution: Mix 10 mL each of Solutions A and B with 20 mL acetic acid and 60 mL water. Add 50 mL 20% $BaCl_2$ solution and store in a brown bottle. For optimal reproducibility, the combined solution should not be more than two days old.

Note: This solution is commercially available, generally without $BaCl_2$.

The test is performed in a small centrifuge tube. One milliliter sample solution is diluted to 4 or 5 mL with 20% acetic acid, combined with an equal volume (4 or 5 mL) of test solution, and shaken vigorously. The presence of nonionic surfactant is shown by the formation of a precipitate. Small quantities of precipitate are more easily detected if the solution is centrifuged. The test may be made quantitative: see Chapter 17, Section II.

Comments: This is a general test for polyoxyethylene and polyoxypropylene groups. Positive tests are therefore also given by polyethylene glycol and anionic or cationic surfactants which contain the polyoxyalkalene functionality. The above test may also be conducted by spotting the sample on a piece of filter paper and spraying with Dragendorff reagent.

REFERENCES

1. Cheronis, N. D., J. B. Entrikin, *Semimicro Qualitative Organic Analysis*, 2nd ed., Interscience, New York, 1957.
2. Rosen, M. J., H. A. Goldsmith, *Systematic Analysis of Surface Active Agents*, 2nd ed., Wiley-Interscience, New York, 1972.
3. Siggia, S., J. G. Hanna, *Quantitative Organic Analysis via Functional Groups*, 4th ed., R. E. Krieger, Malabar, FL, 1988 (first published 1979).
4. Henrich, L. H., Separation and identification of surfactants in commercial cleaners, *J. Planar Chromatogr.—Mod. TLC*, 1992, *5*, 103–117.
5. Matissek, R., TLC identification of surfactants in shampoos, bubble bath formulations, and soaps (in German), *Tenside*, 1982, *19*, 57–66.
6. Dieffenbacher, A., U. Bracco, Analytical techniques in food emulsifiers, *J. Am. Oil Chem. Soc.*, 1978, *55*, 642–646.
7. Desmond, C. T., W. T. Borden, Surface active agents in admixture by TLC, *J. Am. Oil Chem. Soc.*, 1964, *41*, 552–553.
8. Block, H., Toilet bowl additives—surfactant analysis (in German), *Lebensmittelchem., Lebensmittelqual.*, 1985, *9*, 58–70.
9. Bey, K., Detergents (in German), *Tenside, Surfactants, Deterg.*, 1965, *2*, 373–375.
10. Bey, K., TLC analysis of surfactants (in German), *Fette, Seifen, Anstrichm.*, 1965, *67*, 217–221.
11. Armstrong, D. W., G. Y. Stine, Separation and quantitation of anionic, cationic and nonionic surfactants by TLC, *J. Liq. Chromatogr.*, 1983, *6*, 23–33.
12. Bosdorf, V., T. Bluhm, H. Krüssman, TLC determination of adsorbed nonionic surfactants on fabrics (in German), *Melliand Textilber.*, 1994, *75*, 311–312.
13. Bosdorf, V., H. Krüssman, Detergents and cleaning agents with TLC, *World Surfactants Congr., 4th*, 1996, *4*, 92–95.

14. Bare, K. J., H. Read, Fast atom bombardment MS to identify materials separated on high-performance TLC plates, *Analyst*, 1987, *112*, 433–436.

15. Kruse, A., N. Buschmann, K. Cammann, Separation of different types of surfactant by TLC, *J. Planar Chromatogr.—Mod. TLC*, 1994, *7*, 22–24.

16. König, H., Separation of nonionic surfactants by TLC (in German), *Fresenius' Z. Anal. Chem.*, 1970, *251*, 167–171.

17. König, H., E. Walldorf, Skin cleansers and shampoos containing synthetic surfactants (in German), *Fresenius' Z. Anal. Chem.*, 1979, *299*, 1–18.

18. Selden, G. L., J. H. Benedict, Polyoxyalkylene-type nonionic surfactants by paper chromatography, *J. Am. Oil Chem. Soc.*, 1968, *45*, 652–655.

19. Nakamura, K., Y. Morikawa, Surfactant mixtures in shampoos and detergents by HPLC, *J. Am. Oil Chem. Soc.*, 1984, *61*, 1130–1135.

20. Kadono, K., Y. Kitagawa, T. Kohno, HPLC analysis of surfactants present in shampoos and hair conditioners (in Japanese), *J. SCCJ*, 1987, *21*, 5–15. CA 107:223032r.

21. König, H., R. Ryschka, W. Strobel, Separation, identification, and determination of nonionic surfactants using HPLC (in German), *Fresenius' Z. Anal. Chem.*, 1985, *321*, 263–267.

22. Liddicoet, I. H., L. H. Smithson, Surfactants by pyrolysis GC, *J. Am. Oil Chem. Soc.*, 1965, *42*, 1097–1102.

23. Lew, H. L., Acid pyrolysis-capillary chromatographic analysis of anionic and nonionic surfactants, *J. Am. Oil Chem. Soc.*, 1967, *44*, 359–366.

24. Washall, J. W., T. P. Wampler, Pyrolysis of complex multicomponent samples, *J. Chromatogr. Sci.*, 1989, *27*, 144–148.

25. Leyssens, L., V. Martens, E. Royackers, H. Penxten, A. Verheyden, J. Czech, J. Raus, Identication of polyoxyethylene glycols and related compounds by capillary GC, *J. Pharm. Biomed. Anal.*, 1990, *8*, 919–927.

26. Reid, V. W., T. Alston, B. W. Young, Surface-active agents, *Analyst*, 1955, *80*, 682–689.

27. Holness, H., W. R. Stone, Qualitative analysis for anionic surface-active agents, *Analyst*, 1957, *82*, 166–176.

28. Valea Pérez, A., González Arce, M. L., Classification and analysis of surfactant products, *Tec. Lab.*, 1989, *12*, 236–245.

29. Bürger, K., Anionic and cationic surface-active compounds in aqueous solutions (in German), *Fresenius' Z. Anal. Chem.*, 1963, *196*, 15–21.

30. Mandery, K., Surfactant component of an aqueous detergent (in German), *Seifen, Öle, Fette, Wachse*, 1991, *117*, 595–597.

31. Abbott, D. C., Anionic detergents in drinking water, *Analyst*, 1963, *88*, 240–242.

32. Wurzschmitt, B., Substances with capillary activity (in German), *Fresenius' Z. Anal. Chem.*, 1950, *130*, 105–185.

33. Müller, K., D. Noffz, Length and degree of branching of the side chains of surface active alkylbenzene derivatives (in German), *Tenside Surfactants, Deterg.*, 1965, *2*, 68–75.

34. Kunkel, E., G. Peitscher, K. Espeter, Microanalysis of surfactants, *Tenside Surfactants, Deterg.*, 1977, *14*, 199–202.

35. Hummel, D. O., *Analysis of Surfactants: Atlas of FTIR Spectra with Interpretations*, Hanser/Gardner Publications, Inc., Cincinnati, 1996.

36. Nyquist, R. A., Sadtler IR surfactants databases, available from Bio-Rad Laboratories, Hercules, CA.

37. American Society for Testing and Materials, Standard qualitative classification of surfactants by IR absorption, D2357-74 (1995). West Conshohocken, PA 19428.

38. Nettles, J. E., IR spectroscopy for identifying surfactants, *Text. Chem. Color.*, 1969, *1*, 430–441.

39. König, H., Structural elucidation of surfactants by NMR spectrometry (in German), *Tenside*, 1971, *8*, 63–65.

40. Carminati, G., L. Cavalli, F. Buosi, ^{13}C NMR for identification of surfactants in mixture, *J. Am. Oil Chem. Soc.*, 1988, *65*, 669–677.

41.   Ogura, I., D. L. DuVal, S. Kawakami, K. Miyajima, Identification and quantitation of surfactants in consumer products by ion-spray MS, *J. Am. Oil Chem. Soc.*, 1996, *73*, 137–142.
42.   Hunter, K. A., P. S. Liss, Polarographic measurement of surface-active material in natural waters, *Water Res.*, 1981, *15*, 203–215.
43.   Linhart, K., Polarographic determination of surface-active materials in water and wastewater, as well as the determination of their biodegradibility (in German), *Tenside*, 1972, *9*, 241–259.
44.   American Society for Testing and Materials, Standard test method for lignosulfonates (sulfite cellulose) in tanning extracts, D4900-89 (1995). West Conshohocken, PA 19428.
45.   Bürger, K., Surface-active polyoxyethylene compounds and PEG (in German), *Fresenius' Z. Anal. Chem.*, 1963, *196*, 251–259; *ibid.*, 1964, *199*, 434–438.

6
Separation of Surfactants

This chapter summarizes techniques used to separate surfactants in milligram or gram quantities. While many analytical methods are based upon separation, only such applications are covered here which are preparative in nature. The arrangement of the separation techniques in this chapter is inspired by that of Karger, Snyder, and Horvath (1). Procedures for separating individual commercial surfactants into their components are covered in Chapters 1–4, as part of surfactant characterization.

Many comprehensive schemes for separation of commercial products have been published. These attempt the systematic identification of every ingredient in a formulation by a single general approach and are represented by flow charts of as many as a dozen steps. These methods are very valuable in the laboratory of the originator, but are not easily translated to other locations where the nature of the samples and the equipment available to the analyst differ. These general schemes are not covered here. We limit our discussion to separation of only a few components at once, allowing the reader to tailor an approach to the problem at hand.

I. GAS-LIQUID SEPARATIONS

A. Distillation and Gas Chromatography

Distillation is generally not applicable to separation of surfactants. The ionic surfactants have no appreciable volatility unless derivatized. The lower oligomers of the ethoxylated nonionics can sometimes be distilled, but the separation requires high temperature and low pressure, and is not suitable as a routine procedure. Examples include separation of alkyl-phenol ethoxylates by molecular distillation (2) or combined steam distillation/extraction (3) and separation of ethoxylated fatty acids by molecular distillation (4,5). In the latter case, each fraction contained a mixture of mono- and diester, free fatty acid, and PEG.

Molecular distillation of an ethoxylated C_{12}/C_{14} alcohol reaction mixture was possible up to about the 4-mole ethoxylate, but each fraction contained a mixture of different alkyl chain lengths and varying degree of ethoxylation, as well as also containing PEG (6). Similar results were obtained for fatty acid esters of decanol, glycerol, and sorbitan (5). Thus, distillation is only suitable for preliminary sample workup, not for isolation of pure compounds.

Gas chromatography separations are covered in depth in Chapter 8. Preparative gas chromatography is seldom applied to surfactants because of their low volatility.

B. Foam Fractionation and Solvent Sublation

In foam fractionation, gas bubbles are passed through an aqueous solution and the foam which develops is collected from the surface. The foam is enriched 30- to 100-fold in the surfactant compared to the solution, and the surfactant is easily recovered (7,8). This technique has the disadvantage that surface-active materials which are not strong foamers are not recovered, nor are low levels of foamers below the concentration required to generate foam. Mass transfer of surfactant from water to foam is too slow for practical quantitative analysis.

For purposes of chemical analysis, foam fractionation has been supplanted by a somewhat similiar method, a type of adsorptive bubble fractionation called solvent subla-tion. This is a powerful means of separating surface active materials. Trace levels of sur-factants are separated from water by passing bubbles of an inert gas through the solution (9). The surfactant adsorbs to the surface of the bubbles, and is collected in an immiscible solvent floating upon the water (see Fig. 1). This technique is distinct from foam fraction-ation because a stable foam is not formed. It is superior to liquid-liquid extraction because the organic and aqueous phases are not in equilibrium, and thus a much higher concentra-tion of the surfactant can be attained in the organic phase than in the analogous extraction (1).

Wickbold was the first to describe the application of this technique to determination of surfactants (10). He preferred a system where the gas stream was presaturated in ethyl acetate, a refinement given the name "booster bubble separation." Ethyl ether works as well as ethyl acetate to trap the surfactant and the choice of gas is unimportant, though nitrogen is usually most convenient. The sublation must be conducted gently so that there is no turbulence at the water/ethyl acetate interface. Such mixing would permit distribu-tion of the surfactant between the two phases in accord with its relative solubility, prevent-ing the desired concentration enrichment in the organic phase.

Solvent sublation was used a great deal in environmental analysis of surfactants dur-ing the 1970s and 1980s, generally with ethyl acetate as the immiscible solvent. In the case of nonionic surfactants, it has been thought that sublation is effective in isolating materials with an average degree of ethoxylation up to about 30 units. However, careful study shows that this is not a general rule, but must be checked for individual surfactants. Recovery rates for low molecular weight surfactants can be as high as 90% and as low as 50%. Re-coveries for 90-mole ethoxylates are lower than 10% (11). This also is true of highly ethoxylated EO/PO copolymers (12). It is critically important that the water have a high

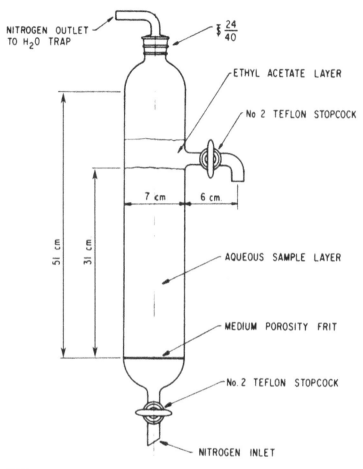

NITROGEN OUTLET
TO H₂O TRAP

$\bar{\xi}\,\frac{24}{40}$

ETHYL ACETATE LAYER

No 2 TEFLON STOPCOCK

7 cm   6 cm.

51 cm   31 cm

AQUEOUS SAMPLE LAYER

MEDIUM POROSITY FRIT

No. 2 TEFLON STOPCOCK

NITROGEN INLET

FIG. 1 Solvent sublation apparatus. (Reprinted with permission from Ref. 231. Copyright 1977 by the American Chemical Society.)

ionic strength, which is accomplished by adding 100 g/L sodium chloride. Otherwise, recoveries of some ionic surfactants can be quite low (13).

Solvent sublation of EO/PO copolymers is a particular problem. The higher MW products are not recovered at all, while the lower MW products give variable yield depending on both molecular composition and on temperature (14). Recovery must be determined for each individual product.

II. LIQUID-LIQUID EXTRACTION

Liquid-liquid extraction techniques are very useful for the routine separation of surfactants. The versatility of ordinary extraction in a separatory funnel makes it possible to develop a procedure which will handle almost any mixture.

Frequently, emulsions will form when a surfactant is partitioned between water and a nonpolar solvent. Emulsions can usually be controlled by limiting the total mass of surfactant in the system, and by increasing the ionic strength by addition of salts such as sodium chloride or sodium sulfate. Alternatively, a co-solvent may be added. For example, if an emulsion forms while extracting a surfactant from water with methylene chloride, ethanol may be added until the emulsion collapses. The extraction is then repeated with methylene chloride until all of the surfactant is extracted. Some investigators suggest placing the vessel in an ultrasonic bath to break emulsions.

The same techniques are applied when the original sample is a water-in-oil or oil-in-water emulsion. Oil-in-water emulsions can sometimes be attacked by adding an inorganic salt to break the emulsion, then extracting with alcohol or methylene chloride. Success depends on the nature of the organic phase as well as the surfactant, since further extraction is probably required to separate the surfactant from other components. Water-in-oil emulsions can sometimes be dispersed in 50:50 ethanol/water, then extracted with petroleum ether. The surfactants may remain behind in the aqueous phase or may need to be recovered from the hydrocarbon phase by washing with additional ethanol/water.

Ionic surfactants are generally insoluble in chloroform or methylene chloride unless the ion pair is formed with an ionic surfactant of opposite charge. This property permits isolation of surfactants from aqueous media with great selectivity. The same procedure, with appropriate changes, can be used as for the two-phase titration of ionic surfactants (Chapter 16).

A general method for isolating hydrophobic materials from wet solids consists in adding 50 mL chloroform and 100 mL methanol to a 10 g sample. After homogenizing, 50 mL chloroform is added and the mixture is homogenized again. After centrifugation, the supernate is removed, and the residue extracted with a further 50 mL chloroform, combining the two extracts. The extract is mixed with 90 mL water and mixed well, adding a little NaCl if an emulsion forms. The layers are separated, the aqueous layer is washed with 50 mL 2:1 chloroform/methanol, and the two organic extracts, containing the hydrophobic material from the sample, are combined. This technique, developed by food technologists, is generally applicable to the isolation of surfactants (15). Further separation is necessary to remove surfactants from nonpolar lipids such as triglycerides.

An interesting approach to separating surfactants from impurities consists of forming a three-phase system, where an oil/water/surfactant/alcohol emulsion is in equilibrium with a layer of water and a layer of organic solvent; salt may also be added. The impurities become more concentrated in the upper and lower layers, while the pure surfactant is more concentrated in the emulsion (16).

A number of liquid-liquid partition systems applied to surfactants are summarized in Table 1. The analyst is warned that the great range in polarity of commercial surfactants makes it impossible to specify a single liquid-liquid extraction technique which will allow isolation of all surfactants from any matrix. Even in the case of a single surfactant such as an ethoxylated nonylphenol, the oligomers with higher degree of ethoxylation are much more hydrophilic than those with less EO, so that extraction from water may change the molecular weight distribution of the sample (17). For quality control applications, ordinary liquid-liquid extraction has serious disadvantages: Typically, the solvents used are flammable, and must be kept out of the laboratory air for industrial hygiene reasons. Extractions require a considerable amount of glassware, which must be cleaned in a manner which does not leave surfactant residues. Thus, solid phase extraction procedures are gradually replacing liquid-liquid extractions in routine analysis.

Centrifugal partition chromatography, a type of countercurrent chromatography, has been demonstrated for automated extraction of a nonionic surfactant into ethyl acetate (18) and for isolation of phosphatidylcholine from natural phospholipids (19). This is an exciting separation technique for preparative scale work, but so far has not been adopted to any great extent by analytical chemists.

Liquid-liquid chromatography has developed into HPLC, which is discussed in Chapter 7. Most HPLC separations can be modified to the preparative or semi-preparative scale. This is a major undertaking for the new practitioner and is beyond the scope of this volume.

Procedure: Isolation of Alkanesulfonate Surfactants by Ion-Pair Extraction (20)

About 200 mg surfactant is dissolved in 70 mL water in a 500-mL separatory funnel. To this are added 30 mL 20% sodium sulfate solution, 0.5 mL 2 M H_2SO_4, and 100 mL methylene chloride. A stoichiometric quantity (determined by a prior titration) of benzethonium chloride solution is added, and the separatory funnel is shaken 1 min. The methylene chloride phase is drawn off and the solvent is evaporated, leaving the benzethonium salt of the anionic.

Note: Chloroform should not be substituted for methylene chloride as the solvent because during the evaporation step small quantities of products such as phosgene can form and react with the amine to form nonvolatile compounds which will give a high weight.

III. LIQUID-SOLID PARTITION

A. Precipitation

1. Reagents

There are a number of reagents which precipitate classes of surfactants. Many anionics are at least partially precipitated as the salts of divalent cations like barium and calcium. The precipitation of ethoxylated nonionics with molybdophosphoric acid in the presence of divalent cations may be used as the basis of a preparative separation from other surfactants (21). Similar reactions are covered in Chapter 17 with gravimetric techniques.

Dodecylsulfate may be separated from aqueous solution by treating the sample with concentrated calcium chloride to precipitate the calcium salt. The solubility of calcium dodecylsulfate is only 0.0003 molal (22). Separation of LAS from water was demonstrated by addition of sodium chloride; about 95% of LAS initially present at concentrations over 0.01% can be removed by making the solution 5% in NaCl (23). The presence of nonionic surfactants will inhibit the precipitation of salts of anionic surfactants due to the formation of mixed micelles (24,25).

2. Solvents

Many methods for isolation of surfactants are based upon the solubility of the materials. For example, the sodium salts of most anionic surfactants are soluble in acetone, while soap is not (26). It is difficult to make general statements in this field because of the many exceptions. For example, the sodium salt of the most common anionic surfactant, LAS, is only partially soluble in acetone. Surfactants which would be insoluble by themselves may become soluble in mixtures with other surfactants. Govindram and Krishnan report for a number of common surfactants that all are soluble in ethanol, but that only nonionics and free fatty acids are soluble in petroleum ether or ethyl acetate, with cocomonoethanolamide being only partially soluble in petroleum ether (26). The solubility of anionics differs depending upon whether they are present as free acids or sodium salts.

TABLE 1 Liquid/Liquid Partition Systems

Separation	Phase 1	Phase 2	In phase 1	In phase 2	Notes	Ref.
LAS from environmental extracts	Water containing 0.001% methylene blue	CHCl₃	Impurities	LAS as the ion pair with methylene blue	Extract is evaporated to dryness, taken up in EtOH, and passed through a cation exchange column to remove the methylene blue.	111
LAS from alkylphenol ether carboxylates	Water, pH 2, low in ionic strength	CH₂Cl₂	LAS and other sulfonates	Carboxylic acids		58
Sodium dodecylsulfate from ointment	1.5:1:1 pentane/ethyl acetate/MeOH	Separates from phase 1	Hydrophobic impurities (upper)	Surfactant (lower)	More solvent is added to the hydrophilic extract, for a total of 3 extractions.	112
Ether sulfate from PEG and PEG sulfate	Water, 59 gpL in NaCl	9:1 ethyl acetate/1-BuOH	PEG and PEG sulfate	Surfactants		113
α-Olefin sulfonate from impurities	3 M HCl	Ethyl ether	Disulfonates and other impurities	Anionic surfactant		114,115
α-Olefin sulfonate from water	3 M HCl, from which monosulfonates have been extracted	n-BuOH or i-BuOH	Impurities	Disulfonates		115
Paraffin monosulfonates from di- and polysulfonates	50:50 EtOH/H₂O, 2 M in HCl	Benzene or petroleum ether	Di- and polysulfonates and Na₂SO₄	Monosulfonates	Continuous extractor required because of unfavorable distribution coefficients.	27,116
Paraffin sulfonate from unsulfonated oil	50:50 EtOH/H₂O, neutral or weakly basic	Pentane	Sulfonates and ionic impurities	Unsulfonated oil		27

Paraffin sulfonate from inorganics	60:40 n-BuOH/acetone	130 g/L aqueous Na$_2$SO$_4$	Mono- and poly-sulfonates	Inorganics	Sodium sulfate decahydrate is allowed to crystallize so that the layers may be separated by filtration and washing.	20
Petroleum sulfonate from oil	50:50 2-PrOH/H$_2$O	Pentane	Sulfonates	Oil	For removal of residual oil, it is necessary to evaporate the isopropanol/water, and extract the surfactant with boiling pentane.	117
Anionic surfactant from benzene-sulfonate or toluene-sulfonate hydrotropes	Water 1 M in HCl and 10% in Na$_2$SO$_4$	Diethyl ether	Hydrotropes	Anionic surfactant		118
AE from emulsion lubricant	Acetonitrile	n-Hexane	AE with EO > 2, some impurities	Oil and methyl oleate; AE with 2 ≥ EO	Surfactants are first extracted from lubricant emulsion (to which brine is added) with CH$_2$Cl$_2$ before partition between hexane and CH$_3$CN.	119
NPE from emulsion lubricant	Acetonitrile	n-Hexane	NPE, water, some impurities	Oil and methyl oleate		120
Sucrose from sucro-glyceride reaction mixture	Water, adjusted to pH 5–6 with H$_3$PO$_4$	n-Butanol	Sucrose and alkali metal phosphate salts	Esters of sucrose and glycerin; free fatty acids		121

(Continued)

TABLE 1 (Continued)

Separation	Phase 1	Phase 2	In phase 1	In phase 2	Notes	Ref.
Nonionics from water	Water, low in ionic strength	Toluene	Ionic surfactants and other materials	Nonionic materials		122
PEG and surfactants from biodegradation tests	CHCl$_3$	Aqueous sample, 32% in NaCl	PEG and surfactants	Other materials		123
Mono- and diethanolamides from fruit drinks	Water	CH$_2$Cl$_2$	Other compounds	Nonionic surfactants	Identification by TLC.	124
Amine oxide from free amines	Water, pH 8	Petroleum ether	Amine oxide	Free amines		125
Ethoxylated nonionics from fruit juice drinks and concentrates	Acetone	Concentrated fruit juice syrup	Nonionics and carotenoids	Other components		126
Ethoxylated nonionics from carotenoids	Carbon disulfide	60:40 MeOH/H$_2$O	Carotenoids	Nonionics		126
Ethoxylated nonionics from aqueous protein solutions	Water (17% MeOH added if necessary to break emulsions)	1,2-dichloroethane	Proteins	Nonionics	Nonethoxylated nonionics removed with much less efficiency.	127
PEG from nonionics	BuOH	Water, 5% in NaHCO$_3$	Nonionic surfactants	PEG	PEG is recovered by neutralizing the aqueous phase to methyl orange with HCl, evaporating to dryness, and extracting the residue with CH$_2$Cl$_2$.	128

PEG from ethoxylates of fatty acids, fatty alcohols, fatty amines, fatty acid amides, and alkylphenols	Methyl ethyl ketone	Water	Ethoxylates	PEG	A 50:50 mixture of 2-butanone/water is prepared. It partitions into an upper phase consisting of 11.4:88.6 water/2-butanone and a lower phase consisting of 77.4:22.6 water/2-butanone. 10 mL of a 1% aqueous solution of the sample is added to a separatory funnel containing 25 mL each of 2-butanone and water. For quantitative separation, the extraction of the lower phase is repeated until no further ethoxylate is recovered.	129
Benzylalkyldimethyl-ammonium chloride from pharmaceuticals	Water containing 0.017% methyl orange and 3% acetic acid	1,2-Dichloroethane	Impurities	Cationic surfactant as the ion pair with methyl orange	Organic phase may be washed with 10% acetic acid or 1 M NaOH to remove impurities.	130
Quaternary amines from other amines	Isoamyl alcohol	Aqueous mineral acid containing inorganic salt	Quaternaries	Other amines		131
C_{12}, C_{14}, and C_{16} alkyltrimethylamines from Bayer liquor	Bayer liquor, acidified to pH 12, to which KI is added	CH_2Cl_2	Other components	Quaternaries as iodide salts		132

Paraffin di- and polysulfonates may be separated from inorganic salts such as sodium sulfate by dissolving in ethanol or isopropanol at neutral pH to precipitate the salts,
which are washed with warm methanol (27).

Emulsions formed using soap and some other carboxylic acid anionic surfactants are
broken by acidification. The surfactant may become water insoluble at low pH and thus
can be separated by precipitation or extraction.

Ethoxylated nonionic surfactants are less soluble in water at higher temperatures.
Thus, at high concentrations, some may be isolated simply by heating an aqueous solution
above the cloud point of the surfactant, which forms a separate layer and can be removed.
The process must be repeated several times for reasonable recovery. The separated phase
will be enriched in the more hydrophobic homologs compared to the original surfactant.
This phase may also act as an extraction medium to concentrate other components of
the solution.

Sucrose and other materials may be separated from sucrose ester surfactants by taking up the mixture in hot methyl ethyl ketone, in which sucrose is insoluble. Acidification
and cooling of the solution will cause the sucrose ester to precipitate, while any fatty acid
remains soluble (28). Another method calls for washing the mixture with acidic water of
high ionic strength. The sucrose esters remain insoluble, while other components of the
reaction mixture are dissolved (29,30).

Phospholipids are isolated from crude lecithin by their solubility properties. For instance, extraction with ethanol yields an ethanol-soluble fraction enriched in phosphatidylcholine and an ethanol-insoluble fraction high in the anionic phosphatides like
phosphatidylserine and phosphatidylinositol (31). Acetone precipitation is often used to
separate phosphatides from other materials, but this step can contribute to hydrolysis of
phosphatidylcholine (31).

B. Ion Exchange

Ion exchange is used extensively for surfactant separation. It is used to separate surfactants
from impurities and byproducts, as well as to separate the classes of surfactants from each
other. It is important to be aware that adsorption occurs on organic ion exchange materials
such that, depending on temperature and the solvent system, the resins may also retard
lipophilic nonionic materials, such as nonionic surfactants. For instance, a weakly acidic
cation exchanger in the H^+ form will strongly adsorb nonionic surfactants, presumably by
a hydrogen-bonding mechanism, from neutral or acidic aqueous solvents (32). With a nonpolar mobile phase, even cationic surfactants may be retarded on an anion exchange column. Depending on the particular compounds, surfactants of all kinds may be easily or
difficultly eluted from the resin because of adsorption. Some amount of experimentation is
required in optimizing the eluents for a particular separation. Adsorption can usually be
inhibited or eliminated by increasing the percentage of organic solvent in aqueous/organic
mixtures.

The chemical properties of the surfactants must not be forgotten during the separation process. For instance, a cation exchange column can catalyze the esterification of
soap with methanol solvent so that fatty acid methyl esters are formed. Alkyl sulfates can
be hydrolyzed to the corresponding alcohols under the strongly acidic conditions used in
some ion exchange separations (33). Micelle formation can interfere with ion exchange
separation of surfactants. This last problem is eliminated by working in solvents with a
strong enough alcohol content (>20%) that micelles do not form (34).

In general, the same ion exchange materials and solvent systems may be used whether the separation occurs in a packed column or is performed in a beaker in batch mode. The particle size of the solid phase and the relative volumes of solid and liquid must be chosen for each case. It is worth noting that many of the ion exchange procedures used in routine analytical laboratories are not optimized as to type and mass of resin, solvent, or acid strength of mobile phase and of desorbing solution. For example, the polystyrene resin most often used contains 8% divinylbenzene for cross-linking and is not as suited for large surfactant molecules as a resin with only 2% cross-linking agent. The 50–100 mesh particle size used in gravity flow columns is only about half as efficient as is the 100–200 mesh particle size designed for use in pressurized systems (33).

1. Ion Exchange Chromatography

(a) Anion exchange chromatography. Applications of anion exchange chromatography to surfactant separation are listed in Table 2. Anionic surfactants are reversibly adsorbed on conventional anion exchange resins. For most anionics, either strongly basic or weakly basic resins can be used, depending on the solvent strength, and the chloride form of the resin is as effective as the hydroxide form. The solvent pH, ionic strength, and water content can be varied to effect the separation of anionics from each other, as well as from other surface-active and non–surface-active compounds.

The use of a weakly basic anionic exchange column in series with a strongly basic column sometimes has advantages (35). The weakly basic column will adsorb only the strongly acidic surfactants, such as the sulfates and sulfonates. The strongly basic column will then adsorb the remaining weakly acid substances, which usually are only carboxylic compounds like fatty acids. This distinction can also be made by using a strongly basic resin in two forms, the chloride or acetate form, which will adsorb the sulfates and sulfonates, followed by the hydroxide form to adsorb the fatty acids (33). Under these conditions, ether carboxylates exhibit behavior similar to the fatty acids, being retained on an anion exchange resin in the OH^- form, but passing through a resin in the Cl^- form (13). Elution of anionic surfactants from ion exchange resin is difficult, but recovery of adsorbed anionics from weakly basic resin is generally easier than from strongly basic resin.

Proprietary anion exchange columns are available for the removal of sodium dodecyl sulfate from protein samples prior to chromatographic analysis (36,37).

(b) Cation exchange chromatography. Cationic surfactants are typically concentrated using cation exchange columns. The conditions must be chosen to suit the particular compound, since some cationics are so strongly retained on conventional resins that they cannot be quantitatively recovered (38). The weaker ion exchange materials are used to avoid this problem. Silica can function as a weak cation exchanger, permitting the separation of cationic surfactants from other materials (39).

Pardue and Williams demonstrated determination of APE in protein solutions by using short anion or (preferably) cation exchange guard columns (40). These held the protein tightly bound, but allowed the nonionic to elute unretarded. An eluent of 0.05–0.1 M potassium phosphate buffer and pH 3–5 with 30% isopropanol was used. This separation is useful in the field of biochemistry and is easily converted to the preparative scale.

A selection of cation exchange chromatography procedures applied to surfactant separation is given in Table 3.

(c) Combined anion/cation exchange chromatography. A number of general schemes for analysis of detergent formulations and other surfactant mixtures make use of two or

TABLE 2 Anion Exchange Separation of Surfactants

Separation performed	Ion exchanger	Procedure	Ref.
LAS from APE	SPE column of strong anion exchanger (Analytichem), 2.8 mL volume	Methanolic soil or sediment extract is passed through the column, 10 to 50 mL in volume, followed by a 10-mL MeOH rinse to remove nonionic material. The anions are then eluted with 2 mL 5:20 conc. HCl/MeOH.	133
Isolation of LAS from pesticide formulations	Rohm & Haas Duolite IMAC HP 555 strongly basic resin, Cl⁻ form, 10 mL	2 mL of the mixture is applied to the column and nonionics and pesticide eluted with 25 mL MeOH. LAS is eluted with 50 mL 80:20 MeOH/0.5 M aqueous HCl.	134
Isolation of alkylarylsulfonate from water	Diaion WA-30 weakly basic anion resin, Cl⁻ form, used in batch mode	0.5 g resin is shaken with 500 mL water sample. The resin is washed with 3% NaOH and water, then the surfactant is removed with 100 mL 1:20 12 M HCl/MeOH.	71,135
Alkyl sulfate from soap and other shampoo components	Amberlite CG-45 Type 2 weakly basic resin, carbonate form, 1.1 × 5 cm	Dissolve sample in EtOH containing HOAc. Pass through column and elute with 99:1 EtOH/HOAc. Wash with H_2O and 1.5 M NH_4OH. Elute the alkyl sulfate from the column with 150 mL of a solution made by mixing 0.8 g $(NH_4)_2CO_3$, 24 mL concentrated NH_4OH, and 200 mL MeOH.	136
Sodium dodecylsulfate from protein solutions prior to analysis of the protein	PolyLC 2SDS Guard Cartridge, 2.1 × 20 mm or 4SDS, 4.6 × 20 mm; hydrophilic imine on silica	Used as a precolumn in front of the C_{18} protein analysis column. SDS is washed from column at CH_3CN/H_2O ratios higher than 70:30.	137
Isolation of C_{12}–C_{15} alcohol 3-mole ethoxysulfate from hand dishwashing detergent	Amberlite IRA-400 CP, OH⁻ form, 1 × 20 cm	An ethanolic solution representing 1 g of original sample is applied to the column and washed with 50 mL EtOH. The AES is eluted with 50 mL 1:4 conc. HCl/EtOH.	138
Paraffin monosulfonates from di- and polysulfonates	DEAE Sephadex A25 weakly basic resin, OH⁻ form	The surfactant mixture is applied in ethanol. Di- and polysulfonates are eluted with 1 M NH_4HCO_3 in H_2O or 70:30 H_2O/MeOH. Monosulfonate is eluted with 0.3 M NH_4HCO_3 in 60:40 PrOH/H_2O.	139
Isolation of alkylnaphthalene sulfonates from pesticide formulations	Waters Sep-Pak QMA quaternary amine type ion exchanger	2 mL of the 0.01 M NaOH-soluble portion of the formulation is applied to the Sep-Pak cartridge. The cartridge is washed with 2 mL H_2O, discarding the eluent. Nonionic material is rinsed from the column with 5 mL H_2O and 6 mL 80:20 MeOH/H_2O. Anionics are eluted with 5 mL MeOH, 5% in HCl.	140

Application	Material	Procedure	Ref.
Isolation of perfluorocarboxylates acids from ground water	Varian SAX disks, 25 mm	55–200 mL sample is passed through the disk and the disk is dried under vacuum. The disk is transferred to an autosampler vial with 1 mL CH_3CN and 0.1 mL CH_3I to elute carboxylates and form methyl esters.	141
Sulfonated anionics from other compounds	Dowex 1-X8, 2 × 10 cm, in Cl⁻ or OH⁻ form	300–500 mg surfactant is dissolved in 96% ethanol and rinsed through the column with 150–200 mL EtOH. The anionics are eluted from the column with 2% NaOH in 50:50 2-PrOH/H_2O. In the case of a fatty alcohol sulfoacetate, the separation must be conducted at 60°C, since at room temperature unesterified fatty alcohol is not completely eluted from the column.	142
Sulfates and sulfonates from soap and other anions	Amberlyst A27, Cl⁻ form, 0.7 g in a column 0.7 cm i.d.	Elution with 100 mL 0.2 M NaCl solution removes soap and inorganics, followed by 100 mL 0.2 M NaCl in 75:25 EtOH/H_2O to remove sulfate and sulfonates.	143
Organic anions from APE	SPE column of anion exchanger (Alltech), 0.5 meq capacity, OH⁻ form	100 mL of a 5% solution of APE is passed through, rinsing with 150 mL of water (or more, until the eluate no longer foams). The anions are eluted with 6 mL 4.5 M H_2SO_4.	144
Isolation of anionics from coolants and lubricants	Strongly basic anion exchange SPE cartridge	After dewatering the emulsion, 50 mg oil sample dissolved in hexane is transferred to the SPE cartridge (conditioned with hexane). The cartridge is washed with 2.5 mL each acetone and MeOH, then anionics are eluted with 7 mL 8:1 MeOH/conc. NH_3.	145
Anionics from amphoterics and cationics	Zerolit M-IP SRA 151, Cl⁻ form	Dissolve sample in MeOH and pass through column. The anionics will be retained. Elute with 3 M NH_4OH in EtOH.	146
Amphoterics from cationics	Bio-Rad AG1-X2, OH⁻ form	Continuation of above. MeOH solution is passed through column, retaining amphoterics. Amphoterics are eluted with 1 M HCl in EtOH.	146
Isolation of ethoxylated surfactants from water	Dowex 1-X8 or Dowex 1-X2 in the tetracyanatocobaltate(II) form	After pumping a large water sample through the column, the ethoxylated materials are eluted with absolute EtOH.	147

TABLE 3 Cation Exchange Chromatographic Separation of Surfactants

Separation performed	Ion exchanger	Procedure	Ref.
Amine oxide from other surfactants	Dowex 50W-X4 strongly acidic cation exchange resin, acid form	Surfactant mixture is dissolved in 50:50 EtOH/H$_2$O and passed through the column, washing with 300 mL EtOH/H$_2$O. The amine oxide behaves as a cationic under acid conditions and remains on the column. The amine oxide is then eluted with 300 mL 1 M ethanolic HCl solution.	109
Isolation of cationics	Bio-Rad AG50W-X8, H$^+$ form	The sample is passed through in MeOH solution. Cationics are retained and are eluted with 1 M HCl in EtOH.	146
Quaternary amines from nonionics	Permutit Zeo-Karb 225 strong cation exchange resin, H$^+$ form, used in batch mode	Contact 50 mL aqueous solution containing not more than 0.005 M quat with 3 mL resin for 1 hr with occasional stirring. Decant, rinsing the resin with six 25-mL portions of water. The solution contains nonionic uncontaminated by quat.	148
Amphoterics from anionic and nonionic surfactants	Dowex 50W-X4, H$^+$ form, 2 × 10 cm	About 300 mg surfactant is dissolved in 10:1 EtOH/H$_2$O and applied to the column. Anionics and nonionics are not retained. Amphoterics are eluted with 50:50 conc. HCl/EtOH. *Note*: Not all amphoterics are completely retained, nor completely eluted again.	149
Anionic from cationic surfactants	Merck Silica Gel 60	The sample is added as a CHCl$_3$ solution to a column containing 1 g silica. Impurities are eluted with 5 mL CHCl$_3$ and 5 mL 95:5 CHCl$_3$/MeOH. Anionics are eluted with 1 mL 3:1 MeOH/2 M NH$_3$ and 5 mL CHCl$_3$. Cationics are eluted with 5 mL 3:1 MeOH/2 M HCl.	39
Isolation of cationics from coolants and lubricants	Strongly acidic cation exchange SPE cartridge	After dewatering the emulsion, 100 mg oil sample dissolved in hexane is transferred to the SPE cartridge (conditioned with hexane). The cartridge is washed with 3 mL each acetone and MeOH, then anionics are eluted with 8 mL 70:30:5 MeOH/H$_2$O/12 M HCl.	145

more ion exchange columns in series. The sample is eluted through the ganged columns, resulting in the adsorption of the ionic components. The nonionic components are recovered from the effluent. The columns are then separated from each other and eluted individually to isolate the various components of the mixture. Generally, the cation exchange column is placed first in the series. This retains the cationic surfactants, while converting the salts of anionic surfactants and fatty acids to the free acid form. Applications are summarized in Table 4. Nonionic impurities originating from anionic surfactants will accompany the nonionic surfactants through the separation sequence.

(d)   Nonexchange separations.   Ion exclusion and ion retardation chromatography have been demonstrated for separation of ionic surfactants from inorganic salts. These approaches have found little application, since the same separation can usually be easily made by, for example, taking up the dried mixture in absolute ethanol.

In salting-out chromatography, an ion exchange resin is used to execute a separation by a mechanism similar to adsorption chromatography. A high concentration of an alcohol is required in the aqueous eluent, so that the surfactant exists as individual molecules, rather than as micelles. The presence of a salt in moderate concentrations in the mobile phase decreases the solubility of the surfactant, thus increasing its affinity for the resin and lengthening its retention time. At higher salt concentrations, the alcohol is also "salted out" of the mobile phase, so that it competes with the surfactant for adsorption sites on the resin, decreasing the retention time of the surfactant (41). Applications of salting-out chromatography are included in the tables of adsorption chromatography references.

2.   Batch Ion Exchange

Batch procedures are often used for separation of a single sample, where there is no incentive to optimize the procedure. The technique requires only a beaker and a coarse filter, and is thus appropriate for preliminary, exploratory investigations. Table 5 summarizes the separation scheme in the classic paper of Rosen, which has been referenced by hundreds of investigators during the four decades since its publication (42).

C. Adsorption

Adsorption techniques are used a great deal with surfactants, since the wide range of polarity of functional groups of the various classes makes adsorption suitable for separation of the surfactants from each other (Tables 6–9). As a general rule, the quantity of surfactant adsorbed from water onto a substrate increases with increasing salt content of the water, and decreases with higher hydrophilicity of the surfactant (43). Most references to adsorption methods for isolating surfactants from water and wastewater are located in the tables of Chapter 18 on environmental analysis.

1.   Adsorption Chromatography and Solid Phase Extraction

Most analytical scale column adsorption chromatography is now performed with so-called solid phase extraction (SPE) columns. These are small, prepacked, single-use columns designed for automated or semi-automated cleanup of samples for analysis by gas or liquid chromatography. A note of caution: recovery is generally less than 100% with any adsorption chromatography procedure. Recovery should be determined as part of the development of the methodology.

Membrane disk adsorption devices are used in the same way as SPE columns, and are listed in the same tables. "Matrix solid phase dispersion" is a workup method in which

TABLE 4 Surfactant Separation by Combined Anion/Cation Exchange Chromatography

Separation performed	Ion exchanger	Procedure	Ref.
Ether carboxylates from crude oil	Connected 15 × 150 mm columns of 10 mL Dowex 50W-X2 cation exchanger, H⁺ form, and 10 mL Merck Lewatit MP 7080 macroporous anion exchanger, OH⁻ form	10 g oil is dissolved in 100 mL 50:25:25 hexane/MeOH/CH₂Cl₂ and added to the columns. The cation exchange column is removed and discarded; the anion exchanger is eluted with 200 mL of the solvent mixture, rinsed free of oil with 100 mL 50:50 hexane/CH₂Cl₂, and finally eluted with 100 mL MeOH, discarding all eluates. The ether carboxylates are removed with 100 mL 0.3 M ammonium bicarbonate in 60:40 2-PrOH/H₂O.	13
AE from ether sulfates and other ionic materials (Wickbold procedure II)	Connected columns of Dowex 50W-X2, H⁺ form, 2 × 20 cm, and Merck Lewatit MP 7080, OH⁻ form, 3 × 17 cm	2.5–3.0 g sulfated alcohol ethoxylate surfactant is dissolved in 100 mL denatured EtOH and applied to the columns, rinsing glassware and columns with 300 mL EtOH. The alcohol ethoxylate is in the eluent, as well as PEG.	150
APE from other detergent components	Connected columns of Dowex 50W-X4, H⁺ form, and Dowex 1-X2, OH⁻ form; 10–15 g of each	5 g detergent is dissolved in 100 mL MeOH, filtered, and applied to the columns. 150–200 mL MeOH is passed through the column; APE is in the eluate.	151
Ethoxylated fatty acid ethanolamide from ethoxylated amine and fatty acid	Connected columns of Bio Rad AG50W-X2 (H⁺ form), 1.8 × 8 cm and Bio Rad AG1-X2 (OH⁻ form), 1.4 × 8 cm	5 g surfactant dissolved in 150 mL 96% EtOH is added to the column, heated to 45°C (if solubility at room temperature is a problem). The columns are rinsed with three 20-mL portions of 96% EtOH. The ethoxylated fatty acid ethanolamide, as well as other nonionic impurities, such as PEG, are in the EtOH fraction. The columns are separated, and ethoxylated ethanolamine is eluted from the cation exchanger with 250 mL 1 M HCl, followed by 1 M HCl in 2-PrOH. Fatty acid is eluted from the anion exchange column with 300 mL 0.2 M KOH in 70:30 EtOH/H₂O.	152,153
Fatty acid ethanolamide from ethanolamine and fatty acid	Connected columns of Bio Rad AG50W-X2 (H⁺ form), 1.8 × 8 cm and Bio Rad AG1-X2 (OH⁻ form), 1.4 × 8 cm	5 g surfactant dissolved in 150 mL 96% EtOH is added to the column, heated to 45°C (if solubility at room temperature is a problem). The columns are rinsed with three 20-mL portions of 96% EtOH. The fatty acid ethanolamide, as well as other nonionic impurities such as fatty acid ester of fatty acid ethanolamide, are in the ethanol fraction. The columns are separated and ethanolamine is eluted from the cation exchanger with 250 mL 1 M HCl. Fatty acid is eluted from the anion exchange column with 300 mL 0.2 M KOH in 70:30 EtOH/H₂O.	152

Nonionics from other surfactants	Connected columns of Dowex 50W-X2, H$^+$ form (first), and Merck Lewatit MP 7080, OH$^-$ form (second), 25 × 120 mm (thermostatted at 50°C to avoid gelling or precipitation of fatty acids and their soaps)	2.5–3.0 g surfactant is dissolved in 150 mL 50:50 2-PrOH/H$_2$O, warmed to 50°C, and passed through the columns, washing with 300 mL warm 2-PrOH/H$_2$O. The nonionic surfactant is in the eluate.	154
Nonionics and PEG from soap components	Connected columns of Dowex 50W-X4, H$^+$ form, and Dowex 1-X2, OH$^-$ form	5–10 g soap product is suspended in 50 mL MeOH, filtered, and applied to the columns. 300 mL MeOH is passed through the column at about 5 mL/min. Nonionic material is in the eluate.	155
Sulfobetaine from anionic and cationic surfactants	Dowex 50W-X4 cation and Dowex 1-X2 anion exchange resins	A solution of the surfactants in 96% EtOH is passed through the column. The sulfobetaine is not retained, but elutes mixed with any nonionic surfactants.	156
Separation of urea, quats, soap, hydrotropes, sulfates/sulfonates, nonionics	Two columns, each 30 × 100 mm, containing Dowex 50W-X4 cation (H$^+$ form) and Dowex 1-X2 anion (OH$^-$ form) exchange resins, respectively; cation exchange column is fitted above the anion exchanger; use fresh resin for each analysis	1 g of the ethanol-soluble portion of the sample is applied to the columns as a 0.5% solution in 96% EtOH. (For surfactants less soluble in EtOH, water can be added to dilute as much as 50:50 EtOH/H$_2$O. However, urea is not quantitatively adsorbed on the cation exchange column from dilute EtOH.) The columns are rinsed with 100 mL solvent. The eluate contains the nonionic surfactants, as well as any unsulfonated material contaminating the anionics. The columns are separated. *Cation column:* Urea is desorbed with 300 mL H$_2$O. Other cations (amines, quats) are eluted with 300 mL 2 M HCl in MeOH, followed by 100 mL EtOH. *Anion column:* Fatty acids are desorbed with 300 mL 1 M HOAc in EtOH, followed by 100 mL EtOH. Hydrotropes like toluene or xylene sulfonate are eluted with 300 mL 1 M HCl in water, followed by 100 mL EtOH. Other anionics are simultaneously desorbed and hydrolyzed by boiling the resin with 2 M methanolic HCl under reflux. The hydrolysis products may be separated, if desired, into anionic and nonionic components by use of another Dowex 1-X2 anion exchange treatment.	157

(Continued)

TABLE 4 *(Continued)*

Separation performed	Ion exchanger	Procedure	Ref.
Separation into the categories of cationics, soaps, other anionics, and nonionics	Three columns in series, each 25 × 150 mm, containing 10 g resin: Dowex 50-X2, H⁺ form; Dowex 1-X2, Cl⁻ or acetate form, Dowex 1-X2, OH⁻ form	A sample of no more than 1 g of the EtOH-soluble portion of a detergent mixture is added to the column train as 100 mL of an EtOH solution. EtOH is added to the columns until all nonionic material is eluted from the last column, as indicated by measuring residue from evaporation of the eluate. The columns are then separated. The cation exchange column is treated with 400 mL 90:10 MeOH/12 M HCl to elute cations. The first anion exchange column is treated with 400 mL 90:10 MeOH/12 M HCl to elute strongly acidic surfactacts. The second anion exchange column is treated with 100 mL 80:20 MeOH/H₂O, 0.5 M in KOH, to elute fatty acids. *Notes:* Partial formation of methyl esters of the fatty acids can occur, especially during passage through column 1. Partial hydrolysis of alkyl sulfates and taurides can occur during elution of column 2 with HCl. This can be avoided by instead eluting with 125 mL 0.3 M NaOAc in 70:30 EtOH/H₂O. The surfactant can be separated from the NaOAc by ion exclusion chromatography or extraction.	33,158,159

TABLE 5 Comprehensive Scheme for Separation of Surfactants by Batch Ion Exchange

Separation performed	Ion exchanger	Procedure	Ref.
Anion exchange			
I. Anionic surfactants from nonionic surfactants	Dowex 1-X4	Dissolve or disperse the surfactant mixture in distilled water. Stir with the resin for 4–5 hr. Filter, washing with 95% EtOH followed by hexane to remove the nonionics.	42
II. Soap from other anionic surfactants	Dowex 1-X4	Stir the resin from I, above, with acidified 50:50 EtOH/H_2O. Filter, washing with 95% ethanol and hexane. The filtrate contains the carboxylic acids.	42
III. Ester-type anionics from other anionic surfactants	Dowex 1-X4	Saponify by refluxing the resin from II with 5% aqueous KOH. Filter and wash with water, ethanol, and hexane. The filtrate contains the alcohols from the original esters. Stir the resin with acidified (to Congo red) 50:50 EtOH/H_2O. Filter and wash with ethanol and hexane to remove the acids from the esters.	42
IV. Sulfates from sulfonates	Dowex 1-X4	Hydrolyze sulfates by refluxing the resin from III with 2 M HCl. Filter and wash with water, alcohol, and hexane to remove the alcohols from the sulfated alcohols. Heat the resin with 50:50 HCl/EtOH, filter, and wash with water, alcohol, and hexane to recover the sulfonates as the free acids.	42
Cation exchange			
I. Cationic surfactants from nonionic surfactants	Dowex 50-X4, acid form	4–5 g of mixture dispersed in 100 mL water is stirred 4–5 hr with 20 g resin, keeping the mixture acid to Congo red by addition of HCl. The resin is washed with 95% EtOH, the filtrate and washings neutralized and evaporated to dryness, then washed with acetone to isolate the nonionic.	42
II. Nonquaternary amines from quats	Dowex 50-X4, acid form	Stir resin from I, above, with 10% diethylamine in benzene. Filter, washing with benzene. The filtrate contains the nonquaternary amines. The extraction of the resin is repeated with 5% diethylamine in methanol, washing with methanol. This yields a small additional quantity of nonquaternary amines, contaminated with quaternaries. The resin is heated with 50:50 HCl/EtOH, filtered, and washed with water and ethanol to remove quats.	42

a solid sample, usually of biological origin, is intimately mixed with, for example, a C_{18} resin, perhaps by grinding in a mortar and pestle. The entire mixture is then transferred to a column for sequential elution (44,45).

(a)   Activated carbon.   In general, all surfactants are adsorbed from aqueous solution onto activated carbon (46). The adsorption of ionic surfactants is affected somewhat by which counterions are in solution, but there is little practical difference (47). While cationics and nonionics are adsorbed under most conditions, adsorption of alkanesulfonate anionics is greatly increased by the presence of inorganic salts in solution, and increases with alkyl chain length (48). Greater adsorption with increasing chain length is also observed with soaps (49). Although activated carbon is perhaps the most effective adsorbent for surfactants, its lack of specificity has made it uninteresting to analytical chemists until the recent commercial development of graphitized carbon black.

(b)   Reversed-phase media.   Surfactants may be isolated from aqueous media by adsorption on silica-based C_8 or C_{18} materials like those used in reversed-phase HPLC or TLC. The concentration is nonselective in that oil and grease are also concentrated. Reversed-phase media are especially effective with nonionics, and in commercial applications can remove about 10% of their weight in APE (50). PEG and ethoxylates may be separated from each other by selective elution from C_{18} media, as indicated in Table 7.

Quantitative isolation of LAS from water is reported using C_{18} columns or extraction discs. In many matrices, LAS is the only anionic material extracted by C_{18}. Best recovery of adsorbed LAS is found with methanol, with incomplete recovery observed with acetonitrile, very incomplete recovery with ethyl acetate, and zero recovery with chloroform (51,52). However, if a large cationic salt like tetrabutylammonium chloride is added to the chloroform, LAS is easily eluted as the ion pair (53). Alkanesulfonates behave similarly, and this technique can be used to add additional selectivity to SPE, since interfering compounds can first be eluted from the resin with pure chloroform, followed by elution of LAS as an ion pair (53,54). Most cationics are not quantitatively isolated from water on C_{18} media. Most investigators use ion exchange media for SPE of cationics.

A particular problem with use of silica-supported reversed-phase media for trace analysis of surfactants is the presence of free Si–OH groups. These bind surfactants tightly, making complete recovery difficult, particularly for cationics (55). Phenyl-modified silica is excellent for SPE of NPE, but recoveries are low for AE, particularly more highly ethoxylated AE (55).

(c)   Synthetic resins.   The Rohm and Haas Amberlite® XAD® series of macroreticular resins (currently available from Supelco) are widely used for separation of surfactants from aqueous solution. Amberlite XAD-2 and XAD-4 are styrene/divinylbenzene copolymers; XAD-7 resin is an acrylic ester copolymer; Amberlite XAD-8 is a methylmethacrylate copolymer.

Anionics.   Anionics are strongly bound to the Amberlite XAD resins. In general, adsorption increases with greater surface area of the resin, with lower temperature, and with more lipophilic surfactants. Thus, LAS is more strongly bound than is alkyl sulfate, and XAD-4 adsorbs more anionic than does XAD-7 resin (56,57). The adsorption of alkanesulfonates is in direct proportion to their alkyl chain lengths (48). LAS is strongly bound to Amberlite XAD-8 under both acidic and basic conditions. On the other hand, the sulfophenylcarboxylate degradation products of LAS are only retained under acidic conditions (58). Adsorption is substantially increased by addition of inorganic salt to the solution (48). In one case, only about 10% of LAS was removed from a solution by XAD-4 in the absence of added salt, but quantitative recovery was obtained by addition of 1% NaCl (55). In general, much more XAD resin is required to isolate an anionic from solution than is the case for carbon (48) or C_{18} media (55).

Szczepaniak prepared a mercurated XAD-2 resin by treating with mercuric acetate. This gave much more rapid adsorption of ordinary anionics, with kinetics resembling ion exchange rather than adsorption. However, recovery of adsorbed surfactant is much easier than with anion exchange resin (59).

Nonionics.   Although activated carbon and silica gel have been used in the past for isolation of nonionics from water, they have now been largely supplanted by copolymer resins (most often styrene/divinylbenzene copolymers). These resins have an affinity for the hydrophobic portion of the nonionic surfactant and are much more effective in adsorbing lower ethoxylates than the more hydrophilic higher ethoxylates. The presence of an inorganic salt encourages more complete adsorption (60). Methanol is an effective solvent for removing the surfactant from the resin. The macroreticular poly(styrene/divinylbenzene) adsorbents of Rohm & Hass are widely used. Because of increased surface area, the adsorptive capacity increases in the order Amberlite® XAD-1 < XAD-2 < XAD-4 (32). Interestingly, Amberlite XAD-7 (a polyacrylic resin with COOR functional groups) as well as weakly acidic cation exchange resins in the H^+ form give even superior performance in adsorbing nonionic surfactants, evidently due to ease of hydrogen bonding with the polyether oxygens (32). However, the effectiveness of the cation exchangers is dependent upon the pH (should be acidic or neutral) and ionic strength (should be low, opposite of the situation with nonpolar adsorbants) of the mobile phase. Sodium hydroxide solution is effective for displacing nonionics from the cation exchange resin (61).

Experimental poly(styrene/divinylbenzene) resins, especially those containing hydroxy groups, have increased capacity for polyether nonionics, since they have an affinity for the hydrophilic as well as the hydrophobic portion of the molecule (62,63).

Fibers coated with silicone/divinylbenzene copolymer have been demonstrated for solid phase microextraction of AE from water (64). Fibers coated with PEG-based polymers have similarly been demonstrated for solid phase microextraction of OPE (65).

Polymer resins are available in the form of beads and membranes from laboratory supply houses such as Bio-Rad and Pierce which are designed specifically to remove ethoxylated nonionic surfactants from protein suspensions. These are very widely used (66).

Cationics.   XAD resin has been successfully used for concentration of cationics from water (55). Kawabata and coworkers have found that a specially prepared resin material of poly(hydroxystyrene) highly crosslinked with divinylbenzene is a superior adsorbent for cationic surfactants (67). The mechanism appears to be acid-base reaction rather than ion exchange, since inorganic cations are not adsorbed. The cationic is recovered and the resin regenerated by elution with methanol. The capacity of this resin for cationics is higher than that of Amberlite XAD resins as well as traditional cationic exchange resins. The presence of salts or alkali in the matrix increases the capacity.

(d) Silica.   Ethoxylated compounds adsorb strongly to silica, with the more highly ethoxylated materials adsorbing more strongly than lower ethoxylates. At high concentrations this behavior can be reversed, since the silica becomes hydrophobic as it is coated with the surfactant (68). Silica is often used to separate ethoxylated materials from nonpolar substances such as grease and oil. It is less effective with more polar substances, such as oxidized grease, which tend to have elution behavior similar to the surfactants.

TABLE 6   Column Adsorption Chromatography: Separation of Anionic Surfactants

Separation performed	Stationary phase	Procedure	Ref.
Isolation of LAS from H$_2$O and from sediment and fish extracts	Analytichem Bond-Elut C$_{18}$ SPE cartridges	Up to 2 L of H$_2$O sample or of MeOH extract diluted with H$_2$O to contain less than 10% MeOH is passed through the column. The column is washed with 2 mL H$_2$O, then the LAS is eluted with 4 mL MeOH.	160
Isolation of LAS from aqueous media	Analytichem C$_2$ SPE cartridge, 500 mg	Sample of less than 50 mL volume is adjusted to pH 3–4 and passed through the MeOH- and H$_2$O-washed cartridge. After rinsing with 5 mL 70:30 H$_2$O/MeOH, LAS is eluted with 10 mL MeOH. Further purification by ion exchange may be performed.	161
Isolation of LAS from wastewater	J. T. Baker C$_{18}$ membrane disks, 0.5 × 47 mm	Up to 500 mL of acidified, filtered sample is pulled through the MeOH-washed disk. The disk is dried by passage of air, then LAS is eluted with 2 10-mL portions of MeOH.	51,52
Isolation of LAS from solutions of environmental materials	SPE column, Analytichem C$_8$ or Waters Sep-Pak C$_{18}$ SPE cartridge	An aqueous solution of 5–100 mL is passed through the column, followed by 5 mL H$_2$O and 3 mL 40:60 MeOH/H$_2$O to remove fatty material. LAS is eluted with 5 mL MeOH. APE remains on the column.	133,162
LAS from branched-chain ABS by salting-out chromatography	Amberlite CG-50, 26 × 500 mm	Elute with 43:57 MeOH/0.5 M (NH$_4$)$_2$SO$_4$ at 50°C, monitoring the eluent by UV absorbance at 254 nm. LAS elutes first, followed by ABS. The column is washed with 43:57 MeOH/H$_2$O.	163
Alkylaryl sulfonate from alkyl sulfate by salting-out chromatography	Amberlite CG-50, 25 × 100 mm, partially H$^+$, partially Na$^+$ form	Elute with 40:60 MeOH/0.5 M NaCl solution, changing to 40:60 MeOH/H$_2$O, at 37°C, monitoring by the methylene blue spectrophotometric method. The alkyl sulfate is eluted first, then the ionic strength is changed to elute the LAS.	164
Isolation of perfluoroalkane-sulfonate from acid etch bath	Dionex IonPac NG1, 4 × 50 mm	Online enrichment for HPLC: 0.1 mL sample loaded onto column, then rinsed with water or 0.02 M NaOH for 20 min at 2 mL/min. Surfactant eluted in reverse direction with LC eluent (55:45 0.011 M NaOH/CH$_3$CN).	165

Isolation of LAS, dodecylsulfate, and dodecylsulfonate from water	Mercurated XAD-2, 5 g in 0.8-cm diameter columns	Water is passed through the column at 5 mL/min. Columns are rinsed with 50 mL water, then the anionics are eluted with 50 mL 1:1 2% aqueous NaOH/2-PrOH.	59
Mono- from disulfonic acids. Applicable to alkyl and aryl sulfonates	Amberlite XAD-2, 1.2 × 25 cm	Use H_2O or 15% NaCl solution to elute H_2SO_4 and disulfonic acids, then MeOH to elute monosulfonic acids.	166
Isolation of LAS and sulfophenylcarboxylate degradation products	Amberlite XAD-8, 500-mL column	After cation exchange treatment to replace cations with H^+, H_2O sample (150 mL) is adjusted to pH 2, applied to the column, and rinsed with 0.01 M HCl. After elution of inorganic salts (detected by conductivity), organic ions are eluted with 75:25 CH_3CN/0.1M HCl.	58
LAS from sulfophenylcarboxylate degradation products	Amberlite XAD-8, 5-mL column	Sample is dissolved in 30 mL 0.01 M NaOH, applied to column, and rinsed with 2 bed volumes of 0.01 M NaOH to elute SPC. LAS is eluted with 75:25 CH_3CN/0.01 M HCl.	58
Sulfated alcohol ethoxylate from PEG, mono- and disulfated PEG, and unsulfated alcohol ethoxylate	Silanized silica gel, Merck No. 7719, 40 g, 2 × 35 cm column	300 mg sample in 5 mL 70:30 2-PrOH/H_2O is placed on the column and eluted with 60 mL 2-PrOH/H_2O to recover PEG, sulfated PEG, and inorganic salts. After rinsing with 10 mL 2-PrOH/H_2O, sulfated ethoxylated alcohol is eluted with 400 mL 2-PrOH/H_2O. Unsulfated ethoxylated alcohol is eluted with 250 mL EtOH. *Note:* Column packing material contaminates the fractions. This can be eliminated by redissolving and filtering.	167
Fractionation of α-olefin sulfonates into mono- and disulfonates	Silanized silica gel, Merck No. 7719, 40 g in a column 2 × 35 cm	0.1 g surfactant is dissolved in 2 mL 30:70 2-PrOH/H_2O, transferred to the column, and eluted with 2-PrOH/H_2O. Disulfonates and sodium sulfate elute in the first 70 mL, the next two 5-mL fractions may contain either mono- or disulfonates, the final 200 mL contains monosulfonates. All fractions are contaminated by column packing, which is removed by successive extractions of dried residue with 90:60:20 n-butanol/acetone/H_2O.	168

(Continued)

TABLE 6 (*Continued*)

Separation performed	Stationary phase	Procedure	Ref.
Fractionation of α-olefin sulfonates (as methyl esters)	Davison 923 silica gel, 3.0% moisture	Methyl esters are prepared by reaction with diazomethane, dissolved in petroleum ether and added to the column. Hydrocarbons are eluted with 150–200 mL petroleum ether. Alkenesulfonates are eluted with 500 mL 90:10 petroleum ether/ethyl ether and hydroxyalkane sulfonate and all disulfonates are eluted with 500 mL ethyl ether. Unesterified material is removed with 50:50 ethyl ether/MeOH.	169
α-Olefin sulfonates: separation of alkene sulfonates from hydroxyalkane sulfonates by salting-out chromatography	Amberlite CG-50, 25 × 400 mm, mixed Na$^+$ and H$^+$ form, 35°C	Elute with 30:70 2-propanol/0.5 M NaCl solution, monitoring the separation by spectrophotometry with methylene blue method. Hydroxyalkane sulfonates elute first, followed by alkene sulfonates.	170
Separation of alkene sulfonates, hydroxyalkane sulfonates, soap, sulfosuccinate, toluenesulfonic acid, alkylsulfate and ABS by salting-out chromatography	Amberlite CG-50, 25 × 400 mm, mixed Na$^+$ and H$^+$ form, 40°C	Elute with 30:70 2-propanol/0.2 M NaCl solution, monitoring the effluent by two-phase titration using bromcresol green indicator. Order of elution: (1) toluenesulfonic acid; (2) the group of alkylsulfate, alkylsulfonate, hydroxyalkane sulfonate, and sulfosuccinate; (3) the group of ABS, alkenesulfonate, and higher MW alkylsulfonate; (4) soap.	171
Petroleum sulfonate from unsulfonated oil	15 g Silica Gel, Davison Grade 62, 22 × 300 mm	10–25 mL CHCl$_3$ solution containing about 2 g of the sodium salt and oil is placed on the column, and the oil is eluted with 210 mL CHCl$_3$, followed by 10 mL denatured EtOH. Elute the sulfonate with 250 mL denatured EtOH.	172
Petroleum sulfonate from unsulfonated oil	200 mL Merck LiChrosorb silica	10 mL CHCl$_3$ containing 4 g of the sodium salt and oil is placed on the column and the oil (wax) is eluted with CHCl$_3$. The mono- and polysulfonates are eluted with MeOH.	173

Alkylarylsulfonates from unsulfonated oil	10 g silica gel	1 g sample is dissolved in $CHCl_3$ and transferred to the column. Neutrals are eluted with $CHCl_3$, then sulfonates are removed with 50:50 $EtOH/CHCl_3$.	174
Anionic and nonionic surfactants from each other and oil and grease	Silica gel, 100 mg	A mixture in 99:1 hexane/MeOH is passed through the column. The oil and grease is not adsorbed. The column is dried, then the nonionic is eluted with two aliquots of 0.5 mL 1,4-dioxane. The anionic is eluted with two aliquots of 0.5 mL MeOH.	175
Alkylbenzenesulfonates and APE from pesticide emulsifiable concentrate containing high-boiling aromatic solvent	Silica gel SPE columns, Waters Sep-Pak	0.1 mL of sample is diluted to 5 mL with 50:50 hexane/CH_2Cl_2 and applied to the column, rinsing with two 1-mL aliquots of hexane/CH_2Cl_2. Aromatic solvent and pesticide are removed by elution with 5 mL 50:50 hexane/CH_2Cl_2, 5 mL CH_2Cl_2, and 5 mL 99:1 (or 98:2) CH_2Cl_2/MeOH. The surfactants are recovered with 5 mL MeOH.	176
Alkylbenzenesulfonate from APE	Alkylamine-modified silica gel SPE column, Waters Sep-Pak	A mixture of 0.2 mg of each surfactant in 5 mL 80:20 CH_2Cl_2/MeOH is applied to the column. Only the APE is eluted. *Note:* This is used in conjunction with the above separation.	176
Bis(2-ethylhexyl)sulfosuccinate from unidentified fluorescent impurities	Bio-Rad C_{18} HL silica, 1 × 25 cm	Prep HPLC: Column equilibrated with 70:30 MeOH/H_2O at 0.5 mL/min; 0.5–1.0 mL 35% solution injected; gradient elution, 70 to 100% MeOH in 40 min with 50 min hold. Monitor UV absorbance at 235 nm.	177
Alkyl sulfate from soap by salting-out chromatography	Amberlite CG-50 weakly acidic cation exchange resin, 25 × 450 mm	Elute with 30:70 2-propanol/water, water initially 0.5 M in NaCl, decreasing to 0 M; 40°C. The alkyl sulfate is eluted first, followed by the soap.	178

TABLE 7 Column Adsorption Chromatography: Separation of Nonionic Surfactants

Separation performed	Stationary phase	Procedure	Ref.
Isolation of pure homologs from commercial AE with alkyl chain = C_{12}; E = 10, 11, 12, 14 isolated, as well as mixtures E = 18/19, 20/21, 22/23	Prep HPLC: for E = 10–14, repetitive fractionation on Shandon Hyperprep HS silica, 4.6 × 250 mm; for higher homologs: repetitive separation first on Merck silica, 8 × 30 cm, then on Shandon Hyperprep HS silica, 5 × 30 cm	E = 10–14: 1-g injections, step gradient at 40 mL/min. 960 mL 90:10 ethyl acetate/MeOH, then 640 mL 85:15, then 640 mL 80:20: 50-mL fractions collected. Fractions were rechromatographed: 92:8 ethyl acetate/MeOH for E = 10; 90:10 for E = 11; 88:12 for E = 12; 85:15 for E = 14. E = 18–23: 20-g injections, elution with 50:50 ethyl acetate/MeOH at 80 mL/min; 100-mL fraction. Fractions were rechromatographed on a 5-cm i.d. column with 70:30 ethyl acetate/MeOH at 40 mL/min, collecting 60-mL fractions. Fractions again rechromatographed at 80:20 ethyl acetate/MeOH, collecting 30-mL fractions.	179
Isolation of AE from water	J. T. Baker Bakerpond Polar Plus C_{18} SPE cartridge, 1 g	250 mL water is passed through cartridge at 5 mL/min. The cartridge is dried, then AE is eluted with 7.5 mL each of MeOH and acetone.	180
AE and NPE from oil and grease	Silica solid phase extraction cartridge, 50-mg capacity	The residue after drying is taken up in CH_2Cl_2 and passed through the SPE column (preconditioned with CH_2Cl_2). Nonpolar material is eluted with CH_2Cl_2, then ethoxylates are eluted with MeOH.	181
NPE (E = 8) from sheepskin grease	Silica	Silica impregnated with the fat paste is eluted with 4:1 n-hexane/ethyl ether to remove fat. NPE is then eluted with MeOH.	182
APE from mineral oil and fatty oils	Alumina, Brockmann activity I, treated sequentially with water, ethanol, and petroleum ether, air-dried, 10 mL in a 50-mL buret	Apply sample (0.01–0.1g APE) to column in 10 mL 50:50 EtOH/10% aqueous $BaCl_2$ solution. Elute with 9 mL 50:50 EtOH/H_2O, then 100 mL H_2O. Filter the eluate containing the nonionics through Whatman No. 42 filter paper	183
Fractionation of extracts containing AE and APE	Silica gel, activated at 300°C 1 hr, 2 × 20 cm	Elute with 2 25-mL portions of 70:30 ethyl acetate-benzene, 30 mL ethyl acetate; 20 mL 10:10 ethyl acetate/acetone, 70 mL acetone, 30 mL 3:2 acetone/H_2O. Most ethoxylates will be in the last two fractions.	184

Fatty alcohol and lower adducts from AE	Alumina, activity IV, 1.3 × 18 cm	300–500 mg surfactant is eluted with water-saturated benzene (about 90:10 benzene/H_2O). 50-mL fractions are collected and identified by TLC analysis. The first fraction contains hydrocarbons and the next 3 contain free alcohol. Subsequent fractions contain 1- and 2-mole ethoxylates.	185
Fractionation of environmental extracts containing AE and APE	Celite 545 loaded with 25% 0.05 M NaOH, 5.3 g in a 1-cm column	Use 30 mL 3:2 benzene/$CHCl_3$ to elute nonionic surfactants, followed by 20 mL MeOH to elute PEG and other materials	184
Isolation of AE and APE from water	Amberlite XAD-4, 1 × 20 cm, prewashed with 1:1 acetone/hexane, acetone, and EtOH	The water sample is pumped through the column, then ethoxylates are eluted with MeOH/H_2O and MeOH. A second fraction is eluted using acetone and acetone/n-hexane.	184,186
Fractionation of NPE and AE into their homologs	Silica, 1.6 × 100 cm	Elute with butanone, monitoring the separation by UV absorbance and TLC/modified Dragendorff reagent.	187
APE from esters, ethoxylated esters, and EO/PO copolymers	Amberlite XAD-4, 1.5 × 25 mm	A dilute aqueous solution of the surfactants is pumped through the column, then APE is eluted with 1 mL EtOH.	188
Isolation of APE from water	Activated carbon	Surfactants are eluted with ethyl acetate.	189
Isolation of APE from water	Alltech graphitized carbon black cartridge, 300 mg/6 mL	100 mL water is passed through pre-extracted cartridge and washed with 20 mL 50:50 MeOH/H_2O and 1 mL MeOH. APE is eluted with 2 mL 80:20 CH_2Cl_2/MeOH.	190
OPE, E = 10; separation from plasma	Various C_8 and C_{18} media, about 22 L to remove 1% OPE from 200 L plasma	Pass plasma through media, monitoring the effluent for breakthrough using HPLC. Only the OPE is retained.	50
OPE, E = 10; separation from protein solutions	Rohm & Haas Amberlite XAD-2, 1.5 × 15 cm	0.5% surfactant solutions at neutral pH are passed through the column, which adsorbs about 0.3 g surfactant/g resin. About 0.01% surfactant remains in solution, but can be removed by passage through fresh XAD-2. Surfactant is quantitatively removed from the resin with 2-PrOH.	191

(Continued)

TABLE 7 (Continued)

Separation performed	Stationary phase	Procedure	Ref.
Isolation of the 9-mole EO adduct from commercial APE	Merck silica gel, 13 × 150 mm	1 g APE mixture dissolved in 10 mL ethyl acetate is applied to the column and eluted with 100 mL each of ethyl acetate, 99.5:0.5 ethyl acetate/MeOH, and 99:1 ethyl acetate/MeOH. The last fraction consists chiefly of the 9-mole adduct.	176
Isolation of pure homologs from commercial APE, E = 10 to 15	Prep HPLC: DuPont Zorbax Sil, 9.4 × 250 mm, 40°C	Gradient: A = 60:40 2-PrOH/n-hexane; B = 80:20 EtOH/H_2O; 10 to 95% B in 45 min.	192
Isolation of pure homologs from commercial NPE, E = 1 to 4	Merck silica gel 60, 0.063–0.200 mm, activated at 200°C, 0.020 × 25 cm	275 mg sample. Eluent is 2-butanone.	193
Fractionation of ethoxylated fatty acids	Silica gel, Davison No. 922 or Machery-Nagel, 20 × 200 mm	A 250-mg sample is dissolved in 10 mL benzene, transferred to the column, and eluted with 30 mL 3:97 MeOH/isopropyl ether (to elute free acid), 12:88 MeOH/isopropyl ether (to elute PEG diester), 10:90 MeOH/$CHCl_3$ (to elute PEG monoester), and 35:65 MeOH/$CHCl_3$ (to elute PEG). See warning on isopropyl ether on p. 173.	4,194
Fatty acid monoethanolamide from the fatty acid ester of fatty acid ethanolamide	Silica gel, activated at 110°C overnight, 2.6 × 25 cm	About 750 mg of the mixture is applied in 3 mL 90:10 $CHCl_3$/n-BuOH, followed by 250 mL of the solvent mixture. This fraction contains the ester. The amide is eluted with 600 mL MeOH. The separation may be run at 45°C if solubility is a problem.	152
PEG from ethoxylated fatty acid ethanolamide	Silanized silica gel, 3.6 × 20 cm	About 1 g surfactant is dissolved in 30:70 2-PrOH/H_2O and applied to the column. The PEG is eluted with 205 mL 2-PrOH/H_2O. The ethoxylated amide is eluted with 500 mL 96% EtOH.	153
Monoethanolamide from fabric softener formulation	Alumina, neutral, activated at 120°C, 8 g	500 g sample is lyophilized, taken up in hexane, and added to the column. Other materials are eluted with 20 mL 3:2 CH_2Cl_2/hexane, then N-(2-hydroxyethyl)alkylamides are eluted with 20 mL 50:50 CH_2Cl_2/MeOH	195

Application	Column	Procedure	Ref.
Mono-, di-, and triglycerides from each other	Silica gel, Fisher grade 923, 5% water content, 30 g in column 19 × 290 mm	*Monoglyceride concentrate:* 1 g sample is dissolved in 15 mL CHCl₃ and applied to the column, rinsing with 5 mL CHCl₃. Triglycerides are eluted with 200 mL benzene; combined diglycerides and free fatty acids are eluted with 200 mL 90:10 benzene/ethyl ether; monoglycerides are eluted with 200 mL ethyl ether. *Shortening:* 5 g sample is dissolved in 10 mL benzene and applied to the column, rinsing with 40 mL benzene. Triglycerides are eluted with 300 mL benzene; combined diglycerides and free fatty acids are eluted with 250 mL 90:10 benzene/ethyl ether; monoglycerides are eluted with 200 mL ethyl ether.	196
Glycerin mono- and diesters from sucrose esters	Silica gel or bleaching clay, 50°C	Use benzene to elute fatty acids and glycerin esters, then anhydrous ethanol to elute sucrose esters.	121
Fractionation of sucrose mono- and diesters	Silica gel, Merck F60, 7 × 70 cm, de-activated with ethyl ether	150 g ester sample is dissolved in 250 mL toluene and applied to the column, rinsing with 100 mL toluene. It is then treated with 5 L ethyl ether to elute glycerides and fatty acids, 7–8 L butanone to elute diesters (initial 600 mL contains higher esters and fatty acids), and 8–10 L butanone saturated in water to elute monoesters (add slowly, since heat is released as the silica gel adsorbs water). *Note:* The separation is monitored by TLC. Overall recovery is only 40%. Most of the residue (a mixture of mono- and diesters) can be removed with MeOH.	197
Sucrose monoester from higher esters	Silica gel, 8 g, in a 1 × 20 cm column	Elute with 20 mL 29.5:70.5 isoamyl alcohol/benzene, 20 mL 25:75 2-PrOH/benzene, and 60 mL 25:75 EtOH/benzene. The higher esters elute first, followed by monoester.	198,199
Fractionation of esters of glycerin or of ethylene glycol	Silica gel, Davison No. 922, 20 × 200 mm	A 250-mg sample is dissolved in 10 mL benzene, transferred to the column, and treated with 60–90 mL 40:60 isopropyl ether/isooctane (to elute triglycerides and diester of EG), 60 mL 70:30 ethyl ether/iso-octane (to elute diglycerides and monoester of EG), and 80 mL 20:80 EtOH/isopropyl ether (to elute monoglycerides). See isopropyl ether warning on p. 173.	194

(Continued)

TABLE 7 (Continued)

Separation performed	Stationary phase	Procedure	Ref.
Esters and ester ethoxylates from nonpolar food lipids	Silica gel, Merck no. 7729, 30 g	1 g lipid extract is dissolved in 10 mL toluene and added to the column. Elution proceeds with 300 mL 7:93 ethyl ether/petroleum ether, 150 mL 2:1 CHCl$_3$/MeOH, and 100 mL MeOH. The nonpolar lipids are in the first fractions, while ester and ethoxylated ester food emulsifiers are in the last fractions.	15
Separation of polysorbates from food	Merck Extrelut 20 (modified large pore silica) and Merck Silica Gel 60 with 10% water	5–10 g chopped food, 10 mL 50:50 CH$_2$Cl$_2$/EtOH, and 8.5 g Extrelut are mixed and let stand 15 min. 5 g Extrelut is added to a column, followed by the food/adsorbent slurry. The polysorbate is eluted with 150 mL 50:50 CH$_2$Cl$_2$/EtOH. The eluate is evaporated to dryness and the residue is taken up in 10 mL THF, then added to a 1.5 × 30 cm column containing 10 g silica gel, prewashed with hexane. 50 mL 80:20 ethyl ether/CHCl$_3$ is added, then 100 mL 75:25 CH$_3$CN/H$_2$O. The eluate contains the polysorbate.	200
Fractionation of sorbitol esters into the mono-, di-, and triesters	Silica gel	Samples of 0.3–1.5 g are eluted successively with 300 mL each of benzene (discard), 90:10 benzene/ethyl ether (triesters and free fatty acid and isosorbide mono esters), ethyl ether (sorbitan mono- and diester), and EtOH (free sorbitol).	201
Sorbitan monostearate from triglycerides and other compounds	Silica gel, containing 5% H$_2$O, 28 × 300 mm	A heptane solution containing about 0.1 g surfactant is applied to the column. 300 mL benzene is run through the column. The surfactant is eluted with 200 mL absolute EtOH.	79
Isolation of sorbitan monolaurate from water	Supelco C$_{18}$ SPE cartridge, 1 g adsorbent	50 mL simulated sea water containing 2.5 mg surfactant was passed through the cartridge; salts were washed out with 6 mL 90:10 H$_2$O/2-PrOH; surfactant was eluted with 5 mL 2-PrOH.	202
Isolation of alkylpolyglucosides from water	Merck LiChrolut EN poly(styrene/divinylbenzene) SPE material, 100 mg	100 mL sample is passed through resin in a glass column. Resin is dried with a stream of nitrogen, then APG is eluted with three 1-mL portions of MeOH.	203

General procedure for separation of nonionics and additives	Silica gel, Davison No. 922; column is a 50-mL buret	Treat with the following: 70 mL CHCl$_3$ (without ethanol stabilizer) to elute nonpolar material such as mineral oil: 100 mL 99:1 CHCl$_3$/ethyl ether to elute triglycerides, fatty acids, and alcohols; 70 mL 50:50 CHCl$_3$/ethyl ether to elute oil-soluble surfactants; 80 mL 50:50 CHCl$_3$/acetone to elute intermediate polarity surfactants; 70 mL 95:5 CHCl$_3$/MeOH to elute more highly ethoxylated surfactants and PEG; 70 mL 90:10 CHCl$_3$/MeOH to elute similar to 95:5 ratio; 70 mL 60:30 CHCl$_3$/MeOH to elute hydrophilic materials like glycerin and high MW PEG.	69,204
Fractionation of nonionics used in shampoos	Silica gel, Davison No. 922, 10 g in a column 0.5 × 24 inches	200 mg nonionic is slurried with 20 mL hot iso-octane and added to 3 g silica gel. After evaporation of the solvent, the silica is added to the column. Elution is performed with 80 mL iso-octane (to elute mineral oil), 100 mL 95:5 iso-octane/ethyl ether (to elute lanolin), 100 mL 90:10 iso-octane/ethyl ether (to elute fatty alcohols), 150 mL 75:25 iso-octane/ethyl ether (to elute glyceryl distearate), 200 mL 50:50 iso-octane/ethyl ether (to elute glyceryl monostearate and lower ethoxylated material), 200 mL ethyl ether (to elute fatty acid diethanolamide and some ethoxylated materials), and 100 mL MeOH (to elute highly polar compounds like glycerin and highly ethoxylated materials).	205
Isolation of nonionics from a metal-working oil	Silica, Isolute Si, International Sorbent Technology, 2 g in a 12-mL SPE cartridge	The cartridge is conditioned with 10:1 cyclohexane/diisopropyl ether solvent. (**WARNING:** diisopropyl ether is notorious for forming an explosive peroxide insoluble in the ether. Do not use without instituting a program to prevent possession of overage reagent.) Add 200 mg of oil dissolved in the solvent and elute oil with 20 mL solvent. Elute surfactants with 10 mL MeOH. (**WARNING:** Check recovery. Some ethoxylated surfactants may not be recovered from the silica.)	145,206

(Continued)

TABLE 7 *(Continued)*

Separation performed	Stationary phase	Procedure	Ref.
Isolation from water of alkoxylated nonionics: AE, NPE, PEG, acid ethoxylates, sorbitan ester ethoxylates, E/P copolymers, E/P adducts of alcohols, PPG	Anion exchange resin in the cobaltithiocyanate form, 15×200 mm	1 L or more of sample is passed through the column; surfactant is eluted with 50 mL MeOH, EtOH, or 2-PrOH.	207
PEG from C_{12}–C_{18} alcohol 5-and 20-mole ethoxylates, C_9 and C_{12} alkylphenol 3- to 20-mole ethoxylates, PEG dioleate, ethoxylated castor oil, sorbitol hexaoleate 40-mole ethoxylate	Merck LiChroprep RP-8 or RP-18	Elute with 85:15 or 80:20 MeOH/H$_2$O. PEG elutes first.	208,209
PEG from lauryl alcohol 4-mole ethoxylate, OP 10-mole ethoxylate, or stearic acid 20-mole ethoxylate	Polygosil 60-4063 C_{18}	Elute PEG with 70:30 MeOH/H$_2$O, then elute ethoxylate with 80:20 MeOH/CHCl$_3$.	210
PEG from ethoxylates; general procedure	Silica gel, prepared in water saturated with n-butanol	Elute with H$_2$O-saturated n-butanol. Surfactant, free of PEG, elutes first.	211
PEG from AE and NPE	Merck LiChroprep RP-18 SPE cartridge	Analytes are adsorbed from water sample. Elution with 2 5-mL portions of 4:1 CH$_2$Cl$_2$/hexane to remove surfactants, then with 2 5-mL portions of 9:1 MeOH/CH$_2$Cl$_2$ to elute PEG. LAS coelutes with PEG.	212

Silica is appropriate with the cationics used in fabric softeners because their highly nonpolar character makes them suitable for normal phase chromatography. Adsorption of cationics on silica is more favored at high pH, usually 9–10. Silica is rarely used in solid phase extraction of surfactants because of the difficulty in removing the surfactant for subsequent analysis. Ethoxylates are particularly tightly bound, but low yields are found also with other surfactants. PEG is most tightly bound of all, and is best separated using C_{18} media.

Rosen devised a very useful general scheme for separation of nonionic surfactant mixtures by chromatography on silica gel (69).

2. Batch Adsorption

The procedures used in column adsorption chromatography may also be used in batch mode, usually with less separation efficiency, but often with some saving of labor. For example, nonionic impurities may be separated from a variety of anionic sulfate and sulfonate surfactants by dissolving the material in a small amount of ethanol/water, mixing with diatomaceous earth, and washing the adsorbant with petroleum ether or 7:93 methylene chloride/petroleum ether. The anionic material remains adsorbed (70). APE may be isolated from water by batchwise separation on XAD-2 resin, 1 g to 0.5–1.0 L of sample. The surfactant is recovered by washing the resin with 100 mL methanol (71).

It has been demonstrated that β-cyclodextrin, polymerized by reaction with epichlorohydrin and formed into beads, can adsorb APE (72), LAS, and tetradecyldimethylbenzylammonium chloride (73) from aqueous solution. Surfactants can be recovered from the β-cyclodextrin by treatment with alcohol/water mixtures.

3. Thin Layer Chromatography

Analytical TLC is covered in Chapter 9. Any of the analytical methods may be used on the preparative scale, and the reader is advised to scan the tables in Chapter 9 to assess the state of the art. A few procedures developed specifically for preparative TLC deserve special mention.

Hellman presents a scheme, developed for environmental samples, for separation of anionic, cationic, and nonionic surfactants from each other by TLC on silica gel (74). Ordinary 20-cm plates are used of Silica Gel F-60, 0.25 mm. The sample is spotted on the plate. Development with 9:1 CHCl₃/MeOH removes the nonpolar impurities. Subsequent development with 4:1 CHCl₃/MeOH moves the nonionics up the plate. Further development with 1:1 CHCl₃/MeOH will move up the anionics, with the cationics still at the origin. The cationics can be removed from the silica gel with 1:1 CHCl₃/MeOH containing 10% water or containing 10% 2 M HCl. Detection is usually made without reagents, by examining the change in transparency of the plate. For low levels, iodine vapor may be used. If anionics are complexed with cationics, a normal situation in environmental samples, they can be separated by a kind of ion exchange on the surface of the thin layer plate. A solvent mixture of 3:1 MeOH/2 M NH₃ is used to move the anionic up the plate, away from the cationic. Conditions are also given to separate the methylene blue or disulfine blue dyes from the surfactant.

Similarly, the three classes of surfactants may be separated readily on reversed-phase TLC plates with 75:25 ethanol/water. Under these conditions, cationics remain near the origin, anionics migrate with the solvent front, and nonionics have intermediate Rf values (75). Interestingly, the same separation is observed on ordinary silica plates, with 80:20 toluene/methanol (76).

TABLE 8 Column Adsorption Chromatography: Separation of Cationic Surfactants

Separation performed	Stationary phase	Procedure	Ref.
C_{12} alkyltrimethylammonium and ditallowdimethylammonium salts from environmental matrices	Alumina "B", Waters SPE cartridge	A methanolic extract of the dried solids is applied to the cartridge and eluted with MeOH. The cationics are in the eluate.	213
C_{10} dialkyldimethylammonium and C_{12} alkyltrimethylammonium salts from beer and water	Waters Sep-Pak C_{18} cartridge	The cartridge is preconditioned with 5 mL MeOH and 10 mL H_2O. 100 mL sample is mixed with 100 mL MeOH and 10 mL 0.05 M benzenesulfonic acid and passed through the cartridge. The ion pair is eluted with 4 mL 95:5 MeOH/H_2O.	214
C_{12} alkyltrimethylammonium and ditallowdimethylammonium salts from environmental matrices	Alumina "B", Waters SPE cartridge	A methanolic extract of the dried solids is applied to the cartridge and eluted with MeOH. The cationics are in the eluate.	213
Isolation of C_{12}–C_{16} benzylalkyldimethylammonium salts from ophthalmic solutions	Waters Sep-Pak C_{18} cartridge or, for use online with column switching, various C_8 and C_{18} packings in 4.6×10 mm columns	*Batch*: 10 mL of aqueous sample is passed through the cartridge, followed by water. Quats are eluted with 3 mL 812:187:2 and 3 mL 150:250:2 THF/H_2O/triethylamine, pH 3. *Online*: 0.2 mL sample injected on precolumn; elute with 1600:400:3 H_2O/THF/H_3PO_4 at 3 mL/min; after 2 min, reverse flow with HPLC eluent (250:150:2 H_2O/THF/triethylamine, adjusted to pH 3.0 with H_3PO_4).	215

Isolation of C_{12}–C_{14} benzylalkyl-dimethylammonium salts from hospital wastewater	Machery-Nagel C_{18} ec cartridge	Cartridge is preconditioned with 3 bed volumes each of MeOH and H_2O. 8 mL aqueous sample is applied and the column is rinsed with 3 bed volumes of H_2O and 2 of ethyl acetate. Benzalkonium chloride is eluted with 2 bed volumes of 1:1 MeOH/ethyl acetate, containing 1% $CaCl_2$.	216
Isolation of benzylalkyldimethyl-ammonium, C_{12} and C_{16} alkyl-trimethylammonium and C_{16} alkylpyridinium salts from water	Poly(hydroxystyrene/divinylbenzene)	The cationic surfactant is selectively adsorbed from aqueous solution. Recovery/regeneration is performed with MeOH.	67
C_{12} alkyltrimethylammonium chloride, C_{12} alkylpyridinium chloride, C_{12} benzylalkyldi-methylammonium chloride; separation from each other and separation by alkyl chain length	Amberlite CG-4B weakly basic anion exchange resin, 25 × 410 mm, OH^- form, 40°C	*Salting-out chromatography*: elute with 45:55 or 55:45 MeOH/0.5 M NaCl, monitoring the effluent by two-phase titration. Order of elution: low numbered alkyl chains before higher homologs; alkyl quats before alkylpyridinium quats before benzylalkonium quats.	217
Fractionation of a quaternized tri-ethanolamine C_{16}/C_{18} ester	Waters Sep-Pak silica cartridge	0.1 g dried ester quat is dried, dissolved in 1 mL $CHCl_3$, and applied to the cartridge. Residual triesteramine is eluted with $CHCl_3$, followed by mixed tri-ester quat, diester quat, and diesteramine eluted with 98:2 $CHCl_3$/MeOH; then diester quat with 95:5 $CHCl_3$/MeOH; followed by monoester quat with 90:10 $CHCl_3$/MeOH. 20 mL total solvent.	218

TABLE 9 Column Adsorption Chromatography: Separation of Lecithin

Separation performed	Stationary phase	Procedure	Ref.
Phosphatidylcholine from other egg phospholipids	Silica gel, 1.5 × 100 cm	Elute with 70:25:5 $CHCl_3$/MeOH/H_2O, monitoring the separation by TLC.	219
Phosphatidylcholine from other egg phospholipids	YMC 120 Å silica gel, 2 × 20 or 5 × 20 cm	Preparative HPLC: elute with 0.005 M NH_4OAc in 80:13:12:5 CH_3CN/2-PrOH/H_2O/MeOH, monitoring the absorbance at 203 nm.	220
Isolation of total phospholipids (phosphatidylcholine, phosphatidylethanolamine, and phosphatidylinositol) from soybean extract	Analtech SPICE C_{18} SPE cartridge	Dissolve an extract in 2.5 mL 100:1 $CHCl_3$/HOAc and pass through the cartridge. Wash out glycerides with 10 mL of the same solvent, then elute phosphatides with 5 mL 10:5:4 MeOH/$CHCl_3$/H_2O.	221
Isolation of phosphatides from total lipids in eggs, meat, and cheese	Bond-Elut C_8 SPE cartridge, 500 mg	Apply 200 mg lipid extract in 0.5 mL 2:1 $CHCl_3$/MeOH to cartridge. Elute phosphatides with 4 mL MeOH, then elute other lipids with 5 mL 3:2 $CHCl_3$/MeOH and 5 mL $CHCl_3$.	222
Phosphatides and other lipids in soy lecithin and beef lipids	Bond-Elut silica SPE cartridge	100 mg sample is applied to the cartridge. Other lipids are eluted with 30 mL 20:80 hexane/ethyl ether, then phospholipids are eluted with 30 mL methanol.	223
Isolation of phosphatides from sunflower oil	Baker diol SPE cartridge, 500 mg	50–150 mg oil is dissolved in chloroform and added to the cartridge. Triglycerides are eluted with 2.5 mL $CHCl_3$, then phosphatides are eluted with 7 mL 100:0.5 MeOH/25% NH_3 solution.	224

Isolation of phosphatidyletha-nolamine, phosphatidic acid, phosphatidylcholine, and phosphatidylinositol from soybean lecithin	Merck silica gel 60, 15–40 mm, 4.6 × 250 mm, or Waters μPorasil, various dimensions	Preparative HPLC of 100 mg soybean lecithin: varying ratios of n-hexane/2-PrOH/H_2O, monitoring UV absorbance at 214 nm.	225,226
Isolation of phosphatidyletha-nolamine, phosphatidylcho-line, phosphatidylinositol, and lysophophatidylcholine from soybean lecithin	Merck silica gel 60, 15–40 mm, 20 × 250 mm	Preparative HPLC of 2 g soybean lecithin: varying ratios of n-hexane/2-PrOH/H_2O, monitoring UV absorbance at 214 nm.	227,228
Separation of soybean leaf phosphatidylcholine accord-ing to acyl groups	Gourley Excello Ultra Pac ODS, 10 × 150 mm	Preparative HPLC: 95:5 MeOH/0.1 M NH_4OAc, pH 7.4, monitoring UV absorbance at 205 nm.	229
Fractionation of phosphatidyl-choline according to degree of unsaturation	Rohm & Haas XN1010 sulfonic acid type cation exchange resin, ground to 15 μm particle size and put in the Ag^+ form, 4.6 × 250 mm	Preparative HPLC: initial hold at 100% MeOH for 5 min, then linear gradient to 75:25 MeOH/CH_3CN in 40 min, monitoring UV absorbance at 206 nm. Elution is in order of increasing number of double bonds per molecule.	230

Preparative TLC can be used for purification of the methylene blue ion pair of anionic surfactants. The ion pair is formed, extracted from water into chloroform, the solvent
is evaporated, and the residue separated on Merck Silica Gel G with methylene chloride
solvent, in which system the ion pair remains near the origin (77).

4. Liquid-Solid Extraction

Surfactants are often isolated from a solid sample matrix by extraction of the dry ground
sample with alcohol. For example, surfactants are removed from detergents by extraction
with successive portions of hot 95% ethanol after neutralization of the detergent to phenolphthalein. The extract may be purified of inorganic salts by redissolving the residue in
50:50 acetone/ethyl ether and filtering (78). Similarly, sorbitan monostearate may be isolated from cake and cake mix by Soxhlet extraction of the dried solids with absolute ethanol for 24 hr. Again, further cleanup of the extract is necessary (79). Emulsifiers are
separated from cosmetics by isopropanol extraction. For additional purification, the dried
isopropanol extract is re-extracted with absolute ethanol at 60°C (80). APE and LAS may
be removed from cleaning products by extraction of the dried solids with 75:25 hexane/
acetone (81). Obviously, the choice of extracting solvent is governed not only by the surfactant, but also by the solubility properties of other components in the matrix.

Residual surfactant may be quantitatively removed from washed fabrics by Soxhlet
extraction with methanol, 9:1 methanol/toluene, or methylene chloride (82,83).

Recovery of anionic surfactants can be improved by adding an ion pairing reagent to
the solvent. For example, tetrabutylammonium bisulfate in methanol effectively removes
LAS and sodium alkanesulfonate from sewage sludge (84).

5. Supercritical Fluid Extraction

Supercritical fluid extraction has advantages over conventional liquid extraction, especially with regard to easy removal of excess extracting agent. Pressure vessels are commercially available for programmed extraction of solids with supercritical carbon dioxide
of varying density. Carbon dioxide has similar extraction properties to ethyl ether, but will
also extract water. The extract is recovered as an oil or, perhaps, an oil floating on water.
The chief advantage of supercritical fluid extraction over conventional extraction is that
selectivity can be improved. Ordinary solvents will dissolve most members of a class of
compounds, with little discrimination. On the other hand, the solvent properties of supercritical fluids can be fine-tuned by changing the density of the fluid; the solvent strength of
CO_2 increases with increasing pressure. The density is easily adjusted by varying the applied pressure.

(a) Anionic surfactants.   Because of its polarity, LAS has very low solubility in pure
CO_2 and pure N_2O at ordinary SFE temperature and pressure. However, modified CO_2 is
quite effective in removing LAS from environmental matrices. Propylene carbonate,
2-methoxyethanol, acetic acid, 1-butanol, and methanol are all effective when mixed with
CO_2, with 40 mole percent methanol giving quantitative recovery. Many other compounds
are extracted from environmental materials by modified CO_2. Thus, it is not clear
whether SFC extraction is more selective than conventional liquid-solid extraction with
pure methanol (85,86).

Supercritical CO_2 by itself may be used to extract sulfonate surfactants if an ion
pairing reagent is added. This approach can be an advantage if two steps of an analytical
scheme can be combined: extraction and derivatization. For example, LAS may be deriva-

tized by adding trimethylphenylammonium hydroxide to the extracting fluid. The extraction apparatus is run in static mode until ion pairing is complete, then in conventional extraction mode. Although it was at first thought that the methyl derivative was formed from the ion pair during the extraction process, further investigation showed that the ion pair survives until it is introduced to the hot GC injection port, where conversion to the methyl derivative of LAS takes place (87). In tests with a number of quaternary salts, Field and coworkers found that tetra-n-butylammonium ion was the most efficient for derivatization of alkanesulfonates in the injection port of a GC, based on the absence of alkanesulfonate residue in the injector. By contrast, only 10% of alkanesulfonate is removed from the injector after the first injection of the trimethylphenylammonium ion pair (84).

(b) Nonionic surfactants. The efficiency of SFE of ethoxylates can be increased at lower pressures by addition of 5% ethanol (88) or 10% methanol (89) to the CO_2 extractant. Otherwise, yields are not quantitative. Extraction of APE from sediment with methanol-modified CO_2 at 450 atmospheres gives recoveries in the 65–85% range (86).

SFE was studied for the removal of free alcohol from dodecanol ethoxylate. Supercritical CO_2, especially CO_2 containing a few percent methanol, was selective for removal of the alcohol and the lower homologs. However, the distribution coefficients were such that an extraction which removed all of the free alcohol also removed 70% of the ethoxylates. Propane and ammonia may be substituted for CO_2/MeOH, with similar results (90).

SFE has been demonstrated for selective separation of AE and APE, first using CO_2 at low density to remove AE from a substrate, then raising the density to remove APE (91). However, this approach is not generally applicable, since the solubility of each homolog differs as a function of CO_2 density and there is no a priori reason that the solubilities of two commercial mixtures should not overlap. Also, a modifier must usually be added to CO_2 for complete extraction of ethoxylates (89). SFE with CO_2 was demonstrated for direct extraction of APE from water using a specially designed cell. However, recoveries were incomplete (89).

Additives in EO/PO copolymers can be selectively extracted with supercritical CO_2 if the product is first adsorbed onto silica and the extract is passed through silica to remove residual copolymer (92). A study of SFC of sorbitan esters adsorbed to silica indicated that extraction with supercritical methanol/CO_2 could separate the linear from the cyclic sorbitan oligomers (92).

Various glycerol and polyglycerol esters were extracted from cornstarch with supercritical CO_2. SFE was more effective than conventional solvent extraction for removing the esters from extruded cornstarch, but recovery was still incomplete (93).

(c) Cationic surfactants. Methanol-modified CO_2 is effective for extraction of ditallowdimethylammonium salts from sediment and sewage sludge. Copper powder is added to the system to avoid plugging due to polymerization of sulfur compounds. Ion pairing reagents like toluene sulfonic acid are helpful in extracting quats from sediment. They are not needed in sludge samples because cationic surfactants are already paired with anionic surfactants (94).

D. Separations on Alumina

Alumina can act as an adsorption substrate, as an anion exchanger, and as a cation exchanger. These properties are determined by the pretreatment of the alumina: whether it is acidic or basic and what its surface water coverage is. Some of the information in the literature is contradictory, indicating that there are differences in acidic, basic, and neutral alumina from various sources.

Alumina can act as an anion exchanger, permitting the separation of anionic and cationic surfactants. Anionics are bound tightly to basic alumina, requiring a strong solvent like methanolic HCl to remove them. Nonionics and cationics are removed more easily and their separation from each other occurs only in a narrow window of solvent composition (55,95).

Cationic surfactants in the form of the complex with Dragendorff reagent can be separated from oil (alkanes) by treating a chloroform or carbon tetrachloride solution with alumina and filtering. The filtrate contains the alkanes. The cationic is recovered by washing the Al_2O_3 with methanol. The stoichiometry of the complex with Dragendorff reagent is changed in this operation (96).

Column chromatographic separations on alumina are included in the tables of adsorption chromatography.

E. Size Exclusion Chromatography

Preparative gel permeation chromatography is sometimes used to separate surfactants into narrow molecular weight fractions. This technique is usually applied to high molecular weight compounds, such as ethoxylated polymers, but low molecular weight sulfonates may also be separated (97). Most published methods describe use of low resolution dextran gels, but high performance media are generally superior. When performing GPC of surfactants at high concentrations in aqueous systems, one must be aware that micelle formation occurs, so that the surfactant molecules behave as very high molecular weight aggregates, rather than as individual molecules weighing 1000 daltons or less.

GPC, usually with the title, "gel filtration" is used to separate surfactants from protein solutions. Most often, the surfactant itself is not recovered, but is discarded with the gel. A gel should be chosen with a pore size such that surfactant molecules are trapped but the higher molecular weight protein is excluded. Separation can be in either batch or column mode, with the column mode being more efficient (98). Gels are sold for this specific purpose by biochemical supply houses and applications literature is available from these vendors. References to preparative GPC procedures are included in the tables in Chapter 7 on HPLC.

IV. OTHER SEPARATION METHODS

A. Membrane Filtration

A number of researchers have studied high-efficiency membrane filtration techniques as they apply to surfacant micelles. This process is called ultrafiltration, microfiltration, and nanofiltration, depending on the pore size of the membranes. These techniques are applied both to isolation of surfactants themselves and, in micelle enhanced ultrafiltration, to separation of other compounds that are trapped in surfactant micelles so that they are too large to permeate the membrane (99).

Ultrafiltration with membranes of typical pore sizes (0.1–20 nm) would not be expected to remove molecules of less than 1000 MW from aqueous solution. Surfactants, however, form micelles of large enough size that ultrafiltration can be effectively used. This was demonstrated for a nonylphenol 9-mole ethoxylate, where more than 90% of the surfactant was removed from solution (100). The separation is incomplete because individual surfactant molecules permeate the ultrafiltration membrane, so that concentrations of surfactant found in the permeate are at or just below the critical micelle concentration (101).

Hydrophobic ultrafiltration membranes allow the above phenomenon to be used in reverse, permitting surfactant to permeate the membrane but holding back soil and oil (102,103). In this case, the surfactant concentration is enriched in the permeate, presumably because adsorption of the surfactant (alcohol ethoxylates) to the membrane aids in permeation. It is important to control the temperature below the cloud point of the surfactant (103).

Nanofiltration has been demonstrated effective on the pilot scale in separating alkanesulfonates from sodium chloride with high selectivity (104). Nanofiltration of a surfactant is a complex process and requires optimization of the membrane and ionic strength of the medium (99).

In a scheme designed to estimate the concentration of biological surfactants in fermentation broth, Lin et al. demonstrated a two-step analysis based on ultrafiltration: (1) The aqueous broth is subjected to ultrafiltration using a membrane of 10,000 MW cutoff; macromolecules, including micelles, do not permeate this membrane. An HPLC chromatogram with RI or low wavelength UV detection is made of the permeate. (2) The solution is adjusted to contain 50% methanol, breaking up the micelles, and the ultrafiltration and HPLC analysis is repeated. Compounds in the filtrate which were absent or at very low concentration in step 1 are assumed to be surfactants (105).

B. Dialysis and Electrodialysis

Dialysis and electrodialysis are techniques which are applicable to the separation of smaller ions and molecules, including ionic surfactants, from larger species. These methods have been used to separate sodium dodecylsulfate from biochemical media (106). Such separations must normally be run at surfactant concentrations below the critical micelle concentration, since the micelles are too large to pass through the dialysis membrane. A general difficulty in application to trace analysis is the loss of surfactant due to adsorption on the surface of the apparatus. Most often, the object of the experiment is the purification of a protein solution, rather than isolation of a pure solution of the surfactant. Dialysis has been used for the removal of hydrocarbon oil solvent from lubricating oil surfactants in micellar form. In this case, n-heptane was used as solvent at reflux temperature (107).

C. Separation After Chemical Reaction

There is a potentially infinite number of separation techniques which can be developed based upon chemical reaction of one component to remove it from consideration. These techniques usually are specific to the particular matrix, rather than to the surfactant. For example, lipids which interfere in the determination of ethoxylated surfactants in tissue samples may be removed by alkaline hydrolysis, followed by extraction of the surfactants into methylene chloride (108). Sulfated anionic surfactants may be separated from sulfonated anionics by hydrolyzing the mixture with 1 M H_2SO_4 or other strong acid at steam bath temperature to convert the sulfates to the corresponding alcohols. These can be removed by CCl_4 extraction from neutral 50:50 ethanol/water solution (109) or by ion exchange. Cationic surfactants may be isolated from clay, if the clay is first destroyed by heating with HF. The silicon is expelled as the volatile silicon tetrafluoride, leaving a soluble residue of the quaternary ammonium fluoride (110).

REFERENCES

1. Karger, B. L., L. R. Snyder, C. Horvath, *An Introduction to Separation Science*, John Wiley & Sons, New York, 1973.
2. Mayhew, R. L., R. C. Hyatt, Effect of mole ratio distribution on the physical properties of a polyoxyethylated alkylphenol, *J. Am. Oil Chem. Soc.*, 1952, *29*, 357–362.
3. Porot, V., NPE in samples from sewage treatment plants (Toulon and Morlaix) by HPLC (in French), *Proc. 2nd World Surfactants Congress*, Paris, May, 1988, *4*, 293–302.
4. Szelag, H., W. Zwierzykowski, Fractionation of ethoxylated fatty acids using molecular distillation, *Tenside. Surfactants, Deterg.*, 1984, *21*, 14–16.
5. Szelag, H., W. Zwierzykowski, Molecular distillation of selected fatty acid derivatives, *SÖFW Journal*, 1995, *121*, 444–448.
6. Szelag, H., W. Zwierzykowski, E. Kobus, Fractionation of ethoxylated fatty alcohols by molecular distillation, *Tenside, Surfactants, Deterg.*, 1992, *29*, 345–348.
7. Tharapiwattananon, N., J. F. Scamehorn, S. Osuwan, J. H. Harwell, K. J. Haller, Surfactant recovery from water using foam fractionation, *Sep. Sci. Technol.*, 1996, *31*, 1233–1258.
8. Kumpabooth, K., J. F. Scamehorn, S. Osuwan, J. H. Harwell, Surfactant recovery from water using foam fractionation: effect of temperature and added salt, *Sep. Sci. Technol.*, 1999, *34*, 157–172.
9. Stefan, R. L., A. J. Szeri, Surfactant scavenging and surface deposition by rising bubbles, *J. Colloid Interface Sci.*, 1999, *212*, 1013.
10. Wickbold, R., Concentration and separation of surfactants from surface waters by transport in the gas/water interface (in German), *Tenside. Surfactants, Deterg.*, 1971, *8*, 61–63.
11. Block, H., W. Gniewkowski, W. Baltes, Concentration of nonionic surfactants for biodegradibility testing (in German), *Tenside*, 1987, *24*, 160–163.
12. Lukaszewski, Z., A. Szymanski, Sources of error in the determination of nonionic surfactants in environmental samples, *Mikrochim. Acta*, 1996, *123*, 185–196.
13. Kunkel, E., Carboxymethylated oxyethylates, *Tenside Detergents*, 1980, *17*, 10–12.
14. Chlebicki, J., W. Garncarz, Nonionic surfactants in water and effluent by atomic absorption spectroscopy, *Tenside Detergents*, 1980, *17*, 13–17.
15. Dieffenbacher, A., U. Bracco, Analytical techniques in food emulsifiers, *J. Am. Oil Chem. Soc.*, 1978, *55*, 642–646.
16. Graciaa, A., J. Lachaise, G. Marion, M. Bourrel, I. Rico, A. Lattes, Purification of surfactants, *Tenside, Surfactants, Deterg.*, 1989, *26*, 384–386.
17. Allan, G. C., J. R. Aston, F. Grieser, T. W. Healy, Partitioning of a polydisperse NPE between water and hexane, *J. Colloid Interface Sci.*, 1989, *128*, 258–274.
18. Menges, R. A., T. S. Menges, G. L. Bertrand, D. W. Armstrong, L. A. Spino, Extraction of nonionic surfactants from waste water using centrifugal partition chromatography, *J. Liq. Chromatogr.*, 1992, *15*, 2909–2925.
19. Baudimant, G., M. Maurice, A. Landrein, G. Durand, P. Durand, Purification of phosphatidylcholine with high content of docosahexaenoic acid from squid by countercurrent chromatography, *J. Liq. Chromatogr. & Rel. Technol.*, 1996, *19*, 1793–1804.
20. Wickbold, R., Alkanesulfonates (in German), *Tenside, Surfactants, Deterg.*, 1971, *8*, 130–134.
21. Zhang, L., X. Li, W. Du, J. Furong, Detection of polyoxyethylene nonionic surfactants and determination of the ethylene oxide content by IR spectroscopy (in Chinese), *Fenxi Huaxue*, 1987, *15*, 810, 823–824. CA 108:188932f.
22. Brant, L. L., K. L. Stellner, J. F. Scamehorn, Recovery of surfactant from surfactant-based separations using a precipitation process, in J. F. Scamehorn and J. H. Harwell, eds., *Surfactant-Based Separation Processes*, Marcel Dekker, New York, 1989.
23. Suri, S. K., P. M. Patel, Recovery of LAS from the effluent stream of surfactant-based industrial processes, *Indian J. Environ. Prot.*, 1996, *16*, 54–58.

24. Stellner, K. L., J. F. Scamehorn, Surfactant precipitation in aqueous solutions containing mixtures of anionic and nonionic surfactants, *J. Am. Oil Chem. Soc.*, 1986, *63*, 566–574.

25. Shiau, B., J. H. Harwell, J. F. Scamehorn, Precipitation of mixtures of anionic and cationic surfactants: effect of added nonionic surfactant, *J. Colloid Interface Sci.*, 1994, *167*, 332–345.

26. Govindram, C. B., V. Krishnan, Complex surfactant systems—classical approach, *Tenside, Surfactants, Deterg.*, 1998, *35*, 104–107.

27. Kupfer, W., J. Jainz, H. Kelker, Alkanesulfonates (in German), *Tenside, Surfactants, Deterg.*, 1969, *6*, 15–21.

28. Osipow, L. I., W. Rosenblatt, Micro-emulsion process for the preparation of sucrose esters, *J. Am. Oil Chem. Soc.*, 1967, *44*, 307–309.

29. Matsumoto, S., Y. Hatakawa, A. Nakajima, Purification of sucrose fatty acid esters by salting out, useful in surfactants, Japanese Patent 02 09,892, 1990.

30. Matsumoto, S., Y. Hatakawa, A. Nakajima, Preparation of powdery sucrose fatty acid esters for use in surfactants, Japanese Patent 02 09,893, 1990.

31. Glonek, T., ^{31}P NMR phospholipid analysis of anionic-enriched lecithins, *J. Am. Oil Chem. Soc.*, 1998, *75*, 569–573.

32. Anielak, P., K. Janio, Adsorption of nonionic surfactants on synthetic adsorbents, *Tenside, Surfactants, Deterg.*, 1990, *27*, 113–117.

33. MacDonald, L. S., B. G. Cooksey, J. M. Ottaway, W. C. Campbell, Automatic detergent analysis, *Anal. Proc.*, 1986, *23*, 448–451.

34. Mandery, K., Qualitative analysis of the surfactant component of an aqueous detergent (in German), *Seifen, Öle, Fette, Wachse*, 1991, *117*, 595–597.

35. König, H., Separation of surfactant mixtures, with special consideration of anionic surfactants (in German), *Fresenius' Z. Anal. Chem.*, 1971, *254*, 337–345.

36. Vissers, J. P. C., J. Chervet, J. Salzmann, Sodium dodecyl sulfate removal from tryptic digest samples for on-line capillary LC-electrospray MS, *J. Mass Spec.*, 1996, *31*, 1021–1027.

37. Vissers, J. P. C., W. P. Hulst, J. Chervet, H. M. J. Snijders, C. A. Cramers, Automated online ionic detergent removal from minute protein/peptide samples prior to LC-electrospray MS, *J. Chromatogr. B: Biomed. Appl.*, 1996, *686*, 119–128.

38. Metcalfe, L. D., Use of cellulosic ion exchangers for determination of quaternary ammonium compounds, *Anal. Chem.*, 1960, *32*, 70–72,380.

39. Hellmann, H., Adsorbants as ion exchange and separation media. Al_2O_3 and SiO_2 in the chromatographic analysis of surfactants (in German), *Fresenius' Z. Anal. Chem.*, 1989, *334*, 126–132.

40. Pardue, K., D. Williams, Nonionic surfactants in protein samples using ion-exchange guard columns, *BioTechniques*, 1993, *14*, 580,582–583.

41. Fudano, S., K. Konishi, Separation mechanism of ionic surfactants in salting-out chromatography: functions of alcohol and sodium chloride in the eluent, *J. Chromatogr.*, 1974, *93*, 467–470.

42. Rosen, M. J., Mixtures of ionic and nonionic surface-active agents: Separation and recovery of components by batch ion exchange, *J. Am. Oil Chem. Soc.*, 1961, *38*, 218–220.

43. Cserhati, T., Adsorption of some nonionic tensides on different carriers. II: Adsorption capacity, *Acta Phytopathol. Entomol. Hung.*, 1993, *28*, 129–135.

44. Tolls, J., P. Kloepper-Sams, D. T. H. M. Sijm, Surfactant bioconcentration—critical review, *Chemosphere*, 1994, *29*, 693–717.

45. Zhao, M., F. van der Wielen, P. de Voogt, Optimization of a matrix solid-phase dispersion method with sequential cleanup for the determination of APE in biological tissues, *J. Chromatogr. A*, 1999, *837*, 129–138.

46. Dusart, O., S. Souabi, M. Mazet, Elimination of surfactants in water treatment by adsorption onto activated carbon, *Environ. Technol.*, 1990, *11*, 721–730.

47. Mazet, M., O. Dusart, P. LaFrance, Adsorption of surfactants on activated carbon in the presence of metallic ions: determination of the adsorption energy parameters, *J. Surf. Sci. Technol.*, 1989, *5*, 345–353.

48. Ihara, Y., Adsorption of anionic surfactants and related compounds from aqueous solution onto activated carbon and synthetic adsorbent, *J. Appl. Polym. Sci.*, 1992, *44*, 1837–1840.

49. Abe, I., K. Hayashi, M. Kitagawa, T. Urahata, Adsorption of fatty acid sodium salts on activated carbons (in Japanese), *Nippon Kagaku Kaishi*, 1978, 1188–1193.

50. Strancar, A., P. Raspor, H. Schwinn, R. Schütz, D. Josic, Extraction of Triton X-100 and its determination in virus-inactivated human plasma by the solvent-detergent method, *J. Chromatogr.*, 1994, *658*, 475–481.

51. Yamini, Y., M. Ashraf-Khorassani, Octadecyl-bonded silica membrane disks for extraction of surfactants from water, *Fresenius' J. Anal. Chem.*, 1994, *348*, 251–252.

52. Yamini, Y., M. Ashraf-Khorassani, Extraction and determination of LAS from the aquatic environment using a membrane disk and GC, *J. High Resolut. Chromatogr.*, 1994, *17*, 634–638.

53. Krueger, C. J., J. A. Field, In-vial C_{18} Empore disk elution coupled with injection port derivatization for the quantitative determination of LAS by GC-FID, *Anal. Chem.*, 1995, *67*, 3363–3366.

54. Field, J. A., T. M. Field, T. Polger, W. Giger, Secondary alkane sulfonates in sewage wastewaters by solid phase extraction and injection port derivatization GC/MS, *Environ. Sci. Technol.*, 1994, *28*, 497–503.

55. Kloster, G., M. Schoester, H. Prast, Concentration of all three surfactant classes from environmental samples (in German), *Tenside, Surfactants, Deterg.*, 1994, *31*, 23–28.

56. García-Delgado, R. A., L. M. Cotoruelo-Minguez, J. J. Rodríguez, Adsorption of anionic surfactant mixtures by polymeric resins, *Sep. Sci. Technol.*, 1992, *27*, 1065–1076.

57. García-Delgado, R. A., L. M. Cotoruelo-Minguez, J. J. Rodríguez, Equilibrium study of single-solute adsorption of anionic surfactants with polymeric XAD resins, *Sep. Sci. Technol.*, 1992, *27*, 975–987.

58. Field, J. A., J. A. Leenheer, K. A. Thorn, L. B. Barber, II, C. Rostad, D. L. Macalady, S. R. Daniel, Persistent anionic surfactant-derived chemicals in sewage effluent and groundwater, *J. Contam. Hydrol.*, 1992, *9*, 55–78.

59. Szczepaniak, W., Sorption and preconcentration of anionic surfactants from aqueous solution onto a mercurated XAD-2 resin, *Chem. Anal.* (Warsaw), 1995, *40*, 281–291.

60. Wolf, F., S. Lindau, Adsorption of nonionic ethylene oxide adducts on porous styrene-divinylbenzene copolymers (in German), *Tenside, Surfactants, Deterg.*, 1977, *14*, 119–122.

61. Saito, S., T. Taniguchi, M. Yukawa, Adsorption of nonionic surfactants by acid-type cation exchangers, *Tenside, Surfactants, Deterg.*, 1975, *12*, 100–103.

62. Fujita, I., Y. Nagano, M. Haratake, K. Harada, M. Nakayama, A. Sugii, Adsorption of nonionic surfactants on chemically modified styrene-divinylbenzene copolymers, *Sep. Sci. Technol.*, 1991, *26*, 1395–1402.

63. Fujita, I., Y. Nagano, K. Harada, M. Nakayama, A. Sugii, Preparation of modified styrene-divinylbenzene copolymers having different physical structures and their adsorption characteristics of nonionic surfactants, *Angew. Makromol. Chem.*, 1992, *200*, 183–192.

64. Aranda, R., R. C. Burk, Nonionic surfactant by solid-phase microextraction coupled with HPLC and on-line derivatization, *J. Chromatogr. A*, 1998, *829*, 401–406.

65. Boyd-Boland, A. A., J. B. Pawliszyn, Solid-phase microextraction coupled with HPLC for APE in water, *Anal. Chem.*, 1996, *68*, 1521–1529.

66. Rigaud, J., D. Levy, G. Mosser, O. Lambert, Detergent removal by nonpolar polystyrene beads. Applications to membrane protein reconstitution and two-dimensional crystallization, *Eur. Biophys. J.*, 1998, *27*, 305–319.

67. Kawabata, N., S. Koichi, M. Tanaka, Selective adsorption of cationic surfactants on cross-linked poly(p-hydroxystyrene), *Ind. Eng. Chem. Res.*, 1990, *29*, 1889–1893.

68. Somasundaran, P., E. D. Snell, Q. Xu, Adsorption behavior of APE on silica, *J. Colloid Interface Sci.*, 1991, *144*, 165–173.

69. Rosen, M. J., Chromatographic separation of nonionic surface-active agents and related materials, *Anal. Chem.*, 1963, *35*, 2074–2077.

70. Cross, C. K., Neutral oil in detergent products by SPE, *J. Am. Oil Chem. Soc.*, 1990, *67*, 142–143.

71. Saito, T., K. Hagiwara, Surfactants in water with anion-exchange resin and polymeric adsorbent, *Fresenius' Z. Anal. Chem.*, 1982, *312*, 533–535.

72. Murai, S., S. Imajo, Y. Maki, K. Takahashi, K. Hattori, Adsorption and recovery of nonionic surfactants by β-cyclodextrin polymer, *J. Colloid Interface Sci.*, 1996, *183*, 118–123.

73. Murai, S., S. Imajo, H. Inumaru, K. Takahashi, K. Hattori, Adsorption and recovery of ionic surfactants by β-cyclodextrin polymer, *J. Colloid Interface Sci.*, 1996, *190*, 488–490.

74. Hellmann, H., Silica gel layers as ion exchangers in surfactant analysis (in German), *Fresenius' Z. Anal. Chem.*, 1983, *315*, 612–617.

75. Armstrong, D. W., G. Y. Stine, Anionic, cationic and nonionic surfactants by TLC, *J. Liq. Chromatogr.*, 1983, *6*, 23–33.

76. Bare, K. J., H. Read, Fast atom bombardment MS to identify materials separated on HPTLC plates, *Analyst*, 1987, *112*, 433–436.

77. McEvoy J., W. Giger, LAS in sewage sludge by GC-MS, *Environ. Sci. Technol.*, 1986, *20*, 376–383.

78. American Society for Testing and Materials, Separation of active ingredient from surfactant and syndet compositions, D2358-82. West Conshohocken, PA 19428.

79. Wetterau, F. P., V. L. Olsanski, C. F. Smullin, Sorbitan monostearate in cake mixes and baked cakes, *J. Am. Oil Chem. Soc.*, 1964, *41*, 791–795.

80. König, H., Emulsifiers of the type orthophosphoric acid esters of fatty alcohols and their ethoxylated derivatives by TLC (in German), *Fresenius' Z. Anal. Chem.*, 1968, *235*, 255–261.

81. Marcomini, A., S. Stelluto, B. Pavoni, LAS and APE in commercial products and marine waters by reversed- and normal-phase HPLC, *Int. J. Environ. Anal. Chem.*, 1989, *35*, 207–218.

82. Bosdorf, V., T. Bluhm, H. Krüssman, Adsorbed nonionic surfactants on fabrics by HPLC (in German), *Text. Prax. Int.*, 1994, *49*, 348–350,353–354.

83. Bosdorf, V., T. Bluhm, H. Krüssman, TLC determination of adsorbed nonionic surfactants on fabrics (in German), *Melliand Textilber.*, 1994, *75*, 311–312.

84. Field, J. A., D. J. Miller, T. M. Field, S. B. Hawthorne, W. Giger, Sulfonated aliphatic and aromatic surfactants in sewage sludge by ion-pair/supercritical fluid extraction and derivatization GC-MS, *Anal. Chem.*, 1992, *64*, 3161–3167.

85. Hawthorne, S. B., D. J. Miller, D. D. Walker, D. E. Whittington, B. L. Moore, Extraction of LAS using supercritical carbon dioxide and a simple device for adding modifiers, *J. Chromatogr.*, 1991, *541*, 185–194.

86. Kreisselmeier, A., H. Dürbeck, Alkylphenols, APE and LAS in sediments by accelerated solvent extraction and supercritical fluid extraction, *J. Chromatogr. A*, 1997, *775*, 187–196.

87. Hawthorne, S. B., D. J. Miller, J. J. Langenfeld, SFE of polar analytes using modified carbon dioxide and in situ chemical derivation, in F. V. Bright and M. E. P. McNally, eds., *Supercritical Fluid Technology: Theoretical and Applied Approaches to Analytical Chemistry*, ACS Symposium Series, No. 488, Washington DC, 1992.

88. Penwell, A. J., M. H. I. Comber, SFE—alternative to traditional techniques in regulatory environmental analysis, *Sample Prep. Biomed. Environ. Anal.* (Proc. Chromatogr. Soc. Int. Symp.), 1991 (Pub. 1994), 203–209.

89. Kane, M., J. R. Dean, S. M. Hitchen, C. J. Dowle, R. L. Tranter, Extraction of surfactants from aqueous media by supercritical fluid extraction, *Analyst*, 1995, *120*, 355–359.

90. Eckert, C. A., M. P. Ekart, B. L. Knutson, K. P. Payne, D. L. Tomasko, C. L. Liotta, N. R. Foster, Supercritical fluid fractionation of a nonionic surfactant, *Ind. Eng. Chem. Res.*, 1992, *31*, 1105–1110.

91. Kane, M., J. R. Dean, S. M. Hitchen, C. J. Dowle, R. L. Tranter, Nonionic surfactants using SPE combined with SFE and SFC, *Anal. Proc.*, 1993, *30*, 399–400.

92. Hunt, T. P., C. J. Dowle, G. Greenway, SFC for the analysis of liquid poly(alkylene glycol) lubricants and sorbitan ester formulations, *Analyst*, 1993, *118*, 17–22.

93. Artz, W. E., M. R. Myers, Supercritical fluid extraction and chromatography of emulsifiers, *J. Am. Oil Chem. Soc.*, 1995, *72*, 219–224.

94. Fernández, P., A. C. Alder, M. J. F. Suter, W. Giger, Ditallowdimethylammonium in digested sludges and marine sediments by supercritical fluid extraction and LC with post-column ion-pair extraction, *Anal. Chem.*, 1996, *68*, 921–929.

95. Heise, S., N. Litz, Extraction of surfactants from solid matrices (in German), *Tenside, Surfactants, Deterg.*, 1999, *36*, 185–191.

96. Hellmann, H., Spectrophotometric determination of cationic surfactants in the presence of anionic and nonionic surfactants (in German), *Fresenius' Z. Anal. Chem.*, 1984, *319*, 272–276.

97. Steuerle, H., Separation and determination of aromatic sulfonic acids by column chromatography with Sephadex G25 (in German), *Fresenius' Z. Anal. Chem.*, 1966, *220*, 413–420.

98. Hall, S. W., S. R. VandenBerg, Solid phase extraction of the zwitterionic detergent CHAPS, *Prep. Biochem.*, 1989, *19*, 1–11.

99. Archer, A. C., A. M. Mendes, R. A. R. Boaventura, Separation of an anionic surfactant by nanofiltration, *Environ. Sci. Technol.*, 1999, *33*, 2758–2764.

100. Paatz, K., J. Rollin, Ultrafiltration of aqueous surfactant solutions (in German), *Wiss. Z.-Tech. Hochsch. Köthen*, 1991, *3*, 95–99.

101. Osborne-Lee, I. W., R. S. Schechter, W. H. Wade, Monomer–micellar equilibrium of aqueous surfactant solutions by the use of ultrafiltration, *J. Colloid Interface Sci.*, 1983, *94*, 179–186.

102. Mahdi, S. M., R. O. Sköld, Reconcentration of a polydisperse nonionic surfactant in aqueous solution by adsorption induced ultrafiltration, *Colloids Surf.*, 1992, *68*, 111–120.

103. Ang, C. C., A. S. Abdul, Ultrafiltration method for surfactant recovery and reuse during washing of contaminated sites, *Ground Water Monit. Rem.*, 1994, *14*, 160–171.

104. Klukas, K., Reclamation of surfactants from NaCl-containing industrial effluents (in German), *Melliand Textilber.*, 1994, *75*, 431–435.

105. Lin, S., Y. Chen, Y. Lin, HPLC for biosurfactant analysis and purification, *J. Chromatogr. A*, 1998, *825*, 149–159.

106. Tuszynski, G. P., L. Warren, Removal of sodium dodecyl sulfate from proteins, *Anal. Biochem.*, 1975, *67*, 55–65.

107. Van de Ven, A. M. C., P. S. Johal, L. Jansen, Synthetic sulfonate and phenate detergents by NMR and IR spectroscopy, *Lubr. Sci.*, 1993, *6*, 3–19.

108. Fung, D., S. Safe, J. F. S. Crocker, Polyethylene glycol ether surfactants in biological tissues, *Toxicol. Environ. Chem.*, 1984, *8*, 109–132.

109. Hoyt, J. L., E. L. Sones, A. J. Sooter, Active ingredients in detergent formulations, *J. Am. Oil Chem. Soc.*, 1979, *56*, 701–703.

110. Spagnolo, F., M. T. Hatcher, B. K. Faulseit, HPLC analysis of commercial alkyl and aryl quaternary ammonium compounds used in organoclay type rheological additives, *J. Chromatogr. Sci.*, 1987, *25*, 399–401.

111. Takada, H., R. Ishiwatari, Linear alkylbenzenes in urban riverine environments in Tokyo: distribution, source, and behavior, *Environ. Sci. Technol.*, 1987, *21*, 875–883.

112. Yamaoka, K., K. Nakajima, H. Moriyama, Y. Saito, T. Sato, Sodium dodecyl sulfate in hydrophilic ointments by TLC with flame ionization detection, *J. Pharm. Sci.*, 1986, *75*, 606–607.

113. International Organization for Standardization, Surface active agents—sulfated ethoxylated alcohols and alkylphenols—estimation of the mean relative molecular mass, ISO 6843:1988. Geneva, Switzerland.

114. Holtzman, S., B. M. Milwidsky, Olefin sulfonation for detergents, *Soap Chem. Spec.*, 1967, *43(11)*, 64,66,68,112–115.

115. Milwidsky, B. M., Continuous liquid/liquid extraction, *Soap Chem. Spec.*, 1969, *45(12)*, 79–80,84,86,88,117–118.

116. International Organization for Standardization, Determination of the mean relative molecular mass of the alkane monosulfonates and the alkane monosulfonate content, ISO 6845:1989, Geneva, Switzerland.

117. Sandvik, E. I., W. W. Gale, M. O. Denekas, Petroleum sulfonates, *Soc. Pet. Eng. J.*, 1977, *17*, 184–192.

118. American Society for Testing and Materials, Sodium toluene sulfonate in detergents, D2023-89. West Conshohocken, PA 19428.

119. Sun, C., M. Baird, H. A. Anderson, D. L. Brydon, Oligomers and homologs of aliphatic AE in textile lubricants and lubricant emulsions by HPLC, *J. Chromatogr. A*, 1997, *771*, 145–154.

120. Sun, C., M. Baird, H. A. Anderson, D. L. Brydon, Separation of broadly distributed NPE and determination of ethylene oxide olgomers in textile lubricants and emulsions by HPLC, *J. Chromatogr. A*, 1996, *731*, 161–169.

121. Zeringue, H. J., R. O. Feuge, Purification of sucrose esters by selective adsorption, *J. Am. Oil Chem. Soc.*, 1976, *53*, 567–571.

122. Inaba, K., Polyoxyethylene-type nonionic surfactants in environmental waters, *Int. J. Environ. Anal. Chem.*, 1987, *31*, 63–66.

123. Anthony, D. H. J., R. S. Tobin, Immiscible solvent extraction scheme for biodegradation testing of polyethoxylate nonionic surfactants, *Anal. Chem.*, 1977, *49*, 398–401.

124. Kröller, E., Emulsifiers in foodstuffs. Part 8 (in German), *Fette, Seifen, Anstrichm.*, 1968, *70*, 431–433.

125. Comité Européen des Agents de Surface et leurs Intermédiaires Organiques, Determination of free amine in amine oxides by TLC, CESIO/AIS Analytical Method 7-91, available from European Chemical Industry Council, Brussels.

126. Kröller, E., Emulsifiers in foodstuffs. Part 5 (in German), *Fette, Seifen, Anstrichm.*, 1964, *66*, 583–586.

127. Terstappen, G. C., M. Kula, Selective extraction and quantitation of polyoxyethylene detergents and its application in protein determination, *Anal. Lett.*, 1990, *23*, 2175–2193.

128. König, H., E. Walldorf, Analysis of toothpaste (in German), *Fresenius' Z. Anal. Chem.*, 1978, *289*, 177–197.

129. Bürger, K., Separation of polyethylene glycols from surface-active ethylene oxide adducts of fatty acids, fatty alcohols, fatty amines, fatty acid amides, and alkylphenols, and quantitative determination, by selective liquid-liquid extraction (in German), *Fresenius' Z. Anal. Chem.*, 1963, *196*, 22–26.

130. Marsh, D. F., L. T. Takahashi, Benzalkonium chloride in the presence of interfering alkaloids and polymeric substrates by reverse-phase HPLC, *J. Pharm. Sci.*, 1983, *72*, 521–525.

131. Lincoln, P. A., C. C. T. Chinnick, Quaternary ammonium compounds as phosphotungstates, *Analyst*, 1956, *81*, 100–104.

132. Hind, A. R., S. K. Bhargava, S. C. Grocott, Alkyltrimethylammonium bromides in Bayer process liquors by GC and GC-MS, *J. Chromatogr. A*, 1997, *765*, 287–293.

133. Matthijs, E., H. De Henau, LAS, *Tenside, Surfactants, Deterg.*, 1987, *24*, 193–199.

134. Bán, T., E. Papp, J. Inczédy, Reversed-phase HPLC chromatography of anionic and ethoxylated nonionic surfactants and pesticides in liquid pesticide formulations, *J. Chromatogr.*, 1992, *593*, 227–231.

135. Saito, T., K. Higashi, K. Hagiwara, LAS by HPLC. Application to water, *Fresenius' Z. Anal. Chem.*, 1982, *313*, 21–23.

136. Newburger, S. H., Analysis of shampoos, *J. Ass. Off. Anal. Chem.*, 1958, *41*, 664–668.

137. Nest Group Ideabook, Nest Group, Southborough, MA 01772, 1993.

138. Rasmussen, H. T., A. M. Pinto, M. W. DeMouth, P. Touretzky, B. P. McPherson, High temperature GC of trimethylsilyl derivatives of alcohol ethoxylates and ethoxysulfates, *J. High Resolut. Chromatogr.*, 1994, *17*, 593–596.

139. Mutter, M., Alkanesulfonates using ion exchangers (in German), *Tenside, Surfactants, Deterg.*, 1968, *5*, 138–140.

140. Schreuder, R. H., A. Martijn, C. van de Kraats, Diisobutyl- and diisopropylnaphthalenesulphonates in pesticide wettable powders and dispersible granules by HPLC, *J. Chromatogr.*, 1989, *467*, 177–184.

141. Moody, C. A., J. A. Field, Perfluorocarboxylates in groundwater impacted by fire-fighting activity, *Environ. Sci. Technol.*, 1999, *33*, 2800–2806.
142. König, H., E. Walldorf, α-Sulfo fatty acid methyl esters and fatty alcohol sulfoacetates (in German), *Fresenius' Z. Anal. Chem.*, 1975, *276*, 365–370.
143. Moody, G. J., J. O. Rutherford, J. D. R. Thomas, Sorption behavior of dodecylsulphate and other anionic surfactants on anion-exchange resins, *Analyst*, 1981, *106*, 537–546.
144. Ianniello, R. M., Organic acids in APE by anion exclusion HPLC, *Anal. Lett.*, 1988, *21*, 87–99.
145. Hülskötter, F., M. Raulf, N. Buschmann, Titration of lipophilic ionic and nonionic surfactants, *World Surfactants Congr., 4th*, 1996, *4*, 96–101.
146. Gabriel, D. M., Cosmetics and toiletries, *J. Soc. Cosmet. Chem.*, 1974, *25*, 33–48.
147. Gorenc, B., D. Gorenc, A. Rošker, Preconcentration of nonionic surfactants by adsorption on ion-exchange resin in the cobaltithiocyanate form, *Vestn. Slov. Kem. Drus.*, 1986, *33*, 467–474.
148. Barber, A., C. C. T. Chinnick, P. A. Lincoln, Mixtures of surface-active quaternary ammonium compounds and polyethylene oxide type of nonionic surface-active agents, *Analyst*, 1956, *81*, 18–25.
149. König, H., Amphoteric surfactants (in German), *Fresenius' Z. Anal. Chem.*, 1970, *251*, 359–368.
150. Wickbold, R., Sulfated AE (in German), *Tenside, Surfactants, Deterg.*, 1976, *13*, 181–184.
151. Bürgi, C., Otz, T., Alkylphenolethoxylates (in German), *Tenside, Surfactants, Deterg.*, 1995, *32*, 22–24.
152. Mutter, M., G. W. van Galen, P. W. Hendrikse, Fatty acid monoethanolamide foam stabilizer (in German), *Tenside, Surfactants, Deterg.*, 1968, *5*, 33–36.
153. Mutter, M., G. W. van Galen, P. W. Hendrikse, Fatty acid monoethanolamide-ethylene oxide condensate nonionic surfactant (in German), *Tenside, Surfactants, Deterg.*, 1968, *5*, 36–39.
154. Wickbold, R., Surfactants in detergents and cleaners (in German), *Tenside, Surfactants, Deterg.*, 1976, *13*, 177–180.
155. König, H., Soaps (in German), *Fresenius' Z. Anal. Chem.*, 1978, *293*, 295–300.
156. König, H., Sulfobetaines (in German), *Fresenius' Z. Anal. Chem.*, 1972, *259*, 191–194.
157. Bey, K. Quantitative ion exchange analysis of surfactants (in German), *Fette, Seifen, Anstrichm.*, 1965, *67*, 25–30.
158. Voogt, P., Ion-exchangers in detergent analysis, *Rec. Trav. Chim.*, 1958, *77*, 889–901.
159. Voogt, P., Ion-exchangers in detergent analysis. Part II, *Rec. Trav. Chim.*, 1959, *78*, 899–912.
160. Kikuchi, M., A. Tokai, T. Yoshida, LAS in the marine environment by HPLC, *Water Res.*, 1986, *20*, 643–650.
161. Castles, M. A., B. L. Moore, S. R. Ward, LAS in aqueous environmental matrices by HPLC with fluorescence detection, *Anal. Chem.*, 1989, *61*, 2534–2540.
162. Garcia Ramon, M. T., I. Ribosa, J. Sanchez Leal, F. Comelles, LAS monitoring in the Tajo river basin, *Tenside, Surfactants, Deterg.*, 1990, *27*, 118–121.
163. Fudano, S., K. Konishi, Separation and determination of linear and branched chain alkylbenzene sulfonates by salting-out chromatography, *J. Chromatogr.*, 1970, *51*, 211–218.
164. Fudano, S., K. Konishi, Separation and determination of LAS and alkylsulfates by salting-out chromatography, *J. Chromatogr.*, 1972, *66*, 153–155.
165. Laikhtman, M., J. S. Rohrer, Fluorochemical surfactants in acid etch baths by IC with on-line matrix elimination, *J. Chromatogr. A*, 1998, *822*, 321–325.
166. Scoggins, M. W., J. W. Miller, Separation technique for mono- and disulfonic acids, *Anal. Chem.*, 1968, *40*, 1155–1157.
167. Puschmann, H., Ether sulfates on silanized silica gel (in German), *Chem., Phys. Chem. Anwendungstech. Grenzflächenaktiven Stoffe, Ber. Int. Kongr., 6th*, 1972, Carl Hanser Verlag, Munich, Band I, 397–406.
168. Puschmann, H., Olefin sulfonates (in German), *Fette, Seifen, Anstrichm.*, 1973, *75*, 434–437.
169. Kuemmel, D. F., S. J. Liggett, Level and position of unsaturation in alpha olefin sulfonates, *J. Am. Oil Chem. Soc.*, 1972, *49*, 656–659.

170. Fudano, S., K. Konishi, Separation and determination of α-olefin sulfonates by salting-out chromatography, *J. Chromatogr.*, 1971, *62*, 467–470.

171. Fudano, S., K. Konishi, Mixtures of anionic surface-active agents by salting-out chromatography, *J. Chromatogr.*, 1973, *77*, 351–355.

172. American Society for Testing and Materials, Oil-soluble petroleum sulfonates by LC, D3712-83. West Conshohocken, PA 19428.

173. Desbène, P. L., C. Rony, B. Desmazières, J. C. Jacquier, Alkylaromatic sulfonates by HPCE, *J. Chromatogr.*, 1992, *608*, 375–383.

174. Grey, R. A., A. F. Chan, Sulfonation studies of monoisomeric di- and trialkylbenzenes, *J. Am. Oil Chem. Soc.*, 1990, *67*, 132–141.

175. Bohne-Matusall, R., K. Otto, P. Wilderer, Analysis of oil emulsions in water (in German), *Z. Wasser Abwasser Forsch.*, 1987, *20*, 35–38.

176. Schreuder, R. H., A. Martijn, LAS and APE in liquid pesticide formulations by HPLC, *J. Chromatogr.*, 1988, *435*, 73–82.

177. Bismuto, E., G. Irace, HPLC purification of sodium bis(2-ethyl-1-hexyl)sulfosuccinate from commercial preparations containing near-UV absorbing and fluorescent impurities, *J. Chromatogr. A*, 1994, *662*, 263–267.

178. Fudano, S., K. Konishi, Alkylsulfate and soap by salting-out chromatography, *J. Chromatogr.*, 1972, *71*, 93–100.

179. Desbène, P. L., F. I. Portet, G. J. Goussot, Surfactant mixtures by reversed-phase HPLC with refractometric detection, *J. Chromatogr. A*, 1996, *730*, 209–218.

180. Stan, H., T. Heberer, P. Billian, AE in drinking water (in German), *Vom Wasser*, 1998, *90*, 93–105.

181. Rothbächer, H., A. Korn, G. Mayer, Nonionic surfactants in cleaning agents for automobile production (in German), *Tenside, Surfactants, Deterg.*, 1993, *30*, 165–173.

182. Marsal, A., J. Cot, M. D. De Castellar, A. Manich, Recovery of natural fat and nonionic surfactant from sheepskin degreasing, *J. Am. Leather Chem. Assoc.*, 1998, *93*, 207–214.

183. Hobson, B. C., R. S. Hartley, Nonionic surface-active agents in oils and solvent extracts from wool, *Analyst*, 1960, *85*, 193–196.

184. Jones, P., G. Nickless, Isolation and examination of polyethoxylated material before and after passage through a sewage plant, *J. Chromatogr.*, 1978, *156*, 99–110.

185. Pollerberg, J., Chromatographic determination of free fatty alcohols in their ethoxylation products (in German), *Fette, Seifen, Anstrichm.*, 1966, *68*, 561–562.

186. Jones, P., G. Nickless, Amberlite XAD-4 resin as an extractant for polyethoxylated material, *J. Chromatogr.*, 1978, *156*, 87–97.

187. Wickbold, R., Chromatographic separation of ethylene oxide adducts into their homologs, and their quantitative determination (in German), *Fette, Seifen, Anstrichm.*, 1968, *70*, 688–692.

188. Leon-Gonzalez, M. E., M. J. Santos-Delgado, L. M. Polo-Diez, Triton-type nonionic surfactants by on-line clean-up and flow injection with spectrophotometric detection, *Analyst*, 1990, *115*, 609–612.

189. Nasonova, S. N., L. F. Kharchenko, N. M. Kondrikova, Determination of OP-7 and OP-10 in colored wastewaters (in Russian), *Tekst. Prom.-St.* (Moscow), 1987, *7*, 62.

190. Mackay, L. G., M. Y. Croft, D. S. Selby, R. J. Wells, NPE and OPE in effluent by LC with fluorescence detection, *J. AOAC Int.*, 1997, *80*, 401–407.

191. Cheetham, P. S. J., Removal of Triton X-100 from aqueous solution using Amberlite XAD-2, *Anal. Biochem.*, 1979, *92*, 447–452.

192. Anghel, D. F., M. Balcan, A. Voicu, M. Elian, Alkylphenol-based nonionic surfactants by HPLC, *J. Chromatogr. A*, 1994, *668*, 375–383.

193. Wahlberg, C., L. Renberg, U. Wideqvist, Nonylphenol and NPE as their pentafluorobenzoates in water, sewage sludge and biota, *Chemosphere*, 1990, *20*, 179–195.

194. Papariello, G. J., S. Chulkaratana, T. Higuchi, J. E. Martin, V. P. Kuceski, Chromatographic analysis of mono-, di-, and triglycerides and the mono- and diesters of ethylene glycol and polyethylene glycol, *J. Am. Oil Chem. Soc.*, 1960, *37*, 396–399.

195. Moldovan, Z., C. Maldonado, J. M. Bayona, Electron ionization and positive ion chemical ionization MS of *N*-(2-hydroxyethyl)alkylamides, *Rapid Commun. Mass Spectrom.*, 1997, *11*, 1077–1082.

196. AOAC International, *Official Methods of Analysis of AOAC International*, 16th ed., Gaithersburg, MD 20877, 1995.

197. Gupta, R. K., K. James, F. J. Smith, Sucrose mono- and diesters prepared from triglycerides containing C_{12}–C_{18} fatty acids, *J. Am. Oil Chem. Soc.*, 1983, *60*, 1908–1913.

198. Mima, H., N. Kitamori, Adsorption chromatography of sucrose palmitates, *J. Am. Oil Chem. Soc.*, 1962, *39*, 546.

199. Mima, H., N. Kitamori, Chromatographic analysis of sucrose esters of long chain fatty acids, *J. Am. Oil Chem. Soc.*, 1964, *41*, 198–200.

200. Kato, H., Y. Nagai, K. Yamamoto, Y. Sakabe, Polysorbates in foods by colorimetry with confirmation by IR spectrophotometry, TLC, and GC, *J. Assoc. Off. Anal. Chem.*, 1989, *72*, 27–29.

201. Sahasrabudhe, M. R., R. K. Chadha, Chromatographic analysis of sorbitan fatty acid esters, *J. Am. Oil Chem. Soc.*, 1969, *46*, 8–12.

202. Wang, Z., M. Fingas, Sorbitan ester surfactants. Capillary SFC, *J. High Resolut. Chromatogr.*, 1994, *17*, 85–90.

203. Eichhorn, P., T. P. Knepper, Metabolism of alkyl polyglucosides and their determination in waste water by LC-electrospray MS, *J. Chromatogr. A*, 1999, *854*, 221–232.

204. Hejna, J. J., D. Daly, Fatty alkanolamides, *J. Soc. Cosmetic Chem.*, 1970, *21*, 107–118.

205. Kirby, D. H., F. D. Barbuscio, W. Metzger, J. Hourihan, Anionic/nonionic shampoo, *Cosmetics and Perfumery*, 1975, *90*(8), 19–23.

206. Buschmann, N., F. Hülskötter, Titration of low ethoxylated nonionic surfactants, *Tenside, Surfactants, Deterg.*, 1997, *34*, 8–11.

207. Kudoh, M., S. Yamaguchi, Selective adsorption of nonionic surfactants using an anion-exchange resin in the cobaltithiocyanate form, *J. Chromatogr.*, 1983, *260*, 483–486.

208. Henke, H., PEG in ethylene oxide adducts using column chromatography (in German), *Tenside, Surfactants, Deterg.*, 1978, *15*, 193–195.

209. Winkle, W., Quantitative analysis in the V_0 zone. Chromatographic approach by coupling HPLC with GPC, *Chromatographia*, 1990, *29*, 530–536.

210. Buschmann, N., SPE of PEG and ethoxylates, *Comun. Jorn. Com. Esp. Deterg.*, 1994, *25*, 333–336.

211. Nakagawa, T., K. Shinoda, Physicochemical studies in aqueous solutions of nonionic surface active agents, in K. Shinoda, T. Nakagawa, B. Tamamushi, T. Isemura, eds., *Colloidal Surfactants*: *Physicochemical Properties*, Academic Press, New York, 1963.

212. Castillo, M., D. Barceló, Polar toxicants in industrial wastewaters using toxicity-based fractionation with LC/MS, *Anal. Chem.*, 1999, *71*, 3769–3776.

213. Simms, J. R., T. Keough, S. R. Ward, B. L. Moore, M. M. Bandurraga, Cationic surfactants in environmental matrices using fast atom bombardment MS, *Anal. Chem.*, 1988, *60*, 2613–2620.

214. Suortti, T., H. Sirvio, Fungistatic quaternary ammonium compounds in beverages and water samples by HPLC, *J. Chromatogr.*, 1990, *507*, 421–425.

215. Elrod, L., T. G. Golich, J. A. Morley, Benzalkonium chloride in eye care products by HPLC and solid-phase extraction or on-line column switching. *J. Chromatogr.*, 1992, *625*, 362–367.

216. Kümmerer, K., A. Eitel, U. Braun, P. Hubner, F. Daschner, G. Mascart, M. Milandri, F. Reinthaler, J. Verhoef, Benzalkonium chloride in the effluent from European hospitals by SPE and HPLC with post-column ion-pairing and fluorescence detection, *J. Chromatogr. A*, 1997, *774*, 281–286.

217. Fudano, S., K. Konishi, Mixtures of cationic surface-active agents by salting-out chromatography, *J. Chromatogr.*, 1973, *87*, 117–124.

218. Wilkes, A. J., C. Jacobs, G. Walraven, J. M. Talbot, Quaternized triethanolamine esters (esterquats) by HPLC, HRCGC, and NMR, *World Surfactants Congr., 4th*, 1996, *1*, 389–412.

219. Richter, H., C. Srey, K. Winter, W. Fuerst, Column chromatographic separation of natural phospholipids, *Pharmazie*, 1977, *32*, 164.
220. Amari, J. V., P. R. Brown, C. M. Grill, J. G. Turcotte, Isolation and purification of lecithin by preparative HPLC, *J. Chromatogr.*, 1990, *517*, 219–228.
221. Dornbos, D. L., R. E. Mullen, E. G. Hammond, Phospholipids of environmentally stressed soybean seeds, *J. Am. Oil Chem. Soc.*, 1989, *66*, 1371–1373.
222. Caboni, M. F., S. Menotta, G. Lercker, Separation and analysis of phospholipids in different foods with a light-scattering detector, *J. Am. Oil Chem. Soc.*, 1996, *73*, 1561–1566.
223. Merton, S. L., Analysis of soybean lecithins and beef phospholipids by HPLC with an evaporative light scattering detector, *J. Am. Oil Chem. Soc.*, 1992, *69*, 784–788.
224. Carelli, A. A., M. I. V. Brevedan, G. H. Crapiste, Phospholipids in sunflower oil, J. Am. Oil *Chem. Soc.*, 1997, *74*, 511–514.
225. Van der Meeren, P., J. Vanderdeelen, M. Huys, L. Baert, Optimization of column loadability for preparative HPLC separation of soybean phospholipids, *J. Am. Oil Chem. Soc.*, 1990, *67*, 815–820.
226. Hanras, C., J. L. Perrin, Preparative HPLC of phospholipids from soybean lecithins, *J. Am. Oil Chem. Soc.*, 1991, *68*, 804–808.
227. De Meulenaer, B., P. Van der Meeren, J. Vanderdeelen, L. Baert, Chromatographic gram-scale preparative fractionation of soybean phospholipids, *Chromatographia*, 1995, *41*, 527–531.
228. De Meulenaer, B., P. Van der Meeren, J. Vanderdeelen, L. Baert, Gram-scale chromatographic purification of soybean phospholipids, *J. Am. Oil Chem. Soc.*, 1995, *72*, 1073–1075.
229. Glass, R. L., Semipreparative HPLC separation of phosphatidylcholine molecular species from soybean leaves, *J. Liq. Chromatogr.*, 1991, *14*, 339–349.
230. Adlof, R., Fractionation of egg and soybean phosphatidylcholines by silver resin chromatography, *J. Chromatogr.*, 1991, *538*, 469–473.
231. Boyer, S. L., K. F. Guin, R. M. Kelley, M. L. Mausner, H. F. Robinson, T. M. Schmitt, C. R. Stahl, E. A. Setzkorn, Nonionic surfactants in laboratory biodegradation and environmental studies, *Environ. Sci. Tech.*, 1977, *11*, 1167–1171.

7
High-Performance Liquid Chromatography

HPLC is used to characterize surfactants according to their molecular composition, as well as to quantitatively determine individual surfactants in mixtures with other materials. This section covers only applications of high-pressure liquid chromatography. Column chromatography, or "flash chromatography," with a driving force of one or two atmospheres, is discussed in Chapter 6.

I. ANIONIC SURFACTANTS

A. Applicability

HPLC is the preferred technique to determine ionic surfactants in mixtures. Anionic surfactants can be determined directly by liquid chromatography, without derivatization and often without preliminary sample workup (1). Because of the polar nature of the surfactants, either ion chromatography or paired-ion chromatography (also called ion interaction chromatography) is generally used. Ion chromatography has the advantage that most common ionic surfactants can be determined using the conductivity detector, while conventional HPLC is most useful for compounds with phenyl rings, where direct UV detection is applicable.

HPLC can be effectively used to determine the degree of sulfonation of anionics such as petroleum sulfonates or paraffin sulfonates, where di- and polysulfonated structures are possible. Another use is determination of the total quantity of a particular surfactant. HPLC may be used to characterize an anionic according to its alkyl chain length, although the resolution of HPLC methods is inferior to that attained with capillary GC (Chapter 8). This is not a disadvantage for routine work, since it is easier to obtain reproducible area measurements from a few well-resolved peaks than it is from dozens of peaks representing isomers of the same compounds.

B. Separation Mechanism

1. Reversed-Phase Chromatography

Most often the analysis of anionics is performed by reversed-phase chromatography. In this mode, all anionics are separated according to increasing hydrophobicity, i.e., according to increasing alkyl chain length. Usually, a salt is added to the mobile phase. If the salt is formed from an organic cation, the mode of separation is called paired-ion chromatography [ion interaction chromatography does not require that the counterion be itself hydrophobic (2)]. If the separation is run under acidic conditions, it is probably based upon ion suppression chromatography.

(a) Paired-ion liquid chromatography. Paired-ion liquid chromatography is the name applied to reversed-phase HPLC when a salt of an organic ion is added to the mobile phase. An ion association occurs between the anionic surfactant and the cation from the salt. The ion pair is retained on the nonpolar packing material longer than the surfactant alone. This technique is generally applicable to anionic surfactants. pH is adjusted to a value where the surfactant exists in the anionic, as opposed to protonated, form. Reversed-phase retention is governed by the total lipophile character of the ion pair, so retention times are longer with higher molecular weight pairing agents. Thus, one group recommends tetraethylammonium ion for analysis of sulfates and sulfonates of chain length C_{12} or less, with tetramethylammonium ion preferred for longer chain lengths (3). Besides common anionics, the important biodegradation products of LAS may be determined by paired ion chromatography, namely the *p*-sulfophenylcarboxylate salts (4).

(b) Ion suppression chromatography. At low pH, anionics are less completely ionized and therefore less polar, so that retention times on a reversed-phase column are longer. If pH adjustment is the main method by which retention is optimized, the system is considered to be one of ion suppression. Since this mechanism is always at work, pH control is important in analysis of ionic surfactants. Ion suppression chromatography is the preferred technique for determination of the *p*-sulfophenylcarboxylate salts (4,5).

(c) Universal HPLC methods. A general system for separating the surfactants commonly found in shampoos consists of two reversed-phase columns in series, a mobile phase of methanol/water containing 0.25 M $NaClO_4$ at apparent pH 2.5, and a refractive index detector. Nonionics and amphoterics, as well as anionics, are separated (6,7). König and Strobel optimized this procedure for determination of the anionics commonly found in toothpaste (8). This approach can be made specific for individual classes of compounds by performing detection by postcolumn reaction. In the case of anionics, methylene blue is used, with the ion pair continuously extracted and its absorbance measured at 630 nm (9).

2. Normal Phase Chromatography

Rarely, the separation of anionics is made by a normal phase mechanism on a silica gel column with a mobile phase of chloroform/ethanol containing a counter ion (10,11). Normal phase conditions may be chosen such that some separation of isomers and oligomers is achieved, with elution in reverse order of alkyl chain length. For simple quantification, a higher ethanol content in the mobile phase will elute all isomers/oligomers as a single peak.

The main use of normal phase chromatograph is in separation of classes of anionics from each other, rather than characterization of a single surfactant. For example, LAS, AOS, paraffin sulfonate, and ester sulfonate were separated from each other after formation of the methyl esters (12). Other sulfates and sulfonates were separated from each other without derivatization by using an ion pairing agent consisting of the crown ether complex of a metal salt (10).

3. Ion Chromatography

Anionic surfactants may be determined by conventional ion chromatography, especially if only one anionic or a simple mixture is present. However, most separations require paired ion chromatography, as described above. One vendor uses the term "mobile-phase ion chromatography" to describe paired-ion HPLC with a conductivity detector and a system to chemically suppress the baseline conductivity of the mobile phase. A number of variations on this theme have been demonstrated, using either polymer backbone or silica backbone reversed-phase columns, varying levels of organic solvent in the mobile phase, and either isocratic or gradient elution conditions (3,13–15).

4. Size Exclusion Chromatography

Anionics may be analyzed on a high-resolution GPC column of the vinyl alcohol copolymer type designed for aqueous solvents. When a mobile phase is applied which is high in water, containing a smaller proportion of acetonitrile, the anionic is separated according to alkyl chain length by a reversed-phase mechanism. When a mobile phase high in methanol with a smaller proportion of water is applied, a single peak is obtained, as would be expected for separation purely by molecular weight. In either case, salt is added to the mobile phase (16).

C. Detection

1. Ultraviolet Absorbance

(a) Direct detection.   The most abundant anionic surfactants, the alkylarylsulfonates, can easily be determined by the UV absorbance detector. The absorbance maximum of the substituted benzene ring is near 225 nm, but there is enough absorbance at the 254-nm mercury doublet that simple fixed-wavelength detectors may be used. The aliphatic surfactants are more difficult to monitor by UV detection. Although they absorb light in the 190–210 nm region, the signal to noise ratio is poor because most solvents and impurities also absorb in this region.

(b) Indirect detection.   Anionics which do not absorb in the UV, like alkyl sulfates and alkanesulfonates, can be characterized by liquid chromatography using indirect photometric detection. An ion exchange column (Whatman Partisil 10 SAX) is used with acetonitrile/water containing a UV absorbing anion (17). Alternatively, a reversed-phase column may be used with water or methanol/water gradient containing an UV absorbing cation, such as N-octylpyridinium bromide. In either case, detection is by the change in UV absorbance of the mobile phase, corresponding to the effect of elution of the surfactant on the concentration of the UV absorbing ions in the eluent. Indirect photometric detection is compatible with gradient elution. An advantage is reduced need for calibration, since the detector response is related to the molar concentration of analytes, not their composition (18,19).

This technique may also be used with less highly absorbing ions, such as nitrate and iodide (20). Sensitivity to low concentrations is less, but this may be offset in particular applications by the larger linear response range and by the fact that the eluent is less likely to cause deposits in the HPLC pumping system. Indirect photometric detection is most sensitive when used with ion exchange columns, rather than with adsorption columns (21).

(c)   Detection of derivatives.   Detection may also be accomplished by derivatizing anionics with a reagent which adds a UV-absorbing end group. The relatively nonpolar compounds formed can then be analyzed by conventional reversed-phase HPLC, without use of ion pairing agents.

2.   Fluorescence

Use of fluorescence gives increased selectivity for detection of LAS in the presence of other compounds, making sample cleanup easier. Calculation is straightforward because all LAS homologs in the usual surfactant range have the same molar absorptivity and fluorescence quantum efficiency (22). Branched-chain alkylbenzenesulfonates (ABS) are reported to give only 22% of the fluorescence detector response of LAS (23). Fluorescence detection allows simultaneous determination of APE and LAS in environmental samples (24).

3.   Conductivity

All common anionic surfactants may be determined with conductivity detection. Both chemically suppressed and electronically suppressed conductivity methods are effective. Indirect conductivity detection has also been applied to determination of anionics, using a highly conductive eluent such as naphthalenedisulfonic acid (25).

4.   Refractive Index

Refractive index detection is generally applicable to determination of surfactants, or indeed any compound, but is relatively insensitive. It is therefore used more often for the characterization of pure surfactants, rather than for their determination at low concentration. A disadvantage of the RI detector when used for characterization is its incompatibility with gradient elution.

5.   Other Detectors

Aliphatic sulfates and sulfonates may be detected with a postcolumn extraction apparatus. As the anions elute from the column, they mix with a stream containing a fluorescent cation. The cation can only be extracted into an organic phase as an ion pair with the anionic surfactant. A fluorescence detector monitors the concentration of the cation extracted. This approach makes possible the analysis of alkylsulfates, alkanesulfonates, ester sulfonates, and ether sulfates by either normal phase or reversed-phase HPLC. Gradient analysis presents no difficulty (10,26,27). Such a system was also used with methylene blue in conjunction with a visible absorbance detector at 630 nm (9). Although methylene blue does not give the sensitivity of fluorescence detection, it is more selective for surfactants. Common fluorescent cations form extractible ion pairs with nonsurfactant anions, giving rise to interference and limiting the choices available for mobile phase modification.

Experimental surfactant-selective electrodes have been demonstrated as detectors for HPLC determination of alkylsulfates (28) and alkyl ether sulfates (29). The column effluents were diluted with water to prevent damage to the membrane from high methanol concentration. Such electrodes are subject to gradual deterioration of response and cannot be recommended for routine use.

The evaporative light-scattering detector has been shown applicable to the analysis of many surfactants. This detector has only limited tolerance for the salts usually added to the mobile phase during analysis of anionic surfactants and is not applicable in many situations. In back-flush mode, the ELS detector can be used to determine total inorganic salts, sulfonated or sulfated surfactant, and unsulfonated or unsulfated material (30).

Mass spectrometers are connected to reversed-phase HPLC systems via a number of interfaces. LC-MS is suitable for the qualitative and semi-quantitative analysis of most surfactants. More information on this topic is found in Chapter 15.

D.   Analysis of Individual Anionic Surfactants

1.   Alkylarylsulfonates

Linear alkylbenzenesulfonates are mixtures which, even when composed of compounds with a single alkyl chain length, contain isomers because the point of attachment of the alkyl chain to the benzene ring varies. Reversed-phase liquid chromatography, usually with sodium perchlorate added to an acetonitrile/water mobile phase, gives complete resolution of homologs of various alkyl chain lengths and partial resolution of the isomers of each LAS homolog. While HPLC does not give complete resolution of isomers by point of attachment of the alkyl chain, it may be optimized to allow quantification of the 2-phenyl isomers to indicate which process was used to prepare the parent alkylbenzenes (31,32). Often, quantitative analysis is simplified by using low resolution, isocratic, conditions, so that the isomer distribution is not observed, but only single peaks for each group of homologs of a single alkyl chain length (Fig. 1). As a rule, C_4 or C_8 columns will give simple chromatograms showing the distribution of LAS only according to alkyl chain length, while C_{18} columns give more complex chromatograms which also differentiate the compounds according to point of attachment of the alkyl chain (12,33,34).

LAS is most often determined by UV detection, with fluorescence used when required to obtain selectivity in environmental samples. In reversed-phase HPLC, retention of dialkylbenzenesulfonates is about the same as the corresponding LAS with a total alkyl carbon content of two carbons less. For example, di-n-octylbenzenesulfonate has a retention time equivalent to that of mono-n-tetradecylbenzenesulfonate (35).

Retention of LAS on a reversed-phase column is increased by increasing concentrations of inorganic salt in the eluent, with multivalent cations giving greater retention than monovalent cations, and lithium having less effect than sodium. Sodium remains the most practical cation, mainly for solubility considerations (36). Publications on HPLC analysis of LAS are summarized in Table 1 [Tables can be found at the end of this chapter].

2.   Alkyl Sulfates and Alkanesulfonates

Most methods for analysis of these compounds are reversed-phase separations on C_8 or C_{18} packings, although anion exchange columns are occasionally used for lower molecular weight products (see Table 2). Polymer columns such as the Hamilton PRP-1 or Dionex MPIC may also be used, with somewhat less resolution. An organic or inorganic salt

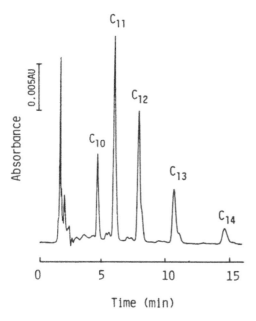

FIG. 1 Reversed-phase HPLC of LAS, with elution according to increasing alkyl chain length. (Reprinted with permission from Ref. 33. Copyright 1991 by Elsevier Science.)

is added to the mobile phase unless the detection system precludes this. Some form of pH control is also required. The salt functions to increase the affinity of the surfactant for the stationary phase. Most efficient separation of complex mixtures is performed with a gradient in which the concentration of the salt decreases while the organic content of the eluent increases (37), although many chromatographers hold the salt content constant. Detection is most readily performed by indirect UV absorbance, with refractive index, low-wavelength direct UV absorbance, and conductivity used less often. LiOH is the preferred salt if membrane-suppressed conductivity detection is used (37).

Reversed-phase HPLC separation is by alkyl chain length, with longer-chain compounds more retained than shorter chains, and alkyl sulfates more strongly retained than sulfonates (38). Purely anion exchange separation results in elution in reverse order of alkyl chain length (17). For good resolution over an extended range of chain lengths, a column packing may be used which combines both anion exchange and hydrophobic properties (21,25,39,40). To prevent micelle formation, the mobile phase must contain an organic solvent. For example, sodium dodecyl sulfate does not form micelles at methanol concentrations above about 30% (41).

Paraffin sulfonates can be analyzed by paired-ion chromatography with tetrabutylammonium sulfate, using refractive index detection (42). Separation is by chain length, but there is enough resolution of the primary sulfonates (more strongly retained) of each chain length from the corresponding secondary sulfonates to allow their quantification. If alkanesulfonates of a single alkyl chain length are analyzed, the HPLC system may be optimized to determine the individual isomers. For instance, three C_{18} columns in series, with a water/acetonitrile gradient and methylpyridinium chloride ion pairing/inverse UV detection system allows baseline separation of all but the two central isomers of dodecane, tetradecane, and octadecanesulfonate (2,19). Matsutani and Endo showed that separation of paraffin sulfonates according to alkyl chain length could be made without an ion pairing agent if the methyl sulfonate derivatives are formed (12).

The evaporative light scattering detector has been demonstrated for analysis of alkanesulfonates using an acetonitrile/aqueous ammonium acetate mobile phase (43). For trace determination of alkyl sulfates, aqueous solutions may be subjected to acid hydrolysis in a steam distillation apparatus which concentrates the resulting fatty alcohols for subsequent HPLC analysis (44,45).

3.   Alkylether Sulfates and Alkylphenolether Sulfates

As described later in this chapter, HPLC of nonionic ethoxylates is easily summarized by the generalization that reversed-phase systems are used to separate on the basis of alkyl chain length, and normal phase systems are chosen to separate by ethoxy chain length. The procedures published for HPLC analysis of sulfated ethoxylates are not as easy to summarize because of the effect of ion pairing reagents and the willingness of many investigators to perform separation according to both hydrophobe and hydrophile distribution in the same chromatogram. This dual separation can be tolerated with ether sulfates because the degree of ethoxylation is much smaller than with commercial nonionics. Reversed-phase systems are always used when a separation on the basis of hydrophobe is required. Either reversed-phase or normal phase chromatography is used for hydrophile separation. Examples of HPLC analysis of ether sulfates are shown in Table 3.

Like the corresponding nonionic compounds, alkyl ether sulfates may be degraded by hydriodic acid cleavage, with the resulting alkyl iodides determined by reversed-phase HPLC (46).

(a)   *Normal phase HPLC.*   Ether sulfates are usually contaminated with some level of unsulfated alkyl or alkylphenol ethoxylate. Like the original ethoxylates, sulfated ethoxylates can be separated according to increasing degree of ethoxylation by normal phase HPLC. In normal phase systems, the sulfated ethoxylate distribution is observed after the oligomers of unsulfated ethoxylates. While the nonionic portion elutes easily, the sulfates are strongly retained and can only be eluted by unusually polar solvents like 2-methoxyethanol (47). Elution is best effected and controlled by addition of a salt or ion pairing reagent to the solvent (48).

(b)   *Reversed-phase HPLC.*   Short-column reversed-phase HPLC with an evaporative light scattering detector allows rapid determination of alkylether sulfate, inorganic salt, and unsulfated material, giving a single peak for each (30). With other detectors, inorganic salt is usually invisible. On reversed-phase columns of the usual length, the unsulfated ethoxylate has a much longer retention time than does the sulfated surfactant (49,50). If a back-flush method is used with a C_{18} column, a single peak can be obtained for unsulfated matter, making quantification easier (51).

The reversed-phase retention order may be changed by addition of a long-chain ion pairing reagent, such as cetyltrimethylammonium bromide, which causes the ether sulfates to be retained longer than the corresponding nonionics. Retention of ether sulfates increases with increasing ion pairing reagent concentration up to about 0.02 M, then begins to decrease (52,53). The elution order is also reversed by replacing the reversed-phase column with a mixed mode anion exchange/reversed-phase column (50,54). Either of these approaches allows determination of the degree of ethoxylation of both unsulfated and sulfated APE.

4. α-Olefin Sulfonates

α-Olefin sulfonates are mixtures of alkenesulfonates and hydroxyalkanesulfonates, typically of a range of hydrocarbon chain lengths, with disulfonates and impurities also present. Under reversed-phase HPLC conditions the hydroxyalkanesulfonates elute first, in order of alkyl chain length and with resolution between isomers that differ in the position of the hydroxyl group. Alkenesulfonates elute later, again in order of alkene chain length, with partial resolution according to position of the double bond. Both compounds elute earlier than would the corresponding alkanesulfonate. Baseline resolution is only obtained in the case of products based on one or two hydrocarbons. In the typical commercial mixed product, the hydroxyalkanesulfonate peaks due to higher molecular weight compounds often elute after the alkenesulfonate peaks of lower molecular weight substances. Disulfonates, if present, will elute prior to the monosulfonate compounds (55,56). Derivatization of the sulfonate groups permits increased retention and improved quantification of the disulfonates (57). Hydrogenation prior to analysis will simplify the chromatogram by converting the alkenesulfonates to alkanesulfonates (12).

Matsutani and Endo demonstrated normal-phase HPLC to separate AOS from other anionics. All sulfonates were first derivatized to the methyl sulfonate compounds (12). Standards for analysis of α-olefin sulfonates are not commercially available and are generally prepared from mono-disperse α-olefins. Published methods are summarized in Table 4.

5. Petroleum Sulfonates

Petroleum sulfonate, after removal of neutral oil by extraction or adsorption (see Chapter 1), may be separated into mono-, di-, and trisulfonated components by HPLC on an anion exchange column, with elution in order of increasing degree of sulfonation (58–60). For relatively simple products, such as nominal diisobutyl- and diisopropylnaphthalene sulfonates, reversed-phase HPLC can be used for characterization according to number and chain length of alkyl groups (Table 5). Some resolution of isomers is also observed (61). More complex products simply give a broad envelope of peaks because of the large number of compounds of varying alkyl and aryl character. Calibration is perhaps the most important consideration for HPLC characterization of petroleum sulfonates, because sulfonates from different sources vary greatly in their UV absorbance.

6. Ether Carboxylates

Carboxylates of alcohol ethoxylates may be characterized by reversed-phase HPLC with an acetonitrile/water gradient containing formic acid and ammonium bicarbonate (Table 6). Doubling the usual column length permits resolution by both alkyl and ethoxy chain length in a single run (62). The carboxylates have shorter retention times than the corresponding nonionic ethoxylates under reversed-phase paired-ion conditions, allowing easy determination of noncarboxylated material (35,62,63). If only the low-resolution determination of unconverted nonionic is required, a simple C_1 or C_8 system can be used without ion pairing agent, in which case the nonionics elute first (64).

7. Other Anionic Surfactants

See Table 7 for published HPLC methods.

(a) Lignin sulfonates. The molecular weight of lignin sulfonates may be determined by SEC in aqueous eluents. As commercially available, lignin sulfonates are complex mixtures of high molecular weight byproducts of paper manufacture. In aqueous SEC sys-

tems, special precautions are necessary to avoid aggregation (micelle formation), ionic effects, and adsorption phenomena, all of which will result in inaccurate molecular weight values (65,66). Nonaqueous SEC of lignin sulfonates is therefore preferable. This can be accomplished by first making the THF-soluble ion pair of the sulfonate with a quaternary amine such as methyltrioctylammonium chloride (67).

(b) Ester sulfonates.   The alkyl chain length distribution of α-sulfo fatty acid esters can be determined by reversed-phase HPLC. The methyl ester components are resolved from the diacids (12,27).

(c) Phosphate esters.   These may be separated into mono- and diester components by reversed-phase HPLC. For example, NPE phosphate will produce peaks for, in order of elution, monoester, diester, and NPE, when analyzed on a C_{18} column with a gradient from 90:10 water/methanol to acetonitrile (68). Alkyl phosphates can be derivatized with 4-diazomethyl-*N,N*-dimethylbenzenesulfonamide to form compounds detectable by UV. These relatively nonpolar derivatives can then be analyzed by conventional gradient elution reversed-phase HPLC (57).

(d) Sulfosuccinate esters.   These are usually analyzed on the same HPLC systems used for alkyl sulfates and alkanesulfonates. Monoalkyl sulfosuccinates are readily separated by paired-ion HPLC according to the length of the alkyl chain. They are separated from each other and from other anionics, such as alkyl sulfates, based on the length of the alkyl chain (69). Dialkyl sulfosuccinate is separated from alkyl sulfates and alkanesulfonates by mixed mode ion exchange/reversed-phase chromatography, again with the separation made according to total alkyl character (25).

(e) N-Acylated amino acids.   Danielson and coworkers demonstrated the separation of a coco sarcosinate according to acyl chain length using either a C_8 hydrocarbon-based or fluorocarbon-based weak anion exchange column with indirect detection (39). Noguchi et al. showed that an acyl glutamate could be separated from other anionics by chromatography on an anion exchange resin, and that the acyl chain distribution could be determined by reversed-phase HPLC (70).

(f) Soap.   Soap can be readily analyzed by HPLC, even though most analysts prefer GC for this purpose. Major manufacturers of HPLC columns have developed systems for fatty acid analysis, usually with refractive index detection. Fatty acid soap can be also be analyzed with ordinary reversed-phase HPLC columns, either at high pH in the presence of an ion pairing agent or in ion suppression mode at low pH (71). A solution of 80:20 or 90:10 MeOH/H_2O containing sodium perchlorate gives good resolution of the fatty acids found in most commercial soap (8). If the acid functionality is derivatized, reversed-phase HPLC conditions without added acid or ion pairing agent may be used (72–74).

Soap may also be determined in mixtures by anion exchange chromatography, without separation by alkyl chain length. Carboxylic acids are not as strongly retained as are sulfates, sulfonates, and dicarboxylic acids (70).

II. NONIONIC SURFACTANTS

A. Modes of Separation

For purposes of characterizing nonionics, the three most frequent HPLC modes are separation on the bases of molecular weight, of hydrophobe chain length, and of hydrophile chain length. These modes are not mutually exclusive. Reviews have appeared summariz-ing HPLC of ethoxylated nonionic surfactants (75–76).

1. Size Exclusion Chromatography

SEC, also called gel permeation chromatography, is very useful for the characterization of the higher molecular weight nonionics. It is particularly useful for supporting production of EO/PO copolymers and PEG diesters. SEC systems are rarely capable of resolving compounds which differ by only one EO unit (44 Daltons). Rather, they are suitable for determining high molecular weight surfactants in formulations and, especially, for detect-ing contamination of neat products. Quantification is possible, with an accuracy of per-haps 5%. SEC has the limitation that, for example, lower ethoxylates of higher MW alcohols overlap with higher ethoxylates of lower MW alcohols. This problem can be par-tially surmounted in simple systems by mathematics (77).

Most manufacturer's literature recommends aqueous SEC for surfactant analysis. Salts are added to the aqueous mobile phase to make polyethoxylates well behaved. In your author's experience, aqueous systems are not suitable for analysis of highly ethoxy-lated materials. Nonaqueous systems with THF as mobile phase give much more reliable results. In the case of lower ethoxylates with 10 or 20 moles of EO, conventional HPLC is preferred to SEC.

A major consideration in the analysis of surfactants by SEC is the formation of mi-celles. If aqueous eluents are used, the surfactant contained in the initial injection into the instrument will be at high enough concentration to form a high molecular weight aggre-gate. These micelles travel along the column faster than the individual molecules and gradually dissociate as the total analyte concentration falls below the critical micelle con-centration. As a result, the shape of the peak and its retention time are a function of the sample size and are not directly related to the molecular weight of the surfactant molecule. Aqueous GPC can, in fact, be used to estimate the extent of micelle formation of a surfac-tant (78). To determine if a particular separation is being influenced by the presence of micelles, the chromatography is repeated at a much different sample concentration. If the separation is not concentration-dependent, micelle formation is probably not occurring.

A study of the Asahipak GS-310 column illustrates some of the peculiarities of SEC of surfactants (16,79). This packing material is a vinyl alcohol copolymer which has about the same volume when swelled with either aqueous or organic solvents, so that it is possi-ble to optimize the solvent for particular surfactants. At low ratios of acetonitrile/water, ethoxylated surfactants are not eluted from the column because the hydrophobic moiety is adsorbed to the column packing. At high acetonitrile/water ratios, apparent size exclusion behavior is observed, with the smaller oligomers eluting after the larger compounds, but without good resolution between oligomers. The best resolution between ethoxylated oli-gomers was obtained at intermediate acetonitrile/water ratios, about 0.3–0.6. The lower the degree of ethoxylation, the higher the percentage of acetonitrile required to elute the compound. Although the elution order is in reverse order of molecular weight, the separa-tion of oligomers appears to proceed according to a reversed-phase mechanism rather than by size exclusion. With increasing column temperature, retention times decreased for oli-gomers with less than 10 moles of EO, and increased for higher degrees of ethoxylation.

2. Separation on the Basis of the Hydrophobe

Conventional reversed-phase chromatography can effect the resolution of most nonionic surfactants on the basis of alkyl chain length. There is normally some discrimination according to extent of branching, with less than baseline resolution of the linear and branched isomers of each chain length. A typical system is a C_{18} column with water/methanol gradient. Usually, each peak represents a single alkyl chain length and contains the whole distribution of ethoxy chain lengths.

When performing reversed-phase separation of nonionics according to alkyl character, undesirable peak broadening may be observed due to partial separation according to ethoxy content, a separation which becomes more pronounced as the column ages. The aging phenomenon is thought to be due to accumulation of Na^+ ions on unreacted silanol sites and can be prevented by adding a low concentration of acid to the eluent (80).

The mobile phase composition must be chosen carefully in reversed-phase HPLC of ethoxylates. Depending on the ratio of, for example, THF/water or acetone/water, the retention time for AE or APE can be proportional to, indifferent to, or even indirectly proportional to ethoxy chain length. Derivatizing agents chosen to enhance detectability will also modify the reversed-phase interactions of the surfactants. Most often, a system is chosen in which ethoxy chain length has no effect on retention time. HPLC conditions under which ethoxy chain length does not affect the separation are sometimes called "critical conditions." For common ethoxylates and EO/PO copolymers and with common reversed-phase media and temperatures, these are in the range of 80:20 to 90:10 methanol/water (81,82) For acetonitrile/water, a ratio of 94.8:5.2 has been reported for naphthylisocyanate derivatives of AE (83), with 98:2 recommended for 3,5-dinitrobenzoyl chloride derivatives (84). Acetonitrile/water at 45:55 (81) or 46:54 (85) is reported for underivatized AE and APE and 42:58 for EO/PO copolymers (86,87). Values of 13:87 are reported for THF/water (81) and 20:80 for isopropanol/water (81). In the case of acetone/water mobile phase, 80:20 and 90:10 ratios are suitable, depending on the stationary phase (83).

HPLC under critical conditions is sometimes used for simple determination of a specific surfactant in a complex matrix, since the smaller number of individual peaks make quantification easier than for normal phase HPLC. Such a system may also be used for qualitative analysis of unknown mixtures of surfactants. For example, a reversed-phase C_{18} column with a mobile phase of 90:10 methanol/water and refractive index detection is suggested for detection of almost every commercial nonionic surfactant. EO/PO copolymers require a C_8 column and 100% methanol. Some sample cleanup is required prior to HPLC (89).

The most common application of critical conditions in reversed-phase HPLC is the determination of PEG in ethoxylates. This is easily carried out with most reversed-phase media, using 95:5 methanol/water and refractive index or evaporative light scattering detection (90,91). PEG elutes prior to the surfactants. PPG can be determined in propoxylates in the same manner (92).

A possible source of error when using reversed-phase systems is the adsorption of surfactants on the surfaces of the HPLC apparatus. For example, if the sample is injected as a purely aqueous solution, surfactant will build up in the metal injection valve depending on the volume of sample used to flush the loop. When the HPLC mobile phase is diverted through the valve, it will remove the adsorbed surfactant along with the measured sample volume. Thus, if varying volumes of sample are used to flush the injector, the peak size observed for a specific surfactant concentration will vary, even with the same loop size (93). This effect is minimized by dissolving the sample in the mobile phase.

3. Separation on the Basis of the Hydrophile

(a) Normal phase chromatography.   Normal phase chromatography on silica gel (or, more often, on NH_2- or CN-modified silica) is commonly used to separate nonionics on

the basis of ethoxy chain distribution. There is usually very little peak-splitting due to varying alkyl chain length. Some other modified silica packings can also give satisfactory separation, particularly the so-called diol packing (94). In this case, there is usually also partial separation according to the hydrophobe. Modified silica packings are preferred to bare silica because equilibration time is long with silica due to its moisture sensitivity.

Bare silica does find use with polar solvents of the type more often applied to reversed-phase separations. Hayes and coworkers demonstrated direct injection and analysis of aqueous AE solutions onto a silica column using an acetonitrile/water gradient where the water content continuously increases. They cleverly used a reversed-phase precolumn to provide an initial concentration of the AE from the sample matrix (95). Ibrahim and Wheals determined the oligomer distribution of APE by chromatography on bare silica, also using an acetonitrile/water gradient (96,97). In each case, performance varied widely between silica columns from different manufacturers.

(b) Reversed-phase LC. Good resolution of oligomers at high degrees of ethoxylation can only be obtained with reversed-phase methods. For lower molecular weight products, reversed-phase systems are generally applied to determination of ethoxy chain distribution only for special purposes, when normal phase eluents are undesirable, as in LC-MS. Because they wished to develop a separation based upon electrochemical detection of derivatives, Desbène and coworkers studied the reversed-phase separation of AE according to degree of ethoxylation (98). C_8 and C_{18} columns, with acetonitrile/water or THF/water, were adequate, although separation by alkyl chain length was of course also observed. For quantification, it was necessary that the alkyl chain length distribution be known from other analyses. Subsequent researchers report that a C_8 column is preferred to the C_{18} column for determination of ethoxy distribution because less discrimination according to alkyl chain length is observed with the C_8 column (99).

Elution of ethoxylates under reversed-phase conditions can be in order of increasing EO content, decreasing EO content, or without regard to EO content, depending on the choice of mobile phase (100). Reversed-phase separation according to ethoxy chain length results in elution in order of increasing ethoxy number at mobile phase solvent ratios below the critical conditions described in the previous subsection. At solvent ratios above the critical conditions, elution is in reverse order of ethoxy number.

Methanol/water mobile phases cannot give adequate resolution for determination of ethoxy distribution with C_8 or C_{18} stationary phases, although success has been demonstrated with C_1 phases (101–103).

(c) Other separation media. Rapid separations with poor resolution can be obtained with a silica-backbone cation exchange column with refractive index detection and normal phase eluents (104). This separation is due to conventional HPLC interactions with the stationary phase matrix, rather than to ion exchange behavior. On the other hand, if the cation exchanger is put in the K^+ form, the interaction of the polyethoxy chains with the potassium ions permits efficient separation of AE and APE oligomers from 6 up to about 30 EO units (105,106). The separation is optimized by adjusting the concentration of the potassium salt and altering the nonaqueous mobile phase, with methanol and methanol/acetone preferred. Temperature programming is very useful for the ion exchange separation (107,108).

Desmazières and coworkers extensively studied the retention of ethoxylates on ion exchange media as a function of the counter ion (108). Best resolution was found with a column of bare silica, which has weak cation exchange properties. Separation of more than 60 oligomers could be obtained with sodium ion in an acetonitrile/water mobile phase, provided that temperature programming was used.

Separation according to degree of ethoxylation can be performed under reversed-phase conditions on a graphitic carbon column. Separation according to alkyl chain length occurs simultaneously (100).

B. Detection

A major concern when applying HPLC to nonionic analysis is the means of detection.

1. Ultraviolet Absorbance

Many publications demonstrate the application of UV absorbance to analysis of APE, usually with a simple mercury lamp detector operating at 254 nm. Depending on the purity of solvents, UV detection is applicable to other nonionics using wavelengths as low as 195 nm, where many compounds show reasonably high absorbance. Use of this detector, rather than refractive index, allows application of gradient elution to resolution of oligomers. Since the mobile phase also absorbs radiation at low UV wavelengths, a flat baseline is not obtained unless steps are taken to match the absorbance of the solvents by adding small quantities of impurities to one or both.

For compounds other than APE, best results are obtained by making UV-absorbing derivatives. Derivatizing agents such as phenyl isocyanate, naphthylisocyanate, and substituted benzoyl chlorides permit use of the UV detector in gradient elution analysis with resolution of up to 50-mole adducts of AE and other compounds. The molar absorptivity of the derivatives is reasonably independent of molecular weight (98). Derivitization changes the polarity of the molecules, and hence their retention time. For example, the phenyl isocyanate derivatives of AE have a longer reversed-phase retention time than underivatized AE (109).

2. Refractive Index

The differential refractive index detector is applicable to most compounds. However, it is less sensitive than other techniques and is incompatible with gradient elution chromatography. Desbène et al. report a detection limit in water of about 100 ppm for a polydisperse AE, or about 0.05 ppm for individual oligomers, with the limit of quantification considerably higher (110). For high accuracy, the RI detector response should be corrected for differences in refractive index between ethoxylated oligomers (111). At first glance, these differences appear small compared to the uncertainties of peak integration. However, depending on the refractive index of the mobile phase, the correction in the value of differential refractive index response can be significant for degree of ethoxylation less than 12. There is little difference in the RI response for different alkyl chain lengths, at least in the case of AE (110).

3. Mass Spectrometry

Mass spectrometric detection allows analysis of most nonionic surfactants without derivatization. Thermospray, electrospray, or atmospheric pressure chemical ionization interfaces permit direct introduction of the effluent of the LC into the MS and make the MS a very selective detector for nonionics. Quasimolecular ions are produced for each discrete compound, so that the HPLC system is not required to separate both by degree of ethoxylation and by alkyl character. A relatively simple HPLC separation, coupled with MS analysis of each peak, is enough to characterize many mixtures of surfactants (112). Quantitative analysis is difficult in any case because of differences in ionization efficiency for similar products. If gradient elution is used, the response of the MS changes during the analysis. Ammonium acetate and methanol/water mobile phase are satisfactory for a reversed-phase system with thermospray LC-MS (101). The commercially available FRIT-FAB interface was demonstrated as a means to obtain mass spectra of AE during LC separation by alkyl chain length (113). However, the electrospray interface is the one most commonly used at the time of this writing.

4. Fluorescence Spectrometry

Fluorescence detection is more selective than absorption spectrophotometry in that both the excitation and emission wavelengths can be tuned for the particular molecule. Fluorescence is useful for trace analysis, where the problem is often not difficulty in detecting the analyte, but rather in interference from other compounds with similar properties. Fluorescence detection gives much improved sensitivity and selectivity for APE, making its environmental determination more reliable (24,114,115). When fluorescence detection is used for APE, the gradient used for the normal phase chromatography must be chosen very carefully, since fluorescence intensity is greatly affected by the matrix, and it is preferable that the same response factor can be used for quantification over the entire chromatogram. Fluorescence response differs slightly between oligomers and is different in different mobile phases (99).

Fluorescence detection can be applied to nonionics other than APE if derivatives are made. Almost all published work concerns AE, although there is no reason other hydroxyl-terminated surfactants could not be treated similarly. A number of reagents have been evaluated for making fluorescent derivatives. 1- and 2-anthranoyl chloride and 1-naphthylisocyanate give excellent results (99). Derivatizing agents profoundly affect the chromatographic behavior as well as detectability.

5. Evaporative Light Scattering

The evaporative light scattering (ELS) detector is generally suitable for analysis of nonionics without the need of forming derivatives. It is not usually applied to trace analysis. A significant advantage of ELS detection over RI or low wavelength UV detection is compatibility with gradient elution (116), especially complex gradient elution (117). The ELS detector may be used effectively with reversed-phase systems, provided that nonvolatile salts are absent from the mobile phase (91). When used for normal phase analysis of ethoxylates, its response to oligomers is linear on a molar basis, rather than a mass basis (30). This linearity of response is lost when gradient elution reversed-phase separations are made, since response varies both with molecular weight and with eluent composition. Thus, standards of individual oligomers are required for calibration (62). Because of the volatility of lower ethoxylates, the temperature of the ELS detector should be set as low as practical, preferably below 50°C (95).

6. Other Detection Methods

For the special case of chromatography of ethoxylates on a cation exchange column in the presence of K^+ ion, inverse conductivity detection may be used (106). Postcolumn reactors are sometimes used for special cases. For example, detection of fatty acid esters, with uniform response according to the molar concentration of the acid portion, can be performed using postcolumn hydrolysis of the ester and formation of the colored 2-nitrophenylhydrazine derivatives of the fatty acids (118)

Flame ionization detectors have been successfully demonstrated for HPLC analysis of surfactants, but so far only prototypes of the hardware have been available. Electrochemical detection is occasionally demonstrated in an academic setting for determination of nonionics. Ethoxylated surfactants can be determined in aqueous eluents by polarography, if the 3,5-dinitrobenzoyl chloride derivatives are first made. Polarographic detection of AE derivatives in a crude oil matrix is said to be more selective and more sensitive than UV detection (119). Alkylpolyglycosides containing free hydroxyl groups can be determined by amperometry at gold electrodes if the HPLC effluent is made basic prior to the detector (120). However, this is inferior to other methods of detection. Infrared absorbance detection is rarely used. It has been demonstrated for analysis of glycerides, using the absorbance band at 1748 cm^{-1}. Careful choice of solvents is required and gradient elution is all but impossible (121).

C. Selectivity

Most methods for analysis of nonionics are based on normal phase chromatography, a technique which separates the various peaks for the ethoxylated oligomers across the entire chromatogram. These methods are intrinsically not suitable for determination of the surfactant concentration in a complex matrix because peaks from many other compounds will appear in the retention time windows for the ethoxylate peaks. Thus, when normal phase HPLC is used in environmental analysis, it is always preceded by a number of cleanup steps aimed at providing a pure concentrate of the nonionic for subsequent chromatography. Reversed-phase chromatography is little more selective, since most methods give a peak for each homolog according to alkyl chain length, and other compounds can easily co-elute and interfere with the identification or quantification of the peaks.

At present, improved chromatographic separation can only be used to increase selectivity if the matrix is well characterized. The best hope for a selective analysis, without false positives, is by use of detectors specific for the surfactant of interest. Other compounds will still be present during chromatography, but they will not be seen.

D. Analysis of Specific Nonionics

1.   Alcohol Ethoxylates

Table 8 lists a number of HPLC systems for AE analysis. Quantitative HPLC characterization of AE is usually performed in two steps. The alkyl chain distribution is determined by reversed-phase HPLC, while the EO chain distribution is determined by normal phase chromatography on any of a number of column packings. Some investigators use reversed-phase conditions which separate according to both homolog and oligomer distribution to give a "fingerprint" (Fig. 2) (122).

(a)   Normal phase HPLC.   Complete resolution of up to 25-mole adducts of AE can be attained by solvent gradients with normal phase HPLC. This is usually performed with UV-absorbing derivatives, since AE has no UV chromophores of its own. Isocratic normal phase HPLC with low wavelength UV or with RI detection can sometimes be used to study the oligomer distribution of underivatized AE, providing that the chromatographer does not require baseline resolution over an extended oligomer range. For gradient elution,

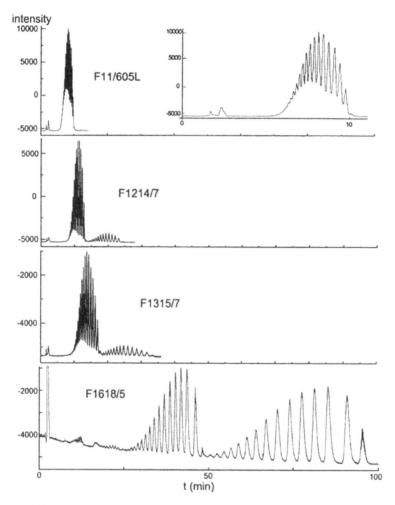

FIG. 2 Reversed-phase HPLC of AE, with separation according to both alkyl and ethoxy chain length. F11/605L–6.5 mole ethoxylate of C_{11} alcohol; F1214/7–7 mole ethoxylate of mixed C_{12}/C_{14} alcohol; F1315/7–7 mole ethoxylate of mixed C_{13}/C_{15} alcohol; F1618/5–5 mole ethoxylate of mixed C_{16}/C_{18} alcohol. ELS detection. (Reprinted with permission from Ref. 122. Copyright 1998 by the American Chemical Society.)

p-nitrophenyl-bonded silica has been proposed as a superior stationary phase, allowing good resolution over a wide range of ethoxy chain length (123). Normal phase HPLC of propoxylated alcohols gives only a single peak (117).

The evaporative light scattering detector is well suited to gradient elution normal phase chromatography and eliminates the need to form derivatives. However, the 1- and 2-mole adducts are generally too volatile for quantification by this method (117,124,125).

(b) Reversed-phase HPLC. As shown in the table, many investigators have separated AE according to alkyl chain length by reversed-phase HPLC. PEG is eluted first, then AE in order of increasing alkyl chain length, with gradient elution rarely necessary. Usually, the system is optimized so that there is no separation by ethoxy chain length, since such

mixed-mode separation causes a loss of sensitivity (83,84). Good separation of oligomers can be obtained for ethoxylates of linear alcohols, but some peak overlap is observed with branched alcohol ethoxylates (84). The branched-chain oligomers elute prior to the straight-chain oligomers of the same molecular weight, and resolution of a complex mixture is best accomplished with MS detection using selected ion monitoring (126).

The most satisfactory results for separation only by alkyl chain length are obtained with methanol/water systems. With acetonitrile/water, it is impossible to completely suppress simultaneous separation according to ethoxy chain length. Elution is in order of decreasing ethoxy chain length at acetonitrile contents below 90%, and in order of increasing ethoxylation above 90%. In acetonitrile/water, very low oligomers (E = 1 to 7) do not always behave in the same way as higher oligomers (127). A problem, especially in environmental analysis, is co-elution of NPE with one or more of the AE homologs (128). This co-elution can be avoided by optimization of the mobile phase (84).

Mixed-mode separations, where both hydrophile and hydrophobe affect the elution volume, are useful in some cases. This is performed by gradient elution HPLC of UV-absorbing derivatives, or by light scattering detection without derivatization. For the case of 4- to 6-mole ethoxylates of even-numbered alcohols, a single chromatogram can show perhaps four series of oligomers corresponding to the ethoxylates of each alcohol component. This resolution is lost if the base alcohol itself is a mixture of isomers (98,129,130).

Fluorescence detection may be applied to AE analysis by using a derivatizing agent, such as 1-anthroylnitrile (114), 1-naphthoyl chloride, or 1-naphthylisocyanate (83,99). Reversed-phase chromatography is then preferred even for determination of ethoxy chain length, since the excess derivatizing reagent interferes with the more common normal phase approach.

An easy approach to determination of AE in the presence of anionic surfactants gives a single peak for AE by a back-flush method. A reversed-phase system is used which causes ionic materials to be eluted first. The direction of flow of the column is then reversed, with the refractive index detector still monitoring the effluent as the AE elutes as a single peak. If other nonionic materials are present, a very short reversed-phase column is added to the system, so that two peaks are registered for nonionic material, one for AE and the other for nonethoxylated compounds (51).

In a determination analogous to the GC method, AE may be reacted with hydriodic acid to yield the alkyl iodide analogs of the starting alcohols. These can be analyzed by HPLC to give the alkyl chain length distribution, with a coefficient of variation of about 2% (46).

If MS detection is used, reversed-phase HPLC may be performed without derivatization to partially separate the AE according to alkyl chain length, with the mass spectrometer determining individual compounds, giving the exact concentration of each individual homolog/oligomer in a single run (126). This is not yet a method suitable for routine analysis, but gives an indication of what the future will bring.

Some groups used reversed-phase HPLC to resolve AE by degree of ethoxylation. The chromatograms are most easily interpreted if these are ethoxylates of a single alcohol (131). Reversed-phase HPLC of propoxylates of a single alcohol or of a single alcohol ethoxylate will resolve oligomers (117). Unfortunately, reversed-phase HPLC of mixed EO/PO adducts of alcohols gives a complex chromatogram with overlapping peaks which can usually not be assigned to specific oligomers. However, the pattern is reproducible and may be used for quantitative analysis (129).

(c) Two-dimensional chromatography. Okada demonstrated complete characterization of an AE as the dinitrobenzoate ester by an initial reversed-phase separation by alkyl chain length, after which the fractions corresponding to individual chain lengths were trapped on short styrene/divinylbenzene copolymer columns (water was mixed with the mobile phase prior to the trapping column to prevent elution of the AE). The trapped fractions were then eluted into a cation exchange column in the K^+ form to give the corresponding ethoxy distributions (132).

Murphy et al. demonstrated two forms of online two-dimensional HPLC for analysis of AE. In each case, the more complex analysis is performed in the first column, with the effluent shunted batchwise to a short second column for a very rapid separation, of the order of one minute. In one case, they used a reversed-phase column first in line for separation by alkyl chain length first, periodically injecting a portion of the effluent onto a SEC column for separation by molecular weight. More information was obtained from a second configuration, where a normal phase column was used first for a separation according to degree of ethoxylation, followed by a very rapid reversed-phase analysis of fractions to determine alkyl chain distribution (133,134).

Two-dimensional HPLC is rarely required in an industrial setting. The information can usually be obtained more easily by LC-MS.

2. Alkylphenol Ethoxylates

APE compounds are very easily analyzed by HPLC because of their UV detectability and their monodisperse hydrophobe component. Each month brings additional publications of HPLC determinations of APE. Alkylphenol ethoxylates can be separated according to the length of the alkyl chain by reversed-phase HPLC, or according to the length of the ethoxy chain, usually by normal phase chromatography. Ultraviolet detection is generally used; APE shows typical aryl absorbance at about 277 nm, with a stronger but less specific band at about 225 nm. The molar UV absorbance of APE oligomers at 277 nm is, for practical purposes, a constant independent of ethoxy chain length (111,115,135,136). This is also true of the 225 nm absorption band (115). Fluorescence detection gives improved detection for APE over UV absorbance, in terms of both sensitivity and selectivity (114,115), although UV detection is applied to trace analysis as well. Fluorescence excitation is effective at either the 225 or 277 nm absorbance bands. The fluorescence emission band is quite broad, with most chromatographers taking their measurements near the 302 nm maximum. Systems used for HPLC analysis of APE are summarized in Table 9.

(a) Normal phase HPLC. Normal phase chromatography is applied only to separation of APE according to the length of the ethoxy chain. Silica columns are generally used for resolution of compounds of shorter ethoxy chain lengths, 10 units or less. NH_2- or CN-bonded phase columns are applied for higher degrees of polymerization because highly ethoxylated materials are strongly retained by bare silica. *p*-Nitrophenyl-bonded silica has been proposed as a superior stationary phase, allowing good resolution over a wide range of ethoxy chain length (123). Isocratic separation is successful over a limited oligomer range, while a binary or ternary gradient allows resolution over an extended range (137–139). Elevated temperature, which reduces the viscosity of the mobile phase, gives increased efficiency for resolution of higher oligomers (123).

Of course, UV or fluorescence detection is not suitable for determining PEG impurity in APE. Differential RI detection is typical. In Zeman's general method for the analysis of ethoxylates on a LiChrosorb DIOL column, the system is optimized for separate determination of free PEG, the ethoxylated surfactant oligomers, or unreacted alkylphenol by adjusting the composition of isocratic hexane/isopropanol/water/acetic acid mobile phase (111,140). A similar approach is suitable for determination of polypropylene glycol in nonylphenol propoxylates (92).

Jandera and coworkers find that peak spacing is irregular and unpredictable in propanol-rich mobile phases on CN- or DIOL-bonded phases, presumably because partition as well as absorption mechanisms play a part in the separation. Because silica- and NH_2-bonded phases are more polar than CN or DIOL phases, such mixed retention mechanisms are usually not observed with these media. A generally useful system for determination of higher APE oligomers consists of a n-heptane/2-propanol gradient with an NH_2-bonded column (137). Quantification of APE oligomers is surprisingly accurate by isocratic HPLC, given that less than baseline resolution is achieved (111).

Technical p-nonylphenol contains about 10% of the ortho isomer, and the same is true of commercial NPE. Normal phase HPLC separates these isomers, so it is thought that normal phase HPLC of NPE is complicated by co-elution of p-NPE_x and o-NPE_{x+1} (141). This effect is not noticed in most analyses and does not apply to OPE.

HPLC is generally not reliable for determination of unethoxylated alkylphenol in APE. The peak corresponding to the retention time of alkylphenol is often much too large, showing the contribution of other unethoxylated materials such as oxidation inhibitors. GC or GC-MS are the preferred techniques for determining alkylphenol.

(b)   *Reversed-phase HPLC.*   Reversed-phase HPLC is used for four purposes:

1. *Determination of alkyl chain distribution*: This is not common, since most commercial APE compounds are monodisperse, containing either tripropylene or diisobutylene structures. It may be of interest for environmental analysis, where the size and form of the sample prevent spectroscopic determination of alkyl chain length. While the diisobutylene chain consists of a single isomeric configuration, the tripropylene group exists in many forms. HPLC cannot resolve NPE according to the alkyl chain isomer composition, instead giving only a single broad peak. Dinonylphenol ethoxylates may contain high amounts of NPE. This determination is readily made by reversed-phase HPLC (142). Similarly, butylphenolethoxylate impurity in OPE can sometimes be detected (143). The presence of a small quantity of *ortho*-isomer in the predominantly *para*-APE may be evident by a small additional peak in the reversed-phase chromatogram (144).

2. *Simplified quantification of total APE*: Analytical results may be more precise if the chromatogram can be simplified by suppressing oligomer separation. For example, nonylphenol 9-mole ethoxylate can be determined at sub-ppm levels in biological fluids using fluorescence detection and nonaqueous reversed-phase HPLC to give a single peak (145).

    For analysis of biological fluids, a simple isocratic procedure may sometimes be used for quantification if a special reversed-phase column is used having a surface network which excludes macromolecules. If the only interfering compounds are these macromolecules, they may be quickly flushed from the column prior to elution of APE (146).

    For increased specificity when using fluorescence detection in environmental matrices, APE may be acetylated. Many interfering compounds lose their fluorescence after acetylation. The difference in retention times after acetylation may be used to confirm peak assignment (147).

3. *Determination of ethoxy chain distribution*: It is sometimes advantageous to have an alternative to normal phase HPLC for characterization of the ethoxy chain, especially for high degrees of ethoxylation. While retention time is affected by both alkyl chain length and by degree of ethoxylation, for practical purposes most products are monodisperse with respect to the alkyl chain and this is not a concern (Fig. 3). Since higher ethoxylates are more polar than lower ethoxylates, APE elutes in order of decreasing ethoxy number under most reversed-phase conditions (148). The theoretical basis of reversed-phase resolution of APE oligomers has been examined in some detail (52).

   A C_1 column with methanol/water mobile phase permits elution of APE in order of increasing ethoxy number. In the case of NPE, there is some peak splitting due to the isomer composition of the nonyl chain, something not observed with OPE (103,104).

4. *Determination of PEG or PPG*: PEG can be determined by the same methodology used for other ethoxylates. Polypropylene glycol can be determined in propoxylated alkylphenol under similar conditions (92).

3. Ethoxylated Acids (Polyethylene Glycol Esters)

Fatty acid ethoxylates usually contain diesters of polyethylene glycols, as well as monoesters. Although these mixtures are easily separated by size exclusion chromatography, most publications describe reversed-phase HPLC methodology. Reversed phase HPLC easily yields the relative amounts of PEG, monoester, and diester, with elution in that order (90,149). Gradient elution chromatography using an evaporative light scattering detector applied to a mixed C_8/C_{10} acid ethoxylate gives peaks, in order of elution, of PEG,

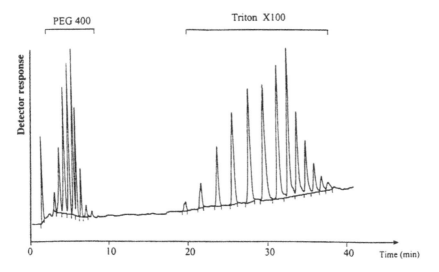

FIG. 3   Gradient elution HPLC analysis of a mixture of 400 MW polyethylene glycol and an octylphenol 10 mole ethoxylate. The column is of porous graphitic carbon and detection is by ELS. (Reprinted with permission from Ref. 100. Copyright 1998 by Elsevier Science.)

PEG monoester of C_8 acid, PEG monoester of C_{10} acid, PEG diester of C_8 acid, PEG diester of C_8/C_{10} acids, and PEG diester of C_{10} acid [Fig. 4] (150).

Zeman's normal phase HPLC method for the analysis of ethoxylates is applicable to ethoxylated acids (140,151). For routine analysis, conditions are used which produce single peaks for monoesters and diesters and a series of peaks showing the PEG distribution. With variation of the mobile phase, the conditions may be optimized to show the oligomer distribution of the monoester or detect the presence of unethoxylated material. These analyses are necessarily isocratic, since refractive index detection is used. If evaporative light scattering detection is available, gradient elution conditions may be chosen which give much more information in a single analysis (117).

Ultraviolet absorbance detection at 210 nm is sometimes used, usually for oleic acid adducts. Absorbance at this wavelength varies greatly for different compounds and is correlated to unsaturation. Thus, there may be twice as much signal for an oleic acid diester as the monoester, and very low response for PEG. HPLC systems for analysis of acid ethoxylates are listed in Table 10.

4. Amides and Ethoxylated Amides

Reversed-phase chromatography is suitable for resolving alkanolamides according to the length of the alkyl chain. For example, homologous series of $C_{10}–C_{18}$ fatty acid monoethanolamides and diethanolamides can be separated on a column of Hitachi Gel 3011, a styrene/divinylbenzene copolymer. A mobile phase of 97:3 methanol/water gives good resolution of series of either the mono- or diethanolamides, with detection by UV absorbance at 215 nm (152). In a similar study, TSK-Gel LS410, with methanol/water/sodium

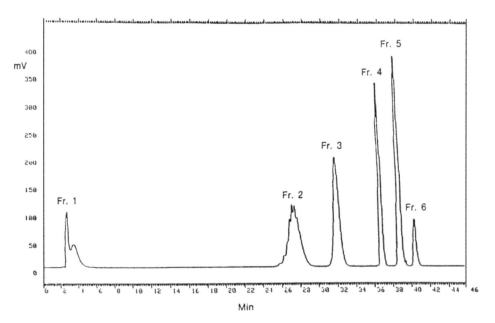

FIG. 4 Reversed-phase HPLC analysis of a mixed C_8/C_{10} acid ethoxylate, using ELS detection. Peak identies: 1, polyethylene glycol (PEG); 2, monoester of C_8 acid and PEG; 3, monoester of C_{10} acid and PEG; 4, C_8 diester; 5, mixed C_8/C_{10} diester; 6, C_{10} diester. (Reprinted with permission from Ref. 150. Copyright 1997 by Wiley-VCH.)

chloride and detection at 210 nm, was used to resolve stearoyl and lauroyl mono- and dial-kanolamides (153). A general method for the determination of surfactants in shampoos separates mono- and dialkanolamides from each other and from other surfactants on two tandem columns of TSK-LS 410 (6,7). A mobile phase of 25:75 water/methanol, 0.25 M in sodium perchlorate and adjusted to pH 2.5, is sufficient for qualitative and quantitative analysis, using a refractive index detector.

In a more demanding application, the synthesis of various fatty acid alkanolamides was followed by quantitative HPLC using a reversed-phase column and a THF/acetoni-trile/water, pH 2.6, mobile phase. Refractive index detection was used. The methyl ester starting materials were resolved, as were free fatty acids; mono-, di-, and triglycerides; dialkanolamides; and amine esters (154). As a rule, GC or TLC methods provide a more complete characterization of reaction products than does HPLC.

C_{12}/C_{14} monoethanolamide, methyl neodecanamide, and N-methyl C_{12}/C_{14} glucosa-mides were determined in cleaning formulations by chromatography on C_{18} with 80:20 or 75:25 methanol/water, using a nitrogen-specific chemiluminescence detector (156).

5. Ethoxylated Amines

Ethoxylated alkylamines can be characterized by HPLC in two steps (157). The alkyl chain distribution is determined on a reversed-phase system with an ion pairing agent. The ethoxy chain distribution is determined with a normal phase system using a NH_2-modified silica column. Isocratic conditions are suitable for separating the rather simple alkyl distri-bution, while a gradient program is required for determination of the ethoxy distribution. These investigators used a sophisticated detection system based on postcolumn extraction in the presence of an ion pairing agent with fluorometric monitoring of the extract. This detection system has the advantage of being specific for ethoxylated amines; ethoxylates of alkylphenols, alcohols, and esters do not interfere.

The ethoxy chain distribution of amine ethoxylates may also be determined by nor-mal phase gradient elution chromatography with an evaporative light scattering detector, with resolution up to about $\varepsilon = 60$ (155).

6. Esters of Polyhydroxy Compounds and Their Ethoxylates

(a) Esters. HPLC analysis of esters may be performed either to determine the concen-tration of the products, often in foodstuffs, or to determine the identity of a particular ester. Normal phase and reversed-phase HPLC, as well as size exclusion chromatography, are used for the analysis of esters. Ultraviolet detection at 220 nm is often used; unsaturated acyl groups disproportionately affect absorbance at shorter wavelengths.

1. *Reversed-phase HPLC*: Under reversed-phase conditions, compounds elute in the order of increasing alkyl character, in the order monoester, diester, triester. Base-line resolution of isomers formed by substitution of acyl groups at the various hy-droxyls of a sugar is usually impossible. It is difficult to separate all compounds in a single isocratic run, so that, for example, mono- and diesters can usually not be resolved in the same analysis that yields discrete peaks for the tetra- and pen-taesters. Residual fatty acids and their methyl esters can also be determined (159).

   All esters can sometimes be separated in a single run if gradient elution is used. The applicability of gradient elution is dependent upon the detector used. For example, a postcolumn reactor gives the freedom to use a complex gradient to resolve mono- through pentaesters of sorbitan and sugar [Fig. 5] (118).

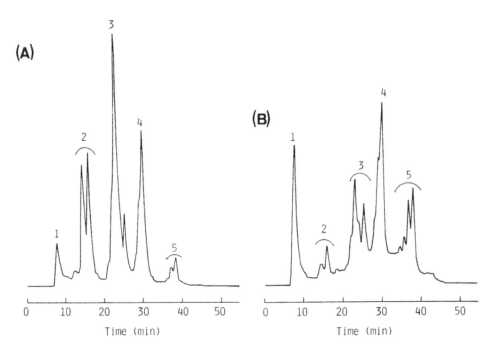

FIG. 5   Reversed-phase gradient elution HPLC analysis of A, sorbitan monooleate and B, sorbitan trioleate. Detection by post-column reaction. Peak assignment: 1, oleic acid; 2, monoester; 3, diester; 4, triester; 5, pentaester. (Reprinted with permission from Ref. 118. Copyright 1991 by Elsevier Science.)

Sorbitol esters can be analyzed using a C_{18} column with various ratios of isopropanol/water as mobile phase (160,161). Mono-, di-, and triesters with palmitic, stearic, oleic, isostearic, and sesquioleic acid were analyzed, including commercial products. Ultraviolet detection at 220 nm is appropriate. Sucrose esters may be analyzed in an analogous manner, with a C_{18} column, methanol/water or methanol/isopropanol mobile phase, and detection by refractive index or UV absorbance at 220 nm (162).

2. *Normal phase HPLC*: Normal phase HPLC of esters is usually performed on a DIOL column. Elution is in order of decreasing acyl substitution, i.e., in order of increasing number of free hydroxyl groups. There is little or no discrimination by chain length, but isomer separation can be achieved, and retention time is sensitive to the presence of unsaturation. Fatty acid methyl esters elute prior to the polyol esters.

Glycerol esters can be analyzed by normal phase chromatography using a LiChrosorb DIOL column and gradient elution with *n*-hexane with low percent levels of isopropanol (163). Polyglycerols are not resolved from glycerol. An isocratic mobile phase of 95:5 isooctane/isopropanol may also be used, with detection at 213 nm and with di-*n*-propyltartrate as internal standard (164). Classwise separation of glycerides and methyl esters of fatty acids may be performed with a silica column and stepwise elution with ethyl ether/*n*-hexane using refractive index detection (165).

3. *Size exclusion chromatography*: Esters of sucrose and other carbohydrates with multiple moles of fatty acid have been determined in biological samples by simple extraction and nonaqueous gel permeation chromatography (166,167). The method has only been demonstrated for high concentrations.

(b)   Ethoxylated esters.   Polysorbate 60 (sorbitan monostearate plus 20 moles EO) and polysorbate 80 (nominally, sorbitan monooleate and 20 moles EO) were separated by gel permeation chromatography, using as detector a postcolumn reactor containing ammonium cobaltithiocyanate immobilized on a solid support. Eluting ethoxylated material acquired the blue color of the polyether–cobaltithiocyanate complex, which was measured at 320 or 620 nm. This approach has not found wide use (168). GPC was demonstrated for detecting mono- and diesters in an ethoxylated glycerol trioleate, but sensitivity was poor. Better results are obtained from reversed-phase HPLC (169).

Reversed-phase HPLC of ethoxylated sorbitan esters has been demonstrated. A number of clusters of peaks are seen, the first cluster of which was identified as due to free PEG (170). HPLC methods applied to analysis of esters are summarized in Table 11.

7.   Alkyl Polyglycosides

Alkylpolyglycosides can be determined by evaporative light scattering and by electrochemical detectors, but these have no clear advantage over refractive index detection unless gradient elution is employed. Detection limits for individual glucosides are reported as approximately 300 ppm for evaporative light scattering, 100 ppm for UV at 190 nm, and 30 ppm for RI (171). Thermospray, electrospray, and atmospheric pressure ionization interfaces have been applied to LC/MS analysis of APG (172–174).

Reversed-phase HPLC gives a separation by alkyl chain length distribution (Table 12). Separation of model compounds occurs by degree of polymerization and isomer composition, but it is difficult to extract this information from the chromatograms of commercial mixtures (Fig. 6).

Reversed-phase HPLC can be used for the analysis of formulations. Because interference from other components is probable, analysis should be based, for example, on separate evaluation of the C_{12} and C_{14} monoglucoside peaks. If these do not lead to the same result for total APG content, then interference is occurring and a preliminary separation of nonionic surfactants must be performed by ethanol extraction and ion exchange treatment (173,175).

Porous graphitized carbon has some advantages over the usual C_{18} stationary phases for reversed-phase chromatography of APG, especially for resolving isomers. For the same alkyl chain length, diglycosides elute before monoglycosides on C_{18} phases, but the order is reversed on graphitized carbon. Good separation of model compounds was demonstrated on graphitized carbon using a gradient from methanol to 70:30 acetonitrile/water and evaporative light scattering detection (176). For trace determination of APG, aqueous solutions may be subjected to acid hydrolysis in a steam distillation apparatus which concentrates the resulting fatty alcohols for subsequent HPLC analysis (44,45).

8.   EO/PO Block Copolymers

Under the proper reversed-phase conditions, copolymers may be separated by the length of the polypropylene oxide chain, with no discrimination by length of the polyethylene oxide chains (86,176). These conditions (approximately 42:58 acetonitrile/water) are se-

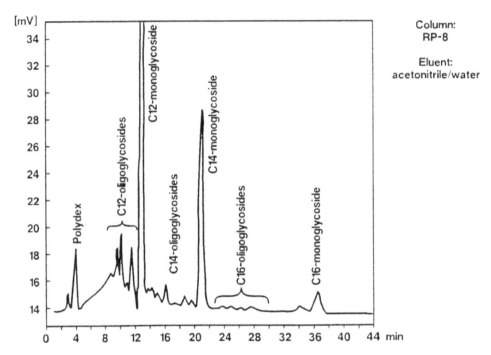

FIG. 6 Reversed-phase HPLC analysis of an alkylpolyglycoside commercial mixture. Peak assignment as indicated. Differential refractive index detection. (Reprinted with the author's permission from Ref. 281.)

lected by chromatographing a series of PEG compounds of approximately the chain length of the hydrophilic portion of the compounds of interest (82).

König and coworkers report that EO/PO copolymers can be qualitatively distinguished from other ethoxylates using a C_8 column and 100% methanol (89).

EO/PO copolymers are most often characterized using SEC (Table 13). Only non-aqueous SEC is reliable for these compounds. Depending on the molecular weight of the compounds and the resolution of the columns used, it is sometimes possible to resolve a minor component of smaller molecular weight from the main component (178). The minor component is thought to contain only a single terminal hydroxyl group and to be formed by a chain termination reaction occurring during polymerization.

9. Amine Oxides

Amine oxides and various amphoteric surfactants can be determined by reversed-phase HPLC using a C_{18} column with a mobile phase of 90:10 or 80:20 methanol/water. Differential refractive index detection is suitable (179)

III. CATIONIC SURFACTANTS

Most modern methods for the analysis of cationics are based upon liquid chromatography. Typically, these give not only the total quantity of surfactant, but also the alkyl chain length distribution (Table 14).

A. Separation Mechanism

1. Reversed-Phase HPLC

HPLC separations of cationics are most often based upon reversed-phase chromatography with a mobile phase containing a salt. Separations are optimized by varying the ionic strength, adjusting pH, or by changing the counter ion, as well as by modifying the solvent ratio. Elution times are long with C_8 or C_{18} packings (180). Most often, cyano-modified columns are used in reversed-phase mode, although other packings have been used, such as polystyrene beads or phenyl-bonded phase. Polymer columns do not provide the high resolution afforded by silica-backbone bonded-phase columns, but the requirements for column conditioning (necessary for reproducibility) are less stringent (181).

For UV detection, a counter ion of low UV absorbance such as perchlorate is chosen. Resolution can be optimized by changing the ratio of two counter ions of different lipophile character, such as perchlorate and methanesulfonate. Resolution is also controlled by changing the pH, usually by addition of phosphoric acid to take advantage of the buffering capability of phosphate ion.

In spite of the good water solubility of many quats, essentially all reversed-phase methods use mobile phases containing a substantial percentage of organic solvent. This is because

1. It is important to inhibit micelle formation. Typical cationic surfactants do not form micelles at methanol concentrations above 20–25% (41).
2. Adsorption losses to apparatus, including glassware, are high unless the solvent is at least partly organic (182).
3. Quats with alkyl chain lengths in the surfactant range are too strongly retained on the column if the eluent contains a high percentage of water.

Reversed-phase methods are summarized in the table. Most use THF/water or acetonitrile/water at low pH on a cyano-bonded phase column. Elution of cationics is in order of increasing total alkyl character. In the case of cationics which have a polyethoxy substituent, it is difficult to choose reversed-phase conditions which do not simultaneously separate the compound according to decreasing ethoxy chain length (183).

Dialkylimidazolinium salts are readily separated according to the total alkyl chain length by reversed-phase HPLC (Fig. 7). Good agreement is reported for alkyl chain length determination by LC and GC, even if some of the alkyl components are unsaturated and thus respond more strongly to low UV detection (184).

2. Normal Phase HPLC

The C_{14}–C_{18} dialkyldimethylammonium salts used in fabric softeners for home laundry have poor water solubility. Normal phase chromatography is most conveniently used for the separation of these compounds, especially in environmental matrices. Most often, detection is by conductivity and the cations are isolated and injected as ion pairs.

Elution of compounds in a normal phase system is in reverse order of alkyl character (Fig. 8). Thus, trialkylmethylammonium compounds elute first, followed by dialkyldimethylammonium compounds, quaternary imidazoline derivatives, alkyldimethylbenzylammonium compounds, and finally monoalkyltrimethylammonium compounds (185,186). Under the proper conditions, partial resolution by chain length can be obtained of the dialkyl dimethyl quaternaries (187) or of the monoalkyltrimethyl compounds (186,188), with the higher carbon number compounds eluting first.

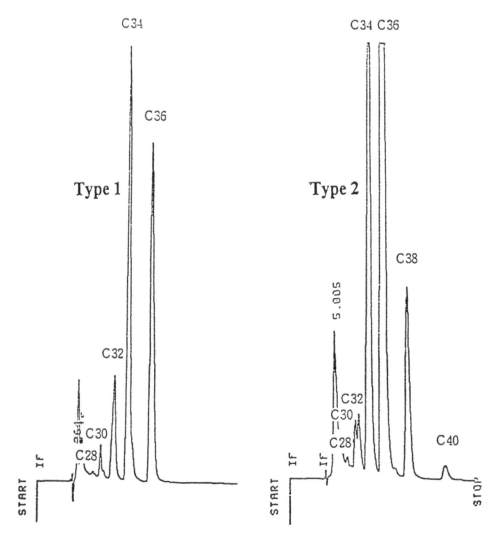

FIG. 7   Reversed-phase HPLC analysis of a quaternary dialkylimidazolinium salt, using UV detection at 230 nm. For the Type 1 compound, the alkyl groups were $C_{14}/C_{16}/C_{18}$. For Type 2, they were $C_{14}/C_{16}/C_{18}/C_{20}$. Peak labels indicate total alkyl character of each dialkyl component. (Reprinted with permission from Ref. 184. Copyright 1998 by the Korean Society of Industrial and Engineering Chemistry.)

    Retention time is controlled by mixing weakly and strongly polar solvents for the mobile phase. Usually, a small percentage of methanol or acetonitrile is added to chloroform. A small amount of an ionic compound greatly improves peak shape, while also decreasing retention time. Acetic acid is usually chosen for this purpose because of its good solubility (189).

    Especially for trace analysis, it is recommended that adsorption losses to the HPLC apparatus be minimized by first making a few injections of high-concentration standards (190).

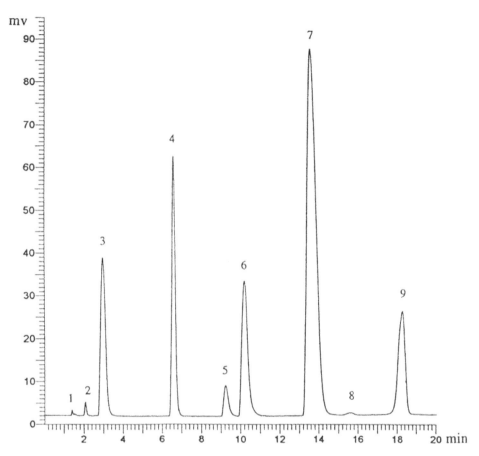

FIG. 8 Normal phase HPLC analysis of a crude ester quat based on quaternized tallow triethano-lamine esters. Peak identies: 1, fatty acid methyl ester; 2, free fatty acid; 3, unquaternized triesteram-ine; 4, tetraoctylammonium bromide internal standard; 5, diesteramine; 6, triesterquat; 7, diesterquat; 8, monoesteramine; 9, monoesterquat. ELS detection. (Reprinted with the author's per-mission from Ref. 288.)

3. Ion Exchange Chromatography

There are fewer applications of ion chromatography than paired-ion chromatography to the analysis of cationics because IC does not allow as much freedom in changing the sepa-ration conditions. However, ion chromatography has a great advantage over reversed-phase HPLC for formulation analysis in that only cationic materials are retained, so that interference is minimized (191).

Ion exchange chromatography analysis is typically based on use of a 25-cm What-man Partisil-10 SCX strong cation exchange column and a mobile phase of methanol or acetonitrile/water. The mobile phase contains a salt, such as ammonium formate, and the pH may be adjusted by addition of acetic acid. Elution is in order of decreasing alkyl char-acter, and retention time is increased by decreasing the concentration of the salt in the elu-ent. Most often, conductivity detection is not used, but rather direct or indirect UV absorbance or refractive index detection.

"Mobile-phase ion chromatography," a technique which uses traditional ion chromatography equipment, a nonpolar poly(styrene/divinylbenzene) separation column, a suppressor column, and conductivity detection, is suitable for analysis of cationic surfactants. Dilute perchloric acid in 70 volume percent acetonitrile is used as eluent (13). The separation mechanism is the same as in the reversed-phase HPLC methods discussed above.

Ion chromatography eluents are somewhat corrosive to stainless steel HPLC systems. With a steel system, it is important that the system be flushed at the end of the day with water to rinse accumulated salts and then stored in pure acetonitrile/water or methanol/water. Because of the nature of detection, the detector response is practically independent of the particular species injected. Sensitivity is of the order of 1 µg (192).

B. Detection

1. Ultraviolet Absorbance

Direct UV detection is suitable for analysis of cationics such as the benzylalkyldimethylammonium salts, providing that the ion pairing agent in the HPLC mobile phase has a comparatively low UV absorptivity. Perchloric or methanesulfonic acids are often used for this purpose. Greatest sensitivity is obtained using detection at 220 nm for benzyl compounds and 260 nm for pyridyl compounds (193). Direct UV detection at about 210 nm is often suitable for determination of compounds which do not show absorbance at higher wavelength, again providing that the proper low-absorbing mobile phase is selected (153).

Indirect photometric detection is used for determination of compounds which are not easily measured by direct UV absorbance. In a typical system, a methanol/water eluent contains *para*-toluenesulfonic acid as counter ion. Ultraviolet detection is made possible by variations in the counter ion concentration as the cationics elute from the column. A similar application in the realm of traditional ion exchange chromatography uses a strong cation exchange column and a mobile phase containing a solution of a strongly UV-absorbing quaternary, such as benzyltrimethylammonium chloride. As the cationics are eluted from the column, negative UV absorbance peaks are observed by the detector, corresponding to the quantity of the benzyltrimethylammonium ion that was required to displace the other surfactants from the column (192).

Benzalkonium chloride may be preconcentrated by liquid-liquid extraction to decrease the UV detection limit. A clever system allows for extraction into acetonitrile, which can be directly analyzed by HPLC. The separate acetonitrile phase is obtained by adding salt to an acetonitrile/water solution. The effect of pH and impurities has not yet been studied sufficiently to recommend this approach for general use (194).

2. Postcolumn Reaction

A postcolumn reactor allows the use of the sensitive and specific reaction of bromphenol blue or other anionic dye which forms an extractible ion pair with cationic surfactants. The dye and organic extractant are added continuously to the effluent of the HPLC column, the streams are mixed and extracted automatically, and the color intensity of the organic phase is monitored. This detection system has the advantages of not imposing constraints on the choice of the HPLC mobile phase and of being applicable to all cationics. Gradient elution techniques can be used without fear of changing the baseline of the detector. These detectors are mainly used with paired ion chromatography because here the freedom to alter the mobile phase can be used to greatest effect (9,183,195). Use of postcolumn reaction systems is not widespread, but they are suitable for use by those who expect to make many measurements, such as in a laboratory performing environmental monitoring of quaternaries. The very sensitive fluorescence detector may be used in conjunction with a postcolumn reactor. The ion pair of the cationic with a fluorescent anion, such as 9,10-dimethoxyanthracene-2-sulfonate, is extracted and continuously measured (189,195,196).

3. Other Detectors

Conductivity detection is often used with cationics. Refractive index detection may be used for analysis of compounds without chromophores, usually by paired-ion chromatography, but sensitivity is not adequate for determination of low ppm levels (197). Mass spectrometry with the newer ionization techniques is well established for HPLC detection, as is evaporative light scattering

IV. AMPHOTERIC SURFACTANTS

A. Detection

Direct HPLC analysis of typical amphoterics is difficult because of their low detectability. Of commonly available equipment, only the refractive index detector and the low wavelength (ca. 200 nm) UV detector are suitable. In the case of lecithin, low wavelength UV detection may be misleading because unsaturated substituents are the predominant absorbing species, and their concentration in lecithin components varies with climate, time of harvest, and other factors. Methods for the determination of a single lecithin component, such as phosphatidylcholine, are therefore usually based on RI detection (198). Because amphoterics are internal salts, they can generally not be made to couple with UV-absorbing ion pairing reagents to improve their detectability. Inverse UV absorbance is, however, a useful method of determining ionic impurities in amphoteric products (199). The detection problem can be circumvented in the specific case of carboxybetaines if the carboxyl group is labeled with a UV-absorbing compound such as 4-bromomethyl-7-methoxycoumarin (200). Similarly, amino phosphatides in lecithin may be determined if the UV-absorbing triphenylmethyl or benzoyl derivatives are made (201). Phosphatides with amino groups may also be detected by fluorescence of their dansyl derivatives (202).

Another approach uses a postcolumn reaction specific for the compounds of interest. For example, amines can be converted to the N-chloramines, and the chloramines treated with iodide to form triiodide, the absorbance of which can be monitored at 355 nm (203). Another postcolumn reaction system uses formation of the ion pair with Orange II, which will occur at low pH, where the amphoteric compound behaves as a cation. The colored complex is continuously extracted, and its absorbance monitored at 484 nm. Nonionic and anionic surfactants do not interfere (9).

A popular device for lecithin analysis is the evaporative light scattering detector. The ELS detector is compatible with gradient elution, and can detect all nonvolatile sample components, although individual calibration curves must be developed. The calibration curves are often nonlinear, although linearity can sometimes be obtained by optimization of the HPLC mobile phase (204). The ELS detector is not as sensitive as the UV detector, and is more useful for normal phase than for reversed-phase chromatography because of its incompatibility with common ion pairing reagents. Even with normal phase separations, it should be noted that calibration curves are affected by the acyl chain length of individual phosphatides (205).

B. Analysis of Specific Compounds

1. Lecithin

Components of lecithin are most often determined by normal phase HPLC (Table 15). Generally, this is sufficient to determine phosphatidyl ethanolamine, phosphatidyl choline, phosphatidyl inositol, and total triglycerides. A collaborative study using normal phase HPLC showed excellent within-laboratory precision for phosphatide quantification (CV = 1–6%), but much poorer agreement between laboratories (CV = 6–39%) (326). Analytical results using different HPLC methods can also differ substantially, especially if different detectors are used (206). Caution must be used, since normal phase LC retention is affected to some degree by the acyl substitution of the phosphatides, with compounds having shorter alkyl chains and unsaturated chains retained slightly longer (205). While separation of phosphatides on β-cyclodextrin–bonded silica resembles normal phase separation on silica, the order of elution is altered (207). The diol column is easier to use for this analysis than silica because conditioning time to equilibrate adsorbed water is less. Light scattering detection allows more facile use of gradient elution and thus resolution of more components (206). Lecithin solutions should be stabilized with 0.05% BHT and analyzed within a day of preparation to prevent changes in composition (206). HPLC separation of phosphatides is the subject of much activity because of the biochemical applications. Much of what has been learned in analysis of cell membrane components can be applied to analysis of lecithin.

The individual phosphatides are themselves mixtures of compounds with various acyl substituents. Separation and quantification of individual components according to acyl substitution is of great interest in biochemistry, and progress is being made rapidly. No HPLC method gives complete resolution in a single chromatogram. If a single class of phosphatide is isolated by preparative normal phase HPLC, it may be further separated according to acyl substitution by reversed-phase HPLC. The preliminary normal phase separation is essential because reversed-phase HPLC will not generally resolve different phosphatides having the same acyl substitution. By coupling reversed-phase and normal phase columns in a single apparatus, the separation may be optimized for particular acyl-substituted phosphatides. This is made most convenient if multi-port valves are used to allow rapid changes in configuration (208).

Complete resolution of individual phosphatides of animal origin may require conversion to other molecular species. For example, the characterization of phosphatidylcholine, after preliminary separation from other phosphatides, is carried out by enzymatic hydrolysis to form the diacylglycerol, which is derivatized to form the benzoate ester. Reversed-phase HPLC is then adequate to completely separate the various diacylbenzoyl-glycerides. Internal standards are recommended, since neither the hydrolysis nor the derivatization is necessarily quantitative (209).

While phosphatidylcholine and phosphatidylethanolamine are truly amphoteric, phosphatides such as phosphatidylinositol are anionic. These are most efficiently separated into their components by reversed-phase HPLC with an ion pairing agent (210,211). This is also true of the phospatide hydrolysis product, phosphatidic acid [$RCOOCH_2CH(OOCR')CH_2OPO_3H$]. Total free phospatidic acid may be determined by normal phase HPLC, but the acid can only be further characterized by reversed-phase chromatography after formation of esters such as the methyl ester (212).

Reversed-phase analysis of phosphatides is much improved if derivatives are made. These enhance the interaction of the compounds with the column stationary phase and give dramatic increases in resolution (212).

Work on LC-MS analysis of phosphatides gives hope of performing the characterization in a single chromatographic run. So far, the limitations of LC-MS ionization techniques have resulted in only reversed-phase HPLC being used for phosphatide analysis (213,214). Thus, the separation is by alkyl character, while the MS further defines each eluting component in terms of its particular species, just the opposite of the way the analysis is performed by other techniques. Information on the MS analysis of phosphatides is given in Chapter 15.

2. Other Amphoteric Surfactants

At low pH, most amphoterics will behave as cationic surfactants. As mentioned above with cationic analysis, ion exchange chromatography is advantageous for formulation analysis in that only cationic materials are retained, reducing interference (191).

Amphoterics of the usual carboxylic acid type are readily analyzed by reversed-phase HPLC, which separates them mainly according to alkyl chain length (Table 16). The pH of the mobile phase is controlled as well as the polarity to give proper resolution

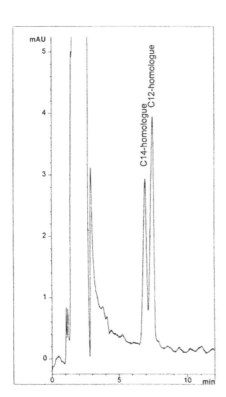 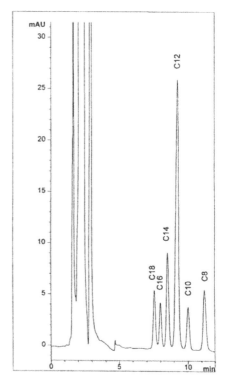

FIG. 9 HPLC analysis of, on the left, an alkylbetaine and, on the right, a cocoamidopropylbetaine, using a cation exchange column and UV detection at 210 nm. Peak labels indicate alkyl chain length. (Reprinted with permission from Ref. 216. Copyright 1995 by Elsevier Science.)

from other compounds. At neutral pH, amphoterics behave as nonpolar molecules with long retention times and can be analyzed under the same sort of HPLC conditions as are used for nonionic surfactants, with cationic and anionic surfactants eluting in the dead volume of the system (179). König and Strobel have developed a system capable of resolving several different types of amphoteric surfactants from each other and from other surfactant classes (179).

Mixed-mode HPLC, where the analytical column has both reversed-phase and cation exchange character, has also been demonstrated for analysis of amphoterics (215). The separation is similar to that obtained with reversed-phase columns.

For formulation analysis, where interference from other long-chain surfactants is a problem, Gmahl and coworkers suggest using a cation exchange column (Fig. 9). Under acidic conditions, the amphoterics are retained while anionic and nonionic materials elute with the dead volume (216).

TABLE 1 HPLC Analysis of Alkylaryl Monosulfonate Anionic Surfactants

Compounds studied	Column	Mobile phase	Detector	Ref.
C_{12} and C_{16} LAS: determination	Unspecified C_{18}, 2×25 mm; $40°C$	Gradient, THF/H_2O	Evaporative light scattering	30
C_{10}–C_{14} LAS: alkyl distribution	GL Science Inertsil ODS-2, 4.6×250 mm	$80:20$ $MeOH/H_2O$	UV, 225 nm	12
C_{10}–C_{14} LAS: determination in pesticide formulations	Merck LiChrosorb 10 RP-8, 4.6×250 mm	Gradient, $MeOH/H_2O$, 0.01 M tetramethylammonium bromide	UV, 225 nm	217
C_{10}–C_{13} LAS: determination in detergents; separation by alkyl chain length and point of attachment	Hitachi Gel 3053 C_{18}, 4.6×150 mm	CH_3CN/H_2O, $45:55$, 0.1 M $NaClO_4$	UV, 225 nm	218
C_{10}–C_{14} LAS: determination in environmental materials; separation by alkyl chain length	Waters μBondapak C18, 3.9×300 mm	Gradient, H_2O/CH_3CN, 0.15 M in $NaClO_4$	UV, 230 nm	219
C_{11}–C_{14} LAS: determination in environmental materials; separation by alkyl chain length	Shim-pack CLC-ODS, 6.0×150 mm	Gradient H_2O/CH_3CN, 0.02 M in $NaClO_4$	UV, 222 nm	220
C_{10}–C_{13} LAS: environmental analysis; separation by alkyl chain length	Shandon ODS or Whatman Partisil 5 ODS 3, 4.6×250 mm	Gradient; H_2O/CH_3CN, both 0.1 M in $NaClO_4$	UV, 225 nm	221
C_{12}–C_{22} LAS: determination in lubricants; separation by alkyl chain length	2 RP-8 columns in series, 4.6×200 mm	Gradient H_2O/CH_3CN to MeOH, all 0.1 M in $NaClO_4$	UV, 225 nm	222
C_9–C_{15} LAS in river water: separation by alkyl chain length	Alltech Spherisorb C18, 4×250 mm	$45:55$ THF/H_2O, 0.1 M in $NaClO_4$	Fluorescence: 225 nm excitation, 290 nm emission	22
C_{10}–C_{14} LAS in river water; separation by alkyl chain length	Wakosil 5C4 C4 silica, 4.6×150 mm	$50:50$ CH_3CN/H_2O, both 0.1 M in $NaClO_4$	UV, 220 nm	33

(Continued)

TABLE 1 (Continued)

Compounds studied	Column	Mobile phase	Detector	Ref.
C_8–C_{18} LAS and ring-substituted LAS; characterization: alkyl distribution, 2-phenyl isomer content, unsulfonated material	Unspecified C_8, 4.6 × 250 for alkyl distribution; C_{18}, 4.6 × 150 mm for 2-phenyl content and unsulfonated material	Alkyl distribution: gradient from 80:20 to 95:5 methanolic 0.5% tetrabutylammonium bisulfate/H_2O in 30 min; 2-phenyl content: gradient 85:15 to 95:5 methanolic 0.5% hexadecyltrimethylammonium bromide/H_2O in 30 min; unsulfonated material: 94:6 MeOH/H_2O, back-flush after 4.5 min	UV, 220 nm	32
C_{10}–C_{15} LAS: determination in various matrices; alkyl distribution and partial isomer separation	Merck LiChrosorb RP-8, 4 × 100 mm or Phase Separations Spherisorb, S3 ODS II, 4 × 250 mm	Gradient CH_3CN/H_2O, 0.02 M $NaClO_4$	UV absorption, 225 nm, or fluorescence: 225 nm excitation, 295 nm emission	24,31
C_9–C_{15} LAS: alkyl distribution and partial isomer separation; study of effect of inorganic salt on separation	DuPont Zorbax ODS, 4.6 × 150 or 4.6 × 250 mm	Isocratic separation by chain length: 3:2 CH_3CN/H_2O, 0.1 M in NaCl; Gradient separation with partial resolution of isomers: A = 60:40 H_2O/CH_3CN, 0.1 M in NaCl, B = 60:40 CH_3CN/H_2O; 0 to 85% B in 30 min	UV, 225 nm	36
C_{10}–C_{14} LAS: alkyl distribution and partial isomer separation	Dionex poly(styrene/divinylbenzene) anion exchanger IonPac AS11 4.6 × 250 mm	30:70 CH_3CN/H_2O containing 0.05 M $MgCl_2$ or 40:60 CH_3CN/H_2O containing 0.0015 M LiOH	UV, 220 or 225 nm	223

C_1–C_{20} LAS and dialkylbenzenesulfonates of total alkyl content C_{12}–C_{26}; separation by alkyl chain length; partial isomer resolution	Alltech RSIL, prep C_{18} precolumn, 4.6 × 32 mm and Dupont Zorbax C_{18} column, 4.6 × 250 mm	Ternary gradient, tetrabutylammonium bisulfate, 0.1 M in $H_2O/H_2O/CH_3CN$; pH 5	UV, 225 nm	35
C_{10}–C_{14} LAS; separation from α-sulfofatty acid methyl esters and α-olefinsulfonates after derivatization	GL Science Inertsil SIL, 4.6 × 250 mm	92:8 n-hexane/ethyl ether	RI	12
C_{12} LAS; determination in sea water in the presence of sulfonated nonylphenol 6-mole ethoxylate; partial homolog separation	Alltech mixed-mode RP4/Anion 100 A 5 μ, 4.6 × 150 mm	CH_3CN/0.05 M KH_2PO_4, pH 4.7: initial hold at 20:80 for 20 min, then gradient to 70:30 in 25 min, holding for 135 min	UV, 223 nm	54
C_{11}–C_{14} LAS; separation by alkyl chain length; demonstration of simultaneous determination with APE	Hichrom TMS C_1, 4.6 × 150 mm	65:35 H_2O/CH_3CN, 0.065 M in NH_4OAc	Fluorescence: 220 nm excitation, 290 nm emission, or API-MS	224
C_{10}–C_{14} LAS; single peak; separation from alkyl sulfates, alkanesulfonates, ether sulfates, and bis(alkyl)sulfosuccinates	Merck LiChrosorb SI 60, 3× 250 mm	90:10 $CHCl_3$/EtOH, 0.0001 M in 18-crown-6 and 0.00015 M in sodium pentanesulfonate	Postcolumn reaction, extraction, and fluorescence detection	10
Di- and trialkylarylsulfonates: C_{11} LAS with one or two methyl, ethyl, or 2-propyl groups on the ring; isomer distribution	Shandon Hypersil C18, 4.6 × 200 mm	CH_3CN/H_2O, 0.01 M tetrabutylammonium chloride	UV, 225 nm	225

TABLE 2 HPLC Analysis of Alkyl Sulfate and Alkanesulfonate Anionic Surfactants

Compounds studied	Column	Mobile phase	Detector	Ref.
C_{12}–C_{18} linear alkylsulfates; separation by chain length	C_{18} column, TSK gel 120T (Toyo Soda), 4.6 × 250 mm	90:10 MeOH/H_2O, 0.05 M NaCl	Ion-selective electrode	28
C_8–C_{14} linear alkylsulfates; separation by alkyl chain length	IBM/Jones Chromatography C_8 or C_{18}; 4.6 × 250 mm	MeOH/H_2O or CH_3CN/H_2O containing $NaNO_3$ or NaI and phosphate buffers	Indirect UV, 242–260 nm	20
C_{12}–C_{18} linear alkylsulfates; separation by alkyl chain length	Toyo Soda TSK-Gel LS-410 C_{18}; 4 × 250 mm	H_2O/MeOH, 0.4 M in NaCl	UV, 210 nm	153
C_{12}–C_{18} linear alkylsulfates; determination in shampoo; separation by alkyl chain length	Toyo Soda TSK Gel LS 410; 2 C_{18} columns in series	MeOH/H_2O, 0.25 M in $NaClO_4$; apparent pH 2.5	Refractive index	6,7
C_{12}–C_{18} linear alkylsulfates after hydrolysis, steam distillation, and formation of fluorescent derivatives; determination in toothpaste and river water; separation by alkyl chain length	Merck LiChrosorb RP8, 4.6 × 125 mm	CH_3CN	Fluorescence: for carbazole-9-carbonyl chloride derivatives, 228 nm excitation, 318 nm emission; for 9-fluorenylmethyl chloroformate derivatives, 260 nm excitation, 310 nm emission	44,45
Octylsulfate; determination in diagnostic reagents; separation from monoalkylsulfosuccinates	Macherey-Nagel Nucleosil C_8, 4 × 250 mm	0.01 M aqueous tetrabutylammonium bisulfate/MeOH. 23:77, pH 3	Refractive index	69
C_2–C_{18} linear alkylsulfates and alkanesulfonates; separation by alkyl chain length	Hamilton poly(styrene/divinylbenzene) PRP-1 4.1 × 50 mm or Zorbax ODS	CH_3CN/H_2O containing various salts	Indirect visible, 510 nm [Fe(II)-phenanthroline added] or conductivity	37,38
C_6–C_{12} linear alkylsulfates and alkanesulfonates; separation by alkyl chain length	Dionex poly(styrene/divinylbenzene) anion exchanger Ion-Pac AS11, 4.6 × 250 mm	40:60 CH_3CN/H_2O containing 0.001 M LiOH	Conductivity (chemically suppressed)	223

Analyte/separation	Column	Mobile phase	Detection	Ref.
C_8–C_{16} linear alkylsulfates and alkanesulfonates; separation by alkyl chain length	Dionex MPIC NS-1, 4 × 250 mm	CH_3CN/H_2O, 0.01 M in ammonia, also containing 0.01 M tetramethyl- or tetraethylammonium hydroxide	Conductivity (chemically suppressed)	3,226
C_1–C_{14} linear alkanesulfonates and C_4–C_{16} alkyl sulfates; separation by alkyl chain length	Waters Novapak C_{18}	Gradient: $A = H_2O$, $B = CH_3CN$, each 0.0005 M in tetrabutylammonium borate; 15% B to 60% B in 10 or 15 min	Conductivity (chemically suppressed)	15
C_{12}–C_{18} linear alkylsulfates; separation by alkyl chain length	Toyo Soda C_{18} TSK-LS410, 6 × 250 mm	85:15 MeOH/H_2O, 1.0 M in $NaClO_4$, pH adjusted to 2.5 with H_3PO_4	Postcolumn reaction with methylene blue	9
C_6–C_{18} linear alkylsulfates and alkanesulfonates; separation by alkyl chain length	Crosslinked amine/Kel-F 800 silica (not commercially available), 4.6 × 250 mm; Alltech C_8 HAE weak anion exchange column, 4.6 × 150 mm, or Alltech mixed-mode RP C_4, RP C_8, or RP Phenyl ion exchange columns, 4.6 × 250 mm (RP Phenyl ion exchange preferred)	Various ratios CH_3CN/H_2O, 0.0002 M in disodium naphthalenedisulfonate, pH 5.5, or 10 μM in trisodium naphthalene trisulfonate, pH 5.5	Direct or inverse conductivity, inverse UV, 285 nm, RI, and inverse fluorescence, 240/660 nm	21,25,39,40
C_{10}–C_{18} alkanesulfonates and C_{12} alkyl sulfates; elution in reverse order of alkane chain length (alkyl sulfate eluted first); separation from bis(alkyl)sulfosuccinates	Merck LiChrosorb SI 60, 3 × 250 mm	90:10 $CHCl_3$/EtOH, 0.0001 M in 18-crown-6 and 0.00015 M in sodium pentanesulfonate	Postcolumn reaction, extraction, and fluorescence detection	10
C_{12} alkanesulfonate; determination	Unspecified C_{18}, 2 × 25 mm; 40°C	Gradient, THF/H_2O	Evaporative light scattering	30

(Continued)

TABLE 2 (*Continued*)

Compounds studied	Column	Mobile phase	Detector	Ref.
C_{10}–C_{18} linear primary and secondary alkanesulfonates; alkyl chain distribution	Shandon reversed-phase, Hypersil SAS or Hypersil ODS, 4.6 × 150 mm	Acetone/H_2O/NaH_2PO_4	Postcolumn reaction, extraction, and fluorescence detection	26
C_2–C_8 linear alkanesulfonates; separation by alkyl chain length	Anion exchange column, Whatman Partisil 10 SAX, 4.6 × 250 mm	CH_3CN/H_2O containing potassium biphthalate, pH 5, sulfosalicylic acid, pH 2.4, or m-sulfobenzoic acid, pH 3.6	Indirect UV, 297 or 320 nm	17
C_{10}–C_{16} alkanesulfonates; alkyl distribution	Shanghai First Reagent Plant ODS, 4.6 × 150 mm	50:50 MeOH/0.001 M aqueous NaH_2PO_4	Conductivity	227
C_{10}–C_{18} alkanesulfonates; alkyl distribution	Merck LiChrosorb RP-8 or Shandon Hypersil ODS 1 or 2 columns, 4.6 × 200 mm 40°C	Gradient, MeOH/H_2O, 0.00025 M in N-methylpyridinium chloride	Indirect UV, 257 nm	18,19
C_{12}–C_{18} alkanesulfonates; separation by position of sulfonation	Shandon Hypersil ODS 3 columns, 4.6 × 200 mm, in series, 27°C	Gradient, H_2O/CH_3CN, 0.0003 M in N-methylpyridinium chloride	Indirect UV, 260 nm	2,19
C_{14} alkanesulfonate; separation by position of sulfonation	Shandon Hypersil ODS 3 columns, 4.6 × 200 mm, in series, 30°C	Gradient, 80:20 to 60:40 H_2O/CH_3CN, each 0.001 M in ammonium acetate	Evaporative light scattering	43

C_{11}–C_{18} alkanesulfonates: separation by chain length and primary sulfonate content	Merck LiChrosorb RP-8 4.6 × 250 mm	MeOH/H_2O, 0.01 M tetrabutylammonium sulfate	Refractive index	42
C_{10}–C_{12} alkanesulfonates: resolution by chain length and separation from LAS	Phase Separations Spherisorb S5-C1, 4 × 40 mm	Gradient: A = H_2O, B = 50:50 CH_3CN/H_2O, both eluents 0.005 M in HCl and 0.01 M in sodium citrate; 20% B to 100% B in 20 min	Postcolumn reaction: ion-pair extraction and fluorescence	228
C_{12}–C_{18} alkanesulfonates: separation from α-sulfofatty acid methyl esters and α-olefinsulfonates after derivatization	GL Science Inertsil SIL, 4.6 × 250 mm	92:8 n-Hexane/ethyl ether	RI	12
C_{10}–C_{12} Alkanesulfonates: resolution by alkyl chain length	Merck LiChrosorb RP-8, 4.6 × 125 mm	80:20 MeOH/H_2O	RI	12
Perfluoroalkanesulfonate: single peak; determination in acid etch bath	Dionex OmniPac PAX-500 mixed mode anion exchanger, 4 × 250 mm	55:45 0.011 M NaOH/CH_3CN	Chemically supressed conductivity	229

TABLE 3 HPLC Analysis of Ether Sulfate Surfactants

Reversed-phase systems

Compounds studied	Column	Mobile phase	Detector	Ref.
C_{12}–C_{18} alkyl 3-mole ethoxysulfates (monodisperse standards); elution in order of increasing alkyl chain length	Phase Separations Spherisorb S5-C1, 4 × 40 mm	Gradient: A = H_2O, B = 50:50 CH_3CN/H_2O, both 0.005 M in HCl and 0.01 M in sodium citrate; 20–100% B in 20 min	Postcolumn reaction: ion-pair extraction and fluorescence	228
C_{12}–C_{15} alkyl 0- to 8-mole ethoxysulfates; determination in environment; elution in order of increasing alkyl chain length	Baker C_8, 4.6 × 250 mm	Gradient: A = 20:80 CH_3CN/H_2O, B = 80:20 CH_3CN/H_2O, both 0.0003 M in NH_4OAc; 20–55% B in 30 min	MS	230
Sulfated C_{10}–C_{14} alkyl 0- to 9-mole ethoxylates; elution in order of increasing alkyl chain and ethoxy chain lengths; determination in shampoo	Alltech Surfactant C_8, 4.6 × 250 mm	45:55 $MeOH/H_2O$, 0.00023 M in NH_4OAc	Conductivity	231
Sulfated C_{12}–C_{15} alkylethoxylates; determination in brine; incomplete resolution by increasing alkyl chain length and increasing ethoxy content; peaks not identified	Mobile phase ion chromatography: Dionex MPIC-NS1, 4.6 × 250 mm	28.5:71.5 CH_3CN/H_2O, 0.01 M in NH_4OH	Refractive index	232
Sulfated lauryldiethoxylate; determination in shampoo; peaks not identified	Dionex MPIC-NS1, 4 × 200 mm	28:72 CH_3CN/H_2O, water 0.01 M in NH_4OH	Conductivity	13
Sulfated C_{12}–C_{14} alkyl and nonylphenol 2- to 5-mole ethoxylates; determination of unsulfated ethoxylates: single peaks for sulfated and unsulfated (elutes last) material	Gasukuro Unisil QC-18, 4.6 × 150 mm and Merck LiChrosorb RP-2, 4.6 × 50 mm	85:15 $MeOH/H_2O$, followed by back-flush to elute nonionics	Postcolumn reaction, extraction, and fluorescence detection	51

Sulfated octadecylphenol 7-mole ethoxylate; APE sulfate elutes first as a single peak, followed by nonionic material	Waters radially compressed C_{18}, 8 × 100 mm	80:20 2-PrOH/H_2O, 0.2% H_3PO_4	UV, 254 nm	49
Sulfated nonylphenol 4-, 5-, and 8-mole ethoxylates; separation from NPE; elution first of NPE then of NPE sulfate in order of decreasing ethoxylation	Lachema Silasorb SPH C_{18}, 3.8 × 300 mm	55:45 H_2O/2-PrOH, 0.04 M in hexadecyltrimethylammonium bromide	UV, 230 or 254 nm	52,53
Sulfated C_{12} 6- to 12-mole ethoxylates and sulfated butyl-hexa(butoxy) 2- and 7-mole ethoxylates; determination of inorganic salt and sulfated and unsulfated organics; single peak for sulfated material; single peak for unsulfated product (elutes last)	Unspecified C_{18}, 2 × 25 mm; 40°C	Step gradient: 90:10 and 40:60 H_2O/THF, followed by back-flush with 100% THF to elute nonionics	Evaporative light scattering	30

Normal phase systems

Sulfated C_{12}–C_{14} alkyl 3-mole ethoxylates and nonylphenol 4-mole ethoxylate; separated peaks not identified; separation from LAS and alkanesulfonate (elute last)	Merck LiChrosorb SI 60, 3 × 250 mm, 50°C	90:10 $CHCl_3$/EtOH, 0.0001 M in 18-crown-6 and 0.00015 M in sodium pentanesulfonate	Postcolumn reaction, extraction, and fluorescence detection	10
Sulfated 2,4,6-tributylphenol 4- to 50-mole ethoxylates; separation from unsulfated material (elutes first); elution in order of increasing ethoxylation	DuPont Zorbax CN, 4.6 × 250 mm	Gradient, hexane to 75:25 2-methoxyethanol/2-PrOH	UV, 255 nm	47

(Continued)

TABLE 3 (*Continued*)

Compounds studied	Column	Mobile phase	Detector	Ref.
Sulfated alkylphenol 6- and 10-mole ethoxylates; elution in order of increasing ethoxylation; nonionics elute first	Shandon Hypersil 5-CPS cyano-propyl, 4.6 × 250 mm	Hexane/EtOH, 70:30, 0.01 M tetramethylammonium chloride	UV, 220 nm	233
Sulfated NPE, E = 4; elution by increasing degree of ethoxylation, unresolved NPE elutes first; or resolution of unsulfated NPE by increasing degree of ethoxylation, sulfated material not seen	Tessek Separon SGX NH$_2$, 3 × 150 mm	Resolution of ether sulfates: 68.6:30:1.4 CH$_3$CN/CH$_2$Cl$_2$/H$_2$O, all 0.04 M in hexadecyltrimethyl-ammonium bromide; resolution of unsulfated material: same, but without salt	UV, 230 nm	234
Sulfated NPE, E = 4; elution by increasing degree of ethoxylation, NPE elutes first, resolved by increasing degree of ethoxylation	Silasorb SPH Nitrile, 4.2 × 300 mm, or Silasorb SPH Amine, 4.2 × 300 mm	CN column: 100:0 to 93:7 n-hep-tane/2-PrOH in 42 min, both 0.02 M in hexadecyltrimethylammonium bromide. NH$_2$ column: 95:5 n-heptane/2-PrOH (no salt) to 80:20 n-heptane/2-PrOH (both 0.02 M in hexadecyltrimethylammonium bromide) in 90 min	UV, 230 nm	48

Other separation modes

Sulfated nonylphenol 6-mole ethoxylate and sulfated octylphenol 3-mole ethoxylate; determination in sea water mixtures; elution in order of decreasing ethoxylation; separation from nonionic matter (elutes first)	Alltech mixed-mode RP8 or RP18/Anion 100 A 5 μ, 4.6 × 150 mm	$CH_3CN/0.02$ M KH_2PO_4, pH 4.7; gradient from 60:40 to 80:20 in 20 min	UV, 223 nm	50
Sulfated nonylphenol 6-mole ethoxylate; determination in sea water; elution in order of decreasing ethoxylation; separation from nonionic matter (elutes first); separation from LAS (elutes last)	Alltech mixed-mode RP4/Anion 100 A 5 μ, 4.6 × 150 mm	$CH_3CN/0.05$ M KH_2PO_4, pH 4.7; initial hold at 20:80 for 20 min, then gradient to 70:30 in 25 min, holding for 135 min	UV, 223 nm	54

TABLE 4 HPLC Analysis of α-Olefin Sulfonate Anionic Surfactants

Compounds studied	Column	Mobile phase	Detector	Ref.
C_{14} and C_{16} α-olefin sulfonates; separation by alkyl chain length; separation of alkenesulfonate from hydroxyalkanesulfonate; isomer separation	Reversed-phase Dupont Zorbax TMS, 4.6×250 mm	75:25 $MeOH/H_2O$, 0.4 M $NaNO_3$	Refractive index	55
C_{16}–C_{30} α-olefin sulfonates; separation of disulfonates; separation by alkyl chain length; separation of isomers; separate peaks for hydroxyalkanesulfonates and alkenesulfonates	Merck LiChrospher 100 CH-18, 4×250 mm	$MeOH/H_2O$, 0.2 M $NaCl$	UV, 210 nm	56
C_{12}–C_{18} α-olefin sulfonates; separation by alkyl chain length; separation of disulfonates; resolution of hydroxyalkanesulfonates from alkenesulfonates; isomer separation	Alltech Ultrasphere C18. 10×250 mm	$H_2O/MeOH/HNO_3$	FID	235,236
C_{16}–C_{18} α-olefinsulfonates after derivatization; separation by alkyl chain length and resolution of hydroxyalkanesulfonates from alkenesulfonates	GL Science Inertsil ODS-2. 4.6×150 mm	85:15 $MeOH/H_2O$	RI	12
C_{14} and C_{16} α-olefin sulfonates, demonstration of mobile phase ion chromatography	Dionex MPIC-NS1, 4.6×250 mm	H_2O/CH_3CN; about 60:40, with the water 0.01 M in NH_3 or 0.002 M in tetramethyl- or tetrapropylammonium hydroxide	Conductivity	13,14
C_{16}–C_{18} α-olefinsulfonates; separation from α-sulfofatty acid methyl esters and alkanesulfonates after derivatization	GL Science Inertsil SIL, 4.6×250 mm	92:8 n-hexane/ethyl ether	RI	12

TABLE 5 HPLC Analysis of Petroleum Sulfonates and Alkylnaphthalene Sulfonates

Compounds studied	Column	Mobile phase	Detector	Ref.
Petroleum sulfonates; separation by degree of sulfonation	DuPont Zorbax SAX, 4.6×250 mm	Complex gradient, 50:50 THF/H_2O or KH_2PO_4 solution, gradually increasing pH from 4.5 to 6.5 and salt from 0.1 to 0.2 M (0.05 to 0.1 in total eluent)	UV, 254 nm	58
Petroleum sulfonate; separation by degree of sulfonation	Whatman Partisil SAX, 4.6×250 mm	Gradient: A = 1:1:1 H_2O/MeOH/CH_3CN, B = 1:1:1 aqueous Na_2HPO_4/MeOH/CH_3CN; 0 to 100% B in 10 min	UV, 254 nm	60
Petroleum sulfonates, apparent total alkyl chain length C_3–C_{28}; determination of approximate alkyl character; homologs not resolved	Alltech RSIL prep C_{18} precolumn, 4.6×32 mm, and Dupont Zorbax 10 C_{18} column, 4.6×250 mm	Ternary gradient, tetrabutylammonium bisulfate, 0.1 M in H_2O/H_2O/CH_3CN, pH 5	UV, 225 or 195 nm	35
Petroleum sulfonate sodium and calcium salts; determination of active agent, salt, and oil; single peak for each	Unspecified C_{18}, 2×25 mm, 40°C	Step gradient: 90:10 and 40:60 H_2O/THF, followed by back-flush with 100% THF to elute oil	Evaporative light scattering	30
Petroleum sulfonates; determination of mono- and disulfonates in oil	C_{18}, 4.6×50 mm, to remove oil, and homemade silica-based anion exchanger, 4.6×150 mm, to separate sulfonates	Gradient: A = 60:40 MeOH/H_2O, B = 60:40 MeOH/0.125 M NaH_2PO_4; 0% B for 8 min, then to 98% B in 2 min, holding for 8 min	UV, 254 nm	59
Petroleum sulfonate; molecular weight distribution	Waters Ultrastyragel 1000 Å SEC column, 7.8×300 mm	THF	RI	237
Alkylnaphthalene sulfonates; determination in formulations; separation by alkyl character	Merck LiChrosorb 10 RP-8, 4.6×250 mm, 40°C	Gradient, MeOH/H_2O, pH adjusted to 2.75 with H_3PO_4	UV, 290 nm	61
Diisobutyl- and diisopropylnaphthalene sulfonates; routine determination in formulations; less complete separation by alkyl character	Chrompack Polygosil 60 DCN 10 cyano, 4.6×250 mm, 40°C	40:60 MeOH/H_2O, pH adjusted to 2.75 with H_3PO_4	UV, 235 nm	61

TABLE 6 HPLC Analysis of Ether Carboxylate Anionic Surfactants

Compounds studied	Column	Mobile phase	Detector	Ref.
Nonylphenol 4- and 6-mole ethoxycarboxylates; C_{12}–C_{14} alkyl 6-mole ethoxycarboxylates; elution in order of increasing ethoxylation; separation of nonionic impurities	DuPont Zorbax C_{18}, 4.6 × 250 mm	0.1 M tetrabutylammonium-bisulfate, pH 5/H_2O/CH_3CN; gradient from 10:50:40 to 10:10:80 in 30 min, then to 0:20:80 in 10 min, holding for 15 min	UV, 225 or 195 nm	35
Nonylphenol 4-, 5-, and 6-mole ethoxycarboxylates; separation from residual NPE	Reversed-phase C_1 or C_8, 4 × 140 mm	MeOH/H_2O or 40:60 CH_3CN/H_2O	UV, 254 nm	64
C_{12} and C_{14} alkylethoxy-carboxylates, with $n = 1$–25; separation by alkyl chain length and in inverse order of ethoxylation; separation from AE	Macherey-Nagel Nucleosil 120-C_{18}, multiple 4 × 250 mm columns in series	1% formic acid/CH_3CN gradient; uncarboxylated AE first trapped on a C_{18} column at neutral pH and low CH_3CN concentration	Evaporative light scattering	62
C_{12} and C_{18} Alkyl 6- and 11-mole ethoxycarboxylates and octyl- and nonylphenol 8- to 20-mole ethoxycarboxylates; separation by hydrophobe, with some differentiation by ethoxy chain length	Merck LiChrospher 100 RP 18, 4 × 125 mm	75:15:10 MeOH/H_2O/CH_3CN, containing 0.004 M tetramethyl-ammonium bisulfate and 0.001 M tetramethylammonium hydroxide	RI and UV, 220 or 276 nm	63

TABLE 7 HPLC Analysis of Other Anionic Surfactants

Compounds studied	Column	Mobile phase	Detector	Ref.
Lignin sulfonates				
Lignin sulfonates; molecular weight distribution	Waters I-250 & I-125 or I-125 & I-60 GPC columns (protein analysis)	0.050 M citric acid/Na$_2$HPO$_4$ buffer, pH 3.0	UV, 280 nm	65
Lignin sulfonates; molecular weight distribution	Toyo Soda TSK G3000SE$_{XL}$, 7.5 × 750 mm and TSK G3000SW$_{XL}$, 7.5 × 300 mm	0.2 M aqueous NaOAc, adjusted to pH 7 with HNO$_3$	UV, 280 nm	66
Lignin sulfonates; molecular weight distribution	Toyo Soda G3000HXL and G4000HXL, each 7.8 × 300 mm, 40°C	THF, 0.02 M in methyltrioctyl-ammonium chloride	UV, 280 nm	67
Ester sulfonates				
C$_{10}$–C$_{18}$ α-sulfofatty acid methyl esters; resolution by alkyl chain length	Macherey & Nagel Nucleosil C$_{18}$, 4 × 250 mm	80:20 or 90:10 MeOH/H$_2$O containing 0.25 M NaClO$_4$	RI	8
C$_{16}$, C$_{18}$ ester sulfonates; resolution by chain length and separation of ester from α-sulfofatty acid	Dionex mobile-phase ion chromatography system, with poly(styrene/divinylbenzene) column and membrane suppressor	Gradient, CH$_3$CN/0.005 M aqueous ammonia	Conductivity	238
C$_{16}$ and C$_{18}$ α-sulfofatty acids and methyl esters; separation from each other and resolution by chain length	Phase Separations Spherisorb S5-C1, 4 × 40 mm	Gradient, citrate buffer/CH$_3$CN, 90:10 to 50:50 in 20 min	Postcolumn reaction: extraction of ion pair, fluroescence	239
C$_{14}$–C$_{18}$ α-sulfofatty acid methyl esters; resolution by alkyl chain length	GL Science Inertsil ODS-2, 4.6 × 250 mm	85:15 MeOH/H$_2$O	RI	12
C$_{14}$–C$_{18}$ α-sulfofatty acid methyl esters; separation from LAS, α-olefinsulfonates, and alkanesulfonates, after derivatization	GL Science Inertsil SIL, 4.6 × 250 mm	97:3, 96:4, or 92:8 *n*-hexane/ethyl ether	RI	12

(Continued)

TABLE 7 *(Continued)*

Compounds studied	Column	Mobile phase	Detector	Ref.
Sulfosuccinate esters				
Bis(2-ethylhexyl)sulfosuccinate; single peak	Crosslinked amine/Kel-F 800 silica (not commercially available), 4.6×250 mm	25:75 CH_3CN/H_2O, 0.0002 M in disodium naphthalenedisulfonate	Inverse conductivity or inverse UV, 285 nm	25
C_6 and C_8 *bis*(alkyl)sulfosuccinates; single peak for each; elution in reverse order of alkyl chain length; separation from alkanesulfonates	Merck LiChrosorb SI 60, 3×250 mm	90:10 $CHCl_3/EtOH$, 0.0001 M in 18-crown-6 and 0.00015 M in sodium pentanesulfonate	Postcolumn reaction, extraction, and fluorescence detection	10
C_{10}–C_{14} monoalkylsulfosuccinates; determination in diagnostic reagents; alkyl distribution; separation from octylsulfate	Macherey-Nagel Nucleosil C_8, 4×250 mm	0.01 M aqueous tetrabutylammonium bisulfate/MeOH, 23:77, pH 3	Refractive index	69
Isethionate esters				
C_8–C_{18} acyl isethionates; alkyl chain length distribution	Macherey & Nagel Nucleosil C_{18}, 4×250 mm	$MeOH/H_2O$, 90:10 for higher homologs and 80:20 for lower, containing 0.25 M $NaClO_4$	RI	8
Acyl taurates				
C_8–C_{18} *N*-acyl taurates and *N*-acyl-*N*-methyl taurates; acyl chain length distribution	Macherey & Nagel Nucleosil C_{18}, 4×250 mm	$MeOH/H_2O$, 90:10 for higher homologs and 80:20 for lower, containing 0.25 M $NaClO_4$	RI	8
N-Acylated amino acids				
C_8–C_{18} *N*-acyl sarcosinates; acyl chain length distribution	Macherey & Nagel Nucleosil C_{18}, 4×250 mm	$MeOH/H_2O$, 90:10 for higher homologs and 80:20 for lower, containing 0.25 M $NaClO_4$	RI	8

Soaps

C8–C18 fatty acid soaps; alkyl chain length	Waters μBondapak C18	MeOH/H2O, 85:15, 0.2% HOAc	RI	240
C8–C18 fatty acid soaps; alkyl chain length distribution	Macherey & Nagel Nucleosil C18, 4 × 250 mm	MeOH/H2O, 90:10, containing 0.25 M NaClO4	RI	8
C12–C22 fatty acid soaps; characterization and determination	Merck LiChrosorb RP-8, 4.7 × 150 mm	MeOH/H2O, with (pH 9.5) or without (pH 3.5) 0.5% cetyltrimethyl-ammonium bromide	RI	71
C12–C20 fatty acids; separation by increasing alkyl chain length; as derivatives with dibromoacetophenone	Beckman RP8, 4.6 × 250 mm	Gradient: A = CH3CN, B = H2O; 20% B for 18 min, then to 15% B in 15 min, then to 0% B in 2 min	UV, 254 nm	241
C10–C22 fatty acid soaps; characterization and determination	Waters μBondapak C18, 2 in series, 38°C	Gradient: A = 60:40 CH3CN/H2O, B = CH3CN; nonlinear gradient from 0 to 100% B in 3 hr	UV absorbance of 4-(bromomethyl)-7-methoxycoumarin derivatives (254 or 280 nm, depending on sensitivity needed	72
C12–C18 fatty acid soaps; characterization and determination	Merck LiChrocart RP-8, 4.6 × 100 mm	Gradient: A = 20:80 MeOH/H2O, B = CH3CN; 60% to 75% B in 10 min	UV absorbance of 2,4-dibromoacetophenone derivatives (254 nm) or fluorescence of 4-(bromomethyl)-7-methoxycoumarin derivatives ($\lambda_{ex}/\lambda_{em}$ = 328/380 nm)	73,74

(Continued)

TABLE 7 *(Continued)*

Compounds studied	Column	Mobile phase	Detector	Ref.
Others				
C_8–C_{16} acyl monoglyceride sulfonates; alkyl chain length distribution	Macherey & Nagel Nucleosil C_{18}, 4×250 mm	MeOH/H_2O, 90:10 or 80:20, containing 0.25 M $NaClO_4$	RI	8
C_{12}–C_{14} alkylsulfoacetates; alkyl chain length distribution	Macherey & Nagel Nucleosil C_{18}, 4×250 mm	80:20 or 90:10 MeOH/H_2O containing 0.25 M $NaClO_4$	RI	8
Glycocholic and taurocholic acids; separation	Alltech mixed mode RP Phenyl ion exchange column, 4.6×250 mm	85:15 CH_3CN/H_2O, 0.0002 M in disodium naphthalenedisulfonate, pH 5.5	Inverse UV, 285 nm	21
C_{16} alkyl disulfonated diphenyl oxide; peaks not identified	Waters Nova-Pak C_8, 3.9×150 mm	Gradient elution: A = 70:20:10 aqueous 0.00075 M $NaHCO_3$ + 0.0022 M Na_2CO_3/CH_3CN/MeOH B = 40:35:25 CH_3CN/ aqueous 0.00075 M $NaHCO_3$ + 0.0022 M Na_2CO_3/ MeOH; 22% B to 35% B in 40 min, then to 50% B in 10 min, 5 min hold, then to 100% B in 5 min	UV, 239 nm	242

TABLE 8 HPLC of Alcohol Ethoxylates

Normal phase systems

Compounds studied	Column	Mobile phase	Detector	Ref.
C_{12} alkyl 5- to 44-mole ethoxylates; ethoxy distribution	Jones Chromatography Apex I silica, 4.6×250 mm, with Shandon Hypersil ODS pre-column, 4.6×7.5 mm	Flow/composition gradient: 0.3 mL/min to 2.0 mL/min in 37 min; 100:0 to 65:35 CH_3CN/H_2O in 43 min	Evaporative light scattering	95
C_8–C_{16} alkyl 1- to 9-mole ethoxylates; ethoxy distribution	Merck LiChrosorb NH_2, 4×250 mm	n-Heptane/THF	RI	243
C_{12}–C_{14} alkyl 5- to 11-mole ethoxylates; ethoxy distribution, with some discrimination by alkyl chain length	Dupont Zorbax NH_2	Hexane/2-PrOH/H_2O gradient from 100:0:0 to 37:60:3	Evaporative light scattering	30
C_{10}–C_{14} alkyl 2- to 16-mole ethoxylates; ethoxy distribution, with resolution also by alkyl chain length	Spherisorb NH_2, 4.6×250 mm	Hexane/$CHCl_3$/MeOH; complex gradient from 76:19:5 to 56:14:30	Evaporative light scattering	117
C_{12}–C_{15} alkyl 1- to 21-mole ethoxylates; ethoxy distribution	DuPont Zorbax NH_2, 4.6×250 mm	Gradient, hexane/THF/(90:10 2-PrOH/H_2O): from 100:0:0 to 80:20:0 in 2 min, then to 10:85:5 in 40 min	Fluorescence of phenyl isocyanate derivatives; 240 nm excitation, 310 nm emission	124
C_8–C_{18} alkyl 1- to 14-mole ethoxylates as 4-nitrobenzoyl derivatives; ethoxy distribution; some resolution by alkyl chain length; analysis of fabric extracts	Spherisorb Amino 3, $20^\circ C$	Gradient: A = hexane, B = 80:20 ethyl acetate/CH_3CN; 5 to 50% B in 30 min, then to 98% B in 15 min; or A = 90:10 hexane/THF, B = MeOH; 3% B for 5 min, then to 18% B in 20 min, then to 50% B in 15 min	UV	244,245

(Continued)

TABLE 8 (*Continued*)

Compounds studied	Column	Mobile phase	Detector	Ref.
C_{12}–C_{14} AE, various products $E = 2$ to $E = 50$; determination in emulsion lubricants as 3,5-dinitrobenzoyl derivatives; ethoxy distribution; some peak-splitting due to alkyl chain length	Phenomonex Spherisorb NH$_2$, 4.6 × 250 mm	Gradient: A = hexane, B = MeOH/CH$_2$Cl$_2$/hexane; 10 to 80% B in 10 min for products with EO = 2–7 and 20 min for EO = 11–50—B ratios are 5:15:80 for EO = 2–3, 8:24:68 for EO = 7, 10:30:60 for EO = 11, 15:35:50 for EO = 20, and 20:40:40 for EO = 50	UV, 276 nm	246
C_{12}–C_{18} AE, E = 1 to 15; study of separation mechanisms	Separon SGX NH$_2$, 3 × 150 mm	CH$_3$CN/CH$_2$Cl$_2$/H$_2$O, various isocratic and gradient combinations	APCI MS	127
C_{12} alkyl 6-mole ethoxylate; ethoxy distribution	LiChrosorb DIOL, 4 × 250 mm	140:60:5:1 or 105:95:10:1 n-hexane/2-PrOH/H$_2$O/HOAc	RI and UV	90,140
C_{16}–C_{18} alkyl 6- to 20-mole ethoxylates; dinitrobenzoyl chloride derivatives; ethoxy distribution	Merck LiChrospher DIOL, 4.6 × 250 mm	n-Heptane/CH$_2$Cl$_2$/MeOH or 2-PrOH; various isocratic and gradient conditions	UV, 254 nm	94,247,248
C_{12}–C_{15} alkyl 1- to 21-mole ethoxylates; ethoxy distribution	Rainin Microsorb CN, 4.6 × 250 mm, 45 °C	Gradient, hexane/THF/(90:10 2-PrOH/H$_2$O): from 100:0:0 to 80:20:0 in 5 min, then to 52:30:18 in 15 min, then to 40:40:20 in 5 min	Evaporative light scattering	124
C_{16}–C_{18} AE from 1- to 80-mole ethoxylates; ethoxy distribution	Macherey-Nagel Nucleosil p-nitrophenyl-bonded silica, 3 × 250 mm; 45°C preferred	Gradient from 90:5:5 to 65:17.5:17.5 n-heptane/CH$_2$Cl$_2$/MeOH in 190 min	Evaporative light scattering or UV, 254 nm, of 3,5-dinitrobenzoyl esters	123

Reversed-phase systems

C_{10}–C_{16} alkyl 50-mole ethoxylates; alkyl distribution after HI cleavage	Waters μBondapak C_{18}, 3.9 × 300 mm	Gradient from 75:25 MeOH/H_2O to 50:50 MeOH/2-PrOH	UV, 252 nm	46
C_{12} and C_{16} alkyl 9-mole ethoxylates, 3,5-dinitrobenzoyl chloride derivatives; elution according to increasing alkyl and increasing ethoxy content	Waters radially compressed C_{18}	Gradient, MeOH/H_2O	UV, 254 nm	249
C_{16}–C_{18} alkyl 6- to 20-mole ethoxylates; dinitrobenzoyl chloride derivatives; elution according to decreasing ethoxy and increasing alkyl character	Macherey-Nagel Nucleosil C_8, 4.6 × 250 mm or Beckman Ultraspher C_{18}, 4.6 × 150 mm	THF/H_2O or CH_3CN/H_2O	UV, 254 nm	98
C_8–C_{18} alkyl 4- to 7-mole ethoxylates, 3,5-dinitrobenzoyl esters; elution according to increasing alkyl and decreasing ethoxy content	Shandon Hypersil ODS, 3 μm, 4 × 120 mm	CH_3CN/H_2O, 60:40 to 100:0 in 15 min, hold for 7 min	UV, 235 nm	129
C_{12}–C_{18} alkyl 10-mole ethoxylates; 3,5-dinitrobenzoyl isocyanate derivatives; optimization study and separation by alkyl chain length; separation from APE	Merck LiChrospher 100 RP-18 endcapped, 4 × 125 mm, 5°C	98:2 CH_3CN/H_2O, 0.0011 M sodium perchlorate	UV, 233 nm	84

(Continued)

TABLE 8 *(Continued)*

Compounds studied	Column	Mobile phase	Detector	Ref.
C_{11}–C_{18} alkyl 1- to 18-mole ethoxylates; separation by increasing alkyl chain length and decreasing degree of ethoxylation	Beckman RP-8, 4.6 × 250 mm	Gradient: A = CH_3CN, B = H_2O; 30 to 0% B in 40 min	UV, 235 nm (phenyl isocyanate derivatives) or evaporative light scattering (no derivatization)	122
C_{10}–C_{12} alkyl 8-mole ethoxylates; 1-anthroylnitrile derivatives; elution according to decreasing ethoxy content	Shandon Hypersil ODS, 6 × 150 mm	70:30 CH_3CN/H_2O	Fluorescence: 395 nm excitation, 450 nm emission	114
C_{12}–C_{18} alkyl 7-mole ethoxylates; 1-naphthoyl chloride derivatives; elution according to increasing alkyl chain length	Merck LiChrospher 100 RP18, 4 × 250 mm	90:10 $CH_3CN/MeOH$	Fluorescence: 228 nm excitation, 365 nm emission	99
C_{12}–C_{18} alkyl 7-mole ethoxylates; 1-naphthoyl chloride derivatives; elution according to decreasing ethoxy content	Merck LiChrosorb RP Select B C_8, 4 × 250 mm	Gradient: A = H_2O with or without 0.05 M tetrabutylammonium dihydrogen phosphate, B = CH_3CN. 60 to 80% B in 20 min, then to 90% B in 20 min, then to 100% B in 10 min	Fluorescence: 228 nm excitation, 365 nm emission	99
C_{12}–C_{18} alkyl 10-mole ethoxylates; 1-naphthyl isocyanate derivatives; optimization study and separation by alkyl chain length	Merck LiChrospher 100 RP-18, 4 × 125 mm	95:5 CH_3CN/H_2O	Fluorescence: 228 nm excitation 368 nm emission	83

C$_{16}$ alkyl 10-mole ethoxylate; single peak for AE; demonstration of solid phase microextraction with online derivatization	Zorbax ODS, 4.6 × 250 mm	A = 70:30 CH$_3$CN/H$_2$O, B = MeOH; 60 to 95% B in 10 min	Fluorescence of 1-naphthoyl chloride derivatives: 228 nm excitation, 366 nm emission	250
C$_{10}$–C$_{16}$ alkyl 2- to 26-mole ethoxylates; elution according to increasing alkyl and increasing ethoxy character	Beckman Ultraspher C$_8$, 4.6 × 250 mm, 21°C	60:40 CH$_3$CN/H$_2$O	RI	110
C$_5$–C$_6$ AE; separation from anionic materials; separation by alkyl chain length	Waters radially compressed C$_{18}$, 8 × 100 mm	40:60 CH$_3$CN/H$_2$O, 0.005 M in tetrabutylammonium phosphate	RI	251
C$_{12}$–C$_{20}$ AE; ethoxy distribution of homologs up to E = 20; elution in order of decreasing ethoxy content	Two columns in series: Merck Multospher RP 18-5 and Superspher 60 RP-8, each 4 × 250 mm, 30°C	CH$_3$CN/H$_2$O, 60:40 for compounds of ≤ C$_{14}$ alkyl chain length; higher, up to 80:20, for chain lengths >C$_{14}$	RI	131
C$_{12}$–C$_{15}$ alkyl 3- to 30-mole ethoxylates; back-flush technique using Unisil column for separation of AE from ionic substances and the Merck column for separation of AE from unethoxylated nonionics; single peak for AE	Gasukuro C18 Unisil QC-18, 4.6 × 150 mm, and Merck LiChrosorb RP-2	85:15 MeOH/H$_2$O	RI	51
C$_{12}$–C$_{18}$ alkyl 6- to 80-mole ethoxylates; PEG determination	Merck LiChrosorb RP-18, 4.6 × 250 mm, or others	95:5 or 80:20 MeOH/H$_2$O	RI	90,252

(Continued)

TABLE 8 (*Continued*)

Compounds studied	Column	Mobile phase	Detector	Ref.
C_{12}–C_{14} alkyl 1 to 25-mole ethoxylates; elution according to increasing alkyl and decreasing ethoxy character	Macherey & Nagel Nucleosil 120-5C_{18}, 4 × 250 mm, two in series	Step gradient, 65:35 CH_3CN/0.1% HOAc for 38 min, then 90:10 for 10 min	Evaporative light scattering	62
C_{10}–C_{16} alkyl 4-mole ethoxylates; elution according to increasing alkyl and decreasing ethoxy character	Kromasil 100 RP-C_{18}, 4.6 × 150 mm	For only alkyl distribution: A = 90:10 MeOH/H_2O, B = MeOH; A to B in 5 min; for both alkyl and ethoxy distribution: A = 65:35 CH_3CN/H_2O, B = CH_3CN; 20 min hold at A, then from A to B in 85 min	Evaporative light scattering	130
C_{12}–C_{18} alkyl 4- to 12-mole ethoxylates; separation by increasing alkyl chain length and increasing degree of ethoxylation	Hypersil Hypercarb-S porous graphitic carbon, 4.6 × 100 mm	Gradient: A = CH_3CN, B = CH_2Cl_2; 20 to 80% B in 30 min	Evaporative light scattering	100
C_{12}–C_{15} alkyl 9-mole ethoxylates; elution according to increasing alkyl and decreasing ethoxy character; separation of linear and branched isomers	Supelco LC$_{18}$, 4.6 × 250 mm for thermospray LC-MS; 3.2 × 250 mm for electrospray	50:50 or 55:45 H_2O/THF; post-column addition of 0.043 M NH_4OAc for thermospray	Thermospray or electrospray MS	126,253
C_{12}–C_{18} AE; elution according to increasing alkyl chain length; MS determination of ethoxy distribution	Alltech C_8, 2.1 × 250 mm	A = H_2O, B = MeOH, each 0.0001 M in trifluoroacetic acid; 80 to 100% B in 20 min	Electrospray MS	80

C_{16}–C_{18} alkyl 6- to 20-mole ethoxylates; dinitrobenzoyl chloride derivatives; elution according to decreasing ethoxy content	Macherey-Nagel Nucleosil C_8, 4.6 × 250 mm	CH_3CN/H_2O, 60:40, 0.01 M in tetrabutylammonium perchlorate	Polarography	119

Other systems

C_{12}–C_{18} alkyl 9- and 23 mole ethoxylates; elution according to increasing ethoxy content	Toyo Soda TSK-gel IC-Cation-SW, 4.6 × 50 mm, K^+ form; temperature programming may be used	MeOH, with and without 0.0004 or 0.0075 M KI	RI, inverse conductivity, or UV (254 nm, of dinitrobenzoyl derivatives)	105–107,132
C_{16}–C_{18} alkyl 25-mole ethoxylates; elution according to increasing ethoxy chain length	Macherey-Nagel Nucleosil Silica, 4.6 × 150 mm; temperature programmed	92:8 CH_3CN/H_2O, both 0.005 M in NaOAc	RI or UV (254 nm of 3.5-dinitrobenzoyl derivatives)	108
C_{12}–C_{18} alkyl 0- to 25-mole ethoxylates; SEC to determine ethoxy chain distribution	Poly(styrene/divinylbenzene), 3 columns in series, 7.8 × 305 mm	THF	RI	77
C_6–C_{18} alkyl 4- to 10-mole ethoxylates; SEC to determine ethoxy chain distribution after preparative RP HPLC to separate by alkyl chain length	Four Phenogel columns in series, 1000 Å, 1000 Å, 500 Å, 500 Å, each 30 cm	$CHCl_3$	Density and RI	254,255
C_{12} AE; oligomer fractionation	Bio-Rad Bio-Beads SX-3, 22 × 250 mm	50:50 CH_2Cl_2/hexane	none specified	256

TABLE 9 HPLC Analysis of Alkylphenol Ethoxylates

Compounds studied	Column	Mobile phase	Detection	Ref.
Normal phase systems				
Determination of NPE9 in pharmaceuticals; separation by ethoxy chain length	Chrom. Sci. Co. S5W silica, 4.6 × 250 mm	50:50 ethyl acetate/MeOH	UV, 280 nm	257
NPE, E = 4–10; ethoxy distribution	Merck LiChrosorb Si 60, 4.6 × 250 mm	70:20:10 n-heptane/CHCl$_3$/MeOH	UV, 276 nm	258
OPE and NPE; system for resolution of homologs where the mean degree of ethoxylation is E = 1 to 10	Hewlett-Packard Si-100, 4.6 × 200 mm, 30°C	Gradient: A = 80/20 n-hexane/ethyl ether, B = 40:30:20:10:1:0.5 dioxane/ethyl ether/n-hexane/2-PrOH/H$_2$O/HOAc; 5 to 95% B in 45 min	UV, 280 nm	144
OPE and NPE; system for resolution of homologs where the mean degree of ethoxylation is E = 1 to 20	Hewlett-Packard Si-100, 4.6 × 200 mm, 30°C	Gradient: A = 80/20 n-hexane/ethyl ether, B = 50:15:10:5:1:0.5 dioxane/ethyl ether/n-hexane/2-PrOH/H$_2$O/HOAc; 10 to 95% B in 40 min	UV, 280 nm	144
OPE and NPE; system for resolution of homologs where the mean degree of ethoxylation is E = 10 to 40	Hewlett-Packard Si-100, 4.6 × 200 mm, 30°C	Gradient: A = 60:40 2-PrOH/n-hexane, B = 80:20 EtOH/H$_2$O: 10 to 95% B in 45 min	UV, 280 nm	144
NPE, E = 4–22; ethoxy distribution	Merck LiChrosorb Si 60, 4.6 × 250 mm, and Adsorbosphere NH$_2$, 4.6 × 250 mm, in series	Gradient: A = 70:15:15 n-heptane/CHCl$_3$/MeOH, B = 25:75 CHCl$_3$/MeOH; 0 to 40% B in 25 min	UV, 270 nm	139
NPE, E = 5–53; ethoxy distribution	Jones Chromatography Apex I silica, 4.6 × 250 mm, with Shandon Hypersil ODS precolumn, 4.6 × 7.5 mm	Flow/composition gradient: 0.3 mL/min to 2.0 mL/min in 37 min; 100:0 to 65:35 CH$_3$CN/H$_2$O in 43 min	Evaporative light scattering and UV, 220 nm	95
NPE, E = 1–19; ethoxy distribution; determination in industrial effluent	Spherisorb silica, 4.5 × 125 mm	Gradient: A = 80:20, B = 50:50 pH 3 phosphate buffer/CH$_3$CN; 0 to 100% B in 12 min	Fluorescence: 230 nm excitation; 302 nm emission	96,97

Nonyl- and tributylphenol ethoxylates, E = 4–22; ethoxy distribution	Adsorbosphere NH₂, 4.6 × 250 mm	Gradient: A = 90:5:5 n-heptane/CHCl₃/MeOH, B = 50:50 CHCl₃/MeOH: 0 to 20% B in 15 min (for NPE), or 0 to 30% B in 20 min (for tributylphenol ethoxylates)	UV, 270 nm	138,139
OPE and NPE, E = 0–18; determination in commercial products and in the environment; ethoxy distribution	Merck LiChrosorb NH₂, 4.6 × 250 mm, or Phase Separations Spherisorb NH₂, 3 × 120 mm, or Shandon Hypersil APS, 4 × 100 mm	Gradient, n-hexane/2-PrOH	UV, 277 nm, or fluorescence: 225 nm excitation, 304 nm emission	24,135,142
OPE, E = 1–9; environmental analysis; separation by ethoxy chain length	DuPont Zorbax NH₂ for higher ethoxylates and Whatman Partisil 5 PAC mixed CN/NH₂ for lower ethoxylates	Gradient: A = methyl t-butyl ether, 0.1% HOAc, B = 95:5 CH₃CN/MeOH and 0.1% acetic acid; 0 to 100% B in 30 min	Fluorescence: 230 nm excitation, 302 nm emission	115
NPE11; ethoxy distribution	Dupont Zorbax NH₂	Hexane/2-PrOH/H₂O gradient, from 100:0:0 to 37:60:3 in 55 min	Evaporative light scattering	30
NPE, E = 0–6; environmental analysis; NP elutes between NP2 and NP3	Shandon Hypersil APS aminosilica, 4.6 × 100 mm	Gradient, n-hexane/2-PrOH	UV, 277 nm	259
NPE, E = 1–20; ethoxy distribution	Lachema Silasorb SPH Amine, 4.2 × 300 mm	n-Heptane/H₂O	UV, 230 nm	148
OPE10 and NPE9; ethoxy distribution	Shandon Hypersil APS aminopropyl-modified silica, 4.6 × 250 mm	Linear gradient: A = 70:30 hexane/THF, B = 90:10 2-PrOH/H₂O: 5 to 50% B in 60 min.	UV, 225 nm	217
NPE, various products NP6 to NP40; determination in emulsion lubricants	Phenomonex Spherisorb NH₂, 4.6 × 250 mm	Gradient: A = CH₃CN; B = 80:20 CH₃CN/H₂O: 2.5 to 40% B in 30 min for NP6, varying up to 35 to 100% B in 60 min for NP40	UV, 276 nm	260

(Continued)

TABLE 9 (Continued)

Compounds studied	Column	Mobile phase	Detection	Ref.
Nonylphenol 5-, 10-, 15-, and 20-mole propoxylates; PPG determination	Merck LiChrosorb DIOL, 4 × 250 mm	75:25:1:0.5 or 60:40:5:1 n-hexane/2-PrOH/H$_2$O/HOAc	RI	92
Dodecylphenol 6-mole ethoxylate; separation by ethoxy chain length; incomplete resolution	Merck LiChrosorb DIOL, 4 × 250 mm	140:60:5:1 or 105:95:10:1 n-hexane/2-PrOH/H$_2$O/HOAc	RI and UV	90,140
NPE, E = 4 to 10: separation of homologs up to E = 18	Merck LiChrosorb DIOL, 4.6 × 250 mm	Gradient from 81.5:15.5:2 to 68:26:6 hexane/CH$_2$Cl$_2$/2-PrOH	UV, 275 nm	136
Dodecylphenol 6-mole ethoxylate, NPE5, NPE6; collaborative study	DIOL columns, various, 4 × 250 mm	n-Hexane/2-PrOH/H$_2$O; various mixtures	RI or UV, various λ	111
NPE, E = 2 to 12: separation of homologs up to E = 20	Merck LiChrospher DIOL-100, 4 × 250 mm	Gradient: A = 94.5:5:0.5 isooctane/2-PrOH/MeOH, B = 80:20 2-PrOH/MeOH; 0 to 20% B in 15 min, then to 50% B in 10 min	UV, 277 nm	261
Tributylphenolethoxylate, E = 4–50; separation by ethoxy chain length	Dupont Zorbax CN, 4.6 × 250 mm	Gradient from hexane to 75:25 2-methoxyethanol/2-PrOH: 2 to 50% B in 50 min	UV, 255 nm	47
Alkylphenol ethoxylates and di-alkylphenol ethoxylates	Dupont Zorbax CN, 4.6 × 250 mm	Gradient from heptane to 50:50 2-methoxyethanol/2-PrOH	UV, 280 nm	142
NPE, E = 1–17, separation by ethoxy chain length	Rainin Microsorb CN, 4.6 × 250 mm	Complex gradient: (80:20 hexane/THF)/(90:10 2-PrOH/H$_2$O); from 99:1 to 58:42 in 22 min	Fluorescence: 230 nm excitation, 310 nm emission	124
NPE, E = 1–20: detailed study of separation by ethoxy number	Various normal phase columns NH$_2$, DIOL, silica, CN	Various isocratic and gradient systems with heptane/PrOH/EtOH	UV, 230 or 254 nm	137
NPE, E = 0–18; ethoxy distribution; environmental analysis	Rainin Microsorb CN, 4.6 × 250 mm	Gradient: A = 80/20 hexane/THF, B = 90:10 2-PrOH/H$_2$O; 1 to 42% B in 20 min	Fluorescence: 229 nm excitation, 310 nm emission	262

NPE; environmental analysis: distribution of homologs for E = 4–14	Waters Resolve CN, 8 × 100 mm	95:5 CH$_3$CN/H$_2$O	UV, 229 nm	263
NPE, E = 12, 40, and 100; ethoxy distribution	Macherey-Nagel Nucleosil p-nitrophenyl-bonded silica, 3 × 250 mm, 45°C	Gradient from 90:5:5 to 65:17.5:17.5 n-heptane/CH$_2$Cl$_2$/MeOH in 190 min	UV, 254 nm	123
NPE, E = 4; ethoxy distribution	Experimental alumina, 4 × 250 mm	Various isocratic mixtures of n-hexane/ethyl acetate	UV, 254 or 275 nm; APCI-MS	264,265
Tris(butyl)phenol ethoxylate, E = 4; apparent separation by both butyl substitution pattern and degree of ethoxylation	Experimental alumina, 4 × 250 mm	Various isocratic mixtures of n-hexane/ethyl acetate; other normal phase solvents also investigated	UV, 275 nm	266,267
Reversed-phase systems				
Environmental analysis: separate single peaks for OPE and NPE; LAS also determined	Merck LiChrosorb RP8, 4 × 100 mm, or Phase Separations Spherisorb S3 ODS II, 4 × 250 mm	Various gradients, CH$_3$CN/H$_2$O/0.02 M NaClO$_4$	UV, 225 nm, or fluorescence: 225 nm excitation, 295 nm emission	24,31,136
Environmental analysis: separate single peaks for OPE and NPE	Alltech Alltima C$_{18}$, 4.6 × 250 mm, 50°C	80:20 MeOH/0.0012 M trifluoroacetic acid	Fluorescence: 225 nm excitation, 295 nm emission, with and without acetylation; confirmation by LC-electrospray-MS	147

(Continued)

TABLE 9 (Continued)

Compounds studied	Column	Mobile phase	Detection	Ref.
NPE9 and OPE10: single peak for NPE or OPE (system optimized for separation of ABS homologs)	Merck LiChrosorb 10 RP8, 4.6 × 250 mm	For OPE: 75:25 MeOH/H$_2$O, 0.005 M in tetramethylammonium bromide; for NPE: linear gradient: A = 65:35 MeOH/H$_2$O, 0.01 M in TMAB, B = 85:15 MeOH/H$_2$O, 0.01 M in TMAB; 40 to 70% B in 20 min	UV, 225 nm	217
OP10 or NP9, ethoxy distribution; demonstration of simultaneous determination with LAS	Hichrom TMS C$_1$, 4.6 × 150 mm	58:42 MeOH/aqueous 0.008 M NH$_4$OAc	Fluorescence: 220 nm excitation, 302 nm emission, or API-MS	224
OPE and NPE; single peaks for each, with minor peaks attributed to presence of *ortho* isomers	Hewlett-Packard RP-18, 4.6 × 200 mm, 40°C	80:20 MeOH/H$_2$O	UV, 280 nm	144
NPE; environmental analysis; single peak	Waters Nova-Pak C$_{18}$, 8 × 100 mm	90:10 CH$_3$OH/H$_2$O	UV, 229 nm	263
NPE in paper mill liquids; single peak for NPE	Waters µBondapak C$_{18}$	85:15 CH$_3$CN/H$_2$O	UV, 276 nm	93,268
NPE; environmental analysis; single peak for NPE; MS determination of ethoxy distribution	Alltech C$_8$, 2.1 × 250 mm	A = H$_2$O, B = MeOH, each 0.0001 M in trifluoroacetic acid; 80 to 100% B in 20 min	Electrospray MS	80
NPE, E = 9, spermicide in biological fluids; single peak	Alltech R-SIL-amine NH$_2$, 4.6 × 250 mm	95:5 THF/CH$_3$CN	Fluorescence: 275 nm excitation, 575 nm emission	145
OPE, E = 10, in virus-inactivated plasma; single peak	Ultrabiosep C$_{18}$, 4.6 × 250 mm, with precolumn 4.6 × 25 mm	3 mL H$_2$O and 3.5 mL 50:50 CH$_3$CN/H$_2$O to remove protein and other materials, then 90:10 CH$_3$CN/H$_2$O for analysis	UV, 225 nm	146

NPE; environmental analysis; distribution of homologs for E = 0–3; elution in order of ethoxylation	Waters Resolve C$_{18}$, 8 × 100 mm	98:2 CH$_3$CN/H$_2$O	UV, 229 nm	263
OPE, E = 10, elution in order of ethoxy number; C$_4$ APE and ortho-OPE impurities detected; OPE, E = 40 analyzed without impurity detection	Neos Fluofix 120N branched fluorinated silica, 4.6 × 250 mm	50:50 MeOH/H$_2$O	UV, 280 nm	143
OPE, E = 10; elution in reverse order of ethoxylation	Knauer RP 100-C18, 4.6 × 250 mm	45:55 isopropanol/0.1% trifluoroacetic acid	UV, 270 nm; identification of collected peaks by MALDI-MS	269
OPE and NPE, E = 1–60; elution in reverse order of ethoxy number	Lachema Silasorb C18 SPH and Silasorb C8 SPH, 3.6 × 300 mm	2-PrOH/H$_2$O, various ratios 40:60 to 65:35	UV, 230 nm	52,148
Phenol 8-mole ethoxylate: elution in order of ethoxy number	Shandon Hypersil 5-SAS C1 phase, 4.6 × 150 mm	Gradient: A = 10:90 MeOH/H$_2$O, B = 95:5 MeOH/H$_2$O (A and B each 0.1 M in NH$_4$OAc); 20 to 65% B in 25 min	UV, 276 nm, and thermospray MS	101
OPE and NPE, E = 1–100: elution in reverse order of ethoxy chain length by mixed size exclusion and reversed-phase mechanisms	Asahipak GS-310 size exclusion packing, 7.6 × 500 mm	Various CH$_3$CN/H$_2$O ratios, 30:70 to 100:0	RI	79
OPE, E = 1–30; NPE, E = 1–22; elution in order of ethoxy number	CSC Spherisorb LC-1 C1 (TMS), 4.6 × 150 mm	Isocratic mixtures of MeOH/0.2% aqueous NH$_4$OAc: 50 to 62% MeOH (higher MeOH % for higher n)	UV, 225 nm	102,103
Nonylphenol 5-, 10-, 15-, and 20-mole propoxylates; single peak for PPG, then elution in order of homolog number	Merck LiChrosorb RP18, 4.6 × 250 mm	90:10 or 99:1 MeOH/H$_2$O	RI (needed for PPG) or UV, 258 nm	92

(Continued)

TABLE 9 (*Continued*)

Compounds studied	Column	Mobile phase	Detection	Ref.
OPE, E = 10; elution in order of increasing ethoxy number; separation from PEG	Hypersil Hypercarb-S porous graphitic carbon, 4.6 × 100 mm	Gradient: A = H$_2$O, B = CH$_3$CN; C = CH$_2$Cl$_2$; 80% A/20% B to 100% B in 15 min, then to 20% B/80% C in 25 min	Evaporative light scattering	100
Tris(butyl)phenol ethoxylate, E = 1 to 10; separation in order of increasing ethoxylation; peaks split due to isomers	Shandon Hypercarb porous graphite, 4.6 × 100 mm	Various isocratic MeOH/H$_2$O ratios, 97.5:2.5 to 80:20	UV, 220 nm	270
PEG determination	Merck LiChrosorb RP18, 4.6 × 250 mm, or others	95:5 MeOH/H$_2$O	RI	90
PEG determination	Merck LiChrosorb RP-18, 4 × 125 mm, and Waters Ultrahydragel 120, 7.8 × 300 mm	80:20 MeOH/H$_2$O	RI	252
Other systems				
NPE, E = 7–20; elution in order of ethoxy number	Toyo Soda TSK-gel IC-Cation-SW, 4.6 × 50 mm, in K$^+$ form;	Gradient from MeOH to MeOH containing 0.005 M KCl, or constant K$^+$ concentration with temperature programming	UV, 280 nm	105–107
OPE, E = 10; determination at 0.01–0.1% level in presence of proteins; single peak: narrow from anion exchanger, broad from cation exchanger	Supelco LC-SCX cation or LC-SAX anion exchange columns, 4.6 × 20 mm	30:70 2-PrOH/0.05 M phosphate buffer; pH adjusted to 3 or 4	UV, 275 nm	271
NP 4- to 85-mole ethoxylates, single broad SEC peak: determination of molecular weight distribution	Polymer Laboratories PLgel mixed-C 10^3–10^7 A, 7.5 × 300 mm and Waters Ultrastyragel 100 A, 7.8 × 300 mm, 40°C	THF	RI	272
OPE, NPE: preparative oligomer fractionation	Bio-Rad Bio-Beads SX-3, 22 × 250 mm	50:50 CH$_2$Cl$_2$/hexane	None specified	256

TABLE 10 HPLC Analysis of Ethoxylated Acids

Compounds studied	Column	Mobile phase	Detector	Ref.
Mixed C_8/C_{10} ethoxylate: separation of PEG, each of the two monoesters, and each of the three diesters; peaks identified by offline MALDI MS	Waters μBondapak C_{18}	MeOH/H_2O gradient	Evaporative light scattering	150
C_{12} acid 6- and 20-mole ethoxylates; percent PEG, mono- and diesters	Merck LiChrosorb RP-18, 4.6×250 mm; other reversed-phase columns also used	95:5 MeOH/H_2O	RI	90
Separation of oleic acid 7 EO, NP 23 EO. and tallowamine 15 EO from each other, single peak for each	Macherey-Nagel Nucleosil C_8, 4×150 mm	80:20 MeOH/0.06 M phosphate buffer, pH 2.7	UV, 220 nm	273
C_9 alkyl ester of PEG: separation of PEG, monoester, diester; single peak for each	Waters Symmetry C18, 4.8×150 mm	Gradient: 50:50 to 90:10 THF/H_2O in 2 min	Electrospray MS	149
Mono- and dioleate of 10-mole ethoxylate of N,N'-bis (2-hydroxyethyl)-5,5-dimethyl-hydantoin; preparative separation of mono- and diester and free ethoxylate	Merck Lobar LiChroprep RP-8, 25×310 mm, six in series	MeOH	RI	274
Mono- and di-heptanoate of tetraethylene glycol: determination of mono- and diester, heptanoic acid, and tetraethylene glycol	Nucleosil S100 silica	For heptanoic acid and diester: 80:20:3 hexane/2-PrOH/HOAc; for tetraethylene glycol and monoester: 97:3 CH_3CN/HOAc	RI	275

(Continued)

TABLE 10 (*Continued*)

Compounds studied	Column	Mobile phase	Detector	Ref.
C_{12}–C_{18} acid 3- to 24-mole ethoxylates; ethoxy distribution of mono- and diester and of free PEG	Merck LiChrosorb DIOL, 4 × 250 mm	105:95:10:1 or 140:60:5:1 *n*-hexane/2-PrOH/H_2O/HOAc	RI (preferred) or UV, 210 nm	90,140,151
Oleic acid 6-, 9-, and 15-mole ethoxylates; resolution of di-ester, monoester, and PEG, with separation of oligomers of monoester and PEG	Spherisorb NH_2, 4.6 × 250 mm	Hexane/$CHCl_3$/MeOH: complex gradient from 75:20:5 to 50:12.5:37.5	Evaporative light scattering	117
Oleic acid ester of PEG 200; preparative separation of mono- and diester, oleic acid, and PEG	Sephadex LH-20, 70 mm × 5 m (three columns)	Acetone	RI	274

TABLE 11 HPLC Analysis of Ester-Type Nonionic Surfactants

Compound	Column	Mobile phase	Detection	Ref.
Normal phase chromatography				
Glycerol and propylene glycol C_{16} and C_{18} esters; separation by number of acyl groups; partial separation of isomers	Merck LiChrosorb DIOL, 4.6 × 250 mm	95:5 isooctane/2-PrOH	UV, 213 nm	164
Glycerol and polyglycerol fatty esters; separation by number of acyl groups; partial isomer separation	Merck LiChrosorb DIOL, 4.6 × 250 mm	Gradient from *n*-hexane to 97:3 *n*-hexane/2-PrOH	UV, 220 nm	163
Glycerol mono-, di-, and tripalmitates; separation by number of acyl groups	Whatman Partisil PXS 10/25 PAC amino/cyano media, 4.6 × 250 mm	Gradient from 60:65 hexane/$CHCl_3$ to 35:25:65 CH_3CN/hexane/$CHCl_3$	IR, 5.72 μ	121
Esters of glycerol, sorbitan, and sucrose; separation by number of acyl groups	Waters μPorasil, 4 × 300 mm	Gradient from 99:1 isooctane/2-PrOH to 2-PrOH or 90:10 EtOH/H_2O	UV, 220 nm	158
Trioleate of glycerol ethoxylate; ethoxy distribution; peaks not identified	Macherey and Nagel Nucleosil NH_2, 4 × 125 mm, 55°C	95:5 to 80:20 hexane/EtOH in 15 min	Evaporative light scattering	169
Reversed-phase chromatography				
Sucrose monoesters; separation by chain length of acyl group	Two in series: Waters μBondapak C_{18}, 3.9 × 150 mm, and Waters Nova-Pak C_{18}, 3.9 × 150 mm	70:30 acetone/H_2O	RI	276
Sucrose C_{16} and C_{18} esters; separation by number and type of acyl groups	Merck LiChrosorb RP-18 4.6 × 250 mm, 40°C	95:5 MeOH/H_2O	RI and UV, 220 nm	162

(Continued)

TABLE 11 (Continued)

Compound	Column	Mobile phase	Detection	Ref.
Sucrose C_{16} and C_{18} mono- and diesters; separation by type of acyl group; partial isomer separation	Phase Separations Spherisorb Octadecyl-1 or Octadecyl-2, 4.6×300 mm	85:15 MeOH/H_2O or 65:25:10 MeOH/ethyl acetate/H_2O	RI	159
Sucrose and sorbitan C_{16} and C_{18} mono- to pentaesters; separation by number of acyl groups and from fatty acids; partial isomer separation	Hatachi Gel 3057 C_{18}, 4.6×150 mm, 50°C	Gradient: $A = 85:15$ MeOH/H_2O, $B = 75:25$ EtOH/MeOH, both A and B 0.0057 M in triethylamine; nonlinear gradient from 0 to 100% B in 25 min	Postcolumn reactor: alkaline hydrolysis of esters and detection of fatty acids by reaction with 2-nitro-phenylhydrazine; absorbance at 550 nm	118
Sorbitan esters; separation by number of acyl groups	Merck LiChrosorb RP-18, 4.6×250 mm	2-PrOH/H_2O	UV, 220 nm	160
Sorbitan esters; separation by number of acyl substituents	Waters Nova-Pak C_{18}, 8×100 mm	85:15 or 90:10 2-PrOH/H_2O	UV, 220 nm	161
Glycerol C_{14}–C_{20} monoesters; separation by type of acyl group	Merck LiChrosorb RP-18	CH_3CN	UV, 204 nm	277

Sample/application	Column	Mobile phase	Detection	Ref.
Trioleate of glycerol ethoxylate; separation of mono-, di-, and trioleates as well as PEG and free oleic acid	Macherey and Nagel Nucleosil 120 C4, 40°C	Acetone for 10 min, then to 60:40 acetone/H_2O in 2 min	Evaporative light scattering	169
Ethoxylated sorbitan esters: demonstration of ELS detector	Polymer Laboratories PLRPS. 4.6 × 150 mm	9:1 H_2O/CH_3CN to 100% CH_3CN in 35 min	Evaporative light scattering detector	170

Size exclusion chromatography

Sample/application	Column	Mobile phase	Detection	Ref.
Sucrose C_{16} and C_{18} hexa-, hepta-, and octaesters; determination in biological samples	Waters μStyragel, 7.6 × 300 mm, two 500 Å and one 1000 Å column in series	THF	RI	166
Esters of sorbitol, sucrose, glucose, and fructose; separation by number of acyl groups	Finepack GEL–201 SEC media, 2 columns, each 2 × 50 cm	THF	RI	167
Trioleate of glycerol ethoxylate; separation of free oleic acid	Waters Ultrastyragel, 500 Å and 100 Å, each 7.8 × 300 mm, 45°C	THF	RI	169

TABLE 12 HPLC Analysis of Alkyl Polyglycosides

Compound	Column	Mobile phase	Detection	Ref.
C_{12}, C_{14}, C_{16} alkylmonoglucosides; alkyl chain length distribution	Unspecified C_{18}	80:20 MeOH/H_2O	RI	278
C_{12}, C_{14}, C_{16} alkylmonoglucosides; alkyl chain length distribution	EnCaPharm 100-RP$_{18}$, 4.6 × 250 mm	95:5 MeOH/H_2O	Evaporative light scattering	279
C_7–C_{17} alkylmonoglucosides; alkyl chain length distribution	Merck LiChrosorb RP8, 4 × 250 mm	A = 40:60, B = 90:10 CH_3CN/H_2O; 6 min initial hold, then to 100% B in 36 min	Evaporative light scattering	280
C_8–C_{12} alkylmonoglucosides; alkyl chain length distribution; resolution of α and β forms of pyranosides and furanosides	Merck Superspher 60 RP-Select B C_8, 2.1 × 125 mm, 35°C	Gradient: A = 5:95 and B = 80:20 CH_3CN/water adjusted to pH 7.9 with NH_3; 0% B for 2 min, then to 25% B in 2 min, then to 50% B in 15 min, then to 70% B in 1 min.	Electrospray MS	174
C_8–C_{12} alkylmonoglucosides; alkyl chain length distribution; AE also detected; formulation analysis	Beckman RP-8, 4.6 × 250 mm	Gradient: A = CH_3CN, B = H_2O: 60 to 20% B in 15 min after 5 min hold	Evaporative light scattering	122
C_6–C_{12} alkylmonoglucosides; alkyl chain length distribution; AE and APE also detected; formulation analysis	Macherey-Nagel Nucleosil RP$_8$, 2 × 250 mm, 55°C	CH_3CN/H_2O, 60/40, 0.001 M in NaOAc; complex CH_3CN/H_2O/NOAc gradient also used	MS	173
C_{12}–C_{16} alkylpolyglucosides; alkyl chain length distribution; resolution of monoglucosides from oligoglucosides	RP-8	CH_3CN/H_2O; ratio not specified	RI	175,281

Application	Column	Mobile phase	Detection	Ref.
C_8–C_{10} alkylpolyglucosides; residual alcohol determination; alcohols resolved by chain length	Waters Nova-Pak C_{18}	66:24:10 $CH_3CN/H_2O/MeOH$	RI: $C_5H_{11}OH$ or $C_{11}H_{23}OH$ internal standard	175
C_6–C_{12} alkylmono- and alkyldiglucosides; model compounds; study of reversed-phase separation	Various reversed-phase columns, 125 × 150 mm	CH_3CN/H_2O, various ratios	ELS	282
C_1–C_{12} alkylmono- and alkyldiglucoside model compounds; demonstration of reversed-phase separation	Macherey-Nagel Nucleosil RP_8, 4 × 250 mm	60:40 CH_3CN/H_2O	RI and UV, 190 nm; nonyl-monoglucoside internal standard	171
C_{12}–C_{16} alkylpolyglycosides; determination of fatty alcohol content after hydrolysis, steam distillation, and formation of fluorescent derivatives	Merck LiChrosorb RP8. 4.6 × 125 mm	CH_3CN	Fluorescence: 228 nm excitation, 318 nm emission for carbazole-9-carbonyl chloride derivatives; 260 nm excitation, 310 nm emission for 9-fluorenylmethyl chloroformate derivatives	44,45
Octyl glucoside; preparative SEC separation from protein and asolectin	Bio-Rad Bio-Beads SM-2, 6 g	pH 8 aqueous buffer	Analysis of fractions	283

TABLE 13 HPLC Analysis of EO/PO Copolymers

Compound	Column	Mobile phase	Detection	Ref.
Compounds not specified	C_8 column	100% MeOH	RI	89
1–11 PO units with 0–40 total EO units; separation by polypropylene oxide chain length	Chrompak C_{18}, 4.6 × 250 mm	42:58 CH_3CN/H_2O	RI	86
1–10 PO units with 0–21 total EO units; separation by polypropylene oxide chain length	Machery-Nagel Nucleosil $5C_{18}$, 4 × 250 mm	43:57 or 42:58 CH_3CN/H_2O	RI; MALDI-MS of fractions	87,177
PO = 34, EO = 24 (MW = 3000, %EO = 30); SEC separation	Polymer Laboratories PLgel 500 A, 7.5 × 300 mm	THF	RI	178
Broad range of products of varying MW and EO/PO ratios; preparative separation into narrow molecular weight fractions	Sephadex LH-20, 2.2 × 115 cm	50:50 2-PrOH/H_2O or 80:20 2-PrOH/n-butanol	RI	284

TABLE 14 HPLC Analysis of Cationic Surfactants

Substance	Stationary phase	Mobile phase	Detection	Ref.
Normal phase methods				
Distearyldimethylammonium chloride, ditalowimidazolinium methylsulfate, N-dodecylpyridinium chloride; separation from each other; determination in water	Whatman Partisil PAC 10, 2 in series	80:20 $CHCl_3$/MeOH	Conductivity	285
C_{12}–C_{14} benzalkonium chloride; environmental analysis	Whatman Partisil PAC 10 (cyanoamino bonded phase), 4.6×250 mm	80:20 $CHCl_3$/MeOH	Postcolumn reaction (ion-pair extraction; fluorescence)	286
C_{12}–C_{18} dialkyldimethylammonium salts; environmental analysis	Whatman Partisil PAC 10 (cyanoamino bonded phase), 4.6×250 mm	Gradient, from 98:2 $CHCl_3$/CH_3CN to 91:7:2 $CHCl_3$/MeOH/CH_3CN to 87:13 $CHCl_3$/MeOH	Postcolumn reaction (ion-pair extraction)	195
Didecyldimethylammonium chloride and didodecyldimethylammonium bromide; determination in wood	YMC PVA-Sil OR S-5, 4.6×250 mm (polyvinylalcohol-coated silica, operated in normal phase mode)	50:250:1:1:700 MeOH/$CHCl_3$/formic acid/triethylamine/hexane; flush with methanol	Evaporative light scattering	287
C_8, C_{12}, and C_{18} monoalkyltrimethyl- and dialkyldimethylammonium chlorides and ester quats; separation and determination	Merck LiChrospher 100 NH$_2$, 4×125 mm, or Spherisorb S5 Aminopropyl, 4×125 mm	Gradient: A = 97.8:1:1:0.2, B = 88.8:20:1:0.2 $CHCl_3$/MeOH/CH_3CN/HOAc; 0 to 60% B in 60 min (and similar programs)	Postcolumn reactor: extraction with fluorescent anion	189,228
C_{14}–C_{18} dialkyldimethylammonium salts; mono- and trialkyl impurities measured	Phase Separations Spherisorb amino, 0.25×250 mm	Gradient: A = 80:20:0.5 $CHCl_3$/n-hexane/HOAc, B = 50:50:0.5 $CHCl_3$/MeOH/HOAc; 5 to 20% B in 18 min	FAB MS	187

(Continued)

TABLE 14 (*Continued*)

Substance	Stationary phase	Mobile phase	Detection	Ref.
Tallow dialkyldimethylammonium salt; single peak for all homologs; environmental analysis	Nucleosil NH$_2$, 4 × 125 mm	Gradient: A = 99:1:0.25, B = 75:25:0.25 CHCl$_3$/MeOH/ HOAc; 5% B for 2 min, then to 25% B in 13 min, then to 40% B in 2 min	Postcolumn reactor: extraction with UV-absorbing or fluorescent anion	190
Tallow dialkyldimethylammonium salt; mono- and trialkyl impurities measured; separated from an imidazolininium compound and from C$_{16}$ alkylbenzyldimethylammonium chloride; monoalkyl compounds resolved by chain length (C$_{14}$–C$_{18}$)	Bio-Rad RSil polyphenol, 4.6 × 250 mm	A = *n*-hexane, B = 3:1 THF/MeOH, all 0.005 M in trifluoroacetic acid; 10% to 90% B in 20 min	Evaporative light scattering	186
Ester quat—Quaternized tallow triethanolamine esters; unquaternized mono-, di-, and triesteramine impurities and fatty acid impurities separated; triesterquat, diesterquat, and monoesterquat resolved; elution according to decreasing alkyl character	Bio-Rad RSil polyphenol, 4.6 × 150 mm, with 3.2 × 50 mm guard column	Gradient: 90:8.5:1.5 to 10:76.5:13.5 *n*-hexane/THF/MeOH (all 0.005 M in trifluoroacetic acid) in 20 min	Evaporative light scattering	288
Distearyldimethylammonium chloride, cocoalkyldimethylbenzylammonium chloride, cetyltrimethylammonium bromide, and a quaternary imidazoline compound; resolved from each other; environmental analysis	Merck DIOL 100	50:50:10 MeOH/THF/H$_2$O, 0.0002 M in HCl	Conductivity	185,289

Reversed-phase methods

Analyte; determination	Column	Mobile phase	Detection	Ref.
Hexadecylpyridinium chloride; determination in pharmaceuticals	Ultrabase C_{18}, 4.6 × 250 mm	10:50:40 0.03 M H_2O/$CHCl_3$/MeOH, 0.01 M in sodium dioctyl sulfosuccinate	UV, 244 or 254 nm	290
Hexadecylpyridinium chloride; determination in lozenges	Shandon Hypersil CPS cyanopropyl, 4.6 × 100 mm	60:40 0.02 M phosphate buffer, pH 2.5/MeOH, 0.03 M in cetyltrimethylammonium bromide	UV, 254 nm	182
Hexadecylpyridinium chloride; determination in mouthwash	Supelco LC-CN, 4.6 × 150 mm	90:10 MeOH/H_2O, buffered at pH 6 with 0.065 M acetate	UV, 254 nm	180
C_{12}–C_{16} alkyldimethylbenzylammonium chlorides; determination in ophthalmic formulations	Spherisorb CN	60:40 H_2O/CH_3CN, 0.1% triethylamine, pH 2.5	UV, 215 nm	291
C_{12}–C_{16} alkyldimethylbenzylammonium chlorides; determination in ophthalmic formulations	DuPont Zorbax Stablebond CN, 4.6 × 150 mm, 40°C	250:150:2 H_2O/THF/triethylamine, adjusted to pH 3.0 with H_3PO_4	UV, 215 nm	292
C_{12}–C_{14} alkyldimethylbenzylammonium chlorides; determination in ophthalmic formulations	Keystone Scientific CPS Hypersil-1 cyano, 4.6 × 150 mm	65:35 CH_3CN/(aqueous 0.05 M sodium propionate adjusted to pH 5.3 with H_2SO_4)	UV, 214 nm	293
C_{12}–C_{16} alkyldimethylbenzylammonium chlorides; determination in ophthalmic formulations	Waters μBondapak Phenyl, 3.9 × 300 mm	65:35 CH_3CN/pH 6.3 buffer (0.05 M KH_2PO_4, 0.057 M sodium hexanesulfonate, pH adjusted with NaOH)	UV, 215 nm	294,295
C_{12}–C_{16} alkyldimethylbenzylammonium chlorides; determination in nasal spray	Beckman C_8, 4 × 250 mm	Gradient, 60:40 to 100:0 CH_3CN/0.02 M HCl in 20 min	UV, 214 nm	296

(Continued)

TABLE 14 (*Continued*)

Substance	Stationary phase	Mobile phase	Detection	Ref.
C_8–C_{16} alkyltrimethyl- and dialkyldimethylammonium salts; determination in the presence of amines and paraffin oil	Dupont Zorbax C_8, 4.6 × 250 mm or Macherey & Nagel Nucleosil CN, 4.6 × 150 mm	25:75 MeOH/H_2O, containing 0.005 M p-toluenesulfonic acid	Inverse photometric detection, 260 nm	297
C_{12}–C_{18} alkylpyridinium, alkyltrimethylammonium, and alkylbenzyldimethylammonium salts; identification in mixtures	Toyo Soda C_{18}; TSK Gel LS 410, 6 × 200 mm, 50°C	85:15 MeOH/H_2O, either 1.0 M in $NaClO_4$, adjusted to pH 2.5 with H_3PO_4, or 0.1 M in $NaClO_4$, adjusted to pH 3.5	Refractive index	6
C_{14} alkyltrimethylammonium chloride, C_{14}–C_{16} alkyldimethylbenzylammonium chlorides, benzethonium chloride; assay	MPIC-NS1 mobile phase ion chromatography column, 4 × 200 mm	70:30 or 75:25 CH_3CN/0.005 M $HClO_4$	Conductivity	13
C_{12}–C_{18} alkylbenzyldimethylammonium chlorides; assay	Waters μBondapak CN, 4. 6 × 300 mm	60:40 CH_3CN/(aqueous 0.1 M NaOAc, adjusted to pH 5 with HOAc)	UV, 254 nm	298
C_{12}–C_{16} alkyltrimethylammonium salts; assay	Polymer Laboratories PLRP-S poly(styrene/divinylbenzene), 4.7 × 250 mm	60:38:2 CH_3CN/H_2O/HOAc, 0.005 M in sodium p-xylenesulfonate	Inverse photometric detection, 262 nm	181
C_{12}–C_{18} alkyltrimethyl-, tetrapropyl-, tetrabutyl-, dioctyldimethyl-, and didecyldimethylammonium salts and cetylpyridinium chloride; determination in water, disinfectant, and mouthwash	Alltech Surfactant/R polydivinylbenzene, 4.6 × 150 mm	Nonlinear gradient from 30:70 to 80:20 CH_3CN/0.002 M nonafluoropentanoic acid	Conductivity after chemical suppression	299
C_{12}, C_{16}, C_{18} alkyltrimethyl-, C_{14} alkyldimethylbenzylammonium salts; separation	Shodex Asahipak GF-310 HQ polyvinyl alcohol, 4.6 × 100 mm	H_2O, 0.0004 M in 4,4'-dipyridyl, 0.0008 M in HCl, and 27% w/v in CH_3CN	Conductivity after chemical suppression	300

C_{14}–C_{18} alkylamidopropyl-N,N-dimethyl-N-(2,3-dihydroxypropyl)ammonium chlorides; determination in shampoo and skin moisturizer	Waters µBondapak CN, two in series, each 3.9 × 150 mm	45:55 CH_3CN/H_2O or 42:57:1 $CH_3CN/H_2O/THF$, each 0.1% in trifluoroacetic acid	Refractive index	301
Didecyldimethylammonium chloride and dodecyltrimethylammonium bromide; determination in beer	Waters Novapak CN RCM cartridge, 8 × 100 mm	80:20 $MeOH/H_2O$, both 0.01 M in benzenesulfonic acid	Refractive index	197
C_{16}–C_{18} dialkyldimethylammonium, dialkylmethyl-2-hydroxyethylammonium, and dialkylmethylpoly(oxyethylene)-ammonium salts; determination in laundry detergent	Shandon Hypersil C8, 4.6 × 150 mm	60:60:5 $MeOH/CH_3CN/H_2O$, all 0.1 M in NH_4OAc	Postcolumn reactor: paired-ion extraction with bromphenol blue, monitor at 605 and 670 nm	183
C_{14}–C_{20} dialkylimidazolium quats; separation by total alkyl chain length	Two columns in series: Phenomenex Bondclone C_{18}, 3.9 × 300 mm, and Waters Novapak C_{18}, 3.9 × 200 mm; 35°C for lower MW imidazolines and 40°C for higher	50:50 CH_3CN/H_2O, 0.1 M in $NaClO_4$	UV, 230 nm	184
Ion-exchange methods				
C_{18} alkyltrimethylammonium chloride; separation from alkylbetaine amphoterics	Macherey-Nagel Nucleosil 5SA cation exchanger, 4.6 × 150 mm, 35°C	95:5 $MeOH/H_2O$, 0.03 M in glycine, 0.02 M in $HClO_4$, apparent pH 2.95	RI	191
C_3–C_6 tetraalkyl-, C_{12} dialkyldimethyl-, C_{16} alkyldimethylethyl-, and C_{14} alkyltrimethylammonium chloride; elution in order of decreasing alkyl content	Whatman Partisil-10 SCX cation exchanger, 4.6 × 250 mm	70:30 CH_3CN/H_2O, 0.010 M in benzyltrimethylammonium chloride and 1% in HOAc, pH 3.7	Inverse photometric detection, 268 nm	192

(Continued)

TABLE 14 (*Continued*)

Substance	Stationary phase	Mobile phase	Detection	Ref.
C_{16} and C_{18} alkyltrimethyl-, dialkyl-dimethyl-, trialkylmethyl-, alkyldibenzylmethyl-, and dialkylbenzylmethylammonium ions, elution in order of decreasing alkyl character	Whatman Partisil-10 SCX cation exchanger, 2.1 × 250 mm	MeOH containing 0.04 M ammo-nium formate	RI and UV, 264 nm	302
C_{13} and C_{16} dialkyldimethylammo-nium chloride, C_{13} dialkylmethyl-N-poly(ethoxy)ammonium chloride, C_{18} alkyldimethylbenzyl-ammonium chloride, single peaks; determination in alkaline hair care formulations	Metachem Spherisorb SCX cation exchanger, 4.6 × 150 mm	MeOH containing 0.06 M ammo-nium formate	Evaporative light scattering	303

TABLE 15 HPLC Analysis of Phosphatides from Egg Yolk and Soy Lecithin

Compounds determined	Column	Mobile phase	Detector	Ref.
Normal phase methods				
Phosphatidylethanolamine (PE), phosphatidylcholine (PC), phosphatidylinositol (PI), lyso PC, free fatty acids, triglycerides, acetylated PE in soy lecithin	Beckman Ultrasphere Si, 4.6 × 250 mm	Gradient, $CHCl_3$ to 14:86 ammonia/MeOH	FID	304
PE, PI, PC, phosphatidic acid (PA), lyso PC in soybean lecithin	Merck LiChrospher Si 60, 125 mm	Gradient, $CHCl_3$/MeOH/30% ammonia/H_2O	Evaporative light scattering	305
PE, PC, PI, phosphatidylserine (PS), PA, lyso PC, lyso PE in soy lecithin	Waters μPorasil silica	Gradient, C_6H_{14}/2-PrOH/H_2O. 6:8:0.5 to 6:8:1.5	UV, 210 nm	306,307
PE, PI, PS, PC, PA in soy lecithin	Merck LiChrosorb Si-60 or Si-100, 4.6 × 250 mm	Gradient elution: from $CHCl_3$/THF to MeOH/NH_3 (or triethyl-amine)/$CHCl_3$, various ratios	Evaporative light scattering	210
PE, PI, PC, in food extracts	Merck LiChrosorb Si-60, 4.6 × 250 mm	A = 80:19.5:0.5 $CHCl_3$/MeOH/NH_4OH, B = 60:34:5.5:0.5 $CHCl_3$/MeOH/H_2O/NH_4OH; 0 to 55% B in 8 min, then to 60% B in 7 min	Evaporative light scattering	308
PC in soy lecithin	Merck LiChrosorb Si-60, 4.6 × 250 mm	10:40:10 n-C_6H_{14}/2-PrOH/H_2O	RI or UV, 210 nm	198
Determination of neutral lipids, PE, PI, PA, PS, and PC in soy lecithin	Spherisorb silica, 3 μm, 4.6 × 100 mm	58:39:3.2 n-C_6H_{14}/2-PrOH/H_2O, switching to 55:44:5 at 5 min, returning to original ratio at 18 min; 1.8 mL/min	Evaporative light scattering	309

(Continued)

TABLE 15 (*Continued*)

Compounds determined	Column	Mobile phase	Detector	Ref.
Determination of PE, total PI + PA, PC, lyso PE and PS in soy lecithin	DuPont Zorbax silica, 4.6 × 250 mm, reconditioned occasionally with THF	Isooctane/THF/2-PrOH/CHCl$_3$/H$_2$O: A = 415:5:446:104:30. B = 216:4:546:154:80: 0 to 100% B in 20 min, then hold for 32 min	Evaporative light scattering	204
Determination of PE, PI, PA, PC in soybean oil	Merck LiChrospher Si-60/II, 4 × 250 mm	A = 75:15 CHCl$_3$/t-butyl methyl ether, B = 92:7:1 MeOH/NH$_4$OH/CHCl$_3$; 0 to 100% B in 30 min, then hold for 10 min	Evaporative light scattering	310
Determination of PE, PI, PA, PC, lyso PC, lyso PE, and PA in soy lecithin and oil	Advanced Separation Technologies β-cyclodextrin-bonded silica, 4.6 × 250 mm	35:32.7:26.8:5.5 C$_6$H$_{14}$/2-PrOH/EtOH/H$_2$O, 0.005 M in tetramethylammonium phosphate, pH 6.3	UV, 208 nm	311
Semi-quant determination of PA, PC, PE, PI, and PS in extracted crude soybean oil	Merck LiChrosorb Si-60, 4.6 × 250 mm	Ternary gradient: 42:56:2 to 51:38:11 2-PrOH/C$_6$H$_{14}$/H$_2$O in 20 min	UV, 206 nm	312
Determination of PC, phosphatidylglycerol, lyso PC, and lyso-phosphatidylglycerol in liposomes	DuPont Zorbax NH$_2$, 4.6 × 250 mm	64:28:8 CH$_3$CN/MeOH/0.01M NH$_4$H$_2$PO$_4$, pH 4.8	RI	313
Reversed-phase methods				
Separation of dioleoyl PC, dipalmitoyl PC, dipalmitoyl PE, dipalmitoyl PA, monopalmitoyl PC isomers, and palmitic acid	Neos Company Fluofix perfluoroalkyl-bonded silica, 4.6 × 250 mm	Various gradients, 50:50 to 100:0 EtOH/H$_2$O	Evaporative light scattering	314
Characterization of PC	Phase Separations Spherisorb ODS-2, 4.6 × 250 mm, 32°C	90:7:3 (60:40 MeOH/EtOH, each 0.020 M in choline chloride)/H$_2$O/CH$_3$CN	UV, 205 and 215 nm	315

Application	Column	Mobile phase	Detection	Ref.
Characterization of PC and PE	Phenomenex Luna C18, 4.6 × 250 mm	87.5:5:3.75:3.75 MeOH/CHCl$_3$/CH$_3$CN/H$_2$O	Evaporative light scattering	316
Characterization of PI (benzoate esters)	Various C$_{18}$ and poly(styrene/divinylbenzene) columns, 250 mm	70:22:8 CH$_3$CN/MeOH/H$_2$O, 0.005 M in quaternary salt	UV, 208 nm	211
Characterization of PA (with and without formation of the methyl ester)	Various C$_{18}$ columns, 22–30 cm long	CH$_3$CN/MeOH/H$_2$O, 70:22:8, containing tetraalkylammonium phosphate ion pairing salts	UV, 208 nm	210
Characterization of PA (without derivatization)	Various C$_{18}$ columns, 15–30 cm long	CH$_3$CN/MeOH/H$_2$O, 49:49:2, 0.05 M in tetramethylammonium phosphate	UV, 208 nm	317
Characterization of PC and PE	Various C$_{18}$ columns, 15–30 cm long	CH$_3$CN/MeOH/H$_2$O, 70:22:8, containing tetraalkylammonium phosphate ion pairing salts	UV, 208 nm	318
Characterization of PC and PE	Alltech RSIL-C18, HL, 4.6 × 250 mm	MeOH/H$_2$O/CHCl$_3$, 20:1:1 for PE; 30:1:1 for PC	Evaporative light scattering	319
Characterization of PS	Various C$_{18}$ columns, 10–30 cm long	CH$_3$CN/MeOH/H$_2$O, 70:22:8, containing tetraalkylammonium phosphate ion pairing salts	UV, 208 nm	320
Characterization of PA, PE, and PC as derivatives	Various C$_{18}$ columns, 25–30 cm long	CH$_3$CN/MeOH/H$_2$O, 70:22:8 or 49:49:2, containing tetraalkylammonium phosphate ion pairing salts	UV, 208 nm	212

TABLE 16 HPLC Analysis of Nonlecithin Amphoteric Surfactants

Compounds	Column	Mobile phase	Detection	Ref.
β-Alanine-types: C_8–C_{18} alkylamino-ethylcarboxylates and dicarboxy-lates; elution in order of increasing alkyl chain length	Nomura Develosil ODS-3. 4.6×150 mm	60:40 CH_3CN/0.2 M $NaClO_4$, adjusted to pH 2.5 with H_3PO_4	Postcolumn reaction for specific detection of amines	203
C_{10}–C_{18} alkylaminopropionate and C_{10}–C_{18} alkyldimethylaminoacetate betaine; elution by increasing alkyl chain length and determination in the presence of other surfactants	Toyo Soda TSK Gel LS 410 C18, 6×250 mm	85:15 MeOH/H_2O, 1.0 M in $NaClO_4$, adjusted to pH 2.5 with H_3PO_4	RI or postcolumn extraction with Orange II at pH 3.4 and detection at 484 nm	6,9
Dodecylaminopropionate, dodecyldi-methylaminoacetic acid betaine, dodecylamidopropyl-N,N-dimethyl-aminoacetic acid betaine, and imidazoline–derived N-undecylami-doethyl-N-2-hydroxyethyl-N-carboxylate; identification in formulations	Toyo Soda TSK Gel LS 410 C18, 6×250 mm; 2 in se-ries at 50°C	75:25 MeOH/H_2O, 1.0 M $NaClO_4$, adjusted to pH 2.5 with H_3PO_4	Refractive index	7
C_{10}–C_{18} alkyldimethylaminoacetate betaine after derivatization with 4-bromomethyl-7-methoxycoumarin; elution in order of increasing alkyl chain length; determination in the presence of other surfactants	Nomura Develosil C8-3 C_{18}, 4.0×250 mm	CH_3CN/H_2O gradient, each containing 0.1 M $NaClO_4$	UV, 325 nm	200
C_8–C_{18} alkylamidopropylbetaine; elu-tion by increasing alkyl chain length; determination in cosmetics and detergents	Alltech mixed-mode RP C_8/ Cation, 4.6×250 mm	75:25 CH_3CN/H_2O	Evaporative light scattering or UV, 214 nm	215

C_{12}–C_{18} alkylamidopropyldimethyl-aminoacetic acid betaines; elution by decreasing alkyl chain length; separation from cationics; determination in shampoo	Macherey-Nagel Nucleosil 5SA cation exchanger, 4.6 × 150 mm, 35°C	95:5 MeOH/H_2O, 0.03 M in glycine, 0.02 M in $HClO_4$, apparent pH 2.95	RI	191,321
C_{12}–C_{14} alkylbetaine and C_8–C_{18} alkyl-amidopropylbetaine; elution by decreasing alkyl chain length; determination in shampoo products	Macherey-Nagel Nucleosil 100-5SA cation exchanger, 4 × 250 mm	70:30 CH_3CN/0.05 M LiOH, adjusted to pH 1.6 with H_3PO_4	UV, 210 nm	216
C_8–C_{18} alkyl imidazoline–derived products made with acrylic acid, ethyl acrylate, or sodium chloroacetate; determination of intermediates and impurities; elution in order of increasing alkyl chain length	Nomura Develosil ODS-3, 4.6 × 150 mm	Various ratios of MeOH/CH_3CN/H_2O, 0.2 or 0.5 M in $NaClO_4$, pH 2.5	UV, 210 nm, or postcolumn reactor	322,323
Lauro- and isostearoamphoglycinate [$RCONHC_2H_4N(C_2H_5OH)CH_2COO$ Na], dihydroxyethyl tallow glycinate [$RN^+(CH_2CH_2OH)_2CH_2COO^-$], cocoamidopropyl betaine, cocoamidopropylhydroxysultaine [$RCONHC_3H_6N^+(CH_3)_2CH_2CHOH$ $CH_2SO_3^-$]; identification in mixtures	Spherisorb ODS, 4.6 × 250 mm, 50°C	75:25 MeOH/H_2O, 0.25 M in $NaClO_4$, adjusted to pH 2.5 with H_3PO_4	UV, 210 nm	324
Coco alkylamidopropyl-N,N-dimethyl-ammonio-2-hydroxy-1-propane-sulfonate; analysis of mixtures with soap	Waters μBondapak C_{18}	85:15 MeOH/H_2O, containing 0.2% HOAc	RI	240

(*Continued*)

TABLE 16 *(Continued)*

Compounds	Column	Mobile phase	Detection	Ref.
A sulfobetaine, alkylamidopropyl-*N*,*N*-dimethylammonio-1-propane-sulfonate, as well as starting materials and intermediates in its synthesis	Waters μBondapak C_{18}	90:10 MeOH/H_2O, containing 0.2% HOAc and 0.0042 M sodium decylbenzenesulfonate, pH 4	RI and inverse UV, 254 nm	199
Alkylimino acids, alkylbetaines, alkylamidobetaines, imidazoline derivatives, and alkyl- and alkylamidosulfobetaines; identification in mixtures	Macherey & Nagel Nucleosil 7 C_{18}, 4 × 250 mm	90:10 or 80:20 MeOH/H_2O	RI	179
3-[(3-cholamidopropyl)-dimethylammonio]-propanesulfonate (CHAPS); preparative SEC separation from proteins	Bio-Rad Bio-Bead SM-4 or Pierce ExtractiGel D	Phosphate-buffered saline solution	UV/vis spectrophotometry of fractions	325

REFERENCES

1. Rasmussen, H. T., B. P. McPherson, Chromatographic analysis of anionic surfactants, in J. Cross, ed., *Anionic Surfactants: Analytical Chemistry*, 2nd ed., Marcel Dekker, New York, 1998.

2. Eppert, G., G. Liebscher, Factors influencing the resolution of positionally isomeric alkane monosulfonates in reversed-phase ion-interaction chromatography with indirect photometric detection, *J. Chromatogr. Sci.*, 1991, *29*, 21–25.

3. Takeda, T., S. Yoshida, T. Ii, Sulfonate- and sulfate-type anionic surfactants by ion chromatography, *Chem. Express*, 1992, *7*, 441–444.

4. Marcomini, A., A. Di Corcia, R. Samperi, S. Capri, Reversed-phase HPLC of LAS, NPE and their carboxylic biotransformation products, *J. Chromatogr.*, 1993, *644*, 59–71.

5. Sarrazin, L., A. Arnoux, P. Rebouillon, HPLC analysis of LAS and its environmental biodegradation metabolites, *J. Chromatogr. A*, 1997, *760*, 285–291.

6. Nakamura, K., Y. Morikawa, Separation of surfactant mixtures and their homologs by HPLC, *J. Am. Oil Chem. Soc.*, 1982, *59*, 64–68.

7. Nakamura, K., Y. Morikawa, Surfactant mixtures in shampoos and detergents by HPLC, *J. Am. Oil Chem. Soc.*, 1984, *61*, 1130–1135.

8. König, H., W. Strobel, Surfactants in toothpastes by HPLC, *Fresenius' Z. Anal. Chem.*, 1988, *331*, 435–438.

9. Kanesato, M., K. Nakamura, O. Nakata, Y. Morikawa, Ionogenic surfactants by HPLC with ion-pair extraction detector, *J. Am. Oil Chem. Soc.*, 1987, *64*, 434–438.

10. Terweij-Groen, C. P., J. C. Kraak, W. M. A. Niessen, J. F. Lawrence, C. E. Werkhoven-Goewie, U. A. T. Brinkman, R. W. Frei, Ion-pair extraction detector for the LC determination of anionic surfactants, *Int. J. Environ. Anal. Chem.*, 1981, *9*, 45–57.

11. Saito, T., K. Higashi, K. Hagiwara, LAS by HPLC: application to water, *Fresenius' Z. Anal. Chem.*, 1982, *313*, 21–23.

12. Matsutani, S., Y. Endo, Sulfonate type anionic surfactants including 2-sulfonatofatty acid methyl ester by methyl ester derivatization and HPLC analysis, *Yugagaku*, 1991, *40*, 566–573.

13. Weiss, J., Ion chromatography (in German), *Tenside. Surfactants. Deterg.*, 1986, *23*, 237–244.

14. Weiss, J., Retention of aliphatic anionic surfactants in ion chromatography, *J. Chromatogr.*, 1986, *353*, 303–307.

15. Li, J. B., P. Jandik, Separation and detection of non-chromophoric, anionic surfactants, *J. Chromatogr.*, 1991, *546*, 395–403.

16. Asahi Chemical Industry Co., Kawasaki, Japan, Asahipak technical literature.

17. Larson, J. R., Alkanesulfonates by LC with indirect photometric detection, *J. Chromatogr.*, 1986, *356*, 379–381.

18. Eppert, G., G. Liebscher, Application of indirect photometric detection with gradient elution. Commercial alkanesulfonates (in German), *J. Chromatogr.*, 1986, *356*, 372–378.

19. Liebscher, G., G. Eppert, H. Oberender, H. Berthold, H. G. Hauthal, HPLC of alkanemonosulfonates (in German), *Tenside*, 1989, *26*, 195–197.

20. Boiani, J. A., Spectator ion indirect photometric detection of aliphatic anionic surfactants separated by reversed-phase HPLC, *Anal. Chem.*, 1987, *59*, 2583–2586.

21. Shamsi, S. A., N. D. Danielson, Mixed-mode LC of aliphatic anionic surfactants with a naphthalenedisulfonate mobile phase, *J. Chromatogr. Sci.*, 1995, *33*, 505–513.

22. Castles, M. A., B. L. Moore, S. R. Ward, LAS in aqueous environmental matrices by LC with fluorescence detection, *Anal. Chem.*, 1989, *61*, 2534–2540.

23. Field, J. A., L. B. Barber, II, E. M. Thurman, B. L. Moore, D. L. Lawrence, D. A. Peake, Fate of alkylbenzenesulfonates and dialkyltetralinsulfonates in sewage-contaminated groundwater, *Environ. Sci. Technol.*, 1992, *26*, 1140–1148.

24. Marcomini, A., W. Giger, Simultaneous determination of LAS, APE, and nonylphenol by HPLC, *Anal. Chem.*, 1987, *59*, 1709–1715.

25. Maki, S. A., J. Wangsa, N. D. Danielson, Aliphatic anionic surfactants using a weak anion exchange column with indirect photometric and indirect conductivity detection. *Anal. Chem.*, 1992, *64*, 583–589.

26. Smedes, F., J. C. Kraak, C. F. Werkhoven-Goewie, U. A. Th. Brinkman, R. W. Frei, HPLC separation and selective detection of anionic surfactants: application to commercial formulations and water samples, *J. Chromatogr.*, 1982, *247*, 123–132.

27. Shoester, M., G. Kloster, Chromatographic post-column ion pair extraction system for the determination of cationic surfactants. *Vom Wasser*, 1991, *77*, 13–20.

28. Masadome, T., T. Imato, N. Ishibashi, Surfactant-selective electrode based on plasticized poly(vinyl chloride) membrane and its application, *Anal. Sci.*, 1987, *3*, 121–124.

29. Gerlache, M., J. M. Kauffmann, PVC-based ion-selective electrode for surfactant detection in microflow systems, *Biomed. Chromatogr.*, 1998, *12*, 147–148.

30. Bear, G. R., Universal detection and quantitation of surfactants by HPLC by means of the evaporative light-scattering detector, *J. Chromatogr.*, 1988, *459*, 91–107.

31. Marcomini, A., S. Stelluto, B. Pavoni, LAS and APE in commercial products and marine waters by reversed- and normal-phase HPLC, *Int. J. Environ. Anal. Chem.*, 1989, *35*, 207–218.

32. Senden, W. A. A., R. Riemersma, Alkylaryl sulphonates by HPLC, *Tenside*, 1990, *27*, 46–51.

33. Yokoyama, Y., H. Sato, Reversed-phase HPLC determination of LAS in river water at ppb levels by precolumn concentration. *J. Chromatogr.*, 1991, *555*, 155–162.

34. Di Corcia, A., M. Marchetti, R. Samperi, A. Marcomini, LC determination of LAS in aqueous environmental samples. *Anal. Chem.*, 1991, *63*, 1179–1182.

35. Bear, G. R., Surfactant characterization by reversed-phase ion pair chromatography, *J. Chromatogr.*, 1986, *371*, 387–402.

36. Chen, S., D. J. Pietrzyk, Reversed-phase HPLC separation of LAS: effect of mobile phase ionic strength, *J. Chromatogr. A*, 1994, *671*, 73–82.

37. Zhou, D., D. J. Pietrzyk, LC separation of alkanesulfonate and alkyl sulfate surfactants: effect of ionic strength, *Anal. Chem.*, 1992, *64*, 1003–1008.

38. Pietrzyk, D. J., P. G. Rigas, D. Yuan, Separation and indirect detection of alkyl sulfonates and sulfates, *J. Chromatogr. Sci*, 1989, *27*, 485–490.

39. Danielson, N. D., S. A. Shamsi, S. A. Maki, Comparison of fluorocarbon and hydrocarbon weak anion exchange columns for the separation of surfactants with indirect detection, *J. High Resolut. Chromatogr.*, 1992, *15*, 343–346.

40. Shamsi, S. A., N. D. Danielson, Mixed-mode LC of aliphatic anionic surfactants with a naphthalenetrisulfonate mobile phase, *Chromatographia*, 1995, *40*, 237–246.

41. Fischer, J., P. Jandera, Chromatographic behavior in reversed-phase HPLC with micellar and submicellar mobile phases: effects of the organic modifier, *J. Chromatogr. B*, 1996, *681*, 3–19.

42. Ullner, H., I. König, C. Sander, U. Schwenk, LC separation of paraffin sulfonates, with special emphasis on primary paraffin sulfonates (in German), *Tenside*, 1980, *17*, 169–170.

43. Hauthal, H. G., Alkanesulfonates, in H. W. Stache, ed., *Anionic Surfactants: Organic Chemistry*, Marcel Dekker, New York, 1996.

44. Meissner, C., H. Engelhardt, Surfactants derived from fatty alcohols—optimization of derivatization, *Chromatographia*, 1999, *49*, 7–11.

45. Meissner, C., H. Engelhardt, Surfactants derived from fatty alcohols—hydrolysis and enrichment, *Chromatographia*, 1999, *49*, 12–16.

46. Benning, M., H. Locke, R. Ianniello, Semi-quantitative identification of hydrophobes in normal alcohol ethoxylates by hydriodic acid cleavage and reversed-phase HPLC analysis with UV detection, *J. Liq. Chromatogr.*, 1989, *12*, 757–770.

47. Pilc, J. A., P. A. Sermon, Ethoxylated nonionic surfactants and their sulfonates using HPLC, *J. Chromatogr.*, 1987, *398*, 375–380.

48. Jandera, P., J. Urbánek, B. Prokeš, H. Blazková-Brúnová, Chromatographic behavior of oligoethylene glycol nonyphenyl ether anionic surfactants in normal phase HPLC, *J. Chromatogr. A*, 1996, *736*, 131–140.

49. Hodgson, P. K. G., N. J. Stewart, Purification of ethoxylated anionic surfactants by preparative HPLC, *J. Chromatogr.*, 1987, *387*, 546–550.

50. Austad, T., I. Fjelde, Chromatographic analysis of commercial products of ethoxylated sulfonates, *Anal. Lett.*, 1992, *25*, 957–971.

51. Yoshimura, H., T. Sugiyama, T. Nagai, HPLC analysis of polyethoxylated nonionics. *J. Am. Oil Chem. Soc.*, 1987, *64*, 550–555.

52. Jandera, P., Selectivity in reversed-phase LC. Retention behaviour of oligomeric series, *J. Chromatogr.*, 1988, *449*, 361–389.

53. Jandera, P., J. Urbánek, Chromatographic behavior of oligoethylene glycol nonyphenyl ether nonionic and anionic surfactants in reversed-phase HPLC, *J. Chromatogr. A*, 1995, *689*, 255–267.

54. Fjelde, I., T. Austad, HPLC analysis of salt tolerant mixtures of ethoxylated and non-ethoxylated sulfonates applicable in enhanced oil recovery, *Colloids Surf. A*, 1994, *82*, 85–90.

55. Johannessen, R. O., W. J. DeWitt, R. S. Smith, M. E. Tuvell, HPLC of α-olefin sulfonates, *J. Am. Oil Chem. Soc.*, 1983, *60*, 858–861.

56. Beranger, A., T. Holt, Middle and heavy α-olefin sulfonates, *Tenside Detergents*, 1986, *23*, 247–254.

57. Kudoh, M., K. Tsuji, HPLC analysis of anionic surfactants by derivatization, *J. Chromatogr.*, 1984, *294*, 456–459.

58. Bear, G. R., C. W. Lawley, R. M. Riddle, Separation of sulfonate and carboxylate mixtures by ion-exchange HPLC, *J. Chromatogr.*, 1984, *302*, 65–78.

59. Shengxiang, J., C. Liren, Petroleum sulfonates in crude oil with multi-dimensional HPLC. *Anal. Sci.*, 1991, *7*, 129–131.

60. Desbène, P. L., C. Rony, B. Desmazières, J. C. Jacquier, Alkylaromatic sulfonates by HPCE, *J. Chromatogr.*, 1992, *608*, 375–383.

61. Schreuder, R. H., A. Martijn, C. van de Kraats, Diisobutyl- and diisopropylnaphthalenesulfonates in pesticide wettable powders and dispersible granules by HPLC, *J. Chromatogr.*, 1989, *467*, 177–184.

62. Mengerink, Y., H. C. J. De Man, Sj. Van der Wal, Evaporative light scattering detector in reversed-phase HPLC of oligomeric surfactants, *J. Chromatogr.*, 1991, *552*, 593–604.

63. König, H., W. Strobel, Alkyl- and alkylphenol ether carboxylates by HPLC. *Fresenius. Z. Anal. Chem.*, 1990, *338*, 728–731.

64. Gerhardt, W., G. Czichocki, H. R. Holzbauer, C. Martens, B. Weiland, Ether carboxylates in reaction mixtures (in German), *Tenside, Surfactants, Deterg.*, 1992, *29*, 285–288.

65. Lewis, N. G., W. Q. Yean, High-performance SEC of lignosulfonates, *J. Chromatogr.*, 1985, *331*, 419–424.

66. Van der Hage, E. R. E., W. M. G. M. van Loon, J. J. Boon, H. Lingeman, U. A. Th. Brinkman, SEC and pyrolysis-GC-MS study of lignosulfonates in pulp mill effluents, *J. Chromatogr.*, 1993, *634*, 263–271.

67. Majcherczyk, A., A. Hüttermann, SEC of lignin as ion-pair complex, *J. Chromatogr. A*, 1997, *764*, 183–191.

68. Frazier, J. D., R. D. Johnson, C. G. Wade, D. J. O'Leary, NPE phosphate anionic surfactants, *Comun. Jorn. Com. Esp. Deterg.*, 1991, *22*, 99–110.

69. Steinbrech, B., D. Neugebauer, G. Zulauf, Reversed-phase ion-pair chromatography of alkylsulfates and alkylsulfosuccinates (in German), *Fresenius' Z. Anal. Chem.*, 1986, *324*, 154–157.

70. Noguchi, H., S. Matsutani, S. Tanaka, Y. Horiguchi, T. Hobo, Alkyl homolog distribution of ionic surfactants by two-dimensional HPLC (in Japanese), *Bunseki Kagaku*, 1998, *47*, 473–479.

71. Thomas, D., J. L. Rocca, HPLC analysis of surface-active agents used in detergent formulations (in French), *Analusis*, 1979, *7*, 386–394.

72. Miller, R. A., N. E. Bussell, C. Ricketts, Quantitation of long chain fatty acids as the methoxyphenacyl esters, *J. Liq. Chromatogr.*, 1978, *1*, 291–304.

73. Moreno, A., J. Bravo, J. Ferrer, C. Bengoechea, Soap in sewage sludge by HPLC, *J. Am. Oil Chem. Soc.*, 1993, *70*, 667–671.

74. Moreno, A., J. Bravo, J. Ferrer, C. Bengoechea, Soap in various environmental matrices (in Spanish), *Comun. Jorn. Com. Esp. Deterg.*, 1993, *24*, 45–57.

75. Miszkiewicz, W., J. Szymanowski, Nonionic surfactants with polyoxyethylene chains by HPLC, *Crit. Rev. Anal. Chem.*, 1996, *25*, 203–246.

76. Rissler, K., HPLC of polyethers and their mono(carboxy)alkyl and -arylalkyl substituted derivatives, *J. Chromatogr. A*, 1996, *742*, 1–54.

77. Mikolajczyk, L., W. Leukefeld, E. Döring, EO distribution of *n*-alcohol EO adducts with high resolution SEC (in German), *Tenside, Surfactants, Deterg.*, 1993, *30*, 34–38.

78. Funasaki, N., S. Hada, S. Neya, Monomer concentrations of nonionic surfactants as deduced with gel filtration chromatography, *J. Phys. Chem.*, 1988, *92*, 7112–7116.

79. Noguchi, K., Y. Yanagihara, M. Kasai, B. Katayama, Chromatographic properties of a vinyl alcohol copolymer gel column for the analysis of nonionic surfactants, *J. Chromatogr.*, 1989, *461*, 365–375.

80. Crescenzi, C., A. Di Corcia, R. Samperi, A. Marcomini, Nonionic polyethoxylate surfactants in environmental waters by LC/electrospray MS, *Anal. Chem.*, 1995, *67*, 1797–1804.

81. Melander, W. R., A. Nahum, C. Horvath, Changes in conformation and retention of oligo(ethylene glycol) derivatives with temperature and eluent composition, *J. Chromatogr.*, 1979, *185*, 129–152.

82. Trathnigg, B., B. Maier, D. Thamer, Polyethers by isocratic HPLC with universal detectors. Study on reproducibility, *J. Liq. Chromatogr.*, 1994, *17*, 4285–4302.

83. Lemr, K., M. Zanette, A. Marcomini, Reversed-phase HPLC separation of 1-naphthyl isocyanate derivatives of linear AE, *J. Chromatogr. A*, 1994, *686*, 219–224.

84. Lemr, K., Homolog separation of linear AE by HPLC, *J. Chromatogr. A*, 1996, *732*, 299–305.

85. Pasch, H., I. Zammert, Chromatographic investigations of macromolecules in the critical range of LC. VIII: Analysis of polyethylene oxides, *J. Liq. Chromatogr.*, 1994, *17*, 3091–3108.

86. Gorshkov, A. V., H. Much, H. Becker, H. Pasch, V. V. Evreinov, S. G. Entelis, Chromatographic investigations of macromolecules in the critical range of LC. I: Functionality type and composition distribution in polyethylene oxide and polypropylene oxide copolymers, *J. Chromatogr.*, 1990, *523*, 91–102.

87. Pasch, H., K. Rode, Matrix-assisted laser desorption/ionization MS for molar mass-sensitive detection in LC of polymers, *J. Chromatogr. A*, 1995, *699*, 21–29.

88. Kudoh, M., Separation of AE by HPLC, *J. Chromatogr.*, 1984, *291*, 327–330.

89. König, H., R. Ryschka, W. Strobel, Separation, identification, and determination of nonionic surfactants using HPLC (in German), *Fresenius' Z. Anal. Chem.*, 1985, *321*, 263–267.

90. Zeman, I., M. Paulovic, Comparison of HPLC determination of free PEG in ethoxylated surfactants with extraction procedure, *Proc. 2nd World Surfactants Congress*, Paris, 1988, 384–398.

91. Brossard, S., M. Lafosse, M. Dreux, Comparison of AE and polyethylene glycols by HPLC and SFC using evaporative light-scattering detection, *J. Chromatogr.*, 1992, *591*, 149–157.

92. Zeman, I., M. Paulovic, HPLC analysis of propylene oxide derivatives of nonylphenol (in Slovak), *Petrochemia*, 1990, *30*, 141–149.

93. Sithole, B. B., B. Zvilichovsky, C. LaPointe, L. H. Allen, Adsorption of aqueous NPE on metal sample loops: effect on quantitation by LC, *J. Assoc. Off. Anal. Chem.*, 1990, *73*, 322–324.

94. Desbène, P. L., B. Desmazières, V. Even, J. J. Basselier, L. Minssieux, Nonionic surfactants used in tertiary oil recovery: optimization of stationary phase in normal phase partition chromatography, *Chromatographia*, 1987, *24*, 857–861.

95. Kibbey, T. C. G., T. P. Yavaraski, K. F. Hayes, HPLC analysis of polydisperse ethoxylated nonionic surfactants in aqueous samples, *J. Chromatogr. A*, 1996, *752*, 155–165.

96. Ibrahim, N. M. A., B. B. Wheals, APE in trade effluents by sublation and HPLC, *Analyst*, 1996, *121*, 239–242.

97. Ibrahim, N. M. A., B. B. Wheals, Oligomeric separation of APE surfactants on silica using aqueous acetonitrile eluents, *J. Chromatogr. A*, 1996, *731*, 171–177.

98. Desbène, P. L., B. Desmazières, J. J. Basselier, A. Desbène-Monvernay, Nonionic surfactants used in enhanced oil recovery: optimization of analytical conditions in reversed-phase partition chromatography, *J. Chromatogr.*, 1989, *461*, 305–313.

99. Marcomini, A., M. Zanette, Derivatization procedures and HPLC separations for environmental analysis of AE, *Riv. Ital. Sostanze Grasse*, 1994, *71*, 203–208.

100. Chaimbault, P., C. Elfakir, M. Lafosse, Comparison of the retention behavior of polyethoxylated alcohols on porous graphitic carbon and polar as well as apolar bonded-silica phases, *J. Chromatogr. A*, 1998, *797*, 83–91.

101. Escott, R. E. A., D. W. Chandler, Ammonium acetate as an ion-pairing electrolyte for ethoxylated surfactant analysis by thermospray LC/MS, *J. Chromatogr. Sci.*, 1989, *27*, 134–138.

102. Wang, Z., M. Fingas, NPE by SFC and HPLC, *J. Chromatogr. Sci.*, 1993, *31*, 509–518.

103. Wang, Z., M. Fingas, Rapid separation of OPE and determination of ethylene oxide oligomer distribution by C1 column reversed-phase LC. *J. Chromatogr.*, 1993, *673*, 145–156.

104. Ernst, J. M., Separation of phenol ethoxylate homologues by LC with an ion exchange column, *Anal. Chem.*, 1984, *56*, 834–835.

105. Okada, T., Chromatographic oligomer separation of poly(oxyethylenes) on K^+-form cation-exchange resin, *Anal. Chem.*, 1990, *62*, 327–331.

106. Okada, T., Indirect conductometric detection of poly(oxyethylenes) after chromatographic separation, *Anal. Chem.*, 1990, *62*, 734–738.

107. Okada, T., Temperature programming for separation of polyoxyethylene oligomers, *Anal. Chem.*, 1991, *63*, 1043–1047.

108. Desmazières, B., F. Portet, P. L. Desbène, Highly condensed nonionic surfactants. Separation on cation exchangers, *Chromatographia*, 1993, *36*, 307–317.

109. Stan, H., T. Heberer, P. Billian, AE in drinking water (in German), *Vom Wasser*, 1998, *90*, 93–105.

110. Desbène, P. L., F. I. Portet, G. J. Goussot, Surfactant mixtures by reversed-phase HPLC with refractometric detection, *J. Chromatogr. A*, 1996, *730*, 209–218.

111. Zeman, I., HPLC of ethoxylated surfactants, *J. Chromatogr.*, 1990, *509*, 201–212.

112. Ott, K. H., W. Wagner-Redeker, W. Winkle, On-line thermospray-LC-MS of nonionic surfactants (in German), *Fett Wiss. Technol.*, 1987, *89*, 208–213.

113. Rockwood, A. L., T. Higuchi, LC/MS analysis of nonionic surfactants using the frit-FAB method, *Tenside, Surfactants, Deterg.*, 1992, *29*, 6–12.

114. Kudoh, M., H. Ozawa, S. Fudano, K. Tsuji, AE and APE by HPLC with fluorimetric detection, *J. Chromatogr.*, 1984, *287*, 337–344.

115. Holt, M. S., E. H. McKerrell, J. Perry, R. J. Watkinson, APE in environmental samples by HPLC coupled to fluorescence detection, *J. Chromatogr.*, 1986, *362*, 419–424.

116. Stockwell, P. B., B. W. King, Light scattering detector for LC, *Am. Lab*, 1991, *23(12)*, 19,20,22–24.

117. Martin, M., Nonionic surfactants by HPLC using evaporative light scattering detection, *J. Liq. Chromatogr.*, 1995, *18*, 1173–1194.

118. Kondoh, Y., A. Yamada, S. Takano, Nonionic surfactants with ester groups by HPLC with post-column derivatization, *J. Chromatogr.* 1991, *541*, 431–441.

119. Desbène, P. L., B. Desmazières, J. J. Basselier, A. Desbène-Monvernay, Polarographic detection of nonionic surfactants analyzed by reversed-phase partition chromatography, *J. Chromatogr.*, 1989, *465*, 69–74.

120. Kruse, A., N. Buschmann, Separation and detection of alkylpolyglucosides in HPLC, *Comun. Jorn. Com. Esp. Deterg.*, 1995, *26*, 209–213.

121. Payne-Wahl, K., G. F. Spencer, R. D. Plattner, R. O. Butterfield, HPLC method for quantitation of free acids, mono-, di-, and triglycerides using an IR detector, *J. Chromatogr.*, 1981, *209*, 61–66.

122. Heinig, K., C. Vogt, G. Werner, Nonionic surfactants by CE and HPLC, *Anal. Chem.*, 1998, *70*, 1885–1892.

123. Desbène, P. L., B. Desmazières, Polyoxyethylene surfactants of high degree of condensation by normal phase LC on *p*-nitrophenyl-bonded silica, *J. Chromatogr. A*, 1994, *661*, 207–213.

124. Dubey, S. T., L. Kravetz, J. P. Salanitro, Nonionic surfactants in bench-scale biotreater samples, *J. Am. Oil Chem. Soc.*, 1995, *72*, 23–30.

125. Miszkiewicz, W., J. Szymanowski, Evaporative light scattering detector used in analysis of AE, *J. Liq. Chromatogr.*, 1996, *19*, 1013–1032.

126. Evans, K. A., S. T. Dubey, L. Kravetz, S. W. Evetts, I. Dzidic, C. C. Dooyema, AE surfactants in environmental samples by electrospray MS, *J. Am. Oil Chem. Soc.*, 1997, *74*, 765–773.

127. Jandera, P., M. Holcapek, G. Theodoridis, AE in normal-phase and reversed-phase systems using HPLC-MS, *J. Chromatogr. A*, 1998, *813*, 299–311.

128. Schmitt, T. M., M. C. Allen, D. K. Brain, K. F. Guin, D. E. Lemmel, Q. W. Osburn, HPLC determination of AE in wastewater, *J. Am. Oil Chem. Soc.*, 1990, *67*, 103–109.

129. Rothbächer, H., A. Korn, G. Mayer, Nonionic surfactants in cleaning agents for automobile production (in German), *Tenside, Surfactants, Deterg.*, 1993, *30*, 165–173.

130. Guerrero, F., J. L. Rocca, RPLC analysis of AE using evaporative light scattering detection, *Chim. Oggi*, 1995, *13(4–5)*, 11–15.

131. Motz, M., B. Fell, W. Meltzow, Z. Xia, GC-HPLC analysis of ethoxylation products of fatty alcohols and Guerbet alcohols (in German), *Tenside, Surfactants, Deterg.*, 1995, *32*, 17–21.

132. Okada, T., Automated analyses of nonionic surfactants in terms of hydrophobic and polyoxyethylated chain lengths, *J. Chromatogr.*, 1992, *609*, 213–218.

133. Murphy, R. E., M. R. Schure, J. P. Foley, Effect of sampling rate on resolution in comprehensive two-dimensional LC, *Anal. Chem.*, 1998, *70*, 1585–1594.

134. Murphy, R. E., M. R. Schure, J. P. Foley, One- and two-dimensional chromatographic analysis of AE, *Anal. Chem.*, 1998, *70*, 4353–4360.

135. Ahel, M., W. Giger, Nonionic surfactants of the APE type by HPLC, *Anal. Chem.*, 1985, *57*, 2584–2590.

136. Zhou, C., A. Bahr, G. Schwedt, Nonionic surfactants of the NPE type by LC, *Anal. Chim. Acta*, 1990, *236*, 273–280.

137. Jandera, P., J. Urbánek, B. Prokeš, J. Churacek, Comparison of various stationary phases for normal-phase HPLC of APE. *Chromatogr.*, 1990, *504*, 297–318.

138. Márquez, N., R. E. Antón, A. Usubillaga, J. L. Salager, HPLC analysis of APE in microemulsion systems. Part II: Gradient mode for extended EON range as found in the analysis of oligomer fractionation, *Sep. Sci. Technol.*, 1993, *28*, 2387–2400.

139. Márquez, N., R. E. Antón, A. Usubillaga, J. L. Salager, HPLC of APE, *J. Liq. Chromatogr.*, 1994, *17*, 1147–1169.

140. Zeman, I., Bonded diol phases for separation of ethoxylated surfactants by HPLC, *J. Chromatogr.*, 1986, *363*, 223–230.

141. Ahel, M., W. Giger, Alkylphenols and alkylphenol mono-and diethoxylates in environmental samples by HPLC, *Anal. Chem.*, 1985, *57*, 1577–1583.

142. Siegel, M. M., R. Tsao, S. Oppenheimer, T. T. Chang, Nonionic surfactants used as exact mass internal standards for the 700–2100 dalton mass range in fast atom bombardment MS, *Anal. Chem.*, 1990, *62*, 322–327.

143. Kamiusuki, T., T. Monde, F. Nemoto, T. Konakahara, Y. Takahashi, OPE by reversed-phase HPLC on branched fluorinated silica gel columns, *J. Chromatogr. A*, 1999, 852, 475–485.

144. Anghel, D. F., M. Balcan, A. Voicu, M. Elian, Alkylphenol-based nonionic surfactants by HPLC, *J. Chromatogr. A*, 1994, 668, 375–383.

145. Beck, G. J., D. Kossak, S. J. Saxena, Spermicide nonoxynol-9 in biological fluids by HPLC, *J. Pharm. Sci.*, 1990, 79, 1029–1031.

146. Strancar, A., P. Raspor, H. Schwinn, R. Schütz, D. Josic, Extraction of Triton X-100 and its determination in virus-inactivated human plasma by the solvent-detergent method, *J. Chromatogr.*, 1994, 658, 475–481.

147. Mackay, L. G., M. Y. Croft, D. S. Selby, R. J. Wells, NPE and OPE in effluent by LC with fluorescence detection, *J. AOAC Int.*, 1997, 80, 401–407.

148. Jandera, P., Mechanism and prediction of retention of oligomers in normal-phase and reversed-phase HPLC, *Chromatographia*, 1988, 26, 417–422.

149. Nielen, M. W. F., F. A. Buijtenhuijs, Polymer analysis by LC/electrospray ionization time-of-flight MS, *Anal. Chem.*, 1999, 71, 1809–1814.

150. Berchter, M., J. Meister, C. Hammes, MALDI-TOF-MS: characterization of products based on renewable raw materials (in German), *Fett/Lipid*, 1997, 99, 384–391.

151. Zeman, I., J. Šilha, M. Bareš, Separation of ethoxylates by HPLC, *Tenside Detergents*, 1986, 23, 181–184.

152. Nakae, A., K. Kunihiro, Homologous fatty acid alkanolamides by HPLC, *J. Chromatogr.*, 1978, 156, 167–172.

153. Nakamura, K., Y. Morikawa, I. Matsumoto, Ionic and nonionic surfactant homologs by HPLC, *J. Am. Oil Chem. Soc.*, 1981, 58, 72–77.

154. Ben-Bassat, A. A., T. Wasserman, A. Basch, Determination of the mono- and diethanolamides of palmitic acid and of soybean oil fatty acids by HPLC, *J. Liq. Chromatogr.*, 1984, 7, 2545–2560. ibid., 1986, 9, 89–101.

155. Lang, R. F., D. Parra-Diaz, D. Jacobs, Ethoxylated fatty amines: determination of molecular weight, *J. Surfactants Deterg.*, 1999, 2, 503–513.

156. Truchan, J., H. T. Rasmussen, N. Omelczenko, B. P. McPherson, Cationic surfactants in household products by HPLC with nitrogen chemiluminescence detection, *J. Liq. Chromatogr. Relat. Technol.*, 1996, 19, 1785–1792.

157. Schreuder, R. H., A. Martijn, H. Poppe, J. C. Kraak, Composition of ethoxylated alkylamines in pesticide formulations by HPLC using ion-pair extraction detection, *J. Chromatogr.*, 1986, 368, 339–350.

158. Brüschweiler, H., Nonionic surface-active compounds (emulsifiers) by HPLC (in German), *Mitt. Gebiete Lebensm. Hyg.*, 1977, 68, 46–63.

159. Jaspers, M. E. A. P., F. F. van Leeuwen, H. J. W. Nieuwenhuis, G. M. Vianen, HPLC separation of sucrose fatty acid esters, *J. Am. Oil Chem. Soc.*, 1987, 64, 1020–1025.

160. Garti, N., E. Wellner, A. Aserin, S. Sarig, Sorbitan fatty acid esters by HPLC, *J. Am. Oil Chem. Soc.*, 1983, 60, 1151–1154.

161. Wang, Z., M. Fingas, Sorbitan ester surfactants. Part I: HPLC, *J. High Resolut. Chromatogr.*, 1994, 17, 15–19.

162. Kaufman, V. R., N. Garti, Sucrose fatty acid esters composition by HPLC, *J. Liq. Chromatogr.*, 1981, 4, 1195–1205.

163. Garti, N., A. Aserin, Analyses of polyglycerol esters of fatty acids using HPLC, *J. Liq. Chromatogr.*, 1981, 4, 1173–1194.

164. Riisom, T., L. Hoffmeyer, HPLC analyses of emulsifiers. I: Quantitative determinations of mono- and diacylglycerols of saturated fatty acids, *J. Am. Oil Chem. Soc.*, 1978, 55, 649–652.

165. Sinsel, J. A., B. M. LaRue, L. D. McGraw, HPLC of glyceride-based lubricants, *Anal. Chem.*, 1975, 47, 1987–1993.

166. Birch, C. G., F. E. Crowe, Sucrose polyesters by high performance GPC, *J. Am. Oil Chem. Soc.*, 1976, 53, 581–583.

167. Seino, H., T. Uchibori, T. Nishitani, S. Inamasu, Enzymatic synthesis of carbohydrate esters of fatty acid. I: Esterification of sucrose, glucose, fructose and sorbitol, *J. Am. Oil Chem. Soc.*, 1984, *61*, 1761–1765.

168. Warner, C. R., S. Selim, D. H. Daniels, Post-column complexation technique for the spectro-photometric detection of poly(oxy-1,2-ethanediyl) oligomers in steric exclusion chromatogra-phy, *J. Chromatogr.*, 1979, *173*, 357–363.

169. Hensel, A., M. Rischer, D. Di Stefano, I. Behr, E. Wolf-Heuss, Chromatographic character-ization of nonionic surfactant polyoxyethylene glycerol trioleate, *Pharm. Acta Helv.*, 1997, *72*, 185–189.

170. Lafosse, M., C. Elfakir, L. Morin-Allory, M. Dreux, Evaporative light scattering detection in pharmaceutical analysis by HPLC and SFC, *J. High Resolut. Chromatogr.*, 1992, *15*, 312–318.

171. Klaffke, H. S., T. Neubert, L. W. Kroh, Alkyl polyglucosides using LC methods, *Tenside, Sur-factants, Deterg.*, 1998, *35*, 108–111.

172. Schröder, H. Fr., Alkyl polyglycosides in the biological waste water treatment process—Degradation behavior by LC/MS and FIA/MS, *World Surfactants Congr., 4th*, 1996, *3*, 121–135.

173. Klaffke, H. S., T. Neubert, L. W. Kroh, Nonionic surfactants by LC/MS using alkyl polyglu-cosides as model substances, *Tenside, Surfactants, Deterg.*, 1999, *36*, 178–184.

174. Eichhorn, P., T. P. Knepper, Metabolism of alkyl polyglucosides and their determination in waste water by LC-electrospray MS, *J. Chromatogr. A*, 1999, *854*, 221–232.

175. Waldhoff, H., J. Scherler, M. Schmitt, J. R. Varvil, Alkyl polyglycosides and determination in consumer products and environmental matrices, in K. Hill, W. von Rybinski, G. Stoll, eds., *Alkyl polyglycosides: Technology, Properties and Applications*, VCH, Weinheim, 1997.

176. Elfakir, C., M. Lafosse, Porous graphitized carbon and octadecyl-silica columns in the sepa-ration of some alkylglycoside detergents, *J. Chromatogr. A.*, 1997, *782*, 191–198.

177. Pasch, H., C. Brinkmann, H. Much, U. Just, Chromatographic investigations of macromolecules in the critical range of LC: polyethylene oxides, *J. Liq. Chromatogr.*, 1994, *17*, 3091–3108.

178. Wigman, L. S., H. Abdel-Kader, G. K. Menon, SEC of poloxalene poloxamers: polyethyene glycol-polypropylene glycol copolymers used to control cattle bloat, *J. Pharm. Biomed. Anal.*, 1994, *12*, 719–722.

179. König, H., W. Strobel, Amphoteric surfactants by HPLC (in German), *Proc. 2nd World Sur-factants Congress*, Paris, 1988, *3*, 108–122.

180. Murawski, D., IC for the analysis of household consumer products, *J. Chromatogr.*, 1991, *546*, 351–367.

181. Dowle, C. J., W. C. Campbell, B. G. Cooksey, Cationic surfactant homologues by HPLC, *Analyst*, 1989, *114*, 883–885.

182. Taylor, R. B., S. Toasaksiri, R. G. Reid, D. Wood, Quaternary ammonium compounds dequal-inium and cetylpyridinium chlorides in candy-based lozenges by HPLC, *Analyst*, 1997, *122*, 973–976.

183. Kawase, J., Y. Takao, K. Tsuji, Non-UV-absorbing dialkyl-type cationic surfactants by LC with an on-line ion-pair extraction detector, *J. Chromatogr.*, 1983, *262*, 293–298.

184. Bak, H., K. Choi, J. Lee, Y. Kim, H. Ahn, Imidazoline type cationic surfactants (in Korean; tables and figures in English), *Kongop Hwahak*, 1998, *9*, 404–406.

185. Emmrich, M., K. Levsen, Cationic surfactants in wastewater and activated sludge (in Ger-man), *Vom Wasser*, 1990, *75*, 343–349.

186. Wilkes, A. J., G. Walraven, J. Talbot, HPLC of quaternary ammonium surfactants with the evaporative light scattering detector. *J. Am. Oil Chem. Soc.*, 1992, *69*, 609–613.

187. Lawrence, D. L., Normal phase LC/MS using coaxial continuous flow FAB, *J. Am. Soc. Mass Spectrom.*, 1992, *3*, 575–581.

188. Matthijs, E., H. De Henau, Monoalkylquaternaries and assessment of their fate in domestic waste waters, river waters and sludges, *Vom Wasser*, 1987, *69*, 73–83.

189. Schoester, M., G. Kloster, Chromatographic post-column ion pair extraction system for the determination of cationic surfactants (in German), *Vom Wasser*, 1991, *77*, 13–20.

190. Fernández, P., A. C. Alder, M. J. F. Suter, W. Giger, Ditallowdimethylammonium in digested sludges and marine sediments by supercritical fluid extraction and LC with post-column ion-pair extraction, *Anal. Chem.*, 1996, *68*, 921–929.

191. Matsuzaki, M., K. Ishii, H. Yoshimura, S. Hashimoto, HPLC of cationic and amphoteric surfactants in cosmetics, *J. SCCJ*, 1993, *27*, 494–497.

192. Larson, J. R., C. D. Pfeiffer, Alkyl quaternary ammonium compounds by LC with indirect photometric detection, *Anal. Chem.*, 1983, *55*, 393–396.

193. Nakae, A., K. Kunihiro, G. Muto, Homologous alkylbenzyldimethylammonium chlorides and alkylpyridinium halides by HPLC, *J. Chromatogr.*, 1977, *134*, 459–466.

194. Parkin, J. E., Salting-out solvent extraction for pre-concentration of benzalkonium chloride prior to HPLC, *J. Chromatogr.*, 1993, *635*, 75–80.

195. De Ruiter, C., J. C. H. F. Hefkens, U. A. T. Brinkman, R. W. Frei, M. Evers, E. Matthijs, J. A. Meijer, LC determination of cationic surfactants in environmental samples using a continuous post-column ion-pair extraction detector with a sandwich phase separator, *Intern. J. Environ. Anal. Chem.*, 1987, *31*, 325–339.

196. Gort, S. M., E. A. Hogendoorn, R. A. Baumann, P. van Zoonen, Ditallowdimethylammonium chloride at the low ppb level in surface water using SPE and normal-phase LC with on-line post-column ion-pair extraction and fluorescence detection, *Int. J. Environ. Anal. Chem.*, 1993, *53*, 289–296.

197. Suortti, T., H. Sirvio, Fungistatic quaternary ammonium compounds in beverages and water samples by HPLC, *J. Chromatogr.*, 1990, *507*, 421–425.

198. Yamagishi, T., H. Akiyama, S. Kimura, M. Toyoda, Phosphatidylcholine by an HPLC-RI system, *J. Am. Oil Chem. Soc.*, 1989, *66*, 1801–1808.

199. Parris, N., Reversed-phase HPLC. Ionic surfactants as UV-absorbing ion pairs, *J. Liq. Chromatog.*, 1980, *3*, 1743–1751.

200. Kondoh, Y., S. Takano, Carboxybetaine amphoteric surfactants in household and cosmetic products by HPLC with prelabeling, *Analytical Sciences*, 1986, *2*, 467–471.

201. Matthees, D. P., Precolumn derivatization of amino phospholipids for LC, *Proc. S. D. Acad. Sci.*, 1980, *59*, 62–64.

202. Caccialanza, G., C. Gandini, M. Kitsos, G. Massolini, LC analysis of phospholipids in washings from rabbit eustachian tube: dipalmitoyl phosphatidylethanolamine content, *J. Pharm. Biomed. Anal.*, 1989, *7*, 1931–1935.

203. Kawase, J., H. Ueno, K. Tsuji, Amphoteric surfactants by LC with post-column detection. I. Mono- and dialanine type surfactants, *J. Chromatogr.*, 1983, *264*, 415–422.

204. Melton, S. L., Soybean lecithins and beef phospholipids by HPLC with an evaporative light scattering detector, *J. Am. Oil Chem. Soc.*, 1992, *69*, 784–788.

205. Van der Meeren, P., J. Vanderdeelen, G. Huyghebaert, L. Baert, Partial resolution of molecular species during LC of soybean phospholipids and effect on quantitation by light-scattering. *Chromatographia*, 1992, *34*, 557–562.

206. Balazs, P. E., P. L. Schmit, B. F. Szuhaj, HPLC separations of soy phospholipids, *J. Am. Oil Chem. Soc.*, 1996, *73*, 193–197.

207. Abidi, S. L., T. L. Mounts, K. A. Rennick, Separations of major soybean phospholipids on β-cyclodextrin-bonded silica, *J. Liq. Chromatogr.*, 1994, *17*, 3705–3725.

208. Bonanno, L. M., B. A. Denizot, P. C. Tchoreloff, F. Puisieux, P. J. Cardot, Determination of phospholipids from pulmonary surfactant using an on-line coupled silica/reversed-phase HPLC system, *Anal. Chem.*, 1992, *64*, 371–379.

209. Cantafora, A., R. Masella, Individual molecular species of phosphatidylcholine in biological samples by HPLC with internal standards, *J. Chromatogr.*, 1992, *593*, 139–146.

210. Abidi, S. L., HPLC of phosphatidic acids and related polar lipids, *J. Chromatogr.*, 1991, *587*, 193–203.

211. Abidi, S. L., T. L. Mounts, K. A. Rennick, Reversed-phase ion-pair HPLC of phosphatidylinositols, *J. Liq. Chromatogr.*, 1991, *14*, 573–588.

212. Abidi, S. L., T. L. Mounts, Reversed-phase HPLC of molecular species of phospholipid derivatives, *J. Chromatogr. A*, 1996, *741*, 213–222.

213. Kim, H., T. L. Wang, Y. Ma, LC-MS of phospholipids using electrospray ionization, *Anal. Chem.*, 1994, *66*, 3977–3982.

214. Li, C., J. A. Yergey, Continuous flow liquid secondary ion MS characterization of phospholipid molecular species, *J. Mass Spectrom.*, 1997, *32*, 314–322.

215. Wilkes, A. J., G. Walraven, J. Talbot, Cocoamidopropylbetaine in raw materials and cosmetic and detergent products by HPLC, *Comun. Jorn. Com. Esp. Deterg.*, 1994, *25*, 209–220.

216. Tegeler, A., W. Ruess, E. Gmahl, Amphoteric surfactants in cosmetic cleansing products by HPLC on a cation exchange column, *J. Chromatogr. A*, 1995, *715*, 195–198.

217. Schreuder, R. H., A. Martijn, Dodecylbenzenesulphonates and APE in liquid pesticide formulations by HPLC, *J. Chromatogr.*, 1988, *435*, 73–82.

218. Nakae, A., K. Tsuji, M. Yamanaka, Alkyl chain distribution of alkylbenzenesulfonates by LC, *Anal. Chem.*, 1981, *53*, 1818–1821.

219. Matthijs, E., H. De Henau, LAS in aqueous samples, sediments, sludges and soils using HPLC, *Tenside Surfactants*, 1987, *24*, 193–199.

220. Inaba, K., K. Amano, HPLC determination of LAS in aquatic environment. Seasonal changes in LAS concentration in polluted lake water and sediment, *Int. J. Environ. Anal. Chem.*, 1988, *34*, 203–213.

221. Kunkel, E., Surfactants in the environment (in German), *Tenside*, 1987, *24*, 280–285.

222. Menez, H. R., C. L. Perez, Alkyl chain distribution in lubricating oil additives of the type alkyl aryl sulfonates by HPLC, *J. High Resolut. Chromatogr.*, 1989, *12*, 562–565.

223. Pan, N., D. J. Pietrzyk, Separation of anionic surfactants on anion exchangers, *J. Chromatogr. A*, 1995, *706*, 327–337.

224. Scullion, S. D., M. R. Clench, M. Cooke, A. E. Ashcroft, Surfactants in surface water by solid-phase extraction, LC, and LC-MS, *J. Chromatogr. A*, 1996, *733*, 207–216.

225. Grey, R. A., A. F. Chan, Sulfonation studies of monoisomeric di- and trialkylbenzenes, *J. Am. Oil Chem. Soc.*, 1990, *67*, 132–141.

226. Hoeft, C. E., R. L. Zollars, Anionic surfactants using ion chromatography, *J. Liq. Chromatogr.*, 1994, *17*, 2691–2704.

227. Jiang, S. X., X. Liu, Reverse phase HPLC analysis of alkyl sulfonates with non-suppression conductivity detection, *J. Liq. Chromatogr. Relat. Technol.*, 1997, *20*, 2053–2061.

228. Kloster, G., M. Schoester, M. J. Schwuger, HPLC analysis of aliphatic ionic surfactants at trace levels, *Comun. Jorn. Com. Esp. Deterg.*, 1993, *24*, 25–33.

229. Laikhtman, M., J. S. Rohrer, Fluorochemical surfactants in acid etch baths by IC with on-line matrix elimination, *J. Chromatogr. A*, 1998, *822*, 321–325.

230. Popenoe, D. D., S. J. Morris, III, P. S. Horn, K. T. Norwood, Determination of alkyl sulfates and alkyl ethoxysulfates in wastewater treatment plant influents and effluents and in river water using LC/ion spray MS, *Anal. Chem.*, 1994, *66*, 1620–1629.

231. Stemp, A., V. A. Boriraj, P. Walling, P. Neill, Ion chromatographic characterization of ethoxylated anionic surfactants, *J. Am. Oil Chem. Soc.*, 1995, *72*, 17–21.

232. Hofman, Y. L., H. P. Angstadt, Enhanced oil recovery formulations, *Chromatographia*, 1987, *24*, 666–670.

233. Escott, R. E. A., S. J. Brinkworth, T. A. Steedman, Ethoxylate oligomer distribution of nonionic and anionic surfactants by HPLC, *J. Chromatogr.*, 1983, *282*, 655–661.

234. Jandera, P., B. Prokeš, Unsulfated and sulfated oligoethyleneglycol nonyphenyl ether surfactants in normal phase LC systems containing water, *Chromatographia*, 1996, *42*, 539–546.

235. Castro, V., J. Canselier, Analysis of α-olefinsulfonates by HPLC, *J. Chromatogr.*, 1985, *325*, 43–51.

236. Castro, V., J. Canselier, Application of the hydrophobic effect in reversed-phase HPLC to the prediction of the critical micelle concentration, *J. Chromatogr.*, 1986, *363*, 139–146.

237. Márquez, N., S. Gonzalez, N. Subero, B. Bravo, G. Chavez, R. Bauza, F. Ysambertt, Isolation and characterization of petroleum sulfonates, *Analyst*, 1998, *123*, 2329–2332.

238. Köhler, M., E. Keck, G. Jaumann, Sulfonated esters (in German), *Fett Wiss. Technol.*, 1988, *90*, 241–243.

239. Schoester, M., G. Kloster, HPLC separation and quantification of anionic surfactants using an automated on-line ion pair extraction system, *Fresenius' Z. Anal. Chem.*, 1993, *345*, 767–772.

240. Parris, N., Surfactant analysis by HPLC: analysis of mixtures of amphoteric surfactants and soap, *J. Am. Oil Chem. Soc.*, 1978, *55*, 675–677.

241. Heinig, K., F. Hissner, S. Martin, C. Vogt, Saturated and unsaturated fatty acids by CE and HPLC, *Amer. Lab.*, 1998, *30(10)*, 24–29.

242. Ye, M., R. Walkup, K. Hill, Alkylated and sulfonated diphenyl oxide surfactant by HPLC, *J. Liq. Chromatogr. Relat. Technol.*, 1996, *19*, 1229–1240.

243. Bogatzki, B. F., H. Lippmann, LC characterization of fatty alcohol-ethylene oxide adducts (in German), *Acta Polymerica*, 1983, *34*, 219–223.

244. Bosdorf, V., T. Bluhm, H. Krüssmann, Retention index system for analysis of surfactants, *Comun. Jorn. Com. Esp. Deterg.*, 1994, *25*, 493–498.

245. Bosdorf, V., T. Bluhm, H. Krüssman, Adsorbed nonionic surfactants on fabrics by HPLC (in German), *Text. Prax. Int.*, 1994, *49*, 348–350,353–354.

246. Sun, C., M. Baird, H. A. Anderson, D. L. Brydon, Separation and determination of oligomers and homologs of aliphatic AE in textile lubricants and lubricant emulsions by HPLC, *J. Chromatogr. A*, 1997, *771*, 145–1545.

247. Desbène, P. L., B. Desmazières, J. J. Basselier, L. Minssieux, HPLC of nonionic surfactants used in tertiary oil recovery, *Chromatographia*, 1987, *24*, 588–592.

248. Desbène, P. L., V. Even, B. Desmazières, J. J. Basselier, L. Minssieux, HPLC on diol-bonded silica gel for the determination of nonionic surfactants used in petroleum recovery (in French), *Analusis*, 1988, *16(7)*, 44–47.

249. Shiraishi, H., A. Otsuki, K. Fuwa, AE in river water at trace levels by field desorption MS, *Bull. Chem. Soc. Jpn.*, 1982, *55*, 1410–1415.

250. Aranda, R., R. C. Burk, Nonionic surfactant by solid-phase microextraction coupled with HPLC and on-line derivatization, *J. Chromatogr. A*, 1998, *829*, 401–406.

251. Hwang, R. J., M. Stauffer, HPLC analysis of cosurfactants used in enhanced oil recovery, *J. Liq. Chromatogr.*, 1987, *10*, 603–615.

252. Winkle, W., Quantitative analysis in the V_0 zone: chromatographic approach by coupling HPLC with GPC, *Chromatographia*, 1990, *29*, 530–536.

253. Evans, K. A., S. T. Dubey, L. Kravetz, I. Dzidic, J. Gumulka, R. Mueller, J. R. Stork, Linear primary AE in environmental samples by thermospray LC/MS, *Anal. Chem.*, 1994, *66*, 699–705.

254. Trathnigg, B., D. Thamer, X. Yan, B. Maier, H. R. Holzbauer, H. Much, Characterization of AE using LC with density and refractive index detection. Quantitative analysis of pure homologous series by size-exclusion chromatography, *J. Chromatogr. A*, 1993, *657*, 365–375.

255. Trathnigg, B., D. Thamer, X. Yan, B. Maier, H. R. Holzbauer, H. Much, Characterization of AE using LC with density and refractive index detection: quantification in LC under critical conditions, *J. Chromatogr. A*, 1994, *665*, 47–53.

256. Jones, F. W., Flame ionization detector relative response factors for oligomers of alkyl and aryl ether polyethoxylates using the effective carbon number concept, *J. Chromatogr. Sci.*, 1998, *36*, 223–226.

257. Black, D. B., B. A. Dawson, G. A. Neville, HPLC system for separation of components in nonoxynol-9 spermicidal agents, *J. Chromatogr.*, 1989, *478*, 244–249.

258. Márquez, N., R. E. Antón, A. Usubillaga, J. L. Salager, HPLC analysis of APE in microemulsion systems: isocratic mode with mixed solvents. *Sep. Sci. Technol.*, 1993, *28*, 1769–1782.

259. Porot, V., NPE from sewage treatment plants (Toulon and Morlaix) by HPLC (in French), *Proc. 2nd World Surfactants Congress*, Paris, May, 1988, *4*, 293–302.

260. Sun, C., M. Baird, H. A. Anderson, D. L. Brydon, Separation of broadly distributed NPE and determination of ethylene oxide olgomers in textile lubricants and emulsions by HPLC, *J. Chromatogr. A*, 1996, *731*, 161–169.

261. Pelizzetti, E., C. Minero, V. Maurino, A. Sciafani, H. Hidaka, N. Serpone, Photocatalytic degradation of NPE, *Environ. Sci. Technol.*, 1989, *23*, 1380–1385.

262. Kubeck, E., C. G. Naylor, Trace analysis of APE, *J. Am. Oil Chem. Soc.*, 1990, *67*, 400–405.

263. Scarlett, M. J., J. A. Fisher, H. Zhang, M. Ronan, NPE in waste waters by gas stripping and isocratic HPLC, *Water Res.*, 1994, *28*, 2109–2116.

264. Forgács, E., T. Cserháti, Retention behavior of NPE oligomers on alumina columns, *Fresenius' Z. Anal. Chem.*, 1995, *351*, 688–689.

265. Kósa, A., A. Dobó, K. Vékey, E. Forgács, NPE oligomers by HPLC with UV and MS detection, *J. Chromatogr. A*, 1998, *819*, 297–302.

266. Forgács, E., T. Cserháti, Retention behavior of tributylphenol ethylene oxide oligomers on an alumina HPLC column, *J. Chromatogr. A*, 1994, *661*, 239–243.

267. Forgács, E., T. Cserháti, Alumina support for the separation of ethoxylated oligomer surfactants according to the length of the ethylene oxide chain, *Anal. Lett.*, 1996, *29*, 321–340.

268. Sithole, B. B., L. H. Allen, NPE in pulp and paper mill process samples by spectrophotometry and LC, *J. Assoc. Off. Anal. Chem.*, 1989, *72*, 273–276.

269. Cumme, G. A., E. Blume, R. Bublitz, H. Hoppe, A. Horn, Detergents of the polyoxyethylene type: comparison of TLC, reverse-phase chromatography, and MALDI MS, *J. Chromatogr. A*, 1997, *791*, 245–253.

270. Németh-Kiss, V., Separation of ethoxylated tributylphenol oligomers on porous graphitic carbon column, *J. Liq. Chromatogr. Relat. Technol.*, 1996, *19*, 217–229.

271. Pardue, K., D. Williams, Nonionic surfactants in protein samples using ion-exchange guard columns, *BioTechniques*, 1993, *14*, 580, 582–583.

272. Ysambertt, F., W. Cabrera, N. Marquez, J. L. Salager, NPE by HPSEC, *J. Liq. Chromatogr.*, 1995, *18*, 1157–1171.

273. Bán, T., E. Papp, J. Inczédy, Reversed-phase HPLC of anionic and ethoxylated nonionic surfactants and pesticides in liquid pesticide formulations, *J. Chromatogr.*, 1992, *593*, 227–231.

274. Henke, H., Preparative low pressure LC (in German), *LaborPraxis*, 1995, *19*, 62,64,66,69.

275. Baudrand, V., Z. Mouloungui, A. Gaset, Synthesis and analysis of tetraethylene glycol mono- and diheptanoate (in French), *Colloq.—Inst. Natl. Rech. Agron.*, 1995, *71*, 183–187.

276. Torres, M. C., M. A. Dean, F. W. Wagner, Chromatographic separations of sucrose monostearate structural isomers, *J. Chromatogr.*, 1990, *522*, 245–253.

277. Sudraud, G., J. M. Coustard, C. Retho, M. Caude, R. Rosset, R. Hagemann, D. Gaudin, H. Virelizier, Analytical and structural study of some food emulsifiers by HPLC and off-line MS, *J. Chromatogr.*, 1981, *204*, 397–406.

278. Spilker, R., B. Menzebach, U. Schneider, I. Venn, Alkylpolyglucosides (in German), *Tenside, Surfactants, Deterg.*, 1996, *33*, 21–25.

279. Buschmann, N., A. Kruse, S. Wodarczak, Alkylpolyglucosides, *Agro-Food-Ind. Hi-Tech*, 1996, *7*, 6–8.

280. Bruns, A., H. Waldhoff, W. Winkle, HPLC with ELSD in fat and carbohydrate chemistry, *Chromatographia*, 1989, *27*, 340–342.

281. Waldhoff, H., J. Scherler, M. Schmitt, Alkyl polyglycosides—analysis of raw material: determination in products and environmental matrices, *World Surfactants Congr., 4th*, 1996, *1*, 507–518.

282. Lafosse, M., P. Marinier, B. Joseph, M. Dreux, Amphiphilic behavior of alkylglycoside surfactants using reversed-phase LC, *J. Chromatogr.*, 1992, *623*, 277–287.

283. Krupinski, J., G. G. Hammes, Phase-lifetime spectrophotometry of deoxycholate-purified bacteriorhodopsin reconstituted into asolectin vesicles, *Biochemistry*, 1985, *24*, 6963–6972.

284. Zgoda, M., S. Petri, Surfactants from copolymers of propylene and ethylene oxides. I: Dispersion of weight-average molecular weights (in Polish), *Chem. Anal.* (Warsaw), 1986, *31*, 577–596.

285. Nitschke, L., R. Müller, G. Metzner, L. Huber, Cationic surfactants in water using HPLC with conductometric detection, *Fresenius' Z. Anal. Chem.*, 1992, *342*, 711–713.

286. Kümmerer, K., A. Eitel, U. Braun, P. Hubner, F. Daschner, G. Mascart, M. Milandri, F. Reinthaler, J. Verhoef, Benzalkonium chloride in the effluent from European hospitals by SPE and HPLC with post-column ion-pairing and fluorescence detection, *J. Chromatogr. A*, 1997, *774*, 281–286.

287. Daniels, C. R., Didecyldimethylammonium chloride on wood surfaces by HPLC with evaporative light scattering detection, *J. Chromatogr. Sci.*, 1992, *30*, 497–499.

288. Wilkes, A. J., C. Jacobs, G. Walraven, J. M. Talbot, Characterization of quaternized triethanolamine esters (esterquats) by HPLC, HRCGC, and NMR, *World Surfactants Congr.*, *4th*, 1996, *1*, 389–412.

289. Levsen, K., M. Emmrich, S. Behnert, Dialkyldimethylammonium compounds and other cationic surfactants in sewage water and activated sludge, *Fresenius' Z. Anal. Chem.*, 1993, *346*, 732–737.

290. Linares, P., M. C. Gutiérrez, F. Lázaro, M. D. Luque de Castro, M. Valcárcel, Benzocaine, dextromethorphan and cetylpyridinium ion by HPLC with UV detection, *J. Chromatogr.*, 1991, *558*, 147–153.

291. Gomez-Gomar, A., M. M. Gonzalez-Aubert, J. Garces-Torrents, J. Costa-Segarra, Benzalkonium chloride in aqueous ophthalmic preparations by HPLC, *J. Pharm. Biomed. Anal.*, 1990, *8*, 871–876.

292. Elrod, L., T. G. Golich, J. A. Morley, Benzalkonium chloride in eye care products by HPLC and solid-phase extraction or on-line column switching, *J. Chromatogr.*, 1992, *625*, 362–367.

293. Parhizkari, G., G. Delker, R. B. Miller, C. Chen, Stability-indicating HPLC method for determination of benzalkonium chloride in Tramadol ophthalmic solution, *Chromatographia*, 1995, *40*, 155–158.

294. Miller, R. B., C. Chen, C. H. Sherwood, HPLC determination of benzalkonium chloride in ophthalmic solution, *J. Liq. Chromatogr.*, 1993, *16*, 3801–3811.

295. Parhizkari, G., R. B. Miller, C. Chen, Stability-indicating HPLC method for determination of benzalkonium chloride in phenylephrine HCl ophthalmic solution, *J. Liq. Chromatogr.*, 1995, *18*, 553–563.

296. Heinig, K., C. Vogt, G. Werner, Cationic surfactants by CE, *Fresenius' Z. Anal. Chem.*, 1997, *358*, 500–505.

297. Huang, C. B., Dialkyldimethylammonium salt in rolling oil, *J. Liq. Chromatogr.*, 1987, *10*, 1103–1125.

298. United States Pharmacopeial Convention, Inc., *United States Pharmacopeia*, 24th Revision/ *National Formulary*, 19th ed. 12601 Twinbrook Parkway, Rockville, MD 20852, 1999.

299. Nair, L. M., R. Saari-Nordhaus, Surfactant analysis by IC, *J. Chromatogr. A*, 1998, *804*, 233–239.

300. Shibukawa, M., R. Eto, A. Kira, F. Miura, K. Oguma, H. Tatsumoto, H. Ogura, A. Uchiumi, Quaternary ammonium compounds by HPLC with a hydrophilic polymer column and conductometric detection, *J. Chromatogr. A*, 1999, *830*, 321–328.

301. Caesar, R., H. Weightman, G. R. Mintz, HPLC determination of alkylamidopropyl-*N,N*-dimethyl-*N*-(2,3-dihydroxypropyl)ammonium chlorides in aqueous solutions and cosmetic formulations, *J. Chromatogr.*, 1989, *478*, 191–203.

302. Spagnolo, F., M. T. Hatcher, B. K. Faulseit, HPLC analysis of commercial alkyl and aryl quaternary ammonium compounds used in organoclay type rheological additives, *J. Chromatogr. Sci.*, 1987, *25*, 399–401.

303. Toomey, A. B., D. M. Dalrymple, J. L. Jasperse, M. M. Manning, M. V. Schulz, Quaternary ammonium compounds by HPLC with ELSD, *J. Liq. Chromatogr. Relat. Technol.*, 1997, *20*, 1037–1047.

304. Grieser, M. D., J. N. Geske, HPLC of phospholipids with flame ionization detection, *J. Am. Oil Chem. Soc.*, 1989, *66*, 1484–1487.

305. Becart, J., C. Chevalier, J. P. Biesse, Analysis of phospholipids by HPLC with a light scattering evaporating detector—application to raw materials for cosmetic use, *J. High Resolut. Chromatogr.*, 1990, *13*, 126–129.

306. Sotirhos, N., C. T. Ho, S. S. Chang, Normal and reverse phase HPLC analysis of soybean phospholipids, *Dev. Food Sci.*, 1986, *12*("Shelf Life of Foods and Beverages"), 601–608.

307. Sotirhos, N., C. T. Ho, S. S. Chang, HPLC of soybean phospholipids, *Fette, Seifen, Anstrichm.*, 1986, *88*, 6–8.

308. Caboni, M. F., S. Menotta, G. Lercker, Separation and analysis of phospholipids in different foods with a light-scattering detector, *J. Am. Oil Chem. Soc.*, 1996, *73*, 1561–1566.

309. Van der Meeren, P., J. Vanderdeelen, M. Huys, L. Baert, Optimization of the column loadability for the preparative HPLC separation of soybean phospholipids, *J. Am. Oil Chem. Soc.*, 1990, *67*, 815–820.

310. Abidi, S. L., T. L. Mounts, T. Finn, Preferred solvent system for HPLC analysis of soybean phospholipids with evaporative light-scattering detection, *J. Am. Oil Chem. Soc.*, 1996, *73*, 535–536.

311. Mounts, T. L., S. L. Abidi, K. A. Rennick, Separations of major soybean phospholipids of β-cyclodextrin-bonded silica, *J. Liq. Chromatogr.*, 1994, *17*, 3705–3725.

312. Mounts, T. L., A. M. Nash, HPLC analysis of phospholipids in crude oil for evaluation of soybean deterioration, *J. Am. Oil Chem. Soc.*, 1990, *67*, 757–760.

313. Grit, M., D. J. A. Crommelin, J. Lang, Determination of phosphatidylcholine, phosphatidylglycerol and their lyso forms from liposome dispersions by HPLC using high-sensitivity refractive index detection, *J. Chromatogr.*, 1991, *585*, 239–246.

314. De Miguel, I., A. Roueche, D. Betbeder, Separation of dipalmitoyl phosphatidyl choline, cholesterol and their degradation products by HPLC on a perfluorinated stationary bonded phase, *J. Chromatogr. A*, 1999, *840*, 31–38.

315. Cantafora, A., M. Cardelli, R. Masella, Separation and determination of molecular species of phosphatidylcholine in biological samples by HPLC, *J. Chromatogr.*, 1990, *507*, 339–349.

316. Wang, T., E. G. Hammond, J. L. Cornette, W. R. Fehr, Fractionation of soybean phospholipids by HPLC with an evaporative light-scattering detector, *J. Am. Oil Chem. Soc.*, 1999, *76*, 1313–1321.

317. Abidi, S. L., T. L. Mounts, Separation of molecular species of phosphatidic acid by HPLC, *J. Chromatogr.*, 1995, *694*, 365–373.

318. Abidi, S. L., T. L. Mounts, HPLC separation of molecular species of neutral phospholipids, *J. Chromatogr.*, 1992, *598*, 209–218.

319. Mounts, T. L., S. L. Abidi, K. A. Rennick, HPLC analysis of phospholipids by evaporative laser light-scattering detection, *J. Am. Oil Chem. Soc.*, 1992, *69*, 438–442.

320. Abidi, S. L., T. L. Mounts, Separation of molecular species of phosphatidylserine by reverse-phase ion-pair HPLC, *J. Liq. Chromatogr.*, 1992, *15*, 2487–2502.

321. Gerhards, R., I. Jussofie, D. Käseborn, S. Keune, R. Schulz, Cocoamidopropyl betaines, *Tenside, Surfactants, Deterg.*, 1996, *33*, 8–14.

322. Kawase, J., K. Tsuji, Y. Yasuda, Amphoteric surfactants by LC with post-column detection. III: Salt-free-type imidazoline amphoteric surfactants, *J. Chromatogr.*, 1983, *267*, 149–166.

323. Kawase, J., K. Tsuji, Y. Yasuda, K. Yashima, Amphoteric surfactants by LC with post-column detection. II: Imidazoline-type amphoteric surfactants derived from sodium chloroacetate, *J. Chromatogr.*, 1983, *267*, 133–148.

324. Cozzoli, O., D. Marini, F. Balestrieri, HPLC for the analysis of mixtures of amphoteric surfactants (in Italian), *Riv. Ital. Sostanze Grasse*, 1989, *66*, 273–277.

325. Hall, S. W., S. R. VandenBerg, Solid phase extraction of the zwitterionic detergent CHAPS, *Prep. Biochem.*, 1989, *19*, 1–11.

326. American Oil Chemists' Society, Determination of lecithin phospholipids by HPLC, Official Method Ja 7b-91. 211 West Bradley Ave., Champaign, IL 61826.

8
Gas Chromatography

I. ANIONIC SURFACTANTS

Gas chromatography provides data on the actual distribution of homologs in anionic surfactants, complementing the volumetric and colorimetric procedures which give a value for total anionic surfactant but do not indicate the compounds present. The sulfated and sulfonated materials which comprise most commercial anionics are not volatile, so GC analysis requires initial treatment to convert them to volatile products. The compatibility of GC with mass spectrometry makes the technique of great interest for nonroutine analysis of unknown products, an area where the application of HPLC is tedious indeed. A review by Rasmussen and McPherson gives additional perspective on the GC analysis of anionic surfactants (1).

A. Derivatization

Most derivatization methods are applied to sulfonates and are based upon formation of either sulfonyl chlorides or of alkyl esters.

Direct analysis of the sulfonyl chlorides formed by reaction of phosphorus pentachloride with LAS is possible (2). The sulfonyl chlorides may also be formed by reaction with thionyl chloride or phosgene (3). Reaction with thionyl chloride is not recommended for LAS because of the superiority of PCl_5 (4). The more volatile sulfonyl fluoride compounds may be produced by further reacting the sulfonyl chlorides with aque-

ous potassium fluoride solution (5). Nagai and coworkers analyzed α-olefin sulfonates after hydrogenation and conversion to the sulfonyl chlorides with thionyl chloride (6).

Volatile methyl sulfonate derivatives are usually prepared by reaction of the free
sulfonic acid with diazomethane. Trimethylorthoformate may also be used (7). Alkyl sulfonate derivatives may also be prepared by reaction of the sulfonyl chlorides with methanol or another alcohol. Dibutylsulfonamide derivatives have been used for determination
of alkylarylsulfonates (8).

A simple reaction of the tetrabutylammonium salt of sulfonates in the hot injection
port of the gas chromatograph to form the butyl esters has been used for quantitative determination of LAS and secondary alkanesulfonates (9,10).

B. Pyrolysis, Cleavage, Desulfonation

If an anionic surfactant is heated anaerobically to 650°C, characteristic fragments are produced. In the case of LAS, the pyrolysis products are linear C_2–C_{10} olefins and paraffins,
benzene, toluene, and xylenes. For alcohol sulfates and ethoxysulfates, C_5–C_{14} α-olefins
are found, as well as the original alcohols. Paraffin sulfonates give primarily C_5–C_{17}
α-olefins (11). This approach can also be used for quantitative analysis, measuring either
the hydrocarbon fragments or the SO_2 evolved. Addition of carbohydrazide to the sample
gives more complete yield of hydrocarbons (12).

Desulfonation is the most useful application of a decomposition reaction to analysis
of anionics. Desulfonation is generally conducted by reaction with phosphoric acid. While
desulfonation and analysis can be conducted within seconds of each other in a pyrolysis
gas chromatograph, there are operational difficulties because of the deleterious effect of
phosphoric acid on quartz (13). Most investigators opt to perform the reaction and chromatography in separate steps. The reaction is often conducted in a reflux apparatus which
permits trapping the resulting hydrocarbons without the need for further cleanup. However, the reaction can be performed in a simple sealed tube (14). The sealed tube is the
basis of the international standard method for analysis of LAS (15). The reaction mixture
must then be extracted to isolate the alkylbenzenes for GC analysis.

Lew (16,17) and Denig (18,19) conducted pyrolysis/desulfonation at relatively low
temperature, with added phosphoric acid. This minimizes "cracking" of the alkyl chain,
while still cleaving the C–S and C–O bonds. LAS and naphthalene sulfonates produce
alkylbenzenes and alkylnaphthalenes identical to those from which they were originally
synthesized. Alkyl sulfates and alkanesulfonates yield olefins. Analysis of alkyl ether sulfates produces chromatograms indistinguishable from the pyrolysis chromatograms of the
alcohol ethoxylates from which they were made. An important advantage of pyrolysis in
the presence of phosphoric acid is that the results are quite reproducible, making quantitative analysis more reliable. This method can even be applied to the analysis of formulations without prior separation.

C. Detection

Almost all GC methods for anionics involve use of the flame ionization detector. For
added selectivity, a sulfur-specific detector may be used. For example, the flame photometric detector (8) and the atomic emission detector in sulfur-selective mode (20) have
been demonstrated for use in quantifying LAS in mixtures.

D. Analysis of Specific Compounds

Since the technology of GC analysis of surfactants has developed slowly, much of the work summarized below was performed using old-fashioned packed columns. The modern chromatographer will surely substitute capillary columns. On the other hand, what has been laboriously discovered in terms of necessary sample pretreatment techniques is still completely applicable.

1. Alkylarylsulfonates

Capillary GC analysis of desulfonated or derivatized LAS gives separation according to alkyl chain length, with the higher chain lengths giving longer retention times (Table 1). Separation of isomers of each homolog is also observed, with the internal isomers eluting first. In other words, for desulfonated dodecylbenzenesulfonate, the order of elution is 6-phenyldodecane, 5-phenyldodecane, 4-phenyldodecane, etc.

Buse and coworkers performed a thorough study of desulfonation and derivatization for GC-MS analysis of LAS. They concluded that desulfonation in glassware allowing simultaneous distillation of the alkylbenzenes gives excellent results and has the advantages of being directly applicable to an aqueous matrix and of being indifferent to the presence of other surfactants. Desulfonation has the disadvantage of requiring about 5 h. Formation of sulfonyl chloride derivatives by reaction with PCl_5 in the presence of DMF occurs without problems and requires only about 2 h once the sample is dried. Reaction with thionyl chloride was not satisfactory and is not recommended. The sulfonyl chloride derivatives can be analyzed directly or be converted to the more volatile sulfonyl fluorides (reaction with KF; requires about 90 min) or to the methyl esters (reaction with methanol; requires about 90 min). The fluoride derivatives elute more quickly as sharper peaks and are better resolved than the chloride derivatives. The methyl esters give little advantage over the sulfonyl chlorides. These authors find desulfonation the preferable method of pretreatment of LAS, especially if GC-MS is used (4).

Many different approaches have been published for desulfonation/GC analysis of LAS. Leidner, Gloor, and Wuhrmann found that the version of the method using distillation for isolation of the desulfonated oil could be applied to laboratory biodegradation studies (21). Waters and Garrigan concentrated the surfactant by the methylene blue method (Chapter 12), purified the extract by passage through an ion exchange column, subjected it to hydrolysis to eliminate alkyl sulfates, again extracted anionic material as the amine ion pair, and finally applied the desulfonation GC procedure with distillation concentration (22). The method of Osburn is similar and is described below. This procedure is fine-tuned specifically for LAS and employs capillary gas chromatography for complete resolution of the alkylbenzenes. Uchiyama and Kawauchi demonstrated that desulfonation could be accomplished in a matter of seconds in a pyrolysis GC, if a trap of KOH-impregnated Celite was used to prevent transfer of phosphoric acid vapors to the column (Fig. 1). Both the trap and the quartz pyrolysis tube were changed after ten analyses (13).

Procedure: Determination of LAS in Sewage Sludge by Derivatization Gas Chromatography (2)

Fifty milligrams of p-(pentadec-3-yl)benzenesulfonate internal standard is added to a 1-mL sample in a separatory funnel. Twenty milliliters of a pH 10 aqueous solution of methylene blue is added, and the solution is extracted repeatedly with 15-mL portions of $CHCl_3$ until no more blue color is extracted. The $CHCl_3$ extracts are combined and evaporated to dryness, then the residue is redissolved in 0.5 mL methylene chloride and purified

TABLE 1 Gas Chromatographic Analysis of Alkylbenzenesulfonate Surfactants

Compound	Preliminary treatment	Column	Temperature	Detector	Ref.
C_{10}–C_{15} LAS; trace analysis; alkyl chain length and isomer separation	Formation of sulfonyl chlorides with PCl_5	SE-54 5% phenyl/1% vinyl dimethylsilicone, 0.31 mm × 19 m	50 to 300°C at 4°C/min	FID	2
C_{10}–C_{14} LAS; trace analysis; alkyl chain length and isomer separation	Formation of methyl sulfonates with PCl_5 and MeOH	1.5% OV-1 dimethylsilicone on Chromosorb W AW-DMCS, 3 mm × 2 m		FID	118
C_9–C_{13} branched ABS; demonstration of alkyl chain length and isomer separation	Formation of methyl sulfonates with diazomethane	OV-1 dimethylsilicone, 320 μm × 25 m, 0.2 μm film thickness	150 to 280°C at 5°C/min	FID	119
C_{10}–C_{14} LAS; trace analysis; alkyl chain length and isomer separation	Formation of trifluoroethyl sulfonates with PCl_5 and trifluoroethanol	Hewlett-Packard Ultra II (95% dimethylsilicone/5% diphenylsilicone), 0.2 mm × 25 m	110 to 300°C at 8°C/min	MS	120
C_{10}–C_{16} LAS/ABS; trace analysis; alkyl chain length and isomer separation	Formation of trifluoroethyl sulfonates with thionyl chloride, pyridine, and trifluoroethanol	Hewlett-Packard HP-5 (95% dimethylsilicone/5% diphenylsilicone), 0.2 mm × 20 m, 0.33 μm film; 0.53 mm × 2 m retention gap	60 to 200°C at 10°C/min, then to 300°C at 6°C/min	MS	121,122
C_{10}–C_{16} LAS; determination of alkyl chain length distribution; isomer separation	Desulfonation with H_3PO_4 to produce alkylbenzenes	OV-101 dimethylsilicone, 0.25 mm × 20 to 50 m	Start: 140–170°C; end: 180–210°C; 0.5–2.0°C/min	FID	15

Analyte	Sample preparation	Column	Temperature program	Detector	Reference
C_{10}-C_{14} LAS; determination of alkyl chain length distribution; isomer separation; trace analysis	Desulfonation with H_3PO_4 to produce alkylbenzenes	Supelco SPB-1 methylsilicone, 0.32 mm × 15 m, 1 μm film	100 to 170°C at 5°C/min	FID	23
C_{10}-C_{13} LAS; determination of alkyl chain length distribution; isomer separation	Desulfonation with H_3PO_4 to produce alkylbenzenes	J&W DB5 5% phenyl dimethylsilicone, 0.25 mm × 60 m, 0.25 μm film; 0.53 mm × 1 m retention gap, H_2 carrier gas	60 to 180°C at 30°C/min, then to 210°C at 2°C/min	FID	7
C_{10}-C_{13} LAS; determination of alkyl chain length distribution; isomer separation	Formation of methyl esters with trimethylorthoformate	J&W DB5 5% phenyl dimethylsilicone, 0.32 mm × 60 m, 0.25 μm film; 0.53 mm × 1 m retention gap, H_2 carrier gas	100 to 200°C at 30°C/min, then to 260°C at 2°C/min	FID	7
C_{10}-C_{14} LAS; determination of alkyl chain length distribution; isomer separation; formulation analysis and trace analysis	Reaction of the tetrabutylammonium salt in the hot injection port to form the butyl ester of LAS	HP-5 5% phenyl dimethylsilicone, 0.2 mm × 20 m, 0.33 μm film, or SE-54, 0.25 mm × 30 m, 0.25 μm film	110 to 220°C at 10°C/min, then to 300°C at 6°C/min	FID or MS	9,10
C_{10}-C_{13} LAS; determination of alkyl chain length distribution; isomer separation; trace analysis	Reaction of the tetrabutylammonium salt in the hot injection port to form the butyl ester of LAS	J&W DB-5MS 5% phenyl dimethylsilicone, 0.25 mm × 30 m, 0.25 μm film	3 min hold at 100°C, then to 300°C at 7°C/min	Ion trap MS	123

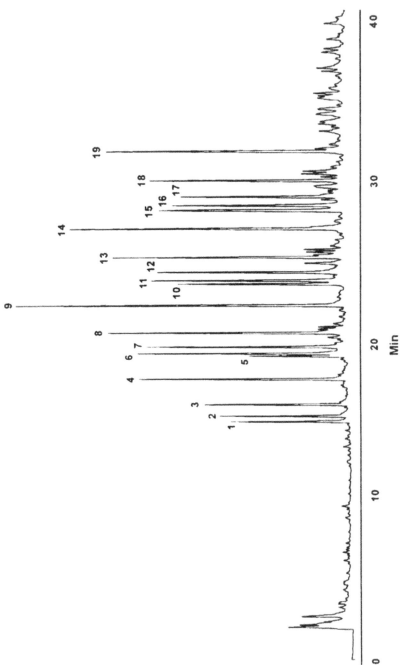

FIG. 1 Phosphoric acid pyrolysis GC analysis of LAS with detection by EI-MS. Alkylbenzene peak assignments: 1, C_{10}-5ϕ; 2, C_{10}-4ϕ; 3, C_{10}-3ϕ; 4, C_{10}-2ϕ; 5, C_{11}-6ϕ; 6, C_{11}-5ϕ; 7, C_{11}-4ϕ; 8, C_{11}-3ϕ; 9, C_{11}-2ϕ; 10, C_{12}-6ϕ; 11, C_{12}-5ϕ; 12, C_{12}-4ϕ; 13, C_{12}-3ϕ; 14, C_{12}-2ϕ; 15, C_{13}-6ϕ:16, C_{13}-5ϕ; 17, C_{13}-4ϕ; 18, C_{13}-3ϕ; 19, C_{13}-2ϕ. (Reprinted with permission from Ref. 13. Copyright 1999 by the American Oil Chemists' Society.)

by preparative TLC on Merck Silica Gel G with CH_2Cl_2. The blue band near the origin is scraped off, and the complex is extracted from the silica with methanol. The MeOH is removed by evaporation, and the residue is dissolved in CH_2Cl_2, transferred to a 3-mL vial, and the solvent is again evaporated. One-half milliliter hexane and 0.05 g PCl_5 are added and the vial is capped tightly and heated at 110°C for 20 min. After cooling, 2 mL of n-hexane is added, then all solvent is evaporated to drive out PCl_5 and the $POCl_3$ reaction product. Since some of the phosphorus compounds remain adsorbed to the walls of the vial, the residue is taken up in n-hexane and transferred to another vial. An aliquot of from 0.5 to 1.0 mL is injected in the splitless mode onto a 19 m × 0.31 mm i.d. capillary column, SE-54 or PS 255, using hydrogen carrier gas and flame ionization detection. The injector is held at 275°C. The column is held at room temperature until the solvent elutes, then heated to 100°C, with temperature programming from 100 to 300°C at 4°C/min.

Procedure: Desulfonation GC Analysis of LAS Recovered from Wastewater (23)

Add to the desulfonation flask 25 mL 85% phosphoric acid and two boiling stones, as well as 1 mg of anionic surfactant and 100 mg of C_9 LAS internal standard. Connect the trap, with stopcock closed, and add water to near the overflow level and 1.0–1.5 mL hexane. After attaching the reflux condenser, the apparatus is lowered into the heating mantle/sand bath. The stopcock is then opened and water removed until the bottom of the hexane layer is about 1 cm above the return tube. The mixture is then heated to boiling. As water is distilled from the flask, it is removed dropwise via the stopcock. The flask is held at the reflux temperature of 215°C for 3 h.

After cooling, most of the water is drained from the trap and discarded. The condenser walls are rinsed with three 1-mL portions of n-pentane, and the remaining water discarded. The solvent is drained into a 2-dram vial containing a bed of about 1 cm anhydrous sodium sulfate. After standing 10 min, the contents are poured into a 3-mL vial stepwise, evaporating most of the solvent before the next transfer. The sodium sulfate is washed with 2 mL pentane, adding the rinsings to the 3-mL vial. The contents of the vial are evaporated to dryness and the residue is taken up in 0.5 mL toluene and analyzed by GC. A Supelco DB-1 fused silica capillary column, 15 m × 0.32 mm i.d., 1.0 micron film thickness, is used with a 25:1 split. Injector and detector (FID) are at 300°C. Analysis is by temperature programming, 100 to 170°C at 5°C/min.

Note: This is a general procedure for workup and analysis of mixed isomers of alkylbenzenesulfonates, such as might be recovered in municipal wastewater. The C_9 standard produces 4 peaks, which provides a check on the matrix effects influencing the separation and on its overall reliability.

2. Alkyl Sulfates and Ethoxysulfates

GC methods are summarized in Tables 2 and 3. Sulfates of alcohols and ethoxylated alcohols are readily decomposed by acid hydrolysis to yield the free alcohols or ethoxylated alcohols. The trimethylsilyl derivatives of these are easily analyzed by gas chromatography (24). Derivatization of the alcohols may be omitted if they have reasonable volatility (25). Alternatively, the surfactant may be decomposed by hydriodic acid, giving alkyl iodide derivatives of the starting alcohols for characterization by gas chromatography (26). Hydrolysis can be combined in the same step with derivatization. Reaction of alkyl sulfates with silylating agents such as BSTFA/1% TMCS results in cleavage of the sulfate and formation of the trimethylsilyl ethers of the alcohols (27). GC or GC-MS analysis will

TABLE 2 Gas Chromatographic Analysis of Alkyl Sulfate Surfactants

Compound	Preliminary treatment	Column	Temperature	Detector	Ref.
C_{12}–C_{18} alcohol sulfates; trace analysis; separation by alkyl chain length	Formation of trimethylsilyl ethers with N,O-bis(trimethylsilyl)trifluoroacetamide (1% trichloromethylsilane) [BSTFA/1% TMCS]	Restek Rtx-1 dimethylsilicone, 0.25 mm × 60 m, 0.25 μ film thickness	50°C for 1 min, then to 215°C at 10°C/min	FID	27
C_{12}–C_{16} alcohol sulfates; determination of alkyl chain length distribution	Acid hydrolysis followed by extraction of alcohols	10% SE-30, 0.375 inch × 12 ft	80 to 256°C at 4°C/min	FID	124
C_{12}–C_{18} alcohol sulfates; semiquantitative determination of chain length distribution	Pyrolysis at 650°C to form olefins	20% SF-96 methylsilicone on Chromosorb W, 0.25 inch × 15 ft	50 to 250°C at 4°C/min	TC	11
C_9–C_{18} alcohol sulfate; determination of alkyl chain length distribution	Pyrolysis at 400°C with P_2O_5 to form alkenes	SF-96 methylsilicone on 0.02 inch × 200 ft steel capillary	120 to 220°C at 16°C/min	FID	16
C_{14}–C_{18} alcohol sulfate; determination of alkyl chain length distribution	Hydrolysis with dilute acid to form alcohols	20% Carbowax 20M polyethylene glycol on Chromosorb W HMDS, 0.25 inch × 10 ft	200°C	FID	16
C_8–C_{16} alcohol sulfate; determination of alkyl chain length distribution	Transesterification with BF_3/MeOH to form free alcohols	5% Carbowax 20M polyethylene glycol, 1% Igepal CO-880 nonylphenol—30 mole ethoxylate on Chromosorb G, 5 ft	150 to 250°C at 12°C/min	TC	125
C_{10}–C_{20} alcohol sulfate; determination of alkyl chain length distribution	Acid hydrolysis to form alcohols; reaction with HI to form alkyl iodides	10% SE-30 dimethylsilicone on Chromosorb W, 2.2 mm × 2 m	150 to 325°C at 10°C/min	TC	24
C_{12} alcohol sulfate; determination in protein extracts	Acid hydrolysis to form decanol	10% SE-30 dimethylsilicone, 2 mm × 3 m	185°C	FID	25

TABLE 3 Gas Chromatographic Analysis of Ether Sulfate Surfactants

Compound	Preliminary treatment	Column	Temperature	Detector	Ref.
Sulfated C_{12}–C_{15} alcohol 3-mole ethoxylate; determination of alkyl chain length distribution	Acid hydrolysis to form alcohol ethoxylates; reaction with HI to form alkyl iodides	10% SE-30 dimethylsilicone on Chromosorb W, 2.2 mm × 2 m	150 to 325°C at 10°C/min	TC	24
Sulfated C_8–C_{18} alcohol 3-mole ethoxylate; determination of alkyl chain length distribution	Reaction with HI to form alkyl iodides	10% PEG adipate on Celite, 5 ft	190°C	TC	26
Sulfated C_{12}–C_{15} alcohol 3-mole ethoxylate; homolog distribution; separation by both alkyl and ethoxy chain length	Acid hydrolysis to form alcohol ethoxylates; derivatization with 2:1:10 1,1,1,3,3,3-hexamethyldisilazane (HMDS)/trimethylchlorosilane (TMCS)/pyridine	10% SE-30 dimethylsilicone on Chromosorb W, 2.2 mm × 2 m	150 to 325°C at 10°C/min	TC	264
Sulfated C_{12}–C_{15} alcohol 3-mole ethoxylate; homolog distribution; separation by both alkyl and ethoxy chain length	Acid hydrolysis to form alcohol ethoxylates; derivatization with 1:1 BSTFA/pyridine	Chrompack High Temperature Sim Dist-CB, 0.1μ silicone film, 0.32 mm × 10 m	50 to 375°C in two temperature ramps	FID	29
Sulfated C_{12}–C_{15} alcohol 3-mole ethoxylate; determination of alkyl chain length distribution	Pyrolysis at 400°C with P_2O_5 to form alkenes	SF-96 methylsilicone on 0.02 inch × 200 ft steel capillary	100 to 190°C at 16°C/min	FID	16
Sulfated C_{12}–C_{14} alcohol ethoxylate; determination of alkyl chain length distribution	Pyrolysis at 500°C with P_2O_5 to form alkenes, acetaldehyde, and 1,4-dioxane	Polyphenylether capillary, 0.8 mm × 100 m	60 to 200°C at 5°C/min	FID	18,19
Sulfated C_{10}–C_{14} alcohol ethoxylate; determination of alkyl chain length distribution; separation of isomers	Reaction with anhydrides of HOAc and p-toluenesulfonic acid to form alkyl acetates and ethylene glycol diacetate	15% Free Fatty Acid Phase (PEG 20 M dinitroterephthalate ester) on Uniport B, 3 mm × 1 m	145°C	TC	53

then permit determination of alkyl chain length distribution. Similarly, reaction of alkyl ethoxysulfates with diazomethane produces the alkyl ethoxylates and their methyl ethers (28).

Rasmussen and coworkers demonstrated analysis of AES by hydrolyzing to the corresponding AE with HCl, then performing high-temperature GC on the BSTFA derivatives (29). Resolution is by both alkyl chain length and ethoxy chain length. Since most commercial products have a reasonably small number of homologues, the chromatogram is easily evaluated.

Sulfated and phosphated alcohol ethoxylates can be decomposed by reaction with the mixed anhydrides of acetic and paratoluenesulfonic acids to give the volatile acetates of the starting alcohols, compounds which are also easily analyzed by GC. The ethoxy units are converted to ethylene glycol diacetate (30).

Pyrolysis GC of alkyl sulfates in the presence of phosphoric acid or P_2O_5 gives olefins of the same length as the original alkyl chain, with the double bond distributed along the chain (16). Pyrolysis without acid yields mainly α-olefins as opposed to internal olefins. Ether sulfates are distinguished from alcohol sulfates by the presence of the characteristic decomposition products acetaldehyde and 1,4-dioxane (19).

Procedure: Determination of Alcohol Sulfate by Gas Chromatography of the Trimethylsilyl Ether Derivative (27)

Convert the AS to free acid form by passage through a strong cationic exchange column preconditioned with 0.1 M HCl. Derivatize by adding 0.5 mL dimethylformamide and 0.5 mL *N,O-bis*(trimethylsilyl)trifluoroacetamide (1% trichloromethylsilane) to a small amount of the AS in a vial. Seal and heat at 80°C for 1 hr. In this reaction, the sulfate group is cleaved and the trimethylsilyl ether of the original alcohol is formed. Analyze by GC-FID on a Restek Rtx-1 dimethylsilicone column, 0.25 mm × 60 m, 0.25 m film thickness, temperature programming from 50°C (1-min hold) to 215°C at 10°C/min. Calibration may be performed with free alcohols derivatized in the same way.

Procedure: Determination of an Alcohol Ether Sulfate in a Hand Dishwashing Detergent (29)

Dilute the sample in 3A alcohol and pass an aliquot containing 1 g of the initial sample through an anion exchange column, Amberlite IRA-400 CP, OH⁻ form, 1 × 20 cm. Wash with 50 mL 3A alcohol, then elute the AES with 50 mL 1:3 con HCl/ethanol. Mix the eluate with an equal volume of aqueous 3 M HCl and boil using a reflux condenser for 90 min. Transfer the solution to a separatory funnel, rinsing the apparatus with ethanol and then water. Extract the solution twice with 5-mL portions of $CHCl_3$. Combine the extracts, treat with 20 g of mixed bed ion exchange resin, H⁺/OH⁻ form, batch mode, to remove impurities. After filtering, treat with 2 mL of 1:1 pyridine/BSTFA, heat at 60°C for 20 min, and dilute 1:5 with $CHCl_3$. The derivatives are analyzed by high-temperature GC using cold on-column injection according to the conditions in Table 3.

3. Alkanesulfonates

Chromatography on, for example, a SE-54 capillary column resolves derivatized or desulfonated alkanesulfonates according to increasing chain length, with individual peaks seen for differing points of attachment of the sulfonate group (see Fig. 2 and Table 4). Alkyl derivatives of alkanesulfonates may be formed by decomposition of their ion pairs with tetrabutylammonium ion in the injection port of the gas chromatograph. Other alkylammonium salts may be used, but tetrabutylammonium ion gives the best conversion (9,31).

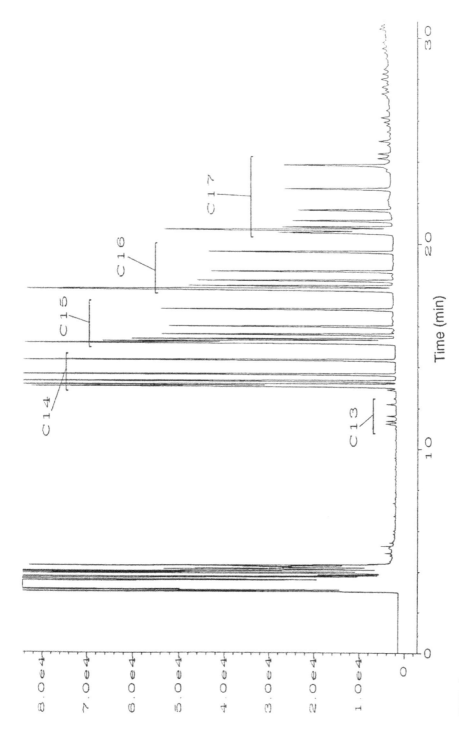

FIG. 2 GC analysis of methyl esters of C$_{13}$–C$_{17}$ alkane sulfonate mixture. Peak assignments denote alkyl chain length. (Reprinted with permission from Ref. 7. Copyright 1995 by Karl Hanser Verlag.)

TABLE 4 Gas Chromatography of Alkane Sulfonate Surfactants

Compound	Preliminary treatment	Column	Temperature	Detector	Ref.
C_{14}–C_{17} branched alkane-sulfonate; demonstration of chain length and isomer separation	Formation of methyl sulfonates with diazomethane	OV-1 dimethylsilicone, 320 μm × 25 m, 0.2 μm film thickness	150 to 280°C at 5°C/min	FID	119
C_{12}–C_{18} n-alkane sulfonates; separation by alkyl chain length; isomer separation	Formation of methyl sulfonates with diazomethane	10% SE-30 dimethylsilicone on Diatoport P, 4 mm × 1.5 m	130 to 250°C at 1°C/min	FID	126
C_{10}–C_{18} alkane sulfonates; determination of alkyl chain length distribution	Pyrolysis at 500°C with P_2O_5 to form alkenes	Polyphenylether capillary, 0.8 mm × 100 m	60 to 200°C at 5°C/min	FID	18,19
C_{12}–C_{18} secondary alkane-sulfonate; separation by chain length; isomer separation	Reaction of the tetrabutylammonium salt in the hot injection port to form the butyl sulfonate ester	Hewlett-Packard HP-5 5% phenyl dimethylsilicone, 0.2 mm × 20 m, 0.33 μm film	110 to 220°C at 10°C/min, then to 300°C at 6°C/min	MS	9,31
C_9–C_{21} secondary alkane sulfonates; determination of alkyl chain length distribution and degree of branching	Fusion with KOH/NaOH at 340°C to form alkenes; optional hydrogenation to form alkanes	Apiezon L hydrocarbon on 50 m capillary column	154°C or 200°C	FID	32
C_{10}–C_{17} paraffin sulfonates; determination of alkyl chain length distribution and isomer distribution	Formation of methyl esters with trimethylorthoformate	J&W Scientific DB5 5% phenyl dimethylsilicone, 0.32 mm × 60 m, 0.25 μm film; 0.53 mm × 1 m retention gap, H_2 carrier gas	100 to 200°C at 30°C/min, then to 225°C at 2°C/min	FID	7

Pyrolysis of alkanesulfonates yields olefins of the same and lower chain length, with unsaturation primarily in the alpha position (11). Chemical desulfonation with acid or base yields olefins of the same chain length with the double bond located randomly (16,32). Thus, compared to derivatization, desulfonation has the disadvantage that information about point of attachment of the sulfonate group is lost. Desulfonation of paraffin sulfonates using phosphoric acid gives poor yields, so that it is suitable only for qualitative analysis. Desulfonation is most completely accomplished by fusion with molten alkali. Since the degree of unsaturation of the desulfonation products varies with the reaction conditions, it is best to perform a subsequent hydrogenation to give the original saturated hydrocarbons. GC analysis will then provide information on the extent of branching as well as the alkyl chain length distribution (32). The fusion is performed in a special apparatus of nickel and quartz with KOH/NaOH in 50:50 mole ratio. The hydrocarbons are distilled and trapped during the reaction. At 330°C, desulfonation requires about 2 h, with 70–80% yield.

4. Other Anionic Surfactants

α-Olefin Sulfonates. Alkene sulfonates can be readily analyzed as the methyl sulfonate derivatives (Fig. 3 and Table 5). The hydroxyalkanesulfonate component is converted during the derivatization reaction to the 1,4-sultone rather than to the methoxyalkanesulfonate (7). Sodium salts must be converted to free acids before methylation. This is performed most elegantly with ion exchange resin. Alternatively, the dried salts may be treated with sulfuric acid, and the liberated acids extracted into ethyl ether (7). An alternative method is derivatization with thionyl chloride after hydrogenation, yielding the alkylsulfonyl chlorides and chloroalkysulfonyl chlorides from the alkenesulfonate and hydroxyalkanesulfonate components, respectively. These alkylsulfonyl chlorides are somewhat unstable at the temperatures required for gas chromatography and partially decompose to form olefin sulfonyl chlorides. By careful standardization of conditions, the extent of decomposition can be made reproducible so that quantitative information may be obtained. The presence of disulfonates compromises the quality of the chromatogram because disulfonates decompose to a greater extent than the monosulfonates. It is best to remove them by extraction (Chapter 1) before analysis (6).

Ester sulfonates. Pyrolysis of α-sulfo fatty acid esters at 400°C in the presence of P_2O_5 results in desulfonation, so that the gas chromatogram shows the fatty acid methyl esters. Free fatty acids are also seen as decomposition products of the esters, as well as those originally present as sulfonated impurities (33).

Lignin sulfonates. Pyrolysis of sodium lignin sulfonate yields SO_2, a product rarely found in pyrolysis of other surfactants. This could possibly be the basis for selective detection of lignin sulfonate (34).

N-*Acyl amino acids.* Both acylsarcosine and free fatty acids can be determined, according to their chain length distribution, by GC of their methyl esters (35)

Ether carboxylates. Ether carboxylates are readily analyzed as the methyl esters. Separation of NPE carboxylate has been demonstrated up to a degree of ethoxylation of about 7 (36).

Fluorinated surfactants. C_6–C_8 perfluorinated fatty acids, analyzed as their methyl esters, require a thick-film column in order to show adequate retention time. The C_4 compounds are too volatile for analysis without applying external cooling (37).

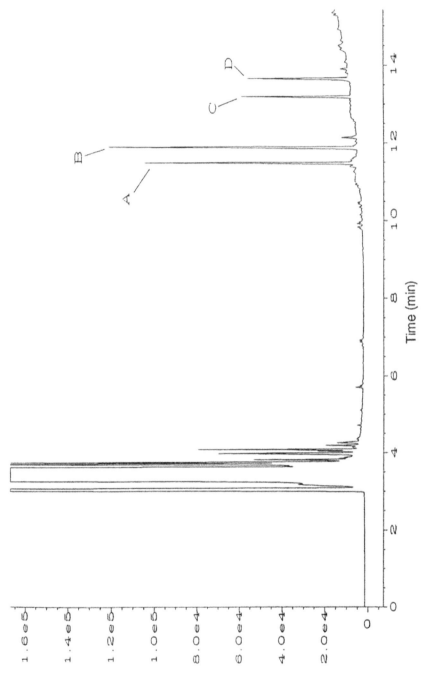

FIG. 3 GC analysis of methyl esters of C_{14}/C_{16} α-olefin sulfonate mixture. Peak assignments: A, methyl 3-tetradecene sulfonate; B, sultone from dehydration of tetradecane hydroxysulfonate; C, methyl 3-hexadecene sulfonate; D, sultone from dehydration of hexadecane hydroxysulfonate. (Reprinted with permission from Ref. 7. Copyright 1995 by Karl Hanser Verlag.)

TABLE 5 Gas Chromatographic Analysis of Other Anionic Surfactants

Compound	Preliminary treatment	Column	Temperature	Detector	Ref.
C_{14}–C_{18} α-olefin sulfonates; separation by chain length and functionality	Formation of alkane sulfonyl chlorides by hydrogenation and reaction with thionyl chloride	30% SE-30 dimethylsilicone on silanized Chromosorb WAW; glass, 4 mm × 2.5 m	200°C	FID	6
C_{16} α-olefin sulfonates; separation of alkenesulfonate isomers	Formation of methyl sulfonates with diazomethane	3% OV-101 dimethylsilicone on Chromosorb W, 0.25 inch × 10 ft	80 to 300°C at 6°C/min	FID	127
C_{14}/C_{16} α-olefin sulfonates; separation by chain length and functionality	Formation of methyl esters with trimethylorthoformate; hydroxyalkanesulfonates are converted to sultones	J&W Scientific DB5 5% phenyl dimethylsilicone, 0.32 mm × 60 m, 0.25 µm film; 0.53 mm × 1 m retention gap, H_2 carrier gas	100 to 200°C at 30°C/min, then to 320°C at 2°C/min	FID or MS	7
Lignin sulfonates; detection in river water	Pyrolysis at 510°C and 610°C to form SO_2, phenols, and other characteristic compounds	CP-SIL-5 (Chrompak dimethylsilicone), 0.32 mm × 50 m, 1 µm film	30 to 300°C at 10°C/min	MS	34
C_{16} and C_{18} α-sulfofatty acid methyl esters; determination of alkyl chain distribution	Pyrolysis with P_2O_5 at 400°C to form the free acids and their methyl esters	DB-5 5% phenyl dimethylsilicone, 30 m capillary	Not given	FID	33
Phosphated C_{12}–C_{18} alcohol ethoxylate; determination of alkyl chain length distribution; separation of isomers	Reaction with anhydrides of HOAc and p-toluenesulfonic acid to form alkyl acetates and ethylene glycol diacetate	15% Free Fatty Acid Phase (PEG 20 M dinitroterephthalate ester) on Uniport B, 3 mm × 1 m	185°C	TC	53

(Continued)

TABLE 5 (*Continued*)

Compound	Preliminary treatment	Column	Temperature	Detector	Ref.
C_{14}–C_{18} acylsarcosines; determination of free fatty acids; separation according to alkyl chain length	Formation of methyl esters by H_2SO_4–catalyzed reaction with MeOH	10% SE-30 dimethylsilicone on Chromosorb W-AW-DMCS, 2 m	320°C	FID	35
NPE carboxylate; E = 5; resolution of homologs to about E = 7; partial resolution of o- and p- isomers	Formation of methyl esters with diazomethane or BF_3/MeOH	10% SE-30 dimethylsilicone on Chromosorb, 50 cm	Not given	MS	36
C_6–C_{12} perfluorocarboxylic acids, resolution by alkyl chain length	Formation of methyl esters with CH_3I	Supelco SPB-1 SULFUR dimethylsilicone, 0.32 mm × 30 m, 4 μm film	60 to 190°C at 6°C/min, then to 270°C at 30°C/min	MS	37
Hydrotropes; determination	Formation of sulfonyl chlorides or methyl esters by reaction with thionyl chloride or diazomethane	20% Dow Corning high vacuum silicone grease on Chromosorb AW 6 mm o.d. × 2 ft, or 20 % SE-76 silicone on Celite 545, 6 mm o.d. × 4 ft	Various	TC	3

II. NONIONIC SURFACTANTS

A. General Considerations

The low volatility of nonionic surfactants makes them challenging subjects for GC analysis. Even with the best of derivatization and high-temperature techniques, oligomers containing more than about 20 EO units will not elute from the GC column. Quantitative GC analysis of nonionics of higher degree of ethoxylation requires chemical reaction to put the sample into a more volatile form.

The most interesting recent work has been in the area of high-temperature (up to about 375°C) GC. Quantitative analysis of derivatized AE up to an average degree of ethoxylation of 12 has been demonstrated, with resolution of oligomers up to the 21-mole adduct (38,39). The short lifetime of GC columns used at high temperature has so far limited the application of this approach. Finke et al. report that longevity of aluminum-clad columns is increased if a reverse program is used to return to starting conditions at the end of a temperature gradient (40).

In HPLC, the separation of commercial mixtures of, for example, AE must normally be performed twice: once with resolution on the basis of the hydrocarbon chain length and once on the basis of the ethoxylate chain length. By contrast, GC analysis separates according to both modes in a single chromatogram. While this makes the instrumentation requirements much simpler, it can make interpretation of the chromatogram difficult (41). In the case of a commercial product made by ethoxylation of a broadly distributed hydrophobe, retention times of lower ethoxylates of higher MW hydrophobes may overlap with those of higher ethoxylates of lower MW hydrophobes such that resolution is impossible.

1. Linearity

Special attention is needed when using GC to determine oligomer distribution. A chromatographer using flame ionization detection to determine concentrations of similar compounds will usually perform quantification from the raw chromatogram by simply calculating area percent of each compound and presume that this is equivalent to weight percent composition. While this is approximately correct for oligomers containing 10 or more EO units, serious errors are seen with the lower adducts. Because the percent oxygen in the free hydrophobe and the lower adducts is significantly less than in the higher adducts, the response of the FID is different for the same mass of each adduct, decreasing with higher chain length. For example, the observed response for underivatized $C_{12}EO_6$ is only 61% of that of dodecanol on a weight basis (42).

In theory, the difference in response for members of an oligomeric series can be corrected by simple calculation: the concept of effective carbon number allows conversion of the FID output to mole percent or weight percent composition, without internal or external calibration. However, the literature on application of this concept is disappointing. Several teams of researchers have published results on analysis of AE (38,42–44), and one on APE (44). All report the same general trends and all report excellent results for quantification within a single laboratory. Unfortunately, the factors diverge significantly between laboratories, as shown in Fig. 4. For precise work, the gas chromatographer must be prepared to check the calibration with pure substances.

Atomic emission detection shows promise for easing the calibration problem. In practice, AED is not completely independent of molecular structure, so calibration should be made with a similar compound. However, calibration standards are required for only one or two members of a homologous series (39).

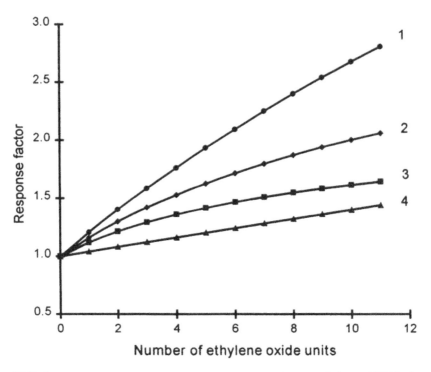

FIG. 4 Response factors for TMS ethers of C_{12} AE relative to dodecanol TMS ether, calculated according to formulas given by different researchers. Curve explanation: 1, Milwidsky and Gabriel, Ref. 147; 2, Scanlon and Willis, Ref. 43; 3, Jones, Ref. 44; 4, McClure, Ref. 148. (Reprinted with permission from Ref. 44. Copyright 1998 by CSIRO.)

B. Direct Analysis

Direct GC analysis, i.e., injection of the surfactant without derivatization or other workup, is attractive to many workers because of the ease and speed with which data can be generated. The analyst must remain aware that such methods discriminate in favor of low molecular weight components.

Packed-column GC. Short (0.5-m) columns are generally used for the direct analysis of alkoxylates to minimize degradation. In a study of analysis of ethoxylated dodecanol by packed-column GC without derivatization, a strategy is given for calculating molecular distribution when it is known that not all the sample elutes from the column (45). Short packed columns are suitable for the direct analysis of AE oligomers up to MW about 500 (corresponding to a 7-mole adduct of a C_{12} alcohol) (42). Low molecular weight compounds of lauric acid and propylene oxide have also been studied, up to tripropylene glycol dilaurate (46).

Capillary column GC. Split injection tends to give higher results for low ethoxylate components, while on-column injection gives better results for the higher oligomers. Attractive chromatograms are obtained up to about 12-mole oligomers, but the distribution is skewed toward the low end compared to HPLC results (47). Programmed-temperature volatilization has been proposed as a method superior to ordinary on-column injection for analysis of ethoxylates (48).

C. Analysis After Derivatization

Derivatization reactions add a terminal group or groups to the molecule, which increases its volatility but otherwise leave it intact. For most nonionic surfactants, capping the polar hydroxyl groups with alkyl functionality greatly increases volatility. In general, derivatization permits the GC analysis of ethoxylates up to oligomers of 16 to 18 EO units, although recovery may not be quantitative. Szymanowski and coworkers have prepared tables of retention indices to help in identifying GC peaks of homologs and oligomers of derivatized PEG, AE, ethoxylated amines, and APE (49).

1. Trimethylsilyl Derivatives

Forming the trimethylsilyl ethers of ethoxylates of moderate MW makes them volatile enough to be analyzed even by packed-column gas chromatography. The complete molecular weight distribution can be calculated on a relative basis, with standards required for definitive work. Compounds also containing oxypropylene moieties usually cannot be distinguished from pure ethoxylates by this technique. Alkanolamides and the various esters of sorbitol, sucrose, and glycerin can be characterized by analysis of the trimethylsilyl derivatives.

A problem with the analysis of surfactants with two hydroxyl groups is the possibility of formation of two different derivatives of each oligomer: the *mono-* and the *bis*trimethylsilyl ether. This leads to more complex chromatograms than for underivatized nonionics. Therefore, for low MW EO adducts of primary amines, analysis without derivatization is sometimes preferred (50).

2. Acetate Esters

Formation of acetate esters is a technique generally applicable to GC analysis of nonionic surfactants. This method is usually applied to simple mixtures, such as ethoxylates of an alcohol of a single alkyl chain length. If the alcohol is a commercial mixture, the gas chromatogram shows dozens of peaks corresponding to the ethoxylates of several alkyl chain lengths, making quantitative analysis difficult or impossible. In the case of acetate esters of NPE, compounds up to an average degree of ethoxylation of 10 can be analyzed. However, due to discrimination against the higher molecular weight compounds, results for ethoxy content were found to be low by 10% for 5-mole ethoxylates and 24% for 9-mole ethoxylates (51).

Care should be taken with acetylation, since, with BF_3 or p-toluenesulfonic acid catalyst, cleavage of the bond between the fatty alcohol and the PEG portions of AE can occur (52,53). This cleavage may be performed deliberately to simplify the gas chromatograms. The alkyl and ethoxy derivatives are separated by extraction and analyzed separately. The PEG diesters can be determined up to a degree of polymerization of about 17, while the alkyl acetates are sometimes saponified and analyzed as the free alcohols (52).

Reaction with acetic anhydride at reflux for 15 min suffices to convert an AE to a volatile acetate derivative, suitable for characterization by packed-column gas chromatography (54). Acetyl chloride may also be used to prepare the derivatives (49). Again, caution is in order, since in the presence of ferric chloride and especially at elevated temperature (150°C), acetyl chloride will cleave ether linkages to give acetates and 2-chloroacetates of the parent alcohol and mono-, di-, and triethylene glycol (55,56). GC analysis of the acetate derivatives of AE is generally performed on a silicone stationary phase with a temperature program from about 100 to as high as 400°C.

Ketene, generated by pyrolysis of acetone, has been suggested as a means of preparing the acetate derivatives. Although the apparatus used is rather elaborate and ketene itself is toxic to humans, the technique has the advantage of not yielding a peak in the chromatogram for excess acetic acid or anhydride (57).

Procedure: Derivatization Procedures for Nonionic Surfactants (Szymanowski, Szewczyk, and coworkers)

Trimethylsilyl derivatives: Add 0.05 g sample and 0.5 g N,O-*bis*(trimethylsilyl) acetamide to a glass micro reaction vial of about 3 mL capacity. Maintain at 40°C for 15 min. For some compounds, higher temperatures and longer reaction times may be required. 60°C and 30 min is recommended for amine derivatives.

Acetate derivatives: A few tenths of a gram of surfactant and a 100 mole% excess of acetic anhydride are refluxed for 15 min in a small reaction vessel under nitrogen purge. The reaction mixture is stripped under reduced pressure at 100°C to remove excess acetic acid and acetic anhydride.

D.   Analysis After Cleavage

In the cleavage reactions, the surfactant is split, generally at the juncture of the hydrophobe and hydrophile groups, and the fragments are put into volatile form.

1.   HBr or HI Fission

Reaction of polyoxyethylene and polyoxypropylene compounds with hydrogen bromide produces mono- and dibromoethane and mono- and dibromopropane, which are easily measured by gas chromatography. Less often, HI is used in place of HBr, giving the corresponding iodoethane and iodopropane compounds. Depending on the product, it may also be possible to determine the hydrophobe component by GC. When coupled with appropriate initial isolation steps, this reaction may even be used for trace analysis, especially if the halogen-selective electron capture detector is used (58–60). A technique for specific determination of AE in wastewater in the parts per billion range is based on GC determination of the alkyl bromides with mass selective detection (61).

If both EO and PO are present in the same molecule, then extra care must be taken for precise analysis by the HBr fission method. It is best to use EO/PO polymers of known composition as bracketing standards. This approach corrects for the fact that formation of byproducts is dependent upon reaction conditions and is not the same for EO and PO (62,63). Generally, yield of halopropane from HBr or HI cleavage of PO units is less complete than haloethane yield from EO units, so that erroneous results may be obtained for EO/PO ratio unless a suitable calibration procedure is selected (64). Organic acids, such as adipic acid, catalyze the formation of propyl iodide from oxypropyl groups, thus minimizing the concentration of propene and other byproducts and allowing more straightforward GC quantification (65).

Procedure: Determination of the Alkyl Chain Distribution of AE by GC After HBr Cleavage (39)

Add to an autosampler vial 100–500 μg AE sample, 20 μg C_{20} alcohol internal standard, and 500 μL HBr as the 30% solution in acetic acid. Seal and heat for 4 hr at 70°C. Cool, add 2 mL water, and extract the alkyl bromides with 600 μL methylene chloride. Analyze by capillary GC on a 15-m J&W Scientific DB5 column.

2. Cleavage with Acetyl Chloride

As noted above, reaction of an ethoxylate with acetyl chloride without catalyst at 60°C for 15 min yields the acetate ester of the intact surfactant. However, reaction with acetyl chloride using ferric chloride catalysis and a temperature of 150°C for 30 min will cleave the ethoxylate units, yielding mainly 2-chloroethyl acetate. Quantification of 2-chloroethyl acetate by GC gives the amount of EO in the original surfactant. Hydrophobe groups are converted to a number of products, including high molecular weight tar and some smaller compounds which may be measured by GC. Most often, chromatographic conditions are chosen which are blind to the hydrophobe, so that quantification of 2-chloroethyl acetate may be performed without distraction (55,66). For higher accuracy, the concentration of byproducts of ethoxylate cleavage may also be determined: 1,2-dichloroethane, *bis*(2-chloroethyl) ether, dioxyethylene glycol diacetate, and 5-chloro-3-oxapentyl acetate (51,56,67).

Acetyl chloride cleavage is also applicable to analysis of block EO/PO copolymers, as well as products containing butylene oxide (68). Propylene oxide groups are converted mainly to 2-chloro-1-methylethyl acetate and 2-chloropropyl acetate, with traces of 1,2-propylene diacetate. Simple GC determination of the ratio of chloroethyl acetate to total chloroethyl and chloropropyl acetates provides an accurate figure for EO content of copolymers up to about 30% EO. A calibration curve must be used for higher ratios (69).

Since this cleavage method is most useful for determining the alkoxy content of surfactants, it is rarely used in facilities that have access to modern NMR instrumentation, which can usually provide the same information with less sample preparation and no need for external calibration.

3. Acetic/Paratoluenesulfonic Acids

Ethoxylates of alcohols, alkylphenols, esters, and alkylamines, as well as EO/PO copolymers, can all be analyzed after cleavage of ether and ester bonds with a mixed anhydride of acetic and paratoluenesulfonic acids (30,53,70). The ethylene oxide units are converted to the diacetate of ethylene glycol. Propylene oxide units react similarly, forming the diacetate of propylene glycol. Alcohol moieties are converted to the volatile acetates. Acid units are released as the free fatty acids. Ethoxylated glycerin, ethylene glycol, and sorbitol esters yield the diacetates of ethylene glycol and isosorbide, and the triacetate of glycerin.

4. Pyrolysis

Pyrolysis gas chromatography can be used by an experienced operator to obtain qualitative data about the nature of nonionics (11,16,18). For example, branched chain APE can be distinguished from linear chain APE, and the alkyl chain length of APE, AE, alkanolamides, and alkylamine oxides can be determined. Polyethylene oxide and polypropylene oxide units are converted to acetaldehyde and propionaldehyde, as well as dioxanes. Ethylene and propylene have also been reported (71). Most often, a cleavage reagent is added prior to pyrolysis. For example, AE is analyzed by pyrolysis in the presence of hydrobromic and acetic acids, which, at 600°C, leads to cleavage of the ethoxy groups, leaving the original alcohols intact (72).

Quantitative analysis is also feasible. A collaborative study has been performed of determination of ethylene and propylene oxide content of surfactants by pyrolysis GC (73). Samples were heated in the presence of phosphoric acid at 500 or 600°C, with the resulting acetaldehyde and propionaldehyde content determined by GC on Porapak Q. For EO/PO copolymers, the EO content could not be calculated directly from the raw data even if the effective carbon number concept was used to correct the response of the FID detector. (This is a substantial correction, since the carbon/oxygen ratio is 2 for acetaldehyde and 3 for propionaldehyde.) Instead, it was necessary to prepare a calibration curve using mixtures of PEG and PPG of molecular weight 600 or more. With empirical calibration curves, excellent accuracy was obtained, as well as good agreement between different laboratories. On the other hand, determination of ethoxy content of nonylphenol ethoxylates was found to be accurate only for an ethoxy content of 66% or more, and EO content of ethoxylated amines or fatty alcohols could not be accurately determined at all.

Pyrolysis GC becomes a much more effective tool if it is combined with mass spectroscopy. Pyrolysis GC-MS is very useful for the characterization of complex copolymers of all sorts and is a valuable supplement to infrared and nuclear magnetic resonance spectrometry for study of surfactants. Although direct pyrolysis can be used, successful pyrolysis GC of ethoxylates generally requires the addition of cleavage reagents to the sample before heating. The distinction between reaction GC and pyrolysis GC is thus blurred.

E. Analysis of Individual Nonionics

1. Alcohol Alkoxylates

Direct-capillary GC analysis of underivatized AE with split injection tends to give higher (i.e., more accurate) results for low ethoxylate components, while on-column injection gives higher results for the higher homologs. In either case, attractive chromatograms are obtained with peaks up to about the 12-mole homologs, but the distribution is skewed toward the low end compared to HPLC (47). Good resolution of isomers of secondary AE is seen, especially for lower homologs (74). The real advantage of GC over HPLC is in detecting volatile impurities in AE.

Silyl derivatives of AE are eluted roughly according to increasing boiling point, with retention times increasing as a function of both the alkyl and ethoxy chain lengths (Fig. 5 and Table 6). Some optimization work is required with mixed ethoxylates, since, for example, $C_{12}E_3$ will elute between $C_{16}E_1$ and $C_{16}E_2$ (Fig. 6). Modern high-temperature capillary chromatography has been shown to give quantitative results for determination of the oligomer distribution of ethoxylates of C_{12}–C_{18} alcohols. Oligomers up to the 20-mole ethoxylate are quantified (29,38–40,75).

Pyrolysis GC of alcohol ethoxylates in the presence of P_2O_5 yields acetaldehyde, 1,4-dioxane, and alkenes corresponding to the initiator alcohols. If the surfactant contained propylene oxide groups, propionaldehyde and dimethyldioxane are also seen. Branched initiators can be distinguished from straight-chain alcohols because the former give many isomeric alkenes after pyrolysis (76).

In the case of AE, HI or HBr fission produces the alkyl iodides or bromides derived from the starting alcohols, as well as the decomposition products from the ethoxy chain described above. The alkyl halides can be readily characterized as to chain length and isomer distribution by gas chromatography (72). An eight-laboratory collaborative study of AE characterization by HI cleavage-GC gave coefficients of variation between laboratories of 1–2% for determination of the major components of a linear alcohol initiator and about twice that for branched alcohols (77). The ethyl and propyl compounds from decomposition of the alkoxy chain may be quantified if the E/P ratio or degree of ethoxylation is to be determined, but such information is more easily available from NMR analysis.

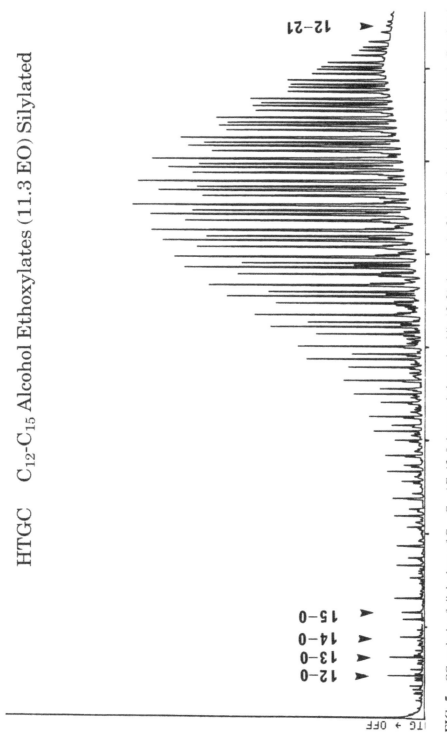

FIG. 5 GC analysis of silyl ethers of C_{12}–C_{15} AE. 12–0 denotes dodecanol, while 12–21 denotes the 21-mole ethoxylate of dodecanol. (Reprinted with permission from Ref. 38. Copyright 1999 by the American Oil Chemists' Society.)

TABLE 6 GC Analysis of Alcohol Ethoxylates

Compound	Preliminary treatment	Column	Temperature	Detector	Ref.
C_{10}–C_{20} AE; resolution of oligomers up to E = 11	None	For 2-mole ethoxylates: Hewlett-Packard Ultra 2, 25 m; for 6-mole ethoxylates: H-P Ultra 2, 12.5 m; for on-column injection of 6-mole ethoxylates: H-P Ultra 1, 0.32 mm × 25 m	80 to 280°C at 15 or 25°C/min	FID	47
C_{12} AE; resolution of oligomers up to E = 4	None	10% SE 301 on Chromosorb W AW-DMCS, 6 mm × 1.4 m	190 to 290°C at 10°C/min	Not specified	45
C_8, C_{12}, C_{18} AE; resolution of oligomers up to E = 6	None	10% SE 30 dimethylsilicone on Chromosorb W AW DMCS, 0.125 in. × 0.5 m	50 to 300°C at 6°C/min	FID	42
AE, alkyl chain not specified; E = 8: resolution of oligomers up to E = 20; determination of average degree of ethoxylation and free alcohol	Derivatization with 2:1:10 1,1,1,3,3,3-hexamethyldisilazane (HMDS)/trimethylchlorosilane (TMCS)/pyridine	SGE AQ5/HT5 (5% phenyl polycarboranesiloxane), 0.53 mm × 25 m, 0.15 μm film	60 to 450°C at 10°C/min; programmed return to 60°	FID	40
C_{12}, C_{14}, C_{16} AE; resolution of oligomers up to E = 6; comparison of various derivatives	With and without derivatization: TMS derivatives with BSA, acetates and trifluoroacetates with the anhydrides, monochloroacetates with monochloroacetyl chloride	WCOT glass capillary with OV-61-OH, 0.7 mm × 12 m	120 to 370°C at 5.5°C/min	FID	128

C_{12}–C_{14} secondary AE: resolution of oligomers up to E = 9; isomer resolution	None	SGE BPX5 5% phenyl dimethylsilicone, 0.32 mm × 12 m, 0.5 mm film	80 to 190°C at 20°C/min, then 190 to 340°C at 2°C/min	FID	74
C_{12}–C_{18} AE; E = 7 to 12; resolution up to E = 21; quantitative analysis	Derivatization with bis(trimethylsilyl)-trifluoroacetamide (BSTFA)	Chrompack SimDist-CB, 0.32 mm × 10 m, 0.1 m film	100 to 375°C at 4° C/min	FID	38
C_{12}–C_{15} AE; E = 0 to 17; separation by alkyl and ethoxy chain length	Derivatization with 1:1 BSTFA/pyridine	Chrompack High Temperature Sim Dist-CB, 0.1m silicone film, 0.32 mm × 10 m	50 to 375°C in two temperature ramps	FID	29
C_{13}/C_{15} AE: demonstration of oligomer separation up to about E = 18	Derivatization with HMDS and TMCS	OV-1 dimethylsilicone, 0.53 mm × 10 m	80 to 180°C at 25°C/min, then to 400°C at 8°C/min	FID	119
C_8–C_{22} AE: resolution of oligomers up to about 13 ethoxy units	Derivatization with N,O-bis(trimethylsilyl)acetamide (BSA) or acetic anhydride	OV-17 50% phenyl di-methylsilicone or OV-101 dimethylsilicone, 3% on Chromosorb G AW DMCS, 2.7 mm × 0.9 m, or Dexsil 400 carborane/methyl phenyl silicone, 1% on Chromosorb G AW DMCS, 2.7 mm × 0.4 m	For the OV-101 column: 100 to 300°C at 8°C/min; for the Dexsil column: 130 to 360°C at 6°C/min	FID	49,129
C_{12} AE: resolution of oligomers up to about 13 ethoxy units	Derivatization with N,O-bis(trimethylsilyl)acetamide (BSA)	SGE BX-5, 0.32 mm × 7.5 m, 0.25 μm film	80 to 390°C at 15°C/min	FID	44

(Continued)

TABLE 6 (*Continued*)

Compound	Preliminary treatment	Column	Temperature	Detector	Ref.
C_{10}–C_{18} linear and branched AE; resolution of oligomers to about E = 17; study of atomic emission detection	Derivatization with 2:3 BSTFA/pyridine or 1:1 acetic anhydride/pyridine	J&W DB5-HT 5% phenyl/95% dimethylsilicone (specially treated), 0.32 mm × 15 m, 0.1 μm film	1 min hold at 50°C, then to 100°C at 20°C/min, then to 375°C at 4°C/min	Atomic emission detection for C, O, and Si	39
C_{10}–C_{18} linear and branched AE; characterization of alkyl composition; collaborative study	HI cleavage	Various nonpolar or low-polarity silicone capillary columns, 0.15–0.30 mm × 20–30 m, 0.1–0.3 μm film	100 to 250°C at 3–5°C/min	FID	77
C_{12}–C_{16} alkyl 3- to 8-mole ethoxylates; resolution of oligomers to up to about 12 ethoxy units	Formation of acetate derivatives	2% SE-30 dimethylsilicone on Chromasorb W-AW, 0.25 inch × 2 ft	125 to 405°C at 5–15°C/min	TC	54
C_{12}–C_{18} AE; determination of alkyl chain distribution; isomers resolved	Cleavage/derivatization with acetic anhydride/p-toluenesulfonic acid	15% Free Fatty Acid Phase (PEG 20 M ester of nitroterephthalic acid) on Uniport B, 3 mm × 1 m	165°C for alkyl distribution; 100 to 140°C at 2°C/min for resolution of C_{11} isomers	TC	53
C_{13}–C_{18} secondary AE; determination of alkyl chain length distribution and ethoxy distribution up to 16 EO units	Cleavage/derivatization with acetic anhydride/BF$_3$/HOAc	Alkyl distribution: 10% PEG 2000 on Embacel 60-AW, 2 mm × 1.5 m; ethoxy distribution: 2% Versamid 900 polyamide resin on Chromosorb G-AW DMCS, 4 mm × 35 cm	Alkyl distribution: 100 to 250°C at 2°C/min; ethoxy distribution: 60 to 350°C at 4°C/min	FID (alkyl distribution) and TC (ethoxy distribution)	52
C_{10}–C_{22} AE, E = 3 to 15; determination of ethoxy content	Cleavage with CH$_3$COCl/FeCl$_3$	12% Carbowax 20M-TPA polyethylene glycol terephthalate on Chromosorb W-AW DMCS, 2.7 mm × 1.6 m	100 to 220°C at 5°C/min	FID	56,66

Application	Sample preparation	Column	Temperature program	Detector	Ref.
C_8–C_{18} alkyl 3- to 14-mole ethoxylates and ethoxy/propoxylates; determination of EO/PO content	Cleavage with HBr/HOAc; extraction	5% OV-210 trifluoropropyl methyl silicone on silanized Chromosorb W, 0.125 inch × 6 ft	50 to 260°C at 20°C/min	FID	130
C_{10}–C_{12} alkyl 4- and 7-mole ethoxylates; determination of alkyl chain lengths	Pyrolysis at 650°C	20% SF-96 methylsilicone on Chromosorb W, 0.25 inch × 15 ft	50 to 250°C at 4°C/min	TC	11
C_{12}–C_{15} alkyl 7-mole ethoxylates; determination of alkyl chain lengths	Pyrolysis at 600°C with HBr/HOAc	SGE BP10 (14% cyanopropylphenyl/86% dimethylsilicone), 0.5 mm film, 0.22 mm × 50 m	70 to 240°C at 12°C/min	FID	72
C_{11}–C_{18} alcohol 9-mole ethoxylates; determination of alkyl chain length distribution	Pyrolysis with P_2O_5 at 400°C to form alkenes	Stainless steel capillary coated with SF-96 methylsilicone, 0.02 inch × 200 ft	100°C	FID	16
C_{10}–C_{18} AE; determination of alkyl chain length distribution and %EO	Pyrolysis with P_2O_5 at 500°C to form alkenes, acetaldehyde, and 1,4-dioxane	Alkene analysis: polyphenylether, 0.8 mm × 100 m; %EO determination: open capillary, 0.25 mm × 50 m	Alkene analysis: 60 to 200°C at 5°C/min; %EO determination: 70°C isothermal	FID	18
C_8–C_{18} alcohol ethoxylates and EO/PO adducts; determination of alkyl chain length distribution, %EO, %PO	Pyrolysis with P_2O_5 at 600°C to form alkenes, acetaldehyde, 1,4-dioxane, propionaldehyde, and dimethyldioxane	SGE HT-5 5% phenyl polycarborane dimethyl silicone, 0.22 mm × 25 m, 0.1 mm film	40°C for 5 min, then to 280°C at 10°C/min	FID	76

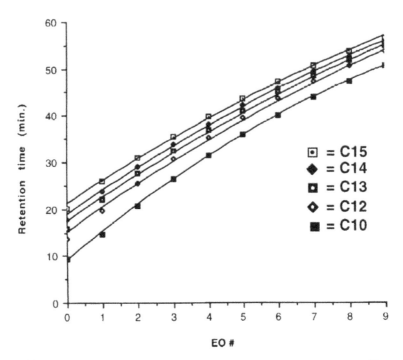

FIG. 6   GC retention times of *bis*(trimethylsilyl)-trifluoroacetamide (BSTFA) derivatives of C_{10}–C_{15} AE. (Reprinted with permission from Ref. 29. Copyright 1994 by Wiley-VCH Verlag.)

For quantitative determination of AE in complex matrices like wastewater, GC-MS analysis of the HBr fission products is preferred to simple GC (61).

Although analysis of AE is almost always performed with FID or MS detection, atomic emission detection eliminates the need for calibration with pure oligomers. Good results under all conditions are obtained by monitoring the carbon emission. There is some nonlinearity in the oxygen trace if silicon is present, so in this case acetate derivatives are preferred to silyl derivatives. If silyl derivatives are made, then the silicon emission may be used for detection (39,75).

2. Alkylphenol Alkoxylates

APE can be analyzed by capillary GC of derivatives (Table 7). The oligomer distribution is seen, along with fine structure due to the small percentage of *ortho*-substituted phenol in the predominantly *para*-alkylphenol from which the surfactant was made.

Packed-column GC analysis without derivitization is not suitable for quantitative analysis of APE, since only the first few oligomers elute and elution is incomplete for them. High-temperature capillary GC of underivatized APE has been demonstrated for resolution of up to about 12 ethoxy groups, but recovery was not determined and is presumably less than quantitative (78,79).

Recovery improves if derivatives are analyzed. In the case of packed-column GC of acetate esters of NPE, compounds up to an average degree of ethoxylation of 10 can be analyzed. However, due to discrimination against the higher molecular weight compounds, results for ethoxy content were found to be low by 10% for 5-mole ethoxylates and 24% for 9-mole ethoxylates (51). Recoveries are much higher for capillary GC, which is most often performed with trimethylsilane derivatives as shown in the table. Linear response is found up to about 18 EO units (44).

As mentioned in Chapter 2, alkylphenol alkoxylates can be fused with molten alkali to decompose ethylene oxide and propylene oxide groups, leaving the starting alkylphenol. This is easily characterized by GC or GC-MS, as well as by spectroscopic methods (80,81).

3. Acid Ethoxylates

Ethoxylated fatty acids can be reacted with paratoluenesulfonic acid in methanol to yield methyl esters of the original acids (30). These are easily extracted from the reaction mixture and analyzed by one of the many procedures supplied by GC column vendors for fatty acid analysis.

4. Amine Ethoxylates

Gas chromatography separates these compounds according to both alkyl and ethoxy chain length (Table 8). Chromatography is possible up to about the 12-ethoxy adduct of octadecylamine, if the BSA derivative is made (41).

Higher ethoxylated amines tend to be unstable at the temperatures required for GC analysis. For the analysis of ethoxylated amines of low degree of ethoxylation, which may contain residual active hydrogen, a drop or two of chlorotrimethylsilane is added to the *bis*(trimethylsilyl)acetamide reagent to replace any amine hydrogen with the trimethylsilyl group (41).

Primary amines will be ethoxylated at each active hydrogen, resulting in the formation of two ethoxy chains, each with a terminal hydroxyl group. This presents a problem when the trimethylsilyl derivatives are formed, since two different derivatives are formed for each oligomer: the mono- and the *bis*trimethylsilyl ether. The proportion of the *bis* derivative decreases with higher oligomers, leading to more complex chromatograms than are observed for underivatized amine ethoxylates. Thus, for low MW EO adducts of primary amines, analysis without derivatization is preferred (50).

5. Fatty Acid Alkanolamides

Fatty acid ethanolamides can also be analyzed in the form of the methyl esters of the fatty acid starting material (82). The ethanolamide is reacted directly with methanol under pressure in the presence of sulfuric acid and chloroform (Table 9).

As mentioned under amine analysis, it is best to react any residual amine hydrogen by addition of chlorotrimethylsilane to the *bis*(trimethylsilyl)acetamide solution (83). Direct GC analysis of underivatized monoethanolamides has been demonstrated using MS detection (84).

6. Esters of Polyhydroxy Compounds

(a) Analysis of decomposition products. These compounds are often determined by cross-esterifying to form the methyl esters of the corresponding fatty acids, which are easily analyzed by GC (Table 10).

Esters of sucrose and fatty acids may be saponified at room temperature with ethanolic NaOH. Derivatives of the saponification products are then analyzed by GC: sucrose is determined as the trimethylsilyl ether, and the acid as the methyl ester (85). Ethoxylated sorbitan esters may be similarly analyzed, with the esters saponified with KOH/ethanol

TABLE 7 GC Analysis of Alkylphenol Ethoxylates

Compound	Preliminary treatment	Column	Temperature	Detector	Ref.
OPE; E = 10; demonstration of direct high temperature GC	None	Specially prepared cross-linked methyl polysiloxane, 0.2 m × 20 mm, 0.1 μm film	130 to 400°C at 15°C/min	FID	79
NPE, E = 6, demonstration of high temperature GC	None	J&W Scientific DB-5HT, 0.25 mm × 15 m	100 to 400°C at 10°C/min	FID	131
C$_4$, C$_5$, C$_9$ alkylphenol EO, PO, and PO/EO adducts; alkoxy distribution to about E = 11	Derivatization with trimethylchlorosilane (TMCS) and hexamethyldisilazane (HMDS) in pyridine	5% SE-52 methylsilicone on silanized Chromosorb G AW, 2 ft	100 to 370°C at 5°C/min	TC	132
NPE, E = 7; determination of ethoxy distribution up to about E = 15	Derivatization with bis(trimethylsilyl)-trifluoroacetamide (BSTFA)	Varian Bonded Phase/1 capillary column, 0.22 mm × 12 m	30 to 340°C at 4°C/min	FID	133
OPE and NPE; resolution of oligomers up to about E = 14	Derivatization with N,O-bis(trimethylsilyl)acetamide (BSA)	OV-17 50% phenyl methylsilicone, 2% on Chromosorb G AW DMCS, 2.7 mm × 0.6 m	130 to 300°C at 4°C/min	FID	49,134
OPE and NPE; resolution of oligomers up to about 16 ethoxy units	Derivatization with N,O-bis(trimethylsilyl)acetamide (BSA)	SGE BX-5, 0.32 mm × 7.5 m, 0.25 μ film	80 to 390°C at 15°C/min	FID	44
OPE, E = 10; demonstration of oligomer separation	Derivatization with HMDS and TMCS	OV-1 vinyl methylsilicone, 0.53 mm × 10 m	80 to 180°C at 25°C/min, then to 400°C at 8°C/min	FID	119

Application	Reaction/Treatment	Column	Temperature	Detector	Ref.
Determination of octylphenol 1- and 2-mole ethoxylates and carboxylated degradation products	Derivatization with BSTFA	ORION SE-54 5% phenyl, 1% vinyl methylsilicone, 0.25 mm × 25 m	50 to 280°C at 3°C/min	MS	135
OPE and NPE, E = 4 to 14; determination of ethoxy content	Cleavage with CH_3COCl/$FeCl_3$	12% Carbowax 20M-TPA poly(ethylene glycol) terephthalate on Chromosorb W-AW DMCS, 2.7 mm × 1.6 m	100 to 220°C at 5°C/min	FID	51,55,67
OPE and NPE; determination of alkyl chain type	Cleavage/derivatization with acetic anhydride/p-toluenesulfonic acid	15% Free Fatty Acid Phase (poly(ethylene glycol) terephthalate) on Uniport B, 3 mm × 1 m	200°C	TC	53
C_8, C_9, and C_{12} alkylphenol ethoxylates; qualitative determination of alkyl chain length and branching	Pyrolysis with H_3PO_4 at 400°C to form alkenes	Stainless steel capillary coated with SF-96 methylsilicone, 0.02 inch × 200 ft	50°C for high resolution; normal analysis; 80°C hold for 10 min, then programmed to 190°C at 16°C/min	FID	16
OPE and NPE; determination of alkyl chain type and %EO	Pyrolysis with P_2O_5 at 500°C to form alkenes, acetaldehyde, and 1,4-dioxane	Uncoated column, 0.25 mm × 50 m	70°C	FID	18
C_4, C_9, and C_{12} alkylphenol ethoxylates; characterization of alkylphenol moiety	Degradation with molten KOH to yield alkylphenol	AK 30000 silicone on Chromosorb G AW-DMCS, 4 mm × 1.5 m	100 to 300°C at 10°C/min	TC	81

TABLE 8 GC Analysis of Amine Ethoxylates

Compound	Preliminary treatment	Column	Temperature	Detector	Ref.
4-Alkylphenylamine ethoxylates; alkyl chain C_1–C_{16}, $E = 1$ to 8; resolution out to about the 13-mole adduct	N,O-bis(trimethylsilyl)acetamide (BSA) derivatization	3% OV-17 50% phenyl methylsilicone on Chromosorb G AW DMCS, 2.7 mm × 0.4, 0.9, or 1.8 m	Start: 80–170°C; end: 320°C; 6°C/min	FID	136
Ethoxylated C_6–C_{18} alkylamines and ethoxylated dibutylamine; ethoxy distribution up to about $E = 17$	None for low EO; BSA derivatization for higher ethoxylates	3% OV-17 50% phenyl methylsilicone on Chromosorb G AW DMCS, 2.7 mm × 0.9 or 1.8 m; 1% OV-17 on a 0.4-m column	Start: 80–170°C; end: 320°C; 4–6°C/min	FID	41,49,50
C_{10}–C_{18} alkyl amine ethoxylates; determination of alkyl chain distribution	Cleavage/derivatization with acetic anhydride/p-toluenesulfonic acid	15% Free Fatty Acid Phase (poly(ethylene glycol) nitroterephthalate) on Uniport B, 3 mm × 1 m	235°C	TC	53

TABLE 9 Gas Chromatographic Analysis of Alkanolamides

Compound	Preliminary treatment	Column	Temperature	Detector	Ref.
C_8–C_{18} alkyldiethanolamides; separation by functionality and by alkyl chain length	Derivatization with N,O-bis(trimethylsilyl)acetamide (BSA) and trimethylchlorosilane (TMCS)	3% SP-2100 dimethylsilicone on Supelcoport, 2 mm × 6 ft, glass	175 to 300°C at 8°C/min	FID	83
C_{10}–C_{18} alkanolamides; determination of alkyl chain length and of impurities	Derivatization with 2:1 1,1,1,3,3,3-hexamethyldisilazane (HMDS)/trimethylchlorosilane (TMCS) in pyridine	10% silicone (OV 1 methylvinylsilicone, OV 101 dimethylsilicone, SE 30 methylsilicone or UCC W-982 methylsilicone) on Chromosorb GAW-DMCS. 2 mm × 1.8 m, glass	120 to 280°C at 6°C/min	FID	137
C_8–C_{18} alkyl mono- and diethanolamides; alkyl chain length distribution	Formation of methyl esters by reaction with CHCl$_3$/MeOH/H$_2$SO$_4$ under pressure at 185°C, followed by extraction of esters	10% PEG adipate on Celite, 4 ft	170°C	TC	82
C_{12} alkyldiethanolamide; determination of impurities	Derivatization with HMDS/TMCS in pyridine	5% SE-52 5% phenyl dimethylsilicone on Anakrom A, 3 ft	100 to 300°C at 20°C/min	TC	138
C_{14}–C_{18} saturated and unsaturated alkyl monoethanolamide; alkyl chain length distribution	None	J&W Scientific DB5, 30 m	90 to 200°C at 12°C/min, then to 310°C at 4°C/min	CI- and EI-MS	84

TABLE 10　GC Analysis of Polyhydroxyesters and Their Ethoxylates

Compound	Preliminary treatment	Column	Temperature	Detector	Ref.
Glycerol monoesters (C$_{14}$–C$_{22}$), separation by alkyl chain length; di- and tri-esters also resolved, as well as glycerol, diglycerol, and free fatty acids	Derivatization with N-methyl-N-trimethylsilyltrifluoroaceta-mide (MSTFA) in pyridine	3% Dexsil 300 GC carborane/methyl silicone on Chromosorb WHP, 2.2 mm × 1 m, or 3% OV 101 dimethyl-silicone on Chromosorb WHP, 2 mm × 1 m	80 to 340°C at 5°C/min	FID	139
Glycerol mono-, di- and tri-esters (C$_{16}$–C$_{18}$), separation by alkyl chain length; iso-mer separation, resolution of free fatty acids	Derivatization with bis(tri-methylsilyl)trifluoroacetamide (BSTFA)	Hewlett Packard HP-5TA (5% diphenyl/95% dimethylarylene siloxane copolymer; low bleed), 0.32 mm × 15 m, 0.1 μm; H$_2$ carrier gas	Cool on column injec-tion; 50°C for 1 min, then to 365°C at 15°C/min	FID	88
Sorbitan laurate; resolution of sorbitol, 1,4-sorbitan, 1,4:3,6-isosorbide, their monolaurates, and free lau-ric acid	Derivatization with 1,1,1,3,3,3-hexamethyldisilazane (HMDS)/trimethylchlorosilane (TMCS) in pyridine	3% SE-30 methylsilicone on Chromosorb G, 5 mm × 2 m	150 to 220°C at 5°C/min	FID	140
C$_{14}$–C$_{18}$ sucrose monoesters; separation by alkyl chain length; partial resolution of isomers	Derivatization with HMDS/TMCS in pyridine	3% OV-17 50% phenyl methylsilicone on Chromosorb W-HP, 0.125 inch × 6 ft	280 to 330°C at 2°C/min	FID	141
Sucrose mono- and diesters (C$_{12}$–C$_{18}$); separation by number of acyl groups and by alkyl chain length	Derivatization with N,O-bis(tri-methylsilyl)acetamide (BSA)/TMCS in pyridine	3% Dexsil 300 GC carborane/methyl silicone on Chromosorb W, 50 cm	70 to 380°C at 16°C/min	FID	142
Sucrose esters (C$_{12}$ and C$_{18}$); separation by number of acyl groups (up to the tristearate and hexalaurate); partial resolution of isomers	Derivatization with trimethyl-silylimidazole (TMSI) in pyridine and MSTFA	Simdis wide bore glass capil-lary coated with PS 264 poly-dimethylsiloxane (Carlo Erba)	4 min hold, then 70 to 400°C at 10°C/min	FID	87

Application	Procedure	Column	Temperature	Detector	Ref.
Sucrose esters (C_{16}–C_{18}); acyl content 60–80%; determination of sucrose and acyl constituents	Saponification; derivatization of sucrose with trimethylchlorosilane (TMCS) and N-trimethylsilylimidazole (TMSI); formation of methyl esters of acids	For sucrose: 2% OV-17 50% phenyl methylsilicone on Chromosorb W AW-DMCS; for acids: 2% DEGS + 0.5% H_3PO_4 on Chromosorb W AW-DMCS, each 3 mm × 225 cm	For sucrose: 230°C; for acids: 165°C	FID	85
Mono- and diesters of glycerol, polyglycerol, sorbitol, and sorbitan (C_{14}–C_{18}); identification of main components	Derivatization with HMDS/TMCS in pyridine	3% JXR on Gas Chrom Q, 3 ft	120 to 325°C at 10°C/min	FID	143–145
Ethoxylated glycerol and sorbitan C_{14}–C_{18} alkyl esters; identification of alcohol and alkyl moieties; determination of alkyl chain distribution	Cleavage/derivatization with acetic anhydride/p-toluenesulfonic acid	15% free fatty acid phase (PEG 20 M dinitroterephthalate) on Uniport B, 3 mm × 1 m	140 to 220°C at 5°C/min	TC	30
Ethoxylated sorbitol esters; determination of alkyl chain distribution (C_{12}–C_{18}); determination in food	Saponification and formation of the methyl esters of the separated fatty acids	5% Advance-DS (Shimadzu) on Chromosorb W AW-DMCS, 3 mm × 0.2 m	150°C	FID	86
Sorbitan and glycerol mono- and diesters (C_{12}–C_{18}); separation by number and type of acyl groups; identification of impurities	Derivatization with HMDS/TMCS in pyridine	5% SE-52 5% phenyl dimethylsilicone on Anakrom A, 3 ft.	100 to 300°C at 20°C/min	TC	138

and the fatty acids separated by extraction, reacted with diazomethane, and determined by GC as the methyl esters (86).

(b)   Analysis of intact esters.   The older literature shows that packed-column GC is adequate for resolution of silyl derivatives of many mono- and diesters, with partial separation of isomers. High-temperature GC on capillary columns extends the analysis to higher esters (87). Karrer and Herberg showed resolution of up to the sucrose hexalaurate and sucrose tristearate. Elution is approximately in order of total carbon number of the derivatized compounds.

Glycerides can be analyzed by high-temperature GC on low-bleed columns (Fig. 7). Care should be taken during sample preparation to avoid transesterification. Derivatization is required for good chromatography of free fatty acid and the mono- and diesters, especially if the column is not new. Triesters are not affected by the derivatization reaction. Elution is in the order derivatized fatty acids, derivatized monoesters, derivatized diesters, and triesters (88).

7.   Alkylpolyglycosides

Injection of alkylpolyglucosides without derivatization allows detection of the alkyl monoglucosides as well as free alcohol (Table 11). The compounds are separated according to alkyl chain length (Fig. 8). There is further separation of the glucofuranoside and glucopyranoside isomers, with fine structure visible due to α and β isomers (89). High-temperature GC of derivatized alkylpolyglucosides allows separaration of clusters of compounds corresponding to increasing degree of polymerization up to about 7 units (89–91).

8.   EO/PO Copolymers

Acetyl chloride cleavage is also applicable to analysis of block EO/PO copolymers, as well as products containing butylene oxide (see Table 12) (68,92). Propylene oxide groups are converted mainly to 2-chloro-1-methylethyl acetate and 2-chloropropyl acetate, with traces of 1,2-propylene diacetate. Simple GC determination of the ratio of chloroethyl acetate to total chloroethyl and chloropropyl acetates provides an accurate figure for EO content of copolymers up to about 30% EO. A calibration curve must be used for higher ratios (69).

9.   Amine Oxides

Amine oxides cannot be analyzed directly by GC, since they will decompose in the injection port at temperatures above 100°C to give a terminal alkene and a tertiary amine (93).

Amine oxides may be reduced with triphenylphosphine to form the tertiary amines, which can be analyzed on a column suitable for amines (Table 13). In the case of an alkylamidopropylamine oxide, the reduction is followed by hydrolysis of the amidoamine and then extraction and separate analyses of the resulting amine and acid (as the methyl ester) (93).

Pyrolysis gas chromatography has been used to characterize amine oxides (11). Neutral or slightly acidic pyrolysis conditions give characteristic alkyldimethylamines. Amine oxides may be analyzed by direct injection of the purified materials. At an injector temperature of 220°C, the amine oxide homologs are converted reproducibly to the olefins. These are separated on a column of Apiezon L with a temperature program of 180 to 280°C at 4°C/min (94).

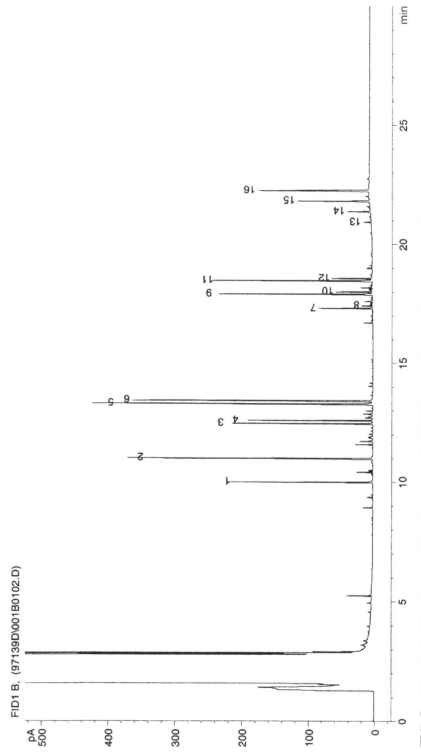

FIG. 7 GC analysis of *bis*(trimethylsilyl)-trifluoroacetamide (BSTFA) derivatives of a glyceride mixture. Peak assignments: 1, palmitic acid; 2, stearic acid; 3 and 4, glycerol monopalmitates; 5 and 6, glycerol monostearates; 7 and 8, glycerol dipalmitates; 9 and 10, mixed diglycerides (palmitate/stearate); 11 and 12, glycerol distearates; 13, glycerol tripalmitate; 14, glycerol dipalmitatemonostearate; 15, glycerol distearatemonopalmitate; 16, glycerol tristearate. (Reprinted with permission from Ref. 29.)

TABLE 11 Gas Chromatographic Analysis of Alkyl Polyglycosides

Compound	Preliminary treatment	Column	Temperature	Detector	Ref.
C_{12}, C_{14}, C_{16} alkylmonogluco-sides; separation by increasing alkyl chain length; separation of α- and β-glucopyranoside and glucofuranoside isomers	None	95% Methylsilicone/5% phenylsilicone, 30 m capillary	175°C to 300°C at 10°C/min	FID	89
C_{12}–C_{16} alkylpolyglucosides; separation of clusters of increasing degree of polymerization (1–5)	Trimethylsilylimidazole (TMSI)	Modified dimethylsilicone glass capillary, 0.53 mm × 10 m	90°C to 400°C at 20°C/min	FID	89
C_8–C_{14} alkylmonoglucosides; separation by increasing alkyl chain length; separation of α- and β-glucopyranoside and glucofuranoside isomers; determination of glucose	TMSI and N-methyl-N-trimethylsilyltrifluoroacetamide (MSTA)	Simulated distillation, manufacturer not specified, glass, 0.53 mm × 10 m	70°C for 4 min, then to 400°C at 10°C/min	FID; pentadecanol internal standard	90,91
C_8–C_{14} alkylmonoglucosides; determination of residual free alcohol	TMSI and MSTA	J&W Scientific DB5 5% phenyl dimethylsilicone, 0.32 mm × 15 m, 0.1 μm film; H_2 carrier gas	45°C for 3 min, then to 300°C at 5°C/min	FID; pentadecanol internal standard	91
C_7–C_{16} alkylmonoglucosides; separation by increasing alkyl chain length; separation of α- and β-glucopyranoside and glucofuranoside isomers	Chlorotrimethylsilane (TMCS), 1,1,1,3,3,3-hexamethyldisilazane (HMDS), pyridine	Hewlett-Packard HP5 95% methylsilicone/5% phenylsilicone, 0.25 mm × 30 m, 0.17 μm film	70°C to 170°C at 50°C/min, then to 300°C at 20°C/min	EI-MS	146

TABLE 12 GC Analysis of EO/PO Copolymers

Compound	Preliminary treatment	Column	Temperature	Detector	Ref.
E/P copolymers, MW = 650–5800, %E = 3–75; determination of ethoxy content	Cleavage with HBr and extraction	10% Apiezon L hydrocarbon on Chromosorb W, 0.25 inch × 9 ft	125°C	FID	63
E/P copolymers, MW in 3500 range, %E = 5–60; determination of ethoxy content	Cleavage with HBr and extraction	30% Silicone Elastomer E301 (ICI Nobel) on acid-washed Celite, 4 mm × 90 cm	65°C	FID	62
E/P copolymers, MW = 2300–4200, %E = 10–50; determination of ethoxy content	Cleavage with CH₃COCl/FeCl₃	15% Diethylene glycol succinate on Chromosorb W-AW, 2.7 mm × 3.3 m	100 to 200°C at 5°C/min	FID	69
E/P copolymers, %E = 14–50; determination of ethoxy content	Cleavage/derivatization with acetic anhydride/p-toluenesulfonic acid	15% Free Fatty Acid Phase (PEG 20M, nitroterephthalate diester) on Uniport B, 3 mm × 1 m	65°C	TC	53
EO/BO copolymers, MW = 1500, 2000, %E = 44; determination of ethoxy content	Cleavage with CH₃COCl/FeCl₃	12% PEG 20 M diterephthalate on Chromosorb W-AW DMCS, 2.7 mm × 1.8 m	100 to 220°C at 5°C/min	FID	68

TABLE 13 Gas Chromatographic Analysis of Amine Oxides

Compound	Preliminary treatment	Column	Temperature	Detector	Ref.
C_{12}–C_{14} alkyl dimethylamine oxide; determination of alkyl chain length	Pyrolysis at 650°C	20% SF-96 methylsilicone on Chromosorb W, 0.25 inch × 15 ft	50 to 250°C at 4°C/min	TC	11
C_{12}–C_{22} alkyl dimethylamine oxides and alkylamidopropyldimethylamine oxides; determination of alkyl chain length distribution	Reduction with triphenylphosphine to form the tertiary amines; amidoamines are hydrolyzed in a second step	Carbowax/KOH, 3 ft	170 to 220°C at 4°C/min	FID	93
C_{12}–C_{16} alkyldimethylamine oxides; determination of alkyl chain length distribution	Pyrolysis at 220°C in GC injection port to form alkenes	20% Apiezon L hydrocarbon on Chromosorb W HMDS, 0.25 inch × 10 ft	180 to 280°C at 4°C/min	TC	94

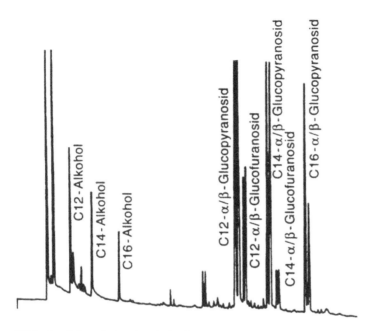

FIG. 8 GC analysis of underivatized alkylmonoglycoside mixture. (Reprinted with permission from Ref. 89. Copyright 1996 by Karl Hanser Verlag.)

III. CATIONIC AND AMPHOTERIC SURFACTANTS

A. Direct Analysis

Because of their lack of volatility, there are few applications published on the use of GC to directly analyze quaternary amines. Only low molecular weight quaternary amines, far below the surfactant range, can be analyzed directly. Higher molecular weight quats can be analyzed only if they are chemically converted to more volatile compounds. At present, amphoterics can generally not be analyzed directly by gas chromatography without decomposition.

B. Derivatization

Alkylbenzyldimethylammonium compounds may be first converted to the corresponding alkyldimethylamines by hydrogenation, then converted to the cyanamide or trichloroethyl carbamate derivatives for gas chromatography analysis (95). The derivatives may be analyzed at relatively low column temperatures, and may be detected with great sensitivity using a nitrogen-selective detector.

Ester quats can be reacted with methanol under acidic conditions to yield the methyl esters of the fatty acids. These are easily analyzed by GC according to standard methods (96).

Some amphoterics containing carboxylic acid functionality, such as the *N*-alkylaminopropylglycines, can be analyzed readily by gas chromatography after formation of the methyl esters (97). Sodium lauroyl sarcosinate is readily determined after extraction from soap products with acidic DMF and subsequent derivatization with *bis*(trimethylsilyl)-trifluoroacetamide/1% trichloromethylsilane (98).

C. Chemical Decomposition Followed by Gas Chromatography

1. Hofmann Elimination

If a quaternary amine hydroxide is heated, it will decompose to give a tertiary amine and an alkene. This method is frequently applied to the analysis of quaternary amines. The amine is reacted with enough alkali to insure that it is in the hydroxide form, and it is heated to give characteristic products volatile enough for conventional gas chromatography analysis.

In a typical application, the surfactants are refluxed with sodium methoxide in *N,N*-dimethylformamide for 1 hr (99). The ether extract of the reaction mixture contains the degradation products. For example, dodecyltrimethylammonium chloride gives 1-dodecene and dimethyldodecylamine. Didodecyldimethylammonium chloride gives 1-dodecene, dimethyldodecylamine, and didodecylmethylamine. Benzyldimethylhexadecylammonium chloride gives benzyldimethylamine, 1-hexadecene, dimethylhexadecylamine, and benzylhexadecylmethylamine. The extracts are analyzed by packed-column gas chromatography, using OV-17 or JXR-Silicone. A program of 100 to 300°C at 10°C/min is adequate, using flame ionization detection. In a variation of the technique, the alkene is separated from the trialkylamine by silica gel chromatography and determined quantitatively using an internal standard. The silica gel treatment also removes interfering compounds from the sample matrix (100).

Similar reaction to form an alkene and amines is observed if a solution of the cationic in methanolic KOH is analyzed directly, with the desired Hofmann degradation occurring in the hot injection port (101). However, injection of untreated sample onto a very alkaline GC column gives less predictable results (102). Suzuki and coworkers studied the degradation reaction, and report that room temperature reaction with potassium *tert*-butoxide favors the formation of alkyldimethylamine from benzalkonium chlorides. Reaction at 80°C favors the formation of α-olefins. A reaction time of 10 min is sufficient (103).

Procedure: Determination of Benzylalkyldimethylammonium Compounds by GC After Alkaline Degradation (103)

The surfactant is separated from the matrix by extraction. An aliquot containing 2–10 mg of the benzalkonium chloride is placed in a glass-stoppered test tube, along with 25 mL 80:20 benzene/DMSO. 100 mg potassium *tert*-butoxide is added and the mixture is shaken vigorously and let stand for 10 min at room temperature. The mixture is transferred to a separatory funnel and shaken with 10 mL 10% aqueous NaCl solution. The organic phase is separated, evaporated to dryness, and dissolved in 5 mL acetone. The acetone extract is analyzed by GC on a glass column of 5% SE-30 on Chomosorb W AW DMCS, 3 mm × 2 m, with a temperature program of 140 to 230°C, at 5°C/min. Calibration is performed by treating benzalkonium compounds of known composition in the same manner.

2. Other Reactions

Alkylbenzyldimethylammonium salts can be hydrogenated to give toluene and alkyldimethylamine. The amine is sufficiently volatile for direct gas chromatographic determination of the alkyl distribution (104). Alternatively, the amine may be derivatized to increase its volatility (95).

Imidazolinium cationics and amphoterics can be hydrolyzed by reflux with alkali, followed by reflux with acid. The fatty acids can then be extracted and analyzed by gas chromatography as their methyl esters (105–108). The amines are converted to the acetate derivatives and also analyzed by GC.

The fatty acid components of the phosphatides in lecithin may be characterized by gas chromatographic analysis. Depending on the objective of the study, either the total sample or isolated individual phosphatides are analyzed. Sample preparation may consist of saponification followed by formation of the methyl esters for GC analysis. More often, decomposition and formation of methyl esters occurs simultaneously in a transesterification reaction with BF_3/methanol, HCl/methanol, or sodium methoxide (109,110).

D. Pyrolysis Gas Chromatography

Denig used a pyrolysis gas chromatograph to perform the Hofmann elimination reaction (19). The quaternary amine is mixed with solid 50:50 KOH/NaOH and heated at 300°C, with the amine and olefin fragments swept into the GC column.

If the halide salt of a quaternary is pyrolyzed in the absence of alkali, Hofmann elimination is only a minor reaction. Pyrolysis may be conducted by simply using a hot injection port, about 310°C, with a conventional GC or GC-MS (111–113). However, a specially designed Curie point pyrolysis unit gives more reliable data (114,115). MS detection is much more useful than conventional GC methodology. In general, quaternary halide salts compounds yield reproducible, predictable, and characteristic fragments consisting of tertiary amines and alkyl halides. In the case of a pyridinic ring compound, secondary reactions occur, so that preliminary experiments with standard compounds are required to determine what are the decomposition products. In the case of hexadecylpyridinium chloride, both hexadecylammonium chloride and pyridine are obtained, and either may be used for quantification. Use of a nitrogen-specific detector makes this approach suitable for analysis of complex matrices (116,117). At higher temperatures, intermolecular reactions of benzyltrialkylammonium chlorides yield 1,2-diphenylethane and stilbene, which are generally not useful for quantification (115).

Takano, Kuzukawa, and Yamanaka showed that direct pyrolysis GC of carboxybetaine and sulfobetaine amphoterics produces mainly nonvolatile compounds, with the main volatile product being alkyldimethylamine (101). However, by injecting a methanolic KOH solution of the surfactant into the heated injection port of the GC, Hofmann elimination can be made to occur, resulting mainly in the formation of volatile products. 2-(Alkyldimethylammonio)-ethane-1-carboxylate, 2-(alkyldimethylammonio)-ethane-1-sulfonate, and 3-(alkyldimethylammonio)-propane-1-sulfonate degrade to give predominantly alkyldimethylamines, with only low levels of α-olefins, while (alkyldimethylammonio)-methane carboxylate and 4-(alkyldimethylammonio)-butane-1-sulfonate give primarily the α-olefins (99).

REFERENCES

1. Rasmussen, H. T., B. P. McPherson, Chromatographic analysis of anionic surfactants, in J. Cross, ed., *Anionic Surfactants: Analytical Chemistry*, 2nd ed., Marcel Dekker, New York, 1998.
2. McEvoy, J., W. Giger, LAS in sewage sludge by high resolution GC-MS, *Environ. Sci. Technol.*, 1986, *20*, 376–383.
3. Kirkland, J. J., Sulfonic acids and salts by GC of volatile derivatives, *Anal. Chem.*, 1960, *32*, 1388–1393.

4. Buse, E., A. Golloch, M. Muzic, Derivatization methods for LAS, *Tenside, Surfactants, Deterg.*, 1996, *33*, 404–409.

5. Parsons, J. S., Sulfonic acids by forming sulfonyl fluoride derivatives, *J. Gas Chromatogr.*, 1967, *5*, 254–256.

6. Nagai, T., S. Hashimoto, I. Yamane, A. Mori, GC analysis for α-olefin sulfonate, *J. Am. Oil Chem. Soc.*, 1970, *47*, 505–509.

7. Louis, D., J. M. Talbot, Chain length distribution of anionic surfactants, *Tenside, Surfactants, Deterg.*, 1995, *32*, 347–350.

8. Okazaki, T., H. Kataoka, N. Muroi, M. Makita, LAS by GC with flame photometric detection (in Japanese), *Bunseki Kagaku*, 1989, *38*, 312–315.

9. Field, J. A., D. J. Miller, T. M. Field, S. B. Hawthorne, W. Giger, Sulfonated aliphatic and aromatic surfactants in sewage sludge by ion-pair/supercritical fluid extraction and derivatization GC-MS, *Anal. Chem.*, 1992, *64*, 3161–3167.

10. Krueger, C. J., J. A. Field, In-vial C_{18} Empore disk elution coupled with injection port derivatization for the quantitative determination of LAS by GC-FID, *Anal. Chem.*, 1995, *67*, 3363–3366.

11. Liddicoet, T. H., L. H. Smithson, Surfactants by pyrolysis-GC, *J. Am. Oil Chem. Soc.*, 1965, *42*, 1097–1102.

12. Siggia, S., L. R. Whitlock, Pyrolysis GC determination of arylsulfonic acids and salts, *Anal. Chem.*, 1970, *42*, 1719–1724.

13. Uchiyama, T., A. Kawauchi, Acid pyrolysis for alkyl homolog and phenyl isomer distribution of LAS, *J. Surfactants Deterg.*, 1999, *2*, 331–335.

14. Zeman, I., Mean molecular mass determination of LAS, *Tenside, Surfactants, Deterg.*, 1982, *19*, 353–356.

15. International Organization for Standardization, Technical straight-chain sodium alkylbenzene-sulfonates—determination of mean relative molecular mass by GC, ISO 6841, Geneva, Switzerland, 1972.

16. Lew, H. Y., Acid pyrolysis-capillary chromatographic analysis of anionic and nonionic surfactants, *J. Am. Oil Chem. Soc.*, 1967, *44*, 359–366

17. Lew, H. Y., Surfactant analysis, *J. Am. Oil Chem. Soc.*, 1972, *49*, 665–670.

18. Denig, R., Surfactant characterization by pyrolysis GC. I: Nonionic and anionic surfactants (in German), *Tenside, Surfactants, Deterg.*, 1973, *10*, 59–63.

19. Denig, R., Characterization of surfactants by pyrolysis GC (in German), *Fette, Seifen, Anstrichm.*, 1974, *76*, 412–416.

20. Hawthorne, S. B., D. J. Miller, J. J. Langenfeld, Supercritical fluid extraction of polar analytes using modified carbon dioxide and *in situ* chemical derivation, in F. V. Bright and M. E. P. McNally, eds., *Supercritical Fluid Technology: Theoretical and Applied Approaches to Analytical Chemistry*. ACS Symposium Series, No. 488, Washington, 1992.

21. Leidner, H., R. Gloor, K. Wuhrmann, Kinetics of degradation of LAS (in German), *Tenside, Surfactants, Deterg.*, 1976, *13*, 122–130.

22. Waters, J., J. T. Garrigan, Microdesulfonation/GC procedure for the determination of LAS in UK Rivers, *Water Res.*, 1983, *171*, 1549–1562.

23. Osburn, Q. W., LAS in waters and wastes, *J. Am. Oil Chem. Soc.*, 1986, *63*, 257–263.

24. Sones, E. L., J. L. Hoyt, A. J. Sooter, Alcohol and ether sulfates and their alkyl carbon distributions in detergent formulations by GC, *J. Am. Oil Chem. Soc.*, 1979, *56*, 689–700.

25. Saraste, M., T. K. Korhonen, Protein-bound dodecyl sulfate by GC, *Anal. Biochem.*, 1979, *92*, 444–446.

26. Lee, S., N. A. Puttnam, GC determination of chain length distribution in AE and sulfated derivatives, *J. Am. Oil Chem. Soc.*, 1966, *43*, 690.

27. Fendinger, N. J., W. M. Begley, D. C. McAvoy, W. S. Eckhoff, Alkyl sulfate surfactants in natural waters, *Environ. Sci. Technol.*, 1992, *26*, 2493–2498.

28. Pinkston, J. D., T. E. Delaney, K. L. Morand, R. G. Cooks, SFC/MS using a quadrupole mass filter/quadrupole ion trap hybrid MS with external ion source, *Anal. Chem.*, 1992, *64*, 1571–1577.

29. Rasmussen, H. T., A. M. Pinto, M. W. DeMouth, P. Touretzky, B. P. McPherson, High temperature GC of trimethylsilyl derivatives of alcohol ethoxylates and ethoxysulfates, *J. High Resolut. Chromatogr.*, 1994, *17*, 593–596.

30. Tsuji, K., K. Konishi, GC analysis of ester-type surfactants by using mixed anhydride reagent, *J. Am. Oil Chem. Soc.*, 1975, *52*, 106–109.

31. Field, J. A., T. M. Field, T. Poiger, W. Giger, Secondary alkane sulfonates in sewage wastewaters by solid phase extraction and injection port derivatization GC/MS, *Environ. Sci. Technol.*, 1994, *28*, 497–503.

32. Pollerberg, J., Homolog composition and the branched-chain content of secondary alkanesulfonates (in German), *Fette, Seifen, Anstrichm.*, 1965, *67*, 927–929.

33. Köhler, M., E. Keck, G. Jaumann, Ester sulfonates (in German), *Fett Wiss. Technol.*, 1988, *90*, 241–243.

34. Van Loon, W. M. G. M., J. J. Boon, B. de Groot, Qualitative analysis of chlorolignins and lignosulfonates in pulp mill effluents entering the river Rhine using pyrolysis-MS and pyrolysis-GC-MS, *J. Anal. Appl. Pyrolysis*, 1991, *20*, 275–302.

35. Daradics, L., J. Pálinkás, Synthesis and analysis of fatty acid sarcosides, *Tenside, Surfactants, Deterg.*, 1994, *31*, 308–313.

36. Gerhardt, W., G. Czichocki, H. R. Holzbauer, C. Martens, B. Weiland, Ether carboxylates: characterization of ether carboxylates in reaction mixtures (in German), *Tenside, Surfactants, Deterg.*, 1992, *29*, 285–288.

37. Moody, C. A., J. A. Field, Perfluorocarboxylates in groundwater impacted by fire-fighting activity, *Environ. Sci. Technol.*, 1999, *33*, 2800–2806.

38. Silver, A. H., H. T. Kalinoski, Comparison of high temperature GC and CO_2 SFC for the analysis of alcohol ethoxylates, *J. Am. Oil Chem. Soc.*, 1992, *69*, 599–608.

39. Asmussen, C., H. J. Stan, AE by HTGC and AED, *J. High Resolut. Chromatogr.*, 1998, *21*, 597–604.

40. Finke, J., U. Kobold, T. Dülffer, S. Pongratz, A. Puhlmann, A. Rudolphi, Quality assurance of ethoxylates using high temperature GC, *Tenside, Surfactants, Deterg.*, 1998, *35*, 478–479.

41. Szewczyk, H., J. Szymanowski, W. Jerzykiewicz, Determination by GC of the composition of some commercial ethoxylated alkylamines, *Tenside, Surfactants, Deterg.*, 1982, *19*, 287–289.

42. Czichocki, G., W. Gerhardt, D. Blumberg, GC separation of AE (in German), *Tenside, Surfactants, Deterg.*, 1988, *25*, 169–173.

43. Scanlon, J. T., D. E. Willis, Calculation of FID relative response factors using the effective carbon number concept, *J. Chromatogr. Sci.*, 1985, *23*, 333–340.

44. Jones, F. W., Flame ionization detector relative response factors for oligomers of alkyl and aryl ether polyethoxylates using the effective carbon number concept, *J. Chromatogr. Sci.*, 1998, *36*, 223–226.

45. Farkas, L., J. Morgós, P. Sallay, I. Rusznák, B. Bartha, G. Veress, Molar mass distribution of AE, *J. Am. Oil Chem. Soc.*, 1981, *58*, 650–655.

46. Chlebicki, J., A. Zwiefka, Polyaddition reaction of propylene oxide to carboxylic acids, *Tenside, Surfactants, Deterg.*, 1990, *27*, 266–269.

47. Motz, M., B. Fell, W. Meltzow, Z. Xia, GC-HPLC analysis of ethoxylation products of fatty alcohols and Guerbet alcohols (in German), *Tenside, Surfactants, Deterg.*, 1995, *32*, 17–21.

48. Van Lieshout, H. P. M., H. G. Janssen, C. A. Cramers, High-temperature PTV injection for HT-CGC, *Am. Lab.*, 1995, *27(12)*, 38,40–44.

49. Szymanowski, J., A. Voelkel, H. Szewczyk, Increments of the arithmetic retention index for nonionic surfactants with a polyoxyethylene chain, *J. Chromatogr.*, 1986, *360*, 43–52.

50. Szymanowski, J., H. Szewczyk, J. Hetper, J. Beger, *N*-oligooxyethylene mono- and dialkylamines, *J. Chromatogr.*, 1986, *351*, 183–193.

51. Szymanowski, J., P. Kusz, E. Dziwinski, Degradation and analysis of commercial APE, *J. Chromatogr.*, 1990, *511*, 325–332.

52. Puschmann, H., Ethylene oxide adducts of secondary alcohols (in German), *Tenside, Surfactants, Deterg.*, 1968, *5*, 207–210.

53. Tsuji, K., K. Konishi, GC of nonionic surfactants by acid cleavage of ether linkages, *J. Am. Oil Chem. Soc.*, 1974, *51*, 55–60.

54. Gildenberg, L., J. R. Trowbridge, GC separation of AE via acetate esters, *J. Am. Oil Chem. Soc.*, 1965, *42*, 69–71.

55. Kusz, P., J. Szymanowski, E. Dziwinski, Analysis of nonionic surfactants after their predegradation with acetyl chloride, *Proc. Second World Surfactants Congress*, Paris, 1988, 239–248.

56. Szymanowski, J., P. Kusz, E. Dziwinski, H. Szewczyk, K. Pyzalski, Degradation and analysis of polyoxyethylene monoalkyl ethers in the presence of acetyl chloride and ferric chloride, *J. Chromatogr.*, 1989, *469*, 197–208.

57. Farkas, L., J. Morgos, P. Sallay, I. Rusznak, GC acetylation, *J. Chromatogr.*, 1979, *168*, 212–215.

58. Kaduji, I. I., J. B. Stead, Determination of polyoxyethylene in nonionic detergents by hydrogen bromide fission followed by GC, *Analyst*, 1976, *101*, 728–731.

59. Smith, R. M., M. Dawson, Polyethylene glycol by hydrogen bromide fission and GC with electron-capture detection, *Analyst*, 1980, *105*, 85–89.

60. Fung, D., S. Safe, J. F. S. Crocker, Polyethylene glycol ether surfactants in biological tissues, *Toxicol. Environ. Chem.*, 1984, *8*, 109–132.

61. Fendinger, N. J., W. M. Begley, D. C. McAvoy, W. S. Eckhoff, AE surfactants in natural waters, *Environ. Sci. Technol.*, 1995, *29*, 856–863.

62. Mathias, A., N. Mellor, Alkylene oxide polymers by NMR spectrometry and by GC, *Anal. Chem.*, 1966, *38*, 472–477.

63. Stead, J. B., A. H. Hindley, EO/PO copolymers by chemical fission and GC, *J. Chromatogr.*, 1969, *42*, 470–475.

64. Cross, J., ed., *Nonionic Surfactants: Chemical Analysis*, Marcel Dekker, New York, 1987.

65. Hodges, K. L., W. E. Kester, D. L. Wiederrich, J. A. Grover, Alkoxyl substitution in cellulose ethers by Zeisel-GC, *Anal. Chem.*, 1979, *51*, 2172–2176.

66. Szymanowski, J., P. Kusz, E. Dziwinski, Cz. Latocha, Degradation and analysis of AE in the presence of acetyl chloride, *J. Chromatogr.*, 1988, *455*, 119–129.

67. Szymanowski, J., P. Kusz, E. Dziwinski, Cz. Latocha, Degradation and analysis of oligooxyethylene glycol mono(4-*tert*-octylphenyl) ethers in the presence of acetyl chloride, *J. Chromatogr.*, 1990, *502*, 407–415.

68. Szymanowski, J., P. Kusz, E. Dziwinski, Chemical degradation and analysis of polyoxyethylene glycols and ethylene oxide-α-butylene oxide block copolymers, *J. Chromatogr.*, 1988, *455*, 131–141.

69. Kusz, P., J. Szymanowski, K. Pyzalski, E. Dziwinski, Degradation and analysis of ethylene oxide—propylene oxide block copolymers in the presence of acetyl chloride, *LC-GC*, 1990, *8*, 48,50.

70. Tsuji, K., K. Konishi, Identification of polyol base compounds in polyurethane polyethers by GC, *Analyst*, 1971, *96*, 457–459.

71. Neumann, E. W., H. G. Nadeau, Polyether and polyolefin polymers by GC determination of the volatile products resulting from controlled pyrolysis, *Anal. Chem.*, 1963, *35*, 1454–1457.

72. Dowle, C. J., W. C. Campbell, Pyrolysis GC analysis of AE with hydrobromic acid cleavage of the ether linkages, *Analyst*, 1988, *113*, 1241–1244.

73. Zeman, I., L. Novak, L. Mitter, J. Stekla, O. Holendova, GC of alkylene oxides in their copolymers, *J. Chromatogr.*, 1976, *119*, 581–589.

74. Sherrard, K. B., P. J. Marriott, R. G. Amiet, R. Colton, M. J. McCormick, G. C. Smith, Photocatalytic degradation of secondary AE: spectroscopic, chromatographic, and MS studies, *Environ. Sci. Technol.*, 1995, *29*, 2235–2242.

75. Asmussen, C., H. Stan, AE—examples of process analysis (in German), *Biol. Abwasserreinig.*, 1998, *10*, 237–251.

76. Rothbächer, H., A. Korn, G. Mayer, Nonionic surfactants in cleaning agents for automobile production (in German), *Tenside. Surfactants. Deterg.*, 1993, *30*, 165–173.

77. Cozzoli, O., C. Ruffo, G. Carrer, F. Pinciroli, E. Faccetti, L. Valtorta, P. Mondani, L. Sedea, Inter-laboratory study of GC determination of alkyl iodides from AE, *Riv. Ital. Sostanze Grasse*, 1997, *74*, 343–348.

78. Lipsky, S. R., M. L. Duffy, High temperature GC: aluminum clad flexible fused silica glass capillary columns coated with thermostable nonpolar phases: Part 2, *Jour. HRC CC*, 1986, *9*, 725–730.

79. Lipsky, S. R., M. L. Duffy, Advances in capillary GC, *LC-GC*, 1986, *4*, 898–906.

80. Müller, K. D. Noffz, Length and the degree of branching of the alkyl chains of surface-active alkylbenzene derivatives (in German), *Tenside. Surfactants. Deterg.*, 1965, *2*, 68–75.

81. Kupfer, W., K. Künzler, Degradation method for the analysis of the alkylphenol component in alkylphenol polyglycol ethers (in German), *Fresenius' Z. Anal. Chem.*, 1973, *267*, 166–169.

82. Lee, S., N. A. Puttnam, GC determination of chain length distribution in fatty acid ethanolamides, *J. Am. Oil Chem. Soc.*, 1965, *42*, 744.

83. O'Connell, A. W., Analysis of coconut oil—diethanolamine condensates by GC, *Anal. Chem.*, 1977, *49*, 835–838.

84. Moldovan, Z., C. Maldonado, J. M. Bayona, Electron ionization and positive ion chemical ionization MS of *N*-(2-hydroxyethyl)alkylamides, *Rapid Commun. Mass Spectrom.*, 1997, *11*, 1077–1082.

85. Tsuda, T., H. Nakanishi, GC determination of sucrose fatty acid esters, *J. Assoc. Off. Anal. Chem.*, 1983, *66*, 1050–1052.

86. Kato, H., Y. Nagai, K. Yamamoto, Y. Sakabe, Polysorbates in foods by colorimetry with confirmation by IR spectrophotometry, TLC, and GC, *J. Assoc. Off. Anal. Chem.*, 1989, *72*, 27–29.

87. Karrer, R., H. Herberg, Sucrose fatty acid esters by high temperature GC, *J. High Resolut. Chromatogr.*, 1992, *15*, 585–589.

88. David, F., D. R. Gere, F. Scanlon, P. Sandra, Elucidation of emulsifiers by high temperature capillary GC, *Proceedings of the Riva 20th International Symposium on Capillary Chromatography*, 1998.

89. Spilker, R., B. Menzebach, U. Schneider, I. Venn, Analysis of alkylpolyglucosides (in German), *Tenside. Surfactants. Deterg.*, 1996, *33*, 21–25.

90. Waldhoff, H., J. Scherler, M. Schmitt, Alkyl polyglycosides—analysis of raw material: determination in products and environmental matrices, *World Surfactants Congr.*, 4th, 1996, *1*, 507–518

91. Waldhoff, H., J. Scherler, M. Schmitt, J. R. Varvil, Alkyl polyglycosides and determination in consumer products and environmental matrices, in K. Hill, W. von Rybinski, G. Stoll, eds., *Alkyl Polyglycosides: Technology, Properties and Applications*, VCH, Weinheim, 1997.

92. Waszeciak, P., H. G. Nadeau, Reaction of ethers with acetyl chloride and the identification of products by GC, *Anal. Chem.*, 1964, *36*, 764–767.

93. Langley, N. A., D. Suddaby, K. Coupland, Carbon chain length of amine oxides, *Int. J. Cosmet. Sci.*, 1988, *10*, 257–261.

94. Lew, H. Y., Detergent mixtures containing amine oxides, *J. Am. Oil Chem. Soc.*, 1964, *41*, 297–300.

95. Abidi, S. L., GC of straight chain homologs of alkylbenzyldimethylammonium compounds, *J. Chromatogr.*, 1980, *200*, 216–220.

96. Wilkes, A. J., C. Jacobs, G. Walraven, J. M. Talbot, Quaternized triethanolamine esters (ester-quats) by HPLC, HRCGC, and NMR, *World Surfactants Congr.*, 4th, 1996, *1*, 389–412.

97. Campeau, D., I. Gruda, Y. Thibeault, F. Legendre, Amphoteric sufactants of the alkylamino-propylglycine type by GC, *J. Chromatogr.*, 1987, *405*, 305–310.

98. Molever, K., Sodium lauroyl sarcosinate by GC, *J. Am. Oil Chem. Soc.*, 1993, *70*, 101–103.

99. Takano, S., C. Takasaki, K. Kunihiro, M. Yamanaka, Homolog distributions by GC on the basis of the Hofmann degradation, *J. Am. Oil Chem. Soc.*, 1977, *54*, 139–143.

100. Suzuki, S., M. Sakai, K. Ikeda, K. Mori, T. Amemiya, Y. Watanabe, Alkyltrimethyl- and dialkyldimethylammonium compounds by GC, *J. Chromatogr.*, 1986, *362*, 227–234.

101. Takano, S., M. Kuzukawa, M. Yamanaka, Homolog distributions by reaction GC on the basis of Hofmann degradation, *J. Am. Oil Chem. Soc.*, 1977, *54*, 484–486.

102. Metcalfe, L. D., Direct GC analysis of long chain quaternary ammonium compounds, *J. Am. Oil Chem. Soc.*, 1963, *40*, 25–27.

103. Suzuki, S., Y. Nakamura, M. Kaneko, K. Mori, Y. Watanabe, Analysis of benzalkonium chlo-rides by GC, *J. Chromatogr.*, 1989, *463*, 188–191.

104. Warrington, H. P., Jr., Homolog distribution of mixed alkylbenzyldimethyl ammonium chlo-rides, *Anal. Chem.*, 1961, *33*, 1898–1900.

105. Takano, S., K. Tsuji, Imidazolinium cationic surfactants, *J. Am. Oil Chem. Soc.*, 1983, *60*, 870–874.

106. Takano, S., K. Tsuji, Amphoteric surfactants obtained by the reaction of 1-(2-hydroxyethyl)-2-alkyl-2-imidazoline with ethyl acrylate, *J. Am. Oil Chem. Soc.*, 1983, *60*, 1798–1806.

107. Takano, S., K. Tsuji, Amphoteric surfactants obtained by the reaction of 1-(2-hydroxyethyl)-2-alkyl-2-imidazoline with sodium monochloroacetate, *J. Am. Oil Chem. Soc.*, 1983, *60*, 1807–1815.

108. Bak, H., K. Choi, J. Lee, Y. Kim, H. Ahn, Imidazoline type cationic surfactants (in Korean; tables and figures in English), *Kongop Hwahak*, 1998, *9*, 404–406.

109. Rezanka T., M. Podojil, Preparative separation of algal polar lipids and of individual molecu-lar species by HPLC and their identification by GC-MS, *J. Chromatogr.*, 1989, *463*, 397–408.

110. Eder, K., A. M. Reichlmayr-Lais, M. Kirchgessner, Studies on the methanolysis of small amounts of purified phospholipids for GC analysis of fatty acid methyl esters, *J. Chromatogr.*, 1992, *607*, 55–67.

111. Grossi, G., R. Vece, GC analysis of primary, secondary and tertiary fatty amines, and of cor-responding quaternary ammonium compounds, *J. Gas Chromatogr.*, 1965, *3*, 170–173.

112. Linhart, K., K. Wrabetz, Composition of quaternary ammonium compounds by GC and MS (in German), *Tenside, Surfactants, Deterg.*, 1978, *15*, 19–30.

113. Hind, A. R., S. K. Bhargava, S. C. Grocott, Alkyltrimethylammonium bromides in Bayer pro-cess liquors by GC and GC-MS, *J. Chromatogr. A*, 1997, *765*, 287–293.

114. Goetz, N., P. Lasserre, G. Kaba, D. Good, Characterization and identification of quaternary ammonium compounds using pyrolysis-GC, *Cosmet. Sci. Technol. Ser.*, 1985, *4*, 105–137.

115. Haskins, N. J., R. Mitchell, Thermal degradation of some benzyltrialkylammonium salts using pyrolysis-GC-MS, *Analyst*, 1991, *116*, 901–903.

116. Christofides, A., W. J. Criddle, Quaternary ammonium compounds by pyrolysis-GC, *Anal. Proc.* (London), 1982, *19*, 314–316.

117. Christofides, A., W. J. Criddle, Hexadecylpyridinium chloride in pharmaceutical preparations using direct injection capillary pyrolysis-GC, *J. Anal. Appl. Pyrolysis*, 1982, *4*, 211–217.

118. Hon-nami, H., T. Hanya, GC-MS determination of alkylbenzenesulfonates in river water, *J. Chromatogr.*, 1978, *161*, 205–212.

119. Sandra, P., F. David, Detergents and lubricants: high temperature capillary GC and capillary SFC, *J. High Resolut. Chromatogr.*, 1990, *13*, 414–417.

120. Tabor, C. F., L. B. Barber, II, Fate of linear alkylbenzene sulfonate in the Mississippi River, *Environ. Sci. Technol.*, 1996, *30*, 161–171.

121. Suter, M. J. F., R. Reiser, W. Giger, Differentiation of linear and branched alkylbenzene-sulfonates by GC/tandem MS, *J. Mass Spectr.*, 1996, *31*, 357–362.

122. Reiser, R., H. O. Toljander, W. Giger, LAS in recent sediments by GC/MS, *Anal. Chem.*, 1997, *69*, 4923–4930.

123. Ding, W., C. Chen, LAS in water by large-volume injection port derivatization and GC-MS, *J. Chromatogr. A*, 1999, *857*, 359–364.

124. Patterson, J. M., S. Kortylewicz, W. T. Smith, Jr., Thermal degradation of sodium dodecyl sulfate, *J. Agric. Food Chem.* 1984, *32*, 782–784.

125. Puchalsky, C. B., Boron trifluoride-methanol transesterification as a means of characterizing alcohol sulfate detergents, *J. Am. Oil Chem. Soc.*, 1970, *42*, 803–804.

126. Kupfer, W., J. Jainz, H. Kelker, Alkylsulfonates (in German), *Tenside, Surfactants, Deterg.*, 1969, *6*, 15–21.

127. Taulli, T. A., Evaluation of isomeric sodium alkenesulfonates via methylation and GC, *J. Chromatogr. Sci.*, 1969, *7*, 671–673.

128. Komárek, K., J. Minár, S. Škvarenina, Capillary GC of higher AE with an even number of carbon atoms in the alkyl group: influence of type of derivatizing agent, alcohol chain length and oxyethylene chain length on the retention indices with a linear temperature increase, *J. Chromatogr. A*, 1996, *727*, 131–138.

129. Szymanowski, J., H. Szewczyk, B. Atamanczuk, Statistical analysis of the content determination of AE as measured by means of GC, *Tenside, Surfactants, Deterg.*, 1984, *21*, 139–143.

130. Luke, B. G., Alkylene oxide-fatty alcohol condensates by GC, *J. Chromatogr.*, 1973, *84*, 43–49.

131. Brumley, W. C., W. J. Jones, A. H. Grange, High temperature capillary GC for environmental analysis, *LC-GC*, 1995, *13*, 228,230,232,234,236.

132. Ludwig, F. J., Ethylene oxide and propylene oxide adducts of alkylphenols or alcohols by NMR, GC, and TLC, *Anal. Chem.*, 1968, *40*, 1620–1627.

133. Allan, G. C., J. R. Aston, F. Grieser, T. W. Healy, Partitioning of NPE between water and hexane, *J. Colloid Interface Sci.*, 1989, *128*, 258–274.

134. Szymanowski, J., H. Szewczyk, J. Hetper, Products obtained in the first stages of the ethoxylation of alkylphenols, *Tenside, Surfactants, Deterg.*, 1981, *18*, 333–338.

135. Stephanou, E., Determination of acidic and neutral residues of APE surfactants using GC/MS analysis of their TMS derivatives, *Comm. Eur. Communities, EUR 1986, EUR 10388, Org. Micropollut. Aquat. Environ.*, 155–156.

136. Wisniewski, M., J. Szymanowski, B. Atamanczuk, Molar mass distribution of polyoxyethylene 4-alkylphenylamines, *J. Chromatogr.*, 1989, *462*, 39–47.

137. Arens, M., H. König, M. Teupel, Working party of the GDR. Analysis of organic surface active compounds (in German), *Fette, Seifen, Anstrichm.*, 1982, *84*, 105–111.

138. Suffis, R., T. J. Sullivan, W. S. Henderson, Identification of surface active agents as trimethyl silyl ether derivatives by GC, *J. Soc. Cosmetic Chemists*, 1965, *16*, 783–794.

139. Soe, J. B., Analyses of monoglycerides and other emulsifiers by GC, *Fette, Seifen, Anstrichm.*, 1983, *85*, 72–76.

140. Giacometti, J., C. Milin, N. Wolf, Monitoring the esterification of sorbitol and fatty acids by GC, *J. Chromatogr. A*, 1995, *704*, 535–539.

141. Torres, M. C., M. A. Dean, F. W. Wagner, Chromatographic separations of sucrose monostearate structural isomers, *J. Chromatogr.*, 1990, *522*, 245–253.

142. Gupta, R. K., K. James, F. J. Smith, Sucrose mono- and diesters prepared from triglycerides containing C_{12}–C_{18} fatty acids, *J. Am. Oil Chem. Soc.*, 1983, *60*, 1908–1913.

143. Sahasrabudhe, M. R., Chromatographic analysis of polyglycerols and their fatty acid esters, *J. Am. Oil Chem. Soc.*, 1967, *44*, 376–378.

144. Sahasrabudhe, M. R., J. J. Legari, GC analysis of mono- and diglycerides, *J. Am. Oil Chem. Soc.*, 1967, *44*, 379–380.

145. Sahasrabudhe, M. R., R. K. Chadha, Chromatographic analysis of sorbitan fatty acid esters, *J. Am. Oil Chem. Soc.*, 1969, *46*, 8–12.
146. Billian, P., H. J. Stan, GC/MS of alkyl polyglucosides as their trimethylsilylethers, *Tenside, Surfactants, Deterg.*, 1998, *35*, 181–184.
147. Milwidsky, M. D., D. M. Gabriel, *Detergent Analysis*: *Handbook for Cost-Effective Quality Control*, Micelle Press, London, 1989.
148. McClure, J. D., Ethylene oxide oligomer distributions in alcohol ethoxylates by HPLC using a rotating disc-flame ionization detector, *J. Am. Oil Chem. Soc.*, 1982, *59*, 364–373.

9
Thin-Layer Chromatography

I. GENERAL CONSIDERATIONS

New publications on surfactant analysis by TLC have all but ceased. TLC is rarely used for quantitative analysis. It is a low technology method, requiring little in the way of capital investment and giving the best results to a careful and experienced operator. It is rare that TLC service is provided by a central analytical services department. It works best if the research or applications chemist performs his/her own TLC analysis to give rapid, semiquantitative, information which the chemist can add to other information already available. To such a careful operator, the color and shape of the spots on the plate, as well as such observations as whether or not bubbles form over the spots when immersed in water, give additional information to help identify surfactants (1,2). Such care and attention are more important than the optimum chromatographic conditions.

TLC is used both for characterization of pure surfactants and for detection of surfactants in complex samples. With modern apparatus, TLC is a reproducible technique for analyzing multiple samples simultaneously and has found a niche in metabolism studies, since it has the advantage that all components of the sample are spread across the surface of a single plate. TLC is often used for the initial separation of surfactants into classes, especially for the separation of nonionics from anionics, as described in Chapter 6.

A number of approaches have been developed, mainly based upon two-dimensional TLC, for qualitative analysis of mixtures of surfactants. These are briefly covered in Chapter 5. TLC spots are commonly identified by removing the silica gel from the plate and extracting the organic material with a polar solvent. With careful technique, subsequent identification can be made by IR spectroscopy, especially by IR microscopy, or by mass spectrometry. Since MS will give the homolog/oligomer distribution of the surfactant, care should be taken not to use a TLC system which separates according to homolog/oligomer distribution; otherwise, the mass spectrum will not represent the original surfactant. Compounds cannot be recovered as easily from reversed-phase TLC plates as from silica plates, since extracts are generally contaminated with fragments of the stationary phase. MALDI-MS is very useful for analysis of TLC spots after scraping from the plate (3). Direct identification of spots while still on the plate has been demonstrated using diffuse reflectance FTIR or secondary ion mass spectroscopy. Relatively high concentrations are required, since the direct method is not as sensitive as more labor-intensive approaches which first remove the spots from the plate (4).

A problem TLC shares with most other analytical techniques is the need to calibrate with standards of the same compounds as those to be determined. This requirement is alleviated somewhat by the Iatroscan technique. This is an instrumental system which substitutes silica-coated rods for TLC plates. The rods are passed through a flame ionization detector to visualize the zones after development. Since the FID response is proportional to carbon content rather than chemical functionality, roughly quantitative work may be performed without external calibration (5).

The interaction of surfactants with silica gel is greatly influenced by the moisture content of the silica. This behavior is even more marked with alumina plates. For reproducible results, it is usually necessary that plates are equilibrated at constant humidity prior to use for analysis.

II. ANIONIC SURFACTANTS

A. General Remarks

Thin-layer chromatography is a rapid, inexpensive method for qualitative and quantitative analysis of mixtures of anionics. Table 1 shows several normal phase systems designed to separate particular surfactant mixtures. For example, LAS, dodecylsulfate, and dodecanesulfonate are completely resolved with a MIBK/propanol/acetic acid/acetonitrile system (6).

As mentioned in other chapters, TLC is frequently used as a cleanup tool to separate anionic surfactants from complex matrices prior to determination by instrumental techniques. Useful for preparative purposes is the behavior of anionics on alumina plates with ethyl acetate/pyridine (7) or chloroform/methanol (8); anionics remain at the origin, while nonionics and cationics migrate far up the plate.

Sulfates and sulfonates typically have low Rf values on silica, independant of the hydrophobe (9). The silica may be deactivated by treatment with ammonium sulfate. By acidification of the mobile phase and use of ammonium sulfate–modified silica, the Rf values can be adjusted so that polysulfonates remain near the origin and di- and monosulfonate and unsulfonated oil have progressively higher Rf (10). Ordinary silica plates with a highly ammoniacal mobile phase will give similar results, again with no discrimination by alkane chain length (11). Normal phase chromatography on silica or alumina is generally not capable of differentiating anionics according to the alkyl chain length. Nonpolar stationary phases are used for this analysis, such as polyamide or C_{18}-modified silica (12). Reversed-phase plates are also used to separate anionics from cationic and nonionic surfactants (13,14).

TABLE 1 TLC Analysis of Mixtures of Anionic Surfactants

Surfactants separated	Stationary phase	Developing system	Visualizer	Ref.
LAS, soap	Merck Silica Gel 60	2-Butanone, H_2O-saturated	Primulin	43
Alkylbenzenesulfonate: paraffin sulfonate; sulfated fatty alcohol ethoxylate: sulfated alkylphenol ethoxylate: sulfated fatty alcohol (some homologs are separated); qualitative	Merck Silica Gel 60	90:10 EtOH/HOAc; 80:20 BuOH/EtOH; 60:10:20:0.5 BuOH/EtOH/H_2O/HOAc	Acid Blue 158	49
Dodecylbenzenesulfonate; dodecyl-sulfate; dodecanesulfonate; quantitative	Silica Gel G with ammonium sulfate, 0.25 mm	20:6:1.6:1 MIBK/n-PrOH/0.1 M HOAc/CH_3CN	Charring at 250°C for 15 min	6
Alkanesulfonate; hydrotropes: fatty alcohol sulfate; alkylbenzene-sulfonate; sulfated fatty alcohol ethoxylate (homologs are separated); soap; α-sulfo fatty acid methyl ester; sulfosuccinic acid diester; qualitative	Merck Silica Gel 60	90:10 acetone/THF; develop 3 times to 10 cm	Pinacryptol yellow; UV light	24
Soap: alkylsulfate; dodecylbenzene-sulfonate; fatty alcohol ethoxylate (homologs are separated); qualitative	Merck Silica Gel 60 G-254	80:10:8:2 $CHCl_3$/MeOH/HOAc/H_2O; 95:5 acetone/HOAc; 80:14:10 ethyl acetate/HOAc/H_2O	Eosin; molybdophosphoric acid; methyl red; phosphoric acid	9
Qualitative identification of various anionic surfactants	Silica Gel G impregnated with dodecyl alcohol	50:50:1 MeOH/NH_4OH/formic acid	Pinacryptol yellow	7
LAS, alkyl sulfates, and alkylether sulfates; detection in formulations	Silica Gel 60, 0.25 mm, Brockmann activity III	90:10 acetone/THF	Pinacryptol yellow; ID by IR analysis	137,138

(Continued)

TABLE 1 (Continued)

Surfactants separated	Stationary phase	Developing system	Visualizer	Ref.
Various anionics; detection and resolution of homologs	Silica Gel 60	80:20:20 n-BuOH/H$_2$O/HOAc	Primulin; Dragendorff reagent for alkyl ether sulfates	19
Isolation of anionics and nonionics for IR analysis	Silica Gel F-60, activated at 80°C for 5 min	CHCl$_3$/MeOH, various ratios	Iodine vapor	139,140
Anionics from nonionics	Merck Silica Gel G	10:10:5:2 PrOH/CHCl$_3$/MeOH/ 10 M NH$_3$	Pinacryptol yellow and modified Dragendorff reagent	7
Anionics from nonionics	Alumina G with 10% Alumina DO	60:40 pyridine/ethyl acetate	Pinacryptol yellow	7

B. Detection

Sulfuric acid charring is the most frequently applied method for detecting anionics. This completely nonspecific operation is usually carried out by spraying the developed plate with dilute sulfuric acid, then heating in an oven to dehydrate. Application of sulfuryl chloride fumes rather than liquid H_2SO_4 to the plates before heating is said to give improved detection limits (15). Improved reproducibility is claimed by heating in a special apparatus in the presence of SO_3 fumes (16).

Modified Dragendorff reagent is useful for detection of anionics containing polyethoxy functionality. Most sulfur-containing anionics react with the cationic dye pinacryptol yellow to give bright orange fluorescence when viewed under long wavelength UV (λ = 366 nm) light, fading to fluorescent blue (14). Sulfates and sulfonates with shorter chain lengths than about C_{12} give yellow rather than orange fluorescence with pinacryptol yellow (17). The anionic dye eosin yellow gives more intense fluorescence with anionics than with other surfactants (9). Other anionic and cationic dyes, such as azur A (18), fuchsin, rhodamine B, primulin (19), and fluorescein derivatives, give color under visible or UV light with anionics, although these are far from specific. The same is true of iodine vapor.

If equipment is available to scan the UV reflectance spectra of the spots, LAS and the aromatic hydrotropes show typical absorption at about 225 and 265 nm. The λ_{max} for alkylnaphthalene sulfonates is somewhat higher. The aromatic compounds can also be detected without a visualizing agent, since they will quench the fluorescence of TLC plates pretreated with a fluorescing agent, appearing as dark spots under UV light.

C. Analysis of Specific Anionic Surfactants

1. Alkylbenzene Sulfonates

LAS is differentiated qualitatively from other surfactants by TLC on silica gel (see Table 2). Resolution according to alkyl chain length is possible by reversed-phase TLC, although this is rarely used. TLC has been used to separate the ion pair of LAS with methylene blue from impurities, but not from other anionic surfactants (20). The complex remains very near the origin in the system silica/methylene chloride.

Argentation TLC was used to separate alkylbenzenes derived from LAS (21). Silica gel plates impregnated with silver nitrate were used with a developing reagent of hexane and visualization with 2,7-dichlorofluorescein. Although the main purpose of the separation was isolation of alkylbenzenes as a group, it was observed that individual isomers are separated from each other.

2. Alkyl Sulfates

Unsulfated alcohol (down to about 0.1%), either from incomplete synthesis or formed by hydrolysis, may be determined in alkylsulfates by TLC on silica, since the alcohol has a high Rf value and is visible after spraying with modified Dragendorff reagent as a light spot on a yellow background (22). TLC is also suitable for detection of sultones and other thermal degradation products (23).

Sulfates may be differentiated from sulfonates by analyzing with and without acid hydrolysis. Sulfates are decomposed by this treatment, while sulfonates are not affected. Methods for analysis of these compounds are summarized in Table 3.

TABLE 2 TLC Analysis of Alkylbenzene Sulfonates

Compounds studied	Stationary phase	Developing system	Visualizer	Ref.
C_{12}–C_{18} alkylbenzene-sulfonate (not differenti-ated); quantitative analysis	Merck Silica Gel G with 5% ammonium sulfate, 0.3 mm	70:30:6 $CHCl_3$/MeOH/0.05 M H_2SO_4	Charring with sulfuryl chloride; 300°C	15
Dodecylbenzenesulfonate; determination in presence of dodecylsulfate and dodecanesulfonate	Silica Gel G with ammonium sulfate, 0.25 mm	20:6:1.6:1 MIBK/n-PrOH/0.1 M HOAc/CH_3CN	Phosphomolybdic acid; 250°C for 15 min	6
C_4–C_{14} alkylbenzene-sulfonates, hydrotropes; separation by alkyl chain length	Polyamide powder B-O (Wako, Osaka, Japan) bound with cellu-lose (Camag)	15:1 0.1 M NH_3/pyridine or 15:15:1 MeOH/0.1 M NH_3/pyridine	Pinacryptol yellow; 254 nm	12
Dodecylbenzenesulfonate (nominal); separation by alkyl chain length	Silica Gel G impregnated with dodecyl alcohol	50:50:1 MeOH/NH_4OH/formic acid	Pinacryptol yellow; UV light	7

TABLE 3 TLC Analysis of Alkyl Sulfate Surfactants

Compounds studied	Stationary phase	Developing system	Visualizer	Ref.
Sodium dodecylsulfate in ointment; determination	Iatroscan Chromarod SII	50:10:2 ethyl acetate/MeOH/NH$_4$OH	FID	87
Free alcohol in dodecanol sulfate: detection	Silica Gel G	80:20 ethyl ether/benzene	Modified Dragendorff reagent	22
Di-n-hexadecylether; di-1-hexadecyl sulfate; hexadecenes; all from thermal degradation of alkyl sulfate; determination	Silica Gel G	1.0–1.5% ethyl ether in hexane; 5°C	50% H$_2$SO$_4$; 180°C overnight	23
Hexadecanols and sultones from thermal degradation of alkyl sulfate	Silica Gel G	60:40 ethyl ether/hexane; 5°C; develop twice	50% H$_2$SO$_4$; 180°C overnight	23
Dodecylsulfate: determination in the presence of decylbenzenesulfonate and dodecanesulfonate	Silica Gel G with ammonium sulfate, 0.25 mm	20:6:1.6:1 MIBK/n-PrOH/0.1 M HOAc/CH$_3$CN	Phosphomolybdic acid; 250°C for 15 min	6
Fatty alcohol sulfate: detection in presence of other surfactants	Merck Silica Gel 60	90:10 acetone/THF; develop three times to 10 cm	Pinacryptol yellow; UV light	24
Dodecyl sulfate: detection in presence of other surfactants	Merck Silica Gel 60 G-254	80:10:8:2 CHCl$_3$/MeOH/HOAc/H$_2$O; 95:5 acetone/HOAc; 80:14:10 ethyl acetate/HOAc/H$_2$O	Eosin or molybdophosphoric acid	9

3. Ether Sulfates

Alkylether sulfates have slightly different Rf values on silica than the corresponding alkyl-sulfates, so mixtures can be resolved (24). The higher the degree of ethoxylation, the lower the Rf value on silica, so these products give a characteristic series of spots. The most common ether sulfates contain an average of 3–4 ethoxy units, so their ethoxy distribution is readily determined by TLC (see Table 4).

With pinacryptol yellow detection, alkyl ether sulfates give light blue fluorescence under UV light, while alkylphenol ether sulfates give a blue-black color (7). Alkyl ether sulfates give more intense colors than alcohol sulfates by about a factor of four (24). Ether sulfate oligomers of more than four EO units react with modified Dragendorff reagent, while the corresponding nonethoxylated compounds will not.

The identification of ether sulfates can be confirmed by analyzing on silica plates with only a low concentration of ammonia in the mobile phase. Under these conditions, alkylether sulfates and alkylphenolether sulfates will give only a single spot near the bottom of the plate. If the sample is boiled with HCl to hydrolyze the sulfate group, then reanalyzed, the resulting nonionic surfactant gives a series of spots (visible with modified Dragendorff reagent) up the plate corresponding to the various oligomers (7).

4. α-Olefin Sulfonates

The components of α-olefin sulfonates, namely alkenesulfonates, hydroxyalkane sulfonates, and the corresponding disulfonates, can be determined by TLC on Silica Gel G, with development with chloroform/methanol/sulfuric acid and visualization by sulfuric acid charring or other means (25,26). Rf values increase in the order disulfonates, hydroxyalkane monosulfonates, alkenesulfonates, and unsulfonated oil, with no differentiation by alkyl chain length (see Table 5).

Determination of alkyl chain distribution has been demonstrated by chromatography on dodecanol-saturated paper, with partial resolution by position of sulfonation (16). Presumably, this separation may also be performed with modern C_{18} reversed-phase media.

5. Alkanesulfonates and Petroleum Sulfonates

Mono-, di-, and polysulfonates are easily separated by chromatography on deactivated silica (10) or ordinary silica with ammoniacal developer (11). Migration in order of increasing Rf is di- and polysulfonates, monosulfonates, and unsulfonated alkylaromatics (27). 4,5-Dichlorofluorescein reacts with the compounds to give yellow spots, or green spots when long wavelength UV light is used (26).

TLC plates prepared from polyamide powder will give a separation according to alkyl chain length, as, presumably, will commercially available reversed-phase plates (12). See Table 6 for a summary of published methods for alkanesulfonates.

6. Sulfosuccinates

On silica, sulfosuccinate diesters have much higher Rf values than the monoesters (Table 7). The opposite is observed under reversed-phase conditions (7). Sulfosuccinate diesters can be resolved by the length of the alkyl chain by reversed-phase TLC, with Rf values decreasing as the alkyl character increases (14). Sulfosuccinates give an intense blue color with pinacryptol yellow under UV light (7).

TABLE 4 TLC Analysis of Ether Sulfate Surfactants

Compounds studied	Stationary phase	Developing system	Visualizer	Ref.
C_{14}–C_{18} alcohol 8-mole ethoxysulfate; tributylphenol 8-mole ethoxysulfate; determination of active agent, PEG, mono- and disulfated PEG, and unsulfated material	Silica gel plate (Merck No. 5715/0025) sprayed with 0.05 M oxalic acid and 2.5% ammonium sulfate solution; dried at 110°C	90:10 $CHCl_3$/MeOH	Modified Dragendorff reagent or SO_3 charring	88
Cocoalcohol 2- to 12-mole ethoxysulfate (oligomers are separated); separation from fatty alcohol sulfate	Merck Silica Gel 60	90:10 acetone/THF; develop three times to 10 cm	Pinacryptol yellow; UV light	24
C_{12} alcohol 3-mole ethoxysulfate; oligomers are separated	Merck Silica Gel 60 G-254	95:5 acetone/HOAc or 80:14:10 ethyl acetate/HOAc/H_2O	Eosin	9
Fatty alcohol ether sulfate and nonylphenol ether sulfate; separation of oligomers after hydrolysis to the corresponding nonionic compounds	Silica Gel G	45:5:2.5 Ethyl acetate/NH_4OH/MeOH	Modified Dragendorff reagent	7

TABLE 5 TLC Analysis of α-Olefinsulfonate Surfactants

Compounds studied	Stationary phase	Developing system	Visualizer	Ref.
C_{14}–C_{18} α-olefin sulfonates: determination of alkene monosulfonate, hydroxyalkane monosulfonate, and disulfonates	Silica Gel G with ammonium sulfate, 0.25 mm	70:32:6 $CHCl_3$/MeOH/0.05 M H_2SO_4	Charring with SO_3 vapor; 20 min at 150°C	25
α-Olefin sulfonate: determination of alkene monosulfonate; hydroxyalkane monosulfonate; disulfonates	Merck Silica Gel G 5715 with ammonium sulfate	80:19:1 $CHCl_3$/MeOH/0.05 M H_2SO_4	50% H_2SO_4; 150°C	26
C_8–C_{18} alkenemonosulfonate and hydroxyalkane monosulfonate; not differentiated; qualitative analysis by chain length	Paper saturated with dodecanol	50:50 or 60:40 CH_3OH/25% NH_3, solution, saturated with dodecanol	Pinacryptol yellow, UV	17
C_{12} and C_{18} alkenesulfonates; C_{16} hydroxyalkanesulfonates; di- and trisulfonates; separation and determination	Merck Silica Gel G with 5% ammonium sulfate, 0.3 mm	70:30:6 $CHCl_3$/MeOH/0.05 M H_2SO_4	Charring with sulfuryl chloride; 300°C	15

TABLE 6 TLC Analysis of Alkanesulfonate and Petroleum Sulfonate Surfactants

Compounds studied	Stationary phase	Developing system	Visualizer	Ref.
C_{11}–C_{20} paraffin sulfonate; determination of alkanesulfonate, disulfonate, polysulfonate; quantitative	Merck Silica Gel GF-254	70:30 n-PrOH/NH_4OH	4,5-Dichlorofluorescein, visible or UV light	11
C_{12}–C_{18} paraffin sulfonate; determination of alkanesulfonate, disulfonate, polysulfonate, and unsulfonated material	Merck Silica Gel G with 5% ammonium sulfate, 0.3 mm	70:30:6 $CHCl_3$/MeOH/0.05 M H_2SO_4	Charring with sulfuryl chloride, 300°C or iodine vapor, 350°C	10,15
Petroleum sulfonate (eq. wt. in 230–620 range); monosulfonate, disulfonate and polysulfonate, unsulfonated material	Silica Gel G with added ammonium sulfate	70:30:6 $CHCl_3$/MeOH/0.05 M H_2SO_4	Fluorescence	27
C_4–C_{18} alkanesulfonates; separation by alkyl chain length	Polyamide powder B-O (Wako, Osaka, Japan) bound with cellulose (Camag)	15:1 0.1 M NH_3/pyridine or 15:15:1 MeOH/0.1 M NH_3/pyridine	Pinacryptol yellow; UV light	12

TABLE 7 TLC Analysis of Other Anionic Surfactants

Compounds studied	Stationary phase	Developing system	Visualizer	Ref.
C_{16} and C_{18} α-sulfo fatty acid ester and α-sulfo fatty acid; determination	Silica Gel 60	2:11:27 0.05 M H_2SO_4/ MeOH/$CHCl_3$	Pinacryptol yellow; fluorescence at 254 and 366 nm	89
Alkylphosphate esters, C_{12} and higher, including phosphate esters of alcohol ethoxylates, E = 4 and 8; identification	Cellulose (S + S) containing 2% corn starch	10:20:40:20:10 benzene/ $CHCl_3$/MeOH/2-PrOH/20% HOAc	Ammonium molybdate/ $HClO_4$; reduction with $SnCl_2$/HCl	28
Sulfosuccinic acid monoesters with ethoxylated lauryl alcohol, ethoxylated coco alkanolamide, and undecyleneth-anolamide; detection	Silica Gel G containing 10% $(NH_4)_2SO_4$	80:19:1 $CHCl_3$/MeOH/0.05 M H_2SO_4	Dichlorofluorescein, 365 nm	90
Monoalkylsulfosuccinates; detection	Silica gel G, 0.25 mm, activated at 100°C	20:20:5 ethyl acetate/MeOH/ NH_4OH	Basic fuchsin	91
Monoalkylsuccinates; detection	Silica gel G, 0.25 mm, activated at 100°C	40:15 toluene/MeOH	Basic fuchsin	91
C_{14}–C_{18} acylsarcosinates; separation from impurities: fatty acids and the ethyl esters of acylsarcosines and fatty acids	Silica Gel 60, 0.2 mm	60:40 n-hexane/ethyl ether	Iodine	33
Carboxymethylated alcohol ethoxylates: C_{12}/C_{14} alcohol 3- and 4,5-mole ethoxyacetates and nonylphenol 3- and 5-mole acetates; qualitative and quantitative analysis; detection of nonionic matter	Silica Gel 60	80:20 $CHCl_3$/MeOH	Dragendorff reagent or 2,7-dichlorofluorescein	29

Carboxymethylated NPE, mean E = 4 to 6: semiquantitative determination of active agent, NPE, PEG, carboxylated PEG	Silica Gel G, 0.5 mm	98:2 $CHCl_3$/MeOH	Dragendorff reagent	30
Carboxymethylated NPE, mean E = 2, 6, and 10; separation from NPE and resolution by ethoxy chain length	Silica gel, activated at 110°C for 1 hr	Two-dimensional separation: 16.25:6.5:5:0.7:0.4 $CHCl_3$/MeOH/1-BuOH/0.1 N EDTA/0.05 M H_2SO_4 to separate NPE from NPE carboxy-methylate, then 9.5:2:2 ethyl acetate/HOAc/H_2O (proportions adjusted depending on degree of ethoxylation) for ethoxy resolution	I_2 vapor	31
C_8–C_{30} fatty acids; separation from each other	Silica Gel G, impregnated with 5% tetradecane	8:3:4:1.3 2-PrOH/EtOH/HOAc/H_2O	Rhodamine B	7
C_{10}–C_{24} fatty acids as phenacyl derivatives; separation from each other	Merck HPTLC RP-18: plate may be reused if washed with 50:50 hexane/ethyl ether	2:1 CH_3CN/2-PrOH	UV, 254 nm	32

7. Phosphates

König showed that a number of phosphate surfactants could be recognized by their behavior on a cellulose plate, with detection by formation of the characteristic molybdenum blue color (28). Henrich gives Rf values for a number of commercial phosphate surfactants on five different TLC systems, but was unable to correlate TLC behavior with structure (14). Although some experienced analysts can recognize individual commercial products from their characteristic spots and color shades, most use other techniques for an exact identification.

8. Ether Carboxylates

Ether carboxylates may be determined by TLC on silica. With a developing solvent of 80:20 chloroform/methanol, the ether carboxylate remains near the origin, while the non-carboxylated nonionic portion rises nearly to the solvent front. Carboxylated PEG has an even lower Rf than the ether carboxylate, while PEG has a somewhat lower Rf than the parent nonionic. 2,7-Dichlorofluorescein is a suitable visualization reagent. A more specific visualizer is modified Dragendorff reagent, which serves to differentiate the ether carboxylate from most other anionics, although low ethoxylates, like nonylphenol-3-mole ethoxylate, cannot be detected with useful sensitivity (29,30). As with ethoxylated non-ionics, the ethoxy chain length distribution of ether carboxylates may be determined by TLC (31).

9. Soap

Soap is readily differentiated from sulfonated surfactants by TLC on unmodified silica with 90:10 acetone/THF. Under these conditions, soap has a high Rf, while sulfates and sulfonates remain near the origin (24).

  TLC is rarely applied to determination of fatty acids, since HPLC and GC (usually of methyl esters) give more precise quantitative information. However, reversed-phase TLC is suitable for separation by acyl chain length, with the higher acids having shorter Rf values. Sensitive detection is a problem, but may be handled by making UV-absorbing derivatives, e.g., by triethylamine-catalyzed reaction with bromoacetophenone (32).

10.   Other Anionic Surfactants

Acylsarcosines can be separated from their main impurities on silica with 60:40 n-hexane/ethyl ether mobile phase. The acylsarcosines stay near the origin, while the impurities, acylsarcosine ethyl ester, free fatty acid, and fatty acid ethyl ester, have progressively higher Rf values (33).

  Isethionates give a light blue color with pinacryptol yellow under UV light (7). Taurides give a light blue color with pinacryptol yellow under UV light (7). Iodine is a suitable visualizer for acylsarcosines and their impurities (33). Acylsarcosines give a bluish black color with pinacryptol yellow under UV light (7).

11.   Hydrotropes

The hydrotropes benzene sulfonate and p-toluene sulfonate are well resolved from anionic surfactants by TLC on dodecanol-impregnated silica (7) or polyamide (12) with an ammoniacal developing solution. Presumably, modern reversed-phase media will give superior performance. Xylene and cumene sulfonates give orange fluorescence with pinacryptol yellow, while p-toluene sulfonate gives a red color (14).

III. NONIONIC SURFACTANTS

TLC is used for qualitative analysis of mixtures of nonionics and determination of individual surfactants. It may also be used for determination of impurities, such as PEG in ethoxylates.

A. General Remarks

1. Normal Phase TLC

Because of their affinity for silica gel, the polyethoxylated compounds that comprise the majority of commercial nonionics can be readily separated according to degree of alkoxylation on ordinary thin-layer plates, using the proper blend of polar and nonpolar solvents to effect resolution. The higher the degree of ethoxylation, the lower the Rf value, with PEG remaining near the origin. Complete resolution of oligomers can be obtained up to a degree of polymerization of about 12. With common developing agents, there is little differentiation between ethoxylates based on different hydrophobes. Thus, TLC on silica gel is completely analogous to normal phase HPLC analysis of ethoxylates. Alumina generally gives better resolution of lower EO adducts than does silica, but resolution of higher adducts is sacrificed (34–36).

Cserháti showed that the Rf values of ethoxylates on silica or alumina could be well described by the equation

$$\log\left(\frac{1}{\text{Rf}} - 1\right) = a + b_a n_e$$

where a is a constant for a particular hydrophobe group, n_e is the degree of ethoxylation, and b_a is a "slope" value showing the retardation effect of each ethoxy unit. The constants a and b_a vary with changes in the composition of the development solvent; b_a is higher on alumina than silica (36).

Often, the analyst is interested only in quantitative determination of a surfactant, not in seeing the oligomer distribution. If ordinary silica plates are used, conditions are sought which will minimize spreading of the ethoxylate, since visualization of a single spot will increase detectability and minimize interference from other compounds. For instance, a developing mixture of 4:1 chloroform/methanol will give a single spot on silica for ethoxylates but a high Rf value, 0.95, which may not separate them from other nonionic material. A 10:1 mixture gives an Rf value of only 0.4, with considerable fractionation according to degree of ethoxylation, while intermediate ratios of chloroform/methanol give higher Rf with progressively less fractionation. One percent ammonia is added in each case to retard anionic material (37). König and Walldorf found that if silica gel is impregnated with oxalic acid, the separation is no longer as strongly controlled by degree of ethoxylation (38,39). By proper choice of solvent, systems may then be developed which separate nonionics according to the hydrophobe group.

2. Reversed-Phase TLC

Reversed-phase TLC is capable of separating nonionics according to the nature of the hydrophobe group, just as in reversed-phase HPLC. In general, the separation is indifferent to degree of ethoxylation, and Rf values are inversely related to alkyl chain length. For example, on C_{18} media with weak developing agents like methanol/water, ethoxylates remain at the origin while PEG has an Rf value of about 0.7. With a strong developer like methanol/chloroform, ethoxylates advance to about Rf 0.5, while PEG moves with the solvent front (40).

Early workers used silica or paper impregnated with long chain alcohols to perform reversed-phase chromatography. Once C_{18}-modified silica plates became available, Armstrong and Stine showed that nonionics (ethoxylated diol, C_9 and C_{12} APE) can be separated from each other with ethanol/water, adding a little sodium tetraphenylborate to suppress streaking (13).

B. Detection

There are few visualizing reagents for nonionic surfactants. The reagents that are available generally fade quickly, so that photography of the spots is required for record-keeping. Only sulfuric acid charring, a completely nonspecific technique, does not suffer from this disadvantage. Because of the lack of a satisfactory visualizing system, some investigators remove the separated spots from the TLC plate, extract the surfactant from the stationary phase, and determine its concentration by spectrophotometric techniques.

The most frequently used visualizer is "modified Dragendorff reagent," a potassium iodobismuthate/barium chloride mixture in concentrated acetic acid, which reacts with some specificity toward polyalkoxy groups, as well as cationic surfactants (41). It is suitable for detection of PEG and most ethoxylated surfactants on silica gel and is generally suitable for qualitative analysis. Quantitative analysis is difficult, since the size and intensity of the spots are dependent upon the percentage of EO in the molecule, as well as the mass of the spot. Dragendorff reagent is sometimes followed by a second spraying with sodium nitrite solution, which gives the spots a purple color. This color fades to yellow, but lasts longer than the color from Dragendorff reagent alone (42). Other researchers suggest that the contrast between spots and background is improved by using only half as much $BaCl_2$ solution as in the original recipe given in Chapter 5 (43).

König reported that equivalent results to those obtained with modified Dragendorff reagent can be obtained by simply spraying the plates with 5% $BaCl_2$ solution (adjusted to pH 2 with HCl), followed by spraying with 0.05 M iodine (presumably also containing KI) (38). These reagents have the advantage of being more stable than the modified Dragendorff reagent, which must be freshly mixed before use.

Since modified Dragendorff reagent is not useful for detection of compounds containing less than about 4 moles of ethylene oxide, primulin is sometimes also used to detect lower oligomers (43).

Flame ionization detection using the Iatroscan apparatus has been demonstrated for analysis of surfactants (5). As predicted by theory, there is a linear relationship between molar detector response and alkyl or ethoxy chain length for homologous series of alcohol ethoxylates (44,45).

Iodine vapor is generally suitable for visualization of nonionics and is compatible with densitometry (14,46,47). The plate must be covered with another glass plate or photographed immediately, since the iodine color fades rapidly in air (48). Acid blue 158 has been shown to be a good, general purpose visualizer for surfactants (49). It was demonstrated for detection of AE and NPE as well as ethoxylated amines and an alkanolamide. Cobalt thiocyanate reagent, commonly used for spectrophotometric determination of nonionic surfactants, has also been used as a visualizer for qualitative TLC analysis (50.51). Formation of the 3,5-dinitrobenzoate ester derivatives allows visualization with UV light of surfactants without chromophores (52).

As a general procedure for detection of food emulsifiers, Dieffenbacher and Bracco recommended using several different visualizers of varying selectivity: dichlorofluorescein is used as a nonspecific visualizer; anisidine/periodate is specific for vicinal hydroxyl groups, as in monoglycerides; Dragendorff reagent is used for detection of ethoxylates; and naphthoresorcinol is recommended for detection of saccharide esters, with still other reagents used for other components (50).

C. Analysis of Specific Surfactants

1. Alcohol Ethoxylates

The molecular weight distribution of AE can be determined on silica or alumina, with unit resolution obtained up to about 13 EO groups (Table 8). Mobile phases which give good resolution of the lower oligomers leave the higher oligomers unresolved near the origin, while solvents which give resolution of the higher oligomers put the low ethoxylates at the solvent front. Hence, double development schemes are preferred for resolution of the largest number of oligomers. The systems can be scaled up for preparative use without difficulty. The same general principles apply to separation of mixed EO/PO alkyl adducts, but alumina rather than silica is used for this analysis; there is usually no oligomer separation of EO/PO adducts on silica (14). Formation of derivatives, such as acetate esters, will prevent streaking of AE on silica gel plates, permitting more precise quantification. Well-defined spots of uniform separation over a broad range of EO chain length are obtained by the technique of programmed multiple development (53). Preparative TLC of dinitrobenzoate esters may be used to isolate pure standards representing a single degree of oxyethylation (52).

Reversed-phase TLC systems will differentiate AE according to the length of the alkyl chain, with the Rf value increasing with decreasing chain length.

Most AE compounds will give blue fluorescence when sprayed with pinacryptol yellow and viewed under long wavelength UV light (14). Primulin is also a suitable reagent, giving fluorescence at 365 nm in the presence of long alkyl chains. For quantitative work, separate calibration curves are required for each alkyl homolog and oligomer, since, with primulin, response on a mass basis increases with alkyl chain length and decreases with length of the ethoxy chain (54).

2. Alkylphenol Ethoxylates

(a) Normal phase TLC.   The molecular weight distribution of APE can be determined on silica or alumina plates or silica rods, with unit resolution obtained up to about 13 EO groups (Table 9). The system can be scaled up for preparative use. The same general principles apply to separation of mixed EO/PO adducts, but alumina rather than silica is used for this analysis. A validation study showed good agreement between TLC and GC results for percent composition represented by the lower molecular weight APE compounds [the only ones for which GC data could be obtained without derivatization] (46,47).

(b) Reversed-phase TLC.   OPE has a slightly higher Rf than NPE on reversed-phase media (14). Thus TLC can be used to distinguish OPE and NPE from each other. Such a method was devised for determining OPE and NPE in cleaning formulations, with a separation system optimized to give a single spot for each (55). The resolution deteriorates as the degree of ethoxylation increases past about 20. It was shown that silicone-impregnated silica gel gave no separation of NPE, while silicone-impregnated alumina and especially

TABLE 8 TLC Analysis of Alcohol Ethoxylate Nonionic Surfactants

Compounds studied	Stationary phase	Developing reagent	Visualization	Ref.
Normal phase TLC				
C_9–C_{18} alkyl 8- to 30-mole ethoxylates; ethoxy distribution; determination of PEG and carboxylate biodegradation products	Silica gel	Quantitative: 40:30:30 ethyl acetate/ HOAc/H_2O (surfactant is a single spot); qualitative: 70:16:15 ethyl acetate/HOAc/H_2O: homologs are separated)	Modified Dragendorff reagent	92,93
C_{12} alkyl 4- to 8-mole ethoxylate; ethoxy distribution	Iatron Iatroscan Chromarod SII	Double development: first, 60:40 benzene/ethyl acetate, 10 cm; second: 80:10:10 ethyl acetate/HOAc/ H_2O, 7 or 9 cm, depending on MW	FID	44
C_{12} alkyl 6-mole ethoxylate; ethoxy distribution	Analabs Anisil HF, 0.25 mm	Two-dimensional TLC: H_2O-saturated methyl ethyl ketone and 50:50:1 benzene/acetone/H_2O	Autoradiography of ^{14}C-labeled products	94
C_{16}–C_{18} alkyl 0- to 12-mole ethoxylates; ethoxy distribution	Silica HPTLC plates	Double development: first with EtOH, then, after drying, with BuOH	Primulin	54
C_{16}–C_{18} alkyl 10-mole ethoxylate; determination of free alcohol and low-mole ethoxylates	Alumina or silica gel G	Methylene chloride for alumina; 95:5 $CHCl_3$/EtOH for silica	Iodine vapor or sulfuric acid charring	34
Oleyl alcohol 8-mole ethoxylate; study of homolog distribution with various media and solvents	Merck Silica Gel 60 and Merck aluminum oxide plates	Various solvent blends: usually alcohols for silica and ethyl acetate, ether, *n*-hexane, and acetone for alumina plates	Modified Dragendorff reagent	36
C_{10}–C_{16} AE: determination of 1- to 12-mole ethoxylates after derivatization with acetic anhydride	Silica Gel G, 0.25 mm, pre-washed in 50:50 $CHCl_3$/EtOH	Programmed multiple development using 70:30, 90:10, and 100:0 $CHCl_3$ (without preservative)/ acetone	Charring with 25% H_2SO_4, 260°C, 30 min	53

C_{12} alkyl 4-mole ethoxylate; identification	Aminopropyl-modified silica	80:20:1 toluene/n-PrOH/25% NH_3	Dragendorff reagent	19
C_{12} alkyl 8-mole ethoxylate; detection in the presence of anionics; single spot	Alumina (90:10 alumina G/Alumina DO)	40:60 ethyl acetate/pyridine	Pinacryptol yellow	7
C_{12} alkyl 8-mole ethoxylate; detection in the presence of anionics; single spot	Silica Gel G	10:10:5:2 PrOH/$CHCl_3$/MeOH/10 M NH_3	Pinacryptol yellow or Dragendorff reagent	7
C_{12}–C_{15} alkyl 12-mole ethoxylates; ethoxy distribution	Haiyang Silica Gel H	24:10:1 ethyl acetate/acetone/H_2O	Iodine vapor	95
C_{16} alkyl 20-mole ethoxylates; determination in the presence of lipids; single spot	Merck Silica Gel 60 HR, 2.5% $MgSO_4$, 0.5 mm, activated at 120°C for 1 hr	100:18:1 $CHCl_3$/MeOH/35% NH_4OH	Spot is removed and measured by oxidation with dichromate and colorimetry	37
C_{10}–C_{12} alkyl 1- to 15-mole ethoxylates; ethoxy distribution	Merck Silica Gel 60 F-254, 0.25 mm, activated at 120°C for 15 min	80:14:10 ethyl acetate/HOAc/H_2O	Eosin; modified Dragendorff reagent; barium chloride/triiodide solution	9
C_{12}–C_{16} alkyl 3- to 14-mole ethoxylates; ethoxy distribution	Silica Gel 60 F254	2-Butanone, saturated with H_2O	Modified Dragendorff reagent	96
Various fatty alcohol 8-mole ethoxylates; ethoxy distribution	Merck Silica Gel G, activated at 120°C for 30 min	2-Butanone, saturated with H_2O	Modified Dragendorff reagent	43
C_{13}–C_{18} alkyl 3- to 11-mole ethoxylates; ethoxy distribution	Silica Gel 60 or Silica Gel 60 F-254, 0.25 mm	2-Butanone or 90:10 EtOH/HOAc or 80:20 BuOH/EtOH or 60:10:20:0.5 BuOH/EtOH/H_2O/HOAc	Acid Blue 158	49
C_{12} alkyl 9-mole ethoxylates; ethoxy distribution; identification and determination in pharmaceuticals	Merck Silica Gel F-254, 0.2 mm	10:1 2-butanone/acetone, water-saturated	$K_2Cr_2O_7$/H_2SO_4; reflectance densitometry, $\lambda = 600$ nm	97

(Continued)

TABLE 8 (Continued)

Compounds studied	Stationary phase	Developing reagent	Visualization	Ref.
C_{12} alkyl 9-mole ethoxylates; incomplete resolution of ethoxy distribution; identification and determination in pharmaceuticals	Merck silica gel	150:20:2 CH_2Cl_2/MeOH/H_2O	Modified Dragendorff reagent	98
C_9–C_{15} alkyl 3- to 8-mole ethoxylates; approximate ethoxy distribution	Merck Silica Gel 60 F_{254}	14:6:1 2-MIBK/PrOH/0.1 M HOAc	Pinacryptol yellow, 366 nm	14
Reversed-phase TLC				
C_8–C_{18} alkyl 6-mole ethoxylates; alkyl chain distribution	RP-8 WF 254 S	88:12 MeOH/H_2O	Modified Dragendorff reagent	96
C_{14} diol 10-mole ethoxylate; determination in mixture with other surfactants; single spot	Whatman KC18F reversed-phase TLC plates	80:20 EtOH/2% aqueous sodium tetraphenylborate solution	Iodine vapor	13
C_9–C_{15} alkyl 3- to 8-mole ethoxylates; separation by average alkyl chain length	Merck RP-18 F_{254}	75:25 EtOH/H_2O	Iodine vapor; pinacryptol yellow, 366 nm	14

TABLE 9 TLC Analysis of Alkylphenol Ethoxylate Nonionic Surfactants

Compounds studied	Stationary phase	Developing reagent	Visualization	Ref.
Normal phase media				
APE, E = 4–16; ethoxy distribution; determination in biodegradation experiments	Silica gel	Quantitative: 40:30:30 ethyl acetate/HOAc/H$_2$O (surfactant is a single spot); qualitative: 70:16:15 ethyl acetate/HOAc/H$_2$O; homologs are separated)	Modified Dragendorff reagent	92,93
OPE, E = 10: ethoxy distribution	Merck Silica Gel 60	13:2:3:10 ethyl acetate/isooctane/HOAc/H$_2$O	Iodine vapor; spots identified by MALDI-MS	3
C$_4$, C$_5$, and C$_9$ alkylphenol 1- to 8-mole ethoxylates, propoxylates, or mixed alkoxylates; alkoxy distribution	Mallinckrodt precoated silicic acid or Brinkmann precoated silica gel	90:5:5 CHCl$_3$/1,4-dioxane/H$_2$O-saturated benzene or 2-butanone, saturated with H$_2$O	Dragendorff reagent	73
C$_4$, C$_5$, and C$_9$ alkylphenol 1- to 8-mole ethoxylates, propoxylates, or mixed alkoxylates; alkoxy distribution	Brinkmann precoated aluminum oxide	42:40.5:17.5 CHCl$_3$/benzene/acetone	Dragendorff reagent	73
APE, E = 3–30; average EO content	Merck Silica Gel F-60	60:10 CHCl$_3$/MeOH	Modified Dragendorff reagent	99
NPE, E = 3–12: preparative isolation by ethoxy chain length	Merck silica gel	80:15:5 ethyl acetate/PrOH/H$_2$O or 50:45:5 benzene/ethyl acetate/MeOH or 60:40:10 hexane/ethyl ether/MeOH or 85:15:3 hexane/CHCl$_3$/MeOH	Iodine vapor or 2,6-dichlorofluorescein or H$_2$SO$_4$	100
NP5, NP6, dodecylphenol 6-mole ethoxylate; ethoxy distribution	Iatroscan Chromarod S II	Double development: 60:40 benzene/ethyl acetate and 80:10:10 ethyl acetate/HOAc/H$_2$O	FID	101

(Continued)

TABLE 9 (*Continued*)

Compounds studied	Stationary phase	Developing reagent	Visualization	Ref.
NPE, E = 5–9; ethoxy distribution	Merck Silica Gel G, 0.2 mm, or circular plates	2-Butanone, saturated with H_2O or 95:5 butanone/H_2O	Iodine vapor or 1% iodine in MeOH	46–48
NP9; ethoxy distribution	Merck Silica Gel G, 0.3 mm	50:50 2-butanone/H_2O; double development	Sulfuric acid charring	102
OPE; ethoxy distribution	Analabs Anasil GF or HF, 0.25 mm	Two-dimensional TLC: 2:1 benzene/acetone and 50:50:1 benzene/acetone/H_2O	Autoradiography of ^{14}C-labeled products	103
Dodecylphenol ethoxylate; ethoxy distribution up to about E = 13	Merck Aluminum Oxide 60	45.9:54.1 CH_3CN/CCl_4	Modified Dragendorff reagent	104
NPE, E = 8, and tributylphenol, E = 8; study of homolog separation with various media and solvents	Merck Silica Gel 60 and Merck aluminum oxide plates	Various solvent blends: usually alcohols for silica, and ethyl acetate, ether, and acetone for alumina plates	Modified Dragendorff reagent	36,105
NPE, E = 4, 5, 6, 8, 9, 11, and 15; study of homolog separation with changes in solvent strength	Merck Aluminum Oxide 60 plates	Various $CHCl_3/CH_3CN$ blends, up to 30% CH_3CN, with higher CH_3CN contents used to separate higher homologs	Reflectance at 275 nm	35
NPE, E = 8; detection in the presence of anionics; single spot	Alumina (90:10 Alumina G/ Alumina DO)	40:60 ethyl acetate/pyridine	Pinacryptol yellow	7
NPE, E = 8; detection in the presence of anionics; single spot	Merck Silica Gel G	10:10:5:2 PrOH/$CHCl_3$/MeOH/ 10 M NH_3	Pinacryptol yellow or Dragendorff reagent	7
OP10; determination in the presence of lipids; single spot	Merck Silica Gel 60 HR, 2.5% $MgSO_4$, 0.5 mm, activated at 120°C for 1 hr	100:18:1 $CHCl_3$/MeOH/35% NH_4OH	Spot is removed and measured by oxidation with dichromate and colorimetry	37
NPE, E = 3 and 8; ethoxy distribution	Merck Silica Gel G	45:2.5:5 ethyl acetate/MeOH/ NH_4OH	Dragendorff reagent	7

NPE, E = 1 to 50: ethoxy distribution	Merck Silica Gel 60 F-254, 0.25 mm, activated at 120°C for 15 min	80:14:10 ethyl acetate/HOAc/H_2O	Eosin; modified Dragendorff reagent; barium chloride/triiodide solution: H_3PO_4	9
C_9 and C_{12} alkylphenol 4- to 23-mole ethoxylates (n-C_9 and tributyl C_{12}); ethoxy distribution; PEG determination	Merck Silica Gel G, activated at 120°C for 30 min	2-Butanone, saturated with H_2O	Modified Dragendorff reagent	41
NPE, E = 7 to 20; ethoxy distribution	Merck Silica Gel 60 or Silica Gel 60 F-254, 0.25 mm	2-Butanone or 90:10 EtOH/HOAc or 80:20 BuOH/EtOH or 60:10:20:0.5 BuOH/EtOH/H_2O/HOAc	Acid Blue 158	49
OPE and NPE, E = 4 to 150: approximate ethoxy distribution	Merck Silica Gel 60 F_{254}	14:6:1 2-MIBK/PrOH/0.1 M HOAc	Iodine vapor; fluorescence quenching; UV reflectance	14
NPE, E = 9: isolation of homologs up to E = 22	Woelm alumina (acid grade), 1.0 mm	2-Butanone	Modified Dragendorff reagent	106

Reversed-phase media

NP5, NP10: determination in mixtures with other surfactants	Whatman KC18F reversed-phase TLC plates	80:20 EtOH/2% aqueous sodium tetraphenylborate solution	Iodine vapor	13
NPE, E = 4 to 30: ethoxy distribution; higher Rf with higher ethoxylation	Merck cellulose impregnated with 1.6% or 5% silicone (30K or 93K molecular weight)	1.26:1 MeOH/H_2O	Iodine vapor	56
C_8, C_9, and C_{12} APE, E = 1 to 48; ethoxy distribution; higher Rf with higher ethoxylation	Whatman chromatography paper grade 1 impregnated with cetyl alcohol	70:20:10 H_2O/EtOH/HCl	Draggendorff reagent	136
OPE and NPE, E = 5 to 20: single spot for each; determination in cleaning products and resolution from each other	Merck RP-8 F254, activated at 120°C	90:9:1 CH_3CN/H_2O/HOAc	UV adsorption at 230 and 278 nm	55

silicone-impregnated cellulose did give separation, with higher Rf values associated with higher degree of ethoxylation. Methanol/water was used as mobile phase in all cases (56).

(c) Detection. APE is a fluorescence quencher because of its UV absorbance and thus can be readily detected without a spray reagent on TLC plates containing a fluorescing agent. With the proper apparatus, it may also be directly detected by UV reflectance measurement of its adsorbance maxima at 228 and 278 nm (14). An unknown should show adsorbance at both wavelengths, otherwise it is not APE (55). While molybdophosphoric acid is a nonspecific visualizer, it gives more intense color with APE than with other surfactants. Use of phosphoric acid results in red spots from APE, which give green fluorescence (9).

3.   Alkanolamides and Their Ethoxylates

As summarized in Table 10, fatty acid ethanolamide and ethoxylated fatty ethanolamide may be quantitatively analyzed by TLC on silica, using iodine vapor for visualization of the ethanolamide and Dragendorff reagent for the ethoxylated compounds, which are separated by chain length (57,58). Either silica gel impregnated with dodecanol or alumina plates allow differentiation of mono- and dialkanolamides (7). Reversed-phase systems will differentiate alkanolamides according to alkyl chain length (14). A 3-plate system allows determination of isopropanolamine and amide, mono-and diethanolamine and amide, and polydiethanolamide in commercial fatty acid alkanolamide surfactant (59). The Iatroscan technique (flame ionization detection) can also be used with multiple development to analyze these compounds (60).

Most alkanolamides can be detected by UV reflectance, with an absorbance band around 240 nm (14). If phosphoric acid is used as the visualizer, amide ethoxylates show as brown spots which give white fluorescence under 366 nm light (9).

Procedure: Analysis of a Fatty Acid Alkanolamide (59)

Sample and various reference standards are spotted on three plates of Merck Silica Gel 60. Plates 1 and 2 are developed with butanone saturated with water; plate 3 is developed with 45:45:8 CHCl$_3$/MeOH/conc. NH$_4$OH. After development, all plates are dried. Plate 1 is not heated, plate 2 is heated at 150°C for 30 min, and plate 3 is heated at 80°C until free of ammonia odor. All three are exposed for 2 min to a chlorine gas atmosphere, stored in a vacuum desiccator until free of excess chlorine, and sprayed with the visualizer, a mixture of 4,4′-methylenebis-(*N,N*-dimethylaniline) (2.5 g dissolved in 10 mL acetic acid, diluted with 50 mL water, and filtered), KI solution (100 mL of a 5% solution), and ninhydrin solution (15 mL of a 0.5% solution in 90:10 H$_2$O/HOAc) The three visualizer components are mixed just before use. The compounds appear as dark blue spots on a light blue background. Plate 1 serves to separate monoethanolamide and isopropanolamide from all other components; plate 2 resolves amides from amines, which remain at the origin; plate 3 differentiates all compounds except diethanolamide and polydiethanolamide.

Procedure: TLC Analysis of Fatty Acid Diethanolamides using FID (60)

The stationary phase is an Iatroscan Chromarod S II, coated with silica gel. Double development is performed, the first 30 min with 60:40 2-butanone (water saturated)/*n*-hexane, followed by scanning of the last 75 mm of the 100 mm length. Development is then repeated with 45:45:8 CHCl$_3$/MeOH/14 M NH$_3$, and the entire rod is scanned with the FID. Commercial products are separated into peaks corresponding to diethanolamides, esters, and, if present, *bis*(2-hydroxyethyl)piperazine.

TABLE 10 TLC Analysis of Alkanolamide and Alkanolamide Ethoxylate Nonionic Surfactants

Compounds studied	Stationary phase	Developing reagent	Visualization	Ref.
Fatty acid mono- and diethanolamides; detection in shampoo	Silica Gel 60, Activity III, 0.25 mm	90:10 acetone/THF	Pinacryptol yellow fluorescence, 366 nm excitation, and modified Dragendorff reagent	137
Fatty acid diethanolamide; determination of main components and impurities, including diethanolpiperazine	Iatroscan Chromarod SII silica gel–coated rod	Double development: first, 60:40 2-butanone (H$_2$O saturated)/n-hexane and 30 min development, and the last 75 mm of the 100 mm length is scanned; development is repeated with 45:45:8 CHCl$_3$/MeOH/25% NH$_3$, and the entire rod is scanned.	FID	60
C$_{10}$–C$_{18}$ fatty acid alkanolamide; determination of isopropanolamine and amide, mono-and diethanolamine and amide, polydiethanolamide	Silica Gel 60	Plates 1 and 2: butanone saturated with H$_2$O; Plate 3: 45:45:8 CHCl$_3$/MeOH/con NH$_4$OH. After development, all plates are dried. Plate 1 is not heated, Plate 2 is heated at 150°C for 30 min, and Plate 3 is heated at 80°C until free of ammonia odor. All three are exposed for 2 min to a Cl$_2$ atmosphere, held in a vacuum desiccator until free of excess Cl$_2$, and sprayed with the visualizer.	Mixture of 4,4′-methylene-*bis*-(N,N-dimethylaniline) (50 mL of a 5% solution in 50:10 H$_2$O/HOAc), KI (100 mL of a 5% solution in 90:10 H$_2$O/HOAc), and ninhydrin (15 mL of a 0.5% solution in 90:10 H$_2$O/HOAc); mix just before use	59
Fatty acid ethanolamide and impurities; ethanolamine and fatty acid ethanolamide ester; identification	Silica Gel G, 0.3 mm, activated at 110°C for 1.5 hr	95:5 CHCl$_3$/96% EtOH	Iodine vapor	57
Fatty acid monoethanolamide ethylene oxide adducts; identification of main components	95:5 Silica Gel G/ammonium sulfate, 0.3 mm	90:10 CHCl$_3$/MeOH	Modified Dragendorff reagent or iodine vapor	58

(Continued)

TABLE 10 (Continued)

Compounds studied	Stationary phase	Developing reagent	Visualization	Ref.
Fatty acid monoethanolamide ethylene oxide adducts; ethoxy distribution	Silica Gel G, activated at 110°C	2-Butanone, H_2O-saturated	Iodine vapor or modified Dragendorff reagent	58
C_{12} acid monoethanolamide 2-mole ethoxylate; detection	Aminopropyl-modified silica	80:20:1 toluene/n-PrOH/25% NH_4OH	Dragendorff reagent	19
Fatty acid mono- and diethanolamide; detection in the presence of anionics	Alumina (90:10 Alumina G/Alumina DO)	BuOH, NH_3-saturated, or 40:60 ethyl acetate/pyridine	Pinacryptol yellow	7
Fatty acid mono- and diethanolamide; detection in presence of other surfactants	Silica Gel G	BuOH, NH_3-saturated, or 10:10:5:2 PrOH/$CHCl_3$/MeOH/10 M NH_3, or 45:2.5:5 ethyl acetate/MeOH/NH_3	Pinacryptol yellow	7
Fatty acid amide 10-mole ethoxylate; detection in presence of anionics	Merck Silica Gel 60 F-254, 0.25 mm, activated at 120°C for 15 min	80:10:8:2 $CHCl_3$/MeOH/HOAc/H_2O	Eosin; modified Dragendorff reagent; barium chloride/triiodide solution	9
Fatty acid amide 10-mole ethoxylate; ethoxy distribution	Merck Silica Gel 60 F-254, 0.25 mm, activated at 120°C for 15 min	80:14:10 ethyl acetate/HOAc/H_2O or 95:5 acetone/HOAc	Eosin; modified Dragendorff reagent; barium chloride/triiodide solution	9
Coco acid mono- and diethanolamide, oleic acid amide 6-mole ethoxylate, coco acid ethanolamide 5-mole ethoxylate; detection in presence of other nonionics	Merck Silica Gel GF-254 impregnated with oxalic acid	90:10 $CHCl_3$/MeOH	Modified Dragendorff reagent; barium chloride/triiodide solution	38
Oleic acid alkanolamide; detection	Merck Silica Gel 60 or Silica Gel 60 F-254, 0.25 mm	2-Butanone or 90:10 EtOH/HOAc or 80:20 BuOH/EtOH or 60:10:20:0.5 BuOH/EtOH/H_2O/HOAc	Acid Blue 158	49

4. Esters and Their Ethoxylates

TLC is used for detection of esters and ethoxylates, especially in food, as well as for differentiating esters according to both degree and position of esterification. In general, TLC on silica can readily separate the products into the categories mono-, di-, and higher esters. Separation of positional isomers is more difficult. No single method can be expected to resolve, for example, the 255 possible ester combinations of a single fatty acid and the eight hydroxyl groups of sucrose (61,62). Published approaches to the analysis are given in Table 11.

A visualizing reagent specific for vicinal hydroxyl groups, as in monoglyceride esters, is anisidine/potassium periodate (50). Plates are sprayed first with 0.1% KIO_4 in water, then with a solution of 2.8 g anisidine in 80 mL 96% ethanol, 70 mL water, 30 mL acetone, and 1.5 mL 1 M HCl (63). Copper phosphate immersion gives excellent results for oleic acid esters (64,65). Naphthoresorcinol solution is a selective reagent for saccharides and their esters (50). Plates are sprayed with a mixture of 10 mL 85% H_3PO_4 and 100 mL 0.2% ethanolic naphthoresorcinol, then heated at 100°C for 5 min (66).

(a) Sucrose esters. TLC on silica is suitable for separation of sucrose monoesters by position of acyl substitution (67). Normal phase TLC is not suitable for separating according to alkyl chain length; reversed-phase HPLC is more often used. Normal phase TLC can separate by number of alkyl chains, since the polarity of the compound changes dramatically with increased substitution.

Diphenylamine/aniline reagent gives a pink or purple color to a sucrose monoester where the fructose moiety is unesterified, or a blue color when the glucose moiety is unesterified (67). Another clue to the identity of individual isomers can be gained by reacting with invertase. This enzyme only catalyzes the hydrolysis of sucrose if the fructose moiety is unsubstituted. Thus, TLC before and after invertase treatment will indicate whether the acyl group is on the fructose or glucose portion of the molecule (67).

(b) Ethoxylated esters. Ethoxylated sorbitol esters of the type used as emulsifiers in foods can be determined by TLC analysis of a chloroform extract of the food. After alumina cleanup of the extract, it is analyzed on silica plates, with visualization by modified Dragendorff reagent. The free PEG present as a byproduct in the surfactant is also seen on the plate (68). Alternatively, the TLC method may be used for qualitative confirmation of identity, after quantitative analysis by spectrophotometry (69). An additional confirmation of identity is performed by reacting the sample with alkali to saponify esters, then subjecting it again to TLC analysis. If nonionic surfactants are still seen, but with different Rf values than observed before saponification, this indicates the original presence of ethoxylated esters (51).

TLC is suitable for showing the ethoxy distribution of an ethoxylated glycerol trioleate, as well as allowing estimation of free oleic acid (64,65).

5. Alkyl Polyglycosides

Applications are summarized in Table 12. APG can be separated according to alkyl chain length by reversed-phase TLC, with the longest alkyl chains exhibiting the shortest Rf values (Fig. 1). C_{18} media are usually applied to this analysis, but C_8 or aminopropyl-modified silica may also be used (70). TLC on silica gel is suitable for differentiating alkylmono-and alkyldiglucosides from each other and from butylmonoglucoside reaction intermediate (71).

TABLE 11 TLC Analysis of Polyhydroxy Esters and Their Ethoxylates

Compounds studied	Stationary phase	Developing reagent	Visualization	Ref.
Esters of fructose, sucrose, sorbitol, glucose; separation by number of acyl groups	Merck Silica Gel 60 F-254, 0.25 mm	81:9:8:2 CHCl$_3$/MeOH/HOAc/H$_2$O	Anisaldehyde/sulfuric acid	107
Sucrose palmitate esters with 1 to 8 acyl groups; separation by number and position of acyl groups; determination of glycerides and free fatty acids	Silica gel containing calcium sulfate or starch binder, activated at 110°C for 1 hr	2:1:1 toluene/ethyl acetate/95% EtOH for resolution of mono-, di-, and triesters; 75:25:1 petroleum ether/ethyl ether/HOAc for resolution of tetra-through octaesters. For qualitative analysis, two-dimensional chromatography of all esters: (1) 2:1:1 toluene/ethyl acetate/95% EtOH; (2) 80:20 CHCl$_3$/MeOH	1 g urea, 4.5 mL 85% H$_3$PO$_4$, 48 mL water-saturated n-butanol, hold at 110°C for 30 min	62,108
Sucrose C$_{12}$–C$_{16}$ esters: determination of sucrose and of mono- and diesters	Mallinckrodt silica gel, 1 mm	25:75 EtOH/benzene or 2-PrOH/benzene	0.2% anthrone in 70:30 H$_2$SO$_4$/water	109,110
Sucrose C$_{16}$–C$_{18}$ esters, separation by number of acyl groups (1 to 3); separation of isomers	Merck Silica Gel G, 0.25 mm, activated at 110°C for 1 hr; plates must be protected from humid air	80:10:8:2 CHCl$_3$/MeOH/HOAc/H$_2$O for separating mono-, di-, and higher esters; 82:17:1 CHCl$_3$/MeOH/HOAc with multiple development for resolution of isomers	50% H$_2$SO$_4$; char 20 min at 150°C	61
Sucrose C$_{16}$ and C$_{18}$ esters; determination of sucrose and of mono- and diesters	Silica gel, Merck No. 7729	10:5:5 toluene/ethyl acetate/95% EtOH	Dichlorofluorescein	111
Sucrose C$_{16}$ and C$_{18}$ esters; separation by type, number (1, 2, 3, and higher), and position of acyl groups	Merck Silica Gel 60 F-254 and Silica Gel 60 HPTLC plates	70:20:2:2 CHCl$_3$/MeOH/HOAc/H$_2$O	50% H$_2$SO$_4$; char at 120°C	112

Application	Sorbent	Solvent system	Detection	Ref.
Sucrose C_{16} and C_{18} esters: separation by number (1, 2, and higher) of acyl groups	Silica gel, Merck type 2357	2-Butanone saturated with water	Sulfuric acid/methanol and charring at 120°C	113
Sucrose C_{14}–C_{18} monoesters; separation of isomers	Whatman K5 silica gel, dipped in 0.2 M potassium phosphate, dried 1 hr at 85°C	Three consecutive developments with 80:20:5 ethyl acetate/pyridine/water, drying between treatments	4 g diphenylamine, 4 mL aniline, 30 mL H_3PO_4, 200 mL acetone; dry, then heat 4 min at 110°C	67
Esters and ethoxylated esters; general method for identification of food emulsifiers	Silica gel 60 F-254, Merck no. 5715	60:40:1 pet ether/ethyl ether/HOAc or 65:25:4 CHCl₃/MeOH/H_2O	Various	50
20-mole ethoxylates of sorbitan mono- and triesters; determination in food	Merck Silica Gel G, 0.25 mm, activated at 120°C for 1 hr; reactivated 10 min	Diethyl ether to separate from fats; then dry, reactivate, and develop with 40:30:30 ethyl acetate/HOAc/H_2O	Modified Dragendorff reagent	68
Sorbitan monostearate 20-mole ethoxylate; determination in food	Silica gel 60, 0.25 mm	CH_2Cl_2, to separate from fats, then 55:20:15:4 CH_2Cl_2/MeOH/acetone/water	Modified Dragendorff reagent followed by sodium nitrite	69
Sorbitan monolaurate, palmitate, stearate 20-mole ethoxylates; PEG determination	Silica Gel G, activated at 110°C for 30 min	50:40:20 MeOH/CHCl₃/HOAc	Modified Dragendorff reagent or starch-iodine	114
Sorbitan monolaurate, palmitate, stearate, oleate 20-mole ethoxylates; confirmation of identity after saponification	Silica Gel 60, precoated plate, Merck No. 5721	68:12:12:3 CHCl₃/MeOH/HOAc/H_2O or 10:10:5:2 n-PrOH/CHCl₃/MeOH/NH_3 or 90:20:13:5:2:5 CHCl₃/acetone/isoamyl acetate/isoamyl alcohol/propionic acid/H_2O	Ammonium cobaltithiocyanate	51
Castor oil 30- and 60-mole ethoxylates, sorbitan monolaurate 20-mole ethoxylate, glycerin cocoate 7-mole ethoxylate; detection	Aminopropyl-modified silica	80:20:1 toluene/n-PrOH/25% NH_3	Dragendorff reagent	19

TABLE 12 TLC Analysis of Alkyl Polyglycoside Surfactants

Compounds studied	Stationary phase	Developing reagent	Visualization	Ref.
Alkyl mono- and diglucosides; separation from each other and from butylmonoglucoside; determination in formulations	Silica gel HPTLC plate	60:40:5 $CHCl_3/EtOH/H_2O$	Lead (IV) acetate/2,7-dichlorofluorescein, dried 3 min at 100°C; fluorescence with 366 nm excitation	71
C_8–C_{14} alkyl glucosides; Rf in inverse order of degree of polymerization; determination in formulations	N60 silica gel	80:20 $CHCl_3/MeOH$	Thymol and H_2SO_4	115,116
C_8–C_{14} alkyl glucosides; separation from each other; Rf in inverse order of alkyl chain length; determination in formulations	C_{18}-modified silica gel	90:10 $MeOH/H_2O$	Thymol and H_2SO_4	115,116
C_8–C_{18} alkyl monoglucosides; separation by alkyl chain length	Macherey-Nagel Nano-SIL C18-100 (on glass) or Alugram RP 18 W/UV$_{254}$ (on aluminum)	90:10 $MeOH/H_2O$	Sulfuric acid charring	72,117
C_1–C_{12} alkyl monoglucosides and C_{10} and C_{12} alkyldiglucosides; separation by alkyl chain length and number of glucose units; determination in formulations	Merck RP$_8$, with fluorescent indicator	30:70 CH_3CN/H_2O; automated multiple development in 8 steps	Thymol/H_2SO_4; nonylglucoside internal standard	118
C_1–C_{12} alkyl monoglucosides and C_{10} and C_{12} alkyldiglucosides; separation by number of glucose units; determination in formulations	Merck silica HPTLC plates without fluorescent indicator	62:27:8.5:2.5 THF/$CHCl_3$/n-butanol/HOAc; automated multiple development 16-step gradient also used	Thymol/H_2SO_4	118

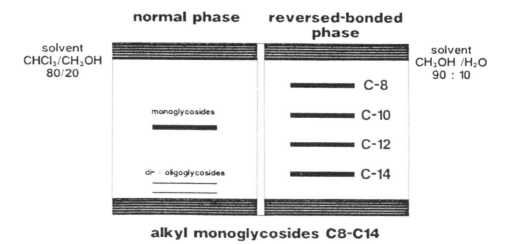

alkyl monoglycosides C8-C14

FIG. 1 TLC analysis of an alkylpolyglycoside mixture on normal phase and reversed-phase media. (Reprinted with permission from Ref. 115.)

Visualization offers a number of choices. Sulfuric acid charring (72), iodine vapor or fluorescence after treatment with lead (IV) acetate/2,7-dichlorofluorescein have all been reported (71). Sulfuric acid and iodine detection have the advantage that other surfactants present will also be visible. Lead acetate/dichlorofluorescein colors monoglucosides more intensely than polyglucosides, so individual calibrations must be performed (71). Also, there is some interference from ethoxylates (70). Vanadium (V)/sulfuric acid gives light spots on a yellowish-green background; anisaldehyde/sulfuric acid gives grey-brown spots only on silica plates; anthrone gives light yellow spots only on reversed-phase plates; lead (IV) acetate/fuchsin gives pink spots; thymol and naphthoresorcinol are generally useful for carbohydrates (70).

Alkylpolyglucosides have Rf values under reversed-phase conditions similiar to those of ethoxylated nonionics. They can be differentiated by visualizing with modified Dragendorff reagent, which colors ethoxylates but not alkylpolyglycosides (72).

6. EO/PO Copolymers

On silica plates EO/PO block copolymers, like other ethoxylates, give a series of spots or streaks corresponding to oligomer distribution, with smaller Rf values for higher EO content (7). Alumina plates provide more information, permitting the MW distribution of the PO chain to be determined, along with an estimate of the EO content (73).

The separation according to degree of ethoxylation can be suppressed by addition of oxalic acid to the silica plates. Under such conditions, it is possible, for example, to distinguish a propylene oxide-initiated EO/PO block copolymer from an ethylene diamine-initiated product (38). Each compound is seen as a single spot. See Table 13 for more information.

7. Other Nonionic Surfactants

TLC analyses of acid ethoxylate, amine ethoxylate, and amine oxide surfactants are summarized in Tables 14–16. Tertiary amine may be separated from amine oxide by TLC on silica or alumina. The amine oxide remains near the start line, while the amine migrates up the plate. Modified Dragendorff reagent is suitable for detection, as are a number of other visualizers.

TABLE 13 TLC Analysis of EO/PO Copolymer Nonionic Surfactants

Compounds studied	Stationary phase	Developing reagent	Visualization	Ref.
M = 1100, %E = 10; detection in the presence of anionics; single spot	Alumina (90:10 Alumina G/Alumina DO)	40:60 ethyl acetate/pyridine	Pinacryptol yellow	7
M = 1100, %E = 10; detection in the presence of anionics; single spot	Silica Gel G	Ammonia-saturated butanol or 10:10:5:2 PrOH/CHCl$_3$/MeOH/10 M NH$_3$	Pinacryptol yellow	7
M = 2900, %E = 40 and ethylene diamine-initiated M = 5500, %E = 40; detection in presence of other nonionics	Merck Silica Gel GF-254, impregnated with oxalic acid	90:10 CHCl$_3$/MeOH	Modified Dragendorff reagent; barium chloride/triiodide solution	38
M = 2500, %E = 20; detection in the presence of other nonionics	Merck Silica Gel 60 F-254, 0.25 mm, activated at 120°C for 15 min	80:10:8:2 CHCl$_3$/MeOH/HOAc/H$_2$O or 80:14:10 ethyl acetate/HOAc/H$_2$O; 95:5 acetone/HOAc	Modified Dragendorff reagent; barium chloride/triiodide solution	9
M = 3530, %E = 8 and M = 1940, %E = 13; in general, for %E < 20; ethoxy content	Brinkmann aluminum oxide	42:40.5:17.5 CHCl$_3$/benzene/acetone	Dragendorff reagent	73
M = 2650, %E = 30; detection	Merck RP-18 F$_{254}$	75:25 EtOH/H$_2$O or 8:2 EtOH/aqueous 2% tetraphenylborate	Iodine vapor or fluorescence quenching	14

TABLE 14 TLC Analysis of Ethoxylated Acid Nonionic Surfactants

Compounds studied	Stationary phase	Developing reagent	Visualization	Ref.
PEG 400 dioleate; homolog separation	Macherey-Nagel silica gel F$_{254}$	Ethyl acetate	Iodine vapor	5
C$_{12}$ acid ethoxylate; detection in presence of other nonionics	Merck Silica Gel GF-254, impregnated with oxalic acid	90:10 CHCl$_3$/MeOH	Modified Dragendorff reagent; barium chloride/triiodide solution	38
Fatty acid 8-mole ethoxylates; ethoxy distribution	Merck Silica Gel G, activated at 120°C for 30 min	2-Butanone, saturated with water	Modified Dragendorff reagent	41
Stearic acid 8-mole ethoxylate: study of homolog distribution with various media and solvents	Merck Silica Gel 60 and Merck aluminum oxide plates	Various solvent blends: usually alcohols for silica and ethyl acetate, ether, n-hexane, and acetone for alumina plates	Modified Dragendorff reagent	36
Mono- and diheptanoate of tetraethylene glycol; determination of mono- and diester, heptanoic acid, and tetraethylene glycol	Iatroscan MK5 silica	80:20:1 ethyl ether/pentane/HOAc	FID	119

TABLE 15 TLC Analysis of Ethoxylated Amine Nonionic Surfactants

Compounds studied	Stationary phase	Developing reagent	Visualization	Ref.
Tallow amine 15-mole ethoxylate; detection in presence of other nonionics	Merck Silica Gel GF-254, impregnated with oxalic acid	90:10 CHCl$_3$/MeOH	Modified Dragendorff reagent; barium chloride/triiodide solution	38
Fatty amine ethoxylates; ethoxy distribution	Merck Silica Gel G, activated at 120°C for 30 min	2-Butanone, saturated with water containing 2.5% NH$_3$	Modified Dragendorff reagent	41
Fatty amine low and 12-mole ethoxylates; detection	Merck Silica Gel 60 and silica gel 60 F-254, 0.25 mm	90:10 ethanol/acetic acid or 80:20 butanol/ethanol or 60:10:20:0.5 butanol/ethanol/water/acetic acid	Acid Blue 158	49

TABLE 16 TLC Analysis of Amine Oxide Surfactants

Compounds studied	Stationary phase	Developing reagent	Visualization	Ref.
Dimethylcocoamines and amine oxides; quantitative determination	Silica Gel G	97:3 CHCl$_3$ (saturated with NH$_4$OH)/MeOH	Dichlorofluorescein	120
Determination of amine impurities in amine oxides	Merck Silica Gel 60	90:10:1 CHCl$_3$/EtOH/NH$_3$	Bromcresol green for identification, then molybdophosphoric acid for quantification	121
Detection of various amines and amine oxides	Merck Silica Gel G and Silica Gel HF	6:1 CHCl$_3$/MeOH	Study of various reagents	122
C$_8$–C$_{10}$ tertiary amines and their oxides; separation from each other	Aluminum Oxide G	98:2 isobutyl acetate/HOAc	Ammonium cobaltithiocyanate	123
Cocoamidopropyldimethylamine oxide and hexadecyldimethylamine oxide; detection in the presence of cationics	Silica Gel 60 or aminopropyl-modified silica	70:20:10 2-BuOH/H$_2$O/HOAc (for silica) or 70:20:10 n-PrOH/2-butanone/25% NH$_3$ (for aminopropyl-modified silica)	Primulin; fluorescence at 365 nm	19

IV. CATIONIC AND AMPHOTERIC SURFACTANTS

A. Quaternary Amines

Thin-layer chromatography is suitable for qualitative and quantitative analysis of cationics, including both quaternaries and simpler amines. Most published applications use ordinary silica gel plates. Visualization is accomplished variously with sulfuric acid charring (74), modified Dragendorff reagent (75), iodine vapor (13), Acid Blue 158 (49), or ninhydrin reagent, ninhydrin being selective for nitrogen compounds (76). Primulin and its derivatives, which are general reagents for the alkyl functionality, may also be used (19).

A number of published methods for analysis of cationic surfactants are listed in Table 17. Generally, quaternary amines can be separated according to class of compound (i.e., by whether they are alkyltrimethylammonium, dialkyldimethylammonium, trialkylmethylammonium, N-alkyl-N-benzyldimethylammonium or N-alkylpyridinium compounds) by normal phase chromatography. Reversed-phase techniques allow resolution by alkyl chain length, with higher chain length compounds being more strongly retained. Often, TLC is used simply to separate cationics and amphoterics from other classes of compounds (77). This is easily performed on silica plates with an ammoniacal mobile phase (78). Däuble describes the separation of a mixture of cationics and of nitrogen-containing nonionics (49). The procedures of Michelson (75) and of Osburn (74) are directed toward confirming the identity of cationic surfactants in environmental samples after isolation and cleanup steps.

On basic alumina, the Rf values of quats are dependent on the identity of the corresponding anion. This is not observed on acidic alumina (79).

Procedure: TLC Measurement of Cationics in Environmental Samples (75)

Dragendorff visualization reagent: Solution A consists of 0.34 g $BiONO_3 \cdot H_2O$ dissolved in 5 mL acetic acid and diluted to 20 mL with water, added to 13 g potassium iodide in 100 mL water. Forty milliliters acetic acid are added and the solution is diluted to 200 mL. Solution A must be prepared fresh daily. Solution B consists only of 290 g $BaCl_2 \cdot H_2O$ in 1 L water. The spray reagent is composed of 2 parts A, 1 part B, 18 parts acetic acid, and 9 parts water. The spray reagent must be prepared fresh daily. (Dragendorff reagent without $BaCl_2$ is commercially available.)

Apply spots of 1–10 µg cationic surfactant to a Silica Gel 60 plate, applying standards at the same time. Condition plates for 20 min at 37% relative humidity and develop with ethyl acetate/acetic acid/water, 4:3:3. Dry at room temperature until the odor of acetic acid is gone. Spray with Dragendorff reagent. Dry in subdued light with a stream of air until the odor of acetic acid is again absent. Cover with glass, and measure on a densitometer by transmission at 525 nm.

B. Amphoteric Surfactants

TLC methods are applied to amphoteric surfactants mainly to identify individual compounds. Only the analysis of phosphatides has been studied extensively. Methods are summarized in Tables 18–19.

1. Lecithin

While the phosphatides in commercial lecithin may be analyzed by normal phase HPLC, the individual peaks represent mixtures of compounds with different hydrocarbon chains. For the most precise identification of peaks, two-dimensional TLC must be used (80).

TABLE 17 TLC Analysis of Cationic Surfactants

Compounds studied	Stationary phase	Developing reagent	Visualization	Ref.
C_{12}–C_{18} alkyltrimethylammonium chlorides; detection in presence of nonsurfactant quaternary amines	Woelm basic alumina	60:30:10 CHCl$_3$/MeOH/NH$_3$	Modified Dragendorff reagent or 2% iodine in MeOH, followed by saturated NaNO$_2$	79
C_{12}–C_{18} alkyltrimethylammonium chlorides; detection in presence of nonsurfactant quaternary amines	Woelm acidic alumina	80:20 or 85:15 CHCl$_3$/MeOH	Modified Dragendorff reagent or 2% iodine in MeOH, followed by saturated NaNO$_2$	79
Cetylpyridinium chloride, cetyltrimethylammonium bromide, distearyldimethylammonium chloride; separation from each other and from primary, secondary, and tertiary amines	Silica gel	Two-dimensional chromatography: (1) 60:15:15 ethyl acetate/MeOH/0.88 M aqueous NH$_3$; (2) 40:40:20:10 n-PrOH/CHCl$_3$/MeOH/0.88 M aqueous NH$_3$	Not given	124
C_{12}–C_{18} alkyltrimethyl-, dialkyldimethyl-, and trialkylmethylamines; separation from each other	Merck Silica Gel G	90:10 acetone/14 M NH$_3$	2′,7′-Dichlorofluorescein, UV light, or charring with K$_2$Cr$_2$O$_7$/H$_2$SO$_4$	125
C_{18} alkyltrimethylamines; determination in C_{18} dialkyldimethylamine	Schleicher and Schuell No. 589 paper	700:1050:1435 MeOH/acetone/H$_2$O	Bromphenol blue, 80 mg in 4 mL 0.1 M NaOH, 10 g Na$_2$CO$_3$, diluted to 100 mL with H$_2$O: 607 nm	126
C_{10}–C_{18} alkylpyridinium, C_{12}–C_{16} alkylisoquinolinium, C_8–C_{18} N-alkyl-N-benzyldimethylammonium chlorides: resolution of each type according to alkyl chain length	Chromatography paper, saturated in lauryl alcohol	50:50 EtOH/1 M aqueous HCl: mixture saturated with lauryl alcohol; 37°C	0.1 g Bi(NO$_3$)$_3$, 1 g KI, per 100 mL H$_2$O	127

(Continued)

TABLE 17 (*Continued*)

Compounds studied	Stationary phase	Developing reagent	Visualization	Ref.
C_{12}–C_{20} alkyltrimethylammonium, C_{16} alkyldimethylethylammonium, N-lauryl-N-methylmorpholinium, and trimethyl(ethyllauramide) ammonium chlorides; resolution of each type according to alkyl chain length	Chromatography paper, saturated in lauryl alcohol	55:45 EtOH/1 M aqueous HCl; mixture saturated with lauryl alcohol; 37°C	0.1 g Bi(NO$_3$)$_3$, 1 g KI, per 100 mL H$_2$O	127
C_{12}–C_{18} alkyltrimethylammonium, C_{12}–C_{18} N-alkyl-N-benzyldimethylammonium, and C_{12}–C_{18} alkylpyridinium chlorides; resolution of each type according to alkyl chain length	Whatman No. 1 paper for chromatography	35:60:5 96% EtOH/H$_2$O/12 M HCl	5:10:40 0.2% rhodamine BS/0.2% Tinopal WG/con NH$_4$OH; UV light	128
C_{12}–C_{18} alkyltrimethylammonium. C_{12}–C_{18} N-alkyl-N-benzyldimethylammonium, and C_{12}–C_{18} alkylpyridinium halides; resolution of each type according to alkyl chain length	Merck Silanized Silica Gel 60, activated at 110°C for 1 hr	65:35:20 MeOH/25% NaOAc · 3H$_2$O (aqueous)/acetone	Dip the oven-dried plate in 1 part 4% KI/2% I$_2$, 19 parts 50:50 MeOH/H$_2$O. Scan in reflection mode at 400 nm	129
C_{16} alkyltrimethylammonium and C_{16} alkylpyridinium chlorides; separation from each other and from tertiary amines	Whatman K6F silica gel	8:1:0.75 CH$_2$Cl$_2$/MeOH/HOAc	Iodine vapor; 405 nm	13

Application	Stationary phase	Mobile phase	Detection	Ref.
Class separation of cationics from anionics and nonionics; cationics remain near the origin	Whatman KC18F reversed-phase plates, activated at 110°C for 2 hr	75:25 EtOH/H$_2$O	Iodine vapor; 405 nm	13
C$_{16}$ alkyltrimethylammonium bromide, C$_{18}$ alkanoylaminophenyl trimethylammonium methylsulfate, and C$_{16}$ alkylpyridinium chloride from pharmaceutical preparations	Silica Gel G, 0.3 mm, activated at 105°C for 30 min	90:10 acetone/NH$_3$	Spectrophotometry after removal of spots	130
Distearyldimethylammonium chloride; determination in environmental matrices	Silica Gel G	75:23:3 CHCl$_3$/MeOH/H$_2$O	Charring with 4.5 M H$_2$SO$_4$	74
Quaternary amines in environmental materials; general method	Merck Silica Gel 60, 0.25 mm	4:3:3 ethyl acetate/HOAc/H$_2$O	Dragendorff reagent	75
Cationic surfactants; separation from anionic and nonionic surfactants	Merck Silica Gel 60, 0.25 mm	3:1 CHCl$_3$/CH$_3$OH to separate nonionics, then 3:1 or 4:1 CH$_3$OH/2M NH$_3$ to separate anionics; cationics remain near origin	Not specified	140
Cationic surfactants; separation from other cationic material, as disulfine blue complexes	Schleicher & Schüll G1600 polyamide	CHCl$_3$ or 9:1 CHCl$_3$/CH$_3$OH	Not required	140

TABLE 18 TLC Analysis of Lecithin

Compounds studied	Stationary phase	Developing reagent	Visualization	Ref.
Separation of phosphatides in lecithin	Silica Gel G, 0.25 mm	Two-dimensional TLC: (1) 130:60:8 CHCl$_3$/MeOH/7 M NH$_4$OH; (2) 170:25:25:6 CHCl$_3$/MeOH/HOAc/H$_2$O	Spots are removed and tested for total P content	131,132
Separation of phosphatides in lecithin	Merck precoated silica gel plates	Two-dimensional TLC: (1) 130:60:8 CHCl$_3$/MeOH/7 M NH$_4$OH; (2) 170:25:25:6 CHCl$_3$/MeOH/HOAc/H$_2$O	CuSO$_4$ solution in 8% H$_3$PO$_4$, then char at 160°C for 20 min; densitometry at 400 nm	133
Phosphatides; separation by class	Analtech preadsorbent silica type HLF plate, prewashed with developer	50:6:6 EtOH/CHCl$_3$/conc. NH$_3$	Charring at 160°C with 10% CuSO$_4$ and 8% H$_3$PO$_4$	134
Phosphatides in lecithin; separation by class	Silica Gel G, 0.2 mm	39:18:4.5 CHCl$_3$/MeOH/40% methylamine or 30:12:6:9:3 CHCl$_3$/acetone/MeOH/HOAc/H$_2$O	I$_2$ vapor or molybdic acid reagent	80
Isolation of phosphatidylcholine from egg lecithin	Silica Gel G, 0.5 mm	25:10:0.75 CHCl$_3$/MeOH/H$_2$O	2,7-Dichlorofluorescein	135
Lecithin in food extract	Silica Gel 60 F-254	65:25:4 CHCl$_3$/MeOH/H$_2$O	Molybdic acid reagent	50
Phosphatides in lecithin	Merck Silica Gel 60, impregnated with H$_3$PO$_4$	65:25:4.3 CHCl$_3$/MeOH/0.2 M acetate buffer, pH 4	Molybdic acid reagent	82

TABLE 19 TLC Analysis of Other Amphoteric Surfactants

Compounds studied	Stationary phase	Developing reagent	Visualization	Ref.
N-Lauryl-β-iminodipropionic acid, C_{12}–C_{14} N-alkyl-dimethylaminoacetic acid, lauroylamidopropyl-N,N-dimethylaminoacetic acid, dicarboxylated imidazoline derivatives based on lauric acid and capric acid; detection	Silica Gel G containing 10% $(NH_4)_2SO_4$	80:19:1 $CHCl_3$/MeOH/0.05 M H_2SO_4	Dichlorofluorescein; 365 nm	84
N-Lauryl-β-iminodipropionic acid, C_{12}–C_{14} N-alkyl-dimethylaminoacetic acid, lauroylamidopropyl-N,N-dimethylaminoacetic acid, dicarboxylated imidazoline derivatives based on lauric acid and capric acid; not separated from each other	Silica Gel GF-254	10:10:5:2 n-PrOH/$CHCl_3$/MeOH/10 M NH_3	Iodine vapor	84
3-(Hexadecyl)dimethylammonium)- propan-1-sulfobetaine; separation from nonionic surfactants; sulfobetaines remain at the origin	Silica Gel G impregnated with oxalic acid	90:10 $CHCl_3$/MeOH	Modified Dragendorff reagent	85
3-Cocoamidopropyl-N,N-dimethylbetaine, cocoamido betaine, N-[3-(cocoamido)-propyl]-N-(2-hydroxy-3-sulfopropyl)-N,N-dimethyl betaine, lauriminodipropionic acid; separation by alkyl chain length	Merck RP-18 F_{254}	80:20 EtOH/aqueous 2% tetraphenylborate	Fluorescence quenching	14
3-Cocoamidopropyl-N,N-dimethylbetaine, cocoamido betaine, N-[3-(cocoamido)-propyl]-N-(2-hydroxy-3-sulfopropyl)-N,N-dimethyl betaine, lauriminodipropionic acid; identification	Merck Silica Gel 60 F_{254}	8:1:0.75 CH_2Cl_2/MeOH/HOAc	Pinacryptol yellow; iodine vapor	14
C_{12}–C_{18} alkylamidobetaine characterization; separation of product from imidazoline intermediate, secondary and tertiary amide intermediates, and from N-hydroxy-ethylethylenediamine	Silica activated 1 hr at 110 °C	20:20:10:8 n-PrOH/$CHCl_3$/MeOH/28% NH_3	Iodine vapor	86

Lecithins are most readily identified by a spray reagent specific for phosphorus compounds, usually molybdic acid reagent (81,82). Synthetic phosphatides may be analyzed by TLC techniques similar to those applied to lecithin (83). The plates should be stored in a controlled atmosphere, since the separation is sensitive to humidity.

Procedure: TLC Analysis of Phosphatides in Vegetable Lecithin (82)

Dittmer-Lester molybdic acid reagent: Dissolve 40 g MoO_3 in 1 L 12.5 M sulfuric acid solution by heating. Separately, dissolve 1.78 g molybenum in 500 mL of the first solution. Mix equal volumes of the two solutions. Stable indefinitely. (A similar solution is commercially available from Sigma, St. Louis, MO).

Immersion solution: Mix the reagent solution with water and ethanol, 1:2:3. Prepare fresh daily.

Merck Silica Gel 60 HPTLC plates, 5 × 5 cm, are dipped in a solution of 0.5% phosphoric acid in 70:30 methanol/water and dried for 20 min at 120°C. Phospholipid mixture (2.5 mg dissolved in 10 mL chloroform) is spotted on the plates, which are developed in an H-chamber with chloroform/methanol/0.2 M pH 4 acetate buffer, 65:25:4.3. Detection is performed by dipping the developed plate in immersion solution and reading the spots by densitometry at 720 nm after 30 min. Colors become more intense for 3 hr. Even very slight variations in the water content of the mobile phase affect the separation.

2.   Other Amphoteric Surfactants

Reversed-phase separation systems will separate homologs of amphoteric surfactants, with compounds having longer alkyl chains exhibiting smaller Rf values (14). Amphoteric surfactants tend to accompany carboxyl-type anionic surfactants in group separations. Amphoterics generally have similar Rf values on silica. They may be distinguished from each other somewhat by using acidic developers (14,84).

In group separations, sulfobetaines travel with the nonionic surfactants. They can be separated from these by further TLC on silica, developing with 9:1 chloroform/methanol (85).

Pinacryptol yellow is not suitable for visualization of most amphoterics. Use of iodine vapor results in intense yellow-brown spots (84).

TLC was used to study the reaction pathway for formation of an alkylamidobetaine. Normal phase chromatography allowed resolution of the main product from the imidazoline intermediate, the secondary and tertiary amide intermediates, and from *N*-hydroxyethylethylenediamine starting material (86).

REFERENCES

1.   Simunic, S., Separation and characterization of surfactants by HPTLC, *J. Liq. Chromatogr. Relat. Technol.*, 1996, *19*, 1139–1149.
2.   Simunic, S., Water as a detecting agent in TLC of surfactants, *J. Liq. Chromatogr. Relat. Technol.*, 1999, *22*, 1247–1256.
3.   Cumme, G. A., E. Blume, R. Bublitz, H. Hoppe, A. Horn, Detergents of the polyoxyethylene type: comparison of TLC, reversed-phase chromatography, and MALDI MS, *J. Chromatogr. A*, 1997, *791*, 245–253.
4.   Buschmann, N., A. Kruse, In-situ TLC-IR and TLC-SIMS for the analysis of surfactants, *Comun. Jorn. Com. Esp. Deterg.*, 1993, *24*, 457–468.
5.   Read, H., Surfactant analysis using HPTLC and the Iatroscan, in R. E. Kaiser, ed., *Proc. Int. Symp. Instrum. High Perform. Thin-Layer Chromatogr.*, 3rd, 1985, 157–171.

6. Yonese, C., T. Shishido, T. Kaneko, K. Maruyama, Anionic surfactants by TLC. I: Mixtures of sodium LAS, sodium dodecyl sulfate and sodium dodecanesulfonate, *J. Am. Oil Chem. Soc.*, 1982, *59*, 112–116.

7. Bey, K., TLC analysis of surfactants (in German), *Fette, Seifen, Anstrichm.*, 1965, *67*, 217–221.

8. Hellmann, H., Adsorbents as ion exchange and separation media: Al_2O_3 and SiO_2 in chromatographic analysis of surfactants (in German), *Fresenius' Z. Anal. Chem.*, 1989, *334*, 126–132.

9. Brüschweiler, H., V. Sieber, H. Weishaupt, TLC analysis of anionic and nonionic surfactants (in German), *Tenside*, 1980, *17*, 126–129.

10. Mutter, M., Alkanesulfonates with ion exchangers (in German), *Tenside*, 1968, *5*, 138–140.

11. Kupfer, W., J. Jainz, H. Kelker, Alkyl sulfonates (in German), *Tenside, Surfactants, Deterg.*, 1969, *6*, 15–21.

12. Takeshita, R., N. Jinnai, H. Yoshida, Detection of sodium alkanesulfonates and alkylbenzene-sulfonates by polyamide TLC, *J. Chromatogr.*, 1976, *123*, 301–307.

13. Armstrong, D. W., G. Y. Stine, Anionic, cationic and nonionic surfactants by TLC, *J. Liq. Chromatogr.*, 1983, *6*, 23–33.

14. Henrich, L. H., Separation and identification of surfactants in commercial cleaners, *J. Planar Chromatogr. Mod. TLC*, 1992, *5*, 103–117.

15. Mutter, M., K. W. Han, TLC of organic sulfonates, *Chromatographia*, 1969, *2*, 172–175.

16. Martin, T. T., M. C. Allen, Charring with sulfur trioxide for the improved visualization and quantitation of TLC, *J. Am. Oil Chem. Soc.*, 1971, *48*, 752–757.

17. Püschel, F., D. Prescher, Paper chromatography of sulfonates and alkanesulfates (in German), *J. Chromatogr.*, 1968, *32*, 337–345.

18. Akinci, S., K. C. Güven, Anionic detergents by TLC, *Pharm. Turc.*, 1992, *34*, 121–126.

19. Hohm, G., TLC methods for surfactant-containing formulations, *Seife, Öle, Fette, Wachse*, 1990, *116*, 273–280.

20. McEvoy, J., W. Giger, LAS in sewage sludge by high resolution GC-MS, *Environ. Sci. Technol.*, 1986, *20*, 376–383.

21. Eganhouse, R. P., E. C. Ruth, I. R. Kaplan, Determination of long-chain alkylbenzenes in environmental samples by argentation TLC/high-resolution GC and GC-MS, *Anal. Chem.*, 1983, *55*, 2120–2126.

22. Czichocki, G., D. Vollhardt, H. Seibt, Preparation and characterization of interfacially-chemically pure sodium dodecylsulfate (in German), *Tenside*, 1981, *18*, 320–327.

23. Roberts, D. W., C. S. Fairclough, J. P. Conroy, Kinetics of the thermal decomposition of alkyl hydrogen sulfates, *J. Am. Oil Chem. Soc.*, 1986, *63*, 799–803.

24. Köhler, M., B. Chalupka, TLC of anionic surfactants (in German), *Fette, Seifen, Anstrichm.*, 1982, *84*, 208–211.

25. Allen, M. C., T. T. Martin, Alkene and hydroxyalkane sulfonates by TLC, *J. Am. Oil Chem. Soc.*, 1971, *48*, 790–793.

26. Puschmann, H., Olefin sulfonates (in German), *Fette, Seifen, Anstrichm.*, 1973, *75*, 434–437.

27. Sandvik, E. I., W. W. Gale, M. O. Denekas, Petroleum sulfonates, *Soc. Pet. Eng. J.*, 1977, *17*, 184–192.

28. König, H., TLC detection of orthophosphoric acid esters of fatty alcohols and AE (in German), *Fresenius' Z. Anal. Chem.*, 1968, *235*, 255–261.

29. Kunkel, E., Carboxymethylated ethoxylates (in German), *Tenside*, 1980, *17*, 10–12.

30. Gerhardt, W., G. Czichocki, H. R. Holzbauer, C. Martens, B. Weiland, Characterization of ether carboxylates in reaction mixtures (in German), *Tenside, Surfactants, Deterg.*, 1992, *29*, 285–288.

31. Yang, J., L. Zhang, X. Li, Ethylene oxide oligomer distribution in ethoxylated alkylphenol carboxymethyl ether salts by TLC, *J. Am. Oil Chem. Soc.*, 1994, *71*, 109–111.

32. Gattavecchia, E., D. Tonelli, G. Bertocchi, Separation and determination of saturated fatty acids by reversed-phase HPTLC, *J. Chromatogr.*, 1983, *260*, 517–521.

33. Daradics, L., J. Pálinkás, Synthesis and analysis of fatty acid sarcosides, *Tenside, Surfactants, Deterg.*, 1994, *31*, 308–313.

34. Pollerberg, J., Free fatty alcohols in their ethoxylation products (in German), *Fette, Seifen, Anstrichm.*, 1966, *68*, 561–562.

35. Cserháti, T., Separation of NPE oligomers according to the length of the ethylene oxide chain, *J. Planar Chromatogr. Mod. TLC*, 1993, *6*, 70–73.

36. Cserháti, T., Solvent strength and selectivity in TLC separation of ethylene oxide oligomers, *J. Chromatogr. Sci.*, 1993, *31*, 220–224.

37. Whitmore, D. A., K. P. Wheeler, Separation and assay of nonionic detergents, *Biochem. Biophys. Methods*, 1980, *2*, 133–138.

38. König, H., Nonionic surfactants by TLC, *Fresenius' Z. Anal. Chem.*, 1970, *251*, 167–171.

39. König, H., E. Walldorf, Skin cleansers and shampoos based on synthetic surfactants (in German), *Fresenius' Z. Anal. Chem.*, 1979, *299*, 1–18.

40. Buschmann, N., Solid-phase extraction separation of polyethylene glycols and ethoxylates, *Comun. Jorn. Com. Esp. Deterg.*, 1994, *25*, 333–336.

41. Bürger, K., TLC determination of the molecular weight distribution and the degree of ethoxylation of polyoxyethylene compounds (in German), *Fresenius' Z. Anal. Chem.*, 1963, *196*, 259–268.

42. Hodda, A. E., Detection and significance of glycols in drug screening, *J. Chromatogr.*, 1976, *124*, 424–425.

43. Bosdorf, V., T. Bluhm, H. Krüssman, TLC determination of adsorbed nonionic surfactants on fabrics, *Melliand Textilber.*, 1994, *75*, 311–312.

44. Sato, T., Y. Saito, I. Anazawa, Polyoxyethylene oligomer distribution of nonionic surfactants, *J. Am. Oil Chem. Soc.*, 1988, *65*, 996–999.

45. Saito, Y., T. Sato, I. Anazawa, Poly(oxyethylene) oligomer distribution in a nonionic surfactant by means of TLC, *Bull. Chem. Soc. Jpn.*, 1989, *62*, 3709–3710.

46. Favretto, L., G. Pertoldi Marletta, L. Favretto Gabrielli, Molecular weight distribution of ethylene glycol oligomers by TLC and their photometric evaluation, *J. Chromatogr.*, 1970, *46*, 255–260.

47. Stancher, B., L. F. Gabrielli, L. Favretto, Polyoxyethylene nonionic surfactants by coupling GC and TLC, *J. Chromatogr.*, 1975, *111*, 459–462.

48. Konishi, K., S. Yamaguchi, Molecular weight distribution of polyoxyethylene-type nonionic surfactants by circular TLC, *Anal. Chem.*, 1966, *38*, 1755–1757.

49. Däuble, M., Identification of surfactants on TLC plates (in German), *Tenside*, 1981, *18*, 7–12.

50. Dieffenbacher, A., U. Bracco, Analytical techniques in food emulsifiers, *J. Am. Oil Chem. Soc.*, 1978, *55*, 642–646.

51. Kato, H., Y. Nagai, K. Yamamoto, Y. Sakabe, Polysorbates in foods by colorimetry with confirmation by IR spectrophotometry, TLC, and GC, *J. Assoc. Off. Anal. Chem.*, 1989, *72*, 27–29.

52. McCoy, R. N., A. B. Bullock, Oxyethylene distribution in primary AE, *J. Am. Oil Chem. Soc.*, 1969, *46*, 289–295.

53. Fischesser, G. J., M. D. Seymour, AE mixtures by programmed multiple development TLC, *J. Chromatogr.*, 1977, *135*, 165–172.

54. Stan, H., T. Heberer, P. Billian, AE in drinking water (in German), *Vom Wasser*, 1998, *90*, 93–105.

55. Bürgi, C., T. Otz, APE (in German), *Tenside, Surfactants, Deterg.*, 1995, *32*, 22–24.

56. Cserháti, T., Z. Illés, Effect of nature of support and impregnating agent on lipophilicity determination for nonionic surfactants by reversed-phase TLC, *Chromatographia*, 1991, *31*, 152–156.

57. Mutter, M., G. W. van Galen, P. W. Hendrikse, Fatty acid monoethanolamide foam stabilizer (in German), *Tenside*, 1968, *5*, 33–36.

58. Mutter, M., G. W. van Galen, P. W. Hendrikse, Fatty acid monoethanolamide-ethylene oxide condensate nonionic surfactant (in German), *Tenside*, 1968, *5*, 36–39

59. Arens, M., H. König, M. Teupel, German Standard Methods for analysis of fats, fatty products, surfactants, and related materials: analysis of organic surface active compounds X (in German), *Fette, Seifen, Anstrichm.*, 1982, *84*, 105–111.

60. Zeman, I., Chromatographic separation of surfactants: fatty acid diethanolamides, *Abh. Akad. Wiss. DDR, Abt. Math., Naturwiss., Tech.*, 1986, 603–608.

61. Wachs, W., K. Gerhardt, Separation of sucrose-fatty acid monoesters by TLC and determination of the capillary active properties of groups of the positional isomers (in German), *Tenside*, 1965, *2*, 6–10.

62. Weiss, T. J., M. Brown, H. J. Zeringue, Jr., R. O. Feuge, Sucrose esters of palmitic acid, *J. Am. Oil Chem. Soc.*, 1971, *48*, 145–148.

63. Halvarson, H., O. Qvist, Monoglyceride content of fats and oils, *J. Am. Oil Chem. Soc.*, 1974, *51*, 162–165.

64. Rischer, M., I. Behr, E. Wolf-Heuss, J. Engel, Characterization of polyoxyethylene glycerol trioleate by TLC, *J. Planar Chromatogr.*, 1995, *8*, 382–387.

65. Hensel, A., M. Rischer, D. Di Stefano, I. Behr, E. Wolf-Heuss, Chromatographic characterization of nonionic surfactant polyoxyethylene glycerol trioleate, *Pharm. Acta Helv.*, 1997, *72*, 185–189.

66. Krebs, K. G., D. Heusser, H. Wimmer, Spray reagents, in E. Stahl, ed., *Thin Layer Chromatography*, 2nd ed., Springer Verlag, New York, 1969.

67. Torres, M. C., M. A. Dean, F. W. Wagner, Chromatographic separations of sucrose monostearate structural isomers, *J. Chromatogr.*, 1990, *522*, 245–253.

68. Murphy, J. M., C. C. Scott, Polyoxyethylene emulsifiers in foods, *Analyst*, 1969, *94*, 481–483.

69. Daniels, D. H., C. R. Warner, S. Selim, Determination of polysorbate 60 in salad dressings by colorimetric and TLC techniques, *J. Assoc. Off. Anal. Chem.*, 1982, *65*, 162–165.

70. Buschmann, N., F. Hülskötter, A. Kruse, S. Wodarczak, Alkylpolyglucosides, *Fett/Lipid*, 1996, *98*, 399–402.

71. Spilker, R., B. Menzebach, U. Schneider, I. Venn, Alkylpolyglucosides (in German), *Tenside, Surfactants, Deterg.*, 1996, *33*, 21–25.

72. Buschmann, N., L. Merschel, S. Wodarczak, Alkyl polyglucosides. Part II: Qualitative determination using TLC and identification by means of in-situ secondary ion MS, *Tenside Surf. Det.*, 1996, *33*, 16–20.

73. Ludwig, F. J., Ethylene oxide and propylene oxide adducts of alkylphenols or alcohols by NMR, GC, and TLC, *Anal. Chem.*, 1968, *40*, 1620–1627.

74. Osburn, Q. W., Cationic fabric softener in waters and wastes, *J. Am. Oil Chem. Soc.*, 1982, *59*, 453–457.

75. Michelson, E. R., Quaternary ammonium bases in water and wastewater by TLC, *Tenside*, 1978, *15*, 169–175.

76. Matissek, R., TLC for the identification of surfactants in shampoos, bubble bath preparations, and soaps (in German), *Tenside*, 1982, *19*, 57–66.

77. Hellmann, H., Extractible cationic surfactants in sewage sludge (in German), *Fresenius' Z. Anal. Chem.*, 1983, *315*, 425–429.

78. Hellmann, H., Application of modified Dragendorff reagent to the determination of cationic surfactants in the aquatic environment (in German), *Fresenius' Z. Anal. Chem.*, 1989, *335*, 265–271.

79. McLean, W. F. H., K. Jewers, TLC of quaternary ammonium salts on alumina layers, *J. Chromatogr.*, 1972, *74*, 297–302.

80. Rivnay, B., Combined analysis of phospholipids by HPLC and TLC. Analysis of phospholipid classes in commercial soybean lecithin, *J. Chromatogr.*, 1984, *294*, 303–315.

81. Dittmer, J. C., R. L. Lester, Spray for the detection of phospholipids on thin-layer chromatograms, *J. Lipid Res.*, 1964, *5*, 126–127.

82. Lendrath, G., A. Nasner, L. Kraus, Behavior of vegetable phospholipids in TLC. Optimization of mobile phase, detection and direct evaluation, *J. Chromatogr.*, 1990, *502*, 385–392.

83. Ranny, M., J. Silhanek, A. Bradikova, R. Seifert, M. Zbirovsky, Synthetic phosphoglycerides. Part 2: Preparative and quantitative TLC (in German), *Tenside*, 1977, *14*, 246–250.

84. König, H., Amphoteric surfactants (in German), *Fresenius' Z. Anal. Chem.*, 1970, *251*, 359–368.

85. König, H., Sulfobetaines (in German), *Fresenius' Z. Anal. Chem.*, 1972, *259*, 191–194.

86. Li, Z., Z. Zhang, Amphoteric imidazoline surfactants, *Tenside, Surfactants, Deterg.*, 1994, *31*, 128–132.

87. Yamaoka, K., K. Nakajima, H. Moriyama, Y. Saito, T. Sato, Sodium dodecyl sulfate in hydrophilic ointments by TLC with flame ionization detection, *J. Pharm. Sci.*, 1986, *75*, 606–607.

88. Puschmann, H., Analysis of ether sulfates on silanized silica gel (in German), *Chem., Phys. Chem. Anwendungstech. Grenzflaechenaktiven Stoffe, Ber. Int. Kongr.*, 6th, 1972, Vol. I, 397–406, Carl Hanser Verlag, Munich.

89. Köhler, M., E. Keck, G. Jaumann, Sulfonated esters (in German), *Fett. Wiss. Technol.*, 1988, *90*, 241–243.

90. König, H., Sulfosuccinic acid half-esters (in German), *Fresenius' Z. Anal. Chem.*, 1971, *254*, 198–209.

91. Malanowska, M., H. Piekacz, E. Kiss, Anionic residues of surface active compounds of the type of salts of sulfosuccinic acid and succinic acid monoesters on the surface of containers in contact with food (in Polish), *Rocz. Panstw. Zakl. Hig.*, 1987, *38*, 539–543.

92. Patterson, S. J., C. C. Scott, K. B. E. Tucker, Nonionic detergent degradation. I: TLC and foaming properties of AE, *J. Am. Oil Chem. Soc.*, 1967, *44*, 407–412.

93. Patterson, S. J., C. C. Scott, K. B. E. Tucker, Nonionic detergent degradation. II: TLC and foaming properties of APE, *J. Am. Oil Chem. Soc.*, 1968, *45*, 528–532.

94. Tanaka, F. S., R. G. Wien, B. L. Hoffer, Photosensitized degradation of a homogeneous nonionic surfactant: hexaethoxylated 2,6,8-trimethyl-4-nonanol, *J. Agric. Food Chem.*, 1986, *34*, 547–551.

95. Zhang, J., L. Lin, Y. Wang, G. Xu, Y. Zhao, TLC comparison of different ethoxylation products in the manufacture of nonionic surfactants, *Chromatographia*, 1998, *47*, 98–100

96. Rothbächer, H., A. Korn, G. Mayer, Nonionic surfactants in cleaning agents for automobile production (in German), *Tenside, Surfactants, Deterg.*, 1993, *30*, 165–173.

97. Krumholz, B., K. Wenz, Polidocanol in a suppository formulation, *J. Planar Chromatogr. Mod. TLC*, 1991, *4*, 370–372.

98. Brahm, R., W. Ziegenbalg, B. Renger, TLC-spectrodensitometric determination of polidocanol (Thesit®) in pharmaceutical preparations, *J. Planar Chromatogr. Mod. TLC*, 1990, *3*, 77–78.

99. Hellmann, H., Monitoring the decomposition of APE in detergents by IR spectrometry/TLC. Part I: Analysis (in German), *Fresenius' Z. Anal. Chem.*, 1985, *321*, 159–162.

100. Cortesi, N., E. Moretti, E. Fedeli, Molecular weight distribution of APE by HPLC (in Italian), *Riv. Ital. Sostanze Grasse*, 1980, *57*, 141–144.

101. Zeman, I., Quantification in HPLC of ethoxylated surfactants, *J. Chromatogr.*, 1990, *509*, 201–212.

102. Hayano, S., T. Nihongi, T. Asahara, TLC analysis of NPE, *Tenside*, 1968, *5*, 80–82.

103. Tanaka, F. S., R. G. Wien, R. G. Zaylskie, Photolytic degradation of a homogeneous Triton X nonionic surfactant: nonaethoxylated *p*-(1,1,3,3,-tetramethylbutyl)phenol, *J. Agric. Food Chem.*, 1991, *39*, 2046–2052.

104. Cserháti, T., A. Somogyi, TLC separation of some tributylphenyl ethylene oxide oligomers according to the length of the ethylene oxide chain, *J. Chromatogr.*, 1988, *446*, 17–22.

105. Szilagyi, A., E. Forgacs, T. Cserhati, Separation of nonionic surfactants according to the length of the ethylene oxide chain on alumina layers, *Toxicol. Environ. Chem.*, 1998, *65*, 95–102.

106. Skelly, N. E., W. B. Crummett, Eighteen to twenty-two mole ethoxymers in nine mole ethylene oxide adduct of *p*-nonylphenol, *J. Chromatogr.*, 1966, *21*, 257–260.

107. Seino, H., T. Uchibori, T. Nishitani, S. Inamasu, Enzymatic synthesis of carbohydrate esters of fatty acid. Esterification of sucrose, glucose, fructose, and sorbitol, *J. Am. Oil Chem. Soc.*, 1984, *61*, 1761–1765.

108. Zeringue, H. J., R. O. Feuge, Purification of sucrose esters by selective adsorption, *J. Am. Oil Chem. Soc.*, 1976, *53*, 567–571.

109. Mima, H., N. Kitamori, Adsorption chromatography of sucrose palmitates, *J. Am. Oil Chem. Soc.*, 1962, *39*, 546.

110. Mima, H., and N. Kitamori, Sucrose esters of long chain fatty acids, *J. Am. Oil Chem. Soc.*, 1964, *41*, 198–200.

111. Gee, M., TLC of sucrose esters and mixtures of raffinose and sucrose, *J. Chromatogr.*, 1962, *9*, 278–282.

112. Jaspers, M. E. A. P., F. F. van Leeuwen, H. J. W. Nieuwenhuis, G. M. Vianen, HPLC separation of sucrose fatty acid esters, *J. Am. Oil Chem. Soc.*, 1987, *64*, 1020–1025.

113. Gupta, R. K., K. James, F. J. Smith, Sucrose mono- and diesters prepared from triglycerides containing C_{12}–C_{18} fatty acids, *J. Am. Oil Chem. Soc.*, 1983, *60*, 1908–1913.

114. Thakkar, A. L., P. B. Kuehn, N. A. Hall, TLC procedure for following partial purification of polysorbates, *Amer. J. Pharm.*, 1967, *139*, 122–125.

115. Waldhoff, H., J. Scherler, M. Schmitt, Alkyl polyglycosides—analysis of raw material; determination in products and environmental matrices, *World Surfactants Congr., 4th*, 1996, *1*, 507–518.

116. Waldhoff, H., J. Scherler, M. Schmitt, J. R. Varvil, Analysis of alkyl polyglycosides and determination in consumer products and environmental matrices, in K. Hill, W. von Rybinski, G. Stoll, eds., *Alkyl polyglycosides. Technology: Properties and Applications*, VCH, Weinheim, 1997.

117. Buschmann, N., S. Wodarczak, Carbohydrate surfactants, *Comun. Jorn. Com. Esp. Deterg.*, 1994, *25*, 203–207.

118. Klaffke, H. S., T. Neubert, L. W. Kroh, Alkyl polyglucosides using LC methods, *Tenside Surfactants, Deterg.*, 1998, *35*, 108–111.

119. Baudrand, V., Z. Mouloungui, A. Gaset, Synthesis and analysis of tetraethylene glycol mono- and diheptanoate (in French), *Colloq. Inst. Natl. Rech. Agron.*, 1995, *71*, 183–187.

120. Pelka, J. R., L. D. Metcalfe, Tertiary amines in long chain amine oxides by TLC, *Anal. Chem*, 1965, *37*, 603–604.

121. Comité Européen des Agents de Surface et leurs Intermédiaires Organiques, Determination of free amine in amine oxides by TLC, CESIO/AIS Analytical Method 7-91, available from European Chemical Industry Council, Brussels.

122. Ross, J. H., TLC of amine oxidation products, *Anal. Chem.*, 1970, *42*, 564–570.

123. Lane, E. S., TLC of long-chain tertiary amines and related compounds, *J. Chromatogr.*, 1965, *18*, 426–430.

124. Gabriel, D. M., Analysis of cosmetics and toiletries, *J. Soc. Cosmet. Chem.*, 1974, *25*, 33–48.

125. Mangold, H. K., R. Kammereck, Industrial aliphatic lipids, *J. Am. Oil Chem. Soc.*, 1962, *39*, 201–206.

126. Moyer, F., J. Nelson, A. Milun, Paper chromatographic determination of trimethyloctadecyl quaternary ammonium chloride in dimethyldioctadecyl quaternary ammonium chloride, *J. Am. Oil Chem. Soc.*, 1960, *37*, 463–466.

127. Cross, J. T., Identification and determination of cationic surface-active agents with sodium tetraphenylboron, *Analyst*, 1965, *90*, 315–324.

128. Holness, H., W. R. Stone, Separation of quaternary halides by paper chromatography, *Analyst*, 1958, *83*, 71–75.

129. Paesen, J., I. Quintens, G. Thoithi, E. Roets, G. Reybrouck, J. Hoogmartens, Quaternary ammonium antiseptics using thin layer densitometry, *J. Chromatogr. A*, 1994, *677*, 377–384.

130. El-Khateeb, S., Z. H. Mohamed, L. El-Sayed, Cationic surfactants, *Tenside, Surfactants, Deterg.*, 1988, 25, 236–239.
131. American Oil Chemists' Society, Official and Tentative Methods, Phospholipids in lecithin concentrates by TLC, Recommended Practice Ja 7-86. Champaign, IL.
132. Erdahl, W. L., A. Stolyhwo, O. S. Privett, Soybean lecithin by thin layer and analytical LC, *J. Am. Oil Chem. Soc.*, 1973, 50, 513–515.
133. Yamamoto, H., K. Nakamura, M. Nakatani, H. Terada, Phospholipids on two-dimensional TLC plates by imaging densitometry, *J. Chromatogr.*, 1991, 543, 201–210.
134. Dugan, E. A., Phospholipids by one-dimensional TLC, *LC Mag.*, 1985, 3, 126–128.
135. Christie, W. W., M. L. Hunter, HPLC analysis of the products of phospholipase: a hydrolysis of phosphatidylcholine, *J. Chromatogr.*, 1984, 294, 489–493.
136. Selden, G. L., J. H. Benedict, Identification of polyoxyalkylene-type nonionic surfactants by paper chromatography, *J. Am. Oil Chem. Soc.*, 1968, 45, 652–655.
137. Matissek, R., E. Hieke, W. Baltes, TLC separation and identification of synthetic surfactants is shampoos and bubble bath preparations (in German), *Fresenius' Z. Anal. Chem.*, 1980, 300, 403–406.
138. Matissek, R., Combination of TLC and IR spectrometry for the identification of ethoxylated and nonethoxylated alkyl sulfate surfactants (in German), *Parfüm. Kosmet.*, 1983, 64, 59–64.
139. Hellmann, H., Nonionic surfactants in water and wastewater by X-ray fluorescence and IR spectrometry (in German), *Fresenius' Z. Anal. Chem.*, 1979, 297, 102–106.
140. Hellmann, H., Cationic surfactants in water and wastewater (in German), *Fresenius' Z. Anal. Chem.*, 1982, 310, 224–229.

10
Supercritical Fluid Chromatography

Supercritical fluid chromatography (SFC) is a technique with considerable power for analysis of low molecular weight polymers, such as those in the surfactant range. Because of the high diffusion coefficients found in the supercritical state and the high solvent strength of many supercritical phases, SFC gives higher resolution and higher speed than HPLC for many analyses, while not requiring that the compounds be volatile, as does GC. An important feature of SFC is that it is compatible with the "universal" gas chromatography detector, the flame ionization detector (FID), for a separation that in other respects resembles HPLC. The FID has a linear response over a wide range of concentrations, and thus calibration is much easier than with the common HPLC detectors based on UV absorbance, differential refractive index, or evaporative light scattering.

I. IONIC SURFACTANTS

SFC is rarely applied to ionic surfactants. Ionic materials have poor solubility in CO_2, which is the usual SFC mobile phase. If another supercritical fluid is substituted for CO_2, then the main advantage of SFC is lost, i.e., the ability to use the flame ionization detector. Applications of SFC to ionic surfactant characterization are listed in Table 1.

Analysis of anionics by SFC is only of academic interest, since other techniques are more useful. Soap may be analyzed as the free fatty acids, with elution on a nonpolar packing in order of boiling point (1). Some anionic surfactants, such as phosphate esters, may be analyzed by SFC of the methyl esters (2). As is also true in gas chromatography,

TABLE 1 Analysis of Ionic and Amphoteric Surfactants by Supercritical Fluid Chromatography

Surfactant	Column	Mobile phase	Detection	Ref.
C_8–C_{24} Fatty acids; separation by alkyl chain length	Ethylvinylbenzene/divinyl-benzene copolymer, 5 µm particles, 750 µm × 15 cm steel	CO_2, pressure programmed, 150°C	FID	1
AE phosphate esters, deriva-tized with diazomethane	SE-54, 0.2 µm, 100 µm × 20 m	CO_2, pressure programmed, 100°C	FID	2
C_{16} mono-, di-, and tri-alkyl quaternary amines, after con-version to tertiary amines	OV-73, 100 µm × 20 m	CO_2, pressure programmed	NP detector	5
Phosphatidyl choline, phosphatidic acid, phos-phatidyl inositol, and phos-phatidyl ethanolamine in soy lecithin	Zorbax Sil, 4.6 × 150 mm, 45°C	78.4:21.6 CO_2/(95:4.95:0.05 MeOH/H_2O/triethylamine)	Evaporative light scattering	32
Phosphatidyl choline (two compounds), phosphatidic acid, phosphatidyl glycerol, and PEG adduct of phos-phatidyl ethanolamine; demonstration of subcritical fluid chromatograhy	Phenomenex Luna Octyl, 4.6 × 250 mm, 70°C	Gradient: A = CO_2; B = 50:50 EtOH/MeOH containing 0.1% tri-fluoroacetic acid; 9% B to 15% B in 14 min, then to 40% in 0.1 min	Evaporative light scattering	33

sulfates must be cleaved before analysis (3). For example, alkyl ethoxysulfates may be reacted with diazomethane, which cleaves the sulfate group and yields the corresponding alkyl ethoxylates as their methyl ethers (4). Quaternary amines may be analyzed by SFC with CO_2 if they are first reacted with phosphoric acid to form the tertiary amines. However, gas chromatography provides better resolution (5).

II. NONIONIC SURFACTANTS

A. Alcohol Ethoxylates

The one niche that SFC fills in the surfactants laboratory is in characterization of AE. SFC using a silicone-coated capillary column with CO_2 mobile phase and flame ionization detection is the most useful of all the chromatographic methods for analysis of AE, including determination of unethoxylated alcohol and determination of molecular weight distribution up to about the 25-mole EO adduct, corresponding to an average degree of ethoxylation of 10–12 (Table 2). Geissler used an empirical equation to correct the response of the FID for the varying ratios of carbon/oxygen in the oligomer series (6), while Silver and Kalinoski used the classical theory of effective carbon number (7).

SFC has also been demonstrated on a packed silica column, but in this case an amine-containing mobile phase is necessary, so that the evaporative light scattering detector must be substituted for the FID (8,9).

Separation of ethoxylates of a single alcohol is simple and quantitative, with results superior to those of GC or HPLC in that derivatization is unnecessary. Ethoxylates of normal and branched-chain alcohols of the same carbon number are readily resolved. Complete resolution of ethoxylates of mixed alcohols is more challenging, because a compound of the form $C_{13}H_{27}O(CH_2CH_2O)_5H$ may elute at the same retention time as the compound $C_{15}H_{31}O(CH_2CH_2O)_4H$. Ordinary ethoxylates seem only to be completely resolvable if no more than two different carbon chain lengths are present, with three carbon chain lengths resolvable only up to about 5 oligomers (4,10). Resolution of ethoxylated alcohol mixtures is improved by derivatization. However, for more complex mixtures where the base alcohol contains several chain lengths, some overlapping of homologs is always observed (7).

B. Alkylphenol Ethoxylates

SFC separation of nonylphenol ethoxylates has been demonstrated on a number of columns. Separation is by ethoxy chain length, with fine structure visible due to ortho/para substitution on the phenyl ring, as well as to isomeric forms of the nonyl chain (11,12). Ultraviolet and FID detection give good results (Table 3).

It has been observed that the appearance of SFC chromatograms of OPE are much better if the surfactant is dissolved in methanol rather than in methylene chloride, perhaps because the higher volatility of methylene chloride is not suited to uniform introduction of sample onto the column (13).

The mass spectrometer can compensate for incomplete separation by the SFC. For example, SFC of an NPE showed resolution only on the basis of ethoxy chain length when viewed with a UV absorbance detector, but the MS detector showed that each peak contained impurities corresponding to a C_7 APE with the same ethoxy distribution (14).

TABLE 2 Analysis of Alcohol Ethoxylate Surfactants by Supercritical Fluid Chromatography

Surfactant	Column	Mobile phase	Detection	Ref.
C_8–C_{16} alkyl 1- to 9-mole ethoxylates; separation by ethoxy chain length; fine structure due to alkyl chain length and branching	Poly(dimethylsiloxane), 100 μm × 20 m, 100°C	CO_2, density programmed	FID	6
C_9–C_{15} alkyl 3- to 8-mole ethoxylates; separation by alkyl and ethoxy chain length	Dionex SB-Methyl 100 or J&W DB-1 (both dimethylpolysiloxane) 50 μm × 10 m, 120°C	CO_2, pressure programmed	FID or MS	18
C_{18} alkyl 12- to 25-mole ethoxylates; ethoxy distribution	Dionex SB-Biphenyl-30 (30% biphenyl; 70% methylpolysiloxane), 50 μm × 10 m, 75°C	CO_2, pressure programmed	FID	19
C_{12}–C_{15} AE, E = 10–12: resolution by alkyl and ethoxy chain length	Dionex SB-Biphenyl-30 (30% biphenyl; 70% methylpolysiloxane), 50 μm × 10 m, programmed 100–200°C	CO_2, density programmed	FID	10
C_{16} alkyl 10- to 40-mole ethoxylates; separation by ethoxy chain length; fine structure visible due to alkyl distribution	Various silica and diol packed HPLC columns	$CO_2/CH_3OH/H_2O/(C_2H_5)_3N$ mixtures and gradients	Evaporative light scattering	8,9

C_{12}–C_{18} alkyl 7- to 12-mole ethoxylates; acetate or trimethylsilyl derivatives; ethoxy distribution up to the 18 ethoxylate	Dionex SB-Biphenyl-30 (30% biphenyl; 70% methylpolysiloxane), 50 μm × 5 m, 125°C	CO_2, density programmed	FID	7
Mixed alkyl 9- to 14-mole ethoxylates; resolution by length and branching of alkyl chain and ethoxy chain length	Dionex SB-Biphenyl-30 (30% biphenyl; 70% methylpolysiloxane), 50 μm × 3 m, 100°C	CO_2, density programmed	FID	20,21
C_{12}–C_{15} alkyl 9-mole ethoxylate, before and after HBr cleavage; separation by alkyl and ethoxy chain length	PS 264 dimethyldiphenylmethylvinylsiloxane, 0.25 μm, 50 μm × 10 m, 140°C	CO_2, density programmed	FID	22
C_{13} alcohol 3-mole ethoxylate; separation by alkyl and ethoxy chain length	ODS-Q3, 0.5 × 800 mm, 260°C	50:50 ethanol/hexane	UV, 200 nm, and MS	14

TABLE 3 Analysis of Alkylphenol Ethoxylate Surfactants by Supercritical Fluid Chromatography

Surfactant	Column	Mobile phase	Detection	Ref.
Octylphenol 10-mole ethoxylate; separation by ethoxy chain length	5% SE-54 crosslinked with azo-t-butane, 1 μm, 100 μm × 30 m	CO_2, pressure programmed, 100°C	MS	23
Octylphenol 10-mole ethoxylate, derivatized with BSTFA, ethoxy distribution	Keystone Scientific Deltabond cyano packed column, 1 × 100 mm	CO_2, pressure programmed 150°C	FID	24
Octylphenol 30-mole ethoxylate, separation by ethoxy chain length	Hamilton PRP-1 poly(styrene/divinylbenzene), 4.6 × 250 mm	CO_2/acetonitrile gradient, 95:5 to 45:55 in 130 min at 135°C and 25 MPa, or CO_2/methanol gradient 95:5 to 64:36 in 130 min at 80°C and 29 MPa	UV, 195 or 280 nm	25
Octylphenol 10-mole ethoxylate; ethoxy distribution	OV-1701, 100 μm × 10 m	CO_2, pressure programmed, 100°C	FID	26
Octylphenol 9-mole ethoxylate; ethoxy distribution	Nomura Develosil ODS-5 C_{18}–bonded silica, 0.7 × 150 mm	Methanol, pressure programmed, 255°C	UV, 215 nm	27
OPE, E = 8 and 16; ethoxy distribution	Macherey-Nagel Nucleosil C_{18}, 2 × 100 mm	CO_2 with MeOH modifier, pressure programmed, 170°C	UV, 278 nm	28
Octylphenol 16-mole ethoxylate; ethoxy distribution	Dionex SB-Biphenyl-30 (30:70 biphenyl/methyl-polysiloxane), 0.25 μm, 50 μm × 10 m	CO_2 with 2-PrOH modifier, pressure programmed, 175°C	UV, 262 nm	28

OPE, E = 1, 3, 5, 7, 10, 15; separation by ethoxy chain length	Dionex SB-Biphenyl-30 (30:70 biphenyl/methyl-polysiloxane). 0.25 μm, 50 μm × 5 m	CO_2, pressure programmed, isothermal at 150°C for lower and 100°C for higher ethoxy-lates; reverse T programming to 88°C used for highest degree of ethoxylation	FID	13
Nonylphenol 4-mole ethoxylate; separation by ethoxy chain length	ODS-Q3, 0.5 × 800 mm	50:50 EtOH/hexane 260°C	UV, 200 nm, and MS	14
NPE, E = 1–25; separation by ethoxy chain length	Dionex SB-Biphenyl-30 (30:70 biphenyl/methyl-polysiloxane). 0.25 μm, 50 μm × 5 m	CO_2, pressure programmed, isothermal at 100–180°C, higher T for higher degree of ethoxylation	FID	11
NPE, E = 2–21; separation by ethoxy chain length	Dionex SB-Biphenyl-30 (30:70 biphenyl/methyl-polysiloxane). 50 μm × 10 m	CO_2, pressure programmed, isothermal at 130°C	FID	12,29

C.   Other Nonionic Surfactants

SFC analyses of other nonionics are summarized in Table 4. Analysis of sorbitol esters gives clusters of peaks corresponding to the mono-, di-, tri-, tetra-, and pentaesters. The clusters in turn consist of compounds of a single degree of esterification which are incompletely resolved according to the chain length of the alkyl moiety (15). Some work has been performed with more complex mixtures, but so far such studies have stopped short of identifying each peak found in the chromatogram (16).

For separation of EO/PO copolymers, it is necessary to perform a preliminary HPLC separation by PO distribution prior to SFC determination of ethoxy distribution (17). Otherwise, overlap of compounds is too severe for resolution.

TABLE 4 Analysis of Other Nonionic Surfactants by Supercritical Fluid Chromatography

Surfactant	Column	Mobile phase	Detection	Ref.
Mixed EO/PO adduct of a fatty alcohol, MW = 1800: demonstration of separation by alkoxy chain length; fine structure due to alkyl chain length	Dionex SB-Methyl-100 (100% methylpolysiloxane), 0.5 µm, 50 µm × 25 m	CO_2, pressure programmed, 160°C	FID	30
C_{17} acid 5-mole ethoxylate, C_{12} alkyldiethanolamide: separation by alkyl and ethoxy chain length	ODS-Q3, 0.5 × 800 µm	50:50 ethanol/hexane, 260°C	UV, 200 nm, and MS	14
Amine ethoxylates, derivatized with trifluoroacetic anhydride, separation by both alkyl and ethoxy chain length	SE-54, 0.1 µm, 50 µm × 10 m	CO_2, pressure programmed, 130°C	FID	2
Sorbitan C_{12}–C_{18} alkyl esters, resolution by number of alkyl groups; partial resolution by alkyl chain length	Dionex SB-Biphenyl-30 (30% biphenyl; 70% methylpolysiloxane), 0.25 µm, 50 µm × 5 m	CO_2, pressure programmed, 160°C	FID	15
Acetylated and lactylated monoglycerides, decaglycerol decaoleate, triglycerol mono/dioleate, hexaglycerol distearate, partial separation of components within each product	Dionex SB-Cyano-25, 0.25 µm, 50 µm × 17 m	CO_2, pressure programmed, 120°C	FID	16
Sorbitan C_{18} alkyl ester 20-mole ethoxylate, resolution by ethoxy chain length (no resolution from free PEG)	Dionex SB-Biphenyl-30 (30% biphenyl; 70% methylpolysiloxane), 0.25 µm, 50 µm × 10 m	CO_2, pressure programmed, 100°C	FID	31
Alkyl glucoside model compounds; demonstration of SFC	DuPont Zorbax CN, 4.6 × 150 mm	93:7 CO_2/MeOH, 40°C	ELS	32
EO/PO copolymers: separation according to ethoxy chain length	Dionex SB-Biphenyl-30 (30% biphenyl; 70% methylpolysiloxane), 50 µm × 10 m	CO_2, pressure and temperature programmed, 130–100°C	FID	17

REFERENCES

1. Liu, Y., F. Yang, C. Pohl, Microbore packed column SFC using a polymer stationary phase, *J. Microcol. Sep.*, 1990, *2*, 245–254.

2. Sandra, P., F. David, Microcolumn chromatography for the analysis of detergents and lubricants. Part 1: High temperature capillary GC and capillary SFC, *J. High Resolut. Chromatogr.*, 1990, *13*, 414–417.

3. Rasmussen, H. T., B. P. McPherson, Chromatographic analysis of anionic surfactants, in J. Cross, ed., *Anionic Surfactants: Analytical Chemistry*, 2nd ed., Marcel Dekker, New York, 1998.

4. Pinkston, J. D., T. E. Delaney, K. L. Morand, R. G. Cooks, SFC/MS using a quadrupole mass filter/quadrupole ion trap hybrid MS with external ion source, *Anal. Chem.*, 1992, *64*, 1571–1577.

5. David, F., P. Sandra, Aliphatic amines and quaternary ammonium salts by capillary SFC, *Jour. HRC. CC.*, 1988, *11*, 897–898.

6. Geissler, P. R., Ethoxylated alcohols by SFC, *J. Am. Oil Chem. Soc.*, 1989, *66*, 685–689.

7. Silver, A. H., H. T. Kalinoski, Comparison of high temperature GC and CO_2 SFC for the analysis of AE, *J. Am. Oil Chem. Soc.*, 1992, *69*, 599–608.

8. Brossard, S., M. Lafosse, M. Dreux, Comparison of AE and PEG by HPLC and SFC using evaporative light-scattering detection, *J. Chromatogr.*, 1992, *591*, 149–157.

9. Lafosse, M., C. Elfakir, L. Morin-Allory, M. Dreux, Evaporative light scattering detection in pharmaceutical analysis by HPLC and SFC, *J. High Resolut. Chromatogr.*, 1992, *15*, 312–318.

10. Murphy, R. E., M. R. Schure, J. P. Foley, One- and two-dimensional chromatographic analysis of AE, *Anal. Chem.*, 1998, *70*, 4353–4360.

11. Wang, Z., M. Fingas, NPE by SFC and HPLC, *J. Chromatogr. Sci.*, 1993, *31*, 509–518.

12. Holzbauer, H.-R., U. Just, Oligomeric ethyene oxide adducts by SFC (in German), *Tenside, Surfactants, Deterg.*, 1994, *31*, 79–82.

13. Wang, Z., M. Fingas, OPE by capillary SFC, *J. Chromatogr.*, 1993, *641*, 125–136.

14. Matsumoto, K., S. Tsuge, Y. Hirata, Nonionic surface active agents by SFC/MS, *Shitsuryo Bunseki*, 1987, *35*, 15–22.

15. Wang, Z., M. Fingas, Sorbitan ester surfactants. Part II: Capillary SFC, *J. High Resolut. Chromatogr.*, 1994, *17*, 85–90.

16. Artz, W. E., M. R. Myers, Supercritical fluid extraction and chromatography of emulsifiers, *J. Am. Oil Chem. Soc.*, 1995, *72*, 219–224.

17. Pasch, H., C. Brinkmann, H. Much, U. Just, Chromatographic investigations of macromolecules in the "critical range" of LC: two-dimensional separations of poly(ethylene oxide-block-propylene oxide), *J. Chromatogr.*, 1992, *623*, 315–322.

18. Pinkston, J. D., D. J. Bowling, T. E. Delaney, SFC-MS involving oligomeric materials of low volatility and thermally labile materials, *J. Chromatogr.*, 1989, *474*, 97–111.

19. Kalinoski, H. T., A. Jensen, Nonionic surfactants using SFC and ^{13}C NMR spectroscopy, *J. Am. Oil Chem. Soc.*, 1989, *66*, 1171–1175.

20. Knowles, D. E., L. Nixon, E. R. Campbell, D. W. Later, B. E. Richter, Industrial applications of SFC: polymer analysis, *Fresenius' Z. Anal. Chem.*, 1988, *330*, 225–228.

21. Crow, J. A., J. P. Foley, Short capillary columns in SFC, *J. High Resolut. Chromatogr.*, 1989, *12*, 467–470.

22. Onuska, F. I., K. A. Terry, SFC of alkylene oxide-fatty alcohol condensates: quantitation in water samples, *Jour. HRC. CC.*, 1988, *11*, 874–877.

23. Smith, R. D., H. R. Udseth, MS interface for microbore and high flow rate capillary SFC with splitless injection, *Anal. Chem.*, 1987, *59*, 13–22.

24. Dean, T. A., C. F. Poole, Solventless injection for packed column SFC, *J. High Resolut. Chromatogr.*, 1989, *12*, 773–778.

25. Gemmel, B., B. Lorenschat, F. P. Schmitz, Oligomers of medium polarity by packed column SFC, *Chromatographia*, 1989, *27*, 605–610.

26. Hensley, J. L., H. M. McNair, Modification of a GC for capillary SFC, *J. Liq. Chromatogr.*, 1986, *9*, 1985–1996.

27. Takeuchi, T., T. Niwa, D. Ishii, Retention behavior in HPLC and SFC using methanol or diethyl ether as mobile phase, *Chromatographia*, 1987, *23*, 929–933.

28. Giorgetti, A., N. Pericles, H. M. Widmer, K. Anton, P. Dätwyler, Mixed mobile phases and pressure programming in packed and capillary column SFC: unified approach, *J. Chromatogr. Sci.*, 1989, *27*, 318–324.

29. Just, U., H.-R. Holzbauer, M. Resch, Molar mass determination of oligomeric ethylene oxide adducts using SFC and matrix-assisted laser desorption-ionization time-of-flight MS, *J. Chromatogr. A*, 1994, *667*, 354–360.

30. Later, D. W., B. E. Richter, M. R. Andersen, Capillary SFC, an emerging technology in perspective, *LC-GC*, 1986, *4*, 992–1003.

31. Ye, M. Y., R. G. Walkup, K. D. Hill, OPE by capillary column SFC and HPLC, *J. Liq. Chromatogr.*, 1995, *18*, 2309–2322.

32. Lafosse, M., P. Rollin, C. Elfakir, L. Morin-Allory, M. Martens, M. Dreux, SFC with light-scattering detection: analysis of polar compounds with packed columns, *J. Chromatogr.*, 1990, *505*, 191–197.

33. Eckard, P. R., L. T. Taylor, G. C. Slack, Phospholipids by subcritical fluid chromatography, *J. Chromatogr. A*, 1998, *826*, 241–247.

11

Capillary Electrophoresis

Most published capillary electrophoresis (CE) methods use buffers containing an organic solvent such as acetonitrile in order to prevent micelle formation. The micelles have different electrophoretic mobility than the isolated surfactant molecules (1). Since individual surfactant molecules are also present, the presence of micelles causes severe tailing of chromatographic peaks (2). Organic eluents also minimize the adsorption of the surfactant to the walls of the capillary.

A number of investigators have demonstrated simultaneous detection of cationic and anionic surfactants by CE. Such analyses require compromises in conditions so that cationics are resolved while anionics elute in a reasonable time. Since anionics and cationics are not formulated together, there is rarely a need for their simultaneous determination and these separations are not discussed here.

Precise quantification is still difficult for CE, mainly because of variability in migration times. CE analysis of surfactants is the subject of a recent review (44).

I. ANIONIC SURFACTANTS

Many of the more hydrophobic anionics have poor solubility in aqueous CE media. This is usually remedied by addition of an organic solvent like acetonitrile or methanol. Selectivity is enhanced by the solvent, and adsorption on the capillary walls is decreased. Because electroosmotic flow is much less with purely organic solvents, polarity is often reversed so that detection occurs at the anode.

402

Migration times of anionics are greatly affected by the presence of Mg^{2+} ion, and resolution of homologs is improved by adding Mg^{2+} to the buffer mixture (3). On the other hand, it is sometimes desirable to have only a single peak to aid quantification of a surfactant. In this case, trace contamination by Mg^{2+} or Ca^{2+} is to be avoided (4).

A. LAS

In the absence of organic solvents, LAS gives a single peak by CE (Table 1). This is often preferred for routine analysis. With the addition of solvent, most often 30% acetonitrile, LAS is resolved according to alkyl chain length, with minor peak-splitting due to partial resolution of isomers. Elution is in reverse order of alkyl chain length. If α-cyclodextrin is added to the mobile phase, more complete resolution of isomers is accomplished, although there is overlap between isomers associated with the various alkyl chain lengths in commercial LAS (5,6). Almost complete resolution of isomers is obtained if sodium dodecylsulfate (SDS) is added to a mobile phase containing 30% acetonitrile (Fig. 1). The high organic solvent content prevents micelle formation (so micellar electrokinetic chromatography is not the mechanism), but the interaction of LAS with SDS changes the elution order so that elution is in order of increasing chain length (7).

B. Alkyl Sulfates and Alkanesulfonates

The electrophoretic mobilities of alkyl sulfates and alkanesulfonates decrease with increasing alkyl chain length (1,8). They can thus be readily separated according to alkyl chain length. Most investigators use cathodic conditions, where elution is in the order of decreasing alkyl chain length (Table 2). The order is reversed if polarity is switched and detection occurs near the anode. This latter approach is typically followed when using entirely organic solvent systems and is best suited to analysis of the higher chain length (lower mobility) surfactants. Under cathodic detection conditions, alkanesulfonates elute prior to alkyl sulfates of the same alkyl chain length, i.e., the alkyl sulfates have higher migration times (3,9).

Partial resolution of isomers may be seen, depending on the buffer system chosen. For indirect UV detection, adenosine monophosphate is a better choice for an electrolyte additive than naphthalene sulfonate because the former has an electrophoretic mobility closer to that of alkyl sulfates and sulfonates in the surfactant range $[C_{12}–C_{18}]$ (10).

C. Ether Sulfates

When analyzed by capillary zone electrophoresis (CZE) with cathodic detection, alkyl ether sulfates elute in reverse order of ethoxy number and also in order of decreasing alkyl chain length (Table 3). This makes resolution of commercial mixtures challenging. Unethoxylated alcohol sulfate impurity is resolved and can be determined by CZE. If a nonaqueous solvent system is used with anodic detection, elution is in order of increasing ethoxy number and increasing alkyl chain length. Capillary zone electrophresis easily separates APE sulfates from nonionic impurities, with resolution of the anionic material by ethoxy chain length. With cathodic detection, the nonionic material (unsulfated APE) elutes first as a single sharp peak. APE sulfates can also be separated by capillary gel electrophoresis (CGE) (11).

TABLE 1 Capillary Electrophoresis Analysis of Alkylarylsulfonate Surfactants

Compound determined	Technique	Column/potential	Buffer	Detection	Ref.
C_{10}–C_{13} LAS, single peak for total LAS	Capillary zone electrophoresis (CZE)	Silica, 75 μm × 50 cm, 25 kV	0.1 M borate buffer, pH 8	Cathodic, UV, 200 nm	7
C_{10}–C_{14} LAS, elution in order of decreasing alkyl chain length	CZE	Silica, 50 μm × 40 cm	3:2 H_2O/CH_3CN, 0.01 M NaOAc buffer, 0.003 M in Mg^{2+} ion, pH 6.0	Cathodic, UV, 220 nm	3
C_2–C_{12} LAS, elution in order of decreasing alkyl chain length	CZE	Silica, 75 μm × 57 cm, 30°C	70:30 H_2O/CH_3CN, 0.0125 M in borate buffer, apparent pH 9	UV, 214 nm	12
C_{10}–C_{13} LAS, elution in order of decreasing alkyl chain length	CZE	Silica, 75 μm × 50 cm, 20 kV	70:30 H_2O/CH_3CN, 0.1 M in phosphate buffer, apparent pH 6.8	Cathodic, UV, 200 nm	6
C_{10}–C_{13} LAS, elution in order of decreasing alkyl chain length; determination in sewage	CZE	Silica, 75 μm × 50 cm, 20 or 25 kV	70:30 0.25 M borate buffer, pH 8/CH_3CN	Cathodic, UV, 200 nm	7
C_{10}–C_{13} LAS, elution in order of decreasing alkyl chain length; separation of isomers	CZE	Silica, 75 μm × 50 cm, 20 kV	80:20 H_2O/CH_3CN, 0.1 M in phosphate buffer and 0.015 M in α-cyclodextrin, apparent pH 6.8	Cathodic, UV, 200 nm	6
C_{10}–C_{13} LAS, elution in order of increasing alkyl chain length; separation of isomers	CZE	Silica, 50 μm × 40 cm, 25 kV	70% 0.01 M phosphate buffer-0.04 M sodium dodecylsulfate, pH 6.8/30% CH_3CN	Cathodic, UV, 200 nm	7
C_{10}–C_{14} LAS, elution in order of increasing alkyl chain length	CZE	Silica, 75 μm × 43 cm, −20 kV	MeOH, 0.01 M in HOAc and 0.02 M in tetramethylammonium hydroxide	Anodic, UV, 214 nm	22
C_{12} LAS, residue determination; single peak	CZE	Silica, 75 μm × 37 cm	0.0025 M borax, 0.005 M Na_2HPO_4, pH 9.5	UV, 200 nm	4

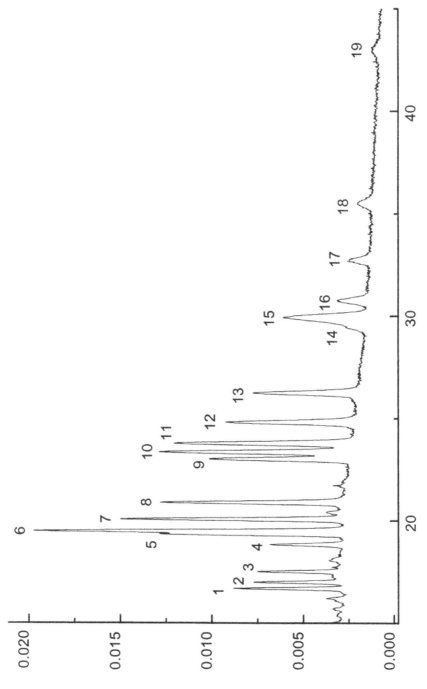

FIG. 1 CE separation of a commercial LAS mixture. Peak identities: 1, C_{10}-5ϕ; 2, C_{10}-4ϕ; 3, C_{10}-3ϕ; 4, C_{10}-2ϕ; 5, C_{11}-6ϕ; 6, C_{11}-5ϕ; 7, C_{11}-4ϕ; 8, C_{11}-3ϕ; 9, C_{11}-2ϕ; 10, C_{12}-6ϕ; 11, C_{12}-5ϕ; 12, C_{12}-4ϕ; 13, C_{12}-3ϕ; 14, C_{13}-7ϕ; 15, C_{13}-6ϕ and C_{12}-2ϕ;16, C_{13}-5ϕ; 17, C_{13}-4ϕ; 18, C_{13}-3ϕ; 19, C_{13}-2ϕ. (Reprinted with permission from Ref. 7. Copyright 1998 by the Royal Society of Chemistry.)

TABLE 2 Capillary Electrophoresis Analysis of Alkyl Sulfate and Alkanesulfonate Surfactants

Compound determined	Technique	Column/potential	Buffer	Detection	Ref.
C_{12} alkyl sulfate; determination in water	Capillary zone electrophoresis (CZE)	Silica, 75 μm × 75 cm, 30°C	95:5 0.005 M 2,4-dihydroxybenzoic acid/MeOH; pH adjusted to 8.1 with NaOH	Cathodic, indirect UV, 250 nm	23
C_{10}–C_{14} alkyl sulfates; separation by decreasing alkyl chain length	CZE	Silica, 100 μm × 80 cm	70:30 H_2O/CH_3CN, 0.001 M in $K_2Cr_2O_7$ and 0.1 in $Na_2B_4O_7$, adjusted to apparent pH 8.0 with H_3BO_3	Cathodic, indirect UV, 265 nm	8
C_4–C_{12} alkanesulfonates; separation by decreasing alkyl chain length	CZE	Silica, 75 μm × 60 cm	30:70 CH_3CN/0.010 M naphthalene sulfonate buffer, pH 10	Cathodic, indirect UV, 254 nm	24
C_4–C_{14} alkyl sulfates; separation by decreasing alkyl chain length	CZE	Silica, 50 μm × 49 cm	0.1 M boric acid/0.005 M borax buffer, 0.005 M in naphthalene sulfonate; pH 8: with or without 30% CH_3CN	Cathodic, indirect UV, 274 nm	25
C_9–C_{13} alkyl sulfates and C_8–C_{18} alkanesulfonates; separation by decreasing alkyl chain length; separation from each other; detection of dodecyl sulfate in shampoo, toothpaste, and facial cleanser	CZE	Silica, 50 μm × 49 cm, 25 kV, or 75 μm × 50 cm, 20 kV	50:50 CH_3CN/H_2O, 0.015 M in p-hydroxybenzoate salt or 30:70 CH_3CN/H_2O, 0.20 M in salicylate ion, pH 6 or 0.005 M dodecylbenzenesulfonate/0.005 M phosphate buffer pH 6.8	Cathodic, indirect UV, 214 nm	6,26
C_9–C_{13} alkyl sulfates and C_8–C_{18} alkanesulfonates; separation by increasing alkyl chain length	CZE	Silica, 50 μm × 49 cm, –30 kV	CH_3OH, 0.01 M in naphthalene sulfonate salt	Anodic, indirect UV, 214 nm	26
C_6–C_{18} alkyl sulfates and C_4–C_{18} alkanesulfonates; separation by decreasing alkyl chain length; separation from each other	CZE	Silica, 50 μm × 50 cm, 30 kV	0.1 M boric acid, 0.005 M in naphthalene sulfonate or adenosine monophosphate; pH 6: 50% CH_3OH	Cathodic, indirect UV, 206 nm for naphthalene sulfonate; 259 nm for adenosine monophosphate	9,10

Application	Technique	Capillary	Electrolyte	Detection	Ref.
C_5–C_{18} alkanesulfonates; separation by decreasing alkyl chain length; simultaneous determination of cationics	CZE	Silica, 50 μm × 50 cm	0.1 M boric acid, 0.005 M in pyridinium p-toluenesulfonate; pH 6.0 or 7.5; 50% CH_3OH	Cathodic, indirect UV, 208 or 220 nm	9
C_9–C_{13} alkyl sulfates; separation by decreasing alkyl chain length	CZE	Silica, 50 μm × 70 cm, 30°C	0.006 or 0.012 M 5,5-diethylbarbituric acid; pH 8.6	Cathodic, indirect UV, 240 nm	27
C_{12} alkyl sulfates; determination in pharmaceutical formulation	CZE	Silica, 75 μm × 30 cm, 30°C, 10 kV	0.008 M 5,5-diethylbarbituric acid; natural pH 9.5	Cathodic, indirect UV, 214 nm	28
C_6–C_{14} alkyl sulfates and C_4–C_{12} alkanesulfonates; separation by decreasing alkyl chain length; separation from each other	CZE	Silica, 50 μm × 40 cm	0.005 M phosphate buffer, 0.05 M in salicylate and 0.001 M in Mg^{2+} ion, pH 7.0	Cathodic, indirect UV, 230 nm	3
C_8–C_{18} alkyl sulfates and C_2–C_{16} alkanesulfonates; separation by increasing alkyl chain length	CZE	Silica, 75 μm × 43 cm, −20 kV	MeOH, 0.01 M in p-toluenesulfonic acid and 0.001 M in sodium p-toluenesulfonate	Anodic, indirect UV, 214 nm	22
C_4–C_{14} alkanesulfonates; separation by decreasing alkyl chain length	CZE	Silica, 50 μm × 49 cm	0.1 M boric acid/0.005 M borax buffer, 0.005 M in naphthalene sulfonate; pH 8: with or without 30% CH_3CN	Cathodic, indirect UV, 274 nm	25
C_8–C_{20} alkyl sulfates and alkanesulfonates; separation by increasing alkyl chain length	Capillary isotachophoresis (CITP)	PTFE, 0.55 mm × 22 cm	Leading electrolyte: 0.01 M L-histidine hydrochloride, apparent pH 4.88, 2% tetraethylene glycol; Terminating electrolyte: 0.01 M picolinic acid: both in 80:20 MeOH/H_2O	Conductivity	1

TABLE 3 Capillary Electrophoresis Analysis of Ether Sulfate Surfactants

Compound determined	Technique	Column/potential	Buffer	Detection	Ref.
Alkyl ether sulfates					
C_{12}–C_{18} alkyl ether sulfates; average E = 1 to 3; partial separation by decreasing alkyl chain length; incomplete resolution	Capillary zone electrophoresis (CZE)	Silica, 50 μm × 49 cm	0.1 M boric acid/0.005 M borax buffer, 0.005 M in naphthalene sulfonate; pH 8; 30% CH_3CN added	Cathodic, indirect UV, 274 nm	25
C_{10}–C_{14} alkyl ether sulfates; E = 0 to 7; elution by decreasing alkyl and decreasing ethoxy chain length	CZE	Silica, 100 μm × 80 cm	70:30 H_2O/CH_3CN, 0.001 M in $K_2Cr_2O_7$ and 0.1 in $Na_2B_4O_7$, adjusted to apparent pH 8.0 with H_3BO_3	Cathodic, indirect UV, 265 nm	8
C_{12} and C_{14} alkyl ether sulfates; E = 0 to 10; elution by decreasing ethoxy chain length; partial resolution by alkyl chain length	CZE	Silica, 50 μm × 49 cm, 25 kV	50:50 CH_3CN/H_2O, 0.015 M in *p*-hydroxybenzoate salt	Cathodic, indirect UV, 214 nm	26
C_{12} and C_{14} Alkyl ether sulfates; E = 0 to 10; elution by increasing ethoxy chain length; resolution by alkyl chain length	CZE	Silica, 50 μm × 49 cm, –30 kV	CH_3OH, 0.01 M in octylbenzene sulfonate salt	Anodic, indirect UV, 214 nm	26
Alkylphenol ether sulfates					
Sulfated alkylphenol 3-mole ethoxylate; elution in reverse order of molecular weight; separation from nonionic material	CZE	Silica, 75 μm × 80 cm, 30 kV	0.006 M $Na_2B_4O_7$, 0.01 M NaH_2PO_4, 20% CH_3CN, 30% CH_3OH, pH 7	Cathodic, UV, 206 nm	11
Sulfated alkylphenol 3-mole ethoxylate; elution in order of increasing molecular weight	CGE	μ-PAGE-3 (3% T/3%C gel) with 7 M urea, 75 μm × 50 cm, –249 V/cm	μ-PAGE (tris-borate, pH 8.3), 7 M urea	Cathodic, UV, 230 nm	11

D. Other Anionic Surfactants

Capillary electrophoresis of other anionic surfactants conditions are summarized in Table 4.

Petroleum sulfonates. Because of the diverse nature of typical petroleum sulfonates, they cannot be resolved according to alkyl chain length or isomer distribution by CZE or MEKC (12).

Sulfosuccinate esters. Sulfosuccinate diesters may be separated according to the length of the acyl chain (3).

Phosphate esters. In the case of phosphate monoesters, pH is critical in controlling migration time because it determines the fraction of the material which exists as a singly charged versus doubly charged anion (13).

Like APE sulfates, APE phosphates can be analyzed by either CZE or CGE (11). The former is adequate for separating material of low degree of ethoxylation and is especially useful for determination of nonionic material, giving a single peak for all nonphosphated APE. The latter is excellent for resolution of higher molecular weight products.

N-*Acyl sarcosinates.* These are eluted in reverse order of alkyl chain length under cathodic conditions. Detection by inverse UV works well (10).

Fatty acid soaps. Carboxylic acids in the range of soaps can be analyzed with buffers containing organic solvents to solubilize the acids. Detection is usually by inverse UV absorbance, with a UV absorber added to the buffer system. The addition of a nonionic surfactant to the buffer system aids in the resolution of certain acids (14).

II. NONIONIC SURFACTANTS

A. General Comments

Nonionics are not ideal candidates for capillary electrophoresis separation, since they are not mobile in an electrostatic field. In general, the separation is inferior to that obtained with HPLC analysis. On the other hand, CE is often easier to perform than HPLC, so there may be reason to apply CE for certain analyses when high precision and high resolution are not required.

Capillary electrophoresis can be applied to nonionics if they are first made ionic by derivatization with, for example, phthalic anhydride (11,15). This has the additional benefit of making them detectable by UV absorbance. Excess derivatizing reagent results in a few minutes of unusable time at the beginning of the electropherogram as the reagent passes the detector. This is minimized if the excess reagent is removed by dialysis or solid phase extraction (16).

Good results for nonionics can be obtained by adding sodium dodecylsulfate to the buffer. While at first this might seem to be micellar electrokinetic chromatography (MEKC), this is not so because the separations are performed in the presence of a high percentage of organic solvent. Thus, "solvophobic association" is a more accurate term for the separation than MEKC (6). Elution is in reverse order of ethoxy content because lower ethoxylated materials are less soluble in the aqueous phase than the ionic micelle phase and thus have higher electrophoretic mobility (Table 5). For good resolution, the aqueous phase must have a higher organic solvent content when analyzing low ethoxylates than higher ethoxylates. For example, Heinig et al. used 40% acetonitrile for separation of NPE of average degree of ethoxylation 5, but required only 20% acetonitrile for separation of a 55-mole ethoxylate (6).

TABLE 4 Capillary Electrophoresis Analysis of Other Anionic Surfactants

Compound determined	Technique	Column/potential	Buffer	Detection	Ref.
LAS, ABS, and petroleum sulfonate; demonstration of MEKC; peaks not identified	Micellar electrokinetic chromatography (MEKC)	Silica, 50 μm × 57 cm, 30°C	70:30 H_2O/CH_3CN, 0.00625 M in borate buffer, apparent pH 9	UV, 214 nm	31
C_4, C_6, and C_8 dialkyl sulfosuccinates; separation by decreasing alkyl chain length	CZE	Silica, 50 μm × 40 cm	0.005 M phosphate buffer, 0.05 M in salicylate and 0.001 M in Mg^{2+} ion, pH 7.0	Cathodic, indirect UV, 230 nm	3
C_1–C_8 phosphate mono- and diesters; elution according to decreasing alkyl chain length; diesters elute prior to monoesters	CZE	Silica, 50 μm × 75 cm, 30°C, 30 kV	0.005 M sodium adenosine monophosphate/0.1 M H_3BO_3, adjusted to pH 7.2	Cathodic, indirect UV, 259 nm	13
Phosphated C_8–C_{10} alkyl 6-mole ethoxylate; elution according to alkyl chain length and degree of ethoxylation; peaks not identified	CZE	Silica, 50 μm × 75 cm, 30°C, 30 kV	0.005 M sodium adenosine monophosphate/0.1 M H_3BO_3, 0.001 M diethylene triamine, adjusted to pH 7.2	Cathodic, indirect UV, 259 nm	13
Phosphated phenol 6-mole ethoxylate; elution according to degree of ethoxylation; peaks not identified	CZE	Silica, 50 μm × 75 cm, 30°C, 30 kV	0.005 M diethylene triamine/0.1 M H_3BO_3, pH 7.2	Cathodic, UV, 220 nm	13
Phosphated alkylphenol 40-mole ethoxylate; separation from non-ionic material	CZE	Silica, 75 μm × 80 cm, 30 kV	0.006 M $Na_2B_4O_7$, 0.01 M NaH_2PO_4, 20% CH_3CN, 30% CH_3OH, pH 7	Cathodic, UV, 206 nm	11
Phosphated alkylphenol 7-mole ethoxylate and 40-mole ethoxylate; separation of oligomers; elution in order of molecular weight	CGE	J&W μ-PAGE-3 (3%T/3%C gel) with or without 7 M urea, 75 μm × 50 cm, −220 or −249 V/cm	μ-PAGE (tris-borate, pH 8.3), with or without 7 M urea	Cathodic, UV, 230 nm	11
Sulfonated castor oil (turkey red oil), determination of sulfonated ricinoleic acid after saponification	CZE	Silica, 50 μm × 57 cm, 35°C	0.03 M tris, 0.0075 M hydroxypropyl-β-cyclodextrin, pH 8	Cathodic, UV, 200 nm	32

Analyte/description	Method	Capillary	Buffer/electrolyte	Detection	Ref.
C_{12}–C_{18} N-alkyl sarcosinates; elution according to decreasing alkyl chain length	CZE	Silica, 50 μm × 50 cm, 30 kV	0.1 M H_3BO_3, 0.01 M sodium adenosine monophosphate or sodium naphthalene sulfonate, pH 6.0, 50% CH_3OH	Cathodic, indirect UV, 259 nm for AMP and 206 nm for NS	10
Hydrotropes: benzene-, toluene-, and xylenesulfonate, elution in reverse order of alkyl character	CZE	Silica, 50 μm × 40 cm	0.01 M phosphate buffer, 0.001 M Mg^{2+} ion, pH 7.0	Cathodic, UV, 220 nm	3
Hydrotropes: toluene-, xylene-, and cumenesulfonate, elution in reverse order of alkyl character	CZE	Silica, 50 μm × 50 cm	0.05 M borate buffer, pH 8.3	Cathodic, UV, 214 nm	33
C_{10}–C_{18} fatty acids; separation by decreasing alkyl chain length	CZE	Silica, 50 μm × 57 cm, 30 kV	60:40 0.005 M $Na_2B_4O_7$ buffer, pH 9.2/EtOH, 10^{-5} M in sodium fluorescein	Cathodic, indirect fluorescence (488 nm excitation, 520 nm emission)	29
C_1–C_{18} fatty acids; separation by increasing alkyl chain length; determination in oxidation studies	Capillary isotachophoresis (CITP)	Fluorinated ethylene/propylene copolymer	Leading electrolyte: 95:5 MeOH/H_2O, 0.002 M in tris(hydroxymethyl) aminomethane hydrochloride, 0.02% in polyvinylalcohol, apparent pH 8.1; Terminating electrolyte: 99:1 MeOH/H_2O, 0.01 M in tris(hydroxymethyl) aminomethane hydrostearate, apparent pH 8.7	Conductivity	30
C_{12}–C_{20} fatty acids; separation by decreasing alkyl chain length	CZE	Silica, 50 μm × 57 cm, 20 kV	50:50 CH_3CN/H_2O: 0.005 M tris, 0.01 M sodium p-hydroxybenzoate; 0.040 M Brij 35 (an AE, $C_{12}E_{23}$); pH 6	Cathodic, indirect UV, 214 nm	14

TABLE 5 Capillary Electrophoresis Analysis of Nonionic Surfactants

Compound determined	Technique	Column	Buffer	Detection	Ref.
C_{12} alkyl ethoxylates, complexes with cations, as 3,5-dinitrobenzoyl derivatives; elution in order of decreasing ethoxy chain length; fine structure due to differences in alkyl chain length; resolution from n = 3 to n = 8	Capillary zone electrophoresis (CZE)	Silica, 50 μm × 30 cm, 20 kV	100% MeOH, 0.005 M KCl, 0.045 M NH_4Cl, 0.045 M triethylamine	UV, 250 nm	17
C_{11}–C_{18} alkyl ethoxylates, as phthalate esters; elution in order of decreasing ethoxy chain length; fine structure due to differences in alkyl chain length; resolution up to about n = 18; detection in formulations; simultaneous detection of LAS	CZE	Silica, 75 μm × 47 cm, 20 kV	50:50 CH_3CN/H_2O, 0.1 M borate buffer, pH 8	Cathodic, UV 200 nm	18
OPE, n = 30; separation of oligomers with elution in order of increasing molecular weight; analyzed after derivatization with phthalic anhydride	Capillary gel electrophoresis (CGE)	J&W μ-PAGE-5 (5%T/5%C gel) without urea, 75 μm × 50 cm, –300 V/cm	μ-PAGE (tris-borate, pH 8.3) without urea	Anodic, UV, 275 nm	11
OPE, n = 40; separation of oligomers with elution in order of increasing molecular weight; separation from PEG; analyzed after derivatization with 1,2,4-benzenetricarboxylic anhydride	CGE	Linear polyacrylamide-coated silica, 75 μm × 45.6 cm, filled with 3% 2,000,000 MW dextran dissolved in buffer, 28 kV	Tris/TAPS, 0.06 M, pH 8.3	Anodic, UV, 210 nm	16
OPE and NPE, n = 5.5 to n = 55; elution by decreasing ethoxy chain length	CZE	Silica, 75 μm × 50 cm, 20 kV	20:80 to 40:60 CH_3CN/H_2O, 0.01 M in phosphate buffer, pH 6.8; 0.07 M sodium dodecylsulfate	Cathodic, UV, 200 nm	6,19
OPE, n = 1 to 46; elution in order of decreasing ethoxy chain length	CZE	Silica, 50 μm × 60 cm, 25 kV	35:65 CH_3CN/H_2O, 0.025 M boric acid, 0.050 M sodium dodecylsulfate, pH 8.6	Cathodic, UV, 200 nm	15

While CZE gives good resolution of derivatized low molecular weight nonionic surfactants, CGE is excellent for the higher molecular weight materials. Wallingford demonstrated excellent resolution of an alkylphenol 30-mole ethoxylate (11).

B. Alcohol Ethoxylates

Okada studied CE of alkyl ethoxylates as their complexes with K^+ and other ions. Separation of AE of low degree of ethoxylation is possible because the effective mobility of complexes increases with increasing degree of ethoxylation up to about $n = 10$ (17).

AE is readily separated by CZE if it is first derivatized to make it anionic. LAS can be determined in the same chromatogram because LAS has higher electrophoretic mobility and hence a longer migration time to the cathode than derivatized AE. Because of the complexity of commercial AE, CZE has been proposed as a method of "fingerprinting" products rather than quantitatively determining the isomer/homolog content (18).

C. Alkylphenol Ethoxylates

OPE can be separated by CGE if it is derivatized to make it anionic. In this case, detection is at the anode and the lower molecular weight materials arrive at the detector first. For the higher molecular weight products, elution times become very long if derivatization is performed with phthalic anhydride, which yields one anionic group per molecule (11). Better results for higher oligomers are obtained if derivatization is performed with benzenetricarboxylic anhydride, which gives two anionic groups per molecule and hence faster migration times (16). With any derivatizing agent, PEG will be derivitized with twice the anionic groups of the APE, and will therefore appear first in the electropherogram (16).

Using CZE, Heinig, Vogt, and Werner showed good separations of APE in buffers containing SDS and a high concentration of organic solvent, typically 40% acetonitrile. 40% isopropanol also gave good results, but with longer retention times than acetonitrile. Retention time of oligomers is inversely proportional to degree of ethoxylation; better resolution of higher oligomers is obtained by decreasing acetonitrile content down to about 20%. OPE is not retained as well an NPE, since, being less hydrophobic, it interacts less with the SDS (19).

III. CATIONIC AND AMPHOTERIC SURFACTANTS

Capillary electrophoresis analyses of cationic and amphoteric surfactants are summarized in Tables 6 and 7.

A. Fatty Quaternary Amines

The electrophoretic mobilities of quaternary amines decrease with increasing alkyl chain length, so that separation of homologs is easily performed, with elution from the column in order of increasing chain length.

Dialkyldimethylammonium salts require a high concentration of organic solvent in the electrolyte for reasonable solubility (10). Even for quats with good water solubility, analysis is performed in buffers with high organic solvent content in order to inhibit micelle formation as well as preventing adsorbance on the wall of the capillary. Since CE instrumentation is normally used only with aqueous buffers and contains plastic parts, the operator is urged to test the compatibility of the equipment with these solvents, particularly THF, before initiating surfactant analysis.

TABLE 6 Capillary Electrophoresis Analysis of Cationic Surfactants

Compound determined	Technique	Column	Buffer	Detection	Ref.
Amine-terminated EO/PO colpolymers, MW 900, 2000, 4000; elution in order of increasing molecular weight; resolution of homologs	Capillary zone electrophoresis (CZE)	Silica, 50 μm × 37 cm, 25 kV	0.03 M creatinine buffer, adjusted to pH 4.8 with HOAc, containing 0.1% poly(ethylene oxide)	Cathodic, indirect UV, 220 nm	15
Amine-terminated EO/PO colpolymers, MW 600 to 2000; 2,3-naphthalenedialdehyde derivatives; elution in order of increasing molecular weight; resolution of homologs	CZE	Silica, 50 μm × 60 cm or 25 μm × 85 cm, 25 or 30 kV, 30°C	20:80 MeOH/H_2O, 0.077 M boric acid, 0.019 M citric acid; 0.011 M Na_3PO_4, pH 4.2	Cathodic, fluorescence, excitation at 442 nm, emission >475 nm	34
C_8–C_{20} alkyltrimethylammonium ions; separation by increasing alkyl chain length	Capillary isotachophoresis (CITP)	PTFE, 0.55 mm × 22 cm	Leading electrolyte: 0.01 M L-histidine hydrochloride, apparent pH 4.88, 2% tetraethylene glycol; terminating electrolyte: 0.01 M picolinic acid; both in 80:20 MeOH/H_2O	Conductivity	1
C_{12} and C_{16} alkylallyldimethylammonium ions; separation from their polymers and decomposition products	CITP	PTFE, 0.55 mm × 22 cm	Leading electrolyte: 0.01 M K^+ ion, 0.0132 M acetate, pH 5.5, 2% tetraethylene glycol; terminating electrolyte: 0.01 M β-alanine; in up to 50% MeOH	Conductivity	35
C_{12}–C_{18} alkyltrimethylammonium ions, separation by increasing alkyl chain length	CZE	Silica, 50 μm × 24 cm	57.5:42.5 THF/H_2O, 0.003 M in benzyldodecyldimethylammonium chloride, 0.003 M in sodium dodecyl sulfate, 0.008 M in NaH_2PO_4	Cathodic, indirect UV, 210 nm	2

Analyte/Description	Mode	Capillary	Buffer	Detection	Ref.
C_4, C_6 Tetraalkylammonium and C_{12} dialkyldimethylammonium ions; separation by increasing alkyl chain length	CZE	Silica, 50 μm × 50 cm	0.1 M boric acid, 0.005 M in benzylamine or l-ephedrine; pH 6.0; 50% CH_3OH	Cathodic, indirect UV. 204 nm	9
C_1–C_6 tetraalkylammonium ions; separation by increasing alkyl chain length	CZE	Silica, 57 μm × 50 cm or 70 μm × 37 cm	0.001 M sodium phosphate buffer containing 0.001 M rhodamine 700, pH 3.6: 50% CH_3OH	Cathodic, indirect absorbance. 254 or 670 nm	36
C_{12}–C_{18} alkyltrimethylammonium bromide, separation by increasing alkyl chain length	CZE	Silica, 75 μm × 50 cm, 20 kV	50:50 THF/H_2O, 0.05 M in phosphate buffer, pH 6.8; 0.005 M benzyltrimethylammonium chloride; 0.003 M sodium dodecylsulfate	Cathodic, indirect UV. 214 nm	6
C_{12}–C_{18} alkyltrimethylammonium bromide and C_{12}–C_{16} dialkyldimethylammonium bromide, separation by increasing alkyl character	CZE	Silica, 75 μm × 50 cm, 20 kV	50:50 THF/H_2O, 0.02 M in phosphate buffer, pH 4.4; 0.005 M benzyldodecyldimethylammonium chloride	Cathodic, indirect UV. 214 nm	37
C_{16} alkyltrimethylammonium bromide, determination in cosmetic product	CZE	Silica, 75 μm × 50 cm, 20 kV	40:60 THF/H_2O, 0.02 M in phosphate buffer; pH 4.4: 0.005 M benzyldodecyldimethylammonium chloride	Cathodic, indirect UV. 214 nm	37
C_{12}–C_{18} dialkyldimethylammonium ions; separation by increasing alkyl chain length	CZE	Silica, 50 μm × 50 cm, 30 kV	0.1 M boric acid, 0.01 M in tetrazolium violet; pH 6.0; 85% CH_3OH	Cathodic, indirect UV, 300 nm	10
C_4 and C_6 Tetraalkylammonium and C_{12} dialkyldimethylammonium ions; separation by increasing alkyl chain length; simultaneous determination of anionics	CZE	Silica, 50 μm × 50 cm	0.1 M boric acid, 0.005 M in pyridinium p-toluenesulfonate: pH 6.0 or 7.5: 50% CH_3OH	Cathodic, indirect UV, 208 or 220 nm	9

(Continued)

TABLE 6 (Continued)

Compound determined	Technique	Column	Buffer	Detection	Ref.
C_{16} alkylpyridinium chloride, determination in mouth wash and throat spray	CZE	Silica, 75 μm × 50 cm, 20 kV	50:50 CH_3CN/H_2O, 0.05 M in phosphate buffer, pH 6.8	Cathodic, UV, 200 nm	6,38
Benzyl-C_{14}–C_{18} alkyldimethylammonium ions, separation by increasing alkyl chain length	CZE	Silica, 50 μm × 24 cm	57.5:42.5 THF/H_2O, 0.044 M in phosphate buffer	Cathodic, UV, 210 nm	2
Benzyl-C_{12}–C_{16}-alkyldimethylammonium chloride, separation by increasing alkyl chain length	CZE	Silica, 75 μm × 50 cm, 20 kV	55:45 THF/H_2O, 0.05 M in phosphate buffer, pH 6.8	Cathodic, UV, 200 nm	6
Benzyl-C_{12}–C_{16}-alkyldimethylammonium chloride, separation by increasing alkyl chain length; demonstration of effect of various buffers	CZE	Silica, 75 μm × 50 cm, 20 or 30 kV	Various solvents	Cathodic, UV, 200 or 214 nm	38
Benzyl-C_{12}–C_{18}-alkyldimethylammonium chloride, separation by increasing alkyl chain length; demonstration of effect of solvents in buffer and in sample solution	CZE	Silica, 75 μm × 37 cm, 15 kV, 25°C	0.02 M phosphate buffer, pH 5.0, 30% CH_3CN or 50% acetone; sample solution contains CH_3OH	Cathodic, UV, 210 nm	20,21
Benzyl-C_{12}–C_{18}-alkyldimethylammonium chloride, elution in order of increasing alkyl chain length	CZE	Silica, 50 μm × 51 cm, 20 kV, 30–35°C	0.5 M phosphate buffer, pH 4, 50% THF	Cathodic, UV, 210 nm	39,40
Benzyl-[2-(4-dodecanoylphenoxy)ethyl]-dimethylammonium chloride; determination in cosmetics	CZE	Silica, 75 μm × 50 cm, 20 kV	55:45 THF/H_2O, 0.05 M in phosphate buffer, pH 6.8	Cathodic, UV, 200 nm	38

TABLE 7 Capillary Electrophoresis Analysis of Amphoteric Surfactants

Compound determined	Technique	Column	Buffer	Detection	Ref.
Phosphatidylcholine (PC), phosphatidylethanolamine (PE), phosphatidylinositol (PI), and phosphatidylserine (PS); determination (after workup) in soybean and other lecithins	Micellar electrokinetic chromatography (MEKC)	Silica, 50 μm × 53 cm, 50°C	0.075 M sodium cholate, 0.010 M Na$_2$HPO$_4$, 0.006 M borax, 30% 2-PrOH, adjusted to apparent pH 8.5 with HCl	UV, 200 nm	41
PC, PE, PI, PS, and phosphatidic acid (PA); determination (after workup) in soybean lecithin; separation according to acyl character observed	MEKC	Silica, 50 μm × 52 cm, 25 kV, 50°C for phospholipid separation; 15°C for acyl distribution	0.035 M sodium deoxycholate, 0.010 M Na$_2$HPO$_4$, 0.006 M borax, 30% 2-PrOH, pH 8.5	UV, 200 nm	42
PC, PE, PS, PA; all with 1-palmitoyl-2-oleyl substitution; demonstration of CE-MS analysis	Capillary zone electrophoresis (CZE)	Silica, 77 μm × 60 cm, 25°C, 30 kV	57:38:5 CH$_3$CN/2-PrOH/ n-hexane, 0.02 M in NH$_4$OAc	Electrospray MS	43

Quats are readily determined by indirect UV detection. Peak tailing can result from the differing electrophoretic mobilies of the surfactants and the UV-absorbing cation. These differences can be minimized by adding an ion pairing agent like dodecylsulfate ion (6).

B. Benzalkonium Compounds

Lin et al. demonstrated excellent resolution of benzylalkyltrimethylammonium salts in buffer containing 30% acetonitrile to avoid micelle formation. Best results are obtained if the sample solution also contains an organic solvent and if sample concentration is minimized (20,21). Addition of cyclodextrins to the mobile phase can be beneficial (45).

C. Amine-Terminated EO/PO Copolymers

These are readily separated by molecular weight. Addition of poly(ethylene oxide) to the buffer reduces the electroosmotic flow and gives better resolution of individual homologs (15).

D. Lecithin

Analogous to normal phase HPLC, CE has been successfully used for separation of the various types of phospholipids: phosphatidylcholine, phosphatidylethanolamine, phosphatidylserine, phosphatidylinositol, and phosphatidic acid. The elution order is affected by buffer strength and composition. Analogous to reversed-phase HPLC, some separation according to the nature of the fatty acid substituents is seen with MEKC.

REFERENCES

1. Tribet, C., R. Gaboriaud, P. Gareil, C_8–C_{20} saturated anionic and cationic surfactant mixtures by capillary isotachophoresis with conductivity detection, *J. Chromatogr.*, 1992, *609*, 381–390.
2. Weiss, C. S., J. S. Hazlett, M. H. Datta, M. H. Danzer, Quaternary ammonium compounds by CE using direct and indirect UV detection, *J. Chromatogr.*, 1992, *608*, 325–332.
3. Chen, S., D. J. Pietrzyk, Separation of sulfonate and sulfate surfactants by CE: effect of buffer cation, *Anal. Chem.*, 1993, *65*, 2770–2775.
4. Altria, K. D., I. Gill, J. S. Howells, C. N. Luscombe, R. Z. Williams, Detergent residues by CE, *Chromatographia*, 1995, *40*, 527–531.
5. Vogt, C., K. Heinig, B. Langer, J. Mattusch, G. Werner, LAS by HPLC and CZE, *Fresenius' J. Anal. Chem.*, 1995, *352*, 508–514.
6. Heinig, K., C. Vogt, G. Werner, Ionic and neutral surfactants by CE and HPLC, *J. Chromatogr. A*, 1996, *745*, 281–292.
7. Heinig, K., C. Vogt, G. Werner, LAS in industrial and environmental samples by CE, *Analyst*, 1998, *123*, 349–353.
8. Goebel, L. K., H. M. McNair, H. T. Rasmussen, B. P. McPherson, Ethoxylated alcohol sulfates by CE using indirect UV detection, *J. Microcolumn Sep.*, 1993, *5*, 47–50.
9. Shamsi, S. A., N. D. Danielson, Individual and simultaneous class separations of cationic and anionic surfactants using CE with indirect photometric detection, *Anal. Chem.*, 1995, *67*, 4210–4216.
10. Shamsi, S. A., N. D. Danielson, CE of cationic surfactants with tetrazolium violet and anionic surfactants with adenosine monophosphate and indirect photometric detection, *J. Chromatogr. A*, 1996, *739*, 405–412.

11. Wallingford, R. A., Oligomeric separation of ionic and nonionic ethoxylated polymers by capillary gel electrophoresis, *Anal. Chem.*, 1996, *68*, 2541–2548.

12. Desbène, P. L., C. Rony, B. Desmazières, J. C. Jacquier, Alkylaromatic sulfonates by HPCE, *J. Chromatogr.*, 1992, *608*, 375–383.

13. Shamsi, S. A., R. M. Weathers, N. D. Danielson, CE of phosphate ester surfactants with adenosine monophosphate and indirect photometric detection, *J. Chromatogr. A*, 1996, *737*, 315–324.

14. Heinig, K., F. Hissner, S. Martin, C. Vogt, Saturated and unsaturated fatty acids by CE and HPLC, *Amer. Lab.*, 1998, *30(10)*, 24–29.

15. Bullock, J., Application of CE to the analysis of the oligomeric distribution of polydisperse polymers, *J. Chromatogr.*, 1993, *645*, 169–177.

16. Barry, J. P., D. R. Radtke, W. J. Carton, R. T. Anselmo, J. V. Evans, Ethoxylated polymers by CE in UV-transparent polymer networks and by MALDI-TOF-MS, *J. Chromatogr. A*, 1998, *800*, 13–19.

17. Okada, T., Nonaqueous CE separation of polyethers and evaluation of weak complex formation, *J. Chromatogr. A*, 1995, *695*, 309–317.

18. Heinig, K., C. Vogt, G. Werner, Nonionic surfactants by CE and HPLC, *Anal. Chem.*, 1998, *70*, 1885–1892.

19. Heinig, K., C. Vogt, G. Werner, Nonionic surfactants of the polyoxyethylene type by CE, *Fresenius' J. Anal. Chem.*, 1997, *357*, 695–700.

20. Lin, C., W. Chiou, W. Lin, CZE separation of alkylbenzyl quaternary ammonium compounds: effect of organic modifier, *J. Chromatogr. A*, 1996, *722*, 345–352.

21. Lin, C., W. Chiou, W. Lin, Separation of alkylbenzyl quaternary ammonium compounds by CZE: effect of organic solvent in sample solution, *J. Chromatogr. A*, 1996, *723*, 189–195.

22. Salimi-Moosavi, H., R. M. Cassidy, Nonaqueous capillary electrophoresis for the separation of long-chain surfactants, *Anal. Chem.*, 1996, *68*, 293–299.

23. Gibbons, J. M., S. H. Hoke, CZE with indirect UV detection: sodium dodecyl sulfate in simulated stream water, *J. High Resolut. Chromatogr.*, 1994, *17*, 665–667.

24. Romano, J., P. Jankik, W. R. Jones, P. E. Jackson, Optimization of inorganic CE for the analysis of anionic solutes, *J. Chromatogr.*, 1991, *546*, 411–421.

25. Shamsi, S. A., N. D. Danielson, Naphthalenesulfonates as electrolytes for CE of inorganic anions, organic acids, and surfactants with indirect photometric detection, *Anal. Chem.*, 1994, *66*, 3757–3764.

26. Heinig, K., C. Vogt, G. Werner, Anionic surfactants using aqueous and nonaqueous CE, *J. Capillary Electrophor.*, 1996, *3*, 261–270.

27. Nielen, M. W. F., Quantitative aspects of indirect UV detection in CZE, *J. Chromatogr.*, 1991, *588*, 321–326.

28. Kelly, M. A., K. D. Altria, B. J. Clark, Sodium dodecyl sulfate by CE, *J. Chromatogr. A*, 1997, *781*, 67–71.

29. Desbène, P. L., C. J. Morin, N. L. Mofaddel, R. S. Groult, Fluorescein sodium salt in laser-induced indirect fluorimetric detection: application to organic anions, *J. Chromatogr. A*, 1995, *716*, 279–290.

30. Koval, M., D. Kaniansky, M. Hutta, R. Lacko, Saturated normal fatty acids in hydrocarbon matrices by capillary isotachophoresis, *J. Chromatogr.*, 1985, *325*, 151–160.

31. Desbène, P. L., C. Rony, High-performance CE of alkylaromatics used as bases of sulfonation in the preparation of industrial surfactants, *J. Chromatogr. A*, 1995, *689*, 107–121.

32. Nawaby, A. V., P. Kruus, E. Dabek-Zlotorzynska, Anionic surfactant turkey red oil by CE with direct UV detection, *J. High Resolut. Chromatogr.*, 1998, *21*, 401–406.

33. Brumley, W. C., Qualitative analysis of environmental samples for aromatic sulfonic acids by HPCE, *J. Chromatogr.*, 1992, *603*, 267–272.

34. Amankwa, L. N., J. Scholl, W. G. Kuhr, Oligomeric dispersion of poly(oxyalkylene)diamine polymers by precolumn derivatization and capillary zone electrophoresis with fluorescence detection, *Anal. Chem.*, 1990, *62*, 2189–2193.

35. Tribet, C., R. Gaboriaud, P. Gareil, Analogy between micelles and polymers of ionic surfactants. Capillary isotachophoresis study of small ionic aggregates in water-organic solutions, *J. Chromatogr.*, 1992, *608*, 131–141.

36. Williams, S. J., E. T. Bergström, D. M. Goodall, H. Kawazumi, K. P. Evans, Diode laser-based indirect absorbance detector for CE, *J. Chromatogr.*, 1993, *636*, 39–45.

37. Heinig, K., C. Vogt, G. Werner, Cationic surfactants by CE with indirect photometric detection, *J. Chromatogr. A*, 1997, *781*, 17–22.

38. Heinig, K., C. Vogt, G. Werner, Cationic surfactants by CE, *Fresenius' Z. Anal. Chem.*, 1997, *358*, 500–505.

39. Piera, E., P. Erra, M. R. Infante, Cationic surfactants by CE, *J. Chromatogr. A*, 1997, *757*, 275–280.

40. Piera, E., P. Erra, M. R. Infante, Cationic surfactants by CE, *World Surfactants Congr., 4th*, 1996, *1*, 403–417.

41. Ingvardsen, L., S. Michaelsen, H. Sørensen, Phospholipids by HPCE, *J. Am. Oil Chem. Soc.*, 1994, *71*, 183–188.

42. Szücs, R., K. Verleysen, G. S. M. J. E. Duchateau, P. Sandra, B. G. M. Vandeginste, Phospholipids in lecithins: comparison between micellar electrokinetic chromatography and HPLC, *J. Chromatogr. A*, 1996, *738*, 25–29.

43. Raith, K., R. Wolf, J. Wagner, R. H. H. Neubert, Phospholipids by nonaqueous CE with electrospray ionization MS, *J. Chromatogr. A*, 1998, *802*, 185–188.

44. Heinig, K., C. Vogt, Surfactants by CE, *Electrophoresis*, 1999, *20*, 3311–3328.

45. So, T. S. K., C. W. Huie, Effects of cyclodextrins and organic solvents on the separation of cationic surfactants by CE, *J. Chromatogr. A*, 2000, *872*, 269–278.

12

Ultraviolet and Visible Spectrophotometry

Spectrophotometric methods have been the standard procedures for determining surfactants in the environment for decades. They are also widely used for determination of surfactants in formulations. Although these methods are more susceptible to interference than chromatographic procedures, there are many situations when they represent the most appropriate technology.

I. ANIONIC SURFACTANTS

A. Ultraviolet Absorbance

Alkylbenzenesulfonates have a characteristic absorbance maximum at about 223 nm, suitable for use in direct analysis. A stronger absorbance at 193 nm is also accessible to sophisticated spectrophotometers (1). LAS homologs with an alkyl chain length longer than 6 carbons have the same molar absorptivity, simplifying calibration (2). Direct determination of LAS by UV absorbance is rarely useful for analysis of mixtures, since many other detergent components, such as the hydrotropes xylene, toluene, and cumene sulfonate (1,3) and the nonionic APE surfactants, also absorb at 223 nm. However, the technique is suitable for matrices in which interfering compounds are absent and is capable of great sensitivity if derivative spectrophotometry is used (4,5). For greatest discrimination, the second derivative is used, with qualitative evaluation at the three wavelength pairs 198/201 nm, 224/227 nm, and 230/233 nm to assure the absence of interference, using 230/233 nm for quantification. (6,7).

Direct measurement of UV absorbance has been used as the basis of a flow injection analysis procedure for determining LAS in spray-dried detergents (8). UV absorption has also been used to monitor the concentration of petroleum sulfonates in tertiary oil recovery studies (9). In this case, errors were minimized by measuring the absorbance at three wavelengths to calculate concentration; $C = 10^{-3}(2.503a_{210} - 1.087a_{235} + 0.528a_{265})$. Sources of error included change in composition of the surfactant due to adsorption on minerals and contamination by crude oil.

LAS may be determined with great sensitivity by a procedure based on extraction of the ion pair with tetraphenylpyridine into isopentyl acetate, followed by measurement of the absorbance of the extract at 305 nm (10). Interferences are similar to those discussed below for the methylene blue procedure.

Procedure: Determination of Sodium Alkylbenzene Sulfonate in Synthetic Detergents by Ultraviolet Absorption (11)

Weigh out triplicate 1-g samples of solid detergent, dissolve in water, and dilute to 500 mL. In the case of slurries, weigh out 20-g samples, add 50 mL 95% ethanol to disperse, and dilute to 1 L with water. Dilute a 50-mL aliquot of the latter solution to 500 mL. Prepare another dilution of the above solutions, 5 mL to 250 mL. Measure the absorbance of the final diluted solutions in 1-cm quartz cuvettes at 224 and 270 nm. In the case of scanning spectrophotometers, obtain the absorbance curve from 210 to 300 nm. The absorbance at 224 nm should lie in the range 0.2 to 0.9 for greatest accuracy. If unsure of the history of the sample, measure the absorbance also at 220 and 228 nm. If the maximal absorbance is not at 224 nm, then LAS is not the only substance responsible for the absorbance, and this method is not valid for the sample. Calculate the average net absorbance ($A_{224} - A_{270}$), and determine the concentration of LAS by comparison with a standard.

Discussion: The test is designed for commercial formulations which do not contain amides. Optical dyes and other typical detergent components do not interfere, but all UV absorbers other than LAS must be absent. For well-characterized samples where a valid standard is available, the main source of error is in the numerous dilutions. Replicate results from a single analyst should agree within 1%, relative. At low concentrations, errors can develop due to adsorption of LAS on the glassware. Addition of dilute potassium dihydrogen phosphate is recommended to limit this adsorption (12). Salt addition is not required in formulated detergents, where the ionic strength is quite high.

B. Visible Spectrophotometry

Like the two-phase titration methods (Chapter 16), almost all of the colorimetric methods rely on the formation of an ion pair by the anionic surfactant and a cation, in this case a dye. The ion pair is extractible into an organic solvent, while the dye alone is not, so the color of the organic phase is directly proportional to the surfactant concentration. This approach is suitable for determination of low concentrations and has been used most often for environmental analysis.

1. Methylene Blue

Because of its simplicity, the methylene blue ion pair method has been a standard for environmental analysis for a half-century. The absorbance of the ion pair is maximum at about 658 nm, depending on the specific anionic surfactant. Maximal color extraction is reached after a minute of agitation, and the color is stable in the organic phase for a half-hour (13). The methylene blue method is susceptible to positive and negative interference, and this interference becomes a greater problem when it is used to measure very low concentrations of anionic surfactants.

(a)   The standard procedure

Procedure: Methylene Blue Method (14,15)

Reagents:

> *Methylene blue reagent*: Dissolve 0.1 g methylene blue, Eastman No. P573, or equivalent, in 100 mL H_2O. Transfer 30 mL to a 1-L volumetric flask. Add 500 mL H_2O, 41 mL 3 M H_2SO_4, and 50 g $NaH_2PO_4 \cdot H_2O$, and shake until dissolved. Dilute to volume with water.
>
> *Wash solution*: Prepare a solution containing 41 mL 3 M H_2SO_4 and 50 g $NaH_2PO_4 \cdot H_2O$ per liter water.
>
> *Surfactant standard solution*: A stock solution of 1 g anionic surfactant/liter is stable for one week; working solutions of 10 mg/liter should be prepared fresh daily.

Prepare a calibration curve by adding to each of a series of separatory funnels 100 mL H_2O spiked to contain anionic surfacant in the range 10–200 μg; include a water blank with no added surfactant. Add samples to other separatory funnels, diluting them so that they fall within the range of the standards. (A sample size of as much as 400 mL may be taken.) If sulfide ion is present, add a few drops of 30% H_2O_2 to prevent decolorization of the methylene blue dye. Neutralize to phenolphthalein by dropwise addition of 1 M NaOH or 0.5 M H_2SO_4. Add to each funnel 10 mL $CHCl_3$ and 25 mL methylene blue reagent. Shake moderately (if necessary, add less than 10 mL isopropanol to break emulsions; treat standards similarly) and carefully draw off the chloroform layer into a second separatory funnel. Repeat the extraction of the aqueous phase twice more with 10-mL portions of $CHCl_3$. If the blue color in the aqueous phase becomes exhausted, the sample size was too large and the extraction must be repeated. Combine the $CHCl_3$ extracts in the second series of funnels, and wash with 50 mL of the wash solution. Draw off the $CHCl_3$ layer through a plug of pre-extracted glass wool into a 100-mL volumetric flask. Extract the wash solution twice with 10-mL aliquots of $CHCl_3$ and add to the flask. The extracts should all be clear. Rinse the glass wool and make the flask up to volume. Measure the absorbance at 652 nm versus a $CHCl_3$ blank, prepare a calibration curve, and calculate the concentration of anionic surfactant in the sample.

This procedure has always been known to be subject to interference in a complex sample matrix (16). Anionic material other than surfactants can form an extractible ion pair with the methylene blue dye, giving high results. In fact, methods based upon methylene blue extraction have been demonstrated for determination of tetrafluoborate, perchlorate, and thiocyanate ions as well as cyano complexes of transition metals. Conversely, other cationic material, if present, will compete with methylene blue and form a colorless ion pair with the anionic surfactant, giving low results.

pH control is important, since the methylene blue loses its cationic character at high pH. At too low a pH, protonation of the anionic surfactant becomes a competitive process to ion pair formation with the titrant. The methylene blue procedure is often used in automated analytical systems. This is discussed in Chapter 17 with flow injection analysis.

(b)   Applicability of the methylene blue procedure.   In general, the materials which can be determined by two-phase titration (Chapter 16) can be determined by the methylene blue spectrophotometric method, and the interferences are similar. Soap is measured at

high pH but not at low pH, where it is insufficiently ionized. The same is true of perfluorinated carboxylic acids, which can be determined in the presence of alkyl sulfate surfactants by performing the analysis at high pH, where both are measured, and at low pH, where only the alkyl sulfate is measured (17). For environmental analysis, the method is said to be suitable for measurement of LAS and alkylether sulfates, but to give only 68% recovery with secondary alkanesulfonates (18). In general, the methylene blue method is applicable to fluorinated surfactants, but each should be checked experimentally (19). Hydrotropes like cumene sulfonate are not hydrophobic enough to form chloroform-soluble ion pairs with methylene blue, and therefore do not interfere with the procedure (1).

(c)   Modifications to the methylene blue procedure.   The methylene blue procedure itself is very simple and rapid, providing one need not worry about interference (20). However, especially for wastewater matrices, interference must be expected. Numerous steps have been proposed to eliminate positive and negative interference. Some are oxidation of sulfide ion with peroxide, careful pH control of the extraction, washing of the extract (16) (these first three steps have been incorporated into the APHA procedure), preliminary cleanup by double extraction of the surfactant with 1-methylheptylamine, acid hydrolysis, use of specially coated glassware (21,22), preliminary isolation of the surfactant by solvent sublation (23), ion exchange treatment to remove cationic surfactants (24), preliminary isolation by adsorption on XAD-2 resin and ion exchange chromatography (25), and oxidation of interfering substances with sulfuric acid/dichromate (26).

Alkyl sulfates can be differentiated from alkane sulfonates and alkylarylsulfonates, because the former are destroyed by hydrolysis with HCl, while the latter are essentially untouched (27). Use of first derivative spectrophotometry at the 631–662 nm range allows increasing the sensitivity tenfold over direct absorbance measurement (4).

An excellent approach to correcting for interference from colored substances is determination of a "blank" value, where the colorless cationic surfactant benzethonium chloride is substituted for the cationic dye (28,29). It is common to substitute other solvents for chloroform for the extraction of the blue ion pair, with methylene chloride being most common. Greater sensitivity is said to be provided by substitution with 1,2-dichloroethane (30). Other modifications to the method have been proposed in the interest of conserving reagents (31).

Procedure: Interference Limited Methylene Blue Method (25)

Prepare two columns for the resin to accommodate a bed of about 11×180 mm. They should have reservoirs of 250–500 mL capacity and stopcocks. A column should be packed with a methanol slurry of Amberlite® XAD-2 resin, then washed with water. The second column should be packed with the anion exchange resin Bio-Rex 9, also as the methanol slurry, and stored under methanol.

A sample volume is selected to contain 0.3–1.0 mg of surfactant. In the case of river water, this will be about 5 L; in the case of feed to a sewage treatment plant, it will be 250–1000 mL. The sample is passed slowly (not faster than one drop per second) through the resin bed, discarding the eluate. The reservoir and column are rinsed with two 25-mL portions of water. Water is purged from the column with ethyl ether, closing the stopcock as the last of the ether reaches the top of the column. Place a clean 250-mL extraction flask below the column and rinse the surfactant from the column with, in order, at a flow of one drop per second, 25 mL methanol, 40 mL 0.05 M methanolic NaOH, 25 mL methanol, 55 mL 25:25:5 methanol/CHCl$_3$/conc. HCl, and 25 mL methanol. The combined eluate is evaporated to dryness and transferred to a 250-mL separatory funnel with a total water volume of about 60 mL. Twenty milliliters concentrated HCl is added, and the funnel mixed well. After cooling, the mixture is extracted twice with 65 mL ethyl ether. The ether extract is evaporated to dryness in an extraction flask, then redissolved in 5 mL chloroform and 10 mL methanol. The solution is passed through the column of Bio-Rex 9 at a rate of one drop per second, rinsing the flask and column with two 10-mL portions of methanol. The resin is washed with 100 mL 0.35 M methanolic acetic acid and two more 10-mL portions of methanol. All washings are discarded. A clean extraction flask is placed under the column and the surfactant is eluted with 120 mL 4 M methanolic HCl, washing with 20 mL methanol. The eluate is evaporated to dryness, purified by re-extraction from water with ethyl ether as described previously, and the residue transferred to a 25-mL volumetric flask and made up to volume with ethyl ether. An aliquot of this solution is analyzed by the methylene blue method according to the APHA procedure.

Discussion: The elimination of interferences is aimed at making the methylene blue method more nearly specific for the chief surfactant component of municipal wastewater: LAS. The acid hydrolysis destroys alcohol sulfate surfactants, for example, which in a specific application may or may not be considered an interference to be eliminated. Acid hydrolysis also aids the recovery of LAS adsorbed onto solids. The main contribution of the anion exchange step is the elimination of cationic materials which would complex with LAS to give low results. The XAD-2 concentration step effectively removes inorganic materials from the system.

2. Other Cationic Dyes

Over the years, many other cationic dyes have been suggested as alternatives to methylene blue for the spectrophotometric determination of anionic surfactants. For example, azure I (oxidized methylene blue) will react in a similar manner to methylene blue (32), as will acridine orange (33), neutral red (34), and *bis*[2-(5-trifluoromethyl-2-pyridylazo)-5-diethylaminophenolato]cobalt(III) (35). Azure A, an analog of methylene blue, is much less sensitive to interference from nitrate and thiocyanate ion than is methylene blue itself (36). A procedure based on dimidium bromide (3,8-diamino-5-methyl-6-phenylphenanthridinium bromide) is more sensitive than the methylene blue procedure, capable of accurate measurements of anionics in the 0.1 ppm range. Even greater sensitivity is attainable if the measurement is made in the UV rather than the visible range, although accuracy is not as good (37).

For systems in which high levels of solvents, for example acetone, are present, methyl green is reported to be superior to methylene blue (38). The color formed by methyl green is stable for 2–3 hr, while the methylene blue color should be read within 30 min (13). A method based on formation of the *bis*(phenanthroline)-Cu(II)-surfactant complex is reported to be faster than the methylene blue method and more sensitive by a factor of two (39). 1-(4-Nitrobenzyl)-4-(4-diethylaminophenylazo)pyridinium bromide is said to be a much more sensitive reagent than methylene blue (40). For automated analysis by the flow injection technique, use of 1-methyl-4-(4-diethylaminophenylazo)pyridinium ion, together with chloroform extraction, was found to give the best results (41).

The iron(II) chelate with 1,10-phenanthroline or bipyridyl has been proposed for determination of anionics (42). This method is not as sensitive as MBAS, but the reagents are available in higher purity.

The triphenylmethane cationic dye, ethyl violet, has advantages over methylene blue (43,44). The ethyl violet ion pair with anionics is more completely extracted than methylene blue, giving an ethyl violet method greater sensitivity to low concentrations and less susceptibility to interfering compounds. Like methylene blue, ethyl violet also forms an extractible ion pair with chloride ion. This can be minimized by back-washing the extract with deionized water. For careful work, the chloride content of the samples and standards must be matched (45). The method has been applied to the analysis of seawater, marine sediments, and other environmental waters (46,47), as well as beverages (48). Liquid-liquid extraction of the ethyl violet-LAS ion pair can be replaced with solvent sublation into just 5 mL 9:1 toluene/MIBK (49).

Choice of the appropriate dye opens the path to a waste minimization program for the laboratory: if a stable cationic dye such as a substituted azopyridine or quinoline is used, the colored organic phase may be washed with water at pH 10 after the analysis, removing the anionic surfactant and leaving a chloroform solution of the dye suitable for use in future analyses (50,51).

It is possible to perform the paired-ion determination without using a water-immiscible solvent. The ion pair formed by an anionic surfactant with many cationic dyes, including methylene blue and rhodamine 6G, is water-insoluble and, with shaking, will adsorb to the walls of the vessel. After discarding the aqueous solution, the ion pair can be washed from the walls with ethanol or methyl cellosolve and measured spectrophotometrically. Best results are obtained with a poly(tetrafluoroethylene) vessel (52). In a similar vein, the ion pair can be filtered out, then dissolved and measured spectrophotometrically. This was demonstrated using a cationic iron complex with a nitrocellulose membrane filter which was soluble in the 2-methoxyethanol used for the spectrophotometric measurement (53).

Procedure: Analysis of River Water by the Ethyl Violet Method (44)

Transfer 100 mL sample to a large separatory funnel, and dilute with 100 mL H_2O. Add 5.0 mL of a 0.2 M, pH 5 acetate buffer solution which is also 1 M in Na_2SO_4 and 0.02 M in disodium EDTA, and 2 mL of 0.001 M ethyl violet solution. Add 5.0 mL toluene and shake well for 10 min. Measure the absorbance of the toluene extract at 615 nm in 1-cm cuvettes against a reagent blank. Determine the sample concentration by comparison to a calibration curve prepared with a suitable surfactant.

3.   Other Methods

Sulfosuccinate esters can be determined by the ferric hydroxamate procedure (54). This is a general test for esters. The esters are reacted with hydroxylamine hydrochloride in the presence of base to form the hydroxamic acids (RCONHOH). The hydroxamic acids are then detected by the color formed when ferric chloride is added to the mixture. The competing reaction, base-catalyzed hydrolysis of the esters to form the carboxylic acid, is inhibited by running the hydroxlaminolysis at ice-bath temperature. Other esters will interfere with this determination, and the various sulfosuccinates can not be distinguished from each other.

A general method for determination of water-soluble compounds having an alkyl chain of length C_8 or higher is based on the displacement of phenolphthalein from its complex with β-cyclodextrin. At pH 10.5, a solution of the complex is colorless. In the presence of surfactants such as dodecyl sulfate or the nonionic surfactant dimethyldodecylamine oxide, phenolphthalein is displaced into the solution and is measured by its absorbance at 550 nm (55).

Some cationic dyes form ion associates with anionic surfactants which give different absorbance spectra in the aqueous phase, making unnecessary the extraction of the ion pair into an organic phase. A specially synthesized dye, 1-(10-bromodecyl)-4-(4-amino-naphthylazo)pyridinium bromide, has been demonstrated effective for determination of fractional ppm levels of LAS, alkyl sulfate, and soap in water. A nonionic surfactant is added to stabilize the solution (56). Ethyl violet has also been proposed for this analysis, with the absorbance determined at two wavelengths to correct for the absorbance of excess reagent (57).

If certain anionic dyes, including bromcresol purple, methyl orange, and methyl orange derivatives, are mixed with a solution of a cationic surfactant, the wavelength of maximal light absorption shifts, giving a decrease of color intensity at the former λ_{max}. If an anionic surfactant is added, it binds the cationic, allowing the absorbance to again increase. This phenomenon can form the basis of a spectrophotometric method of determining anionics in the low parts per million range, and has the advantage over the methylene blue procedure that no organic solvent layer is required. Usually it is necessary to add a stabilizer such as a nonionic surfactant or polyvinyl alcohol to prevent precipitation of the ion pair (58–62). A number of triphenylmethane cationic dyes (malachite green, brilliant green, methyl green, iodine green, methyl violet, ethyl violet, crystal violet) can be used in similar manner, but without addition of cationic surfactant. The dye color changes due to the presence of the anionic surfactant (63,64). These methods appear to be more susceptible to interference than the methylene blue procedure, but may be suitable for systems where the matrix is well understood. They have been proposed for use in flow injection analysis. Such systems have also been adapted to optodes, but so far have not attracted practical interest (65).

One method proposed is formation of a complex of a cationic surfactant with a colored or fluorescent dye, like methyl orange or fluorescein, which is adsorbed onto silica gel. If the silica gel is stirred with a water sample containing an anionic surfactant, a proportional amount of the dye is displaced into the water (66).

A sensitive indirect method for determination of anionics consists of extracting the ion pair with bis(ethylenediamine) copper(II), then determining the copper in the extract by one of the hundreds of published methods, such as that based on the colored complex with diethylamine and 1-(2-pyridylazo)-2-naphthol (67).

II. NONIONIC SURFACTANTS

A. Nonionics Containing Alkylene Oxide Groups

As is also true for titration methods, most spectrophotometric methods for determination of ethoxylates rely on the ability of polyethers to form complexes with large cations such as K^+ or Ba^{2+} (68). For analytical purposes, the complex is formed as an ion pair with a large lipophilic anion which is measurable by its UV or visible absorbance spectrum. The ethoxy chain must be at least four units in length to form the complex.

Calibration is more complicated for nonionic than ionic surfactants. Most ionic surfactants have only one functional group per molecule. Thus calibration can be performed with any well-characterized standard compound, and the results for an unknown will still be fairly accurate. With nonionics, the color development is not only related to molar concentration, but is also a function of the ethoxy chain length and the particular hydrophobe. Thus, all of the spectrophotometric methods suffer from the disadvantage that, if an arbitrary standard must be chosen on which to base concentration measurements, large errors may creep in, since the standard will not yield the same molar calibration curve as the sample.

1. Direct Ultraviolet Absorbance

The commercial nonionic surfactants do not absorb radiation in the visible spectrum. The simplest form of spectrophotometric analysis of nonionics is the direct measurement of the UV absorbance of the sample. The ethoxylated alkylphenols are the only compounds which can be readily determined by this method, with a maximal absorbance at about 223 nm and another peak at 276 nm (69). Ethoxylated amides may be determined in model systems by their absorption of light at 202 nm (70), but many other compounds found in typical samples also have absorbance in this region. Because of the sensitivity of direct UV analysis to interference, it can only be used in well-defined situations. Interferences often encountered in nonionic surfactants are oxidation inhibitors like butylated hydroxytoluene.

For routine quality control of well-characterized alkylphenolethoxylates, the absorbance due to the aromatic ring at 275 nm may be monitored. The specific absorptivity of the pure compound is linearly related, inversely, to the degree of ethoxylation (71–73).

2. Cobalt Thiocyanate Methods

The most commonly used spectrophotometric method for determination of nonionic surfactants is the cobalt thiocyanate method. This is based upon formation of a tetrathiocyanatocobaltate(II) complex with materials containing polyether linkages which is extractable from water into organic solvents. The cation associated with this complex is usually ammonium ion, although there are hints in the literature that the potassium ion might lead to more complete extraction. Most common nonionic surfactants can be determined, providing that they have more than about 4 moles of ethylene oxide per molecule. The complex exhibits maximal absorbance in the ultraviolet region at 318 nm. However, most analysts use the somewhat weaker absorbance in the visible region at 620 nm, either because the instrumentation is less expensive or to avoid interference from other materials.

A thorough study of the cobaltithiocyanate method has been published (74). Response factors for the various surfactants vary by a factor of two or more, so the calibration substance must be chosen carefully. Anionic surfactants give enhanced absorbance but, if their concentration is approximately known, compensation may be made. Cationic surfactants must be absent, since they also form an extractable ion pair with tetracyanatocobaltate. The method is only applicable to alkoxylates, but a slight response is observed with fatty acid diethanolamides, and, to a lesser extent, with diisopropanolamides (but not the monoalkanolamides). In a study of the cobaltithiocyanate method using pure oligomers of dodecyl ethoxylates rather than the commercially available mixtures, it was concluded that Beer's law was not always obeyed (75). These investigators confirmed the observation of others that at least 2.5 moles of ethoxylated material are required to complex 1 mole of cobalt compound. No general rules could be developed, other than that a calibration curve should be prepared with the exact compound to be determined. The molar absorptivity of the complex, in the case of ethoxylated octylamides, is practically the same for the range of 10 to 20 moles of EO. It falls off somewhat below this range, with the 5-mole EO compound showing less than half the absorptivity of the 10-mole adduct (76).

Results are less predictable with high molecular weight EO/PO copolymers and with PEG, because the solubility of the ion pair in methylene chloride diminishes. For some copolymers, the response falls off to zero (77). A modified procedure works well with an 8400 MW, 80% EO copolymer: ethyl acetate is substituted for methylene chloride (78). The cobaltithiocyanate/surfactant complex formed is soluble in neither the aqueous nor the ethyl acetate phase, but can be centrifuged out as a precipitate. The precipitate is washed free of excess reagent with ethyl acetate, then dissolved in acetone and measured spectrophotometrically (79).

The cobaltithiocyanate method is lacking in sensitivity, so it is generally used in conjuction with a concentration step if low levels of surfactant must be measured. While halogenated solvents are generally used, toluene can also give acceptable results (80). If concentration and separation steps are included in the procedure, a great deal of sensitivity and specificity is attained (81–83).

In an alternative spectrophotometric analysis capable of greater sensitivity to low levels of surfactant, the cation–surfactant–cobaltithiocyanate complex may be broken up and the extracted cobalt measured by a highly sensitive reaction with 1-nitrosonaphth-2-ol-3,6-disulphonic acid (also called Nitroso-R salt), (84,85) or 4-(2-pyridylazo)-resorcinol (80).

Procedure: Determination of Cobalt Thiocyanate Active Substance in Water (14)

Ion exchange column: Prepare a glass column of dimensions about 1×30 cm. Slurry pack with Bio-Rad AG1-X2 anion exchange resin, 50–100 mesh, hydroxide form and Bio-Rad AG 50W-X8 cation exchange resin, 50–100 mesh, acid form, each to about 10 cm depth. One column may be used about six times, depending on sample impurities.

Preconcentrate the sample, if necessary, by solvent sublation or adsorption on XAD resin (Chapter 6). Purify by passage through the mixed bed ion exchange column, dissolved in methanol. Elute with methanol at a one drop per second flow rate until about 125 mL is collected in a 150-mL extraction flask. Evaporate the methanol to dryness.

Prepare a calibration curve by adding to a series of 150-mL extraction flasks, 0.0, 0.5, 1.0, 2.0, and 3.0 mg of nonionic surfactant calibration standard. Evaporate to dryness. Prepare a series of 125-mL separatory funnels. Add to each 5 mL ammonium cobalto-thiocyanate reagent [30 g $Co(NO_3)_2 \cdot 6H_2O$ and 200 g NH_4SCN per liter water]. Add 10.00 mL methylene chloride to each of the extraction flasks (containing both the sample and standards) and swirl to dissolve the residue. Take care not to experience excessive evaporation of the CH_2Cl_2 during this step. Transfer the CH_2Cl_2 solution, without rinsing, to the separatory funnel. Shake the funnels well and let the layers separate. Run the lower phase through a plug of glass wool into a 2.0-cm cuvette and determine the absorbance at 620 nm against a blank of methylene chloride. Plot a calibration curve and determine the milligrams of nonionic surfactant in the samples in the usual way. Greater sensitivity, by a factor of about four, can be obtained by making absorbance measurements at 320 nm, rather than 620 nm (86), but with a loss of selectivity.

3. Barium Iodobismuthate Methods

Another colorimetric method is a modification of Bürger's sedimetric method for nonion-ics (Chapter 17). The cationic complex formed by ethoxylates with barium ion is precipi-tated by tetraiodobismuthate ion in the presence of acetic acid. In order to avoid the inaccuracies inherent in measuring the small volume or mass of the precipitate, the precip-itate is separated and dissolved, and the bismuth ion is determined spectrophotometrically. An example is the determination of bismuth by measuring the absorptivity of its EDTA complex (87). A similar modification is the spectrophotometric determination of bismuth in the dissolved precipitate by measuring the absorbance of the diethyldithiocarbamate complex (88). A disadvantage of the iodobismuthate reagent is its poor stability. On stand-ing, free iodine is evolved, which perturbs the determination because the triiodide ion competes with the tetraiodobismuthate ion in combining with the barium complex of the nonionic. Cationic surfactants, if present, will give positive interference with iodobismuth-ate methods.

Procedure: Spectrophotometric Determination of Nonionic Surfactants After Precipitation with Modified Dragendorff Reagent (87)

Reagents:

> *KBiI₄ solution*: Dissolve 1.7g basic bismuth nitrate or bismuth nitrate pentahydrate in 20 mL glacial acetic acid and 80 mL distilled water. Mix with a solution of 65 g potassium iodide in 200 mL H_2O, add 200 mL acetic acid, and dilute to 1 L with water.
>
> *Precipitation reagent*: Add 200 mL $KBiI_4$ solution to 100 mL of barium chloride solution (290 g $BaCl_2 \cdot 2H_2O$ in 1 L water). Store in a brown bottle and discard after a week.
>
> *Ammonium tartrate solution*: 1.77 g/100 mL, freshly prepared.
>
> *Bromcresol purple indicator*: 0.1% in methanol. *EDTA solution*: 0.02 M, 7.44 g/L reagent grade (ethylenedinitrilo)tetraacetic acid disodium salt dihydrate.

Prepare a calibration curve by adding to a series of 250-mL beakers, 0.00, 0.20, 0.40, 0.60, 0.80, and 1.00 mg of the appropriate nonionic surfactant standard, along with about 40 mL distilled water. Put the surfactant samples in similar beakers, choosing the sample size to fall within the range of the standards and adding water if needed to reach at least the 40 mL mark. If the sample is dissolved in a few milliliters methanol, add the same amount of methanol to the standard solutions. Add a few drops of bromcresol purple indicator solution, then adjust with 0.2 M HCl as needed to obtain the color change to yellow. Add 30 mL precipitation reagent and stir occasionally for 10 min, then let stand for 10 min or more. Filter with suction through a fine-porosity sintered glass crucible, and wash with three 10-mL portions of glacial acetic acid. Transfer the filtering crucible to a clean crucible holder fitted into a clean filter flask. Dissolve the precipitate by pouring three 15-mL aliquots of hot ammonium tartrate solution through the crucible, using suction, followed by 10 mL water. Pour the contents of the filter flask back into the original 250-mL beaker, rinsing with two 10-mL aliquots of water. Add 4 mL EDTA solution, mix, and transfer to a 100-mL volumetric flask. Make up to volume with water and determine the absorbance at 264 nm versus water in a 2-cm silica cuvette. Determine the mg of nonionic in the sample beakers by comparison to the calibration curve, and calculate the original concentration.

> *Note*: Filtering crucibles may be cleaned by soaking in 1% acid permanganate solution, followed by soaking in 10% hydrogen peroxide.

4. Potassium Picrate Methods

Favretto and coworkers demonstrated a spectrophotometric method based upon the extraction of picrate ion from water into an organic solvent in association with potassium ion complexes of polyoxyethylene chains (89–92). This procedure, due to the greater molar absorptivity of the picrate ion, is about eight times more sensitive than the cobaltithiocyanate method (at 620 nm; when the cobaltithiocyanate method is used at 320 nm, the sensitivity of the two methods is similar). The barium–surfactant–picrate complex is reported to be extracted more readily, and have a higher apparent absorptivity, than the potassium complex (93). The procedure is more sensitive to AE and APE than to ethoxylated esters or to EO/PO copolymers (94). The method has been successfully applied to environmental analysis after concentration of nonionic by solvent sublation and further purification by anion/cation exchange (95).

The procedure typically consists of mixing 5 mL of 0.02 M potassium picrate solution (in 10^{-4} M KOH) with 20.0 mL 2.5 M KNO$_3$ solution and the aqueous sample in a 50-mL volumetric flask and diluting to volume. The entire mixture is transferred to a separatory funnel and extracted with 5 mL 1,2-dichloroethane, the absorbance of which is measured at 378 nm versus a reagent blank, with the concentration determined from a calibration curve prepared by adding aliquots of surfactant standard solution in the range of 0.5 to 50 μg to 50-mL volumetric flasks. For increased sensitivity, the nonionic surfactant may first be concentrated from a large aqueous sample by extraction with methylene chloride, followed by purification of the extract by washing sequentially with acidic and alkaline water (92).

The potassium picrate method is susceptible to positive interference from cationic surfactants and, to a lesser degree, negative interference from anionic surfactants. If the quantity of cationic surfactant is not too large, the interference is eliminated by back-washing the organic extract with plain water. In the absence of excess potassium picrate in the aqueous phase, the nonionic surfactant complex is destroyed. Thus, the difference in absorbance of the organic phase before and after water-washing is due to the concentration of the nonionic surfactant (92).

Modifications have been suggested to the picrate method (96). These consist of addition of 1 mL ethanol to speed the extraction and adjustment of pH to 12 to inhibit the extraction of free picric acid and obtain a lower blank value. The linear range is from approximately 0.04–1.0 ppm, depending on the surfactant.

Despite the advantages of this method, it is rarely used in North America, where most older chemists have had the experience of watching police or military bomb squads coming to their workplaces to dispose of aged picric acid. Picrate compounds will detonate when handled in the dry state. Picric acid must always be stored with 10% or more water and the inventory must be carefully controlled to avoid future problems.

5. Heteropoly Acid Methods

A family of methods once popular in the pharmaceutical industry is based on the ability of barium complexes of polyethylene oxide compounds to form precipitates with heteropoly acid anions, such as tungstosilicic, molybdophosphoric, and tungstophosphoric acid. The precipitate is separated, dissolved, and the tungsten or molybdenum determined by one of many colorimetric procedures (97–99). This approach is too tedious for use in the modern quality assurance laboratory. It is possible to streamline the procedure by replacing the gravimetric separation of the complex with a liquid-liquid extraction (100). The heteropoly acid methods are generally avoided because these reagents give positive tests with many substances, including nitrogen compounds and cellulose.

6. Iodine Methods

Polyether compounds form stable adducts with iodine, a property which has led to their use as iodophores for bactericides on human and animal skin. For medical applications, iodide concentration is kept low, since the triiodide ion has little disinfecting activity compared to iodine. By contrast, iodine-based analytical methods for surfactant determination are usually conducted in the presence of a large excess of potassium iodide. The likely mechanism consists of the K$^+$–polyether complex associating with the large, lipophilic, triiodide anion. The absorbance maximum of the complex is shifted from that of triiodide

alone, allowing quantitative determination of most nonionics, including the propylene ox-
ide/ethylene oxide copolymers (101,102). This behavior is dependent upon the structure of
the compound, the ethoxy chain length, and concentration (103,104).

In the past, methods based upon triiodide were seldom used, since they were neither
sensitive nor selective. This area has been revisited in the last decade by Boyd-Boland and
Eckert, who added an extraction step to the determination, making the method suitable for
analysis in the 0.1 ppm range. In their study with an OPE, they found that a 1:1:1 complex
of OPE/potassium/triodide was extracted into 1,1,1-trichloroethylene, where it gave a sensitive absorption peak at 380 nm (due to triiodide ion), well removed from the 510 nm
peak of free I_2. The method was demonstrated for analysis of sea water containing low
levels of anionic surfactant and worked well at the 1 ppm OPE level. High levels of anionics would be expected to interfere by competing with the triiodide ion to form the ion pair
with the K^+–nonionic adduct (105).

7. Decomposition Methods

Other procedures are based upon determination of the decomposition products of polyoxyalkalene compounds. By reaction with phosphoric acid, ethylene oxide and propylene
oxide moieties are converted, with 80–90% yield, to acetaldehyde and propionaldehyde,
which may be measured spectrophotometrically (106). By more drastic decomposition
with perchloric acid and periodate ion, the oxyalkalene compounds are converted to formaldehyde, with 55–65% yield, which may be distilled and determined colorimetrically
(107). Neither of these methods has been much used.

As discussed in Chapter 8, ferric chloride–catalyzed reaction of alkoxylates with
acetyl chloride results in decomposition of the polyether structure, converting ethylene oxide units to mainly 2-chloroethyl acetate. The reaction mixture develops a red color which
is proportional to the amount of polyether originally present. It has not been determined
exactly which reaction product produces the colored complex with ferric ion, but the phenomenon has been demonstrated as suitable for quantitatively determining low levels of
polyether (108). Since the procedure cannot be used with aqueous samples, its application
is limited.

8. Other Approaches

Conditions can often be found under which the absorbance of a dye is affected by the presence of a nonionic surfactant, usually APE. For example, the anionic dye dichlorofluorescein has been proposed for the determination of APE (109). The surfactant lowers the
absorptivity of a dichlorofluorescein solution, while other nonionic surfactants and low
amounts of anionic surfactants have little effect. Cationic surfactants interfere with the
analysis. Similarly, the anionic dye alizarin fluorine blue is proposed for determination of
APE because of the increase in intensity and change in wavelength of the dye's absorbance. Ionic and other nonionic surfactants must be absent (110).

A colorimetric method is based upon the extraction of potassium ion into an organic
phase by ethoxylates in a system where the counter ion, also extracted, is the anionic
dye tetrabromophenolphthalein, ethyl ester (111). Anionic surfactants are a serious
interference.

An indirect method is based upon the extraction of the potassium adduct of the nonionic with tetrathiocyanatozincate(II) ion, with subsequent determination of the zinc in the
extract by the 1-(2-pyridylazo)-2-naphthol method (112).

An extraction-spectrophotometric method has been developed based upon ferric thiocyanate which is generally similar to the cobalt thiocyanate method (113). It has the advantage of giving a higher specific absorbance value than the cobalt method, making the technique more applicable to determination of low concentrations of surfactant.

B. Nonethoxylated Nonionics

Esters and amine oxides may be decomposed and determined by methodology appropiate for analysis of the polyol, fatty acid, or fatty amine starting materials. For example, sucrose esters may be determined using the same procedures as are used for determination of sucrose itself, such as anthrone (9,10-dihydro-9-oxoanthracene) reagent (114). Alkylpolyglycosides can also be determined by reaction with anthrone, which gives a general test for carbohydrates. Hydrolysis with strong acid is part of the procedure, so that the polyglucosides are converted to glucose, which forms a green complex with anthrone. Cellulose derivatives like carboxymethylcellulose interfere, as do reducing sugars and, to a minor extent, propylene glycol (115,116). The method is, however, applicable in many circumstances, such as for determination of alkylpolyglycosides in shampoo formulations.

III. CATIONIC AND AMPHOTERIC SURFACTANTS

Some cationic and amphoteric surfactants contain aryl groups, giving them strong UV absorptivity. Imidazoline surfactants have a strong absorbance at about 235 nm (117). However, this is a function of pH, since the unstabilized imidazoline ring opens at high pH (118). Direct UV determination of surfactants is practical only in special cases because of the abundance of interfering compounds.

Like anionic surfactants, cationics can be determined by formation of ion pairs. This is the principle of all spectrophotometric methods in general use. A colored, water-soluble, anionic species is chosen that is only extracted into an organic phase in the presence of cationic surfactants. Many such extraction methods have been published. They generally suffer from negative interference from anionic surfactants and positive interference from certain nonsurfactant cationic substances. The most widely accepted method is based upon disulfine blue (119). Ion exchange chromatography has long been used to avoid the drawbacks of the direct ion pair analysis, namely the insensitivity of the methods to low concentration and interference with the extraction step by fatty matter in the sample (120). The cationic is concentrated by ion exchange and put in the form of the chloride salt or some other well-defined species.

While most of the reagents discussed here will give adequate results over a wide range of pH with most cationics, some are more sensitive than others for determination of individual surfactants. If a method is developed to determine a specific cationic surfactant, it is desirable to optimize the pH for maximal extraction and color development, and to choose the dye which gives the best performance (121,122). Difunctional anionic dyes usually have more narrow optimal pH ranges than do monofunctional dyes. In general, amphoteric surfactants have cationic character at low pH and can be determined by the same dye transfer methods applicable to cationics (123).

A. Disulfine Blue

Disulfine blue is the dye most often used for determination of cationic surfactants, especially in environmental matrices. The disulfine blue method is made suitable for the analysis of materials containing other surfactants and interfering substances by the addition of an ion exchange step to isolate the cationic (124). The disulfine blue method has the disadvantage that a calibration curve must be prepared for each cationic surfactant, and even for each technician, since color formation is influenced by the vigor of mixing the phases, as well as the pH of the aqueous phase (4). Inorganic ions such as perchlorate and iodide will compete with disulfine blue in forming extractible complexes with quaternary amines, but this interference is only significant at relatively high concentrations of the anions (125).

Procedure: Determination of Cationic Surfactants by Reaction with Disulfine Blue (119)

Reagents:

> *Disulphine blue VN 150 solution* is prepared by dissolving 0.16 g of the dye in 250 mL 90:10 water/ethanol.
> *Acetate buffer* is prepared by dissolving 115 g anhydrous sodium acetate in about 500 mL H_2O and adjusting the pH to 5.0 by addition of glacial acetic acid, diluting to a final volume of 1 L.

Fill an anion exchange column (1.3×17 cm, equipped with a stopcock and a reservoir) with Bio-Rad AG 1-X2, chloride form. Wash with methanol and precondition by passing 10 mL of a 0.1% solution of a cationic surfactant through the column at 2 mL/min. Wash with 600 mL methanol, at 2–3 mL/min. Pass through the ion exchange column 10 mL methanol containing 50 to 500 µg cationic surfactant, using a flow rate of 1 mL/min. Collect the eluate in a 250-mL extraction flask. Wash the column with 100 mL methanol, collecting the washings in the same flask. Evaporate to dryness on a steam bath. Redissolve the residue in 20 mL boiling $CHCl_3$ and transfer to a 100-mL volumetric flask. Rinse the extraction flask with additional $CHCl_3$, adding the rinsings to the volumetric flask. Make up to volume. To a 40-mL conical centrifuge tube, add 2.5 mL acetate buffer solution, 1.0 mL disulphine blue solution, and 15 mL water. Add 10.0 mL of the $CHCl_3$ sample solution, stir vigorously, then centrifuge at 2000 rpm for 20 sec for complete separation of the layers. Use a siphon to transfer a portion of the lower, organic phase into a 1-cm cuvette and measure the absorbance at 628 nm versus $CHCl_3$. Compare to a calibration curve prepared by adding a reagent blank and cationic surfactant standards in the 0–50 µg range to centrifuge tubes. The reagent blank, but not the standards, is also subjected to the ion exchange cleanup step.

B. Picric Acid

Picric acid, also a reagent for spectrophotometric determination of nonionic surfactants, is useful for determination of cationics (126–128). The ion pair of the quaternary amine and picric acid is extracted from the aqueous sample solution into chloroform or 1,2-dichloroethane, and the UV absorbance of the phenol read at 365 or 375 nm. The pH range 4–12 is suitable. Nonionic surfactants interfere positively because of their ability to form complexes with inorganic cations. Most such interference can be eliminated by back-washing the extract with pure water before measuring the absorbance (92). Anionic surfactants give negative interference. As noted above, because of the ability of picric acid to detonate when dry, it is not used for routine analysis in North America.

C. Tetraiodobismuthate

This reagent is often used in the form of modified Dragendorff reagent, a solution containing barium chloride as well as potassium tetraiodobismuthate and acetic acid. As such, it is applied most often to precipitation of nonionic surfactants and visualization of TLC plates. If used for spectrophotometric determination of cationics, barium ion is not necessary, since it is the ion pair of the cationic with tetraiodomismuthate which is extracted (129). The method can be made reasonably specific for quaternary alkylammonium compounds; the iodobismuthate complexes of quaternary imidazolines and benzylalkyldimethyl-ammonium compounds are not soluble if CCl_4 is used for the extraction (4,130,131). Nonionics form a precipitate with Dragendorff reagent which is not extracted; if barium is omitted, only a slight turbidity is encountered. Anionics are reported to not interfere with the determination of distearyldimethylammonium chloride. Thus, the ion exchange pretreatment step required for the disulfine blue method may sometimes be omitted for the Dragendorff method. If freshly prepared reagent is not used, the free iodine which forms will extract into the organic phase and interfere with the measurement, although the free iodine may be stripped off if desired. Positive interference sometimes occurs with environmental samples. This can be eliminated by passing the extract through an alumina column to remove the Dragendorff reagent, then performing the disulfine blue determination. Alternatively, the original environmental extract may be cleaned up by ion exchange chromatography on silica or alumina prior to reaction of the cationic fraction with tetraiodobismuthate reagent. If the extract has a yellow color prior to reaction, the final spectrophotometric reading is made versus the "specimen blank." The disulfine blue method is 11 times more sensitive than the Dragendorff method (131). However, the Dragendorff method has adequate sensitivity for most applications if first-derivative spectrophotometry is used (4).

D. Orange II

Orange II, sodium *para*-(2-hydroxy-1-naphthylazo)benzenesulfonate, is used for determination of a variety of amphoteric and cationic surfactants (132). The cationic–Orange II complex is extracted into $CHCl_3$, and the absorbance determined at 485 nm. This technique has been combined with flow injection analysis to develop an automated method (133). Amines and amphoteric surfactants can be distinguished from quaternaries by performing the analysis at low pH (total amines, amphoterics, and quats) and at neutral pH (quats only). The method was demonstrated for analysis of *N*-lauryl-3-aminopropionic acid and 2-heptadecyl-1-methyl-1-[(2-stearoylamido) ethyl] imidazolinium methylsulfate. At pH 3.6, both amphoterics and cationics are determined, while at pH 9.2 only cationics are measured (132). Gerhards and Schulz demonstrated linear calibration curves down to 50 ppb of cocoamidopropyl betaine, alkyl betaine, and amphoglycinate amphoterics. A minimum alkyl chain length of about ten carbons is required for extraction of the ion pair (134).

E. Other Reagents

Chromotrope 2B (1,8-dihydroxy-2-(*p*-nitrophenylazo)naphthalene-3,6-disulfonic acid) is said to be a superior ion pairing reagent for colorimetric determination of cationics because its two sulfonate groups make it less lipophilic than other dyes (135). This property makes it incapable of forming an extractible complex with a nonionic surfactant, even at high cation concentrations. It has the additional advantage that the free reagent cannot be extracted, even at low pH. Only the presence of a hydrophobic cation allows it to be extracted. In the presence of nonionic surfactants, best results are obtained if the solution is 0.1 M in sulfuric acid. Anions, both anionic surfactants and high concentrations of inorganic ions such as chloride and perchlorate, introduce negative interference which may be eliminated by exchanging them for sulfate ions via an ion exchange resin.

Quaternary amines will form an extractible complex with tetrathiocyanatocobaltate ion, the same reagent used for determination of nonionic surfactants (136). This is the basis for a colorimetric method suitable for relatively high concentrations of cationics. Nonionic surfactants do not interfere in the absence of potassium ion. Methods have also been written based on use of bromophenol blue (137) and methyl orange (122).

Tetrabromophenolphthalein ethyl ester has the useful property of "thermochromism," i.e., colored associates of amines and the reagent become colorless at elevated temperature, while the ion pairs with quats remain colored. This permits convenient elimination of interference from amines (138).

Sakai proposed increasing the selectivity of the dye transfer methods by forming a ternary complex of multiprotic dyes such as bromcresol green and bromophenol blue with quinine or quinidine and the quaternary amine of interest. Such complexes form over a relatively narrow pH range and the methods suffer from less interference from amines and other cationic compounds (139–141).

Extraction with bromothymol blue is proposed for determination of quaternary amines. A yellow complex is formed in aqueous solution, which is extractible into an immiscible solvent. The optimal pH varies for each surfactant, but is in the range 7–8. If the extract is made alkaline with triethanolamine, an intense blue color is liberated (69). Bengal red (sodium tetraiodotetrachlorofluorescein) has also been used (142). If the anionic surfactant LAS is used as a reagent, the UV absorbance of the organic phase is proportional to the amount of cationic surfactant present (4).

Quercetin, a natural flavonol, oxidized with N-bromosuccinimide, has been demonstrated as a colored extraction reagent for the determination of cetylpyridinium chloride and other amines (143).

Spectrophotometric determination of cationics without extraction can be performed by forming a ternary complex of iron (III), chrome azurol S, and the surfactant. The method is susceptible to interference and has not yet been proven in environmental analysis (144).

The change in absorbance wavelength of the anionic dye pyrocatechol violet in the presence of germanium and cationic surfactants has been used as a method for determination of imidazoline quats in the 10 ppm range in acid metal plating baths (145).

A novel method has been demonstrated for determination of cationics based on formation of a hydrophobic ion pair with bis[4-hydroxy-3-(8-quinolyl)azo-1-naphthalene-sulfonato]cobaltate(III). The ion pair is filtered from aqueous solution on a transparent membrane filter, and the color of the filter is measured directly in a spectrophotometer (146).

F. Indirect Determination

So far, a reagent has not been found for cationics which has the advantages of methylene blue for anionics: a well-defined, stoichiometric complex, rapid kinetics of ion pair formation, and good sensitivity. These advantages can be obtained if the methylene blue/LAS ion pair is used for an indirect determination of cationics. The cationic surfactant pairs with LAS, displacing the methylene blue, so that the color of the organic phase is indirectly proportional to cationic surfactant concentration. The same principle may be used with quantification of unextracted LAS by UV absorbance measurement of the aqueous phase, if a known quantity of cationic surfactant such as distearyldimethylammonium chloride is used in place of methylene blue (4).

Addition of a cationic surfactant to a solution of an anionic dye results in a change in the wavelength of maximum light absorbance, observed as a decrease in intensity of the dye color at the normal λ_{max}. This phenomenon may be used as a means of determining low levels of cationics which does not require extraction of ion pairs into an organic solvent. Methodology has been developed based on bromothymol blue (147), azo dyes of the methyl orange series (61,62), and bromcresol purple (59). Precipitation is avoided by adding a nonionic surfactant to solubilize the ion pair in micelles in the aqueous phase. This approach has been proposed for flow injection analysis. In practice, the blue shift phenomenon is subject to interference from a number of ionic and nonionic compounds, so that it is useful only in well understood sample matrices.

REFERENCES

1. Hellmann, H., Cumene sulfonate in the presence of LAS by spectroscopic methods (UV, fluorescence, IR) (in German), *Tenside, Surfactants, Deterg.*, 1994, *31*, 200–206.
2. Weber, W. J., Jr., J. C. Morris, W. Stumm, Alkylbenzenesulfonates by UV spectrophotometry, *Anal. Chem.*, 1962, *34*, 1844–1845.
3. Setzkorn, E. A., R. L. Huddleston, UV spectroscopic analysis for following the biodegradation of hydrotropes, *J. Am. Oil Chem. Soc.*, 1965, *42*, 1081–1084.
4. Hellmann, H., Cationic and anionic (LAS) surfactants in sediments, suspended matter, and sludges (in German), Z. *Wasser Abwasser Forsch.*, 1989, *22*, 131–137.
5. Theraulaz, F., L. Djellal, O. Thomas, LAS in sewage using advanced UV spectrophotometry, *Tenside, Surfactants, Deterg.*, 1996, *33*, 447–451.
6. Hellmann, H., UV spectroscopy of LAS in relevant environmental matrices—fundamentals and LAS determination under relatively problem-free conditions (in German), Z. *Wasser Abwasser Forsch.*, 1990, *23*, 62–69.
7. Hellmann, H., Anionic surfactants (LAS) in German streams using second derivative UV spectroscopy (in German), Z. *Wasser Abwasser Forsch.*, 1991, *24*, 178–187.
8. Brandli, E. H., R. M. Kelley, Automated determination of alkyl aryl sulfonates in spray-dried detergents by UV absorption, *J. Am. Oil Chem. Soc.*, 1970, *47*, 200–202.
9. Sandvik, E. I., W. W. Gale, M. O. Denekas, Petroleum sulfonates, *Soc. Pet. Eng. J.*, 1977, *17*, 184–192.
10. Ortuño, J. A., C. Sánchez-Pedreño, M. C. Torrecillas, Extraction-spectrophotometric determination and design of a selective electrode for determination of anionic surfactants based on the ion pair with 1,2,4,6-tetraphenylpyridine (in Spanish), *Quim. Anal.* (Barcelona), 1990, *9*, 255–268.
11. American Society for Testing and Materials, Sodium alkylbenzene sulfonate in synthetic detergents by UV absorption, D1768. West Conshohocken, PA 19428.
12. Weber, R., K. Levsen, G. J. Louter, A. J. H. Boerboom, J. Haverkamp, Direct mixture analysis of surfactants by combined field desorption/collisionally activated dissociation MS with simultaneous ion detection, *Anal. Chem.*, 1982, *54*, 1458–1466.
13. Ströhl, G. W., D. Kurzak, Absorption spectra and stability of complexes of anionic surfactants with methyene blue and methyl green (in German), *Tenside, Surfactants, Deterg.*, 1969, *6*, 74–76.
14. Clesceri, L. S., Greenberg, A. E., A. D. Eaton, eds., *Standard Methods for the Examination of Water and Wastewater*, 20th ed., American Public Health Association, Washington, DC, 1998.
15. International Organization for Standardization, Water Quality—determination of surfactants—Part 1: Determination of anionic surfactants by measurement of the methylene blue index (MBAS), ISO 7875/1:1984. Geneva, Switzerland.

16. Longwell, J., W. D. Maniece, Anionic detergents in sewage, sewage effluents and river waters, *Analyst*, 1955, *80*, 167–171.

17. Sharma, R., R. Pyter, P. Mukerjee, Spectrophotometric determination of perfluoro carboxylic acids (heptanoic to decanoic) and sodium perfluorooctanoate and decyl sulfate in mixtures by dye-extraction, *Anal. Lett.*, 1989, *22*, 999–1007.

18. Schröder, H. Fr., Surfactants—comparison of class-specific methods with LC-MS substance-specific analysis (in German), *Vom Wasser*, 1992, *79*, 193–209.

19. Kissa, E., Anionic fluorinated surfactants, in J. Cross, ed., *Anionic Surfactants: Analytical Chemistry*, 2nd ed., Marcel Dekker, New York, 1998.

20. Koga, M., Y. Yamamichi, Y. Nomoto, M. Irie, T. Tanimura, T. Yoshinaga, Anionic surfactant by the methylene blue spectrophotometric method, *Analytical Sciences*, 1999, *15*, 563–568.

21. Fairing, J. D., F. R. Short, Spectrophotometric determination of alkyl benzenesulfonate detergents in surface water and sewage, *Anal. Chem.*, 1956, *28*, 1827–34.

22. American Society for Testing and Materials, Methylene blue active substances, D2330. West Conshohocken, PA 19428.

23. Wickbold, R., ABS in surface waters by the Longwell-Maniece method (in German), *Tenside, Surfactants, Deterg.*, 1976, *13*, 32–34.

24. Ströhl, G. W., Anionic and cationic surfactants in wastewater (in German), *Tenside, Surfactants, Deterg.*, 1969, *6*, 78–79.

25. Osburn, Q. W., LAS in waters and wastes, *J. Am. Oil Chem. Soc.*, 1986, *63*, 257–263.

26. Hill, W. H., M. A. Shapiro, Y. Kobayashi, Alkylbenzene sulfonate in water, *J. Amer. Water Works Ass.*, 1962, *54*, 409–416.

27. Ströhl, G. W., D. Kurzak, Sulfate and sulfonate type anionic surfactants, *Talanta*, 1969, *16*, 135–137.

28. Wiskerchen, J., Collaborative study of a method for sodium lauryl sulfate in egg whites, *J. Assoc. Off. Anal. Chem.*, 1968, *51*, 540–542.

29. Cunniff, P., ed., *Official Methods of Analysis of AOAC International*, 16th ed., AOAC International, Arlington, VA, 1999.

30. Ho, W., Spectrophotometric determination of anionic surfactant with methylene blue, *Anal. Sci.*, 1991, *7(Suppl. Proc. Int. Congr. Anal. Sci., 1991)*, 61–64.

31. Chitikela, S., S. K. Dentel, H. E. Allen, Modified method for the analysis of anionic surfactants as methylene blue active substances, *Analyst*, 1995, *120*, 2001–2004.

32. Kalinichenko, K. P., Anionic surfactants in natural waters using Azure I (in Russian), *Gidrobiol. Zh.*, 1987, *23*, 107–110.

33. Canoves, A. M. F., F. B. Serrat, Colorimetric determination of anionic surfactants with acridine orange by ion pair extraction (in Spanish), *Afinidad*, 1987, *44*, 483–486.

34. Kuban, V., J. Jurasova, Extraction spectrophotometric and two-phase titrimetric determination of ionic surfactants with neutral red, *Scr. Fac. Sci. Nat. Univ. Purkynianae Brun.*, 1988, *18(3–4)*, 159–166.

35. Kasahara, I., K. Hashimoto, T. Kawabe, A. Kunita, K. Magawa, N. Hata, S. Taguchi, K. Goto, Spectrophotometric determination of anionic surfactants in sea water based on ion-pair extraction with bis[2-(5-trifluoromethyl-2-pyridylazo)-5-diethylaminophenolato]cobalt(III), *Analyst*, 1995, *120*, 1803–1807.

36. Van Steveninck, J., J. C. Riemersma, Long-chain alkyl sulfates as chloroform-soluble Azure A salts, *Anal. Chem.*, 1966, *38*, 1250–1251.

37. Orthgiess, E., B. Dobias, Colorimetric determination of anionic surfactants, *Tenside, Surfactants, Deterg.*, 1990, *27*, 226–228.

38. Gould, E., Extractant for microamounts of anionic surfactant bound to large amounts of protein, with subsequent spectrophotometric determination, *Anal. Chem.*, 1962, *34*, 567–571.

39. Bailey, B. W., J. M. Rankin, R. Weinbloom, Anionic surfactants in water, *Intern. J. Environ. Anal. Chem.*, 1971, *1*, 3–9.

40. Higuchi, K., Y. Shimoishi, H. Miyata, K. Toei, T. Hayami, Spectrophotometric determination of anionic surfactants in river waters using 1-(4-nitrobenzyl)-4-(4-diethylaminophenylazo)-pyridinium bromide, *Analyst*, 1980, *105*, 768–773.

41. Motomizu, S., Y. Hazaki, M. Oshima, K. Toei, Spectrophotometric determination of anionic surfactants in river water with cationic azo dye by solvent extraction-flow injection analysis, *Anal. Sci.*, 1987, *3*, 265–269.

42. Taylor, C. G., B. Fryer, Determination of anionic detergents with iron(II) chelates: application to sewage and sewage effluents, *Analyst*, 1969, *94*, 1106–1116.

43. Motomizu, S., S. Fujiwara, K. Toei, Liquid-liquid distribution behavior of ion pairs of triphenylmethane dye cations and their analytical applications, *Anal. Chim. Acta*, 1981, *128*, 185–194.

44. Motomizu, S., S. Fujiwara, A. Fujiwara, K. Toei, Solvent extraction-spectrophotometric determination of anionic surfactants with ethyl violet, *Anal. Chem.*, 1982, *54*, 392–397.

45. Yamamoto, K., S. Motomizu, Solvent extraction-spectrophotometric determination of anionic surfactants in sea water, *Analyst*, 1987, *112*, 1405–1408.

46. Wu, J., J. Li, Y. Li, H. Song, Anionic surfactants in the estuary, *Kexue Tongbao* (foreign lang. ed.), 1987, *32*, 863–864.

47. Adam, B., B. Gorenc, D. Gorenc, Triphenylmethane dyes as reagents for extractive-spectrophotometric determination of anionic surfactants, *Vestn. Slov. Kem. Drus.*, 1988, *35*, 339–350.

48. Gorenc, D., B. Adam, B. Gorenc, Anionic surfactants in mineral waters and soft drinks by spectrophotometry, *Mikrochim. Acta*, 1991, *I*, 311–315.

49. Ueno, K., E. Kobayashi, T. Hobo, S. Suzuki, Solvent sublation/spectrophotometric determination of anionic surfactants, *Bunseki Kagaku*, 1987, *36*, 740–744.

50. Kubota, H., M. Idei, S. Motomizu, Liquid-liquid distribution of ion associates of anions with 4-[4-alkyl(aryl)aminophenylazo]pyridines and their use as spectrophotometric reagents for anionic surfactants, *Analyst*, 1990, *115*, 1109–1115.

51. Kubota, H., M. Katsuki, S. Motomizu, Batchwise and flow-injection methods for the spectrophotometric determination of anionic surfactants with 4-(4-*N*,*N*-dimethylaminophenylazo)-2-methylquinoline, *Anal. Sci.*, 1990, 6, 705–709.

52. Kamaya, M., Y. Tomizawa, K. Nagashima, Spectrophotometric method for the determination of an anionic surfactant without liquid-liquid extraction, *Anal. Chim. Acta*, 1998, *362*, 157–161.

53. Chen, Y., S. Wang, R. Wu, D. Qi, T. Zhou, Spectrophotometric determination of SDS and LAS in water after preconcentration on an organic solvent-soluble membrane filter, *Anal. Lett.*, 1998, *31*, 691–701.

54. Feldman, J. A., S. G. Frank, T. J. Holmes, Di-(2-ethylhexyl) sodium sulfosuccinate and related dialkyl esters in aqueous solutions by the ferric hydroxamate procedure, *J. Pharm. Sci.*, 1971, *60*, 920–921.

55. Sasaki, K. J., S. D. Christian, E. E. Tucker, Visible spectral displacement method to determine the concentration of surfactants in aqueous solution, *J. Colloid Interface Sci.*, 1990, *134*, 412–416.

56. Shimoishi, Y., H. Miyata, Spectrophotometric determination of anionic surfactants in tap and river waters with 1-(10-bromodecyl)-4-(4-aminonaphthylazo)pyridinium bromide, *Fresenius' Z. Anal. Chem.*, 1990, *338*, 46–49.

57. Gao, H. W., Dual-wavelength β-correction spectrophotometry for directly determining anionic detergents in environmental water, *Anal. Proc.*, 1995, *32*, 197–200.

58. Yamamoto, K., K. Ikehara, S. Motomizu, Spectrophotometric method for anionic surfactants on the basis of color change of bromocresol purple with a quaternary ammonium ion (in Japanese), *Bunseki Kagaku*, 1990, *39*, 393–397.

59. Yamamoto, K., S. Motomizu, Spectrophotometric method for ionic surfactants by flow injection analysis with acidic dyes, *Anal. Chim. Acta*, 1991, *246*, 333–339.

60. Motomizu, S., M. Oshima, Y. Gao, Flow injection spectrophotometric determination of anionic surfactants with an anionic azo dye and a quaternary ammonium ion, *Anal. Sci.*, 1991, 7(*Supp.*), 301–304.

61. Motomizu, S., M. Oshima, Y. Hosoi, Spectrophotometric determination of cationic and anionic surfactants with anionic dyes in the presence of nonionic surfactants. Part I: General aspect, *Mikrochim. Acta*, 1992, *106*, 57–66.

62. Motomizu, S., M. Oshima, Y. Hosoi, Spectrophotometric determination of cationic and anionic surfactants with anionic dyes in the presence of nonionic surfactants. Part II: Development of batch and flow injection methods, *Mikrochim. Acta*, 1992, *106*, 67–74.

63. Shaopu, L., H. Zhigui, J. Fangying, Color reaction of triphenylmethane basic dyes with anionic surfactants in aqueous solution, *J. Surf. Sci. Technol.*, 1990, 6, 223–229.

64. Oshima, M., S. Motomizu, H. Doi, Interaction of hydrophobic anions with cationic dyes and its application to the spectrophotometric determination of anionic surfactants, *Analyst*, 1992, *117*, 1643–1646.

65. Chan, W. H., A. W. M. Lee, J. Lu, Optode for the specific determination of anionic surfactants, *Anal. Chim. Acta*, 1998, *361*, 55–61.

66. Zaporozhets, O. A., O. Y. Nadzhafova, V. V. Verba, S. A. Dolenko, T. Y. Keda, V. V. Sukhan, Solid phase reagents for the determination of anionic surfactants in water, *Analyst*, 1998, *123*, 1583–1586.

67. Bhat, S. R., P. T. Crisp, J. M. Eckert, N. A. Gibson, Spectrophotometric method for anionic surfactants, *Anal. Chim. Acta*, 1980, *116*, 191–193.

68. Okada, T., Complexation of poly(oxyethylene) in analytical chemistry, *Analyst*, 1993, *118*, 959–971.

69. El-Khateeb, S., S. M. Hassan, A. K. S. Ahmad, M. M. Amer, Spectrophotometric determination of some nonionic surfactants, *Tenside, Surfactants, Deterg.*, 1979, *16*, 27–29.

70. Lin, J. T., D. G. Cornell, Correction method for UV spectrophotometry of turbid systems: determination of *N*-polyethoxylated alkyl amide in clay supernatant, *Anal. Chem.*, 1986, *58*, 830–833.

71. De la Guardia, M., J. L. Carrion, J. Medina, UV-visible spectophotometry in the determination of average properties of NPE, *Anal. Chim. Acta*, 1983, *155*, 113–121.

72. Carrion, J. L., S. Sagrado, M. De la Guardia, Characterization of ethylene oxide/*tert*-octylphenol condensates by UV and IR spectrometry, *Analyt. Chim. Acta*, 1986, *185*, 101–107.

73. De la Guardia, M., J. E. Tronch, J. L. Carrion, A. Aucejo, Mathematical models in the characterization of ethylene oxide condensate surfactants by IR and UV spectroscopy, *Analusis*, 1988, *16*, 124–130.

74. Milwidsky, B. M., Production control of the nonionic component in synthetic detergents, *Analyst*, 1969, *94*, 377–386.

75. Nozawa, A., T. Ohnuma, T. Sekine, Nonionic surfactants that contain polyoxyethylene chains by the method involving solvent extraction of the thiocyanatocobaltate(II) complex, *Analyst*, 1976, *101*, 543–548.

76. Lin, J. T., D. G. Cornell, T. J. Micich, Cobaltothiocyanate colorimetric analysis for homologous polyoxyethylated alkyl amides, *J. Am. Oil Chem. Soc.*, 1986, *63*, 1575–1579.

77. Andrew, B. E., Nonionic surfactants in waste water by direct extraction with FTIR detection, *Analyst*, 1993, *118*, 153–155.

78. Tercyak, A. M., T. E. Felker, Colorimetric assay for Pluronic F-68 in isolated rat liver perfusion systems, *Anal. Biochem.*, 1990, *187*, 54–55.

79. Ghebeh, H., A. Handa-Corrigan, M. Butler, Surfactant Pluronic® F-68 in mammalian cell culture medium, *Anal. Biochem.*, 1998, *262*, 39–44.

80. Inaba, K., Polyoxyethylene-type nonionic surfactants in environmental waters, *Int. J. Environ. Anal. Chem.*, 1987, *31*, 63–66.

81. Boyer, S. L., K. F. Guin, R. M. Kelley, M. L. Mausner, H. F. Robinson, T. M. Schmitt, C. R. Stahl, E. A. Setzkorn, Nonionic surfactants in laboratory biodegradation and environmental studies, *Env. Sci. Tech.*, 1977, *11*, 1167–1171.

82. Daniels, D. H., C. R. Warner, S. Selim, Polysorbate 60 in salad dressings by colorimetric and TLC techniques, *J. Assoc. Off. Anal. Chem.*, 1982, *65*, 162–165.

83. Kato, H., Y. Nagai, K. Yamamoto, Y. Sakabe, Polysorbates in foods by colorimetry with confirmation by IR spectrophotometry, TLC, and GC, *J. Assoc. Off. Anal. Chem.*, 1989, *72*, 27–29.

84. Anderson, N. H., J. Girling, Polyoxyethylene nonionic surfactants, *Analyst*, 1982, *107*, 836–838.

85. Paz Antolin, I., M. Paz Castro, S. Vicente Perez, Polyoxyethylene nonionic surfactants (in Spanish), *Afinidad*, 1990, *47*, 201–206.

86. Warner, C. R., S. Selim, D. H. Daniels, Post-column complexation technique for the spectrophotometric detection of poly(oxy-1,2-ethanediyl) oligomers in steric exclusion chromatography, *J. Chromatogr.*, 1979, *173*, 357–363.

87. Waters, J., G. F. Longman, Colorimetric modification of the Wickbold procedure for the determination of nonionic surfactants in biodegradation test liquors, *Anal. Chim. Acta*, 1977, *93*, 341–344.

88. Polyak, E. A., G. A. Nudel, Synthetic nonionic surfactants in solutions of complex composition (in Russian), *Zh. Anal. Khim.*, 1981, *36*, 1155–1159.

89. Favretto, L., F. Tunis, APE in water, *Analyst*, 1976, *101*, 198–202.

90. Favretto, L, B. Stancher, F. Tunis, Extraction and determination of AE in water, *Analyst*, 1978, *103*, 955–962.

91. Favretto, L., B. Stancher, F. Tunis, Reaction mechanisms in the determination of nonionic surfactants in waters as potassium picrate active substances, *Analyst*, 1979, *104*, 241–247.

92. Favretto, L., B. Stancher, F. Tunis, Spectrophotometric determination of polyoxyethylene nonionic surfactants in waters as potassium picrate active substances in presence of cationic surfactants, *Int. J. Environ. Anal. Chem.*, 1983, *14*, 201–214.

93. Qian, X., J. Wang, Z. Hu, Nonionic surfactants in water: barium picrate complex extraction spectrophotometric method (in Chinese), *Fenxi Huaxue*, 1987, *15*, 97–100.

94. Leon Gonzalez, M. E., M. L. Merino Teillet, M. J. Santos Delgado, L. M. Polo-Diez, Extractive-spectrophotometric determination of nonionic surfactants in milk products, based on the surfactant–K^+–picrate system, *An. Quim., Ser. B*, 1989, *85*, 135–139.

95. Cozzoli, O., C. Ruffo, G. Carrer, F. Pinciroli, M. Ciambella, M. C. Grillo, N. Cogliati, M. Fiorito, A. Gellera, N. Giannelli, V. Zamboni, Ethoxylated nonionic surfactants: comparison of the precision of the iodobismuthate method and other methods, *Riv. Ital. Sostanze Grasse*, 1993, *70*, 199–203.

96. Merino-Teillet, M. de la Luz, M. E. Leon-Gonzalez, M. J. Santos-Delgado, L. M. Polo-Diez, Picrate method for the spectrophotometric determination of nonionic surfactants, *Analyst*, 1987, *112*, 1323–1325.

97. Shaffer, C. B., F. H. Critchfield, Solid PEG in biological materials, *Anal. Chem.*, 1947, *19*, 32–34.

98. Stevenson, D. G., Absorptiometric determination of a nonionic detergent, *Analyst*, 1954, *79*, 504–507.

99. Pitter, P., Colorimetric method for nonionic polyoxyethylene-type surfactants, *Chem. Ind.*, Oct. 20, 1962, 1832–1833.

100. Burttschell, R. H., Ethylene oxide based nonionic detergents in sewage, *J. Am. Oil Chem. Soc.*, 1966, *43*, 366–370.

101. Heron M. W., B. C. Paton, Measuring a nonionic surface-active agent (Pluronic F-68) in biological fluids, *Anal. Biochem.*, 1968, *24*, 491–495.

102. Baleux, B., Colorimetry of ethoxylated nonionic surfactants using iodide-iodine solution (in French), *Compt. Rend.*, 1972, *274*, 1617–1620.

103. Chang, J. H., M. Ohno, K. Esumi, K. Meguro, Interaction of iodine with nonionic surfactant and PEG in aqueous potassium iodide solution, *J. Am. Oil Chem. Soc.*, 1988, *65*, 1664–1668.

104. Nowaczyk, F. J., R. L. Schnaare, R. J. Wigent, C. M. Ofner, III, Charge-transfer complexes of iodine and nonionic surfactants: interpretation and use in the Winkler method, *J. Pharm. Biomed. Anal.*, 1993, *11*, 835–842.

105. Boyd-Boland, A. A., J. M. Eckert, Nonionic surfactants by spectrophotometry after extraction with potassium triiodide, *Anal. Chim. Acta*, 1993, *271*, 311–314.

106. Williams, J. L., H. D. Graham, Microdetermination of polyoxyethylene and polyoxypropylene surface active agents, *Anal. Chem.*, 1964, *36*, 1345–1349.

107. Huber, W., E. Fröhlke, Ethoxylates (in German), *Tenside*, 1975, *12*, 39–40.

108. Aminuddin, M., H. D. C. Rapson, Colorimetric assay procedure for polyoxyethylene nonionic surfactants, *J. Chem. Soc. Pak.*, 1991, *13*, 192–196.

109. Stewart, R. G., NPE, *Analyst*, 1963, *88*, 468–470.

110. Leon-Gonzalez, M. E., M. J. Santos-Delgado, L. M. Polo-Diez, Selective determination of Triton-type nonionic surfactants by on-line clean-up and flow injection with spectrophotometric detection, *Analyst*, 1990, *115*, 609–612.

111. Toei, K., S. Motomizu, T. Umano, Extractive spectrophotometric determination of nonionic surfactants in water, *Talanta*, 1982, *29*, 103–106.

112. Crisp, P. T., J. M. Eckert, N. A. Gibson, I. J. Webster, Extraction-spectrophotometric method for the determination of nonionic surfactant, *Anal. Chim. Acta*, 1981, *123*, 355–357.

113. Kautzner, B., R. H. Laby, Colorimetric determination of nonionic polyethylene oxide surfactants, *GIT Fachz. Lab.*, 1988, *32*, 527,529–532.

114. Mima, H., N. Kitamori, Chromatographic analysis of sucrose esters of long chain fatty acids, *J. Am. Oil Chem. Soc.*, 1964, *41*, 198–200.

115. Buschmann, N., S. Wodarczak, Carbohydrate surfactants, *Comun. Jorn. Com. Esp. Deterg.*, 1994, *25*, 203–207.

116. Buschmann, N., A. Kruse, S. Wodarczak, Alkylpolyglucosides, *Agro-Food-Ind. Hi-Tech*, 1996, *7*, 6–8.

117. Bistline, R. G., Jr., J. W. Hampson, W. M. Linfield, Synthesis and properties of fatty imidazolines and their *N*-(2-aminoethyl) derivatives, *J. Amer. Oil Chem. Soc.*, 1983, *60*, 823–828.

118. Baumung, H., L. Huber, D. Vetter, Ecological impact of cationic surfactants in laundry softeners (in German), *Seifen, Öle, Fette, Wachse*, 1991, *117*, 287–292.

119. Waters, J., W. Kupfer, Cationic surfactants in the presence of anionic surfactant in biodegradation test liquors, *Anal. Chim. Acta*, 1976, *85*, 241–251.

120. Metcalfe, L. D., Cellulosic ion exchangers for determination of quaternary ammonium compounds, *Anal. Chem.*, 1960, *32*, 70–72.

121. Chatten, L. G., K. O. Okamura, Quaternary ammonium compounds in various dosage forms by acid-dye method, *J. Pharm. Sci.*, 1973, *62*, 1328–1332.

122. Bonilla Simon, M. M., A. De Elvira Cozar, L. Maria Polo Diez, Spectrophotometric determination of cationic surfactants in frozen and fresh squid by ion-pair formation with methyl orange, *Analyst*, 1990, *115*, 337–339.

123. Ruf, E., Photometric determination of cationic surfactants and cation-active amphoteric surfactants, (in German) *Fresenius' Z. Anal. Chem.*, 1964, *204*, 344–355.

124. Osburn, Q. W., Cationic fabric softener in waters and wastes, *J. Am. Oil Chem. Soc.*, 1982, *59*, 453–457.

125. Biswas, H. K., B. M. Mandal, Extraction of anions into chloroform by surfactant cations: relevance to dye extraction method of analysis of long chain amines, *Anal. Chem.*, 1972, *44*, 1636–1640.

126. Sloneker, J. H., J. B. Mooberry, P. R. Schmidt, J. E. Pittsley, P. R. Watson, A. Jeanes, Spectrophotometric determination of high molecular weight quaternary ammonium cations with picric acid: application to residual amounts in polysaccharides, *Anal. Chem.*, 1965, *37*, 243–246.

127. Chin, T. F., J. L. Lach, Spectrophotometric determination of some quaternary compounds, *J. Pharm. Sci.*, 1965, *54*, 1550–1551.

128. Sheiham, I., T. A. Pinfold, Spectrophotometric determination of cationic surfactants with picric acid, *Analyst*, 1969, *94*, 387–388.

129. Hellmann, H., Application of the modified Dragendorff reagent to the determination of cationic surfactants in the aquatic environment (in German), *Fresenius' Z. Anal. Chem.*, 1989, *335*, 265–271.

130. Hellmann, H., Spectrophotometric determination of cationic surfactants in the presence of anionic and nonionic surfactants (in German), *Fresenius' Z. Anal. Chem.*, 1984, *319*, 272–276.

131. Hellmann, H., Spectrophotometric determination of cationic surfactants (in German), *Fresenius' Z. Anal. Chem.*, 1986, *323*, 29–32.

132. Scott, G. V., Spectrophotometric determination of cationic surfactants with Orange II, *Anal. Chem.*, 1968, *40*, 768–773.

133. Kawase, J., M. Yamanaka, Continuous solvent-extraction method for the spectrophotometric determination of cationic surfactants, *Analyst*, 1979, *104*, 750–755.

134. Gerhards, R. Schulz, Amphoteric surfactants in water, *Tenside, Surfactants, Deterg.*, 1999, *36*, 300–302,304–307.

135. Pesavento, M., A. Profumo, E. Medaina, Selective spectrophotometric determination of long-chain quaternary ammonium salts in the presence of nonionic surfactants, *Ann. Chim.* (Rome), 1989, *79*, 243–258.

136. Rosen, M. J., H. A. Goldsmith, *Systematic Analysis of Surface Active Agents*, 2nd Ed., Wiley, New York, 1972.

137. Hellmann, H., Problems in the determination of cationic surfactants in water and wastewater (in German), *Fresenius' Z. Anal. Chem.*, 1982, *310*, 224–229.

138. Sakai, T., Spectrophotometric determination of quaternary ammonium salts by a flow injection method coupled with thermochromism of ion associates, *Analyst*, 1992, *117*, 211–214.

139. Sakai, T., Extraction-spectrophotometric determination of benzethonium and benzalkonium salts with bromocresol green and quinine, *Anal. Chim. Acta*, 1983, *147*, 331–337.

140. Sakai, T., Spectrophotometric determination of trace amounts of quaternary ammonium salts in drugs by ion-pair extraction with bromophenol blue and quinine, *Analyst*, 1983, *108*, 608–614.

141. Sakai, T., N. Ohno, T. Kamoto, H. Sasaki, Formation of ternary ion associates using diprotic acid dyes and its application to determination of cationic surfactants, *Mikrochim. Acta*, 1992, *106*, 45–55.

142. Lengyel, J., J. Krtil, Spectrophotometric determination of Septonex in waters using Bengal red (in Czech), *Vodni Hospod. B*, 1987, *37*, 244–246.

143. Mohamed, F. A., A. I. Mohamed, H. A. Mohamed, S. A. Hussein, Quercetin for determination of some tertiary amine and quaternary ammonium salts, *Talanta*, 1996, *43*, 1931–1939.

144. Song, M., S. Liang, Spectrophotometric determination of cationic surfactants by formation of ternary complexes with Fe(III) and chrome azurol S, *Chin. J. Chem.*, 1996, *14*, 228–234.

145. Chernova, R. K., T. D. Smirnova, V. V. Krut, I. V. Konovalova, Spectrophotometric determination of cationic surfactants in strong acid media, *J. Anal. Chem.* (Transl. of *Zh. Anal. Khim.*), 1997, *52*, 288–291.

146. Taguchi, S., K. Morisaku, Y. Sengoku, I. Kasahara, Transparent membrane filter for the solid-phase spectrophotometric determination of cationic surfactant in water, *Analyst*, 1999, *124*, 1489–1492.

147. El-Khateeb, S. Z., E. M. Abdel-Moety, Quaternary ammonium surfactants in pharmaceutical formulations by the hypochromic effect, *Talanta*, 1988, *35*, 813–815.

13
Infrared and Near Infrared Spectroscopy

I. GENERAL CONSIDERATIONS

Infrared (IR) spectroscopy is most often used for qualitative identification of chemical compounds. Comprehensive works on the application of IR spectroscopy to identification of surfactants are available elsewhere. The volume of Hummel is an essential reference for experts as well as a teaching aid for novices (1). The Sadtler database is invaluable (2). A number of shorter works provide an introduction to surfactants for the beginner (3–6). With experience, the spectroscopist will find it possible to identify not only pure compounds, but also mixtures of surfactants. Modern computerized instruments aid by permitting subtraction of spectra of known compounds from the spectrum of the mixture. IR spectroscopy is widely used for detailed examination of purified fractions prepared by extraction or ion exchange chromatography. The small sample size requirement makes it possible to identify compounds collected from the eluent of a liquid chromatograph, especially if techniques like diffuse reflectance Fourier transform IR are used.

Although IR spectroscopy is most useful for qualitative analysis, many quantitative tests have been developed to satisfy special needs. Fuller and coworkers describe a technique suitable for quality control analysis of formulated liquid detergents by FTIR using a trough liquid attenuated total reflectance cell (7–9). The method has the capability of simultaneously measuring the concentration of the surfactant and of several solvents and adjuvants. ATR spectroscopy, with FT or dispersive instrumentation, allows direct, qualitative analysis of surfactant-containing samples, such as toothpaste, bar soap, and shampoos, and minimizes interference from water (10). Reflectance IR spectroscopy has even been used to determine residues of soap on human skin as a function of rinsing and Ca^{2+} concentration in the rinse water (11).

Spectroscopy in the near infrared region is widely applied to quality control analysis of surfactants. Absorbance peaks in this region are not as well defined as in conventional IR, nor are they as easy to assign to specific functional groups. Near IR is used as a semi-empirical method, with calibration curves prepared from the same matrix as the sample. In modern instruments, calibration is performed automatically using dozens of wavelengths. This method is preferred because, for example, there is interference between the hydroxyl group and the water absorption in polyether nonionics, which make simple calibration at two or three wavelengths unreliable. Using an instrument designed for process control, NIR with a flow-through transmission cell permits simultaneous monitoring of several components of a detergent formulation (8).

Reflectance NIR cells allow rapid analysis of many different materials, but with some loss in precision over transmission measurement. Walling and Dabney have demonstrated the use of a conventional reflectance instrument to measure active agent content and functionality of an alcohol sulfate, as well as concentration of an additive (12). For a well-defined system, the determination of several parameters requires only a very few minutes. A reflectance NIR instrument is also suitable for quality control monitoring of surfactants and other components of powdered detergent formulations (13). NIR reflectance has even been demonstrated for detecting residues of an unspecified detergent on a stainless steel surface, with a detection limit in the 1 $\mu g/cm^2$ range and reasonable linearity down to about the 15 $\mu g/cm^2$ level (14). The patent literature describes the use of NIR reflectance to determine the amount of water film left behind on a quartz tube after a surfactant solution has contacted the exterior of the tube. Within limits, the mass of the film is related to the surfactant concentration of the solution (15).

II. IONIC SURFACTANTS

A. Qualitative Analysis

Infrared indicates that a sulfonate is a sulfonate, whether or not it is aromatic, and can show whether the alkyl portion has appreciable unsaturation. Infrared spectroscopy can distinguish between branched and linear alkyl chains in alkylbenzenesulfonates (16). The two types can be distinguished by the relative strengths of the bands for methyl groups at 1379 and 1361 cm^{-1} and the band for $(CH_2)_n$ groups at 719 cm^{-1}. Similarly, short chain alkylaryl sulfonates like hydrotropes can be distinguished because they lack the parafin absorbance at 719 or 720 cm^{-1} (17). The IR spectra of the alkylbenzene, after cleavage of the sulfonate group, is easier to interpret. In this case, the bands at 1379–1361 and at 719 cm^{-1} are very well resolved, making it easy to distinguish the extent of branching of the alkyl chain. In the case of pure materials, the position of the absorbance band at 750–840 cm^{-1} indicates whether the compound has *ortho*, *meta*, or *para* substitution, but this is of less value in analysis of commercial mixtures (18). IR analysis is suitable for identifying sultones in α-olefin sulfonates subjected to acid dehydration of hydroxyalkanesulfonates (19).

Although this is no longer common, anionic surfactants may be identified in wastewater after concentration by methylene blue extraction, as described elsewhere in this volume. A note of caution: the IR spectrum of a surfactant after formation and breaking of the methylene blue ion pair is not identical to the original spectrum. The sulfonate absorption decreases and a carbonyl band becomes noticeable (20).

Phosphate esters are readily identified by the phosphate bands at 2700, 2300, 1250–1300, and 900–1100 cm^{-1}. The spectrum is adequate to detect aliphatic, aromatic, and polyether character, but is normally insufficient to discriminate among similar products (21).

IR spectroscopy can be used to follow the formation of carboxylated NPE from the nitrile intermediate. In this case, the hydroxyl band at 3400 cm^{-1}, the CN band at 2250 cm^{-1}, and the carbonyl band at 1710 cm^{-1} are monitored. Residual intermediate can be detected down to about 5% (22).

Infrared spectra of the most common cationic surfactant, distearyldimethylammonium chloride are very simple, little different from hydrocarbon spectra. The two methyl groups and the N–C group cannot be distinguished from the overpowering –CH$_2$– bands, at least not by ordinary dispersive IR (23). Cross published information on the infrared spectra of the tetraphenylborate salts of common cationics (24). He finds IR spectroscopy suitable for the classification (he reserved the term "identification" for a technique which would give the exact length of the alkyl chain) of surfactants isolated by precipitation from complex samples.

The cyclization of an imidazoline derivative can be followed by observing the disappearance of the bands at 1560 and 1638 cm^{-1} and the appearance of the band at 1600 cm^{-1} (25). Hydrolysis of the ring structure to produce the amide can be followed by reversing the process.

B. Quantitative Analysis

While infrared spectrophotometry is most useful for the qualitative analysis of surfactants, various quantitative methods have been developed for well-characterized systems. For example, an attenuated total reflectance cell with a ZnSe crystal is useful for direct analysis of aqueous anionic surfactant solutions by FTIR, while avoiding the deleterious effects of water on the usual transmission cells. In this case, the sulfonate absorbance at 1175 cm^{-1}, or the sulfate absorbance at 1206–1215 cm^{-1}, is used for quantification (10,26). In another application, the weak absorption bands in the 1429–1333 cm^{-1} region are used to measure the relative amounts of linear and branched chain alkylbenzene sulfonates extracted from environmental waters (27). This is the one advantage of the infrared technique over those that have supplanted it for wastewater analysis: its ability to differentiate the straight and branched chain compounds (28). No procedure will be given here, since the cleanup prior to IR analysis can be handled adequately by the method for LAS analysis by desulfonation/gas chromatography, described in Chapter 8.

Another approach to infrared quantification after methylene blue extraction is preparation of the sulfonyl chloride derivatives, followed by measurement in the far infrared (29). By determining the absorption at three wavelengths, the ratio of LAS (640 cm^{-1}), ABS (618 cm^{-1}), and alkanesulfonates (524 cm^{-1}) can be estimated. Depending on the matrix, there may be some advantage in replacing the methylene blue visible spectrophotometric procedure with one based on a cationic dye which absorbs light in the near infrared region. This is because sensitivity may be increased, and interference is less of a problem (30).

Diffuse reflectance IR has been proposed for the determination of didecyldimethylammonium chloride on softwood, using the absorbance region 2500–3000 cm^{-1}. It is necessary to subtract the background spectrum due to the wood matrix (31).

III. NONIONIC SURFACTANTS

Rauscher has reviewed the use of infrared spectroscopy in characterization of nonionic surfactants (32).

A. Qualitative Analysis

The spectra of highly ethoxylated surfactants are mainly influenced by the presence of the polyether groups; further characterization is difficult. If the FTIR spectra of an ethoxylated surfactant and PEG are taken under carefully controlled conditions, it is often possible to determine by difference what was the starting material, i.e., to determine whether the product is AE, APE, or another compound (33).

B. Quantitative Analysis

1. Determination of Nonionic Surfactant

A number of infrared spectroscopy procedures have been developed to characterize environmental samples. Frazee and coworkers isolated AE and APE from wastewaters by various means, analyzing the extracts qualitatively and quantitatively by IR (28). Andrew simply extracted process water samples with CH_2Cl_2 and measured the ether absorbance at 1110 cm^{-1}, making a blank measurement at 990 cm^{-1} (34). Soap and other materials interfered.

Cassidy and Niro eliminated interferences by purifying the extract by column chromatography on a poly(styrene-divinylbenzene) column of the type used for gel permeation chromatography (35). They report a detection limit of 10 µg of ethoxylated acid surfactant using the ester carbonyl absorbance at 1740 cm^{-1}. Hellman determined nonionic surfactants in sediments by separating them by extraction and then by precipitation with Dragendorff reagent and partitioning on silica gel, finally measuring the IR absorbance at the $-CH_2-$ regions of 2920 and 2926 cm^{-1} (36,37). In this last case, specificity is provided by the separation technique, not by the IR measurement.

2. Determination of Hydroxyl End Groups

The number-average molecular weight of most nonionic surfactants can be determined by measurement of the concentration of the hydroxyl groups which terminate the molecules. The standard methods for this determination are based upon esterification and titration. IR spectroscopy has been used as an alternative procedure, especially for quality control purposes where the matrix is well characterized. (In recent times, mid-IR methods for hydroxyl number determination have been displaced by near IR technology.) The IR determination of hydroxyl number requires that a calibration curve be prepared from compounds having the same chemical structure as that of the sample. Control of temperature and of water content is critical (38,39). Normally, the OH band chosen for IR determination is due to hydrogen bonding, and is thus very matrix dependent. For EO and PO units, some of these matrix effects can be eliminated by using an intramolecular hydrogen bonding wavelength, at 3520 cm^{-1} (40). Another approach is dilution of the polymer in a solvent, such as THF, and use of the OH–solvent bond absorbance. This latter method is less sensitive to temperature effects and polymer impurities, although water interference remains a problem (41).

While the mid-IR region remains most suitable for qualitative work, near infrared spectroscopy has advantages for quantitative work such as the determination of hydroxyl content. In the near IR region only hydrogen vibrations are observed, which means that there are fewer interferences than in the IR region. The –OH band in the NIR suffers from less interference from water than does the corresponding band in the IR. Also, NIR bands are much weaker than IR, allowing used of longer cell pathlength, so that quantitative measurements are more convenient. For example, quartz cells of 0.5 cm pathlength can be used. Often, dilution is unnecessary. For hydroxyl number measurement in the NIR, one calibration curve can be used for a homologous series of surfactants. If a copolymer is involved, a calibration curve must be prepared for each monomer ratio. The weaker band at 1450 nm or the stronger band at 2050 nm may be used for monitoring the hydroxyl content by NIR (42).

Jouan-Rimbaud and coworkers used NIR determination of OH number to demonstrate a calibration strategy. They found it advantageous to use wavelengths with low correlation to hydroxyl content as well as those highly correlated and those correlated to water content. This gives a more robust calibration than otherwise obtained (43).

3. Determination of Polyether Content

Infrared spectrophotometry can be used to determine the ethylene oxide content of ethoxylated surfactants (44). The main absorbance is found at 2295 cm^{-1} for the –CH$_2$– groups, and at 2485 cm^{-1} for the oxygen. The latter absorption is the more useful. For quantification, the sample and an appropriate standard are dissolved and their specific absorptivities determined at 2485 cm^{-1}.

Procedure: Near IR Determination of EO Content of an Ethoxylated Surfactant (44)

Use for calibration a polyethylene glycol of 20–30 average degree of polymerization and known water content. Dissolve 1 g, weighed exactly, in CCl$_4$ in a 25-mL volumetric flask. Record the spectrum of this solution in a quartz cuvette of 1.0-cm pathlength from 2200–2750 nm, using CCl$_4$ as reference. Prepare a solution of the sample and measure its absorbance in the same way. Draw a baseline on the spectrum, connecting the absorbance minima at about 2360 and 2605 nm. Determine the net absorbance at 2485 nm, corrected for baseline. The EO content of the standard is calculated:

$$\%EO = \frac{44n(100 - W_a)}{44n + 18}$$

where n is the average number of EO units and W_a is the percent water as obtained from Karl Fischer titration. The specific extinction, k_c, is then calculated:

$$k_c = \frac{a_c}{G_c}$$

where G_c is the weight of the standard and a_c is the absorbance measured. The specific extinction of the sample, k_s, is calculated in the same manner, and the %EO calculated:

$$\%EO \text{ in sample} = \frac{(\%EO \text{ in std.})k_s}{k_c}$$

4. Specific Applications

Many quantitative applications of IR analysis have been developed by individual analysts throughout the world to control the production of certain well-defined mixtures.

(a) Alcohol ethoxylates.   The ratio of the ether and methylene absorbances at 720 and 843 cm^{-1} is proposed as the basis for monitoring the ethylene oxide content of straight-chain alcohol ethoxylates (45)

NIR is proposed for routine process control of the ethoxylation of fatty alcohol or alkylphenol (46). Normally, a computerized instrument is used which chooses its own calibration wavelengths. These wavelengths correspond to the regions of absorbance of the –OH (2080 nm), –CH$_2$– (2272, 2304, 1732, and 1750 nm), aromatic (2136 and 2160 nm), and –CH$_2$CH$_2$O– (2488 and 2498 nm) groups. The calibration curve is, of course, different for each family of surfactants. NIR analysis is also appropriate for online determination of water content (42).

(b) Alkylphenol ethoxylates.   An example is a correlation between the structure and surfactant properties of ethoxylated octyl- and nonylphenols and the height ratio of absorbance bands at 960 and 840 cm^{-1} (47–49). The ratio of the absorbance in the aromatic region (1500 cm^{-1}) to that in the –CH$_2$O– region (1350 cm^{-1}) can be used, with proper calibration, to give the approximate degree of ethoxylation of an APE surfactant (50). Similarly, Meszlényi and coworkers proposed use of the ratio of absorbances at 1350 cm^{-1} and 1610 cm^{-1} (substituted benzene) for determination of degree of ethoxylation of nonylphenols (51). Wavelengths of 1350 cm^{-1} and 1205 cm^{-1} (aromatic C-H) are recommended for tri-*t*-butylphenol ethoxylates (52). Such measurements may be used as an identity check by the surfactant customer, but IR is a precise enough tool that surfactant manufacturers sometimes rely on it to control the degree of ethoxylation.

(c) EO/PO copolymers.   Turley and Pietrantonio have described the development of a NIR reflectance procedure for quality control of polyether compounds (53). While the particular application was to polyethers as raw materials for urethane foam, the same conclusions are applicable to surfactants. By using automatic measurement at multiple wavelengths and computerized data manipulation, a rapid method for determining hydroxyl number, water content, and relative concentration of EO and PO units was obtained. Calibration requires a data set of dozens of well-characterized compounds. This is because many of the NIR bands represent the absorbance of "combinations," i.e., their intensities vary with the concentration of more than one functional group.

In the case of EO/PO copolymers, IR spectroscopy can give the concentration ratio of the hydrophobe/hydrophile units (39,54). (This information can also be obtained from NMR spectrometry, with the added advantage that no external calibration is necessary.) A calibration curve is prepared from mixtures of poly(propylene glycol) and poly(ethylene glycol) dissolved in chloroform. The infrared spectrum is scanned from 1667 to 1250 cm^{-1} and a baseline drawn connecting the minima at 1420 and 1318 cm^{-1}. The ratio of the absorbance at 1379 cm^{-1}, corresponding to the methyl group of the PO moiety, and 1350 cm^{-1}, corresponding to the CH$_2$–O absorption, is measured and plotted versus known %PO. Alternatively, the infrared stretching vibrations at 2880 cm^{-1} and 2975–2980 cm^{-1} (55) or the near infrared overtone bands at 8503 and 8271 cm^{-1} (39) may be used, with each pair corresponding to methyl and methylene absorption, respectively. In the near IR, the methyl absorbance bands at 1680 and 2260 nm are appropriate for determining the PO content of an EO/PO copolymer (42). Both the IR and NMR methods suffer from the disadvantage that they measure only the average properties of the sample. If a mixture of compounds is present, the results will have limited meaning. Generally, infrared spectrometry is accurate only over a limited range of EO/PO ratios, while the NMR method is more widely applicable.

(d) *Alkanolamides.* The ester content of alkanolamides may be determined by measuring the absorbance at 1740 cm^{-1} and comparing to a suitable calibration curve (56).

REFERENCES

1. Hummel, D. O., *Analysis of Surfactants: Atlas of FTIR Spectra with Interpretations*, Hanser/Gardner Publications, Inc., Cincinnati, 1996.
2. Nyquist, R. A., Sadtler IR Surfactants Databases, available from Bio-Rad Laboratories, Hercules, CA.
3. Nettles, J. E., IR spectroscopy for identifying surfactants, *Text. Chem. Color.*, 1969, *1*, 430–441.
4. American Society for Testing and Materials, Qualitative classification of surfactants by IR absorption, D2357. West Conshohocken, PA 19428.
5. Mandery, K., Surfactant component of an aqueous detergent (in German), *Seifen, Öle, Fette, Wachse*, 1991, *117*, 595–597.
6. Hancewicz, T. M., IR and Raman spectroscopy, in J. Cross, ed., *Anionic Surfactants: Analytical Chemistry*, 2nd ed., Marcel Dekker, New York, 1998.
7. Fuller, M. P., G. L. Ritter, C. S. Draper, Partial least-squares quantitative analysis of IR spectroscopic data. Part II: Application to detergent analysis, *App. Spectrosc.*, 1988, *42*, 228–236.
8. Fuller, M. P., M. E. Meyers, On-line FT-MIR/NIR process analysis, *Mikrochim. Acta*, 1988, *1*, 31–34.
9. Fuller, M. P., M. C. Garry, Z. Stanek, FTIR liquid analyzer, *Am. Lab.*, 1990, *22(15)*, 58,60–62,64–69.
10. Ferrer, N., M. Roura, M. Baucells, Alkylbenzenesulfonates in water solution by ATR/FT-IR Spectroscopy, *Mikrochim. Acta, Suppl.*, 1997, *14*, 297–299.
11. Fujiwara, N., I. Toyooka, K. Ohnishi, E. Onohara, Adsorption residue of fatty acid soap on human skin: analysis of human skin surface *in situ* using IR spectroscopy, *J. SCCJ*, 1992, *26*, 107–112.
12. Walling, P. L., J. M. Dabney, Application of near IR reflectance spectroscopy to the quality assurance of surfactants, *J. Soc. Cosmet. Chem.*, 1986, *37*, 445–459.
13. Bernardini, M., D. Baroni, E. Fedeli, Dishwashing machine powdered detergents by NIR reflectance (in Italian), *Riv. Ital. Sostanze Grasse*, 1986, *63*, 155–161.
14. Biwald, C. E., W. K. Gavlick, Total organic carbon analysis and FTIR to determine residues of cleaning agents on surfaces, *J. AOAC Int.*, 1997, *80*, 1078–1083.
15. Eberl, R., J. Wilke, Surfactants in aqueous solutions, German Patent DE 19740266, March 18, 1999.
16. Müller, K., D. Noffz, Length and the degree of branching of the alkyl chains in surface active alkylbenzene derivatives (in German), *Tenside, Surfactants, Deterg.*, 1965, *2*, 68–75.
17. Hellmann, H., Cumene sulfonate in the presence of LAS by spectroscopic methods (UV, fluorescence, IR) (in German), *Tenside, Surfactants, Deterg.*, 1994, *31*, 200–206.
18. El-Emary, M., L. O. Morgan, Structural features of LAS as observed in ^{13}C magnetic resonance spectra, *J. Am. Oil Chem. Soc.*, 1978, *55*, 593–599.
19. Ranky, W. O., G. T. Battaglini, Alpha olefin sulfonates, *Soap Chem. Spec.*, 1968(*4*), 36–39, 78–86.
20. Hellmann, H., Transformation of anionic surfactants under the conditions of analysis—a comparison of colorimetry and IR spectrometry (in German), *Fresenius' Z. Anal. Chem.*, 1979, *294*, 379–384.

21. Frazier, J. D., R. D. Johnson, C. G. Wade, D. J. O'Leary, Composition of a variety of nonylphenol poly(ethylene oxide) phosphate anionic surfactants, *Comun. Jorn. Com. Esp. Deterg.*, 1991, *22*, 99–110.

22. Gerhardt, W., G. Czichocki, H. R. Holzbauer, C. Martens, B. Weiland, Characterization of ethercarboxylic acids during synthesis (in German), *Tenside, Surfactants, Deterg.*, 1992, *29*, 285–288.

23. Hellmann, H., Spectrophotometric determination of cationic surfactants (in German), *Fresenius' Z. Anal. Chem.*, 1986, *323*, 29–32.

24. Cross, J. T., Identification and determination of cationic surface-active agents with sodium tetraphenylboron, *Analyst*, 1965, *90*, 315–324.

25. Koeber, A., W. Melloh, M. Bloch, Ampholytic cycloimidinium surfactants, *Soap Cosmet. Chem. Spec.*, 1972, *48(5)*, 86,88,173.

26. Sabo, M., J. Gross, I. E. Rosenberg, Anionic surfactants in aqueous systems via Fourier transform IR spectroscopy, *J. Soc. Cosmet. Chem.*, 1984, *35*, 207–220.

27. Maehler, C. Z., J. M. Cripps, A. E. Greenberg, Differentiation of LAS and ABS in water, *J. Water Poll. Cont. Fed.*, 1967, *39*, R92–R98.

28. Frazee, C. D., Q. W. Osburn, R. O. Crisler, Application of IR spectroscopy to surfactant degradation studies, *J. Am. Oil Chem. Soc.*, 1964, *41*, 808–812.

29. Oba, K., K. Miura, H. Sekiguchi, R. Yagi, A. Mori, Anionic surfactants in waste water by IR spectroscopy, *Water Res.*, 1976, *10*, 149–155.

30. Roberson, M. A., D. Andrews-Wilberforce, D. C. Norris, G. Patonay, Surfactants using ion pair extraction of near-IR laser dyes, *Anal. Lett.*, 1990, *23*, 719–734.

31. Manville, J. F., J. R. Nault, Antisapstain chemical didecyldimethylammonium chloride on wood surfaces by diffuse reflectance FTIR, *Appl. Spectrosc.*, 1997, *51*, 721–724.

32. Rauscher, G., IR spectroscopy and MS of nonionic surfactants, in J. Cross, ed., *Nonionic Surfactants: Chemical Analysis*, Marcel Dekker, New York, 1987.

33. Hummel, D. O., FTIR spectrometry of surfactants, *GIT Fachz. Lab.*, 1994, *38*, 439–440, 443–444, 446–447.

34. Andrew, B. E., Nonionic surfactants in waste water by direct extraction with FTIR detection, *Analyst*, 1993, *118*, 153–155.

35. Cassidy, R. M., C. M. Niro, HPLC analysis of polyoxyethylene surfactants and their decomposition products in industrial process waters, *J. Chromatogr.*, 1976, *126*, 787–794.

36. Hellmann, H., Nonionic surfactants in water and wastewater by X-ray fluorescence and IR spectrometry (in German), *Fresenius' Z. Anal. Chem.*, 1979, *297*, 102–106.

37. Hellmann, H., Nonionic surfactants in sewage and other sludges (in German), *Fresenius' Z. Anal. Chem.*, 1980, *300*, 44–47.

38. Hilton, C. L., Hydroxyl numbers by near-IR absorption, *Anal. Chem.*, 1959, *31*, 1610–1612.

39. Weis, G., Propylene oxide/ethylene oxide block copolymers by IR spectroscopy (in German), *Fette, Seifen, Anstrichm.*, 1968, *70*, 355–359.

40. Burns, E. A., R. F. Muraca, Hydroxyl concentration in polypropylene glycols by IR spectroscopy, *Anal. Chem.*, 1959, *31*, 397–399.

41. Kim, C. S. Y., A. L. Dodge, S. Lau, A. Kawasaki, Hydroxyl concentrations in prepolymers from the IR absorption band of tetrahydrofuran-associated hydroxyl groups, *Anal. Chem.*, 1982, *54*, 232–238.

42. Schirmer, R. E., A. G. Gargus, Monitoring polymer processing through fiber optics, *Amer. Lab.*, 1988(11), 37–43.

43. Jouan-Rimbaud, D., D. Massart, R. Leardi, O. E. De Noord, Genetic algorithms as a tool for wavelength selection in multivariate calibration, *Anal. Chem.*, 1995, *67*, 4295–4301.

44. Voogt, P., Ethylene oxide content of nonionic detergents (in German), *Fette, Seifen, Anstrichm.*, 1963, *65*, 964–970.

45. Das, S., V. V. Kumar, FTIR spectroscopic determination of ethylene oxide content of nonionic surfactants, *Indian J. Chem., Sect. A: Inorg., Bioinorg., Phys., Theor. Anal. Chem.*, 1993, *32A*, 1004–1005.

46. Bernardini, M., D. Baroni, E. Fedeli, G. Gaia, Simultaneous determination on line of the polymerization parameters of ethoxylated alcohols and alkylphenols by NIR reflectance (in Italian), *Riv. Ital. Sostanze Grasse*, 1987, *64*, 57–61.

47. De la Guardia, M., J. L. Carrion, J. Medina, UV-visible spectophotometry in the determination of average properties of nonylphenol ethylene oxide condensates, *Anal. Chim. Acta*, 1983, *155*, 113–121.

48. Carrion, J. L., S. Sagrado, M. De la Guardia, OPE by UV and IR spectrometry, *Analyt. Chim. Acta*, 1986, *185*, 101–107.

49. De la Guardia, M., J. E. Tronch, J. L. Carrion, A. Aucejo, Mathematical models in the characterization of ethylene oxide condensate surfactants by IR and UV spectroscopy, *Analusis*, 1988, *16*, 124–130.

50. Zhang, L., X. Li, W. Du, J. Furong, Detection of polyoxyethylene nonionic surfactants and determination of the ethylene oxide content by IR spectroscopy (in Chinese), *Fenxi Huaxue*, 1987, *15*, 810, 823–824.

51. Meszlényi, G., G. Körtvélyessy, É. Juhász, M. Eros-Lelkes, Molecular mass and composition of NPE by IR spectroscopy, *Acta Chim. Hung.*, 1991 *128*, 179–181.

52. Meszlényi, G., É. Juhász, M. Lelkes, 2,4,6-Tri-*tert*-butylphenol polyethylene glycol ethers, *Tenside, Surfactants, Deterg.*, 1994, *31*, 83–85.

53. Turley, P. A., A. Pietrantonio, Rapid hydroxyl number determination by near IR reflectance analysis, *J. Cell. Plast.*, 1984, *20*, 274–278.

54. Meszlényi, G., M. Sipos, É. Juhász, M. Lelkes, IR spectroscopic determination of the average composition of derivatives of ethylene oxide-propylene oxide block polymers, *Acta Chim. Hung.*, 1990, *127*, 495–499.

55. Zgoda, M. M., Surfactants in the group of propylene and ethylene oxide copolymers. II: Application of IR and proton NMR spectroscopy to the determination of the hydrophilic-lipophilic balance in polyetherdiols (in Polish), *Acta Pol. Pharm.*, 1988, *45*, 63–70.

56. Cosmetic, Toiletry and Fragrance Association, Total esters in alkylolamides, CTFA Method G 7-1. 1110 Vermont Ave. NW, Washington, DC 20005.

14

Nuclear Magnetic Resonance Spectroscopy

NMR is typically used to characterize pure materials, as opposed to determining their concentration in mixtures. An overview of the use of ^{13}C NMR spectroscopy to identify surfactants has been published (1,2). König has published proton NMR spectra of a number of surfactants (3), as well as a table of chemical shifts of functional groups found in common surfactants (4). This allows use of NMR to identify components of commercial products when the range of possible structures is limited. We will not discuss the qualitative identification of surfactants here. Neither will we discuss the use of NMR in the study of micelle formation.

I. ANIONIC SURFACTANTS

The NMR spectra of anionic surfactants are discussed with examples in a companion publication (5). Micelles cause peak broadening, so these should be eliminated prior to NMR analysis. This is not usually a problem when analyzing neat surfactants. In the case of "overbased" detergent additives for lubricating oil, the sulfonates will be mixed with a salt such as calcium carbonate, which leads to the formation of reverse micelles with the sulfonate groups clustered together with the salts. These can be destroyed by boiling with HCl and then extracting the surfactant from the salt solution using ethyl ether (6).

A. Alkylarylsulfonates

A few publications have demonstrated application of NMR to routine analysis. The most common use in environmental work is determination of the degree of branching of the alkyl chain of an alkylbenzenesulfonate. NMR is much more straightforward for this analysis than is IR spectroscopy because ^{1}H NMR can clearly distinguish methyl from methylene protons. ^{13}C NMR may also be used for this purpose, since the aryl carbon peaks are affected by the degree of branching of the side chain. Differences are also seen in the alkyl carbon peaks, but these are more difficult to quantify (7). A field strength as low as 100 MHz (25.16 MHz for ^{13}C) is adequate for this analysis. In a typical environmental application, Thurman and coworkers used ^{13}C NMR to determine whether alkylbenzenesulfonate concentrated from groundwater had a branched alkyl chain or was a straight-chain product of more recent origin (8). The approximate composition of mixtures could be measured by this technique. Use of attached proton–test software permitted identification of methylene and methyl moieties in the molecule. ^{13}C NMR is not generally superior to ^{1}H NMR for this purpose. Trace analysis is accomplished by simply running the instrument for a longer time, collecting more individual spectra to give a better signal/noise ratio. These investigators used a 300-MHz instrument (75.4 MHz for ^{13}C).

Proton and ^{13}C chemical shifts in the aromatic region are suitable for determining the position of sulfonation in monoalkylarylsulfonates, but are not useful for studying sulfonation of di- and trialkylbenzenes (9). These compounds are best studied by comparing the aliphatic ^{13}C shifts to those of the original alkylbenzenes. Significant changes are seen in the benzylic carbon shifts *ortho* to the sulfonate group, while much smaller shifts are seen for *meta* substitution. NMR is considered accurate to ±5% for determination of isomer distribution (10).

^{13}C NMR permits determination of individual isomers of LAS, indicating which of the alkyl carbons is attached to the phenyl ring (1,2). This allows determination of the source of the parent alkylbenzene, whether from alkylation of benzene with *n*-olefins (about 18% 2-phenyl isomer), or alkylation with *n*-chloroparaffins (about 30% 2-phenyl isomer). Since NMR spectroscopy can define only the average chemical properties of a mixture, in terms of alkyl chain length and substitution, it is not as useful as gas chromatography for characterization of commercial products.

B. Alkyl Sulfates

^{13}C NMR of linear alkyl sulfates shows individual peaks for the two or three carbons at each end of the chain, with overlapping peaks for most of the methylene carbons. The chemical shift for the carbon alpha to the sulfate group is the same for each compound, but the location of the methyl group differs enough so that C_{12}, C_{14}, and C_{16} alkyl sulfates can sometimes be determined in the presence of each other (7). ^{13}C NMR can determine whether the parent alcohol of an alkyl sulfate was a linear natural product or a branched oxo alcohol (1,2).

C. α-Olefin Sulfonates

Proton and ^{13}C NMR were used to characterize α-olefin sulfonates and the intermediate sultones (11). While the spectra of purified fractions were studied, this technique is potentially applicable to the characterization of commercial mixtures. Characterization of sodium α-sulfo fatty acid methyl ester products has also been described (12). A ^{13}C NMR study (75 MHz for C) of *n*-octyl and *n*-decyl sulfonates showed resolution of resonance signals for each of the methylene carbons, although the central carbon resonances were almost equivalent (13)

The sulfonation of internal olefins has been studied by NMR (14,15). Proton and ^{13}C signals were assigned for alkenesulfonates, hydroxyalkanesulfonates, alkenes, and β- and γ-sultones.

D. Phosphate Esters

^{31}P NMR is suitable for determination of monoester, diester, and phosphoric acid in phosphate esters. Peaks for these compounds are well separated from each other and easily identified. Quantitative results show good agreement with HPLC data (16).

II. NONIONIC SURFACTANTS

NMR is quite suitable for the characterization of nonionic surfactants and even of mixtures of surfactants. NMR is used to determine the concentration ratios of various functionalities in a sample which is concentrated and reasonably pure. If a mixture of materials is present, such as a blend of polyethylene glycol and ethoxylated surfactant, this must be known in advance or false qualitative and quantitative information will be developed. A review of NMR analysis of nonionic surfactants appears in a companion volume of the Surfactant Science Series (17).

While most nonionics are soluble in deuterated water, this is not necessarily the solvent which provides the most information for proton NMR work, partly because of the micelle formation which occurs in water. A great deal of routine work is performed in deuterated chloroform. Deuterated acetone was reported to provide the richest proton spectra for APE, with deuterated DMSO being somewhat less useful (18).

A. Alcohol Alkoxylates

In the case of an alcohol ethoxylate, ^{1}H NMR measurement of the ratio of methyl protons to methylene protons will allow calculation of the chain length of a linear alkyl chain or the degree of branching of a chain of known length. The ratio of chain lengths of the alkyl and ethoxy portions of the molecule can easily be determined (19,20), but finding an "internal standard" to give the absolute length of either chain is more difficult than for APE. (Of course, if the alkyl chain length is known by GC or HPLC, the rest is easy.) By proton NMR, the signal for the –OH of the terminal alcohol functionality cannot normally be resolved from the signal for the –CH$_2$CH$_2$O– units. Besides, low level contamination by water or ethanol would have a large effect on the results. Formation of the trimethylsilyl derivative of the AE solves this problem, giving a strong signal of nine protons per terminal hydroxy group, well-resolved at near 0 ppm (21). Trifluoroacetic acid may be added instead to separate the signal of the terminal hydroxyl from the ether signals. In general, proton NMR is sufficient for determining the nature of the surfactant and its relative content of EO and PO groups. ^{13}C NMR is required for more exact information (22).

^{13}C NMR is capable of distinguishing the methylene carbon adjacent to the terminal hydroxyl from the methylene carbons in the ethoxy chain, thus permitting calculation of ethoxy chain length. Additional information may be garnered as to the nature of the alkyl chain. For example, it is easy to determine whether the starting alcohol was an oxo alcohol (some branching) or a fatty alcohol (no branching) (23). In the case of mixed alkoxylates, the sequence of addition of EO and PO can be determined (22,24). Shift reagents allow the direct determination of the PEG content of ethoxylated acids and alcohols by NMR (25).

Using a 500-MHz instrument (125.8 MHz for ^{13}C), Heatley and coworkers demonstrated that it was possible to directly determine the concentration of the first three oligomers of a C_{14} alcohol ethoxylate, with the higher oligomers indistinguishable. The NMR spectrum also contains detailed information on the various branched alkyl moieties (26).

Low resolution NMR instruments are available for quality control use by relatively unskilled operators. Typically, they give a signal proportional to the total hydrogen content of the sample. Differentiation between types of hydrogen is not performed. The use of such an instrument has been demonstrated for quality control of ethoxylated alcohols (27). The greater the degree of ethoxylation on the alkyl chain, the lower the percent hydrogen measured by NMR. The technique could also be applied to propoxylated alcohols, but not to alcohols to which were added both ethylene oxide and propylene oxide.

Application of LC–^1H NMR to analysis of AE has been demonstrated (28).

B. Alkylphenol Ethoxylates

In the case of alkylphenol ethoxylates, proton NMR can give the ratio of alkyl to ethoxy chain length, thus giving the molecular weight of the material if either chain length is known from other techniques. The absolute length of each chain can be determined by ratio of the signal to that of the four aryl protons. This measurement is most accurate for short chain lengths, where the ratio of two signals of about equal strength is determined. NMR of the starting alkylphenol will give the degree of branching of the alkyl chain and the position of the ring substitution (29). High field proton NMR permits obtaining this same information from the ethoxylated surfactant (30).

While proton NMR does not allow resolution of the signals from the polyoxyethylene chain, ^{13}C NMR gives separate lines for each of the seven types of carbon in the chain, as well as four lines for the five types of alkyl carbon in the octyl chain, and four lines for the four types of phenyl carbon (31).

In a demonstration of proton NMR detection for HPLC, impurity peaks in a C_8-APE were identified as PEG and C_4-APE. In this case, a mobile phase of acetonitrile/D_2O was used for reversed-phase chromatography (32). A 500-MHz instrument and special solvent-suppression techniques gave excellent results.

C. Ethoxylated Acids

Proton NMR (250 MHz) can be used to determine relative content of monoester and diester in acid ethoxylates, since only the monoester contains free hydroxyl. Derivatization is necessary to resolve the hydroxyl signal; trichloroacetylisocyanate was used in this case. The spectrum is run twice, with and without derivatization. It was necessary to remove free polyethylene glycol by extraction. Moisture must also be absent, to avoid side reactions with the derivatizing agent (33).

D. Ethoxylated Alkanolamides

NMR analysis of an ethoxylated amide indicates whether the amide is secondary or tertiary, depending on the relative strength of the signals from hydrogens associated with the nitrogen. Examination of the ethoxylated ethanolamine impurity present in the amide can tell what proportion of the EO added to the amine nitrogen, as opposed to the alcohol group (34).

E. Alkyl Polyglycosides

Proton NMR of alkylpolyglucosides gives peaks for protons corresponding to hydroxyl and alkyl groups, so that an estimate can be made of the number of glucose units per molecule, providing that free alcohol and glucose have been removed (35). ^{13}C NMR allows this determination to be made with greater precision and also permits resolution of the α- and β-isomers (35).

F. Ethylene Oxide/Propylene Oxide Copolymers

The NMR analysis of EO/PO block copolymers has been studied extensively, since they are used in great volume as polyurethane raw materials as well as surfactants. EO/PO ratio is easily determined by proton NMR, often in deuterated chloroform solvent (36–38). This measurement is based on the presence of methyl protons in the PO units, but only methylene protons in the EO units. The technique can not, of course, differentiate between block or random copolymers. NMR is very often used to determine the ratio of primary to secondary hydroxyl in copolymers of ethylene oxide and propylene oxide. (Only secondary hydroxyl is observed from propylene oxide–terminated molecules.) Derivatization is required before determination of primary/secondary hydroxyl groups by proton NMR. This technique also allows quantitative analysis of mixtures of, say, a polyethylene glycol and a polypropylene glycol. A small amount of trifluoroacetic acid added to an EO/PO copolymer can prevent overlap of the signal from the hydroxy group with that of the CHO and CH_2O groups (39).

Much more information is available from ^{13}C magnetic resonance. Not only can ^{13}C NMR measure the EO/PO distribution along the chain, and determine the ratio of primary/secondary terminal hydroxyl groups (40), but ^{13}C NMR is also suitable for indicating whether the structure of a copolymer is block or heteric. Differentiation can be made between PO–EO–PO block copolymers and EO–PO–EO blocks (41). The degree of heteric structure can be determined by measuring the occurrence of specific triads in the molecule (22,42). While ^{1}H NMR cannot differentiate the protons of some initiators (e.g., glycerin, trimethylolpropane) from those of EO, ^{13}C spectroscopy resolves these resonances (38). ^{19}F NMR can also be used to give the primary/secondary hydroxyl ratio if the trifluoroacetate derivatives are made (43).

Two routine methods are suitable for determining if the end groups of a copolymer are primary or secondary hydroxyls (44). Method A makes use of ^{19}F NMR. The surfactant is reacted with trifluoroacetic anhydride to form the trifluoroacetate diester. High resolution ^{19}F NMR shows well-resolved resonance peaks for the esters of the primary and secondary alcohols. Integration gives the ratio of primary to secondary hydroxyl, and thus the ratio of terminal EO groups to terminal PO groups. This method cannot be applied to surfactants with high EO content, since they will be extracted into the aqueous solution used to wash the derivatives. Method B is based on direct ^{13}C analysis of the surfactant, which yields well-resolved bands for the two types of hydroxyl groups. Deuterated chloroform or acetone is suitable as solvent.

^{13}C NMR can determine the relative abundance of EO or PO units near the ends of the molecule, thus indicating differences in the rate of incorporation of the monomers during synthesis (22). Determination of the molecular weight of the surfactant by NMR is possible by first forming the carbamate derivative with trichloroacetyl isocyanate (45).

III. CATIONIC AND AMPHOTERIC SURFACTANTS

NMR spectroscopy of cationic surfactants has been well summarized by Mozayeni (46).

^{13}C NMR is effective for qualitative analysis of ditallowdimethylammonium salts, as well as for determination of alkyltrimethyl- and trialkylmethylammonium impurities. Methyl and methylene carbons adjacent to the nitrogen are well resolved, and the chemical shift values differ measurably for methyl and methylene carbons of the various compounds (47). This method was demonstrated for impurities higher than about 2% (48).

Proton NMR of ester quats (made by quaternization of a mixture of mono-, di-, and triesters of triethanolamine) with a 300-MHz instrument allows determination of unsaturation, ester number, molecular weight, and degree of quaternization; all of these, of course, are average values (49).

^{15}N NMR has been demonstrated for differentiation of quaternary amines; monoalkyltrimethyl-, dialkyldimethyl-, and tetramethylammonium chlorides gave well-resolved individual peaks (46).

Proton NMR of betaines shows a betaine-specific signal at 3.3 ppm versus trimethylsilylpropionate. This signal can be used for quantitative analysis of betaines by comparison to the trimethylsilylpropionate internal standard (50).

^{31}P NMR has been demonstrated for direct determination of phospholipids in lecithin. It may even prove to be superior to normal phase HPLC for this purpose. Clearly separated signals are obtained for the major components and their hydrolysis products, but without differentiation by alkyl chain length (51). Metal cations must first be removed by extraction with EDTA solution, since they interfere with the analysis (51).

REFERENCES

1. Carminati, G., L. Cavalli, F. Buosi, ^{13}C NMR for identification of surfactants in mixture, *J. Am. Oil Chem. Soc.*, 1988, 65, 669–677.
2. Carminati, G., L. Cavalli, Detergents analysis. ^{13}C NMR spectroscopy (in Italian), *Riv. Ital. Sostanze Grasse*, 1989, 66, 341–347.
3. König, H., Surfactants using NMR spectrometry (in German), *Fresenius' Z. Anal. Chem.*, 1970, 251, 225–262.
4. König, H., Surfactants by NMR spectrometry (in German), *Tenside, Surfactants, Deterg.*, 1971, 8, 63–65.
5. Van Gorkom, L. C. M., A. Jensen, NMR Spectroscopy, in J. Cross, ed., *Anionic Surfactants: Analytical Chemistry*, 2nd ed., Marcel Dekker, New York, 1998.
6. Van de Ven, A. M. C., P. S. Johal, L. Jansen, Synthetic sulfonate and phenate detergents by NMR and IR spectroscopy, *Lubr. Sci.*, 1993, 6, 3–19.
7. Kosugi, Y., Y. Yoshida, T. Takeuchi, Alkylbenzenesulfonates and alkylsulfates as anionic surfactants by carbon-13 NMR spectrometry, *Anal. Chem.*, 1979, 51, 951–953.
8. Thurman, E. M., T. Willoughby, L. B. Barber, Jr., K. A. Thorn, Alkylbenzenesulfonate surfactants in groundwater using macroreticular resins and ^{13}C NMR spectrometry, *Anal. Chem.*, 1987, 59, 1798–1802.
9. El-Emary, M., L. O. Morgan, Structural features of LAS as observed in ^{13}C magnetic resonance spectra, *J. Am. Oil Chem. Soc.*, 1978, 55, 593–599.

10. Grey, R. A., A. F. Chan, Sulfonation studies of monoisomeric di- and trialkylbenzenes, *J. Am. Oil Chem. Soc.*, 1990, *67*, 132–141.

11. Boyer, J. L., J. P. Canselier, V. Castro, SO_3-sulfonation products of 1-alkenes by spectrometric methods, *J. Am. Oil Chem. Soc.*, 1982, *59*, 458–464.

12. Hashimoto, S., T. Nagai, Sodium α-sulpho fatty acid methyl ester by NMR spectrometry, *Tenside, Surfactants, Deterg.*, 1977, *14*, 271–272.

13. Herke, R., K. Rasheed, Addition of bisulfite to α-olefins: synthesis of *n*-alkane sulfonates and characterization of intermediates, *J. Am. Oil Chem. Soc.*, 1992, *69*, 47–51.

14. Yoshimura, H., E. Yoshihisa, S. Hashimoto, NMR study of sulfonation of internal olefins, *J. Am. Oil Chem. Soc.*, 1991, *68*, 623–628.

15. Radici, P., L. Cavalli, C. Maraschin, Internal *n*-olefin sulfonates, *Comun. Jorn. Com. Esp. Deterg.*, 1992, *23*, 205–218.

16. Frazier, J. D., R. D. Johnson, C. G. Wade, D. J. O'Leary, Composition of a variety of nonylphenol poly(ethylene oxide) phosphate anionic surfactants. *Comun. Jorn. Com. Esp. Deterg.*, 1991, *22*, 99–110.

17. Montana, A. J., NMR spectrometry of nonionic surfactants, in J. Cross, ed., *Nonionic Surfactants: Chemical Analysis*, Marcel Dekker, New York, 1987.

18. Black, D. B., B. A. Dawson, G. A. Neville, HPLC system for separation of components in nonoxynol-9 spermicidal agents, *J. Chromatogr.*, 1989, *478*, 244–249.

19. Cosmetic, Toiletry and Fragrance Association, Mole ratio of ethylene oxide in PEG alkyl ethers, CTFA Method J 3-1. 1110 Vermont Ave. NW, Washington, DC 20005.

20. Hammond, C. E., D. K. Kubik, Ethylene oxide content of *n*-alcohol ethoxylates by proton NMR spectroscopy, *J. Am. Oil Chem. Soc.*, 1994, *71*, 113–115.

21. Cross, C. K., A. C. Mackay, AE by NMR, *J. Am. Oil Chem. Soc.*, 1973, *50*, 249–250.

22. Gronski, W., G. Hellmann, A. Wilsch-Irrgang, ^{13}C NMR characterization of ethylene oxide/propylene oxide adducts, *Makromol. Chem.*, 1991, *192*, 591–601.

23. Auf der Heyde, W., ^{13}C NMR analysis of oxo alcohol ethoxylates (in German), *Tenside, Surfactants, Deterg.*, 1981, *18*, 265–268.

24. Kalinoski, H. T., A. Jensen, Nonionic surfactants using SFC and carbon-13 NMR spectroscopy, *J. Am. Oil Chem. Soc.*, 1989, *66*, 1171–1175.

25. Perov, P. A., V. L. Laptev, E. E. Zaev, L. A. Bez'yazychnaya, Free PEG in ethoxylated alcohols, acids, and alkylphenols by NMR spectroscopy (in Russian), *Zh. Anal. Khim.*, 1987, *42*, 2088–2092.

26. Yang, L., F. Heatley, T. G. Blease, R. I. G. Thompson, Oligomer distribution in ethoxylated linear and branched alkanols using ^{13}C-NMR, *Eur. Polym. J.*, 1997, *33*, 143–151.

27. Hearmon, R. A., M. P. Rhodes, Hydroxyl value determination by NMR, *Tenside, Surfactants, Deterg.*, 1986, *23*, 245–246.

28. Schlotterbeck, G., H. Pasch, K. Albert, Online HPLC ^{1}H NMR coupling for the analysis of AE, *Polym. Bull.* (Berlin), 1997, *38*, 673–679.

29. Szymanowski, J., H. Szewczyk, J. Hetper, Products obtained in the first stages of the ethoxylation of alkylphenols, *Tenside, Surfactants, Deterg.*, 1981, *18*, 333–338.

30. Podo, F., A. Ray, G. Nemethy, Structure and hydration of nonionic detergent micelles: high resolution NMR study, *J. Amer. Chem. Soc.*, 1973, *95*, 6164–6171.

31. Ribeiro, A. A., E. A. Dennis, Carbon-13 spin-lattice relaxation study on *p-tert*-OPE, *J. Phys. Chem.*, 1976, *80*, 1746–1753.

32. Pasch, H., W. Hiller, Technical poly(ethylene oxide) by on-line HPLC/^{1}H-NMR, *Macromolecules*, 1996, *29*, 6556–6559.

33. Stefanova, R., D. Rankoff, S. Panayotova, S. L. Spassov, Proton NMR determination of linoleic acid mono- and diesters of polyethyleneglycols via reaction with trichloroacetyl isocyanate, *J. Am. Oil Chem. Soc.*, 1988, *65*, 1516–1518.

34. Krusche, G., Amides in surfactants (in German), *Tenside, Surfactants, Deterg.*, 1973, *10*, 182–185.

35. Spilker, R., B. Menzebach, U. Schneider, I. Venn, Alkylpolyglucosides (in German), *Tenside, Surfactants, Deterg.*, 1996, *33*, 21–25.
36. Gabriel, D. M., Cosmetics and toiletries, *J. Soc. Cosmet. Chem.*, 1974, *25*, 33–48.
37. Zgoda, M. M., Surfactants in the group of propylene and ethylene oxide copolymers. II: Application of IR and proton NMR spectroscopy to the determination of the hydrophilic-lipophilic balance in polyetherdiols (in Polish), *Acta Pol. Pharm.*, 1988, *45*, 63–70.
38. American Society for Testing and Materials, Polymerized ethylene oxide content of polyether polyols, D4875. West Conshohocken, PA 19428.
39. Mathias, A., N. Mellor, Alkylene oxide polymers by NMR spectrometry and by GC, *Anal. Chem.*, 1966, *38*, 472–477.
40. Naylor, C. G., Nonionic surfactants containing propylene oxide, *J. Am. Oil Chem. Soc.*, 1986, *63*, 1201–1208.
41. Vonk, H. J., A. J. van Wely, L. G. J. van der Ven, A. J. J. de Breet, F. P. B. van der Maeden, M. E. F. Biemond, A. Venema, W. G. B. Huysmans, Analytical methods for ethoxylated surfactants, *Tr.-Mezhdunar. Kongr. Poverkhn.-Akt. Veshchestvam, 7th*, 1976, *1*, 435–449.
42. Heatley, F., Y. Luo, J. Ding, R. H. Mobbs, C. Booth, ^{13}C NMR study of the triad sequence structure of block and statistical copolymers of EO and PO, *Macromolecules*, 1988, *21*, 2713–2721.
43. Lebas, C. L., P. A. Turley, Primary hydroxyl content in polyols-evaluation of two NMR methods, *J. Cell. Plastics*, 1984, *20*, 194–199.
44. American Society for Testing and Materials, Primary hydroxyl contents of polyether polyols, D4273. West Conshohocken, PA 19428.
45. Carrion, J. L., M. De la Guardia, NMR characterization of ethylene oxide condensates with propylene oxide (in Spanish), *Quim. Anal.* (Barcelona), 1987, *6*, 76–82.
46. Mozayeni, F., Molecular spectroscopy of cationic surfactants, in J. Cross and E. J. Singer, eds., *Cationic Surfactants: Analytical and Biological Evaluation*, Marcel Dekker, New York, 1994.
47. Mozayeni, F., C. Plank, L. Gray, Mixture analysis of fatty amines and their derivatives by ^{13}C NMR, *Appl. Spectrosc.*, 1984, *38*, 518–521.
48. Fairchild, E. H., ^{13}C NMR spectroscopy of fatty quaternary amines, *J. Am. Oil Chem. Soc.*, 1982, *59*, 305–308.
49. Wilkes, A. J., C. Jacobs, G. Walraven, J. M. Talbot, Quaternized triethanolamine esters (esterquats) by HPLC, HRCGC, and NMR, *World Surfactants Congr., 4th*, 1996, *1*, 389–412.
50. Gerhards, R., I. Jussofie, D. Käseborn, S. Keune, R. Schulz, Cocoamidopropyl betaines, *Tenside, Surfactants, Deterg.*, 1996, *33*, 8–14.
51. Glonek, T., ^{31}P NMR phospholipid analysis of anionic-enriched lecithins, *J. Am. Oil Chem. Soc.*, 1998, *75*, 569–573.

15
Mass Spectrometry

I. GENERAL CONSIDERATIONS

Mass spectrometry has become the tool most widely used in industrial laboratories for characterization of organic compounds. It is inherently a sensitive and specific method of analysis, applicable to mixtures. The parent ions give an estimate of the molecular weight distribution of surfactant oligomers. Fragmentation patterns allow determination of isomer distribution and other chemical information, such as location of unsaturation, location of side chains, and degree of branching. As a general rule, direct MS analysis of surfactants, i.e., MS without prior separation, is not suitable for precise determination of homolog or oligomer distribution. This is because soft ionization techniques vary in efficiency according to the composition of the individual molecules.

The soft ionization techniques have greatly increased the accessibility of surfactants to MS analysis. Ionization and volatilization of molecules occur simultaneously, without subjecting the entire sample to high temperatures. These ionization techniques usually require that a large excess of the surfactant is subjected to the process, only a tiny portion of which is volatilized and ionized. True quantitative analysis is generally not possible. If possible, the system is tuned so that the mass spectrum qualitatively gives the correct molecular weight distribution for those surfactants which are mixtures of oligomers. Often, true quantification can only be performed by the use of internal standards of isotope-labeled compounds of the precise surfactants to be determined, too expensive an undertaking for most work.

461

It is a challenge to analyze surfactants by the most common configuration of MS: GC-MS. As discussed in Chapter 8, most surfactants are not sufficiently volatile to pass through the GC unless they are first derivatized or degraded to form volatile products. Liquid chromatography–MS presents problems of its own, since there is no one LC-MS interface suitable for all compounds. The use of LC-MS in characterizing surfactants in the environment has been reviewed (1). Electrospray ionization gives better results for surfactants than thermospray, with not much published on particle beam or atmospheric pressure ionization of surfactants.

II. ANIONIC SURFACTANTS

A review of MS of anionic surfactants, with special emphasis on the newer ionization techniques, is part of a companion volume of the Surfactant Science Series (2). In general, greater sensitivity and selectivity toward anionics is obtained by operating in the negative ion detection mode.

A. Alkylarylsulfonates

Linear alkylbenzenesulfonate is a mixture of components, with the alkyl chain length of most commercial products falling in the range C_9–C_{15}. The aryl ring may be attached to the alkyl chain anywhere along its length but at the C_1 position. Mass spectrometry by itself is an effective technique for unequivocally identifying homologs, with GC-MS allowing identification of isomers. High resolution capillary chromatography is essential to separate the isomers, which can be detected using the mass spectrometer in either the electron impact or the chemical ionization (with methane) mode. The parent ion is predominant in the CI spectrum. Desulfonation was for many years the preferred technique of sample pretreatment for GC-MS, but since the 1980s most published work is based on formation of sulfonyl chloride, sulfonyl fluoride, and methyl sulfonate derivatives (see Chapter 8). Derivatization with PCl_5 will form the sulfonyl chlorides (3) or, after subsequent reaction with methanol, the methyl sulfonate esters (4). More often, the methyl esters are formed directly from the free sulfonic acids using diazomethane. Alternatively, the tetraalkylammonium salt is made. This decomposes at the temperature of the GC injection port or the MS direct probe to yield the alkyl ester and tertiary amine [Fig. 1] (5).

Electron impact ionization. EI mass spectra of LAS show the tropylium ion, $C_7H_7^+$, at m/z 91 and a series of ions of the form $C_nH_{2n}C_6H_4SO_3H^+$ (6). Parent ions are detectable, whether analysis is performed of the free sulfonic acids or of methyl, trifluoroethyl, and sulfonyl chloride derivatives, with fragmentation patterns analogous to those of alkylbenzenes. Characteristic positive daughter ions result from cleavage of the alkyl chain at the point of attachment to the benzylic carbon. Thus, abundant ions correspond to the loss of both the sulfonate group and the longer of the alkyl side chains, giving ions of m/z 104, 118, 132, 146, 160, and 174 for dodecylbenzenesulfonate with phenyl substitution in the C_2, C_3, C_4, C_5, C_6, and C_7 positions, respectively. For derivatized LAS, even more abundant and characteristic positive ions correspond to the loss of the longer alkyl chain but without loss of the sulfonate group. The mass of these fragments depends on which derivatizing agent was used (3,4,7). At least in positive ion mode, parent ions are not usually observed for higher alkyl derivatives of LAS like the butyl esters; the highest m/z peak is due to loss of the derivatizing group (5).

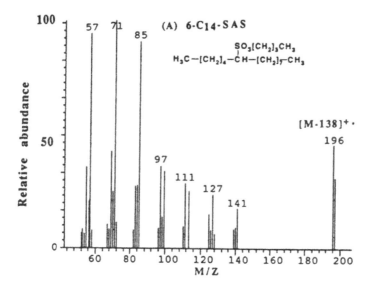

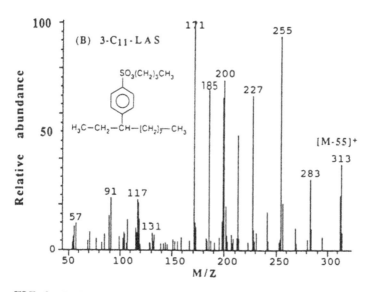

FIG. 1 Positive ion electron impact MS of butyl esters of an alkanesulfonate component and a LAS component. Esters were formed by injecting the tetrabutylammonium salts of the surfactants into the GC. (Reprinted with permission from Ref. 5. Copyright 1992 by the American Chemical Society.)

Agazzino and coworkers found that, by direct introduction of sodium alkylarylsulfonates mixed with $KHSO_4$, good electron impact spectra of the acids could be obtained without resorting to FAB or other special techniques. Severe fragmentation occurred with ionization at 75 eV, but at 15 eV good abundance of the parent ions (positive mode) was observed, with fragmentation characteristic of the degree of branching and the alkyl chain length (6). However, most investigators use electron impact with GC-MS rather than direct introduction, so derivatives are almost always made.

Chemical ionization. Yield of molecular ions of trifluoroethyl derivatives is much higher under positive CI conditions than negative CI conditions, making positive CI more useful for trace analysis than either EI or negative CI (7,8). Negative ion CI-MS of trifluoroethanol derivatives of LAS and dialkyltetralinsulfonate shows predominantly the parent ion, [parent – 100]⁻, due to loss of trifluoroethanol, and m/z 163 due to $O_3SCH_2CF_3^-$ (9). The intensity of the rather uninformative 163 ion is increased by increasing the source temperature.

Collisionally activated dissociation. LAS can be readily differentiated from the branched chain tetrapropylene-based alkylbenzenesulfonates (ABS) by the different fragmentation patterns to which the parent ions give rise under conditions of MS-MS, using the instrument in negative ion mode. Most characteristic for LAS is the ion $O_3S\phi CH=CH_2^-$, m/z 183, while the corresponding ion for ABS is $O_3S\phi C(CH_3)=CH_2^-$, m/z 197. Both fragments show further loss of SO_2 to form ions at m/z 119 and 133. Similar behavior is seen when initial ionization is by FAB or by CI of trifluoroethyl derivatives. Positive CI-CAD does not allow differentiation of LAS and ABS (10,11). If trifluoroethanol derivatives are used, the diagnostic peaks are at $O_2S\phi CH=CH_2^-$, m/z 167, and $O_2S\phi C(CH_3)=CH_2^-$, m/z 181, corresponding to loss of CF_3CH_2O (11).

Fast atom bombardment. FAB has been demonstrated as a means of determining LAS and ABS in environmental samples. Good abundance of the molecular anions is seen. Calibration with a single internal standard yields inaccurate results for determination of homolog distribution of LAS because of differences in surface activity, and hence FAB ionization, between homologs. Three internal standards of various alkyl chain lengths are required for accurate quantification (12). When subjected to CAD, the molecular ions produce the characteristic daughter ions mentioned above, at m/z 183 for LAS and 197 for ABS (10,13).

Matrix-assisted laser desorption introduction. MALDI of LAS was rather uninformative, showing little more than the peak for the parent ion; much more information was developed analyzing desulfonated LAS. The authors concluded that good MALDI analysis of LAS must necessarily await discovery of a suitable matrix (14).

Electrospray ionization. This process is suitable for use in direct MS or LC-MS analysis of anionics, including LAS, using negative ion mode (15). LC-MS analysis with electrospray ionization in the negative ion mode allows very thorough determination of LAS and its coproducts and metabolites in wastewater. The sample extract is first scanned at a high cone voltage (–80 V) to detect the characteristic ion $O_3S\phi CH=CH_2^-$, m/z 183. A second run is made as necessary with a lower cone voltage (–20 V) to detect the parent ions, [M – H]⁻, of the individual LAS homologs corresponding to C_{10}–C_{13} alkyl substitution, at m/z 297, 311, 325, and 339 (16–20).

Atmospheric pressure chemical ionization. LC-MS analysis with an APCI interface can be used in either the positive or negative ion mode. The positive ion mode is much more sensitive, but negative ion mode gives less interference from nonionic surfactants. Pseudomolecular ions of Na⁺ adducts predominate in the positive ion spectra, with molecular ions predominant in the negative ion spectra (21).

Field desorption. FD-MS of LAS, used in either the positive or negative ion mode, gives quasimolecular ions free of fragments and ionization products (22–24). Use of FD in the negative ion mode has the advantage that the surfactant need not be separated from the sample matrix prior to analysis, since the technique is very selective for sulfonates. Even if cationic surfactants are present in mixture with the anionics, the anionic spectrum is not affected. A disadvantage of the FD technique is that no structural information can be obtained. To obtain structural information, the ions are subjected to CAD (25,26).

Laser ionization. This procedure followed by MS of the positive or negative ions, produces intense ions corresponding to the molecular ion or the dimer, plus a cation (such as sodium). Alkyl sulfates and alkylarylsulfonates can be analyzed by this technique (27). Accurate quantification requires an internal standard, generally the deuterated analog, for each homolog. The technique has been demonstrated for direct analysis of cloth, but this approach is clearly only appropriate for very special problems.

Procedure: Determination of LAS in Sewage Sludge by Gas Chromatography–Mass Spectrometry (3)

50 µg of p-(pentadec-3-yl)benzenesulfonate internal standard is added to a 1-mL sample in a separatory funnel. Twenty milliliters of a pH 10 aqueous solution of methylene blue is added, and the solution is extracted repetitively with 15 mL portions of $CHCl_3$ until no more blue color is removed. The chloroform extracts are combined and evaporated to dryness, then the residue is redissolved in 0.5 mL methylene chloride and purified by preparative TLC on Merck Silica Gel G with CH_2Cl_2. The blue band near the origin is scraped off and the LAS eluted from the silica with methanol. The methanol is evaporated from the extract, which is then rinsed into a 3-mL vial with CH_2Cl_2. The solvent is again evaporated. One-half milliliter hexane and 0.05 g PCl_5 are added and the vial is capped tightly and heated at 110°C for 20 min. After cooling, 2 mL of n-hexane is added, then all solvent is evaporated to drive out unreacted PCl_5 and the reaction product, $POCl_3$. Since some of the phosphorus compounds adsorb to the walls of the vial, the residue is taken up in n-hexane and transferred to another vial. An aliquot of from 0.5 to 1.0 µL is injected in the splitless mode on a 0.31 mm × 19 m capillary column, SE-54 or PS 255. The injector is held at 275°C. The end of the column is connected directly to the ion source of the MS by a fused silica capillary. The temperature program is run from 60 to 270°C at 4°C/min. For EI, an ion source temperature of 250°C is used with 40 eV ionizing energy and filament emission current of 350 µA. For CI, methane is the reagent gas, the ionization temperature is 150°C, the ionizing energy is 70 eV, and the pressure 0.37 torr.

B. Alkyl Sulfates and Alkanesulfonates

Alkanesulfonates are usually analyzed by GC-MS of their alkyl esters. Alternatively, the tetraalkylammonium salt is made. This decomposes at the temperature of the GC injection port or the MS direct probe to yield the alkyl ester and tertiary amine [Fig. 1] (5).

Electron impact ionization. Under electron impact conditions, the parent ion of derivatized alkanesulfonate has very low abundance. Intense ions, suitable for quantitative analysis, are observed for the alkyl ions resulting from loss of the derivatized sulfonate group. A series of alkyl ions of lower m/z is also observed. Much smaller quantities of RO^+ ions are formed by rearrangement (5).

Collisionally activated dissociation. CAD spectra of FAB-generated negative molecular ions from alkyl sulfate gives ions beginning with m/z 96 and increasing in increments of 14 out to the molecular anion itself. Alkanesulfonates appear similar, but begin with a peak at 80 (SO_3^-) rather than 96 (SO_4^-) (15,28).

Laser ionization. Followed by MS of the positive or negative ions, laser ionization produces intense ions corresponding to the molecular ion or the dimer, plus a cation (such as sodium). Alkyl sulfates and alkylarylsulfonates can be analyzed by this technique (27).

MALDI. A strong parent ion for alkyl sulfate is provided by MALDI, but little other information can be derived (14).

Fast atom bombardment. In the case of alkyl sulfates, positive ion FAB shows positive ions from clusters of the molecular ions with sodium ions in the general formula $[Na_{n+1}AS_n]^+$. Negative ion FAB shows a molecular anion and analogous clusters (28,29).

Atmospheric pressure chemical ionization. This technique is compatible with the effluent from a chemically suppressed ion chromatography system (30). In this case, field assisted ion evaporation, operated in the negative ion mode, gave easily identified molecular ions for alkyl sulfates and sulfonates. By using a tandem mass spectrometer, fragmentation data were also obtained, corresponding chiefly to loss of a SO_3^- or HSO_4^- ion.

Electrospray ionization. ESI-MS in the negative-ion mode is effective for MS of alkyl sulfates in the surfactant range of C_{12}–C_{15} chain length (16). The parent ion by itself is not sufficient for unequivocal identification, but this is provided by coupling with LC (31) or by CAD analysis of the parent ion (15).

C. Ether Sulfates

Ether sulfates are not analyzed by GC-MS. Analysis is successful by soft ionization techniques, including LC-MS. MS analysis alone is not suitable for determining the level of unsulfated alcohol or ethoxylated alcohol in an alkyl ether sulfate (32).

Electrospray ionization. Direct electrospray analysis of alkylethoxysulfate in positive ion mode gives two major ions, one corresponding to $[R(OCH_2CH_2)_nOSO_3Na + Na]^+$ and the other to $[R(OCH_2CH_2)_nOH + Na]^+$. In the absence of a prior HPLC separation, these positive ion spectra are difficult to interpret because of overlap of, for example, $[C_{13}H_{27}(OCH_2CH_2)_3OH + Na]^+$ and $[C_{12}H_{25}(OCH_2CH_2)OSO_3Na + Na]^+$ (32). In negative ion mode, only the molecular anion, $[R(OCH_2CH_2)_nOSO_3]^-$, is seen (32). An electrospray interface was demonstrated for LC-MS of alkyl sulfates and ethoxysulfates of C_{12}–C_{15} alkyl chain length and 0–8 ethoxy groups. The LC gave partial separation by alkyl chain length, with quantification of individual homologs accomplished by MS in the negative ion mode (31).

Atmospheric pressure chemical ionization. APCI-MS analysis in positive ion mode gives only one major ion for each homolog/oligomer, corresponding to loss of the sulfate group $[R(OCH_2CH_2)_nOH + H]^+$ (32).

Collisionally activated dissociation. CAD of ions produced by ESI or APCI: CAD of $[R(OCH_2CH_2)_nOSO_3Na_2]^+$ gives only Na^+ and $[Na_2HSO_4]^+$, while CAD of $[R(OCH_2CH_2)_nOH + Na]^+$ yields only Na^+. CAD of anions is similarly uninformative, with $[R(OCH_2CH_2)_nOSO_3]^-$ yielding HSO_4^-, m/z 97; SO_3^-, m/z 80; and a "tiny" amount of $[CH_2{=}CHOSO_3]^-$, m/z 123, regardless of the degree of ethoxylation. CAD of the ion produced by APCI, $[R(OCH_2CH_2)_nOH + H]^+$, gives more information. An important fragment is seen corresponding to loss of the alkyl group, namely $[H(OCH_2CH_2)_nOH + H]^+$. Thus, the length of the alkyl group can be calculated from difference in m/z between the parent ion and this fragment. Also seen is a series of ions of the form $[C_2H_4(OCH_2CH_2)_nOH]^+$, with n having values 1–4, but not higher than the value of n in the parent ion (32).

In contrast to the above, Lyon et al. report considerable structural information by CAD of the anions formed by FAB. Fragments of the parent anion, $[R(OCH_2CH_2)_nOSO_3]^-$, give many peaks corresponding to many different combinations of losses of alkyl and ethoxy fragments (28).

MALDI. These spectra of alkyl ether sulfates are complex because of the overlapping distributions of ethoxylates of individual alkyl chain lengths. Only pseudomolecular ions are seen, which consist of monosodium and disodium adducts of the main components, as well as monosodium adducts of PEG (33).

Fast atom bombardment. Sulfated alcohol ethoxylates give abundant negative ions corresponding to deprotonation of the main components and of clusters. Positive clusters with sodium and potassium ions are also seen (29).

Thermospray LC-MS. This process can be used with reversed-phase HPLC to qualitatively analyze sulfonated phenol ethoxylates (34).

D. α-Olefin Sulfonates

Olefin sulfonates may be analyzed by GC-MS of derivatives or by applying the newer ionization methods to underivatized material.

Electron impact ionization. When hexadecenesulfonate is analyzed as the methyl sulfonate, the highest m/z ion observed corresponds to $[M - OCH_3]^+$. Other ions correspond to $C_{16}H_{31}^+$, $C_{16}H_{29}^+$, SO_2^+, $CH_3SO_2^+$, and the like (35). EI-MS of the methyl derivative of 2-hydroxyalkanesulfonate also shows no parent ion. The characteristic peak is at 153, corresponding to the fragment $CH_3O^+=CHCH_2SO_3CH_3$ (36). A peak at 121 corresponds to loss of methanol from the 153 peak.

Electrospray ionization. ESI-MS in the negative ion mode has been demonstrated for obtaining molecular ions of olefin sulfonate. Low concentrations were detected by introducing the compounds as solutions in 50:50 benzene/methanol or carbon tetrachloride/methanol (16).

Collisionally activated dissociation. Hydroxyalkane sulfonates cannot be differentiated from alkyl sulfates by simple electrospray MS in negative ion mode because the parent ions are similar. However, CAD shows a daughter ion at m/z 80, indicating a sulfonate, and neutral loss of a water, characteristic of the hydroxyalkane compound (15).

Fast atom bombardment. FAB analysis of olefin sulfonates shows clusters in both positive and negative ion mode. CAD of the negative ions can give rise to considerable fragmentation, with a peak due to SO_3^-, m/z 80, and many peaks corresponding to $C_nH_{2n-2}SO_3^-$ (28).

E. Sulfosuccinate Esters

MALDI. In the positive ion mode MALDI-MS gives only a weak signal of pseudomolecular ions which contain two sodium ions, one of which is associated with the sulfonate group. The signal is much stronger in the negative ion mode, with molecular ions corresponding to the whole range of homologs. This difference in response eases the detection of succinate esters, as well as most other anionics, in mixtures (37).

Electrospray ionization. ESI-MS in the negative-ion mode has been demonstrated for obtaining molecular ions of *bis*(2-ethylhexyl)sulfosuccinate. Low concentrations were detected by introducing the compounds as solutions in 50:50 benzene/methanol or carbon tetrachloride/methanol (16).

Fast atom bombardment. FAB shows a molecular anion (m/z 421 in the case of the 2-ethylhexyl diester) as well as cations in the form of clusters with sodium. Monoesters as well as diesters can be detected. CAD of the 421 anion gives a fragment at 80 (for SO_3^-), as well as one at 309 corresponding to loss of a 2-ethylhexyl group (28).

F. Other Anionics

1. Fluorinated Anionics

Schröder examined these by thermospray MS and found negative ion MS to be much more valuable than positive ion spectra. Negative ion spectra of perfluorinated alkyl compounds are dominated by peaks at intervals of $\Delta m/z$ 100, corresponding to loss of the C_2F_4 group (38). Moody and Field analyzed perfluorinated fatty acids using GC-MS of the methyl esters. In positive ion EI mode, molecular ions were not seen, but rather fragments differing by CF_2, m/z 50. Characteristic ions were $C_3F_5^+$, m/z 131, $C_3F_7^+$, m/z 169, and $C_4F_9^+$, m/z 219. Electron capture negative ionization gave negative molecular ions for each of the methyl esters (39).

2. *N*-Acylated Amino Acids

FAB of an *N*-laurylsarcosinate shows positive clusters with sodium and a negative molecular anion, as well as a negative dimer ion. CAD of the negative molecular ion gives important ions due to the loss of CO_2 and to loss of the alkyl group (28).

3. Fatty Acids

Electrospray ionization.   ESI-MS in the negative-ion mode has been demonstrated for obtaining molecular ions of fatty acids (15). Low concentrations were detected by introducing the compounds as solutions in 50:50 benzene/methanol or carbon tetrachloride/methanol (16).

Fast atom bombardment.   Negative ion FAB of fatty acids gives a molecular anion. CAD analysis of the anion shows a peak due to loss of CO_2 as well as a series on ions differing by 14 units, characteristic of alkyl chains (28).

III. NONIONIC SURFACTANTS

MS has found most use in qualitative rather than quantitative analysis of nonionic surfactants. This is partly due to their lack of volatility, but also because the ionization efficiency varies for different homologs and oligomers, so that the average degree of ethoxylation found by MS usually differs from that found by other methods. Nonionics are usually analyzed in the positive ion mode. Few negative ions are formed; hydroxyl groups are not deprotonated in the mass spectrometer. Rauscher has compiled a table of fragments found in the MS analysis of common nonionic surfactants (40).

LC-MS using either the electrospray interface or the thermospray interface gives the best results for MS analysis of nonionics. Most often, an HPLC system is used which separates the homologs according to the length of the alkyl chain. MS analysis of each peak gives the information on the EO chain and can often unambiguously identify each compound. The peaks observed are mainly the quasimolecular ions of $[M + NH_4]^+$ (Ammonium acetate solution is mixed with the HPLC mobile phase downstream from the column). Little fragmentation information is developed with electrospray or thermospray MS, but it has the advantage of providing mass spectra for compounds which are not ordinarily amenable to MS analysis. It functions as a specific detector, allowing differentiation of mixtures of nonionics, even if they are not completely resolved by LC. As of this writing, the electrospray interface is most popular for LC-MS of nonionics and is replacing the thermospray interface.

A. Alcohol Ethoxylates

Direct MS analysis is handicapped because the ethoxylates of commercial interest are not volatile unless derivatized, and even then the higher ethoxylates cannot be analyzed.

Electron impact ionization. With experience, the composition of AE can be deduced from interpretation of the EI spectrum alone (41). While the EI spectrum is dominated by nonspecific ions corresponding to polyethylene glycols (42), substantial structural information can be extracted, especially if the direct inlet probe is temperature-programmed. It was observed that if the alkyl chain is unsaturated, the $[M + 1]^+$ ion is absent, while the $[M - 1]^+$ ion is much more abundant (43). However, not all workers report this phenomenon. Surfactants containing propylene oxide groups are easily distinguished by their EI fragmentation from compounds containing only ethylene oxide.

GC-MS determination of the alkyl bromide decomposition products of AE was successfully used in a wastewater monitoring study. EI products of alkyl bromides are similar, so SIM was used to monitor the ions at 135 and 137, corresponding to $C_4H_8Br^+$ ions containing the ^{79}Br and ^{81}Br isotopes (44).

Chemical ionization. Desorption chemical ionization of nonionics, as with most other substances, gives abundant quasimolecular ions. Commercial ethoxylates are therefore easily recognized by the pattern of ions differing by 44 mass units. The lack of fragments compared with other techniques makes CI less valuable for extracting structural information. This shortcoming can be improved somewhat by the choice of reactant gas. Direct probe distillation coupled with chemical ionization with ammonia and ammonia/methane mixture was used for characterization of commercial AE mixtures (45). AE shows mainly the $[M + 1]^+$ and $[M - 1]^+$ ions by GC-CI-MS (46). Optimized DCI conditions also give predominantly the protonated molecular ions. In the case of a C_{12} alkyl 23-mole ethoxylate, the oligomer distribution found was consistent with that found by other techniques (47). Low energy CAD of the AE parent ion results in cleavage of the ether bonds to give fragments corresponding to the loss of 1, 2, 3, and 4 EO groups (ions at m/z of 45, 89, 133, and 177, with others at $[M + H]^+$, $[M + H - 44]^+$, and $[M + H - 88]^+$) (48). At higher energy, low abundance ions are seen representing the alkyl chain (47).

The carboxylates of AE (formed by biological degradation of AE) can be analyzed by GC-CI-MS of the *n*-propyl esters with methane reagent gas. The most abundant ions correspond to $[M + 1]^+$, with M^+ and $[M - H]^+$ ions also observed, the relative amounts varying depending mainly on the alkyl chain length. Diagnostically important fragments result from loss of the alkyl chain by the $[M + H]^+$ ion (olefin displacement) thus giving information on the length of both the alkyl and ethoxy chain length. For example, a C_{12} AE will give a fragment of $[MH - 168]^+$ (49).

Thermospray MS. Analysis of AE by this method shows almost exclusively the molecular and quasimolecular ions of the homologs (50–52). Some lower MW compounds show an $[M + 1]^+$ ion. Response varies among the homologs, so use of an internal standard is preferred.

Fast atom bombardment. FAB of AE gives quasimolecular ions for each homolog (29,53–55). CAD analysis of individual peaks from FAB analysis gives mainly nonspecific polyethylene glycol peaks ($m/z = [44n + 1]^+$), as for EI-MS (55). However, other fragments give information on the alkyl chain composition (56,57).

Electrospray MS. Applied to an acidic solution of an average 9-mole ethoxylate, electrospray MS gives $[M + 1]^+$ ions representing the true molecular weight distribution (58). This means that the main peaks are at intervals of $m/z = 44$, with peaks of $m/z = 14$,

displaced in either direction corresponding to other alkyl chain lengths. Presence of sodium or potassium salts in the mobile phase causes a shift in the mass distribution corresponding to formation of the corresponding adducts ($[M + 23]^+$ or $[M + 38]^+$). Similar
results are seen with ammonium salt buffers, with the $[M + NH_4]^+$ adducts dominant (15).
Electrospray gives good results in LC-MS [Fig. 2] (59–61).

If CAD is applied to ions produced by electrospray MS, fragmentation of secondary
AE occurs more easily than that of primary. With higher energy CAD of secondary AE
compounds, the PEG ions become dominant and extend to the higher mass range (58).

Atmospheric pressure chemical ionization. Solutions of AE give molecular or
pseudomolecular positive ion peaks for all homologs in the sample, limited only by the
mass range of the spectrometer. There is some indication that the technique is more sensitive to PEG than to AE, so care must be used in quantification (17,62). Positive ions characteristic of AE are of the form $[14n + 44x + 18]^+$, corresponding to the molecular ions
(18). Environmental samples will also give ions for PEG, $[44x + 18]^+$, and mono- and dicarboxylated PEG, $[44x + 76]^+$ and $[44x + 134]^+$ (18). APCI in negative ion mode also
gives pseudomolecular peaks. There is a definite discrimination against free alcohol and
low ethoxylates, with corresponding higher sensitivity to higher oligomers (63).

MALDI. Analysis of AE by MALDI works well, giving molecular ions of the form
$[M + Na]^+$ and/or $[M + K]^+$. It is undesirable to have both sodium and potassium present
because, for example, the potassium adduct of a C_{14} alcohol 4-ethoxylate has the same m/z (at unit resolution) as the sodium adduct of a C_{12} alcohol 5-ethoxylate (37). For this reason, some investigators use ion exchange to insure that only a single cation, either Na^+ or
K^+, is present (33). MALDI is applicable to high ethoxylates which cannot be quantitatively analyzed by other MS techniques because of their lack of volatility. Good results are
obtained for average degree of ethoxylation of 30 or even higher (37,64). MALDI is a
good technique for the characterization of LC and TLC fractions (65,66). MALDI of low
ethoxylates must be used carefully because the efficiency of ionization increases as the
degree of ethoxylation increases, making MALDI unsuitable for determination of average
degree of ethoxylation (Fig. 3). This effect is not observed above about 12 EO units (33).

Supercritical fluid chromatography. Kalinoski and Hargiss report that SFC-MS in
CI mode of AE yields abundant fragments because of the presence of the carrier gas. Addition of dichlorodifluoromethane suppresses this effect, giving primarily $(M + Cl)^-$ ions
(67). Fragmentation seems not to occur with hexane/ethanol mobile phase (68). SFC-MS,
using CO_2 as mobile phase and chemical ionization MS detection, was shown very suitable for characterization of commercial AE, giving information on the presence of
branched alkyl chains, as well as the molecular weight of the homologs (69).

B. Alkylphenol Ethoxylates

Analysis of higher oligomers by thermospray or other LC-MS techniques is problematic
due to the MW-dependance of both the mass spectrometer response and the ionization process. Typically, MS of APE shows trace levels of impurities of dialkylphenol ethoxylates
and of PEG. APE is best detected in positive ion mode, while alkylphenol itself is detected
with greatest sensitivity in negative ion mode (70)

Electron impact ionization. A small molecular ion may or may not be observed, as
well as a higher intensity fragment corresponding to loss of pentyl or hexyl radical to give
$^+C(CH_3)_2C_6H_4O(CH_2CH_2O)_nH$. Other ions correspond to fragments of the alkyl and

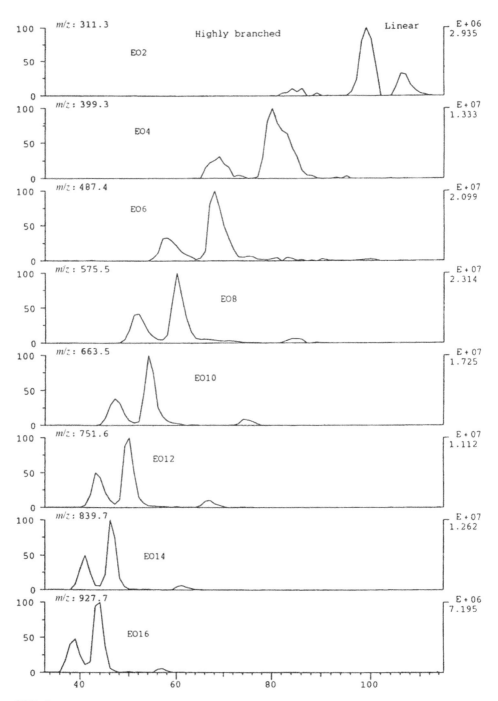

FIG. 2 AE analysis by electrospray MS interfaced to reversed-phase HPLC. Shown are selected ion chromatograms for individual $C_{13}H_{27}O(CH_2CH_2O)_nHNa^+$ oligomers. Note that because of the presence of isomers, HPLC alone would not be able to resolve the mixture. (Reprinted with permission from Ref. 61. Copyright 1997 by the American Oil Chemists' Society.)

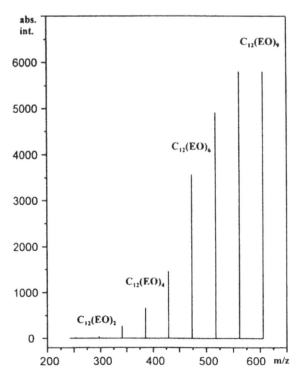

FIG. 3 Positive ion MALDI spectrum of a mixture of equimolar amounts of dodecanol ethoxy-late oligomers containing 1–9 moles of EO. Note the poor response from the lower oligomers. (Reprinted with permission from Ref. 33. Copyright 1998 by Karl Hanser Verlag.)

ethoxy chains (42,71). Similar fragmentation is observed with halogenated OPE and trim-ethylsilyl derivatives of APE (72).

   Chemical ionization.   The lower ethoxylates of alkylphenol can be analyzed by di-rect probe CI-MS or by GC-CI-MS. Characteristic of the 1- and 2-mole ethoxylates of nonylphenol is an ion corresponding to loss of a hexyl or heptyl radical, depending on the branching of the alkyl chain (73). Important fragments observed with CI of APE oligom-ers of up to 6 EO units correspond to the loss of nonene or octene, i.e., the entire alkyl chain. With oligomers higher than 6, this fragmentation becomes unimportant and a strong ion is observed corresponding to $RC_6H_4OCH_2CH_2OCH_2CH_2^+$ (demonstrated up to 16 EO units) (42). Analysis without derivatization is applicable up to a maximum degree of ethoxylation of about 5 for APE. Higher oligomers are not sufficiently volatile to pass through the GC. Derivatization with diazomethane or methanol/BF_3 is used for analysis of higher oligomers of APE, as well as for carboxylates of APE and halogenated APE (54,72).

   Laser desorption Fourier transform ion cyclotron resonance spectrometry.   Laser desorption FTICR-MS looks very promising for the analysis of high ethoxylates, but this technique is still the province of research universities. It has been demonstrated for deter-mination of oligomer distribution of OPE of average degree of ethoxylation of 59 (74).

   Desorption chemical ionization.   Vincenti and coworkers showed that desorption chemical ionization could be optimized with small sample size, cool source temperature, high reagent gas pressure, and very rapid sample heating and m/z scanning (75). Such conditions give a mass spectrum consisting predominantly of molecular ions. Under these conditions, the molecular weight distributions of NPE with an average degree of ethoxylation of 12 show good agreement between DCI and other methods (47).

Electrospray MS. This method is commonly used for LC detection (59). Usually, the sodium adducts are seen. Electrospray MS interfaced to a SEC system worked well for analysis of OPE of $n = 16$ or 44. A small amount of a sodium salt was either added to the tetrahydrofuran eluent or used in the sheath liquid. Quasimolecular ions were obtained for all oligomers (76,77). Electrospray MS of an acidic solution of a nonylphenol 9-mole ethoxylate gives $[M + 1]^+$ ions representing the true molecular weight distribution, with the main peaks at intervals of $m/z = 44$. Addition of sodium or potassium salts to the mobile phase causes the appearance of other peaks corresponding to the adducts $[M + 22]^+$ or $[M + 38]^+$ (58). Application of MS-MS shows daughter ions corresponding to neutral loss of CH_2CH_2O (44), $(CH_2CH_2O)_2$ (88), and C_9H_{18} (126) (78). MS-MS can be used in the absence of a preliminary LC separation, monitoring the product ion $\phi O(H^+)CH{=}CH_2$, m/z 121, for 1–4 ethoxylates of nonylphenol, and $CH_2{=}CHO(H^+)CH_2CH_2OCH_2CH_2OH$, m/z 133, for higher ethoxylates (79).

Ikonomou and coworkers investigated quantitative LC-ES-MS using normal phase conditions, with injection of an aqueous solution before the interface. They noted that the signal was dispersed among several ions ($[M + H]^+$, $[M + NH_3]^+$, $[M + H_3O]^+$, $[M + Na]^+$, and $[M + K]^+$) and that the signal strength was much different for different oligomers, particularly at degrees of ethoxylation below about 10. Best results were obtained by adding a sodium acetate solution, so that exclusively sodium adducts were produced. They report the unexplained observation that the intensity of the 1- and 2-mole ethoxylate adducts is greatly increased if a NPE/sodium acetate solution is allowed to stand 1–2 days before analysis. Above 12 EO units, disodium adducts are also seen. They report the optimum cone voltage range as 15–37 V. Below 15 V, cluster formation is favored, while above 37 V fragmentation begins to weaken the signal of the quasimolecular ion. Reasonably good reproducibility was obtained with careful and frequent optimization of conditions and use of an internal standard (70).

LC-ES-MS was used for detection of carboxylated degradation products of NPE. Because of prior purification, protonated molecular ions were seen rather than sodium or potassium adducts. Fragmentation spectra can be obtained by applying in-source CAD (80).

Thermospray MS. Analysis of NPE shows only quasimolecular ions corresponding to the individual homologs of NPE and PEG (34,50).

Fast atom bombardment. FAB gives protonated molecular ions, but these are of low abundance, especially for higher oligomers (81). Much greater sensitivity is obtained by adding sodium or potassium salts to give the $[M + Na]^+$ or $[M + K]^+$ quasimolecular ions (82). This has been demonstrated up to about 50 EO units (83). Compounds containing both EO and PO react similarly (53). Under certain FAB conditions, intense fragmentation occurs to give typical fragments of the ethoxy and alkyl chains (84). Ions are also seen corresponding to products of benzylic cleavage of the alkyl chain (85).

Field desorption. APEs give very simple mass spectra by field desorption, showing almost exclusively the molecular or $[M + 1]$ ions, demonstrated to a degree of ethoxylation of about 16 (22,82,86). FD can be used for quantitative determination of NPE in environmental samples (87,88). Field desorption mass spectroscopy was used to characterize fractions collected after HPLC separation of APE (86). The HPLC system separated only by alkyl chain length, giving three fractions corresponding to all ethoxylates of C_8,

C_9, and C_{12} alkylphenol. Tables were developed to identify the ethoxylates on the basis of their field desorption mass spectra.

Collisionally activated dissociation. APE ions are rather resistant to CAD at low energy, showing very little fragmentation to form PEG ions (58). CAD of the molecular ions at higher energy shows the quasimolecular ion and fragments corresponding to loss of portions of both the alkyl chain and the ethoxy chain (47,79).

MALDI. MALDI-MS analysis of APE works well, giving pseudomolecular ions of the form $[M + Na]^+$ and $[M + K]^+$ (Fig. 4). It is a good technique for the characterization of LC and TLC fractions (64,65). For quantification, addition of LiCl is recommended, rather than Na^+ or K^+ salts. Under these conditions, there is little discrimination among oligomers, so that the measured distribution is similar to the true distribution. Fine structure is seen in the mass spectra, indicating that, in the case of NPE, traces of octyl-, decyl-, and undecylphenolethoxylates are present (14,89).

Atmospheric pressure chemical ionization. LC-MS with acetonitrile/water/ammonium acetate or even columnless injection of acetonitrile solutions of APE gives molecular positive ion peaks or pseudomolecular adducts for all oligomers in the sample, limited only by the mass range of the spectrometer. Identification and quantification of individual peaks is difficult and requires experience to differentiate PEG from the surfactant (61,62). Positive molecular ions characteristic of NPE are of the form $[44x + 220]^+$, while those for PEG are $[44x + 18]^+$ (18). Plomley et al. report that APCI is superior to electrospray ionization for determination of APE in the environment, because fewer interfering compounds are ionized (79). APCI is also more suitable for use with normal phase LC.

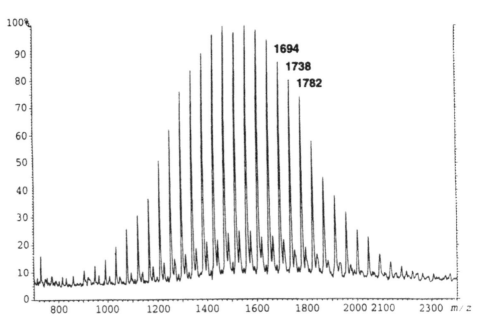

FIG. 4 Positive ion MALDI spectrum of a commercial nonylphenol ethoxylate. The main peaks are quasimolecular adducts with Na^+; the small peaks are K^+ adducts. (Reprinted with permission from Ref. 14. Copyright 1995 by the American Oil Chemists' Society.)

Supercritical fluid chromatography. SFC-MS of NPE using experimental inter-faces has been demonstrated. Investigators have used CO_2 (90) and hexane/ethanol (68). No clear advantage has been demonstrated over other MS techniques.

C. Acid Ethoxylates

Electron impact ionization. EI-MS of these compounds shows molecular ions which are more intense in the case of unsaturated than saturated acids. $[M - H_2O]^+$, $[M + 1]^+$, and $[M - 1]^-$ ions are also seen. Typical aliphatic chain fragments are common: $[C_xH_{2x+1}]^+$ and $[C_xH_{2x}]^+$. The length of the alkyl chain can be determined by the presence of the fragments $RC\equiv O^+$ and $RCOOCH_2CH_2^+$ (43).

MALDI. Analysis of ethoxylates of mixed acids gives very complex spectra due to the quasimolecular ions of the many homologs. Excellent results are obtained if the sam-ple is first fractionated into monoesters, diesters, etc., by HPLC (37).

Electrospray ionization. Analysis was demonstrated using electrospray ionization of HPLC effluent. Efficiency of ionization varies between PEG, monoester, and diester, so the MS signal is not suitable for use in quantification (91).

Supercritical fluid chromatography. SFC-MS has been demonstrated for analysis of acid ethoxylates, identifying peaks due to the active agent and PEG up to about $n = 6$, the limitation being the mass range of the instrument. Only molecular ions are seen (70).

D. Fatty Acid Alkanolamides

There is little reported information on these compounds, and part of what is reported is contradictory, indicating that useful mass spectra are dependent on finding optimal condi-tions for the instrument.

Electron impact ionization. EI-MS of fatty diethanolamides shows an M^+ ion for unsaturated and $[M + 1]^+$ ion for saturated alkyl compounds. $[M - 18]^+$ ions are important in the case of unsaturated compounds. The length of the alkyl chain may be determined by the $RC\equiv O^+$ and $RCON^+(\equiv CH_2)CH_2CH_2OH$ fragments. It must be noted that unsatur-ated homologs give more intense ions than the saturated homologs. A number of ions are independent of the alkyl chain: McLafferty rearrangement gives rise to the ion $CH_2=C(OH^+)N(CH_2CH_2OH)_2$ of m/z 147. Also seen are $^+N(CH_2CH_2OH)_2$, m/z 104, $^+CH_2=NHCH_2CH_2OH$, m/z 74, and $^+O\equiv CN(CH=CH_2)CH_2CH_2OH$, m/z 114 (43).

For EI of monoethanolamides, molecular ions are not seen, but there are low abun-dance $[M - 18]^+$ ions. The location of double bonds can be deduced from the fragmenta-tion pattern. For EI of the trimethylsilyl ether derivative, peaks diagnostic of the molecular weight are seen at $[M - CH_3]^+$ and $[M - (CH_3)_3SiOH]^+$. Ions independent of alkyl chain length: $CH_2=CHNHCO^+CH_2CH_2$, m/z 98, is the base peak for unsaturated compounds and an important peak for saturated compounds. McLafferty rearrangement gives rise to the ion $CH_2=C(OH^+)NHCH=CH_2$ of m/z 85, which is the base peak for saturated com-pounds and a large peak in the unsaturated homologs (92).

Chemical ionization. For CI of monoethanolamides, with methane as reactant gas, molecular ions are not seen, but rather the quasimolecular ions $[M - H_2O + H]^+$ and $[M - H_2O - H]^+$, as well as $[M - H_2O + 29]^+$ and $[M - H_2O + 41]^+$. The relative intensity of the $[M - H_2O + H]^+$ and $[M - H_2O - H]^+$ ions depends on the degree of unsaturation as well as the length of the alkyl group (92).

Thermospray MS. Coconut acid diethanolamide showed mainly quasimolecular ions, as well as cluster ions of the form $[2M + 1]^+$ (50).

Fast atom bombardment.   FAB mass spectrometry of fatty acid alkanolamides shows only protonated and cationized molecular ions in the positive ion mode. Fatty acid impurities are seen in the negative ion mode (29,85).

Supercritical fluid chromatography.   SFC-MS shows only molecular ions of alkyl mono- and diethanolamides (68).

E.   Amine Ethoxylates

Electron impact ionization.   Trimethylsilyl derivatives of ethoxylated amines are readily analyzed by conventional GC-EI-MS. The parent ion is absent or at very low abundance, but characteristic fragments permit identification of the compounds (93).

Direct EI or FD-CAD spectra of ethoxylated primary amines show the molecular ion and many fragments corresponding to loss of C_2H_4O groups and to α-cleavage of the alkyl chain, the latter reaction giving ions of the form $CH_2=N^+(C_2H_4O)_2$, with weights of 74, 118, 162, etc. Ions of the form $^+CH_2CH_2NH(CH_2CH_2O)_n$ are also seen. Nitrogen-containing fragments are easily identified, since they have even numbers for m/z. FD is reported to give a higher abundance molecular ion than does EI. $[M - 31]^+$ ions, corresponding to loss of methanol, are seen with both techniques (25,43).

Chemical ionization.   CI spectra allow determination of the alkyl chain from the ion $RN^+(=CH_2)(CH_2CH_2O)_nH$, formed by α-cleavage of one ethoxy chain (82).

Fast atom bombardment.   FAB gives quasimolecular ions and fragments from α-cleavage of the alkyl chain and from cleavage of ethoxy chains (82).

F.   Esters of Polyhydroxy Compounds

Chemical ionization.   Heated probe hydroxyl negative ion MS of polysorbate esters showed weak $[M - H]^-$ ions. Ions corresponding to the fatty acid moieties, RCO_2^-, were stronger, as were ions at 145 and 163 corresponding to sorbitol and dianhydrosorbitol (94). Direct probe CI-MS was demonstrated for characterization of trimethylsilyl derivatives of sucrose esters. $[M + NH_4]^+$ ions were seen for compounds up to the hexalaurate ester (95).

MALDI.   MALDI-MS of polyglyceride esters shows pseudomolecular ions (Na^+ adducts) corresponding to a degree of polymerization up to about five and mono-, di-, tri-, and tetraesters (37).

Field desorption.   FD-MS gives molecular ions for sorbitan esters. The spectra are complex, because of the complex nature of the product itself, although mainly pseudomolecular ions are seen (22).

G.   Ester Ethoxylates

Chemical ionization.   Heated probe hydroxyl negative ion MS of ethoxylated polysorbate esters shows a series of $[M - H]^-$ ions from free PEG present as a byproduct in the esters, with higher oligomers appearing as the probe temperature is raised. A second series of ions is seen corresponding to the homologs of ethoxylated dianhydrosorbitol. Ions corresponding to the fatty acid moieties are seen, RCO_2^-, but molecular ions corresponding to the esters are not observed. Cl⁻ attachment spectra are analogous to the OH⁻ spectra. Confirmation of the presence of polysorbate 60 in salad dressing was made after TLC separation. (94).

MALDI.   MALDI-MS of ethoxylated polyglyceride esters shows a complicated spectrum due to the pseudomolecular sodium adduct ions of all the various possible combinations. For example, peaks are seen corresponding to the distribution of glycerin distearate ethoxylates, glycerin tristearate ethoxylates, diglycerindistearate ethoxylates, diglycerintristearate ethoxylates, and so on (37). MALDI of ethoxylated sorbitan esters is straightforward, permitting accurate determination of the molecular weight distribution (64).

H.   Alkyl Polyglycosides

Electron impact ionization.   EI of underivatized glycosides provides fragments characteristic of the sugar, but only fragments of the alkyl chain and no molecular ion. Thus, the identity of the alkyl chain cannot be deduced, nor is there information on whether the product is a mono-, di-, or higher ester (97). Analysis of trimethylsilyl ethers gives superior results (Fig. 5). In the case of APG, there is a base peak for all pyranosides at m/z 204, corresponding to $(CH_3)_3SiOCHC^+HOSi(CH_3)_3$ and a base peak for all furanosides at m/z 217, corresponding to $(CH_3)_3SiOCH=CHC^+HOSi(CH_3)_3$. Although a molecular ion is not seen, the ions at $[M-105]^+$ allow identification of the homologous series of alkyl substituents (96).

Chemical ionization.   CI-MS complements EI by giving the molecular weight of the compound. Direct CI of underivatized dodecyl maltoside with ammonia showed strong quasimolecular ions and weak fragments corresponding to scission of the molecule into its sugar and alkyl constituents (97).

Thin layer chromatography.   SIMS analysis of spots on a TLC plate was demonstrated using alkyl polyglycosides (98). In this method, a thin, porous layer of silver is vapor deposited on the developed TLC plate. Individual compounds on the plate are concentrated in the upper silver layer by placing a drop of methanol or DMSO on the spot and allowing the solvent to evaporate. Subsequent SIMS analysis results in mainly quasimolecular ions of $[M + Ag]^+$, with the ions showing the expected silver isotope ratio of $^{107}Ag/^{109}Ag = 51{:}49$. Quasimolecular ions are obtained representing $[M + Na]^+$ and $[M + Ag]^+$. This suffices to identify homologs differing by degree of polymerization and chain length of the alkyl substituent. The experimental time-of-flight unit used in this demonstration showed quasimolecular ions out to a degree of polymerization of 12. The technique is not suitable for determining the average degree of polymerization since ionization efficiencies differ among the homologs.

Atmospheric pressure ionization.   API has been demonstrated for direct analysis of APG and esters of APG (99), as well as for LC-MS analysis (100). It is expected in API that ionic compounds are detected as $[M + H]^+$ or $[M – H]^-$ ions and nonionic compounds are detected as positively charged clusters with cations such as sodium or ammonium. For the analysis of APG itself, cationized APG was detectable up to a degree of polymerization of four, with higher m/z compounds visible due to formation of mixed clusters of the form $[2M + Na]^+$. Cluster formation is minimized by using smaller sample size, for example, by using microbore LC (100). APG is also detectable with less sensitivity in the negative ion mode, giving $[M – H]^-$ ions.

Maleate esters of APG are intermediates in the preparation of sulfosuccinate APG esters. Positive ion spectra show unreacted APG, as discussed above, and $[M + Na]^+$ and $[M + NH_4]^+$ ions of the esters up to a degree of APG polymerization of 3. $[2M + Na]^+$ and $[2M + NH_4]^+$ ions are also visible, where M represents mono- or dimaleates of APG with varying alkyl chain lengths and with degree of APG polymerization up to 3. $[M + H]^+$ ions

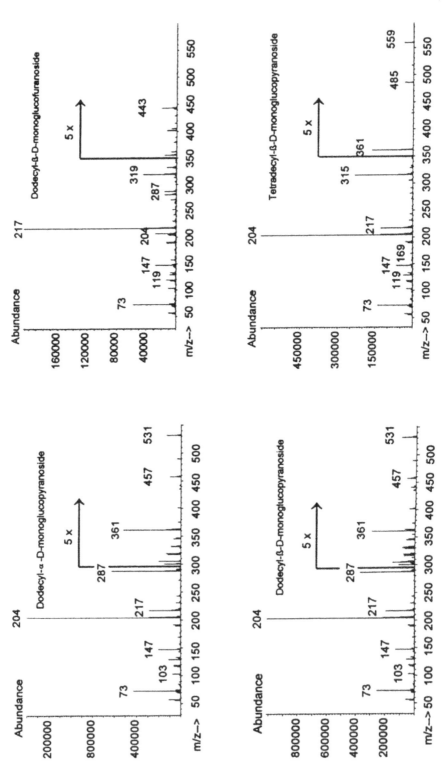

FIG. 5 Positive ion electron impact MS detection of trimethylsilyl derivatives of a dodecylmonoglucoside after separation by GC. Identities as labeled. (Reprinted with permission from Ref. 96. Copyright 1998 by Karl Hanser Verlag.)

of byproduct alkyl maleates are seen at lower m/z, with clusters of the same type also visible. As with APG, negative ion spectra are simpler, showing only $[M - H]^-$ ions. Positive ion spectra of APG sulfosuccinate esters show $[M + 2Na]^+$ and $[M + 3Na]^+$ ions. Negative ion spectra show also peaks at half the expected m/z, because of the possibility of formation of $[M - 2H]^{2-}$ ions. In general, spectra of citrate and tartrate esters of APG have the same form as the maleate spectra (99).

Thermospray MS.   Thermospray MS has been applied to LC analysis as well as direct introduction analysis of environmental materials. MS in the positive ion mode with addition of ammonium acetate gives ions of the form $[M+NH_4]^+$ corresponding to the monoglucosides (and clusters) of the various alkyl chain lengths. Diglucosides are detectable. Negative ion MS is more selective but less sensitive, giving ions, including clusters, of the form $[M + CH_3CO_2]^-$. CAD can be applied to give fragment ions. The daughter ions of APG are characteristic of monosaccharides (101).

Electrospray MS.   Analysis of APG in an ammonium acetate buffer under positive ion conditions shows ammonium adduct parent ions for the alkylmono- and alkyldiglycosides (15). Electrospray MS is also appropriate for LC detection. Negative ion mode is preferred for quantitative work because of more uniform response. Good results are obtained in single ion monitoring mode using the $[M - H]^-$ ions. In positive ion mode, response varies among isomers because of varying affinities of the isomers for cations. For example, the α-glucofuranoside form of dodecyl monoglucoside has a much higher affinity to sodium ion than does the β-glucofuranoside form (102).

MALDI.   MALDI-MS of APG gives clean pseudomolecular ions of the form $[M + Na]^+$. With a repeating unit of 162, the oligomers are widely separated and it is a simple matter to determine the alkyl chain length. Constituents of the commercial product are seen out to a degree of polymerization of 7. Mono-, di-, and triesters of APG can be similarly analyzed with excellent results (37).

I. EO/PO Copolymers

Fast atom bombardment.   FAB of these copolymers in the presence of Li^+ gives $[M + Li]^+$ ions.

Collisionally activated dissociation.   CAD of these ions gives abundant product ions, of which $[M + Li – 88]^+$, corresponding to loss of 1,4-dioxane, is quite intense. There is no corresponding loss of two moles PO or of EO + PO. Examination of the series of ions allows deduction of whether the compound is a block or random copolymer (103).

Laser desorption FTMS.   Analysis of amine-terminated EO/PO copolymers with molecular weight in the 600–2000 range effectively gives the complete homolog distribution, including the number of moles of EO and PO in each homolog. Both $[M + H]^+$ and $[M + K]^+$ ions are seen, presumably because potassium salts often remain in the product as neutralized alkoxylation catalyst (104).

MALDI.   MALDI-MS analysis of LC fractions after addition of 2,5-dihydroxybenzoic acid matrix and LiCl gives $[M + Li]^+$ peaks representative of the homolog distribution of the fractions, both for the copolymers and for PEG impurity (66). MALDI-FTICR-MS has been demonstrated for analysis of an EO/PO copolymer, giving a very precise molecular weight distribution. Again, 2,5-dihydroxybenzoic matrix was used and in this case the spectrum obtained was of $[M + Na]^+$ adduct ions. A mathematical technique was used to correct for the expected presence of ^{13}C, ^{17}O, and ^{18}O (105).

J. Amine Oxides

FAB of amine oxides gives positive ions corresponding to the protonated molecular ion and the dimer ion (85). No negative ion spectrum is seen. Tandem mass spectrometry with collisionally activated dissociation gives structural information (28).

IV. CATIONIC AND AMPHOTERIC SURFACTANTS

Mass spectrometry is not routinely applied to analysis of cationic and amphoteric surfactants. The most common cationic surfactants, the quaternary amine salts, cannot be analyzed directly by conventional GC-MS or by direct electron impact MS because of their low volatility. Under EI or CI conditions, quaternary amines normally decompose to tertiary amines, which can become ionized or protonated for MS detection. Thus, ordinary MS shows only fragments and rearrangement products, and its use is quite limited in the analysis of materials of unknown composition.

A surprising number of publications continue to appear describing the application of exotic ionization techniques to the MS analysis of quaternary amines. This is because their thermal instability and lack of volatility make their analysis an interesting test case for new techniques. Kalinoski has written a review of mass spectroscopy of cationic surfactants, including detailed explanations of the various instrumental techniques (106).

In the case of amphoterics, cleavage usually occurs before volatilization, so that molecular ions are not obtained with ordinary EI or CI techniques. Soft ionization techniques must be used to create the vaporized ions which can be measured by mass spectrometry.

Calibration. A high molecular weight ethoxylated quaternary amine is proposed for mass value calibration of a FAB instrument, with perfluorobutyrate esters used to increase sensitivity at the high molecular weight end of the distribution (107). A fluorinated alkyl betaine, DuPont's Zonyl FSB, was used to calibrate high resolution FAB mass spectometry measurements. A series of positive peaks were observed corresponding to the $[M + H]^+$ ion and two important fragments, all of various fluoroalkyl chain lengths (108).

The use of an internal standard, a deuterated analog of the particular compound determined, allows the quantitative application of FAB to environmental analysis. This has an advantage over other instrumental techniques in that very little sample work-up is necessary and is suitable for quantitative analysis below the 10 ppb range (109). Deuterated standards of alkyltrimethylammonium salts can be prepared relatively easily by reaction of perdeuterated trimethylamine with the corresponding alkyl bromide (110).

A. MS Analysis of Quaternary Amines

Direct GC-MS. If GC-MS is used with a high injection port temperature (about 310°C), quaternary salts are decomposed to tertiary amines and alkyl halides, which are readily separated and identified (111). Good results are obtained with chloride and iodide salts, but not bromides (112). Quantitation of the resulting tertiary amines can be performed using the fragment at m/z 58 due to the $CH_2\!=\!N(CH_3)_2^+$ ion (112).

Desorption chemical ionization. In desorption CI (also called direct CI), the sample is deposited on a wire placed in the source. The wire is heated rapidly, volatilizing molecules which are immediately ionized. This technique allows formation of the molecular ions, as well as the much more abundant decomposition ions resulting from loss of an alkyl or benzyl group. Quantitative determination of relative percentage of the alkyl groups is most reliably based on the decomposition ions (81,113).

Fast atom bombardment and the related technique of liquid secondary ion mass spectrometry (LSIMS). Analysis of alkyltrimethyl and alkylpyridinium salts in glycerol matrix gives abundant molecular ions. At low quat concentrations, ions are also seen at $[M + 30]^+$, $[M + 60]^+$, and $[M + 90]^+$, corresponding to adducts of glycerol and glycerol fragments with the molecular ion (114). Glycerol dimer and trimer ions are also seen. At higher quat concentrations, the glycerol adducts diminish but dimer clusters of the form M_2Br^+ become significant (115). Chen and Shiea report that the addition of sodium dodecylsulfate to the glycerol matrix has the effect of increasing the intensity of the molecular ions of the shorter chain length quats relative to those of higher chain length (116).

Ignoring dimer clusters and the glycerol adducts, linear alkyltrimethylammonium compounds give mainly molecular ions, with other ions seen at 2–5% of the molecular ion. These minor ions correspond to a series of compounds given by neutral alkane loss (CH_4, C_2H_6, ...) down to m/z 114, corresponding to $H_2C=CHC_2H_4N^+(CH_3)_3$. Other minor ions correspond to $C_6H_{13}^+$, $C_5H_{11}^+$, $C_3H_7^+$, $C_3H_5^+$, $N(CH_3)_3^+$, and $CH_2=N(CH_3)_2^+$ (110,115,117). It is likely that what appear to be fragment ions are often only impurities (29). Clusters of two molecular ions with one halide ion may also be detected (84).

Continuous flow FAB has been demonstrated for HPLC detection of the components of ditallowdimethylammonium chloride fabric softener. In this technique, the matrix substance is added to the HPLC column effluent prior to introduction to the MS. MS-MS was demonstrated for confirmation of identity. While continuous flow FAB is typically performed with aqueous HPLC systems, the solubility of dialkyl quats is such that normal phase HPLC with nonpolar solvents is preferred. Lawrence was able to adapt a FAB system to this analysis, using a matrix solution of 75:25 glycerol/methanol (118). The use of FAB for analysis of surfactants has dropped off sharply with the availability of electrospray and thermospray interfaces for HPLC-MS.

Tandem mass spectrometry can be used to give more power to the FAB technique (119). In the case of dioctadecyldimethylammonium ion, further fragmentation of the molecular cation of m/z 550 yields a parent ion and a strong ion at m/z 296, corresponding to loss of neutral octadecane. A number of weaker ions are seen at intermediate m/z, reflecting chain scission. Similarly, CAD fragmentation of the molecular cation of octadecylhexadecyldimethylammonium ion gives two main daughter ions, corresponding to loss of neutral octadecane and hexadecane (118). By contrast, the main fragment from alkyltrimethylamine salts is $(CH_3)_3NH^+$, m/z 60, along with very short alkyl fragments (117).

MALDI. Analysis of quaternaries was demonstrated using 2,5-dihydroxybenzoic acid as matrix. This is suitable for showing the parent ions (14).

Electrospray MS. Analysis of methanol solutions in the positive ion mode gives molecular ions of distearyldimethylammonium chloride. Strong ions are seen for the C_{18}/C_{18}, C_{18}/C_{16}, and C_{16}/C_{16} homologs, as well as some minor components (15,120). It must be noted that the response factors differ sharply for similar compounds. For example, the response factor for hexadecyltrimethylammonium bromide is only 45% of that for dodecyltrimethylammonium bromide (117). The same observations on calibration with deuterated analogs and fragments for MS-MS analysis made above for FAB also apply to electrospray (117). Electrospray MS was demonstrated for detection of benzalkonium chloride on human skin, using a triple quadrupole instrument in multiple reaction monitoring mode (121).

Atmospheric pressure ionization. API is compatible with the effluent from a chemically suppressed ion chromatography system (30). Field assisted ion evaporation in the positive ion mode with atmospheric pressure ionization gave the molecular ion of dioctyldimethylamine, while CAD in the next stage of the triple quadrapole instrument gave fragments consistent with loss of alkyl groups.

Thermospray MS. This soft ionization technique is used in conjunction with HPLC. Under thermospray conditions, benzyldimethyldodecylammonium chloride and dodecylpyridinium chloride give a parent ion as well as protonated amines formed from cleavage of C–N bonds. The alkyl fragments are only seen with difficulty (122).

B. Fluorinated Cationics

As is the case with other fluoro compounds, negative ion MS of fluorinated surfactants is more informative and less subject to interference than is positive ion MS. Negative ion spectra generated by thermospray analysis of perfluorinated alkyl compounds are dominated by peaks at intervals of $\Delta m/z$ 50 or 100, corresponding to loss of the CF_2 and C_2F_4 groups (38).

C. Tertiary Amines

Electrospray MS. In positive ion mode electrospray MS gives parent ions for the amines. Subsequent CAD analysis gives fragmentation, with the main fragments corresponding to the protonated alkyl chains cleaved at the tertiary nitrogen (15).

D. Alkyldimethylbetaines

Direct GC-MS. Like the quaternary amines, alkyldimethylbetaines react in the hot injection port of the GC to give alkyldimethylamines, which are readily identified by MS (123).

E. Alkylamidobetaines

Fast atom bombardment. Amidobetaines give only positive ions which mainly correspond to protonated and "cationized" (addition of sodium or potassium ion) molecular ions, with a small amount of fragments (29,85).

F. Amino Acids

Fast atom bombardment. FAB gives intense molecular ions from the amino acid *N*-coco-β-aminopropionate, which by the CAD technique show alpha cleavage at the nitrogen resulting in loss of $C_{11}H_{24}$ and $C_2H_3O_2$ (119).

G. Phosphatides

Direct GC-MS. Phosphatides isolated from lecithin may be analyzed after isolation and enzymatic hydrolysis of the phosphorus-containing substituent. The diglycerides are analyzed by GC-MS of the TMS derivatives (124).

Fast atom bombardment. Intact phosphatides are not volatile enough to be analyzed by conventional MS. Field desorption CI MS will give a quasimolecular ion and some fragments. However, most work is reported using FAB. In the case of mixtures, neither positive nor negative ion FAB-MS is enough to characterize the phosphatides present. Tandem MS, however, gives enough information for semiquantitative analysis (125–127).

Low energy CAD of protonated phospholipids give a few typical fragments. In the positive ion mode, the main reaction is loss of a neutral phosphate ester, leaving the charge on the diacylglycerol fragment. The exception is phosphatidylcholine, where the choline phosphate fragment is charged. In negative ion mode, one of the acyl groups is cleaved as a charged fragment, leaving the remaining fragment neutral (125). CAD of protonated phosphatidylcholine ions produced by FAB gives, as the most abundant ions, those containing the phosphate group at m/z 184 and 224, corresponding to $[(OH)_2P(O)OCH_2CH_2N(CH_3)_3]^+$ and $[CH_2\!\!=\!\!CHCH_2OP(OH)(O)OCH_2CH_2N(CH_3)_3]^+$. Deacylated ions are weak, especially for higher molecular weight compounds (127).

The FAB negative ion mass spectrum of phosphatidylcholines shows characteristic ions at $[M - 15]^-$, $[M - 60]^-$, and $[M - 86]^-$, corresponding to loss of CH_3, $HN(CH_3)_3$, and $CH_2\!\!=\!\!CHN(CH_3)_3$, respectively, with only $[M - 15]^-$ having an even m/z number (126). CAD of the $[M - 15]^-$ ion shows abundant ions corresponding to phosphorylcholine and glycerophosphocholine, with deacylation ions corresponding to $[M - 15 - RCO]^-$ also visible (127). The most abundant ions are those due to carboxylate anions, with the ion from the 2 position generally being more abundant than that from the 1 position. Thus negative ion MS-MS allows characterization of the molecule. On sophisticated instruments, the carboxylate ions can be subjected to a further MS stage to determine location of unsaturation (127).

Electrospray MS.   Analysis of phosphatidylcholine gives negative ions in the presence of acetate or chloride ions. The same observations hold true as mentioned above for FAB-MS of phosphatidylcholine (128).

MALDI.   MALDI-MS has been demonstrated for the characterization of individual phosphatides. In general, molecular or pseudomolecular ions are formed and some fragmentation is observed. Phosphatidylcholine gives a strong M^+ ion and no negative ions, while phosphatidylinositol will only give a $[M + Na]^+$ cation. $[M - H]^-$ anions are formed in all cases except for phosphatidycholine. MALDI allowed determination of the acyl substituents of individual phosphatides without undue difficuly (129).

LC-MS can be performed using Frit-FAB-MS, also called LSIMS (130). If the system is stable enough, interference from the matrix can be subtracted out. The phosphatides are separated according to alkyl character on a reversed-phase column, with the MS identifying the components making up each peak. Identification is straightforward, with the positive ion mode showing abundant molecular ions and characteristic fragments for phosphatidylcholine, phosphatidylethanolamine, and phosphatidylinositol, as well as ions representing loss of an acyl chain, while the negative ion mode further aids in identifying the acyl substituents (130).

REFERENCES

1. Di Corcia, A., Surfactants and their biointermediates by LC-MS, *J. Chromatogr. A*, 1998, *794*, 165–185.
2. Kalinoski, H. T., MS of anionic surfactants, in J. Cross, ed., *Anionic Surfactants: Analytical Chemistry*, 2nd ed., Marcel Dekker, New York, 1998.
3. McEvoy, J., W. Giger, LAS in sewage sludge by high-resolution GC-MS, *Environ. Sci. Technol.*, 1986, *20*, 376–383.
4. Hon-nami, H., T. Hanya, GC-MS determination of alkylbenzenesulfonates in river water, *J. Chromatogr.*, 1978, *161*, 205–212.
5. Field, J. A., D. J. Miller, T. M. Field, S. B. Hawthorne, W. Giger, Sulfonated aliphatic and aromatic surfactants in sewage sludge by ion-pair/supercritical fluid extraction and derivatization GC/MS, *Anal. Chem.*, 1992, *64*, 3161–3167.

6. Agozzino, P., L. Ceraulo, M. Ferrugia, E. Caponetti, F. Intravaia, R. Triolo, Sodium *p*-alkyl-benzenesulfonates: characterization by electron impact MS, *J. Colloid Interface Sci.*, 1986, *114*, 26–31.

7. Ding, W. H., J. H. Lo, S. H. Tzing, LAS and degradation products in water samples by GC with ion-trap MS, *J. Chromatogr. A*, 1998, *818*, 270–279.

8. Reiser, R., H. O. Toljander, W. Giger, Alkylbenzenesulfonates in recent sediments by GC/MS, *Anal. Chem.*, 1997, *69*, 4923–4930.

9. Trehy, M. L., W. E. Gledhill, R. G. Orth, LAS and dialkyltetralinsulfonates in water and sediment by GC-MS, *Anal. Chem.*, 1990, *62*, 2581–2586.

10. Borgerding, A. J., R. A. Hites, Alkylbenzenesulfonate surfactants using continuous-flow fast atom bombardment spectrometry, *Anal. Chem.*, 1992, *64*, 1449–1454.

11. Suter, M. J. F., R. Reiser, W. Giger, Differentiation of linear and branched alkylbenzene-sulfonates by GC/tandem MS, *J. Mass Spectr.*, 1996, *31*, 357–362.

12. Wernery, J. D., D. A. Peake, Effect of surface activity on quantification of linear alkylbenzenesulfonates by FAB-MS, *Rapid Commun. Mass Spectrom*, 1989, *3*, 396–399.

13. Field, J. A., L. B. Barber, II, E. M. Thurman, B. L. Moore, D. L. Lawrence, D. A. Peake, Fate of alkylbenzenesulfonates and dialkyltetralinsulfonates in sewage-contaminated groundwater, *Environ. Sci. Technol.*, 1982, *26*, 1140–1148.

14. Thompson, B., Z. Wang, A. Paine, A. Rudin, G. Lajoie, Surfactant analysis by matrix-assisted laser desorption time-of-flight MS, *J. Am. Oil Chem. Soc.*, 1995, *72*, 11–15.

15. Ogura, I., D. L. DuVal, S. Kawakami, K. Miyajima, Surfactants in consumer products by ion-spray MS, *J. Am. Oil Chem. Soc.*, 1996, *73*, 137–142.

16. Hiraoka, K., I. Kudaka, Negative-mode electrospray MS using nonaqueous solvents, *Rapid Commun. Mass Spectrom.*, 1992, *6*, 265–268.

17. Castillo, M., M. C. Alonso, J. Riu, D. Barcelo, Polar, ionic, and highly water soluble organic pollutants in untreated industrial wastewaters, *Environ. Sci. Technol.*, 1999, *33*, 1300–1306.

18. Castillo, M., D. Barceló, Polar toxicants in industrial wastewaters using toxicity-based fractionation with LC/MS, *Anal. Chem.*, 1999, *71*, 3769–3776.

19. Di Corcia, A., F. Casassa, C. Crescenzi, A. Marcomini, R. Samperi, Fate of LAS and coproducts in a laboratory biodegradation test using LC/MS, *Environ. Sci. Technol.*, 1999, *33*, 4112–4118.

20. Di Corcia, A., L. Capuani, F. Casassa, A. Marcomini, R. Samperi, Fate of LAS, coproducts, and their metabolites in sewage treatment plants and in receiving river waters, *Environ. Sci. Technol.*, 1999, *33*, 4119–4125.

21. Scullion, S. D., M. R. Clench, M. Cooke, A. E. Ashcroft, Surfactants in surface water by SPE, LC, and LC-MS, *J. Chromatogr. A*, 1996, *733*, 207–216.

22. Shiraishi, H., A. Otsuki, K. Fuwa, Potentialities of field desorption MS using emitter current programmer for direct analysis of multicomponents, *Bull. Chem. Soc. Jpn.*, 1979, *52*, 2903–2907.

23. Dähling, P., F. W. Röllgen, J. J. Zwinselman, R. H. Fokkens, N. M. M. Nibbering, Negative ion field desorption MS of anionic surfactants, *Fresenius' Z. Anal. Chem.*, 1982, *312*, 335–337.

24. Schneider, E., K. Levsen, P. Dähling, F. W. Röllgen, Analysis of surfactants by newer mass spectrometric techniques. Part II: Anionic surfactants, *Fresenius' Z. Anal. Chem.*, 1983, *316*, 488–492.

25. Weber, R., K. Levsen, G. J. Louter, A. J. H. Boerboom, J. Haverkamp, Direct mixture analysis of surfactants by combined field desorption/collisionally activated dissociation MS with simultaneous ion detection, *Anal. Chem.*, 1982, *54*, 1458–1466.

26. Schneider, E., K. Levsen, A. J. H. Boerboom, P. Kistemaker, S. A. McLuckey, M. Przybylski, Identification of cationic and anionic surfactants in surface water by combined field desorption–collisionally activated decomposition MS, *Anal. Chem.*, 1984, *56*, 1987–1988.

27. Chiarelli, M. P., M. L. Gross, D. A. Peake, Surfactants on textiles by laser desorption–Fourier transform MS, *Anal. Chim. Acta*, 1990, *228*, 169–176.

28. Lyon, P. A., W. L. Stebbings, F. W. Crow, K. B. Tomer, K. L. Lippstreu, M. L. Gross, Anionic surfactants by MS/MS with FAB, *Anal. Chem.*, 1984, *56*, 8–13.

29. Facino, R. M., M. Carini, P. Minghetti, G. Moneti, E. Arlandini, S. Melis, Surfactants in raw materials and in finished detergent formulations by FAB-MS, *Biomed. Environ. Mass Spectrom.*, 1989, *18*, 673–689.

30. Conboy, J. J., J. D. Henion, M. W. Martin, J. A. Zweigenbaum, Ion chromatography/MS for the determination of organic ammonium and sulfate compounds, *Anal. Chem.*, 1990, *62*, 800–807.

31. Popenoe, D. D., S. J. Morris, III, P. S. Horn, K. T. Norwood, Alkyl sulfates and alkyl ethoxysulfates in wastewater treatment plant influents and effluents and in river water using LC/ion spray MS, *Anal. Chem.*, 1994, *66*, 1620–1629.

32. Jewett, B. N., L. Ramaley, J. C. T. Kwak, API-MS for alkylethoxysulfate, *J. Am. Soc. Mass Spectrom.*, 1999, *10*, 529–536.

33. Bartsch, H., M. Strassner, U. Hintze, Ethoxylated surfactants by MALDI-MS, *Tenside, Surfactants, Deterg.*, 1998, *35*, 94–102.

34. Escott, R. E. A., D. W. Chandler, Ammonium acetate as an ion-pairing electrolyte for ethoxylated surfactant analysis by thermospray LC/MS, *J. Chromatogr. Sci.*, 1989, *27*, 134–138.

35. Taulli, T. A., Isomeric sodium alkenesulfonates via methylation and GC, *J. Chromatogr. Sci.*, 1969, *7*, 671–673.

36. Boyer, J. L., J. P. Canselier, V. Castro, SO_3-sulfonation products of 1-alkenes by spectrometric methods, *J. Am. Oil Chem. Soc.*, 1982, *59*, 458–464.

37. Berchter, M., J. Meister, C. Hammes, MALDI-TOF-MS: characterization of products based on renewable raw materials (in German), *Fett/Lipid*, 1997, *99*, 384–391.

38. Schröder, H. Fr., Fluorine-containing surfactants—another challenge to the environment? Part 1: Anionic and cationic surfactants (in German), *Vom Wasser*, 1991, *77*, 277–290.

39. Moody, C. A., J. A. Field, Perfluorocarboxylates in groundwater impacted by fire-fighting activity, *Environ. Sci. Technol.*, 1999, *33*, 2800–2806.

40. Rauscher, G., IR spectroscopy and MS of nonionic surfactants, in J. Cross, ed., *Nonionic Surfactants: Chemical Analysis*, Marcel Dekker, New York, 1987.

41. Vettori, U., S. Issa, R. M. Facino, M. Carini, AE in raw materials by capillary column GC/electron impact MS, *Biomed. Environ. Mass Spectrom.*, 1988, *17*, 193–204.

42. Levsen, K., W. Wagner-Redeker, K. H. Schäfer, P. Dobberstein, On-line LC-MS of nonionic surfactants, *J. Chromatogr.*, 1985, *323*, 135–141.

43. Julià-Danés, E., A. M. Casanovas, MS of nonionic surfactants, *Tenside, Surfactants, Deterg.*, 1979, *16*, 317–323.

44. Fendinger, N. J., W. M. Begley, D. C. McAvoy, W. S. Eckhoff, AE in natural waters, *Environ. Sci. Technol.*, 1995, *29*, 856–863.

45. Rudewicz, P., B. Munson, AE by probe distillation/chemical ionization MS, *Anal. Chem.*, 1986, *58*, 674–679.

46. Stephanou, E., Nonionic detergents by GC/CI-MS, *Chemosphere*, 1984, *13*, 43–51.

47. Vincenti, M., C. Minero, E. Pelizzetti, Nonionic surfactants in spiked tap water by desorption chemical ionization and tandem MS, *Ann. Chim.* (Rome), 1993, *83*, 381–396.

48. Lin, H. Y., A. Rockwood, M. S. B. Munson, D. P. Ridge, Proton affinity and collision-induced decomposition of AE, *Anal. Chem.*, 1993, *65*, 2119–2124.

49. Ding, W., Y. Fujita, M. Reinhard, CI mass spectra of linear alcohol polyethoxy carboxylates and polyethylene glycol dicarboxylates, *Rapid Commun. Mass Spectrom.*, 1994, *8*, 1016–1020.

50. Ott, K. H., W. Wagner-Redeker, W. Winkle, On-line thermospray–LC-MS of nonionic surfactants (in German), *Fett Wiss. Technol.*, 1987, *89*, 208–213.

51. Evans, K. A., S. T. Dubey, L. Kravetz, I. Dzidic, J. Gumulka, R. Mueller, J. R. Stork, Linear primary AE in environmental samples by thermospray LC/MS, *Anal. Chem.*, 1994, *66*, 699–705.

52. Cassani, G., M. Comber, A. Guarini, J. Lux, M. Hetheridge, R. Wolf, AE in environmental samples by thermospray LC/MS, *World Surfactants Congr., 4th*, 1996, *4*, 436–447.

53. Rivera, J., J. Caixach, A. Figueras, D. Fraisse, F. Ventura, Application of FAB and tandem MS to the identification of organic pollutants in water, *Biomed. Environ. Mass Spectrom.*, 1988, *16*, 403–408.

54. Ventura, F., A. Figueras, J. Caixach, I. Espadaler, J. Romero, J. Guardiola, J. Rivera, Characterization of polyethoxylated surfactants and their brominated derivatives formed at the water treatment plant of Barcelona by GC-MS and FAB-MS, *Water Res.*, 1988, *22*, 1211–1217.

55. Rockwood, A. L., T. Higuchi, LC-MS analysis of nonionic surfactants using the FRIT-FAB method, *Tenside, Surfactants, Deterg.*, 1992, *29*, 6–12.

56. Ventura, F., D. Fraisse, J. Caixach, J. Rivera, Identification of [(alkyloxy)polyethoxy]carboxylates in raw and drinking water by MS-MS and mass determination using FAB and nonionic surfactants as internal standards, *Anal. Chem.*, 1991, *63*, 2095–2099.

57. Lattimer, R. P., Tandem MS of lithium-attachment ions from polyglycols, *J. Am. Soc. Mass Spectrom.*, 1992, *3*, 225–234.

58. Sherrard, K. B., P. J. Marriott, M. J. McCormick, R. Colton, G. Smith, Electrospray MS analysis and photocatalytic degradation of polyethoxylate surfactants used in wool scouring, *Anal. Chem.*, 1994, *66*, 3394–3399.

59. Crescenzi, C., A. Di Corcia, R. Samperi, A. Marcomini, Nonionic polyethoxylate surfactants in environmental waters by LC/electrospray MS, *Anal. Chem.*, 1995, *67*, 1797–1804.

60. Dubey, S. T., K. A. Evans, L. Kravetz, I. Dzidic, LC/MS analyses and sample preparation of AE in environmental samples, *World Surfactants Congr., 4th*, 1996, *4*, 395–401.

61. Evans, K. A., S. T. Dubey, L. Kravetz, S. W. Evetts, I. Dzidic, C. C. Dooyema, AE in environmental samples by electrospray MS, *J. Am. Oil Chem. Soc.*, 1997, *74*, 765–773.

62. Pattanaargsorn, S., P. Sangvanich, A. Petsom, S. Roengsumran, Oligomer distribution of APE and AE by positive ion atmospheric pressure chemical ionization MS, *Analyst*, 1995, *120*, 1573–1576.

63. Asmussen, C., H. Stan, AE—examples of process analysis (in German), *Biol. Abwasserreinig.*, 1998, *10*, 237–251.

64. Cumme, G. A., E. Blume, R. Bublitz, H. Hoppe, A. Horn, Detergents of the polyoxyethylene type: comparison of TLC, reversed-phase chromatography, and MALDI MS, *J. Chromatogr. A*, 1997, *791*, 245–253.

65. Pasch, H., I. Zammert, Chromatographic investigations of macromolecules in the critical range of LC. VIII: Analysis of polyethylene oxides, *J. Liq. Chromatogr.*, 1994, *17*, 3091–3108.

66. Pasch, H., K. Rode, MALDI MS for molar mass-sensitive detection in LC of polymers, *J. Chromatogr. A*, 1995, *699*, 21–29.

67. Kalinoski, H. T., L. O. Hargiss, SFC-MS of nonionic surfactant materials using chloride-attachment negative ion chemical ionization, *J. Chromatogr.*, 1990, *505*, 199–213.

68. Matsumoto, K., S. Tsuge, Y. Hirata, Nonionic surface active agents by SFC/MS, *Shitsuryo Bunseki*, 1987, *35*, 15–22.

69. Pinkston, J. D., D. J. Bowling, T. E. Delaney, Industrial applications of SFC-MS involving oligomeric materials of low volatility and thermally labile materials, *J. Chromatogr.*, 1989, *474*, 97–111.

70. Shang, D. Y., M. G. Ikonomou, R. W. Macdonald, NPE surfactants in marine sediment using normal phase LC-ECMS, *J. Chromatogr. A*, 1999, *849*, 467–482.

71. Ding, W. H., S. H. Tzing, NPE and degradation products in river water and sewage effluent by GC-ion trap (tandem) MS with EI and CI, *J. Chromatogr. A*, 1998, *824*, 79–90.

72. Stephanou, E., M. Reinhard, H. A. Ball, Halogenated and non-halogenated OPE residues by GC-MS using electron chemical ionization, *Biomed. Environ. Mass Spectrom.*, 1988, *15*, 275–282.

73. Jandera, P., J. Urbánek, B. Prokeš, J. Churáček, Comparison of various stationary phases for normal-phase HPLC of APE, *J. Chromatogr.*, 1990, *504*, 297–318.

74. Liang, Z., A. G. Marshall, D. G. Westmoreland, Molecular weight distributions of *tert*-OPE by laser desorption Fourier transform ion cyclotron resonance MS and HPLC, *Anal. Chem.*, 1991, *63*, 815–818.

75. Vincenti, M., E. Pelizzetti, A. Guarini, S. Costanzi, MW distributions of polymers by desorption chemical ionization MS, *Anal. Chem.*, 1992, *64*, 1879–1884.
76. Prokai, L., W. J. Somonsick, Jr., Electrospray ionization MS coupled with SEC, *Rapid Commun. Mass Spectrom.*, 1993, *7*, 853–856.
77. Prokai, L., D. J. Aaserud, W. J. Somonsick, Jr., Microcolumn SEC coupled with electrospray ionization MS, *J. Chromatogr. A*, 1999, *835*, 121–126.
78. Mackay, L. G., M. Y. Croft, D. S. Selby, R. J. Wells, NPE and OPE in effluent by LC with fluorescence detection, *J. AOAC Int.*, 1997, *80*, 401–407.
79. Plomley, J. B., P. W. Crozier, V. Y. Taguchi, NPE in sewage treatment plants by combined precursor ion scanning and multiple reaction monitoring, *J. Chromatogr. A*, 1999, *854*, 245–257.
80. Di Corcia, A., A. Costantino, C. Crescenzi, E. Marinoni, R. Samperi, Recalcitrant intermediates from biotransformation of the branched alkyl side chain of NPE, *Environ. Sci. Technol.*, 1998, *32*, 2401–2409.
81. Takeuchi, T., S. Watanabe, N. Kondo, M. Goto, D. Ishii, LC-FAB-MS of nonionic detergents, *Chromatographia*, 1988, *25*, 523–525.
82. Schneider, E., K. Levsen, P. Dähling, F. W. Röllgen, Surfactants by newer MS techniques. Part I: Cationic and non-ionic surfactants, *Fresenius' Z. Anal. Chem.*, 1983, *316*, 277–285.
83. Siegel, M. M., R. Tsao, S. Oppenheimer, T. T. Chang, Nonionic surfactants used as exact mass internal standards for the 700–2100 Dalton mass range in FAB-MS, *Anal. Chem.*, 1990, *62*, 322–327.
84. Kamiusuki, T., T. Monde, F. Nemoto, T. Konakahara, Y. Takahashi, OPE by reversed-phase HPLC on branched fluorinated silica gel columns, *J. Chromatogr. A*, 1999, *852*, 475–485.
85. Ventura, F., J. Caixach, A. Figueras, I. Espadaler, D. Fraisse, J. Rivera, Identification of surfactants in water by FAB-MS, *Water Res.*, 1989, *23*, 1191–1203.
86. Otsuki, A., H. Shiraishi, APE in water at trace levels by reversed phase adsorption LC and field desorption MS, *Anal. Chem.*, 1979, *51*, 2329–2332.
87. Schneider, E., K. Levsen, Identification of surfactants and study of their degradation in surface water by MS, *Comm. Eur. Communities*, 1986, EUR 10388, *Org. Micropollut. Aquat. Environ.*, 14–25.
88. Schneider, E., K. Levsen, Monitoring the biological degradation of surfactants with field desorption MS (in German), *Fresenius' Z. Anal. Chem.*, 1987, *326*, 43–48.
89. Just, U., H.-R. Holzbauer, M. Resch, Molar mass determination of oligomeric ethylene oxide adducts using SFC and MALDI time-of-flight MS, *J. Chromatogr. A*, 1994, *667*, 354–360.
90. Smith, R. D., H. R. Udseth, MS interface for microbore and high flow rate capillary SFC with splitless injection, *Anal. Chem.*, 1987, *59*, 13–22.
91. Nielen, M. W. F., F. A. Buijtenhuijs, Polymer analysis by LC/electrospray ionization time-of-flight MS, *Anal. Chem.*, 1999, *71*, 1809–1814.
92. Moldovan, Z., C. Maldonado, J. M. Bayona, Electron ionization and positive ion chemical ionization MS of N-(2-hydroxyethyl)alkylamides, *Rapid Commun. Mass Spectrom.*, 1997, *11*, 1077–1082.
93. Szymanowski, J., H. Szewczyk, J. Hetper, J. Beger, N-Oligooxyethylene mono- and dialkylamines, *J. Chromatogr.*, 1986, *351*, 183–193.
94. Brumley, W. C., C. R. Warner, D. H. Daniels, D. Andrzejewski, K. D. White, Z. Min, J. Y. T. Chen, J. A. Sphon, Polysorbates by OH^- negative ion CI-MS, *J. Agric. Food Chem.*, 1985, *33*, 368–372.
95. Karrer, R., H. Herberg, Sucrose fatty acid esters by high temperature GC, *J. High Resolut. Chromatogr.*, 1992, *15*, 585–589.
96. Billian, P., H. J. Stan, GC/MS of alkyl polyglucosides as their trimethylsilylethers, *Tenside, Surfactants, Deterg.*, 1998, *35*, 181–184.
97. Ott, K. H., Soft ionization methods for mass spectrometric analysis of surfactants (in German), *Fette, Seifen, Anstrichm.*, 1985, *87*, 377–382.

98. Buschmann, N., L. Merschel, S. Wodarczak, Alkyl polyglucosides: qualitative determination using TLC and identification by means of in-situ secondary ion MS, *Tenside, Surfactants, Deterg.*, 1996, *33*, 16–20.

99. Facino, R. M., M. Carini, G. Depta, P. Bernardi, B. Casetta, Atmospheric pressure ionization MS analysis of new alkylpolyglucoside esters, *J. Am. Oil Chem. Soc.*, 1995, *72*, 1–9.

100. Klaffke, H. S., T. Neubert, L. W. Kroh, Nonionic surfactants by LC/MS using alkyl polyglucosides as model substances, *Tenside, Surfactants, Deterg.*, 1999, *36*, 178–184.

101. Schröder, H. Fr., Alkyl polyglycosides in the biological waste water treatment process—degradation behavior by LC/MS and FIA/MS, *World Surfactants Congr., 4th*, 1996, *3*, 121–135.

102. Eichhorn, P., T. P. Knepper, Metabolism of alkyl polyglucosides and their determination in waste water by LC-electrospray MS, *J. Chromatogr. A*, 1999, *854*, 221–232.

103. Lattimer, R. P., Tandem MS of PEG lithium-attachment ions, *J. Am. Soc. Mass Spectrom.*, 1994, *53*, 1072–1080.

104. Nuwaysir, L. M., C. L. Wilkins, W. J. Simonsick, Jr., Copolymers by laser desorption FTMS, *J. Am. Soc. Mass Spectrom.*, 1990, *1*, 66–71.

105. Van Rooij, G. J., M. C. Duursma, C. G. de Koster, R. M. A. Heeren, J. J. Boon, P. J. W. Schuyl, E. R. E. van der Hage, Block length distributions of poly(oxypropylene) and poly(oxyethylene) block copolymers by MALDI-FTICR MS, *Anal. Chem.*, 1998, *70*, 843–850.

106. Kalinoski, H. T., MS of cationic surfactants, in J. Cross and E. J. Singer, eds., *Cationic Surfactants: Analytical and Biological Evaluation*, Marcel Dekker, New York, 1994.

107. DeStefano, A. J., T. Keough, Quaternary ammonium salts as calibration compounds for FAB-MS, *Anal. Chem.*, 1984, *56*, 1846–1849.

108. Gilliam, J. M., P. W. Landis, J. L. Occolowitz, Mass measurement in FAB-MS, *Anal. Chem.*, 1983, *55*, 1531–1533.

109. Simms, J. R., T. Keough, S. R. Ward, B. L. Moore, M. M. Bandurraga, Cationic surfactants in environmental matrices using FAB-MS, *Anal. Chem.*, 1988, *60*, 2613–2620.

110. Bambagiotti-Alberti, M., S. A. Coran, F. Benvenuti, P. LoNostro, S. Catinella, D. Favretto, P. Traldi, Decomposition channels alternative to remote-charge fragmentations observed in MS of alkyltrimethylammonium cations, *J. Mass Spectrom.*, 1995, *30*, 1742–1746.

111. Linhart, K., K. Wrabetz, Composition of quaternary ammonium compounds by GC and MS (in German), *Tenside, Surfactants, Deterg.*, 1978, *15*, 19–30.

112. Hind, A. R., S. K. Bhargava, S. C. Grocott, Alkyltrimethylammonium bromides in Bayer process liquors by GC and GC-MS, *J. Chromatogr. A*, 1997, *765*, 287–293.

113. Cotter, R. J., G. Hansen, T. R. Jones, MS determination of long-chain quaterary amines in mixtures, *Anal. Chem. Acta*, 1982, *136*, 135–142.

114. Tuinman, A. A., K. D. Cook, FAB-induced condensation of glycerol with ammonium surfactants. I: Regioselectivity of the adduct formation, *J. Am. Soc. Mass Spectrom.*, 1992, *3*, 318–325.

115. Paul, G. J. C., I. Marcotte, J. Anastassopoulou, T. Theophanides, M. Arkas, C. M. Paleos, M. J. Bertrand, Clustering processes occurring in liquid secondary ion mass spectrometry for alkyl quaternary ammonium bromides, *J. Mass Spectr.*, 1996, *31*, 95–100.

116. Chen, Y. C., J. Shiea, Selective enhancement of ion signal of charged surfactant in FAB MS, *J. Mass Spectrom.*, 1995, *30*, 1435–1440.

117. Coran, S. A., M. Bambagiotti-Alberti, V. Giannellini, G. Moneti, G. Pieraccini, A. Raffaelli, Continuous flow FAB and ion spray ionization techniques for the simultaneous determination of alkyltrimethylammonium surfactants by MS, *Rapid Commun. Mass Spectrom.*, 1998, *12*, 281–284.

118. Lawrence, D. L., Normal phase LC/MS by using coaxial continuous flow fast atom bombardment, *J. Am. Soc. Mass Spectrom.*, 1992, *3*, 575–581.

119. Lyon, P. A., F. W. Crow, K. B. Tomer, M. L. Gross, Cationic surfactants by MS/MS with FAB, *Anal. Chem.*, 1984, *56*, 2278–2284.

120. Fernández, P., A. C. Alder, M. J. F. Suter, W. Giger, Ditallowdimethylammonium in digested sludges and marine sediments by supercritical fluid extraction and LC with post-column ion-pair extraction, *Anal. Chem.*, 1996, *68*, 921–929.
121. Kawakami, S., R. H. Callicott, N. Zhang, Benzalkonium chloride on skin by flow injection ionspray MS-MS, *Analyst*, 1998, *123*, 489–491.
122. Vicchio, D., A. L. Yergey, Thermospray LC/MS of quaternary ammonium salts, *Org. Mass Spectrom.*, 1989, *24*, 1060–1064.
123. Tegeler, A., W. Ruess, E. Gmahl, Amphoteric surfactants in cosmetic cleansing products by HPLC on a cation exchange column, *J. Chromatogr. A*, 1995, *715*, 195–198.
124. Rezanka, T., M. Podojil, Preparative separation of algal polar lipids and of individual molecular species by HPLC and their identification by GC-MS, *J. Chromatogr.*, 1989, *463*, 397–408.
125. Cole, M. J., C. G. Enke, Phospholipid structures in microorganisms by FAB triple quadrupole MS, *Anal. Chem.*, 1991, *63*, 1032–1038.
126. Hayashi, A., T. Matsubara, M. Morita, T. Kinoshita, T. Nakamura, Choline phospholipids by FAB-MS and tandem MS, *J. Biochem.* (Tokyo), 1989, *106*, 264–269.
127. Bryant, D. K., R. C. Orlando, C. Fenselau, R. C. Sowder, L. E. Henderson, Four-sector tandem MS analysis of complex mixtures of phosphatidylcholines present in a human immuno-deficiency virus preparation, *Anal. Chem.*, 1991, *63*, 1110–1114.
128. Harrison, K. A., R. C. Murphy, Negative electrospray ionization of glycerophosphocholine lipids: formation of [M-15]$^-$ ions occurs via collisional decomposition of adduct anions, *J. Mass Spectrom.*, 1995, *30*, 1772–1773.
129. Harvey, D. J., Matrix-assisted laser desorption/ionization MS of phospholipids, *J. Mass Spectrom.*, 1995, *30*, 1333–1346.
130. Li, C., J. A. Yergey, Continuous flow liquid secondary ion MS characterization of phospholipid molecular species, *J. Mass Spectrom.*, 1997, *32*, 314–322.

16
Titration of Surfactants

The most useful methods for volumetric determination of ionic surfactants are based on titration of an anionic surfactant with a cationic surfactant or the titration of a cationic surfactant with an anionic surfactant. In general, equivalent results are obtained by pouring the sample into the buret and using it to titrate a precisely known quantity of a standard solution of opposite charge. Thus, the methods given below for determination of anionics and determination of cationics are to a great extent interchangeable with each other; the reader should study both sections to appreciate the choices available for the analysis.

Reiner Schulz has written a wonderfully clear book on surfactant titrations. Anyone able to read German is urged to study that work, which contains a wealth of practical knowledge (1).

I. ANIONIC SURFACTANTS

A. Introduction

Titrimetric methods were the first procedures to be widely applied to the analysis of anionic surfactants and they remain very useful for assay purposes. Their speed and low cost make them especially suitable for quality control. Cross has written a thorough and recent discussion of the titration of anionic surfactants (2)

1. Acid-Base Titration

Most anionic surfactants are salts of moderately strong acids. As such, they can be titrated directly with base, provided that a suitable solvent system and visual or instrumental end point can be found. However, since anionic surfactants are generally found in the company of other ionic materials, some of them also acidic, direct acid-base titration is not used for most applications. For example, acid-base titration for assay of alkylarylsulfonate would risk high results because of titration of byproduct sulfate or other ions. Acid-base titration of LAS in a detergent formulation suffers from interference from such buffering compounds as sodium silicate and sodium tripolyphosphate. Titration with alkali is therefore limited to cases where the anionic surfactant can be isolated in pure form.

Simple acid-base titration may be used for the characterization of pure anionics (3). This approach is most often applied to nonaqueous titration of fatty acid soap (4). Colored indicators may be used, although potentiometric titration is generally preferred. In this approach, the free fatty acids are solubilized in acetone with excess HCl. Conventional nonaqueous titration with tetrabutylammonium hydroxide solution neutralizes first the excess HCl, then the total fatty acid. Care is required to characterize the sample well enough to assure that interference from other detergent components does not occur.

Soap can be determined in mixture with oil or ester-type nonionic emulsifiers by nonaqueous titration. The solvent system is 15:5 acetic acid/benzene, with crystal violet indicator. The soap is titrated with HBr in glacial acetic acid. If compounds are present which react with HBr, such as epoxides, the alkali metal cations are first extracted into acetic acid from amyl acetate/*n*-butanol solution and then titrated with HBr. In either case, the procedure is not specific for soap; any alkaline component can be expected to interfere (5).

2. Precipitation Titration

Titration is usually performed with a cationic surfactant to form a water-insoluble ion pair. This technique has the advantage of being selective for ionic molecules with nonpolar "tails," i.e., those with classical surfactant structure. A few years ago, the most popular method for determination of anionic surfactants was two-phase titration with benzethonium chloride. This has very rapidly been replaced by one-phase titration with potentiometric end point. Several vendors provide PVC membrane electrodes suitable for potentiometric titration of ionic surfactants.

B. Two-Phase Titration

Two-phase titration with a cationic surfactant was the most common method for determination of anionics from about 1950–1990. The technique remains in wide use and is incorporated into many national and international standards.

1. The Standard Method

The titration is based on the reaction of anionic surfactants with cations, normally large cationic surfactants, to form an ion pair. With the ionic charges neutralized, the ion pair has net nonpolar character, much more lipophilic than either free ion. In a two-phase system the ion pair is therefore extracted continually into the organic phase as it is formed. The reaction is monitored by addition of a water-soluble cationic dye, dimidium bromide, and a water-soluble anionic dye, disulphine blue. The cationic dye forms an extractible ion pair with the anionic surfactant, coloring the organic phase a pink hue. At the end point, the cationic titrant displaces the dye and the pink color leaves the organic phase. Further addition of titrant carries the blue anionic dye into the organic phase. Thus, the end point is indicated by a color change from pink to blue in the organic phase. This version of the two-phase titration has become the standard method, a thorough study of which has been written by Reid, Longman, and Heinerth (6–9).

The procedure is suitable for determination of anionic surfactants in concentrates and in detergent formulations and is used for assay of alkyl aryl sulfonates, alkyl sulfates, sulfated APE and AE, and sulfosuccinates. As described in Chapter 1, alkane monosulfonates require special precautions. Phosphate surfactants will generally not be determined. Since the analysis is conducted at pH 2, soap is not determined. Hydrotropes in low concentrations do not interfere, nor do most other detergent constituents.

Procedure: Assay of an Anionic Surfactant by Two-Phase Titration (6)

Reagents:

> *Benzethonium chloride* (Hyamine 1622) is available from chemical supply houses. A 0.004 M solution is prepared in water and standardized against sodium lauryl sulfate.
>
> *Mixed indicator solution* is prepared by dissolving separately 0.5 g dimidium bromide and 0.25 g disulphine blue in warm 10:90 ethanol/water. Mix and make up to 250 mL with 10:90 ethanol/water. Transfer 20 mL of this solution to a 500-mL volumetric flask, add 50 mL 1 N H_2SO_4, and dilute to volume with water.

Weigh a sample sufficient to give about 1 meq of anionic surfactant. For an unknown, an equivalent weight of 360 may be assumed. Dissolve in about 100 mL water and neutralize to phenolphthalein with 1 N NaOH or H_2SO_4, as required. Transfer to a 250-mL volumetric flask, dilute to volume, and mix well. Transfer a 20-mL aliquot to the titration vessel, which is a 100-mL graduated cylinder fitted with a ground glass stopper. Add 10 mL water, 15 mL methylene chloride, and 10 mL indicator solution. Titrate with the benzethonium chloride solution, shaking well after each addition of titrant. The end point occurs when the pink color is completely discharged from the methylene chloride layer, which is then a faint blue.

Notes: (a) In letters to several journals during 1995–1996, Professor Norbert Buschmann suggested that dimidium bromide may have mutagenic properties and that its solutions should therefore be handled with all appropriate precautions. (b) Mettler has developed titration accessories which allow this analysis to be performed automatically (10).

2. Modifications to the Standard Procedure

The precipitation titration has been studied extensively, particularly by Buschmann (11). Sources of error include solubility of the ions at the equivalence point, which is influenced by formation of micelles. Micelle formation is influenced by the presence of solvents and salts in the system. A vast number of modifications have been suggested by various authors, based on greater or lesser understanding of the principles at work.

(a) Indicator. Many ionic dyes have been proposed as indicators for the titration. The first widely-used version of the two-phase titration method, sometimes called the "Epton method," employed as indicator a single water-soluble cationic dye, methylene blue. At the start of the titration, the methylene blue salt of the anionic surfactant is extracted into the organic phase, where it remains until the methylene blue ion is gradually displaced from the complex by the cationic titrant. The end point has been demonstrated to be at the

point where the blue color of the aqueous and organic phases is of equal intensity. (12,13) This end point is somewhat indistinct and the methylene blue method is sensitive to interference from chloride and other anions. The cationic dye dimethyl yellow is said to give a sharper end point than the standard mixed indicator. Since the dye has low water solubility, the aqueous phase remains nearly colorless, and the organic phase changes sharply from red to yellow at the end point, as the dye responds to the change in its environment (14,15). The cationic dye safranine T is used for assay of alkylethercarboxylates at pH 8.6, taking as the end point the titrant volume when the color intensity of each phase is the same (16).

Replacement of the acidic titration medium with a basic medium has been suggested as a way to minimize the interference of weakly cationic surfactants, such as amine oxides, which are not ionized at higher pH (17). Titration under these conditions allows determination of soap (18,19). Since cationic dyes do not generally retain their ionic character at high pH, an anionic dye, such as bromcresol green or bromophenol blue, is used as indicator. With such a system, the color remains in the aqueous phase until the end point, when it is extracted into the organic layer. (The dimidium bromide/disulfine blue mixed indicator may also be used in alkaline conditions, but the color change is different and less sharp than when used in acid solution.)

If interference from soap is not a problem, anionic dyes have certain advantages over the cationic dyes as indicators. With tetrabromophenolphthalein, ethyl ester, the need for color-matching of the phases is avoided. During the titration, the aqueous phase remains colorless, while the organic phase changes sharply from yellow to blue at the end point (20). Similarly, thymolphthalein is recommended for the titration of acylsarcosines at high pH (21). Phenol red gives quantitative results for titration of tetrahydronaphthalene sulfonate as well as surfactants, while with bromcresol green, only the surfactants are titrated (22). Use of phenol red in conjuction with titration at pH 9–10 permits determination of both carboxylate and sulfonate groups in an unesterified α-sulfo fatty acid, while use of methylene blue at lower pH allows titration of the sulfonate group alone (23); with bromcresol green only the α-sulfo fatty acid ester is measured (24).

(b) Other end points. Instead of shaking the mixture after each addition of titrant, a special stirred glass vessel resembling a Soxhlet extractor may be used for the titration. Values obtained are comparable to the shaking method, with much less operator fatigue (25).

A turbidimetric end point has been found to make the two-phase titration easily automated (26). The indicator is omitted, and the titration is followed by monitoring the turbidity of the aqueous phase, which suddenly increases at the end point. The titration is, in other respects, similar to the standard method, although methanol is added to the system to optimize end point detection. Hendry and Hockings found that the organic phase could be omitted altogether, as described below under one-phase titration (27). Pinazo and Domingo performed a similar titration with only a small amount of chloroform finely dispersed by stirring. In this case, the end point corresponds to an inflection point on the turbidity curve (28).

An automated two-phase titration may be conducted to a spectrophotometric end point. Various researchers have demonstrated this technique with specially made glass vessels or by using membranes to separate the aqueous and organic phases. Nowadays, the manufacturers of automatic titrators supply equipment and optimized conditions to permit this type of titration (1,29). Usually, the vessel is stirred after each increment of titrant, then the phases are allowed to separate and the transmittance of the organic phase is determined. Spectrophotometric methods are not used for most surfactant titrations, since the single-phase potentiometric titration works quite well. They are used for special cases, such as determination of polyelectrolytes.

As described below, the two-phase titration has been largely supplanted by aqueous titration with a potentiometric end point. The aqueous titration fails for certain samples, such as those containing oily material which fouls the electrode or those containing surfactants which do not form an insoluble ion pair with the titrant. Such problems can be solved by going to a two-phase system but keeping the potentiometric end point detection in the aqueous phase. Oil is extracted into the organic layer, as well as the ion pair formed during the titration. At the end point, the electrode responds to the increase in concentration of the titrant in the organic phase. It is essential that the electrode used be resistant to organic solvents, a property not enjoyed by most commercial PVC-membrane surfactant-selective electrodes. The best design is a coated-wire electrode made with plasticized PVC containing a polymeric ionophore based on silicone (1).

(c) Titrant. Although most cationic surfactants will function as titrants, a compound is usually chosen which consists of a single pure compound, rather than containing a range of alkyl chain lengths like typical surfactant quaternary amines. The most frequently used material is the disinfectant benzethonium chloride (often referred to by the Rohm & Haas brand name, Hyamine® 1622). TEGO®trant A100 (1,3,-didecyl-2-methylimidazolium chloride, available from Metrohm) is a commercially available alternative titrant which is more hydrophobic than benzethonium chloride. This allows the formation of hydrophobic ion pairs with anionics of shorter alkyl chain length than can be titrated with benzethonium chloride. It also leads to sharper end points when using an electrode for end point detection as discussed below.

In eastern Europe, a well-characterized disinfectant, Septonex (carbethoxypentadecyltrimethylammonium bromide), is sometimes used as titrant (14,15). Results are comparable with benzethonium chloride. Benzyldimethyldodecylammonium bromide is also suitable (30).

(d) Sample pretreatment by hydrolysis. Some classes of anionic surfactants are readily hydrolyzed under acidic or basic conditions, while others are not. Procedures have therefore been developed to analyze surfactant mixtures by titration before and after hydrolysis. Acid hydrolysis for 3 hours with 1 M sulfuric acid will decompose alkyl sulfates, alkyl hydroxysulfates, alcohol ether sulfates, and alkylphenolether sulfates so that they no longer respond to titration, while sulfonates are not affected (7,31–33). Alkaline hydrolysis for 30 minutes with 2 M sodium hydroxide will saponify dialkylsulfosuccinates and fatty acid glyceride sulfates so that they are no longer surface active and are not titrated (34).

Saponification for four hours will decompose α-sulfo fatty acid methyl esters so that they can no longer be titrated by the standard two-phase procedure, either in acidic or basic media. A one-hour saponification will cleave the ester, but will remove only about 20% of the sulfonate group, so that these compounds are still titrated. Reflux with HCl has no effect. On the other hand, reflux with either acid or base will decompose fatty alcohol sulfoacetates so that they are no longer titratible (35).

3. Extending the Application of Two-Phase Titration

The standard method is effective for determination of most detergent sulfonates and sulfates, as well as dialkylsulfosuccinates. Interference is not usually encountered from inorganic anions, chelates, soap, or common hydrotropes (8). Other anionics, such as phosphate esters, must be tested individually for applicability. Li and Rosen reported that the standard method gives low recoveries (as low as 89%) for anionic surfactants with an alkyl chain length less than C_{12} (36). Complete recovery can be obtained by changing the extraction solvent from chloroform to 40:60 chloroform/1-nitropropane, and by drawing off the organic layer after the end point is reached, replacing it with fresh solvent, and titrating to a second end point.

While the two-phase titration will not determine alkanedisulfonates if monosulfonates are absent, the presence of disulfonate impurities in commercial alkanesulfonate products will cause a variable interference (37). The response of the two-phase titration to molecules containing more than one sulfate or sulfonate group must be determined for each individual case. Lew reports stoichiometric reaction with each functional group for a C_{22} surfactant with two anionic sulfate groups separated by four carbon atoms (38). Wickbold suggests addition of 10 mL 20% sodium sulfate solution to eliminate interference from disulfonates, and this was adopted in the ISO procedure (39). Alternatively, modification of the organic solvent to a 93:7 chloroform/1-hexanol mixture will allow complete recovery of disulfonates, regardless of pH (40).

Two-phase titration is effective in determination of diesters of sulfosuccinic acid, whether the titration is conducted in acidic or basic medium. Results with monoesters are variable due to their hydrophilic nature. Each product must be tested individually. Generally, half esters cannot be titrated in basic media and can be titrated in acid medium only if they have good solubility in organic solvents (41).

By performing the titration twice, once at acidic and once at basic pH, soap can be differentiated from synthetic anionic surfactants (18,40). Under acid conditions, only the sulfate and sulfonate type surfactants are determined, while under basic conditions both soap and synthetic surfactants are measured.

Most cationic surfactants interfere quantitatively with the titration of the anionics because of their similarity in properties to the titrant. Their presence leads, therefore, to low values. This is not generally a serious problem, because the antagonism of anionic and cationic surfactants insures that they are rarely formulated together. If the level of anionic surfactant is lower than that of the cationic, both components may be determined by titration of the cationic with tetraphenylborate (42). In this procedure, the total cationic is titrated at pH 12.5, using 1,2-dichloroethane as the second phase and potassium tetrabromophenolphthalein, ethyl ester, as indicator. A similar titration is performed at pH 6.0, giving a value for that amount of cationic which is not neutralized by the anionic surfactant (which quantitatively interferes at the lower pH value). This procedure was developed specifically for analysis of a hospital wastewater stream which contained anionic surfactant as well as cationic biocides. A similar approach is used if the level of the cationic is lower than that of the anionic (43). The sample is first titrated at pH 9.0 with cationic surfactant mixture, using ethylene dichloride as second phase and Victoria Blue B as indicator. This value corresponds to the anionic surfactant not neutralized with cationic surfactant. Another sample aliquot is then titrated at high pH with tetraphenylborate solution to give the total cationic concentration. The anionic concentration is calculated from the sum of the two titrations.

Alkylbetaine amphoteric surfactants behave as cationics at pH 1, at which the standard two-phase titration of anionics is performed. Their presence thus leads to low results for anionics (1).

C. One-Phase Titration

It is possible to eliminate the second phase of an organic solvent during the titration. This change will usually permit the titration to be performed in less time and ease automation of the procedure. Properly, surfactant titrations conducted in an aqueous medium are not truly one-phase reactions, since the ion pair of the cationic and anionic surfactant is water-insoluble and constitutes a second dispersed phase of its own. The titration can be monitored by following the removal of the surfactant ion from the bulk aqueous phase or by observing the formation of the hydrophobic ion pair.

1. Potentiometric Titration

Because of its reproducibility and ease of use, one-phase potentiometric titration has largely supplanted the two-phase titration with indicator end point. This has been made possible in the last decade by the easy availability of electrodes which respond to surfactant concentration. The titration is performed in a stirred one-phase aqueous system with no need to wait for phase separation after each addition of titrant. Concern about operator-sensitive color matching at the end point, as well as consideration of different colored forms of the dye at various pH values, is eliminated.

(a) Applicability. The ASTM method reproduced below has been validated for determination of alkylbenzenesulfonates, α-olefin sulfonates, alcohol sulfates and alcohol ether sulfates. It has been extended to analysis of α-sulfo methyl tallowate (23). In the latter case, the pH of the titration medium is important: At pH 3, only the sulfonate group is titrated; at pH 10, both the sulfonate and the carboxylate groups are determined. In general, anionics are titrated at pH 3 except for sulfosuccinates, which are titrated at pH 6 (10). Potentiometric titration, unlike two-phase titration, is suitable for determination of both sulfosuccinate diesters and monoesters. pH control is important for the sulfosuccinate monoesters, since at low pH only the sulfonate group is titrated, at high pH both the carboxylate and sulfonate groups are titrated, and at pH 5 no potententiometric end point is seen (1). Fluorinated anionic surfactants are generally susceptible to measurement using potentiometric titration (44).

(b) Equivalence of potentiometric and two-phase methods. Both the one-phase and two-phase titrations are precipitation titrations relying on formation of an insoluble ion pair. Both one-phase and two-phase titrations have limitations, but each titration is limited in its own way (45,46): (a) at low surfactant concentrations, a correspondingly high titrant concentration is required to exceed the solubility product; (b) micelle formation is competitive to ion pair formation, and components which influence micelle formation (salts, alcohols, nonionic surfactants) will influence the results. Thus, while purified surfactants will usually give the same result by either procedure, commercial surfactants consisting of mixtures may have some components which respond to one type of titration but not the other. An example is commercial alkanesulfonate, which contains about 10% di- and polysulfonate impurities. These impurities are measured to varying extents when an alkanesulfonate is titrated potentiometrically or by two-phase titration (1,11,45,46).

(c) Titrant. When used in potentiometric titration, hexadecylpyridinium chloride, hexadecyltrimethylammonium chloride, and trioctylmethylammonium chloride all give higher potential jumps than does benzethonium chloride. The highest potential jump is obtained with 1,3-didecyl-2-methylimidazolium chloride (47), and this compound is therefore widely used for potentiometric surfactant titration. It is commercially available

from Metrohm under the TEGOtrant® trademark. A lower standard deviation is found for replicate titrations with the imidazolium compound compared to benzethonium chloride. The superiority of 1,3-didecyl-2-methylimidazolium ion is due to its having a lower solubility with some anionics than does benzethonium chloride (45,46). This leads to more complete reaction at the end point. It also means that more hydrophilic anions can be titrated. For example, octylsulfate can be potentiometrically titrated with the imidazolium compound, but not with benzethonium chloride (48,49)

(d) *Minimum quantities.* It has been reported that titration of small quantities of anionics leads to low results (48). It has also been reported that titration of small quantities leads to high results (45,46). When developing a titration method it is prudent to make determinations at several concentrations to determine the effect of sample size.

(e) *Mixtures.* If two anionics of somewhat different functionality are present, it is possible to obtain a double inflection of the potentiometric curve during titration. This results from different solubilities of the ion pair formed with the titrant and may permit determination of two formulation components by a single titration (50). Unfortunately, such quantification often becomes difficult in commercial formulations (30).

(f) *Electrodes.* Electrodes for determining the end point of surfactant titrations are described in Section V of this chapter.

(g) *Interference.* Especially in the case of home-made electrodes, the shape of the potentiometric curve deteriorates in the presence of salts and of other materials, such as nonionic surfactants (51). This interference can be largely circumvented by using the more versatile electrodes sold by the titrator manufacturers and by carefully optimizing the titration conditions. In special cases, a two-phase titration using a special solvent-resistant electrode is preferred (1,52).

(h) *Other remarks.* Since the polymer membrane electrodes are susceptible to fouling by hydrophobic substances, special precautions are needed in the case of, for example, determination of surfactants in crude oil. One approach is conducting the titration in the presence of a small amount of an upper organic phase which will trap the oil but allow anionic surfactant to be extracted into the titration solution (53).

Some researchers have suggested adding a nonionic surfactant to the sample to keep the precipitate suspended and the electrode free from fouling. This has the undesirable effect of making the end point less sharp. A better strategy is the addition of 5% methanol to prevent fouling (1).

Surfactant titrations are based on slower reactions than other analytical titrations and the results are more dependent on variables like titration speed, stir rate, etc. Evaluation of the titration curves is more susceptible to error (1). The best sources for surfactant titration methodology are the manufacturers of the titrators and electrodes.

Procedure: Potentiometric Titration for Assay of Anionic Surfactants (54)
Reagents:

> *Benzethonium chloride* (Hyamine 1622, diisobutylphenoxyethoxyethyl dimethylbenzyl ammonium chloride) is dissolved in water containing 0.4 mL of 50% NaOH solution. Eighteen grams are dissolved in 1 L water and standardized against sodium lauryl sulfate solution.

Sodium lauryl sulfate, $NaC_{12}H_{25}O_4S$, MW 288.38, may be obtained from Gallard Schlesinger Chemical Mfg. Corp. A 0.04 M solution is prepared by dissolving about 11.5 g in 1 L H_2O.

Add to a 150-mL beaker a sample consisting of about 0.15 mmoles anionic surfactant in a volume of about 50 mL H_2O. Titrate potentiometrically with benzethonium chloride solution at a rate of 0.5 mL/min until well past the inflection point of the curve. An automatic titrator with 5-mL buret assembly is used, equipped with a nitrate-selective electrode (HNU Systems model ISE-20-31-00, Orion model 93-07, or equivalent) and a Ag/AgCl reference electrode with ground-glass sleeve.

2. Miscellaneous One-Phase Titration Methods

(a) Indicator methods. The titration of an anionic surfactant with a quaternary amine can be conducted in a one-phase aqueous system using thiocyanatocobalt(II) as indicator. At the end point, excess amine titrant forms a complex with the indicator, causing a color change from pink to violet. The procedure was demonstrated by titration of sodium lauryl sulphate, sodium lauryl ether sulfate, and alkylbenzene sulfonate with cetrimide, a mixture of C_{12}-, C_{14}-, and C_{16}-alkyltrimethylammonium bromides (55).

   If a nonionic surfactant is added to the aqueous solution, the ion pair of the anionic and the titrant will be solubilized in the micelles (56), making unnecessary the separate organic solvent layer used in conventional two-phase indicator titrations. This approach has been demonstrated using tetrabromophenolphthalein ethyl ester as indicator. At the end point, a color change from yellow to blue is observed, which may be monitored with a fiber optic colorimeter (57). As mentioned above, potentiometric titration using a few percent methanol in place of the nonionic surfactant will generally give superior results.

(b) Other electrochemical titrations. A more specialized electrochemical method for end point detection makes use of the general effect of surfactants on the capacitance of a mercury drop electrode (58). Since the anionic and cationic surfactants interfere with each other's effect on the capacitance, the change in capacitance can be used to follow the titration of one with the other.

(c) Turbidimetric titration. Hendry and Hockings found that the organic phase and indicator could be omitted altogether from the standard two-phase titration procedure, with the end point of the titration taken as the point of maximum turbidity of an aqueous solution during the benzethonium chloride titration (27). The turbidity of the single, aqueous phase reaches a maximum at the equivalence point, then decreases again as the titrant concentration increases. Both Metrohm and Mettler market devices that allows automatic titration to the turbidity end point (10). This approach was demonstrated for the automatic titration of alcohol sulfate, sulfosuccinate, alkylaryl sulfonate, ether sulfate, and sulfonated amide salt. This approach does not have clear advantages over the potentiometric end point (1). Interference with the end point is experienced with high levels of salts or of nonionic surfactants. The point of maximum turbidity of the titration medium is not necessarily the stoichiometric end point (59). This is because excess anionic surfactant tends to solubilize the ion pair prior to the end point, while excess cationic solubilizes it after the end point, and these phenomena need not be exactly symmetric. The end point does seem to be reproducible, so good results can be obtained if the titrant is standardized in the same way and under the same conditions as the analysis is performed.

A commercial device allows titration to a so-called refractive index end point. The response of this device is actually more related to turbidity than refractive index, and end point detection is inferior to that provided by photometric devices optimized to respond to turbidity (1).

II. NONIONIC SURFACTANTS

A. Nonionics Containing Alkylene Oxide Groups

The basis of most titration methods for nonionics is the ability of polyethers to form weak complexes with large cations such as Ba(II). The formation of the complex is governed by the presence of lipophilic anions, and the titrimetric finish usually involves the anion. Tetraphenylborate is the most commonly used anion, although substituted tetraphenylborates, ferrocyanide, tetraiodobismuthate, and heteropoly acids are also used. Thorough discussions of these reactions have appeared. Generally, individual homologs containing more than four alkylene oxide units are titratable (60,61).

1. Titration with Tetraphenylborate

Polyalkylene glycol adducts can be titrated directly, making use of their ability to form an insoluble complex with divalent cations and tetraphenylborate ions. An excess of a divalent salt is added to the sample, and the complex is titrated potentiometrically with sodium tetraphenylborate solution, using a metal indicator electrode or, more often, one of the PVC-membrane surfactant electrodes. The method is simple and rapid, although it is necessary to standardize with the particular surfactant to be determined, and there are many interfering species. (In fact, the titrant is often standardized against potassium or thallium nitrate solutions.) Of common cations, barium ion gives the largest change in potential during titration and is therefore always used (30). The average calibration factor is 5.2 moles oxyethylene per mole tetraphenylborate, or 10.4 moles per mole barium ion. Compounds containing 1 to 4 moles oxyethylene are not titrated, so empirical calibration factors are generally required for commercial products containing a range of oxyethylated homologs.

Concentration of alcohol in the titration medium must be kept to an absolute minimum, since it solubilizes the complex and hence raises the detection limit. Ammonium and potassium ions give insoluble tetraphenylborate salts and so should be avoided, especially in the filling solutions of reference electrodes. Anionic surfactants containing polyethoxy chains interfere with this method and must be removed (62). Cationic surfactants generally react quantitatively with tetraphenylborate. Their interference can be corrected by performing the titration without added barium chloride solution to determine cationics, then subtracting this value from the titration with added barium chloride (62). The same technique is used if interference from ammonium or potassium ions is suspected, although in practice this interference is often negligible (1).

Procedure: Titration of Ethoxylated Surfactant with Sodium Tetraphenylborate (62)

A sample containing 5–50 mg ethoxylate in about 90 mL water is mixed with 10 mL of 0.1 M barium chloride solution and titrated with 0.01 M aqueous sodium tetraphenylborate solution (prepared with 10 g polyvinyl alcohol and 10 mL 0.1 M NaOH per liter of titrant). The titration is made potentiometrically, using the Metrohm surfactant indicating electrode and a silver/silver chloride reference electrode (NaCl filling solution). The electrode must be rinsed free of precipitate after every few titrations using methanol. An electrode equilibration time of 30 sec should be allowed prior to beginning each titration. Calibration is performed versus surfactants of known composition. End point detection is best in the first derivative mode.

This method is effective for determination of OPE, AE, and PO-containing surfactants (62–64). Ionic surfactants interfere quantitatively, but can be removed by precipitation in a prior anionic/cationic titration (64). PEG is determined along with surface active polyoxyethylene adducts. In the case of EO/PO copolymers, Schulz states that the polypropylene oxide portion of the molecule takes no part in the titration. He says that the polyethylene oxide portion must be at least 25% for a successful titration (1).

2. Iodobismuthate Methods

There are several procedures based on the formation of an insoluble complex between the iodobismuthate ion and polyether linkages in the presence of barium ion (65). The first widely used procedure quantified the precipitate by titration of the incorporated bismuth. This method is described in Chapter 17 under gravimetric procedures.

3. Two-Phase Titration

A significant drawback of direct tetraphenylborate titration is interference from anionic surfactants, which form an ion pair with the barium–nonionic surfactant complex in competition with tetraphenylborate. This interference can be eliminated by resorting to two-phase titration. Tetrakis(4-chlorophenyl)borate or tetrakis(4-fluorophenyl)borate do not precipitate with potassium ion alone, as does tetraphenylborate. The complex formed between potassium ion and a nonionic surfactant is extracted as an ion pair with the tetrakis(halophenyl)borate from an aqueous into an organic phase, while the corresponding ion pair with an anionic surfactant does not extract under the same conditions (66–68). When the nonionic surfactant is exhausted, excess tetrakis(halophenyl)borate titrant serves to change the color of a cationic dye in the organic phase, Victoria Blue B, allowing visualization of the end point. At high pH, anionic surfactant does not interfere with the end point color change. This procedure works well with most ethoxylated and propoxylated surfactants, including the EO/PO copolymers, giving an approximately constant ratio of moles titrant per alkoxylate unit for various compounds. The method has even been found suitable for environmental monitoring, provided that the nonionic is first concentrated by solvent sublation and purified by anion/cation exchange (69).

Unlike the two-phase titration of anionic and cationic surfactants, the titration of nonionics with tetrakis(halophenyl)borate does not occur on a mole to mole basis. The stoichiometry of the reaction is dependent on the number of EO units per mole of the particular surfactant. O'Connell reports an average value of 4.1 mL titrant per milligram surfactant for common ethoxylated alcohols, while Tsubouchi and coworkers report a value of 3.6 mL for most types of nonionics, and Hei and Janisch report about 3 mL/mg. It is important to run a blank determination because the color change at the end point is gradual, making the titration operator-sensitive. Cationic surfactants interfere with the titration at a molar ratio of about 1.3 to 1. The interference is reproducible so that for quality control purposes, where the amount and nature of the cationic is known, this procedure will still be useful. Typical anionic surfactants do not interfere, except to the extent that they cause emulsions at high concentration. This can be readily solved by dilution of the sample. Anionics of the type sulfated alcohol ethoxylate would be expected to interfere, depending on the length of the EO chain. For complex detergent matrices, calibration should be performed in the same matrix to minimize problems in duplicating the end point color.

In another version of this procedure the nonionic surfactant was first extracted batch-wise with sodium tetraphenylborate into 1,2-dichloroethane. The tetraphenylborate in the isolated organic phase was then titrated with a cationic surfactant, using Victoria Blue B as indicator (70). This titration can also be performed to an electrochemically detected end point. In this version, an excess of anionic surfactant is added to the cationic complex formed by the ethoxylated nonionic surfactant and potassium ion. The ion pair is extracted into dichloroethane, separated from the initial aqueous phase, then titrated with cationic surfactant in the presence of additional water. The ion pair of the anionic surfactant and $Fe(II)(1,10\text{-phenanthroline})_3$ is added as indicator. The end point of the titration is indicated when the last of the anionic surfactant is complexed by the cationic titrant, causing the iron-phenanthroline cation to migrate to the aqueous phase, where it is detected as a change in potential at a platinum electrode (71).

Procedure: Two-Phase Titration of Nonionic Surfactants with Potassium Tetrakis(4-chlorophenyl)borate (67)

Reagents:

> *Potassium tetrakis(4-chlorophenyl)borate titrant*, 0.0005 M, CAS 25776-12-9. Prepare a 0.025% aqueous solution by vigorous boiling and stirring.
> *Victoria Blue B indicator solution*, 0.04% in ethanol, CAS 2580-56-5, $C_{33}H_{32}N_3 \cdot$ Cl, CI 44045.

Use a stoppered, 50-mL, conical-bottom centrifuge tube as the titration vessel. Add 10.0 mL aqueous sample solution, 5 mL 6 M KOH solution, 50 mL indicator, and 5.0 mL 1,2-dichloroethane. Add titrant solution, shaking well after each increment. For some detergent formulations, it will be necessary to centrifuge the titration vessel when near the end point to produce a clear organic layer. The end point is taken at the color change in the organic phase from pink to purple, with the further change to blue indicating overtitration. The aqueous phase remains colorless throughout the titration, because of the insolubility of the dye in alkali. The titrant is standardized against the same surfactant as is being determined.

4. HI Cleavage

If the identity of the nonionic surfactant is known and it can be separated from interfering compounds, it may be determined by the method of HI cleavage (72). The material is refluxed with hydriodic acid in the absence of oxygen, leading to decomposition of the ethoxylate chain and the formation of free iodine, which is titrated in the usual manner with sodium thiosulfate.

Various modifications of this procedure have been proposed. For example, the reproducibility problems inherent in using the unstable HI can be sidestepped by instead using a mixture of KI and H_3PO_4 (73). The HI cleavage method is discussed in more detail in Chapter 2.

5. Potassium Ferrocyanide Method

Ferrocyanide ion will form an insoluble complex with the potassium salt of a polyether. After isolation of the precipitate, the ferrocyanide ion remaining in the filtrate can be determined by titration (74).

6. Heteropoly Acid Titration

Potentiometric titration of Ba^{2+} or Ca^{2+} adducts of ethoxylates with tungstophosphoric acid has been demonstrated, using a tungsten wire electrode for end point detection. This is proposed for samples containing oil or solvents which would attack the PVC membrane of a surfactant-selective electrode (75).

B. Nonionics Containing Hydroxyl Groups

Nonionics containing hydroxyl groups can be readily converted to the corresponding sulfates and then be determined according to the method for potentiometric titration of anionic surfactants given above (76,77). The reaction can be performed in a few minutes using the dimethylformamide/SO_3 complex formed by reaction of DMF with chlorosulfonic acid. The reagent is also commercially available. This approach has utility for special situations. The method is useful for alcohol and alkylphenol ethoxylates having only a few moles of ethylene oxide and thus poorly titrated by the tetraphenylborate methods. It is also suitable for use with alkylpolyglucosides, N-acylglucamides, diethanolamides, ethoxylated monoethanolamides, and partial esters of sorbitan and glycerol. Amine ethoxylates can be determined in alkaline solution, but only by two-phase titration (76). Nonionics of degree of ethoxylation higher than about 10 cannot be determined by this procedure because they become too water-soluble to react with the titrant. This is also true of ethoxylated sorbitan esters. Cationic and anionic surfactants cause quantitative interference.

C. Amine Oxides

Long chain amine oxides are sufficiently basic that they may be titrated directly with HCl in isopropanol. Any residual tertiary amine will also be titrated under these conditions, but modern potentiometric titration apparatus allows differentiation of the amine and amine oxide in a single determination (Chapter 2). Alternatively, reaction with methyl iodide to form the quaternary amine from the tertiary amine will eliminate interference (78). Amine oxides which cannot be determined in isopropanol can generally be titrated potentiometrically with perchloric acid in acetic acid/acetic anhydride solvent. Acetic acid alone is not a suitable solvent for this titration (79,80).

Amine oxides have cationic properties at low pH and so can be titrated with anionic reagents according to the procedures developed for cationic surfactants. Titrations with tetraphenylborate (81) and with dodecylsulfate (82) have been demonstrated, using a potentiometric end point. A turbidimetric end point has also been demonstrated (28). An amperometric end point may be used in a two-phase system with dodecylsulfate titrant, if a suitable cationic indicator is added such as the iron(II) 1,10-phenanthroline complex (83). Long-chain amines, including any unreacted amine from synthesis of the amine oxide, will quantitatively interfere.

A two-phase titration method for determining amine oxides in the presence of anionic surfactants calls for titration of the anionics first with benzethonium chloride at pH 9.5 using bromcresol green as indicator. A second aliquot of the sample is analyzed by adding exactly the amount of benzethonium chloride needed to complex the anionic surfactant. The amine oxide, with the ion pair of the anionic surfactant, is then extracted into chloroform at pH 9.5. The chloroform solution is mixed with water and titrated with alkylbenzene sulfonate solution at pH 2 using methylene blue indicator (17). Another approach is to perform potentiometric titration with a cationic surfactant at pH 7 or higher, which will determine only the anionic. The titration is repeated at pH 2, at which pH the amine oxide will react with an equivalent amount of the anionic. The amount by which the apparent anionic content is lowered is equivalent to the amine oxide concentration (82). Depending on the relative levels of amine oxide and anionic, it may be necessary to spike in a known additional amount of anionic surfactant (28).

   Amine oxides can also be determined by redox titration. The compounds are reduced with stannous chloride to the corresponding amines, and the excess stannous chloride is determined by titration with ferric ammonium sulfate. Potassium indigo sulfate is a suitable indicator (84). Interferences are abundant, including atmospheric oxygen.

D.   Mixed Micelle Formation

The determination of nonionic surfactants has been demonstrated using mixed micelle formation. This relies on the principle that, to a great extent, micelle formation is additive, i.e., if two surfactants are present, each will be present in the micelles, and the critical micelle concentration (CMC) is the sum of the concentrations of the two surfactants. A dye is chosen which has an absorbance which changes if surfactant micelles are present in solution. A "blank" solution is titrated with a surfactant, usually an anionic, to determine the CMC. A solution of the unknown surfactant is then titrated, giving an apparent CMC lower than the blank. The difference in the two values represents the contribution of the unknown surfactant, a contribution which is additive over a useful concentration range (85).

III.   CATIONIC SURFACTANTS

Many cationic surfactants can be determined by direct acid-base titration in nonaqueous solvents according to general methods for determination of amines. Usually, perchloric acid titrant is used in a solvent of glacial acetic acid. Since such titrations are in no way specific for surfactants, they can only be used for solutions in which there are no unknown components. Details may be found in standard texts on nonaqueous titration.

   The most generally useful methods for titration of cationics are based on formation of ion pairs as described below. For an in-depth discussion of volumetric analysis of cationic surfactants, the reader is urged to consult the thorough review of Cross (86).

A.   Two-Phase Titration

In general, the same two-phase titration methods based on formation of water-insoluble ion pairs can be used for determination of cationic surfactants as are used in the determination of anionic surfactants, described earlier in this chapter. This means that most methods have the disadvantage that anionic surfactants interfere with cationic determination. This is not a serious problem in formulations, since cationics are not blended with anionics. However, in environmental samples anionics are usually present at much higher levels than cationics. All of the following methods suffer from anionic interference.

   The minimum "effective chain length" allowing amines to be quantitatively titrated by dodecylsulfate is 12 carbons, or 10 in the case of alkyldimethylamines titrated with tetradecylsulfate. This is true of both two-phase indicator titration and one-phase potentiometric titration. If the titrant is tetraphenylborate rather than an alkyl sulfate, the effective chain length need only be 8 for tertiary symmetric amines, while all quaternaries are titrated (87). The effective chain length concept corrects for overlapping of alkyl chains, so that tetrabutylammonium bromide, for example, has an effective chain length of 10.2, rather than 16.

In the special case of titration of cationics in the presence of an anionic known to be an alkylsulfate, the sample may be subjected to acid hydrolysis to destroy the anionic prior to analysis (31).

1. Titration with Tetraphenylborate

A widely accepted titrant for two-phase titration of cationics is tetraphenylborate ion (88). The titration depends upon the surfactants maintaining their ionic character, and therefore the pH is critical. All common cationic surfactants can be titrated at pH 3, using methyl orange as indicator. At pH 10, all quaternaries can be titrated, but other amines are not sufficiently ionized. At pH 13, only certain quaternaries can be determined, but these include the commercially most abundant products: alkyltrimethyl- and dialkyldimethylammonium and benzyltrialkylammonium salts. Bromophenol blue is a suitable indicator for the titrations at pH 10 and 13.

In another version of the two-phase titration, also suffering from interference by anionic surfactants, the cationic is first extracted from water into chloroform with excess methyl orange. The two phases are left in the separatory funnel and titrated with sodium tetraphenylborate solution, with shaking, until the yellow color is expelled from the organic phase (89).

Procedure: Determination of Cationic Surfactants by Two-Phase Titration with a Hydrophobic Indicator (90)

> *Phosphate buffer solution* is prepared by dropwise addition of 5 M sulfuric acid to a 0.3 M solution of Na_2HPO_4 until pH 6.0 is reached.
> *Indicator solution* is 0.2% potassium salt of tetrabromophenolphthalein, ethyl ester, in ethanol.
> *Tetraphenylborate solution*, 0.01 M, is prepared by dissolving 3.42 g sodium tetraphenylborate, dried at 80°C, in 1 L 0.001 M NaOH solution. This is diluted to prepare the 5×10^{-5} M titrant.

A sample of 15 mL volume, containing 5×10^{-8} to 5×10^{-7} M of cationic surfactant, is placed in a 200-mL conical flask with 5 mL of phosphate buffer solution, 1 drop of indicator solution, and 0.5-1.5 mL 1,2-dichloroethane. Titrate with 5×10^{-5} M sodium tetraphenylborate solution from a 50-mL buret, shaking after each addition of titrant as the end point is reached. The 1,2-dichloroethane phase is initially blue, turns green in the vicinity of the end point, and becomes yellow when one drop of excess titrant is added.

2. Titration with Anionic Surfactant

The numerous indicator dyes described under the analysis of anionics are generally suitable also for titration of cationics. The Epton procedure with methylene blue indicator is still widely used, even though the end point is difficult to match, since the color remains the same throughout the titration, simply changing in its distribution between phases.

Cross gives the general rule that the titration of a cationic surfactant with an anionic is valid if each surfactant's alkyl chain has a length of at least C_{12}, if standardization is frequent and with the same class of compound as being determined, and if nonionic surfactants are absent (91). Accuracy varies with the choice of titrant and indicator (87). If a quality control method is being developed, these parameters must be carefully evaluated.

Sodium dodecylsulfate is available in a high purity pharmaceutical grade, and is therefore the most popular anionic surfactant for titration. Many other compounds have been proposed, including the heteropoly compound octapotassium 1-cobalto(II)[1-aqua]-1-molybdo-16-tungsto-2-phosphate, $K_8[P_2CoMoW_{16}O_{61}(H_2O)] \cdot 19H_2O$ (92).

The same pH effect as noted above is observed when lauryl sulfate is employed as titrant, so that even primary amines may be determined at low pH (93). In the case of a strongly alkaline detergent formulation containing a low level of cationic, pH control must be assured by adding the proper quantity of a suitable buffer.

The ISO method for determination of cationics gives separate procedures for determination of high and low molecular weight compounds (94,95). Low MW compounds (MW less than about 500) such as alkyltrimethylammonium salts are simply dissolved and titrated as described above. Higher MW compounds, such as dialkyldimethylammonium fabric softeners have poor water solubility and are therefore taken up in isopropanol, diluted with water, and used to titrate a known quantity of anionic surfactant.

3. Photometric Titration

Two-phase titration may be conducted to a spectrophotometric end point, observing the change in absorbance of either the aqueous or organic phase (40,96–98). For example, cationic surfactants can be titrated with picrate ion, while monitoring the UV absorbance ($\lambda = 400$ nm) of either phase. As titrant is added, the absorbance of the organic phase increases linearly from zero to a maximum. It remains at the maximum value after the end point, which is determined by extrapolation of the two linear portions of the titration plot. If the absorbance of the aqueous phase is monitored, its absorbance increases linearly after the end point, as additional picrate ion is added but not extracted into the organic phase.

A stirred cuvette can be used for the spectrophotometric titration of cationics with sodium dodecylsulfate titrant and the cationic dye neutral red as indicator. The end point is taken as the midpoint of the inflection when the color of the organic phase changes from a maximum to a minimum value (99).

4. Indirect Determination

A modification of two-phase titration calls for preparing only one titration system for both cationics and anionics: the benzethonium chloride, disulfine blue/dimidium bromide method which is standard for anionics. For the determination of cationics, an excess of standard sodium dodecylsulfate is added, and the excess determined by titration with benzethonium chloride (31).

In a wastewater stream, cationics are often present together with anionic surfactants. In such cases, only the excess of one or the other is available for determination by the standard method; the balance can be considered to already be titrated. Prior to analysis, the sample must be treated with ion exchange resin to remove either anionic or cationic surfactants. If, for example, cationics are removed and two titrations of anionics are performed, one with and one without cationic removal, the difference is equivalent to the concentration of cationics (100).

B. One-Phase Titration

The two-phase titration has disadvantages, the most serious being that the method is resistant to automation and the end point is operator sensitive. Foaming can be a problem in some systems, and some materials, such as polymeric surfactants or protein-based surfactants, do not form extractible ion pairs.

The titration of cationics may be conducted without extracting the ion pair into an organic solvent. Since the ion pair is insoluble in the aqueous titration medium, the methods below do not generally involve true one-phase titrations, but simply substitute the solid phase of a precipitate for the second liquid phase as a means to separate the ion pair.

1. Indicator Titration

Cationics may be titrated with tetraphenylborate in a single aqueous phase using 2,7-dichlorofluorescein as indicator (101). Dichlorofluorescein forms a complex with quaternary compounds, which dissociates, with a color change from pink to yellow, when the end point of the titration is reached. Since this is a precipitation reaction, alcohols and other water-miscible organic solvents must be avoided since they partially dissolve the precipitate and spoil the end point. By adjustment of pH as described above, the procedure may be used for the determination of amines other than quaternaries.

Quaternary amines may be titrated directly with stabilized polyvinyl sulfate, using the cationic dye toluidine blue O as indicator. A color change from purple to blue occurs at the end point (102).

Bromophenol blue indicator has maximum absorbance at 613 nm as an ion pair with a cationic surfactant, while the λ_{max} shifts to 592 nm for the free ion. Using a special apparatus to measure the difference between the two absorbances, the true titration end point is found at the maximum difference. This procedure is not affected by the turbidity of the solution, which rises to a maximum prior to the true end point (59).

If the precipitation of the ion pair is inhibited by addition of a nonionic surfactant, a one-phase titration is possible using as end point the color change of an anionic dye, tetrabromophenolphthalein ethyl ester (103). The indicator color change is from blue (ion pair with a cationic) to yellow (free acid). Since more hydrophobic cationics are slow in breaking the complex with the indicator and forming one with the anionic titrant, this determination is best made as a back-titration, where an excess of anionic surfactant is added and the excess determined by titration with a cationic. An octylphenol 10-mole ethoxylate at 0.02% gave good results in keeping the solution clear enough for photometric determination of the end point at 630 nm. pH is buffered at 3.2–4.0 because of the pK_a of the indicator. The method fails with cationics with alkyl chain lengths less than C_{14}.

2. Potentiometric Titration

Potentiometric titration of cationics is well established, using either tetraphenylborate or anionic surfactants as titrants (104). Tetraphenylborate generally gives sharper end points because of the larger potential jump (105). However, it is more subject to interference.

(a) Titration with tetraphenylborate. Tetraphenylborate titration can be conducted potentiometrically using a metal electrode. A potentiometric end point makes the procedure amenable to automation and eliminates consideration of the color of solution. The biggest advantage is that the titration now occurs in a single phase. Two metal electrodes have been described, made from silver and platinum. The polymer membrane electrodes designated as surfactant-selective electrodes are also used for this purpose (47).

The silver electrode responds to the reaction of excess tetraphenylborate with the equilibrium amount of silver ion in solution (106). Pinzauti and La Porta describe an electrode consisting of a spiral prepared from silver wire, 1 × 300 mm. The reference was a mercury(I) sulfate electrode. These workers found it necessary to clean the metal electrode with nitric acid followed by KCN solution between titrations. For analysis, a quantity of sample containing about 0.01 meq cationic surfactant was adjusted to the proper pH and diluted to a volume of about 50 mL. Titration was conducted with 0.001 M aqueous sodium tetraphenylborate solution, using an automatic recording potentiometric titrator. It is recommended that the titration be conducted at pH 3.4 for benzethonium chloride, methylbenzethonium chloride, and benzyldodecylbis(2-hydroxyethyl) ammonium chloride; at neutral pH (sodium acetate solution) for cetylpyridinium chloride; and at low pH (1 mL acetic acid added) for benzyldimethylmyristylammonium chloride, benzalkonium chloride, cetyltrimethylammonium bromide, and dequalinium chloride. Bromide salts give smaller potential breaks at the end point than chloride, due to the lower solubility of silver bromide. Buschmann and coworkers report that the most accurate results are obtained when the tetraphenylborate solution is titrated with the cationic, rather than vice versa (47).

Conventional platinum ring, disk, or foil electrodes are used to monitor the end point of the tetraphenylborate titration (107,108). The reference electrode may be either a platinum wire in the buret tip immersed in the titration solution or a Ag/AgCl or similar commercial electrode. In the latter case, the reference should be isolated from the titration solution by a double junction arrangement. The water-insoluble ion pair resulting from the titration tends to foul the electrodes, so Donkerbroek and Wang recommend vigorous stirring and maintenance of a temperature above 70°C. The electrode and buret tip should be cleaned with acetone between titrations. They observe that a source of error in the titration is a non-zero "blank" value. This error is eliminated by spiking each sample with the same volume of a standard quaternary solution (benzethonium chloride, of a concentration to give a 2-mL titration), so that the blank titration value can be accurately measured and a correction made. A major advantage of this platinum electrode system is that dialkyldimethylammonium quats may be determined directly, rather than by the back-titration required with membrane electrodes (108). A collaborative study of potentiometric tetraphenylborate titration for assay of quats showed a coefficient of variation of less than 1% (109).

Some specialty cationics which are not sufficiently hydrophobic for direct potentiometric titration with tetraphenylborate can be determined indirectly. An excess of sodium tetraphenylborate is added to precipitate the surfactant, which is removed by filtration. Excess tetraphenylborate ion in the filtrate is then determined by titration with thallium(I) nitrate (110). This last procedure was used for determination of the amount of a cationic surfactant fixed to cotton fibers in a textile dying bath.

(b) Titration with anionic surfactants. Polymer membrane electrodes are suitable for monitoring the titration of a cationic with an anionic surfactant (30). The electrodes are discussed later in this chapter. The ASTM procedure for determination of quats in fabric softeners is based on titration with lauryl sulfate, using the Orion nitrate-selective electrode (a polymer membrane electrode) for end point detection (111). With this method, diamidoamine-based quats may be titrated directly. However, dialkyldimethylammonium quats must be determined indirectly because their low solubility makes membrane electrodes unsuitable for end point detection by direct titration (108). An excess of lauryl sulfate is added, with the excess determined by back-titration with benzethonium chloride (111). Many practitioners add a little nonionic surfactant to suspend the precipitate so that the electrode does not become fouled. Shulz and Gerhards recommend instead adding 5% methanol to the titration medium to reduce electrode fouling (48).

Ester quats may also be titrated, although their comparatively high water solubility leads to less sharply defined end points (48). In general, cationics are titrated at pH 10 except for the ester quats, which are titrated at pH 6 to avoid hydrolysis of the esters (10).

(c)   Titration with other compounds.   Other anionic compounds, such as congo red, methyl orange, and sodium picrate, form water-insoluble complexes with cationic surfactants and have been proposed as reagents for their titration (112). In most cases, these offer no advantage over the titrants mentioned above.

The heteropoly compound octapotassium 1-cobalto(II)[1-aqua]-1-molybdo-16-tungsto-2-phosphate, $K_8[P_2CoMoW_{16}O_{61}(H_2O)] \cdot 19H_2O$ is recommended as a superior titrant because it complexes cationics more strongly that do other compounds (92). For example, very high concentrations of phosphate ion obscure the end point during determination of cationics using conventional titrants, requiring that the phosphate be precipitated with lead ion prior to analysis, which is conducted with high stirring to keep the precipitate suspended (113). The need to precipitate the phosphate is eliminated if the heteropoly titrant is used (92).

3.   Turbidimetric Titration

See the discussion in Section I of this chapter on titration of anionic surfactants.

IV.   AMPHOTERIC SURFACTANTS

Acid-base potentiometric titration is applicable to amphoterics. Typically, either an excess of acid or of base is added, and the titration is conducted with base or acid, respectively. The first inflection is due to the excess acid or base and subsequent inflections are due to end points associated with the surfactant. Hydrolysis and consequent blurring of the end point can be minimized by using a highly alcoholic solvent (114). Since other acidic or basic components will interfere, this approach is only applicable to concentrated surfactant solutions, not to formulations. In fact, impurities in the concentrated surfactant will interfere, which sometimes limits the usefulness of titration for assay. For determination of amphoterics in formulations, ion-pair titration is often used.

A.   Two-Phase Titration

Many amphoterics behave as cationic surfactants at acid pH, allowing them to be determined by the standard two-phase titration method. The mixed indicator method for the analysis of ionic surfactants, as described by Reid, Longman, and Heinerth, is adequate, but emulsions form in the vicinity of the end point so that precise analysis is difficult (115). Sometimes these problems can be minimized by reversing the titration. For example, a solution of sodium lauryl sulfate, buffer, and indicator is titrated with an aqueous sample containing a quaternary or amphoteric surfactant. The same end point is observed as described under the titration of anionics (94,95). Further improvement can be made by carefully adjusting the amount of ethanol added to the titration flask, depending on the individual surfactant being analyzed (116).

Some amphoterics can be titrated at high pH as anionic surfactants, using benzethonium chloride titrant. Betaines do not lose their cationic nature at high pH and are not measured.

An amperometric end point may be used with the two-phase titration of amphoterics at low pH with dodecylsulfate. Iron(II) 1,10-phenanthroline is added as the redox indicator. At the end point of the titration, the cationic indicator is extracted from the aqueous phase, resulting in the amperometric change. Sulfobetaines must be determined by back-titration, since they do not form ion pairs with anionics until the indicator is consumed (83). Taurides and sarcosine derivatives behave as amphoterics, rather than anionics, under these conditions and are likewise determined.

A word of caution: most paired-ion titrations of amphoterics will also measure amine impurities (117). It is difficult to generalize about amphoterics; some cannot be determined by titration with anionics (48,114).

Procedure: Two-phase Titration of Zwitterionics (116)

Acid mixed indicator solution: To a 250-mL volumetric flask add 0.050 g dimidium bromide dissolved in 15 mL hot 10:90 EtOH/H_2O and 0.050 g disulphine blue V which has been dissolved in 10 mL hot 10:90 EtOH/H_2O. Add about 100 mL water and 10 mL 2.5 M H_2SO_4, then dilute to volume with water. Store protected from light.

An aqueous solution of the sample is prepared at about 0.001 M. Ten milliliters are added to a glass-stoppered conical flask, together with 10 mL acid mixed indicator solution, 0.235 mL concentrated H_2SO_4, 15 mL $CHCl_3$, and 5.0 mL 95% ethanol. Titrate with 0.001 M dodecanesulfonate solution, shaking vigorously after each addition of titrant. At the end point, the chloroform phase is light pink.

Discussion: The end point of the titration is very dependent on the ethanol concentration. For each new zwitterionic analyzed, it is necessary to calibrate the method with a substance of known concentration. If the end point comes too soon, giving low values, the amount of ethanol added must be increased. Conversely, if the end point is too late, giving high values, the ethanol concentration must be decreased. Similarly, fatty alcohol impurities lead to high results, so the level of ethanol required should be determined for the commercial product, not on a specially purified material.

B. One-Phase Titration

All that was said above of two-phase titration of amphoterics is generally true of one-phase titration using a surfactant-selective electrode. Amphoteric surfactants containing carboxylate groups, including cocoamidopropylbetaine, dimethyllaurylammonium betaine, cocodimethylbetaine, dihydroxyethyl tallow glycinate, and some imidazolium compounds, can be successfully titrated at low pH with tetraphenylborate using a membrane electrode (81,105) or a coated-wire electrode (110) for end point detection. Because of the differences in solubility products of the tetraphenylborate salts, it is sometimes possible to simultaneously titrate a cationic and an amphoteric (Fig. 1). Two breaks are observed, the first corresponding to the cationic and the second to the amphoteric (81). A surfactant usually considered to be anionic, *N*-lauroylsarcosine, responds as an amphoteric in this system and can be readily determined.

Betaines are best titrated with tetraphenylborate, since the solubility of the complex is much lower than is the case with an anionic surfactant titrant (118). Of course, one has to remember that potassium and ammonium ions interfere with the tetraphenylborate titration. Tetraphenylborate titration can be used with alkylamidobetaines of alkyl chain length C_8 at concentrations higher than 0.2%, this being determined by the solubility of the ion pair. For the less soluble, higher chain length homologs, the detection limit is correspondingly lower (118). A protective colloid like gum arabic inhibits precipitation onto the electrode and reduces electronic noise during potentiometric titrations.

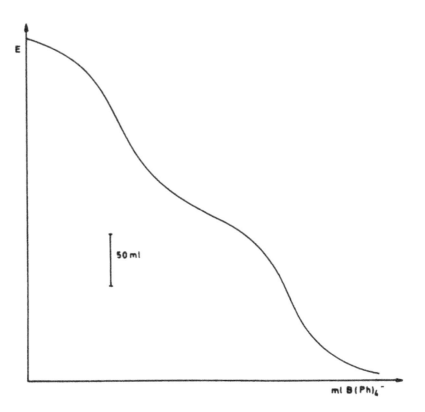

FIG. 1   Potentiometric titration with sodium tetraphenylborate of a mixture of a cationic (cetylpy-ridinium chloride) and an amphoteric (cocoamidopropylbetaine) surfactant, followed with a spe-cially made PVC-membrane electrode. The first inflection is due to titration of the cationic and the second to titration of the amphoteric surfactant. (Reprinted with permission from Ref. 81. Copyright 1992 by Karl Hanser Verlag.)

V.   ELECTRODES TO MONITOR TITRATIONS

A.   Electrode Design

1.   Physical Form

An electrode sensitive to both cationic and anionic surfactants can be made from a glass tube sealed at the bottom with a membrane of poly(vinylchloride) containing a high per-centage of plasticizer, 40% or more. Dissolved in the plasticized PVC is a surfactant ion pair, such as hexadecyltrimethylammonium dodecylsulfate. The tube is filled with a dilute solution of anionic surfactant with a small amount of a chloride salt, into which a silver/silver chloride electrode is inserted. For potential measurements in solutions, the circuit is completed with a reference electrode such as saturated calomel. Commercial surfactant-selective electrodes based on PVC membranes have been available since approximately 1990.

The PVC-membrane electrode can be made in different forms. A particularly rugged version is the coated-wire electrode, in which the PVC covers a length of metal or graphite rod (30,119). A similar form of solid state electrode is made by applying the PVC mem-brane to a substrate of conductive epoxy resin (120,121). A flow-through electrode suitable for flow injection analysis can be made by coating the membrane on the inside of a tube of carbon or carbon-containing polymer (122,123).

An ion-selective field effect transistor device has been invented to serve as an anionic or cationic surfactant-selective electrode. In this case, a plasticized PVC membrane is made incorporating a surfactant salt. This membrane is cast over the gates area of a field effect transistor, the whole (except the sensing surface) is encapsulated in epoxy resin, and a suitable measuring circuit is connected. Application is similar to the polymer membrane electrodes described above (124–126).

A specially made electrode is now commercially available which is much more tolerant of solvents than the conventional PVC-membrane electrodes. It is in the form of a coated-wire electrode, made with the ion pair incorporated into a silicone polymer by chemical reaction (1,127). This permits use in titration media where high organic solvent concentrations are necessary, as for example when the sample matrix is oily or when an ionic surfactant must be determined in the presence of a very high concentration of non-ionic surfactant. The electrode can even be used if two phases are present, analogous to the two-phase titration discussed at the beginning of this chapter. As of this writing, the electrode is only available from Metrohm, but presumably competitive products will be available.

2. Membranes

(a) Ionic surfactants. PVC membranes for surfactant electrodes are made by slowly evaporating the solvent from a solution containing PVC resin, plasticizer, ion pair, and perhaps other additives. Even without incorporation of an ion pair, ordinary plasticized PVC is suitable for monitoring the titration of cationics and anionics with each other. This is because the lipophilic ion pair formed in the titration will migrate into the membrane, attaining an equilibrium value. Thus, after the membrane is conditioned by a few preliminary titrations, it performs exactly as one to which an ion pair has been deliberately added. Because of this, the first electrodes widely used for surfactant analysis were commercial ion-selective electrodes developed for other purposes entirely. These included the nitrate ion-selective electrode, the calcium and potassium selective electrodes, the fluoroborate electrode, and the perchlorate electrode. These commercial electrodes were all polymer-membrane electrodes based on Ag/AgCl internals, as described above, and were used for titration end point detection after conditioning in a solution of the surfactant to be analyzed. Since their manufacturers do not test them or warrant them for surfactant determination, their use is becoming uncommon now that special electrodes are marketed specifically for surfactant applications.

Many papers have been published over the past two decades demonstrating that electrodes can be made more selective to particular surfactants by incorporating additives into the membane. For example, relatively more selectivity is attained to anionic or cationic surfactants by incorporating into the membrane a salt of a short chain length anion or cation with a long chain, comparatively less mobile, cationic or anionic counter ion, respectively. A typical anion-selective membrane might incorporate tetrabutylammonium dodecylsulfate, while a cation-selective membrane might contain hexadecyltrimethylammonium 1-pentanesulfonate (128). Other counter ions for cationic determination are reineckate and phosphotungstate (126,129). One group recommends incorporating into the membrane the salt of sodium lauryl sulfate and the Metrohm TEGOtrant®, 1,3-didecyl-2-methylimidazolium chloride (130).

(b) Nonionic surfactants. Since polyethoxylate chains can complex with certain cations, ion-selective electrodes may be devised to respond to ethoxylate concentration. An example is an electrode based on a PVC membrane incorporating the tetraphenylborate salt of a barium/ethoxylated nonylphenol complex. This electrode responds to the concentration of other ethoxylated surfactants. A comprehensive description has been published of the fabrication and use of PVC-membrane electrodes for nonionic surfactant analysis (131,132).

These experimental electrodes with surfactants incorporated into the membranes can be used to monitor the titration of nonionics (as their Ba^{2+} complexes) with tetraphenylborate (133). Ivanov and Pravshin report that an electrode suitable for determination of nonionics can be made from a membrane of plasticized PVC containing either cetylpyridinium dodecylsulfonate or the complex of barium and tetraphenylborate ions with a nonionic (134). These electrodes also respond to ionic surfactants and respond much better to tetraphenylborate than to nonionic surfactants. Thus, they are most useful for monitoring the end point of a titration of the barium complexes of nonionics with tetraphenylborate. Ionic surfactants interfere in the titration.

The Orion PVC-membrane surfactant electrode can be used for end point detection in the titration of ethoxylated nonionics (in the presence of barium ion) with tetraphenylborate (63,64), as described earlier in this chapter.

(c) General remarks. Most surfactant titrations are performed using automatic titrators. Please note that titrator settings, particularly the rate of titrant addition near the end point, must be optimized for the particular surfactant. This is because response of the surfactant electrode is neither as rapid nor as large as with a pH or redox electrode. This means that an automatic titrator may very well add titrant too rapidly in the end point region because the internal logic does not recognize that the gradual response change indeed corresponds to the region of the end point (1).

Surfactant-selective electrodes from different manufactures contain different additives in the membranes and necessarily differ from each other in their response to particular compounds. In particular, they differ in their response time and in the potential difference registered during a titration. This means that titration conditions must be optimized for each type of electrode, as well as for each surfactant (1). Care must be taken with other aspects of the titration also. The electrode should be rinsed with methanol and then water after each titration if it has a tendency to become coated with the ion pair.

Since commercial surfactant electrodes are available, there is no reason for the analyst to prepare membranes and construct electrodes. Considerable know-how is involved in producing durable electrodes which behave reliably, so we should leave this to the instrument manufacturers. To illustrate this point, we may mention a study in which six electrodes were compared for use in end point detection of a titration of a cationic with an anionic. Two commercial surfactant-selective electrodes, two tetrafluoroborate selective electrodes, a nitrate selective electrode, and a homemade PVC-membrane electrode incorporating a tetraphenylborate salt were tested (135). Only one electrode, one of the commercial models, was truly suited for routine use, giving smooth potentiometric curves without reconditioning for over 100 titrations. The relative standard deviation of the end point was about 0.5%, while that for the other electrodes was 1.3–2.3%. The standard deviation of the end point potential was 3 mV for this electrode, compared to 8 mV or more for the other electrodes. Besides this, those electrodes not designed as surfactant electrodes required reconditioning (i.e., soaking in a dispersion of a surfactant ion pair) after 25 titrations in order to remain usable.

Even PVC-membrane electrodes designed for surfactant use must be "conditioned" by performing a few preliminary titrations before their response is suitable for use in quantitative titration. PVC electrodes are best used in water; solvents can swell the polymer and remove plasticizer and ion pairs. Schulz recommends no more than 10% alcohols in the titration mixture, with 15% ethanol or 30% methanol permitted (1).

B. Electrode Applications

1. Direct Measurement of Surfactant Concentration

The literature contains many studies of experimental PVC-membrane electrodes containing novel ionic materials to enhance sensitivity to surfactants. Quite properly, these studies demonstrate the quality of the final electrode by showing the millivolt response as a function of surfactant concentration, and many show Nernstian or near-Nernstian response. This might lead one to think that surfactant concentration could be determined as easily as pH, or at least as easily as certain ions for which ion-selective electrodes are available: the calibrated electrode would be dipped into the test solution, and the electrode potential would be recorded and converted directly to concentration by the millivolt/pH meter.

This approach will not be practical for some time to come. The fundamental properties of surfactants (micelle formation, enrichment at interfaces) mean that the activity of a surfactant will usually differ from its absolute concentration (1). Just as serious is the technical problem that current surfactant-selective electrodes suffer from response which varies with their past and recent history; they are also sensitive to the concentration of nonsurfactant ions. The result is that quantitative applications use electrodes not in direct measurements relating potential to concentration, but as indicators of the end point of a titration. In this latter application, it is not important that the electrode potential be exactly reproducible, but only that the potential change sharply as the surfactant concentration changes. For the titration of an anionic surfactant with a cationic surfactant, the electrode used for end point detection can be chosen to respond to either surfactant. Because of the drift in electrode potential, titrations must be conducted to an inflection in the titration curve rather than to a specific millivolt value. Details of the potentiometric titration methods can be found earlier in this chapter. The electrodes have also been demonstrated as detectors for flow injection analysis.

The user should remain aware that surfactant-selective electrodes respond in nonlinear fashion in the range of the critical micelle concentration. The activity of individual surfactant molecules may actually decrease as the total concentration is raised to just over the CMC. If an organic solvent is used in place of water, the formation of micelles may not occur or may occur at higher concentration, but dimers and other aggregates may form. Even for titrations, electrodes are most reliably used below the critical micelle concentration of the surfactant.

The literature of direct electrode measurement of surfactant concentration has been reviewed elsewhere (136,137).

2. Flow Injection Analysis

An automated version of potentiometric titration can be used with flow injection analysis. In this application, termed pseudotitration (138), the mobile phase of the flow injection analysis system contains a constant concentration of the titrant, which can be either an anionic or cationic surfactant. The sample, containing either a cationic or anionic surfactant, is passed through an exponential dilution chamber prior to mixing with the mobile phase. The detection system is a flow-through electrode sensitive to the ionic surfactant to be determined. The peak width, rather than the peak height, of the signal is proportional to the logarithm of the surfactant concentration. The system must be calibrated with the specific surfactant being determined, and electrode life currently is 2–4 weeks (122).

3. HPLC Detection

Experimental surfactant-selective electrodes have been demonstrated as detectors for HPLC determination of alkylsulfates (139) and alkyl ether sulfates (140). The column effluents were diluted with water to prevent damage to the membrane from high methanol concentration. Such electrodes are subject to gradual deterioration of response and cannot be recommended for routine use.

REFERENCES

1. Schulz, R., *Titration of Surfactants and Pharmaceuticals* (in German), Verlag für chemische Industrie, H. Ziolkowsky GmbH, Augsburg, 1996.
2. Cross, J., Volumetric analysis of anionic surfactants, in J. Cross, ed., *Anionic Surfactants: Analytical Chemistry*, 2nd ed., Marcel Dekker, Inc., New York, 1998.
3. Griffin, E. H., E. W. Albaugh, α-olefin-derived sodium sulfonates by ion exchange and potentiometric titrimetry, *Anal. Chem.*, 1966, *38*, 921–923.
4. Bares, M., Soap in detergents, *Tenside*, 1969, *6*, 21–24.
5. Zarins, Z. M., J. L. White, R. O. Feuge, Hydrogen bromide titration for soaps in fat products, *J. Am. Oil Chem. Soc.*, 1983, *60*, 1109–1111.
6. Reid, V. W., G. F. Longman, E. Heinerth, Anionic-active detergents by two-phase titration, *Tenside*, 1967, *4*, 292–304.
7. Reid, V. W., G. F. Longman, E. Heinerth, Anionic detergents by two-phase titration (II), *Tenside*, 1968, *5*, 90–96.
8. American Society for Testing and Materials, Synthetic anionic ingredient by cationic titration, D3049. West Conshohocken, PA 19428.
9. International Organization for Standardization, Surface active agents—detergents—determination of anionic-active matter by manual or mechanical direct two-phase titration procedure, ISO 2271:1989. Available from national ISO affiliates.
10. Walter, C. M., C. A. De Caro, Tenside titration, *World Surfactants Congr., 4th*, 1996, *1*, 432–445.
11. Buschmann, N., H. Starp, Titrimetric determination of paraffin sulfonates, *Tenside, Surfactants, Deterg.*, 1997, *34*, 84–94.
12. Epton, S. R., Surface-active agents, *Nature*, 1947, *160*, 795–796.
13. Epton, S. R., Titrimetric analysis of sodium alkyl sulfates and related compounds, *Trans. Faraday Soc.*, 1948, *44*, 226–230.
14. Eppert, G., G. Liebscher, Two-phase titration for determination of anionic surfactants (in German), *Z. Chem.*, 1978, *18(5)*, 188–189.
15. Markó-Monostory, B., S. Börzsönyi, Oil-soluble sulphonic acids and sulphonates by two-phase titration with carbethoxypentadecyltrimethylammonium bromide (Septonex), *Tenside*, 1985, *22*, 265–268.
16. Spilker, R., Hüls AG, Marl, Germany. Personal communication, 1993.
17. Lew, H. Y., Detergent mixtures containing amine oxides, *J. Am. Oil Chem. Soc.*, 1964, *41*, 297–300.
18. Milwidsky, B. M., S. Holtzman, Soaps/syndets, *Soap Chem. Spec.*, 1966, *42(5)*, 83–86,154–158.
19. Cullum, D. C., P. Platt, Analysis of formulated products, in M. R. Porter, ed., *Recent Developments in the Analysis of Surfactants*, Elsevier, New York, 1991.
20. Tsubouchi, M., N. Yamasaki, K. Matsuoka, Anionic surfactants by two-phase titration with tetrabromophenolphthalein ethyl ester as indicator, *J. Am. Oil Chem. Soc.*, 1979, *56*, 921–923.

21. Daradics, L., J. Pálinkás, Fatty acid sarcosides, *Tenside, Surfactants, Deterg.*, 1994, *31*, 308–313.

22. Han, K. W., Two-phase titration method for determination of anionic surfactants (in German), *Tenside*, 1966, *3*, 265–269.

23. Battaglini, G. T., J. L. Larsen-Zobus, T. G. Baker, Alpha sulfo methyl tallowate, *J. Am. Oil Chem. Soc.*, 1986, *63*, 1073–1077.

24. Marcinkiewicz-Salmonowicz, J., A. Górska, W. Zwierzykowski, α-Sulfonated fatty acid methyl esters, *Tenside, Surfactants, Deterg.*, 1995, *32*, 240–242.

25. Hoffmann, A. R., W. W. Böer, G. W. G. Schwarz, Two-phase titration of anionic surfactants using an improved stirring technique (in German), *Fette, Seifen, Anstrichm.*, 1976, *78*, 367–368.

26. Seguran, P., Turbidimetric determination of anionic surfactants, *Tenside*, 1985, *22*, 67–70.

27. Hendry, J. B. M., A. J. Hockings, Photometric determination of sulfated and sulfonated anionic surfactants, *Analyst*, 1986, *111*, 1431–1433.

28. Pinazo, A., X. Domingo, Turbidimetric analysis of amine oxides and amine oxide–anionic surfactant mixtures, *J. Am. Oil Chem. Soc.*, 1996, *73*, 143–147.

29. De Caro, C. A., Automated surfactant titration: a review, *Riv. Ital. Sostanze Grasse*, 1998, *75*, 197–205.

30. Vytras, K., Coated wire electrodes in the analysis of surfactants, *Electroanalysis*, 1991, *3*, 343–347.

31. Lerschmacher, P., U. Altenhofen, H. Zahn, Alkylsulfonates and quaternary ammonium compounds on wool (in German), *Tenside*, 1981, *18*, 306–308.

32. International Organization for Standardization, Surface active agents—detergents—anionic-active matter stable to acid hydrolysis—determination of trace amounts, ISO 2868:1973. Geneva, Switzerland.

33. International Organization for Standardization, Surface active agents—detergents—determination of anionic-active matter hydrolyzable and non-hydrolyzable under acid conditions, ISO 2870:1986. Geneva, Switzerland.

34. International Organization for Standardization, Surface active agents—detergents—anionic-active matter hydrolyzable under alkaline conditions—determination of hydrolyzable and non-hydrolyzable anionic-active matter, ISO 2869:1973. Geneva, Switzerland.

35. König, H., E. Walldorf, α-Sulfofatty acid methyl esters and fatty alcohol sulfoacetates (in German), *Fresenius' Z. Anal. Chem.*, 1975, *276*, 365–370.

36. Li, Z., M. J. Rosen, Two-phase mixed indicator titration method for the determination of anionic surfactants, *Anal. Chem.*, 1981, *53*, 1516–1519.

37. Wickbold, R., Alkanesulfonates (in German), *Tenside*, 1971, *8*, 130–134.

38. Lew, H. Y., Surfactant analysis, *J. Am. Oil Chem. Soc.*, 1972, *49*, 665–670.

39. International Organization for Standardization, Surface active agents—technical alkane sulfonates—determination of alkane monosulfonates content by direct two-phase titration, ISO 6121:1988. Geneva, Switzerland.

40. Takano, S., A. Hasegawa, H. Ohotsuka, Determination of soap in anionic surfactant mixtures and amine and quaternary ammonium salts in cationic surfactant mixtures by using an automated photometric titrator (in Japanese), *Bunseki Kagaku*, 1988, *37(3)*, 137–141.

41. König, H., Sulfosuccinic acid half-esters (in German), *Fresenius' Z. Anal. Chem.*, 1971, *254*, 198–209.

42. Tsubouchi, M., J. H. Mallory, Differential determination of cationic and anionic surfactants in mixtures by two-phase titration, *Analyst*, 1983, *108*, 636–639.

43. Tsubouchi, M., Y. Yamamoto, Anionic surfactants in presence of cationic surfactants by two-phase titration, *Anal. Chem.*, 1983, *55*, 583–584.

44. Kissa, E., Anionic fluorinated surfactants, in J. Cross, ed., *Anionic Surfactants: Analytical Chemistry*, 2nd ed., Marcel Dekker, New York, 1998.

45. Buschmann, N., Accuracy of potentiometric titration of anionic surfactants, *Comun. Jorn. Com. Esp. Deterg.*, 1995, *26*, 215–225.

46. Buschmann, N., Titrimetric determination of ionic surfactants (in German), *Tenside, Surfactants, Deterg.*, 1995, *32*, 504–506.

47. Buschmann, N., U. Görs, R. Schulz, Comparison of different cationic titrants for the potentiometric determination of anionic surfactants, *Comun. Jorn. Com. Esp. Deterg.*, 1993, *24*, 469–476.

48. Schulz, R., R. Gerhards, Potentiometric titration of ionic detergents, *Amer. Lab.*, 1994, *26(11)*, 40–44.

49. Schulz, R., R. Gerhards, Titrant for the potentiometric titration of anionic surfactants, *Tenside, Surfactants, Deterg.*, 1995, *32*, 6–11.

50. Makowezkaja, L. I., I. L. Majslina, S. P. Wolosowitzsch, Ion-selective electrodes with anionic surfactant functionality: use in analysis of laundry products (in German), *Seifen, Öle, Fette, Wachse*, 1991, *117*, 565–571.

51. Knittel, D., W. Schollmeyer, Influence of additives on the response of detergent electrodes in titrations of anionic surfactants, *SOFW J.*, 1998, *124*, 338,340–342.

52. Schulz, R., Potentiometric two-phase titration: influence of cosurfactants (in German), *Parfuem. Kosmet.*, 1998, *79(4)*, 20,22–23.

53. Gronsveld, J., M. J. Faber, Surfactant analysis in oil-containing fluids, *Tenside, Surfactants, Deterg.*, 1990, *27*, 231–232.

54. American Society for Testing and Materials, Active matter in anionic surfactants by potentiometric titration, D4251. West Conshohocken, PA 19428.

55. Zarapkar, S. S., K. R. Athalye, Titrimetric method for determination of sodium lauryl sulphate, sodium lauryl ether sulphate, triethanolamine lauryl sulphate and alkylbenzene sulphonate by titration with centrimide, domiphen bromide and oxyphenonium bromide using cobalt thiocyanate indicator, *Soaps, Deterg. Toiletries Rev.*, 1987, *17(10)*, 21–23.

56. Shiau, B., J. H. Harwell, J. F. Scamehorn, Precipitation of mixtures of anionic and cationic surfactants. III: Effect of added nonionic surfactant, *J. Colloid Interface Sci.*, 1994, *167*, 332–345.

57. Motomizu, S., M. Oshima, Y. Gao, S. Ishihara, K. Uemura, Indicator system for the photometric titration of ionic surfactants in an aqueous medium: determination of anionic surfactants with distearyldimethylammonium chloride as titrant and tetrabromophenolphthalein ethyl ester as indicator, *Analyst*, 1992, *117*, 1775–1780.

58. Bos, M., Tensammetric titration of ionic detergents, *Anal. Chim. Acta*, 1982, *135*, 249–261.

59. Majer, H. J., R. Schulz, G. Hägele, R. Gerhards, Photometric one-phase surfactant titration using the PHOTO T concept, *Tenside, Surfactants, Deterg.*, 1994, *31*, 74–78.

60. Cross, J, ed., *Nonionic Surfactants: Chemical Analysis*, Marcel Dekker, New York, 1987.

61. Vytras, K., V. Dvorakova, I. Zeman, Titrations of nonionic surfactants with sodium tetraphenylborate using simple potentiometric sensors, *Analyst*, 1989, *114*, 1435–1441.

62. Schulz, R., Potentiometric titration of nonionic surfactants with the surfactant-sensitive electrode (in German), *SÖFW-Jour.*, 1996, *122*, 1022,1024–1028.

63. Gallegos, R. D., Titrations of nonionic surfactants with sodium tetraphenylborate using the Orion surfactant electrode, *Analyst*, 1993, *118*, 1137–1141.

64. Dammer, H. U., Ion-selective electrode for determination of nonionic surfactants (in German), *Tenside, Surfactants, Deterg.*, 1995, *32*, 12–16.

65. Bürger, K., Detection and determination of surface active polyoxyethylene compounds and of polyethylene glycol (in German), *Fresenius' Z. Anal. Chem.*, 1963, *196*, 251–259; 1964, *199*, 434–438.

66. Tsubouchi, M., N. Yamasaki, K. Yanagisawa, Two-phase titration of poly(oxyethylene) nonionic surfactants with tetrakis(4-fluorophenyl)borate, *Anal. Chem.*, 1985, *57*, 783–784.

67. O'Connell, A. W., Titration of nonionic surfactants with potassium tetrakis(4-chlorophenyl)borate, *Anal. Chem.*, 1986, *58*, 669–670.

68. Hei, R. D., N. M. Janisch, Nonionic surfactants, *Tenside*, 1989, *26*, 288–290.

69. Cozzoli, O., C. Ruffo, G. Carrer, F. Pinciroli, M. Ciambella, M. C. Grillo, N. Cogliati, M. Fiorito, A. Gellera, N. Giannelli, V. Zamboni, Ethoxylated nonionic surfactants: comparison of the precision of the iodobismuthate method and other methods, *Riv. Ital. Sostanze Grasse*, 1993, *70*, 199–203.

70. Tsubouchi, M., Y. Tanaka, Poly(oxyethylene) non-ionic surfactants by two-phase titration, *Talanta*, 1984, *31*, 633–634.

71. Buschmann, N., Nonionic surfactants by displacement titration (in German), *Fresenius' Z. Anal. Chem.*, 1988, *330*, 700–703.

72. American Society for Testing and Materials, Ethylene oxide content of polyethoxylated nonionic surfactants, D2959. West Conshohocken, PA 19428.

73. Obruba, K., E. Kucerová, M. Jurecek, Determination of the oxyethylene group by cleavage with nascent hydriodic acid (in German), *Mikrochim. Ichnoanal. Acta*, 1964(1), 44–48.

74. Siggia, S., J. G. Hanna, *Quantitative Organic Analysis via Functional Groups*, 4th ed., Robert E. Krieger, Malabar, FL, 1988.

75. Reifler, M., P. A. Bruttel, Automatic titration of nonionic surfactants, *World Surfactants Congr., 4th*, 1996, *1*, 489–496.

76. Buschmann, N., F. Hülskötter, Titration of low ethoxylated nonionic surfactants, *Tenside, Surfactants, Deterg.*, 1997, *34*, 8–11.

77. Buschmann, N., F. Hülskötter, Potentiometric titration of carbohydrate surfactants: alkylpolyglucosides, *N*-acylglucamides, sorbitan esters, *Comun. Jorn. Com. Esp. Deterg.*, 1997, *27*, 419–425.

78. Metcalfe, L. D., Potentiometric titration of long chain amine oxides using alkyl halide to remove tertiary amine interference, *Anal. Chem.*, 1962, *34*, 1849.

79. Muth, C. W., R. S. Darlak, W. H. English, A. T. Hamner, Titration of aromatic *N*-oxides in acetic anhydride with perchloric acid, *Anal. Chem.*, 1962, *34*, 1163–1164.

80. Wimer, D. C., Titrimetric determination of some amine oxides and phosphine oxides in acetic anhydride, *Anal. Chem.*, 1962, *34*, 873–874.

81. Buschmann, N., R. Schulz, Cationic and zwitterionic surfactants using ion selective electrodes, *Tenside, Surfactants, Deterg.*, 1992, *29*, 128–130.

82. Walton, B., Potentiometric titration of long chain tertiary amine oxides and tertiary amines using a poly(vinyl chloride) membrane electrode, *Analyst*, 1994, *119*, 2202–2203.

83. Buschmann, N., Zwitterionic surfactants by electrochemically indicated amphimetry, *Tenside, Surfactants, Deterg.*, 1991, *28*, 329–332.

84. Glynn, E., Reductometric determination of the sulphoxide and amine oxide groups, *Analyst*, 1947, *72*, 248–250.

85. Sicilia, D., S. Rubio, D. Pérez-Bendito, Determination of surfactants based on mixed-micelle formation, *Anal. Chem.*, 1995, *67*, 1872–1880.

86. Cross, J., Volumetric analysis of cationic surfactants, in J. Cross and E. J. Singer, eds., *Cationic Surfactants: Analytical and Biological Evaluation*, Marcel Dekker, New York, 1994.

87. Morak, M., Titration of alkylammonium salts (in German), *Tenside*, 1989, *26*, 215–221.

88. Cross, J. T., Identification and determination of cationic surface-active agents with sodium tetraphenylboron, *Analyst*, 1965, *90*, 315–324.

89. Wang, L. K., Anionic or cationic surfactants in industrial water, *J. Am. Oil Chem. Soc.*, 1975, *52*, 339–344.

90. Tsubouchi, M., H. Mitsushio, N. Yamasaki, Cationic surfactants by two-phase titration, *Anal. Chem.*, 1981, *53*, 1957–1959.

91. Cross, J. T., Identification and determination of cationic surfactants, in E. Jungermann, ed., *Cationic Surfactants*, Marcel Dekker, New York, 1977.

92. Lowy, D. A., A. Patrut, M. E. Walter, Reagent for the potentiometry of cationic surfactants, *Process Control Qual.*, 1993, *4*, 125–137.

93. Jansson, S. O., R. Modin, G. Schill, Two-phase titration of organic ammonium ions with lauryl sulphate and methyl yellow as indicator, *Talanta*, 1974, *21*, 905–918.

94. International Organization for Standardization, Surface active agents—detergents—determination of cationic-active matter content. Part 1: High-molecular-mass cationic-active matter, ISO 2871-1:1988. Available from national ISO affiliates.

95. International Organization for Standardization, Surface active agents—detergents—determination of cationic-active matter content. Part 2: Cationic-active matter of low molecular mass (between 200 and 500), ISO 2871-2:1990. Available from national ISO affiliates.

96. Cantwell, F. F., H. Y. Mohammed, Photometric acid-base titrations in the presence of an immiscible solvent, *Anal. Chem.*, 1979, *51*, 218–223.

97. Mohammed, H. Y., F. F. Cantwell, Photometric ion-pair titrations in the presence of an immiscible solvent and their application to drug analysis, *Anal. Chem.*, 1979, *51*, 1006–1012.

98. Mohammed, H. Y., F. F. Cantwell, Two-phase photometric ion-pair titrations of drugs and surfactants, *Anal. Chem.*, 1980, *52*, 553–557.

99. Kubán, V., J. Jurasová, Extraction spectrophotometric and two-phase titrimetric determination of ionic surfactants with neutral red, *Scr. Fac. Sci. Nat. Univ. Purkynianae Brun.*, 1988, *18*, 159–166.

100. Ströhl, G. W., Simultaneous determination of anionic and cationic surfactants in wastewater (in German), *Tenside, Surfactants, Deterg.*, 1969, *6*, 78–79.

101. Metcalfe, L. D., R. J. Martin, A. A. Schmitz, Titration of long-chain quaternary ammonium compounds using tetraphenylboron, *J. Am. Oil Chem. Soc.*, 1966, *43*, 355–357.

102. *Product Catalog*, CHEMetrics. Route 28, Calverton, VA 22018.

103. Motomizu, S., Y. Gao, K. Uemura, S. Ishihara, Photometric titration of cationic surfactants in an aqueous medium, *Analyst*, 1994, *119*, 473–477.

104. American Society for Testing and Materials, Disinfectant quaternary ammonium salts by potentiometric titration, D5806. West Conshohocken, PA 19428.

105. Oei, H. H. Y., I. Mai, D. C. Toro, Anionic or cationic surfactants using a surfactant electrode, *J. Soc. Cosmet. Chem.*, 1991, *42*, 309–316.

106. Pinzauti, S., E. La Porta, Silver electrode in the potentiometric determination of quaternary ammonium compounds, *Analyst*, 1977, *102*, 938–942.

107. Donkerbroek, J. J., C. N. Wang, Cationic surfactant analysis, *Proc. 2nd World Surfactants Cong.*, Paris, 1988, *3*, 134–152.

108. Wang, C. N., L. D. Metcalfe, J. J. Donkerbroek, A. H. M. Cosijn, Potentiometric titration of long chain quaternary ammonium compounds using sodium tetraphenylborate, *J. Am. Oil Chem. Soc.*, 1989, *66*, 1831–1833.

109. Arens, M., J. G. A. Kooijman, H. P. Wingen, Potentiometric titration of surfactants—Determination of the active agent in quaternary fatty alkyl ammonium salts—Collaboration of the DGF. Communication 155: German standard methods for study of fats, fatty products, surfactants and related materials. Communication 121: Analysis of surface active materials XXVIII (in German), *Fett/Lipid*, 1996, *98*, 128–130.

110. Vytras, K., J. Kalous, J. Symerský, Ampholytic and cationic surfactants by potentiometric titration based on ion pair formation, *Anal. Chim. Acta*, 1985, *177*, 219–223.

111. American Society for Testing and Materials, Synthetic quaternary ammonium salts in fabric softeners by potentiometric titrations, D5070. West Conshohocken, PA 19428.

112. Tyagi, V., A. K. Jain, Estimation of cationic surfactants using liquid membrane electrodes, *Indian J. Chem., Sect. A: Inorg., Bio-Inorg., Phys., Theor. Anal. Chem.*, 1991, *30A*, 844–848.

113. Oniciu, L., D. A. Lowy, I. A. Silberg, D. F. Anghel, Potentiometric determination of cationic surfactants used in adiponitrile electrosynthesis, *Analusis*, 1986, *14*, 456–461.

114. Lomax, E. G., ed., *Amphoteric Surfactants*, 2nd ed., Marcel Dekker, New York, 1996.

115. König, H., Amphoteric surfactants (in German), *Fresenius' Z. Anal. Chem.*, 1970, *251*, 359–368.

116. Rosen, M. J., F. Zhao, D. S. Murphy, Two-phase mixed indicator method for the determination of zwitterionic surfactants, *J. Am. Oil Chem. Soc.*, 1987, *64*, 439–441.

117. Plantinga, J. M., J. J. Donkerbroek, R. J. Mulder, Betaine and free amine in alkyldimethyl betaine by potentiometric titrations, *J. Am. Oil Chem. Soc.*, 1993, *70*, 97–99.

118. Gerhards, R., I. Jussofie, D. Käseborn, S. Keune, R. Schulz, Analysis of cocoamidopropyl betaines, *Tenside, Surfactants, Deterg.*, 1996, *33*, 8–14.

119. Leuchte, H. W., H. D. Kahleyss, J. Faltermann, Ion-selective electrodes in surfactant analysis (in German), *Fett Wiss. Technol.*, 1992, *94*, 64–66.

120. Baró-Romà, J., J. Sánchez, M. del Valle, J. Alonso, J. Bartrolí, Construction and development of ion-selective electrodes responsive to anionic surfactants, *Sens. Actuators, B*, 1993, *15*, 179–183.

121. Alegret, S., J. Alonso, J. Bartrolí, J. Baró-Romà, J. Sánchez, M. del Valle, Solid-state ion-selective electrode for automated titration of anionic surfactants, *Analyst*, 1994, *119*, 2319–2322.

122. Dowle, C. J., B. G. Cooksey, J. M. Ottaway, W. C. Campbell, Ionic surfactants by flow injection pseudotitration, *Analyst*, 1988, *113*, 117–119.

123. Alonso, J., J. Baró, J. Bartrolí, J. Sànchez, M. del Valle, Flow-through tubular ion selective electrodes responsive to anionic surfactants for flow injection analysis, *Anal. Chim. Acta*, 1995, *308*, 115–121.

124. Campanella, L., M. Battilotti, A. Borraccino, C. Colapicchioni, M. Tomassetti, G. Visco, ISFET device responsive to anionic detergents, *Sens. Actuators, B*, 1994, *19*, 321–328.

125. Campanella, L., C. Colapicchioni, L. Aiello, M. Tomassetti, Solid state sensors (ISFETs) for environmental analysis, *Sens. Microsyts., Proc. Ital. Conf.*, *1st*, 1996, 157–162.

126. Campanella, L., L. Aiello, C. Colapicchioni, M. Tomassetti, Cationic surfactants in environmental aqueous matrices by new ISFET devices, *Anal. Lett.*, 1997, *30*, 1611–1629.

127. Schulz, R., J. Thiede, Titration of ionic surfactants in cooling lubricants, *Amer. Lab.*, 1999, *31(21)*, 12,14.

128. Dowle, C. J., B. G. Cooksey, J. M. Ottaway, W. C. Campbell, Ion-selective electrodes for use in the titration of ionic surfactants in mixed solvent systems, *Analyst*, 1987, *112*, 1299–1302.

129. Campanella, L., L. Aiello, C. Colapicchioni, M. Tomassetti, Ion-selective polymeric membranes for cationic surfactant analysis, *Analusis*, 1996, *24*, 387–391.

130. Gerlache, M., Z. Sentürk, J. C. Viré, J. M. Kauffmann, Potentiometric analysis of ionic surfactants by a new type of ion-selective electrode, *Anal. Chim. Acta*, 1997, *349*, 59–65.

131. Moody, G. J., J. D. R. Thomas, Potentiometry of oxyalkylates, in *Nonionic Surfactants: Chemical Analysis*, J. Cross, ed., Marcel Dekker, New York, 1987.

132. Alexander, P. H. V., G. J. Moody, J. D. R. Thomas, Electrode membrane and solvent extraction parameters relating to the potentiometry of polyalkoxylates, *Analyst*, 1987, *112*, 113–120.

133. Sugawara, M., S. Nagasawa, N. Ohashi, Potentiometric sensors based on tetraphenylborates of calcium and barium complexes with poly(oxyethylene)mono(6-methylheptyl)phenyl ether for the determination of nonionic surfactants, *J. Electroanal. Chem.*, 1984, *176*, 183–194.

134. Ivanov, V. N., Y. S. Pravshin, Nonionic surfactants by using ion-selective electrodes, *Zh. Anal. Khim*, 1986, *41*, 360–364.

135. Buschmann, N., R. Schulz, Comparison of ion selective electrodes for the titrimetric determination of ionic surfactants, *Tenside, Surfactants, Deterg.*, 1993, *30*, 18–23.

136. Moody, G. J., J. D. R. Thomas, Potentiometry of cationic surfactants, in J. Cross and E. J. Singer, eds., *Cationic Surfactants: Analytical and Biological Evaluation*, Marcel Dekker, New York, 1994.

137. Gerlache, M., J. M. Kauffmann, G. Quarin, J. C. Vire, G. A. Bryant, J. M. Talbot, Electrochemical analysis of surfactants: an overview, *Talanta*, 1996, *43*, 507–519.

138. Stewart, K. K., A. G. Rosenfeld, Exponential dilution chambers for scale expansion in flow injection analysis, *Anal. Chem.*, 1982, *54*, 2368–2372.

139. Masadome, T., T. Imato, N. Ishibashi, Surfactant-selective electrode based on plasticized poly(vinyl chloride) membrane and its application, *Anal. Sci.*, 1987, *3*, 121–124.

140. Gerlache, M., J. M. Kauffmann, PVC-based ion-selective electrode for surfactant detection in microflow systems, *Biomed. Chromatogr.*, 1998, *12*, 147–148.

17
Miscellaneous
Analytical Methods

I. POLAROGRAPHY

Electrochemical measurement of surfactant concentration by polarography and ion-selective electrodes is the subject of a recent review (1). While direct measurement does not promise to have practical application except perhaps for HPLC detection, it is briefly described here for the sake of completeness.

Surfactants are commonly added to polarographic cells to suppress the overpotential associated with dissolved oxygen or other species. This phenomenon is periodically proposed as a general quantitative method for determination of surfactants in environmental samples (2–4). The approach is not useful for determining which surfactant is present. In natural waters, naturally occurring materials like humic acids are the predominant surface active materials detected by polarographic methods (5).

The effect of surfactants on the capacitance of an electrode may also be exploited, the exploitation sometimes being called "tensammetry." A mercury or mercury-coated electrode is most often used. The electrode response is generally a nonlinear function of the concentration of the adsorbing species (6–11). Through sophisticated electronics, this capacitance may be monitored during the titration of an anionic with a cationic, or vice versa, to observe the end point without use of a two-phase system or colored indicator (12). By control of the pH, soap can be differentiated from other anionic surfactants in this titration (13). A suitable setup for general analysis consists of a flow-through, mercury-coated gold electrode associated with a flow injection system. The electrode responds to the concentration of both ionic and nonionic surfactants (14).

520

Another universal approach is based on "stripping analysis." In the technique of adsorptive stripping voltammetry, metal ions are first concentrated from aqueous solution by adsorption on a hanging mercury drop, then determined by a very rapid polarographic scan. If a constant level of the Ni(II) dimethylglyoxime complex is added to a water sample, the amount of current measured from the complex will be inversely proportional to the surfactant concentration due to interference from the surfactant in the preconcentration of the nickel complex on the surface of the mercury drop. Anionic, cationic, and nonionic surfactants can all be determined in this way at parts per billion concentration, but they cannot be distinguished from each other (15,16). An organic substance such as ethyl acetate can be substituted for the metal ion in this general technique of indirect tensammetry (17–19). One disadvantage of this technique is that the signals arising from surfactant mixtures do not represent the simple sum of the signals from the constituents (17); another disadvantage is that polyethylene glycol interferes. Surfactants can also be determined directly, without addition of another substance, although interpretation of the data becomes more complicated (20–23).

Ethoxylated nonylphenols (24) and ethoxylated alcohols (25) can be determined with some selectivity by alternating current polarography, measuring the amount of surfactant adsorbed at the mercury surface by the effect on the double-layer capacitance.

Because of the lack of specificity, polarographic techniques are only useful for analysis of "real world" samples if they are coupled with separation procedures. For example, one team has demonstrated that the BiAS procedure for trace analysis of nonionics can be improved by using an electrochemical procedure for surfactant determination after first precipitating and isolating the potassium iodobismuthate–nonionic complex from the sample (18,26). They prevent interference of hydrocarbons by washing the precipitate with isooctane (27).

The polarographic methods tend to be studied only in academic settings, where they are applied to measurement of very low levels of surfactants—in the 100 ppb range. This is an awkward measurement range because there are significant adsorption losses to the laboratory apparatus. For example, aluminum and most plastics cannot be used in the polarographic cell, and the ceramic frit of the salt bridge must be isolated. Only quartz is suitable, and even with quartz there are some losses (20). Of course, this problem is not unique to polarographic measurements.

II. GRAVIMETRIC METHODS

Little has been published in recent times on gravimetric determination of surfactants. Gravimetric techniques are rather more labor-intensive than other methods, and generally are susceptible to both positive and negative interference. These disadvantages are partially offset by the low cost of the apparatus and the excellent reproducibility which can be attained. They are discussed in this volume mainly because some of the precipitation methods are readily converted to qualitative methods of the "go/no-go" type for use by plant operators or in the field.

A. Anionic Surfactants

1. Extraction Methods

(a) Alcohol extraction.   One of the oldest gravimetric methods consists of extracting a dry detergent formulation with hot alcohol or alcohol/water mixture to selectively isolate

the surfactant and leave behind inorganics and such compounds as carboxymethylcellulose. The alcohol extract is heated to remove solvent, weighed, and identified by infrared spectroscopy or other techniques.

(b) Amine salts.   Greater specificity for anionic surfactants is attained if they are extracted into an organic phase as an amine salt. The p-toluidine hydrochloride method is an approach to analysis of anionic surfactants, consisting of a gravimetric method to which is appended a titrimetric finish (28,29). An alcohol/water solution containing the sample is extracted with carbon tetrachloride to remove nonpolar organics, and the extract is discarded. p-Toluidine hydrochloride is then added, forming the amine salt of the anionic. This ion pair is extracted into carbon tetrachloride, dried, and weighed. The residue is dissolved in alcohol/water and titrated with NaOH solution to determine the amine, and with AgNO$_3$ solution to correct for amine hydrochloride which may have contaminated the precipitate. This approach allows determination of the concentration and the average equivalent weight of the anionic surfactant. Hydrotropes do not interfere.

2. Precipitation Methods

Gravimetric techniques applied to anionics are necessarily nonspecific. There are no reagents which precipitate only particular sulfates or sulfonates. Many quality control laboratories use gravimetric techniques which are effective for certain mixtures. For example, certain anionics can be precipitated quantitatively as their barium salts, while many others cannot (30). The precipate is ashed and the resulting BaSO$_4$ weighed for the most precise results.

A number of inorganic cations form salts with anionic surfactants which have sufficiently low solubility to be useful for gravimetric determination. Generally, multivalent cations are more useful than monovalent cations, allowing the solubility of particular anionics to be reduced below the millimolar range (31). For example, the calcium salt of dodecylsulfate can be separated with better than 99% recovery by precipitation from aqueous solution.

α-Sulfo fatty acid esters and fatty alcohol sulfoacetates are precipitated quantitatively by BaCl$_2$ from neutral solution at 40–50°C. Determining the sulfated ash content of the precipitate allows calculation of the equivalent weight of the compounds (32).

A method of determining LAS and AS is proposed based on precipitating the calcium salts. Rather than bringing the precipitates to constant weight, quantification is performed by dissolving the filter paper and precipitate by microwave digestion with nitric acid, then determining total sulfur by ICP. The sample is analyzed with and without preliminary acid hydrolysis to determine the contribution of AS to the result (33). The final sulfur determination can also be performed by X-ray fluorescence (34).

By detailed study, the above methods can be made more selective for individual surfactants. For example, while the p-toluidine salts of most anionics can be extracted into ethyl ether, that of the methyl ester of α-sulfopalmitic acid cannot, but instead forms a precipitate in the aqueous phase. This can be quantitatively removed by filtration (32). Sulfosuccinic acid diesters can generally be determined by the p-toluidine method, but many sulfosuccinic acid monoesters cannot, because of their lack of solubility in organic solvents (35).

B. Nonionic Surfactants

1. Precipitation with Heteropoly Acids

One of the oldest methods for determination of nonionic surfactants involves forming the insoluble precipitate of a cation complex of the surfactant with a heteropoly acid. Shaffer and Critchfield (36,37) described the use of tungstosilicic acid or molybdophosphoric acid, together with barium ion, for gravimetric determination of PEG. This method found later use for determination of surfactants containing polyoxyethylene moieties, such as the EO/PO copolymers, APE, and AE (38–43). Tungstophosphoric acid seems to give better results than does molybdophosphoric acid (44). Since the same heteropoly acids, without added barium ion, are effective for determination of quaternary amines and amphoterics, some preliminary separation may be necessary prior to the gravimetric finish. Ion exchange methods are suitable for this separation (44). The heteropoly acid approach has been modified, even for trace analysis, by atomic absorption spectroscopic determination of excess reagent in the supernatant phase (45,46). Proteins interfere with the heteropoly acid methods and must be removed. Sulfate will interfere in any determination involving barium ion.

Procedure: Gravimetric Determination of Ethoxylated Sorbitol Ester in Shortening (43,47)

Weigh about 9 g shortening (containing 10–40 mg emulsifier) into a 300-mL alkali-resistant conical flask. Add 50 mL 1 M alcoholic KOH solution, attach a reflux condenser, and boil for 45 min on a hot plate. Add 10 mL H_2O and reflux an additional 45 min. Transfer to a 250-mL separatory funnel, rinsing with about 50 mL warm water. Add 4.5 mL concentrated HCl, let cool, then add 50 mL hexane. Shake well, and transfer the lower, aqueous phase to a second separatory funnel. Repeat the hexane extraction. Transfer the aqueous layer to an addition funnel or separatory funnel. Combine the hexane extracts in one of the used separatory funnels and wash twice with 20-mL portions of 50:50 EtOH/H_2O, adding the wash solution to the addition funnel with the aqueous phase. Purify the aqueous phase by passage through a mixed-bed ion exchange column, 2.8 cm in diameter, 30-cm bed depth, at a flow of about 2 mL/min. Wash with 200 mL water, collecting the eluate in a 600-mL beaker. Evaporate the eluate to a volume of about 300 mL. While still hot, add 2.0 mL 3M HCl, 4.0 mL 10% $BaCl_2 \cdot 2H_2O$ solution, and 4.0 mL 10% molybdophosphoric acid. Stir, then let stand overnight. Filter through a tared Gooch crucible with glass fiber mat. Wash with 50 mL water, dry 1 hr at 110°C, cool, and weigh. Determine the gravimetric factor by analyzing a known amount of the emulsifier to be determined.

   Discussion: In this test, the sorbitol ester itself is not determined, but rather the polyethylene glycol obtained upon hydrolysis. A collaborative study gave recoveries of 98–103% for polysorbate 60 in shortening, with the gravimetric factor varying from 2.43–2.83 (43).

2. Precipitation with Iodobismuthate Ion

(a) Volumetric finish. There are several colorimetric and turbidimetric procedures based on the formation of an insoluble complex between the iodobismuthate ion and polyether linkages in the presence of barium ion. Wickbold adapted this method to a titrimetric finish, determining bismuth ion by potentiometric titration with 1-pyrrolidinecarbodithio-

ate (48,49). In its original scope, the procedure was designed for the analysis of contaminated water samples containing parts per million levels of nonionic surfactants, after concentration by solvent sublation. More concentrated samples may be analyzed after appropriate dilution. This approach has been shown suitable for analysis of APE, AE, ethoxylated acids and amines, EO/PO block copolymers, ethoxylates of esters of polyhydroxy compounds, PEG, and PPG (18,50,51). The iodobismuthate precipitation with volumetric determination of bismuth has been adopted as an ISO standard for wastewater analysis, with determination of bismuth by atomic absorption or UV spectrophotometry specified as approved modifications (52).

Researchers report various figures for the number of ethoxy groups required for a surfactant to be detected by this procedure, typically stating that 4 polyether groups represent the lower limit of what can be detected and that the response of the BiAS method increases with increasing degree of ethoxylation to about 8, after which the increase is smaller (53). An explanation of this phenomenon is that 1- and 2-mole ethoxylates are not precipitated at all, while higher homologs are completely precipitated (18). Thus, recoveries for low-mole ethoxylates are incomplete to the extent that 1- and 2-mole ethoxylates are present in the initial sample. Analyzing environmental samples, Schröder reports recoveries of 88–115% for a heptanol 6-mole propoxylate, 1–2% for a perfluorinated hexanol 8-mole ethoxylate, and 24% for a coconut fatty acid diethanolamide (54). The iodobismuthate methods will also give a positive response to ether carboxylates (55). These may be separated by ion exchange prior to the analysis.

The main source of error in the procedure is the washing of the precipitate with acetic acid. This was found to give a variable loss of 5–15% for samples containing 1 mg nonionic. Losses due to incomplete precipitation and adsorptive losses are lower, perhaps accounting for another 5%. Thus, the higher the surfactant concentration, the higher will be the accuracy of the BiAS procedure (56). Some improvement can be obtained by filtering with a finer porosity sintered glass crucible, the European size G5, corresponding to the US ultrafine (57). A proposed modification to the BiAS procedure calls for determining the surfactant in the precipitate by electrochemical means. Since excess iodobismuthate reagent does not interfere with the electrochemical method, the washing step can be omitted, eliminating this source of error (18).

Procedure: Determination of Nonionic Surfactant in Wastewater Using Modified Dragendorff Reagent (48)

Reagents:

> $KBiI_4$ *solution*: Dissolve 1.7 g basic bismuth nitrate or bismuth nitrate pentahydrate in 20 mL glacial acetic acid and 80 mL distilled water. Mix with a solution of 65 g potassium iodide in 200 mL H_2O, add 200 mL acetic acid, and dilute to 1 L with distilled water.
>
> *Precipitation reagent*: Add 200 mL $KBiI_4$ solution to 100 mL of barium chloride solution (290 g $BaCl_2 \cdot 2H_2O$ in 1 L distilled water). Store in a brown bottle and discard after a week.
>
> *Ammonium tartrate solution*: 12.4 g tartaric acid, $C_4H_6O_6$, and 18 mL concentrated ammonium hydroxide in 1 L water.
>
> *Acetate buffer solution*: Cautiously mix 40 g sodium hydroxide and 120 mL glacial acetic acid to prepare 1 L aqueous solution.

Sodium pyrrolidinecarbodithioate titrant, 0.001 M: Dissolve exactly 0.206 g of the reagent in 1 L distilled water containing 10 mL *n*-amyl alcohol and 0.5 g sodium bicarbonate as preservatives. Discard after one week.

Bromcresol purple indicator: 0.1% in methanol.

Prepare a calibration curve by adding, to a series of 250-mL beakers, 0.00, 0.10, 0.20, 0.30, 0.40, and 0.50 mg of the appropriate nonionic surfactant standard, along with about 50 mL distilled water. Put the aqueous samples in similar beakers, choosing the sample size to fall within the range of the standards, and adding water if needed to reach at least the 50-mL mark. Add a few drops of bromcresol purple indicator solution and add 0.5 M HCl as needed to obtain the color change to yellow. Add 30 mL precipitation reagent, and stir rapidly for 10 min, then let stand for 5 min or more. Filter with suction through a fine-porosity sintered-glass crucible, and wash with 175 mL glacial acetic acid. Transfer the filtering crucible to a clean filtering flask, fitted with an adapter, if necessary, to prevent loss or contamination of the filtrate. Dissolve the precipitate by pouring three 10-mL aliquots of hot ammonium tartrate solution through the crucible, using suction. Use two more 10-mL aliquots of ammonium tartrate solution to rinse down the walls of the original 250-mL beaker, leaving the rinsings in the beaker. Use distilled water to rinse the filtering crucible and the crucible holder into the 250-mL beaker. Add a few drops of bromcresol purple indicator, then add 0.5 M NH_3 solution dropwise until the color changes to purple, then add 10 mL acetate buffer solution. Titrate bismuth ion with 0.001 M sodium pyrrolidinecarbodithioate solution using an automatic titrator with a combination electrode consisting of a platinum indicator and saturated calomel reference. A span of 100 mV and a 20-mL buret are appropriate. Determine the end point of each titration by the extrapolation method, and plot the calibration curve on linear graph paper, milligrams surfactant versus milliliters titrant to the end point, drawing the best straight line through the points. Determine mg of nonionic in the sample beakers by comparison to the calibration curve, and calculate the original concentration.

Note: It does no harm to let the solutions stand overnight after the initial precipitation or immediately prior to the titration.

(b) Sedimetric finish.   A simpler version of the iodobismuthate method incorporates sedimetry (i.e., the volume of the precipitate is measured, rather than its mass) with barium tetraiodobismuthate (58). As described above, a reddish orange precipitate forms in the presence of polyether compounds, down to a concentration of 0.1 ppm in water, provided that more than about 5 polyether units are present in the molecule.

Procedure: Sedimetric Procedure for Determination of Nonionic Surfactants (58)

Reagent: Dissolve 1.7 g basic bismuth nitrate or bismuth nitrate pentahydrate with 20 mL glacial acetic acid and dilute with water. Separately, dissolve 40 g KI in water. Combine the two solutions in a 1-L volumetric flask, add 200 mL glacial acetic acid, and dilute to the mark with water. This is the stock solution, which is stable for only 2 days. Prepare a mixed reagent shortly before beginning the analysis by mixing 100 mL of this stock solution with 50 mL 20% $BaCl_2$ solution. Store in an amber bottle.

Separate the surfactant from the sample by preliminary extraction. Redissolve in water, and transfer a 6.0-mL aliquot containing not more than 60 mg nonionic surfactant to a specially designed centrifuge tube, such as Corning No. 8360, with a narrow, graduated tip. Add 6.0 mL mixed reagent solution and shake well. Precipitation of the complex will begin in 0–5 min, depending on the surfactant concentration. When precipitation begins, centrifuge 3 min at 3000 rpm. If some of the precipitate failed to fall into the bottom of the tube, dislodge it from the tube wall with a wire and re-centrifuge. Measure the height of the sediment column in the tube, using a magnifying lens and suitable equipment to attain an accuracy of 0.1 mm. Determine the surfactant content by comparison with a calibration curve made from analysis of standards of the same type of surfactant, analyzed under precisely the same conditions.

Discussion: If the nonionic surfactant is contaminated by the presence of polyethylene glycols, these should be separated by extraction (Chapter 6) and determined separately. This same method serves for determination of PEG. The pH of the system is critical. If the final solution is diluted, as, for example, by use of a larger sample size than specified, the white precipitate of hydrated bismuth hydroxide will form. The procedure has been tested for ethoxylation products of fatty acids, fatty alcohols, amines, alkylphenols, and amides, as well as the polyethyene and polypropylene glycols. In the case of surfactants with low degree of ethoxylation, it may be necessary to dissolve the sample in isopropanol, rather than water, and dilute to a final concentration of 60:40 2-PrOH/H_2O.

(c) Turbidimetric finish. A rapid, simple test for determining levels of ethoxylated nonionic surfactants in process streams and other applications is provided by the reaction with potassium tetraiodomercurate to form an insoluble complex (37). The time of reading the turbidity is critical, since the particle size of the precipitate changes rapidly with time. For process samples, the matrix of the standards should be matched as closely as possible with the expected sample composition.

A similar colloidal complex can be formed with barium ion and triiodide (59). The ethoxylated material is mixed with barium chloride solution and iodine/iodide, and the amount of light scattered read with a laser nephelometer at 600 nm. The method has been demonstrated with EO/PO copolymer, OPE, and PEG 1000 monostearate, as well as PEG. Both of these tests are based on the ability of alkoxylate chains to form complexes with potassium or barium ion in the presence of large, lipophilic anions.

Procedure: Turbidimetric Determination of Nonionic Surfactants (37)

Reagent: Dissolve 250 g potassium iodide in water. Add 68 g mercuric chloride, $HgCl_2$, stir to dissolve, and dilute to 1.00 L.

Transfer 20 mL of a sample containing not more than 4 mg nonionic surfactant to a beaker. Add 20 mL 8% NaCl solution. Rapidly add 10.0 mL potassium tetraiodomercurate reagent and let stand exactly 15 min. Read the turbidity promptly with a nephelometer and compare to a calibration curve prepared with standards of the same composition as the sample.

3. Other Reagents

Ethoxylated surfactants form water-insoluble complexes with polycarboxylic compounds such as polyacrylic or polymethacrylic acid (60,61). This seems to be based upon hydrogen bonding between the hydrophilic ethoxy groups of the surfactant and carboxylic groups of the polymer, producing a net hydrophobic effect for the complex. The presence of inorganic salts markedly decreases the solubility of the complex. Although this has been proposed as the basis of a gravimetric method of surfactant determination, practical details have yet to be developed since the complex is soluble if the ratio of the polycarboxylic acid and the ethoxylate concentrations differ too much from ideal stoichiometry.

In general, EO/PO copolymers are precipitated by the reagents used to precipitate proteins. For example, poloxamer 407 is reported to be precipitated by deoxycholate/ trichloracetic acid reagent (62). Because gravimetric methods are little used today, this phenomenon remains more a laboratory curiousity than the basis for quantitative analytical methods.

C. Cationic and Amphoteric Surfactants

A number of ions have been used to precipitate quaternary amines. Quantitative determination follows by weighing the precipitate, titrating the residual reagent, or titrating the isolated precipitate. Di- and multivalent anions generally have lower solubility products with quaternaries than do monovalent ions. Various anions have been used: ferro- and ferricyanide, dichromate, reineckate, and heteropoly compounds. These reagents are not selective for quats.

The reagent with widest use is tungstophosphoric acid (63). The quaternary is precipitated from acid solution as the tungstophosphate, filtered, and dried to constant weight, first at 105°C, then again at red heat. The compound at the lower temperature corresponds to $3Quat \cdot PO_4 \cdot 12WO_3$, while that at higher temperature is $HPO_3 \cdot 12WO_3$. Thus, the gravimetric determination may be made on an unknown quaternary, without knowing its equivalent weight. However, Banick and Valentine reported that quaternaries containing the amido group precipitate with the formula $2Quat \cdot HPO_4 \cdot 12WO_3$ (64). Other amines interfere in this procedure, so quaternaries must first be separated from amines by selective extraction. In acid solution, most amphoteric surfactants have cationic character. They can be determined by the same reaction with heteropoly acids which is effective for cationics. This method has been demonstrated for alkyl- and alkylamidobetaines (65,66).

Most cationics form salts of low solubility with tetraphenylborate ion. Although this phenomenon may be used as the basis of a gravimetric determination, the reagent is more often applied as a means of isolating the cationic for subsequent spectroscopic identification (67).

Procedure: Gravimetric Determination of an Amphoteric Surfactant (65)

Five milliliters of an aqueous solution of the amphoteric, at a concentration of about 0.05– 1%, is mixed with 5 mL 1.5 M H_2SO_4 in a 100-mL conical flask and cooled in ice water for 3 min. Five milliliters of a 4% solution of molybdophosphoric acid ($H_3[P(Mo_3O_{10})_4] \cdot 14H_2O$) is added while swirling the flask, and the solution is mixed well. The flask is let stand in a freezer for 30 min, then the precipitate is filtered with a fine-porosity sintered-glass crucible which was also cooled in the freezer. The precipitate is washed with 100 mL ice water, dried 3 hr in a vacuum desiccator over sulfuric acid, and weighed. For determination of the gravimetric constant, a quaternary amine of known purity is analyzed, and that factor corrected for the molecular weight of the amphoteric.

III. FLOW INJECTION ANALYSIS

For many years, flow injection analysis has been used for automated spectrophotometric determination of anionics. Although UV absorbance at 224 nm has been reported for FIA of LAS (68), FIA is normally based upon the more robust methylene blue visible spectrophotometric method (69–71). Other cationic dyes may be used in place of methylene blue (72,73). The methylene blue method requires a two-phase system, a feature which is a continuing target of optimization experiments. Most commercial systems rely on gravity separation of the aqueous phase before reading the color of the organic phase. Teflon membranes allow facile separation of the organic phase for absorbance measurements (74–76), although methods based on membrane separation are not as sensitive or rugged as traditional approaches. More recent publications focus on eliminating the phase separation altogether by reading the absorbance of the small segments of the organic solvent as they travel through the instrument extraction coil (77). A so-called chromatomembrane cell, made from porous teflon, is the latest in the separation technologies applied to the methylene blue method (78).

In a study optimizing the FIA method for the determination of parts per billion levels of anionics in river water and sewage plant streams, it was observed that greater sensitivity is obtained if the sample, rather than being injected into a water stream, is pumped undiluted into the system, mixing only with a reagent stream of pH buffer and methylene blue before extraction. Operation at pH 2 inhibits fouling of the fluorocarbon membrane separator by proteins. Addition of 10% methanol or ethanol to the chloroform extractant aids in isolation of the ion pair. The presence of high levels of nonionic surfactants will inhibit the extraction of the colored species, but this can be alleviated somewhat by increasing the time of mixing of the two phases (79).

A number of other FIA approaches for anionic determination have been proposed, although it is doubtful that they are much used. For example, determination of high concentrations of LAS can be performed based on the solvatochromism of p-diphenylaminoazobenzene sulfonate (80). Another method for automated analysis of anionic surfactants involves flow injection analysis with continuous formation and extraction of the ion pair with copper(II)-1,10-phenanthroline. The extract is introduced into an atomic absorption spectrophotometer for determination of the copper (81). Cationics can be determined similarly, using adduct formation with tetrathiocyanatocobaltate(II) (82). The last system is tolerant of interference by ethoxylated nonionic surfactants, since potassium ion is not added.

FIA of anionics without a phase separation may be based on the quenching effect of surfactants on the fluorescence intensity of 8-anilino-1-naphthalenesulfonic acid coupled with bovine serum albumin (83). Another method of avoiding a phase separation is to use a surfactant-selective electrode. Since interference is a problem, the surfactant-selective electrode approach is best coupled with online concentration with a reversed-phase column. Such a system has been demonstrated for trace analysis of sodium dodecylsulfate (84).

The ion-pair extraction and colorimetric determination of cationics described in Chapter 12 has been automated using flow injection analysis. Orange II is a suitable anionic dye. A fluorocarbon membrane permits the separation of the organic from the aqueous phase for color measurement. Addition of methanol to the chloroform/water system aids in the efficiency of the extraction, so that nearly identical molar responses are obtained for a variety of cationic surfactants. As in the manual method, use of a neutral pH allows determination of only quaternary surfactants, while an acid pH permits determination of both quats and fatty amines (85,86). A flow injection method based on tetrabromophenolphthalein ethyl ester eliminates interference from amines by measuring the absorbance at 45°C, at which temperature the ion associates with amines are colorless (87). The amine interference may also be eliminated by conducting the determination at high pH (88).

Other spectrophotometric methods have been proposed for FIA of ionic surfactants based on the change in absorbance of an anionic dye in the presence of a cationic surfactant, which change is attenuated by the presence of an anionic surfactant in the system

(89,90). Alternatively, the decrease in absorbance of a cationic dye in the presence of an anionic surfactant can be monitored (91). These approaches, which do not require phase separation, are not widespread so it is not yet clear whether they have advantages over the older methods.

Still other methods are proposed periodically. For example, high concentrations of all surfactants without discrimination can be determined by an FIA titration approach using a fluorescent probe responding to the presence of micelles (92). Cationics can be determined in an FIA system containing the fluorescent material 3,6-*bis*(dimethylamino)-10-dodecylacridinium bromide and LAS. The LAS quenches the fluorescence of the acridinium compound. Cationics combine with the LAS, reversing the quenching effect (93)

Nonionic surfactants may be determined by FIA using the spectrophotometric procedure based on extraction of the ion pair of the potassium adduct of polyethoxylates with the anionic dye tetrabromophenolphthalein ethyl ester (94). The utility of this method is dependent upon the particular matrix to be analyzed. An FIA procedure for determining APE is based on the change in wavelength and intensity of the absorbance of a solution of alizarin fluorine blue (95). Since other surfactants interfere with this phenomenon, they are removed by online ion exchange or adsorption chromatography. Online ion exchange cleanup may also be used in FIA determination of APE by the potassium picrate extraction/spectrophotometric procedure (96).

IV. ELISA

Enzyme-linked immmunosorbent assay (ELISA) has been demonstrated for determination of LAS (97), octylphenolethoxylate (98), and for monoalkyl and dialkyl quaternary amines (99–101). The OPE method was shown to be highly specific, being selective to both the C_8 alkyl chain and the average degree of ethoxylation of a particular product (9–10 moles). The quat methods are reasonably selective, but have some sensitivity to other quaternary compounds with alkyl substituents differing from the target molecules, as well as having sensitivity to some other compounds.

ELISA method development is complex, requiring as a first step that a surfactant standard be covalently linked to a protein molecule and injected into a rabbit or mouse to develop anti-surfactant antibodies. The procedure would not be used for occasional analysis, but is of great interest for the routine analysis of thousands of samples.

V. SURFACE TENSION MEASUREMENT

A device has been demonstrated for continuous monitoring of surface tension, and hence surfactant concentration. It consists of two capillaries of different diameters dipped into the sample stream. A stream of air passes through each, and the difference in backpressure between the two is measured. This can be readily correlated with the surface tension of the sample, and allows optimization of the amount of surfactant added to, for example, a textile treatment bath (102).

Another device is based on optical equipment which measures the drop size of a liquid stream exiting a capillary into air. In the presence of a surfactant, the drop size is smaller. This was demonstrated for sodium dodecylsulfate, and it was shown that the effect was accentuated by the presence of an ion pairing reagent such as tetrabutylammonium ion. The instrument is proposed as a detector for flow injection analysis or HPLC

(103). This technique has also been proposed, without preliminary separation, for continuous monitoring of total surfactant content of metal cleaning and electroplating baths (104).

VI. MICROBIAL SENSORS

Microbial sensors have been demonstrated for determination of LAS (105) and alkyl sulfate (106). A strain of bacteria is selected for ability to degrade the surfactant in question, then placed in an oxygen-limited cell or made part of an oxygen-measuring electrode. The oxygen concentration is monitored as sample solution is introduced, and the decrease in oxygen related to the surfactant concentration. The systems are hardly selective. In the case of the AS-sensitive device, selectivity was measured to be 100% for AS, 82% for alkanesulfonate, 56% for ethanol, 36% for decylbenzenesulfonate, 27% for acetate, and 20% for alkylmethyltaurine (106). Such systems are of course also sensitive to temperature and ionic strength.

A biosensor has been demonstrated based on acrylodan-labeled bovine serum albumin immobilized on a silica optical fiber. The fluorescence behavior of the sensor changes in the environment of ionic surfactants. It was shown to be effective in the 5–60 micromolar surfactant range (107).

VII. FOAM MEASUREMENT

A device has been developed for continuously monitoring the surfactant concentration of an aqueous stream by conducting it into a glass cell through which air is blown. The foam height is measured optically (108). So far, there are no commercial applications of such measurements.

VIII. X-RAY FLUORESCENCE

X-ray fluorescence can measure the concentration of total sulfur in a sample, and thus can be applied to the quality control of anionic surfactants and to quality control of detergents containing anionics (34). This technique is suitable for quality control of well-characterized materials, and can also be used for semi-quantitative analysis of unknown materials. Because of interelement interference with, for example, silicate and phosphorus, quantitative analysis of unknowns is more difficult.

X-ray fluorescence can be used for trace analysis of nonionic surfactants. In this case, the analysis is performed according to the titrimetric method using tetraiodobismuthate reagent, but the final measurement of bismuth concentration in the precipitate is made not by potentiometric titration, but by XRF (109).

IX. ATOMIC ABSORPTION SPECTROSCOPY

Atomic absorption spectroscopy can sometimes be used for indirect determination of surfactants.

A. Anionics

Anionics may be extracted from water into an organic solvent as the ion pair with bis(ethylenediamine)copper(II). The copper is then determined by graphite furnace atomic absorption spectrometry and related to the original concentration of the surfactant (110,111). Similiar procedures consist of extracting the complex of LAS, copper, and α,α'-bipyridine into methylisobutylketone, determining copper ion in the extract by AAS (112) or of extracting the copper–phenanthroline–anionic surfactant complex and quantifying by atomic absorption or induction coupled plasma atomic emission spectroscopy (113). A similar method consists of extracting the ion pair of sodium dodecylsulfate and cobalt(III) *bis*(2-benzoylpyridine thiosemicarbazone) and determining cobalt (114)

B. Cationics

Cationics can be determined by forming the ion pair with the tetrathiocyanatocobalt(II) or tetrathiocyanatocopper(II) anion and extracting into an organic solvent. The extract is analyzed by atomic absorption spectroscopy, and the cobalt (115) or copper (116) content related to the original concentration of cationic surfactants.

C. Nonionics

If the nonionic surfactant is extracted from water into an organic solvent as its potassium tetrathiocyanatozincate(II) complex, its original concentration can be related to the concentration of zinc in the extract, as determined by atomic absorption spectrometry (117) or visible spectrophotometry (118). The gravimetric barium chloride/molybdophosphoric acid method for determination of nonionics has also been adapted to an atomic absorption finish, with the residual molybdenum being determined in the supernate after centrifugation (45). Similarly, the bismuth in the barium/ethoxylated surfactant/tetraiodobismuthate precipitate can be determined by AAS (52). This procedure is discussed with gravimetric analysis.

X. FLUOROMETRY

Because fluorometry is subject to many interferences, its main application to surfactant analysis has been use as an HPLC detector for quantification of APE or LAS. There has been some use as a "stand-alone" technique.

A. Direct Analysis

LAS has relatively weak fluorescence compared to multi-ring compounds, so that there is no advantage to using fluorescence rather than UV absorbance in terms of sensitivity. However, fluorescence may be used for increased selectivity.

Direct fluorescence measurement has been used to determine the approximate level of lignin sulfonate in natural waters polluted by paper mill waste. In this case, the choice of excitation and emission wavelengths is made to minimize interference from naturally occurring fluorescent compounds, such as humic acids (119,120).

B. Ion-Pair Formation

Anionic surfactants can be determined at low concentrations by extraction of the ion pair with a cationic fluorescent molecule, such as Rhodamine B or Safranine-T (121–123). This method is entirely analogous to the ion-pair extractions, such as the methylene blue method, discussed in Chapter 12. The fluorescent method has no greater selectivity, but is somewhat more sensitive. Because of fluorescence quenching effects, this approach can be predicted to be somewhat more susceptible to negative interference than is the methylene blue method. By choice of the proper dye, the fluorescence measurement may be performed in the near infrared region rather than at UV/visible wavelengths. This has been demonstrated with the cationic dyes 3,3´-diethyl-2,2´-(4,5;4´,5´-dibenzo)thiatricarbocyanine iodide [DDTC] (124) and rhodamine 800 (125). NIR measurement may be advantageous in some circumstances because of interference at other wavelength ranges.

If certain anionic fluorescent molecules are immobilized in a matrix such as cross-linked polyvinylalcohol or a modified glass surface, they show an increase in fluorescent intensity and a decrease in emission wavelength in the presence of a cationic surfactant. Although not suitable for trace analysis, this phenomenon is proportional to concentration over a range of about 30, and is insensitive to anionic and nonionic surfactants (126).

The intensity of fluorescence of the anionic 8-octadecyloxypyrene-1,3,6-trisulfonate is decreased by the presence of cationic surfactants in the 40–400 ppb range, presumably because of ion-pair formation (127). Although calibration curves based on this phenomenon are linear, different slopes are observed with different cationics or even with the same cationic depending on whether it is added as the chloride or bromide salt. It seems reasonable that this could be the basis of a very sensitive analytical technique, provided that proper cleanup steps are added to the procedure to eliminate interferences. Anionic surfactants interfere.

XI. CHEMILUMINESCENCE

An unusual indirect method for determination of anionics consists of extracting the ion pair with dimethylbiacrydil cation. The complex is broken up by back-extraction with 0.1 M KOH and the cation is determined in the aqueous phase by its chemiluminescence when reacted with hydrogen peroxide (128).

A flow injection analysis method for determining low parts per million levels of an amine ethoxylate was based on the chemiluminescence reaction of tertiary amines with hypochlorite ion (129). Because of the low intensity of the chemiluminescence, Rhodamine B is added as a sensitizer.

XII. OPTRODE METHODS

Kawabata and coworkers demonstrated an optrode sensitive to the presence of surfactants (130). The optrode was an optical fiber, one end of which was coated with a plasticized PVC membrane containing valinomycin, sodium tetrakis[3,5-bis(trifluoromethyl)phenyl]borate. The cationic fluorescent dye hexadecylacridine orange was attached to the membrane by ion exchange with the sodium ions (accomplished by simply dipping the optrode in a solution of the bromide salt of the dye). By using suitable optics connected to the optical fiber, the fluorescence of the dye could be observed when the optrode was immersed in a solution. This device was sensitive to the presence of potassium ion, which would be transported into the membrane as the valinomycin complex, paired with the boron compound. The dye would be partially displaced by an ion exchange mechanism, changing the observed fluorescence. The above response to potassium ion has been found to be affected by the presence of anionic, cationic, and nonionic surfactants. At present, this phenomenon is of only academic interest.

REFERENCES

1. Gerlache, M., J. M. Kauffmann, G. Quarin, J. C. Vire, G. A. Bryant, J. M. Talbot, Electrochemical analysis of surfactants: an overview, *Talanta*, 1996, *43*, 507–519.

2. Linhart, K., Polarographic determination of surface active materials in water and wastewater, and the determination of their degradibility (in German), *Tenside*, 1972, *9*, 241–259.

3. Bednarkiewicz, E., Z. Kublik, Surfactants by suppression of the polarographic maximum of the first kind, *J. Electroanal. Chem. Interfacial Electrochem.*, 1987, *218*, 93–106.

4. Bednarkiewicz, E., Surfactants in fresh waters based on suppression of polarographic oxygen maximum, *Electroanalysis* (N.Y.), 1991, *3*, 839–845.

5. Cosovic, B., V. Vojvodic, Surface active substances in natural waters, *Mar. Chem.*, 1987, *22*, 363–373.

6. Batina, N., B. Cosovic, D. Tezak, Surface-active compounds in precipitation studies by AC polarography, *Anal. Chim. Acta*, 1987, *199*, 177–180.

7. Matysik, J., H. Kroszka, A. Persona, Cationic surface-active agents using DME and its relevance to their adsorption, *Adsorpt. Sci. Technol.*, 1987, *4*, 53–57.

8. Bos, M., P. Van Marion, W. E. Van der Linden, Multivariate calibration procedure for the tensammetric determination of detergents, *Anal. Chim. Acta*, 1989, *223*, 387–393.

9. Emons, H., T. Schmidt, K. Stulik, Tensammetric determination of phospholipids in batch and flow injection systems, *Analyst*, 1989, *114*, 1593–1596.

10. Lohse, H., Electrochemical measuring system for determination of surface-active substances in aqueous solutions, *Anal. Chim. Acta*, 1995, *305*, 269–272.

11. Sander, S., G. Henze, AC-voltammetric determination of the total concentration of nonionic and anionic surfactants in aqueous systems, *Electroanalysis*, 1997, *9*, 243–246.

12. Bos, M., Tensammetric titration of ionic detergents, *Anal. Chim. Acta*, 1982, *135*, 249–261.

13. Sak-Bosnar, M., M. S. Jovanovic, Electrocapillary potentiometric titration of soaps in the presence of synthetic anionic surfactants in detergents, *J. Serb. Chem. Soc.*, 1987, *52*, 159–165.

14. Bos, M., J. H. H. G. Van Willigen, W. E. Van der Linden, Flow injection analyis with tensammetric detection for the determination of detergents, *Anal. Chim. Acta*, 1984, *156*, 71–76.

15. Pihlar, B., B. Gorenc, D. Petric, Indirect determination of surfactants by adsorptive stripping voltammetry, *Anal. Chim. Acta*, 1986, *189*, 229–236.

16. Adeloju, S. B., S. J. Shaw, Indirect determination of surfactants by adsorptive voltammetry. Part 1: Sodium dodecylbenzene sulfonate, *Electroanalysis*, 1994, *6*, 639–644.

17. Szymanski, A., Z. Lukaszewski, Indirect tensammetric method for the determination of nonionic surfactants. Part 1: General properties of the analytical signal. Part 2: Investigation and improvement of tolerance to manmade anionic surfactants. Part 3: Properties of the analytical signal of mixtures of nonionic surfactants, *Anal. Chim. Acta*, 1992, *260*, 25–34; 1993, *273*, 313–321; 1994, *293*, 77–86.

18. Wyrwas, B., A. Szymanski, Z. Lukaszewski, Tensammetric determination of nonionic surfactants combined with BiAS separation procedure: precipitation of different ethoxylates with modified Dragendorff reagent in the proposed and classical BiAS procedures, *Talanta*, 1994, *41*, 1529–1535.

19. Szymanski, A., B. Wyrwas, Z. Lukaszewski, Indirect tensammetric method for the determination of nonionic surfactants in surface water, *Anal. Chim. Acta*, 1995, *305*, 256–264.

20. Szymanski, A., Z. Lukaszewski, Tensammetry with accumulation on the hanging mercury drop electrode: errors of determination caused by adsorption of nonionic surfactants on the material of the measuring cell, *Anal. Chim. Acta*, 1990, *231*, 77–84.

21. Szymanski, A., Z. Lukaszewski, Adsorptive stripping tensammetry of commercially available oxyethylated alcohols: surfactants having C_{16}–C_{18} hydrophobe, *Electroanalysis*, 1991, *3*, 963–972.

22. Szymanski, A., Z. Lukaszewski, Adsorptive stripping tensammetry of oxyethylene-oxypropylene block copolymers, *Electroanalysis*, 1994, *6*, 1094–1102.

23. Szymanski, A., Z. Lukaszewski, Adsorptive stripping tensammetry of oxyethylated isooctylphenols of the Triton X series, *Electroanalysis*, 1995, *7*, 114–119.
24. Jehring, H., A. Weiss, Alternating current polarography: absorption behavior of polyethyleneglycol ethers at the mercury/electrolyte boundary (in German), *Tenside*, 1969, *6*, 251–257.
25. Rosen, M. J., X. Hua, P. Bratin, A. W. Cohen, Tensammetric determination of AE with low oxyethylene content, *Anal. Chem.*, 1981, *53*, 232–236.
26. Wyrwas, B., A. Szymanski, Z. Lukaszewski, Tensammetric determination of nonionic surfactants combined with BiAS separation procedure: optimization of the precipitation and investigation of interferences, *Talanta*, 1995, *42*, 1251–1258.
27. Wyrwas, B., A. Szymanski, Z. Lukaszewski, Tensammetric determination of nonionic surfactants combined with BiAS separation procedure: determination in the presence of hydrocarbons, *Talanta*, 1998, *47*, 325–333.
28. Stüpel, H., A. von Segesser, Anionic colloid electrolytes with *p*-toluidine (in German), *Fette und Seifen*, 1951, *53*, 260–264.
29. American Society for Testing and Materials, Anionic detergent by *para*-toluidine hydrochloride, D4224. West Conshohocken, PA 19428.
30. Jenkins, J. W., K. O. Kellenbach, Identification of anionic surface active agents by infrared absorption of the barium salts, *Anal. Chem.*, 1959, *31*, 1056–1059.
31. Brant, L. L., K. L. Stellner, J. F. Scamehorn, Recovery of surfactant from surfactant-based separations using a precipitation process, in J. F. Scamehorn and J. H. Harwell, eds., *Surfactant-Based Separation Processes*, Marcel Dekker, New York, 1989.
32. König, H., E. Walldorf, Analysis of α-sulfofatty acid methyl esters and fatty alcohol sulfoacetates (in German), *Fresenius' Z. Anal. Chem.*, 1975, *276*, 365–370.
33. Kawauchi, A., Nonsolvent quantitation of anionic surfactants and inorganic ingredients in laundry detergent products, *J. Am. Oil Chem. Soc.*, 1997, *74*, 787–792.
34. Kawauchi, A., Nonsolvent quantitation of anionic surfactants and other inorganic ingredients in laundry detergent products by X-ray fluorescence spectrometry, *J. Surfactants Deterg.*, 1999, *2*, 79–83.
35. König, H., Sulfosuccinic acid half-esters (in German), *Fresenius' Z. Anal. Chem.*, 1971, *254*, 198–209.
36. Shaffer, C. B., F. H. Critchfield, Solid polyethylene glycols (Carbowax compounds): determination in biological materials, *Anal. Chem.*, 1947, *19*, 32–34.
37. Siggia, S., J. G. Hanna, *Quantitative Organic Analysis via Functional Groups*, 4th ed., Robert E. Krieger, Malabar, FL, 1988.
38. Oliver, J., C. Preston, Nonionic detergents, *Nature*, 1949, *164*, 242–243.
39. Stuffins, C. B., Nonionic detergent in soap mixtures, *Soap, Perfumery, Cosmetics*, 1958, *31*, 369–370.
40. Kimura, W., T. Harada, Nonionic surface-active compounds, *Fette, Seifen, Anstrichm.*, 1959, *61*, 930.
41. Hobson, B. C., R. S. Hartley, Surface active agents in oils and solvent extracts from wool, *Analyst*, 1960, *85*, 193–196.
42. Smullin, C. F., F. P. Wetterau, V. L. Olsanski, Polysorbate 60 in foods, *J. Am. Oil Chem. Soc.*, 1971, *48*, 18–20.
43. Smullin, C. F., Polysorbate 60 in shortening: collaborative study, *J. Ass. Off. Anal. Chem.*, 1974, *57*, 62–64.
44. Barber, A., C. C. T. Chinnick, P. A. Lincoln, Mixtures of surface-active quaternary ammonium compounds and polyethylene oxide type of nonionic surface-active agents, *Analyst*, 1956, *81*, 18–25.
45. Sheridan, J. C., E. P. K. Lau, B. Z, Senkowski, Nonionic surfactants by atomic absorption spectrophotometry, *Anal. Chem.*, 1969, *41*, 247–250.
46. Chlebicki, J., W. Garncarz, Nonionic surfactants in water and effluent by atomic absorption spectroscopy, *Tenside, Surfactants, Deterg.*, 1980, *17*, 13–17.

47. AOAC International, *Official Methods of Analysis of AOAC International*, 16th ed., Gaithersburg, MD 20877, 1999.

48. Wickbold, R., Nonionic surfactants in river and wastewater (in German), *Tenside*, 1972, *9*, 173–177.

49. Wickbold, R., Nonionic surfactants (in German), *Tenside*, 1973, *10*, 179–182.

50. Chlebicki, J., W. Garncarz, Block copolymers of ethylene oxide and propylene oxide in water, *Tenside*, 1978, *15*, 187–189.

51. Anthony, D. H. J., R. S. Tobin, Immiscible solvent extraction scheme for biodegradation testing of polyethoxylate nonionic surfactants, *Anal. Chem.*, 1977, *49*, 398–401.

52. International Organization for Standardization, Water quality—determination of surfactants—Part 2: Determination of nonionic surfactants using Dragendorff reagent, 7875/2-1984. Available from national ISO affiliates.

53. Brüschweiler, H., H. Gämperle, F. Schwager, Primary degradation, complete degradation, and degradation intermediates of APE (in German), *Tenside, Surfactants, Deterg.*, 1983, *20*, 317–324.

54. Schröder, H. Fr., Determination of surfactants—comparison of class-specific methods with LC/MS substance-specific analysis (in German), *Vom Wasser*, 1992, *79*, 193–209.

55. Kunkel, E., Carboxymethylated oxyethylates, *Tenside, Surfactants, Deterg.*, 1980, *17*, 10–12.

56. Wyrwas, B., A. Szymanski, Z. Lukaszewski, Sources of error in precipitation of nonionic surfactants with modified Dragendorff reagent, *Anal. Chim. Acta*, 1993, *278*, 197–203.

57. Lukaszewski, Z., A. Szymanski, Sources of error in the determination of nonionic surfactants in environmental samples, *Mikrochim. Acta*, 1996, *123*, 185–196.

58. Bürger, K., Detection and determination of surface active polyoxyethylene compounds and of polyethylene glycol (in German), *Fresenius' Z. Anal. Chem.*, 1963, *196*, 251–259; 1964, *199*, 434–438.

59. Cole, S. C., G. A. Christensen, W. P. Olson, PEG quantitation by laser nephelometry, *Anal. Biochem.*, 1983, *134*, 368–373.

60. Saito, S., Taniguchi, T., Matsuyama, H., Precipitation of nonionic surfactants by polymeric acids: additive effects of inorganic salts and acids (in German), *Colloid Polym. Sci.*, 1976, *254*, 882–889.

61. Saito, S., Precipitation of nonionic surfactants by polymeric acids: precipitation limit of surfactant concentration, *Tenside, Surfactants, Deterg.*, 1977, *14*, 113–116.

62. Pec, E. A., Z. G. Wout, T. P. Johnston, Biological activity of urease formulated in poloxamer 407 after intraperitoneal injection in the rat, *J. Pharmaceut. Sci.*, 1992, *281*, 626–630.

63. Lincoln, P. A., C. C. T. Chinnick, Quaternary ammonium compounds as phosphotungstates, *Analyst*, 1956, *81*, 100–104.

64. Banick, W. M., Jr., J. R. Valentine, Quaternary ammonium compounds as phosphotungstates, *Analyst*, 1964, *89*, 435–436.

65. König, H., Amphoteric surfactants (in German), *Fresenius' Z. Anal. Chem.*, 1970, *251*, 359–368.

66. Govindram, C. B., V. Krishnan, Complex surfactant systems—classical approach, *Tenside, Surfactants, Deterg.*, 1998, *35*, 104–107.

67. Cross, J. T., Identification and determination of cationic surfactants, in E. Jungermann, ed., *Cationic Surfactants*, Marcel Dekker, New York, 1977.

68. Brandli, E. H., R. M. Kelley, Automated determination of alkyl aryl sulfonates in spray-dried detergents by UV absorption, *J. Am. Oil Chem. Soc.*, 1970, *47*, 200–202.

69. Motomizu, S., M. Oshima, T. Kuroda, Spectrophotometric determination of anionic surfactants in water after solvent extraction coupled with flow injection, *Analyst*, 1988, *113*, 747–753.

70. Canete, F., A. Rios, M. D. L. de Castro, M. Valcarcel, Liquid-liquid extraction in continuous flow systems without phase separation, *Anal. Chem.*, 1988, *60*, 2354–2357.

71. Fan, S., Z. Fang, Two-step solvent extraction flow injection system for the determination of anionic surfactants by spectrophotometry, *Fresenius' Z. Anal. Chem.*, 1997, *357*, 416–419.

72. Motomizu, S., Y. Hazaki, M. Oshima, K. Toei, Spectrophotometric determination of anionic surfactants in river water with cationic azo dye by solvent extraction–flow injection analysis, *Anal. Sci.*, 1987, *3*, 265–269.

73. Sakai, T., H. Harada, X. Liu, N. Ura, K. Takeyoshi, K. Sugimoto, Phase separator for extraction–spectrophotometric determination of anionic surfactants with malachite green by FIA, *Talanta*, 1998, *45*, 543–548.

74. Kawase, J., A. Nakae, M. Yamanaka, Anionic surfactants by flow injection analysis based on ion-pair extraction, *Anal. Chem.*, 1979, *51*, 1640–1643.

75. Sahlestrom, Y., B. Karlberg, Unsegmented extraction system for flow injecton analysis, *Anal. Chim. Acta*, 1986, *179*, 315–323.

76. Sahlestrom, Y., B. Karlberg, Flow-injection extraction with a microvolume module based on integrated conduits, *Anal. Chim. Acta*, 1986, *185*, 259–269.

77. Liu, H., P. K. Dasgupta, Flow injection extraction without phase separation based on dual wavelength spectrophotometry, *Anal. Chim. Acta*, 1994, *288*, 237–245.

78. Moskvin, L. N., J. Simon, P. Löffler, N. V. Michailova, D. N. Nicolaevna, Photometric determination of anionic surfactants with a flow-injection analyzer that includes a chromatomembrane cell for sample preconcentration by liquid-liquid solvent extraction, *Talanta*, 1996, *43*, 819–824.

79. del Valle, M., J. Alonso, J. Bartroli, I. Marti, Spectrophotometric determination of anionic surfactants in water by solvent extraction in a flow injection system, *Analyst*, 1988, *113*, 1677–1681.

80. Liu, J., Flow injection determination of anionic surfactants based on the solvatochromism of *p*-diphenylaminoazobenzene sulfonate, *Anal. Chim. Acta*, 1997, *343*, 33–37.

81. Gallego, M., M. Silva, M. Valcarcel, Indirect atomic absorption determination of anionic surfactants in wastewaters by flow injection continuous liquid-liquid extraction, *Anal. Chem.*, 1986, *58*, 2265–2269.

82. Martinez Jimenez, P., M. Gallego, M. Valcarcel, Indirect atomic absorption spectrometric determination of some cationic surfactants by continuous liquid/liquid extraction with tetrathiocyanatocobaltate, *Anal. Chim. Acta*, 1988, *215*, 233–240.

83. Recalde Ruiz, D. L., A. L. Carvhalo Torres, E. Andrés García, M. E. Díaz García, Fluorimetric flow-injection method for anionic surfactants based on protein–surfactant interactions, *Analyst*, 1998, *123*, 2257–2261.

84. Martínez-Barrachina, S., J. Alonso, L. Matia, R. Prats, M. del Valle, Anionic surfactants in river water and wastewater by a FIA system with on-line preconcentration and potentiometric detection, *Anal. Chem.*, 1999, *71*, 3684–3691.

85. Kawase, J., M. Yamanaka, Continuous solvent extraction–spectrophotometric determination of cationic surfactants, *Analyst*, 1979, *104*, 750–755.

86. Kawase, J., Automated determination of cationic surfactants by flow injection analysis based on ion-pair extraction, *Anal. Chem.*, 1980, *52*, 2124–2127.

87. Sakai, T., Spectrophotometric determination of quaternary ammonium salts by a flow injection method coupled with thermochromism of ion associates, *Analyst*, 1992, *117*, 211–214.

88. Sakai, T., Solvent extraction–spectrophotometric determination of berberine and benzethonium in drugs with tetrabromophenolphthalein ethyl ester by batchwise and flow injection methods, *Analyst*, 1991, *116*, 187–190.

89. Motomizu, S., M. Oshima, Y. Hosoi, Spectrophotometric determination of cationic and anionic surfactants with anionic dyes in the presence of nonionic surfactants: general aspect, *Mikrochim. Acta*, 1992, *106*, 57–66.

90. Motomizu, S., M. Oshima, Y. Hosoi, Spectrophotometric determination of cationic and anionic surfactants with anionic dyes in the presence of nonionic surfactants: development of batch and flow injection methods, *Mikrochim. Acta*, 1992, *106*, 67–74.

91. Patel, R., K. S. Patel, FIA determination of anionic surfactants with cationic dyes in water bodies of central India, *Analyst*, 1998, *123*, 1691–1695.

92. Lucy, C. A., J. S. W. Tsang, Surfactant concentration using micellar enhanced fluorescence and flow injection titration, *Talanta*, 2000, *50*, 1283–1289.

93. Masadome, T., Flow injection fluorometric determination of cationic surfactants using 3,6-*bis*(dimethylamino)-10-dodecylacridinium bromide, *Anal. Lett.*, 1998, *31*, 1071–1079.

94. Whitaker, M. J., Spectrophotometric determination of nonionic surfactants by flow injection analysis utilizing ion-pair extraction and an improved phase separator, *Anal. Chim. Acta*, 1986, *179*, 459–462.

95. Leon-Gonzalez, M. E., M. J. Santos-Delgado, L. M. Polo-Diez, Triton-type nonionic surfactants by on-line clean-up and flow injection with spectrophotometric detection, *Analyst*, 1990, *115*, 609–612.

96. Leon-Gonzalez, M. E., M. J. Santos-Delgado, L. M. Polo-Diez, Triton-type nonionic surfactants in different samples by on-line clean-up and FIA, *Fresenius' Z. Anal. Chem*, 1990, *337*, 389–392.

97. Fujita, M., M. Ike, Y. Goda, S. Fujimoto, Y. Toyoda, K. Miyagawa, ELISA for detection of LAS: development and field studies, *Environ. Sci. Technol.*, 1998, *32*, 1143–1146.

98. Wie, S. I., B. D. Hammock, Use of enzyme-linked immunosorbent assays (ELISA) for the determination of Triton X nonionic detergents, *Anal. Biochem.*, 1982, *125*, 168–176.

99. Chen, T., C. Dwyre-Gygax, R. S. Smith, C. Breuil, ELISA for didecyldimethylammonium chloride, a fungicide used by the forest products industry, *J. Agric. Food Chem.*, 1995, *43*, 1400–1406.

100. Bull, J. P., A. N. Serreqi, H. R. Gamboa, C. Breuil, ELISA to detect didecyldimethylammonium chloride, a quaternary ammonium compound, *J. Agric. Food Chem.*, 1998, *46*, 4779–4786.

101. Bull, J. P., A. N. Serriqi, T. Chen, C. Breuil, Immunoassay for benzyldimethyldodecylammonium chloride, *Water Res.*, 1998, *32*, 3621–3630.

102. Müller-Kirschbaum, E. J. Smulders, On-line tensiometer: sensor for continuous monitoring of laundry detergent concentration and simultaneous correlation with washing performance (in German), *Seifen, Öle, Fette, Wachse*, 1992, *118*, 427–434.

103. Young, T. E., R. E. Synovec, Enhanced surfactant determination by ion-pair formation using flow-injection analysis and dynamic surface tension detection, *Talanta*, 1996, *43*, 889–899.

104. Janocha, B., D. Renzow, J. Matheis, Process control of electroplating and cleaning baths: automated tensiometric determination of surfactant content (in German), *Galvanotechnik*, 1997, *88*, 3265–3268.

105. Nomura, Y., K. Ikebukuro, K. Yokoyama, T. Takeuchi, Y. Arikawa, S. Ohno, I. Karube, Novel microbial sensor for anionic surfactant determination, *Anal. Lett.*, 1994, *27*, 3095–3108.

106. Reshetilov, A. N., I. N. Semenchuk, P. V. Iliasov, L. A. Taranova, Amperometric biosensor for detection of sodium dodecyl sulfate, *Anal. Chim. Acta*, 1997, *347*, 19–26.

107. Lundgren, J. S., F. V. Bright, Biosensor for the nonspecific determination of ionic surfactants, *Anal. Chem.*, 1996, *68*, 3377–3381.

108. Soran, P. D., E. E. Neal, B. Smith, K. I. Mullen, On-line surfactant monitoring by foam generation, *J. Chem. Educ.*, 1996, *73*, 819–821.

109. Hellmann, H., Nonionic surfactants in water and wastewater by X-ray fluorescence and IR spectrometry (in German), *Fresenius' Z. Anal. Chem.*, 1979, *297*, 102–106.

110. Crisp, P. T., J. M. Eckert, N. A. Gibson, G. F. Kirkbright, T. S. West, Anionic detergents at ppb levels by graphite furnace atomic absorption spectrometry, *Anal. Chim. Acta*, 1976, *87*, 97–101.

111. Gagnon, M. J., Anionic detergents in natural waters at the ppb level, *Water Res.*, 1979, *13*, 53–56.

112. Xu, T., B. Xu, W. Tong, Y. Fang, Indirect determination of trace anionic surfactants in water using atomic absorption spectrophotometry (in Chinese), *Huanjing Kexue*, 1987, *8(3)*, 76–78.

113. Pressouyre, B., Anionic detergents by ICP emission spectroscopy, *Analusis*, 1989, *17*, 346–354.

114. Chattaraj, S., A. K. Das, Indirect AAS determination of anionic surfactants by formation of an ion-pair with *bis*(2-benzoylpyridine thiosemicarbazone) cobalt(III), *Indian J. Chem. Technol.*, 1994, *1*, 98–102.

115. Le Bihan, A., J. Courtot-Coupez, Cationic surfactants using flameless atomic absorption spectrophotometry, *Analusis*, 1976, *4*, 58–64.

116. Martinez Gonzales, P., C. Camara Rica, L. Polo Diez, Cationic surfactants in frozen squid by flame and electrothermal atomization atomic absorption spectrometry, *J. Anal. At. Spectrom.*, 1987, *2*, 809–811.

117. Crisp, P. T., J. M. Eckert, N. A. Gibson, Atomic absorption spectrometric method for the determination of nonionic surfactants, *Anal. Chim. Acta*, 1979, *104*, 93–98.

118. Crisp, P. T., J. M. Eckert, N. A. Gibson, I. J. Webster, Extraction–spectrophotometric method for the determination of nonionic surfactant, *Anal. Chim. Acta*, 1981, *123*, 355–357.

119. Thruston, A. D., Jr., Fluorometric method for the determination of lignin sulfonates in natural waters, *J. Water Pollut. Contr. Fed.*, 1970, *42*, 1551–1555.

120. Wilander, A., H. Kvarnas, T. Lindell, Fluorometric method for measurement of lignin sulfonates and its *in situ* application in natural waters, *Water Res.*, 1974, *8*, 1037–1045.

121. Rubio-Barroso, S., M. Gomez-Rodriguez, L. M. Polo-Diez, Fluorometric determination of anionic surfactants by extraction as the rhodamine-B ion pair, *Microchem. J.*, 1988, *37*, 93–98.

122. Rubio-Barroso, S., L. M. Polo-Diez, V. Rodriguez-Gamonal, M. Gomez-Rodriguez, Comparative study of spectrofluorometric and spectrophotometric methods for anionic surfactants determination in wastewater, *An. Qhim., Ser. B*, 1988, *84*, 361–365.

123. Rubio-Barroso, S., V. Rodriguez-Gamonal, L. M. Polo-Diez, Fluorometric determination of anionic surfactants by extraction as safranine-T ion-pairs, *Anal. Chim. Acta*, 1988, *206*, 351–355.

124. Imasaka, T., A. Yoshitake, N. Ishibashi, Semiconductor laser fluorimetry in the near-IR region, *Anal. Chem.*, 1984, *56*, 1077–1079.

125. Hindocha, R. K., J. N. Miller, N. J. Seare, On-line liquid-liquid extraction with very near-IR fluorescence detection, *Anal. Proc.*, 1993, *30*, 129–131.

126. Shakhsher, Z. M., W. R. Seitz, Optical detection of cationic surfactants based on ion pairing with an environment-sensitive fluorophor, *Anal. Chem.*, 1990, *62*, 1758–1762.

127. Marhold, S., E. Koller, I. Meyer, O. S. Wolfbeis, Fluorimetric assay for cationic surfactants, *Fresenius' Z. Anal. Chem.*, 1990, *336*, 111–113.

128. Pyatnitskii, I. V., A. Yu. Nazarenko, O. A. Zaporozhets, Chemiluminescence determination of anionic surfactants (in Russian), *Izv. Vyssh. Uchebn. Zaved., Khim. Khim. Tekhnol.*, 1988, *31*(6), 127–128.

129. Lancaster, J. S., P. J. Worsfold, A. Lynes, Nonionic surfactant in aqueous environmental samples by flow injection analysis with chemiluminescence detection, *Anal. Chim. Acta*, 1990, *239*, 189–194.

130. Kawabata, Y., T. Yamamoto, T. Imasaka, Theoretical evaluation of optical response to cations and cationic surfactant for an optrode using hexadecyl acridine orange attached on a plasticized PVC membrane, *Sens. Actuators, B*, 1993, *11*, 341–346.

18
Environmental Analysis

I. GENERAL CONSIDERATIONS

The most challenging applications of surfactant analysis are in determination of low levels of surfactants and their metabolites in environmental materials. The techniques used in environmental analysis are sufficiently different from those used in general analysis that we discuss them in a separate chapter.

A. Strategy

1. Purpose

Measurements of surfactants in the environment are made for a variety of reasons. Sometimes, the surfactant is a convenient tracer that allows estimation of the extent of contamination of groundwater by sewage and septic tank effluents. In this case, a simple test for total man-made anionic surfactants may be appropriate. At other times, the environmental impact of a specific commercial surfactant is studied. In this case, specific tests are needed to measure the concentration of the surfactant through sewage treatment and ultimate disposal. A feature of such a study may be determination of the concentrations of individual homologs and oligomers of the surfactant to decide whether, for example, longer alkyl chains or ethoxy chains are degraded more quickly than are shorter chains. This is especially true when biological effects differ among the homologs. Sometimes compounds such as polyethylene glycol or alkylbenzenes are measured. These are not themselves sur-

factants, but are impurities in commercial surfactants or products of surfactant biodegradation, and hence are present in the environment as a consequence of surfactant use.

Surfactants are often added to agricultural products to increase solubility or otherwise enhance the efficacy of the active agents. Depending on the toxicity of the surfactants
and the volumes used, it may be necessary to determine pathways and yields of their photodegradation.

2. Chemical and Biochemical Degradation

The fate of man-made compounds in the environment is explained in terms of "primary"
and "ultimate" degradation. In the case of surfactants, primary degradation has occurred
when the molecule is fragmented in such a way that it is no longer surface-active. Ultimate
degradation, also called mineralization, is complete when the organic compound has decomposed to CO_2, H_2O, SO_4^{2-}, and NO_3^-. Studies of the environmental impact of individual surfactants require knowledge of the degradation pathways and the relative rates of the
reactions. While society does not require that commercial surfactants undergo complete
and immediate mineralization (there are few organic compounds, synthetic or natural,
which do not give long-lived residues), it is necessary to identify stable intermediates and
determine that they will not have harmful effects.

The rate of primary degradation is determined by measuring the level of intact surfactant in the sample as a function of time. Such an experiment is sometimes called a "die-
away test." For a system in which the surfactant provides the only nutrient source, as in a
laboratory test cell, ultimate degradation can be determined by monitoring the level of total organic carbon. If the TOC concentration stabilizes at a level above zero, this indicates
the presence of stable metabolites or impurities. Carbon dioxide evolution may be quantitatively monitored to give the same information.

Isotopically labeled compounds are used to determine the ultimate fate of specific
substances in model systems. Usually a ^{14}C-labeled analog of the compound of interest is
introduced to the lab-scale biotreatment system. Over time, the evolved gas is trapped to
determine $^{14}CO_2$, an indicator of ultimate biodegradation. This approach has the great advantage that nutrients and other organic matter can be freely added to the test cell without
confounding the interpretation of evolved CO_2 data. Tritium labeling is also used in metabolite studies. The liquid and solid phases from the model system are separated to produce fractions which, together with evolved gases, account for all of the radioactivity
originally introduced. The fractions are characterized to determine the specific compounds, or at least the functionality of the compounds, that result from degradation of the
original surfactant (1).

Most commercial surfactants consist of a series of homologs or oligomers. Depending on the object of the biodegradation experiments, the radioactive label may be applied
to a series of compounds representing the commercial product or to a single compound.
For example, tagged ethoxylates may be prepared in the laboratory by reaction of an alcohol or alkylphenol with ^{14}C-tagged ethylene oxide to give oligomers with the label distributed randomly along the hydrophile chain. Alternatively, the alcohol or alkylphenol may
be reacted through suitable intermediate steps with, e.g., pentaethylene glycol, to prepare a
homogeneous tagged compound (2,3).

Degradation of surfactants is the subject of full-length books (4,5), as are other environmental aspects (6). The literature of bioaccumulation of surfactants was reviewed in
1994 (7).

3. General Analytical Approach

The analytical techniques used in environmental studies are basically the same as those used in other applications, as described in previous chapters. A number of review articles have appeared on the subject (4,8,9). Only rarely can quantitative methods be applied directly to determination of low levels of surfactants in the environment; it is first necessary to isolate the materials of interest from the matrix.

The normal sequence in an environmental analysis is (a) first a concentration step to isolate surfactants from the bulk of the matrix; (b) then a separation of the compound of interest from similar materials; (c) and finally quantification by a suitable method.

All environmental matrices contain interfering compounds, some of which will accompany the synthetic surfactants through all the isolation steps. Whenever possible, the purity of the final extract should be cross-checked by, for example, infrared or TLC analysis to make sure that the effects of positive interference are not too severe (10)

B. Isolation Techniques

Sampling and sample preservation is of paramount importance. Samples must be chosen with complete knowledge of the system to be characterized. For example, a sewage treatment plant can operate very differently during the rainy season than during the dry season. Surfactant input to a treatment plant will vary in daily and weekly cycles depending on the household routines and commercial work schedules of the community. Care must be taken that the samples are representative of whatever situation the experiment is to model.

Environmental samples are inherently unstable. It is common practice to protect liquid samples against further microbiological activity by addition of formaldehyde at the 0.3% level (addition of 1% formalin, which is a 37% formaldehyde solution), most often in combination with refrigeration at 4°C. Solid samples are stored well below 0°C. Most investigators agree that refrigeration alone is not effective for preservation of water samples, while formaldehyde alone is effective (11). Glass is preferred for sample containers. Loss of surfactant due to adsorption on the container walls may be minimized by addition to the sample of another surfactant, selected so as not to interfere in the analysis, at a comparatively high level. If PEG is to be determined, one group recommends preserving with 20 ppm mercuric chloride rather than formaldehyde because formaldehyde can react with glycols (12).

In environmental studies of natural waters or wastewaters, sample preparation must be carefully thought out. All surfactants preferentially adsorb to solids in aqueous systems. For example, they will be found enriched in the sludge phase of sewage treatment systems and in the sediment of rivers. This effect is stronger with the more hydrophobic members of a homologous series. Filtration of the sample to remove particulates will result in some loss of surfactants from solution, especially since surfactants also adhere preferentially to filter media and glassware. This is most pronounced for cationics and less so for nonionics, with anionics being the least strongly adsorbed (13).

1. Solvent Sublation

Solvent sublation was used very extensively in the 1970s and 1980s for environmental analysis of surfactants, but has now been largely replaced by solid phase extraction. The advantages and problems of the technique are discussed in Chapter 6. Surfactant degradation products and impurities which are not themselves surface-active are not removed by sublation. These include low molecular weight EO adducts and carboxylated LAS. Carboxylated APE retains hydrophile/lipophile properties and can be isolated by sublation at pH 2.

While sublation is not necessarily more efficient than liquid-liquid extraction for isolation of individual surfactants, it is preferred in analyzing sewage because emulsions are less likely to form. Sublation is less likely to concentrate impurities than is either liquid-liquid or solid phase extraction. A significant advantage of sublation over solid phase extraction is the ease with which large sample volumes can be handled.

2. Extraction

(a) Liquid-liquid extraction. Direct extraction is not generally useful in determination of surfactants in the environment because the amount of surfactant to be found in a liter or two of wastewater is not high enough to provide the mass required for specific detection. Use of an automated continuous extractor can provide a large concentration factor. For example, a unit which extracted 200 L of effluent over a period of 3 days with only 150 mL hexane was used in sample preparation for GC-MS determination of 1- and 2-mole EO adducts of alkylphenol (14). However, the greatest use of liquid-liquid extraction is in fractionation of material initially isolated by other means, such as solid phase extraction or evaporation. The most common approach is to extract sample portions at different pH, giving, for example, an acidic and a basic-plus-neutral fraction.

(b) Liquid-solid extraction. Extraction of a solid phase, such as river sediment or sewage sludge, is a prerequisite for some environmental analyses. Especially in the case of sewage sludge, the challenge is to choose a solvent system which will solubilize the surfactants without forming an emulsion or co-extracting too many of the other sludge components. Most often, solid soils or sludges are subjected to exhaustive methanol extraction to remove surfactant. This can be performed by Soxhlet extraction, accelerated solvent extraction (i.e., extraction at elevated pressure and temperature), or by microwave-assisted solvent extraction (15). The material isolated in this initial extract must almost always be further purified before analysis.

To provide a suitable concentration factor, water samples are sometimes evaporated to dryness and the residue then extracted. A polar solvent such as methanol or ethanol is used which will, in addition to surfactants, also remove such materials as PEG and organic acids.

A combination steam distillation/solvent extraction apparatus is effective for removing APE metabolites: nonylphenol and its mono-and diethoxylates, from sewage sludge (16,17). This technique has also been demonstrated for isolation of alkylbenzene and other materials from sludge (17).

(c) Solid phase extraction. Nonpolar materials in general can be concentrated from aqueous matrices by passage through a column of Amberlite® XAD® resin (Rohm & Haas trademarks, available from Supelco), which has a strong affinity for the hydrophobe portion of most surfactants. This approach is used extensively in surfactant analysis. Resins used include XAD-2 (18), XAD-4 (19), and XAD-8. Substantial enrichment is obtained, since, for example, 300 L of water may be pumped through 1.2 L of resin in the concentration step (20). The surfactants can be quantitatively eluted with a suitable solvent prior to analysis or further separation.

A column containing reversed-phase HPLC packing material will effectively concentrate anionic and nonionic surfactants from water. For a typical application, 100 mL river water is passed through a C_8 column of 40 micron particle size and 2.8 mL volume (21). No difference was seen in the performance of C_{18} SPE columns from 3 manufacturers (Waters, Supelco, Analytichem), each containing about 300 mg of packing material. The addition of 8% NaCl is recommended to encourage adsorption of the surfactants from the sample. Sodium dodecylsulfate may be used for the same purpose, but less effectively (22,23). Desorption of the surfactants from the cartridge has been accomplished with many different solvents, as described in the tables found later in this chapter.

Activated carbon will effectively concentrate surfactants, including fluorine-containing surfactants (24), from dilute aqueous samples (25). However, many other organic compounds are also extracted, so that extensive further purification may be required.

A very significant advantage of using SPE is that the adsorption can be performed at the sampling site, eliminating the need to preserve and transport large water samples to the laboratory. Only the small quantity of resin need be stabilized by refrigeration and shipped.

Procedure: Isolation of LAS, APE, and Their Carboxylated Degradation Products from Wastewater by SPE on Graphitized Carbon Black (26)

Ten milliliters of sewage influent or 100 mL sewage effluent are acidified to about pH 3 to insure retention of carboxylates, then passed through a SPE cartridge containing 1 g Alltech graphitized carbon black. The sample container is rinsed with 10–15 mL deionized water, which is also passed through the SPE cartridge. The cartridge is rinsed with an additional 7 mL water, then air is pulled through the cartridge. Two milliliters MeOH is added to dry the cartridge, followed by more air. APE and nonylphenol are eluted with 7 mL 70:30 CH_2Cl_2/MeOH. Carboxylated APE is removed with 7 mL 0.025 M formic acid in CH_2Cl_2. LAS and its carboxylated degradation products are removed by elution with 7 mL 0.01 M tetramethylammonium hydroxide in 90:10 CH_2Cl_2/MeOH.

Notes: Care should be exercised if the first eluate is evaporated to avoid losing volatile nonylphenol. The basic eluate should be neutralized prior to evaporation in order to prevent formation of methyl esters.

(d) Supercritical fluid extraction. At the present state of knowledge, there is little reason to choose supercritical fluid extraction over conventional liquid extraction. This is particularly true in complex matrices, since the efficiency of extraction of polar substances by CO_2 is greatly affected by the presence of other materials in the sample.

C. General Test Methods

Useful and reproducible determinations of surfactant concentration are based on such methods as measurement of surface tension or the height of a foam that can be generated on recirculating the solution. These analyses are especially useful in monitoring laboratory bench-scale studies of primary biodegradation, which usually only involve a single surfactant. Of course, such measurements are only related to the concentration of the intact surfactant; they do not detect degradation products without surface-active properties. Since naturally occurring substances are not differentiated from man-made compounds, these techniques are of no value for natural waters not grossly polluted. These methods are reviewed by Swisher (4).

For initial screening tests of biodegradability, conducted on the bench scale, it is often possible to monitor results by simply measuring total organic carbon. This test is also used as a gross measure of purity of water. The TOC instrument is designed specifically so that it does not discriminate as to the nature of the carbon compounds.

Methods for determining fluorine-containing surfactants are not yet well developed. One approach is to perform a combustion and total fluorine determination on activated carbon used to adsorb surfactants from aqueous solution (24).

II. ANIONIC SURFACTANTS

A. Surfactants and Their Metabolites

Reviews have appeared describing the degradation pathways of many anionics (4,27). These should be consulted before beginning any investigation.

1. LAS and ABS

Most published work deals with the environmental determination of the dominant anionic surfactant, linear alkylbenzenesulfonate (LAS). The same methods are applicable to the less common branched-chain analog, ABS. Thorough environmental studies not only track the amount of "intact" LAS in various matrices, but also follow the alkyl chain distribution in order to reach an understanding of which homologs are dominant in the matrix due to the selective biodegradation of other homologs.

Commercial LAS contains significant concentrations of 19–26 different homologs and isomers, because of the many different positions at which the aryl ring can be attached to the alkyl chain. Also, there is a small amount of a compound present which consists of LAS with a single methyl branch on the otherwise linear alkyl chain; this is sometimes called "iso-LAS" (28,29). The different homologs vary widely in aqueous solubility and propensity to adsorb to sediment, with hydrophilicity increasing with shorter alkyl chain length and as the point of aryl attachment approaches the center of the chain. The homologs and isomers can be determined individually by gas chromatography after desulfonation or derivatization. Initial biodegradation gives the corresponding sulfophenyl carboxylic acids, with the alkyl chain length becoming progressively shorter as degradation proceeds. These intermediates may rearrange to form indanones and tetralones. Primary biodegradation, as followed by tests for the concentration of intact LAS, occurs very quickly. Ultimate degradation, as followed by decrease of UV absorption or TOC, is less rapid. Biodegradation is more rapid for longer alkyl chains and for isomers having the sulfophenyl group near the end of the alkyl chain. A study of a European river showed that about 1% of the soluble TOC content could be attributed to LAS and its sulfophenylcarboxylate degradation products (30).

A partially degraded LAS sample will contain sulfophenyl carboxylate compounds and their cyclized desulfonation products, 1-alkyl indanones and 1-alkyl tetralones. The predominant sulfophenyl carboxylate homologs found are 3-(4-sulfophenyl)-pentanedioic acid, 4-(4-sulfophenyl)-heptanedioic acid, 3-(4-sulfophenyl)-hexanoic acid, and 3-(4-sulfophenyl)-heptanoic acid. The monocarboxylic acids are more abundant than the dicarboxylates (31). The most important product of fish metabolism of LAS is 3-(4-sulfophenyl)-butanoic acid (32,33).

It is generally agreed that alkylbenzenes in natural waters originate as unsulfonated hydrocarbon impurities in commercial LAS and are not formed from biodegradation of LAS. They may be determined by GC or GC-MS after extensive work-up of the sample.

Dialkylindanesulfonates and dialkyltetralinsulfonates are also introduced to the environment as impurities in commercial LAS (34,35).

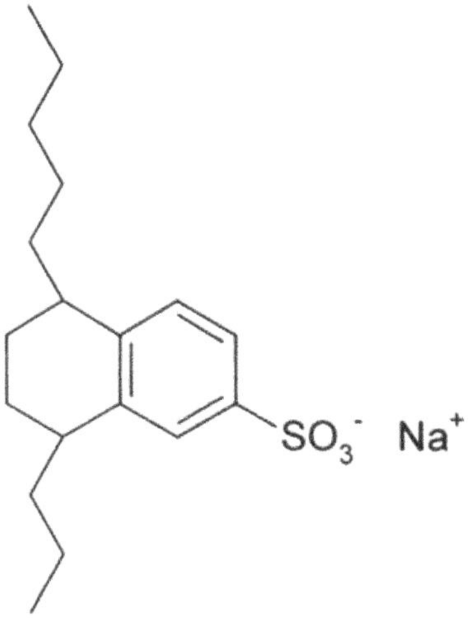

SO_3^- Na^+

Dialkyltetralinsulfonate

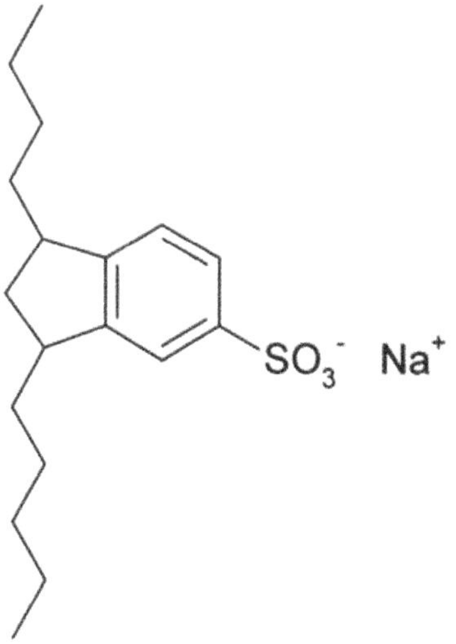

SO_3^- Na^+

Dialkylindanesulfonate

Early work on the environmental fate of ABS and LAS focused on the amount of intact surfactant which found its way into natural waters from sewage treatment plants. This analytical methodology is now well developed, as are methods for determining the chief metabolites in water. Current areas of method development are in determination of the surfactants and their metabolites in the solid phase: sewage sludge, river sediment, and in soil which has been fertilized with digested sewage sludge. Generally speaking, LAS is degraded to a great extent in properly run activated sludge wastewater treatment plants, and is degraded to a lesser extent in trickling filter treatment plants. A significant portion of the surfactant removed from water by activated sludge treatment is not actually biodegraded, but is simply adsorbed to the sludge. As long as anaerobic conditions are maintained in the sludge, LAS does not further biodegrade and as much as 1% of the dried weight of some sewage sludges consists of intact LAS. LAS concentration in sludge de-creases rapidly under conditions of aerobic digestion or composting (36).

In sewage sludge, the anionic surfactants are present as complexes or salts with cat-ionic materials. An ion exchange procedure is usually necessary to put the anionic mate-rial into a soluble form which can be measured (37,38). While much LAS leaves the sewage treatment plant as part of the sludge, the sulfophenylcarboxylic acid biodegrada-tion intermediates are more hydrophilic and are not generally found in the sludge (39).

Typical values for LAS are 3–7 ppm in the influent to a sewage treatment plant, 0.07 ppm or less in the effluent, 0.01 ppm or less in the receiving river (40,41). Values in the 0.2–3.4 ppm range are reported for lake and river sediments (42).

2. Other Anionic Surfactants

Secondary alkanesulfonate biodegrades to form alcohols as intermediates. The longer chain homologs are more likely to adsorb to particulates and sludge and are slower to de-grade. Alkanesulfonate has been reported in European sludges in the 0.01–0.1% concen-tration range (43,44).

A partially degraded sample of an alcohol sulfate will also contain alcohols and sul-fated carboxylic acids. Fish metabolites of ether sulfates include 4-polyethoxybutyric ac-ids, both sulfated and unsulfated (32).

Determination of soap is problematic because of other sources of natural fatty acids, particularly from partial biodegradation of fats and oils. Soap is only soluble at the sub–part per million range in sewage because of the insolubility of calcium salts. It is therefore associated with sludge and other solid fractions (45).

B. Isolation Techniques

Nondiscriminating measurements of anionic concentration, notably the spectrophotomet-ric methylene blue method (MBAS), can generally be performed without preliminary frac-tionation of an aqueous sample. Interference from other materials, including other surfactants, is minimal. This is easily confirmed by application of the "interference-limited" MBAS method, to determine whether the result is much different. However, if additional structural information about the anionic is required, complete separation of the anionic surfactant from other sample components is desirable to provide a concentrate suitable for characterization by spectroscopic or chromatographic methods.

In the case of LAS, 57% of that in a system was found to be adsorbed to river sedi-ment, 31% irreversibly adsorbed (46). Losses of LAS to laboratory apparatus can be re-duced or eliminated by addition of 3% NaCl or 0.003 M sodium dodecylsulfate (47).

1. Liquid-Solid Extraction and Supercritical Fluid Extraction

(a) Conventional solvents.   LAS is found adsorbed to sediment, soil, and sludge, with the more hydrophobic homologs being more strongly adsorbed (48). For desorption, most investigators use pure methanol, with either Soxhlet extraction or accelerated solvent ex-traction (ASE) giving recovery in the 80–90% range. Chloroform may be used for extrac-tion of LAS, but recovery is not as good as with methanol; results are in the 45% range (15). ASE gives better reproducibility than Soxhlet extraction, and is much faster (15). Solubilization with methanol in an ultrasonic bath has also been suggested to remove LAS from sediment (49). Kornecki et al. recommend 90:10 acetone/0.1 M NaCl for extraction of LAS and sodium lauryl sulfate from soil (50).

LAS as well as its sulfophenylcarboxylic acid degradation products may be concentrated from wastewater by first evaporating to near dryness, then taking up the residue in methanol, 0.05 M in sodium dodecylsulfate, with the help of an ultrasonic bath (51).

(b) Supercritical solvents. While supercritical CO_2 is a poor solvent for anionics, supercritical CO_2 saturated in methanol (about 40 mole%) gives quantitative recovery of LAS from soil and sludge (52–54). Supercritical CO_2 containing an ion pairing reagent such as tetrabutylammonium bisulfate gives complete recovery of LAS and alkane-sulfonate from sludge in much less time than is required for conventional liquid-solid extraction (43).

2.   Liquid-Liquid Extraction

Simple liquid-liquid extraction is not a good method for isolating anionic surfactants from water. At low pH and low ionic strength, LAS and its carboxylated degradation products are not extracted from water into nonpolar solvents (35). A preliminary concentration of LAS from aqueous samples can be made by extraction from 0.25 M salt solution with methylisobutylketone (MIBK). For greater selectivity, the LAS is back-extracted from the MIBK into water by first diluting the organic phase with five times its volume of hexane. Recovery from this back-extraction is reproducible but varies according to the length of the alkyl chain (49). Hellmann finds that use of *n*-hexane or cyclohexane leads to emulsion formation in environmental samples, and recommends chloroform or, preferably, 4:1 trichlorotrifluoroethane/chloroform (55).

A more efficient method of isolating anionic surfactants is extraction as part of an ion pair (33). An inorganic salt is added to decrease the solubility of the ion pair in the aqueous phase. Sometimes, the methylene blue spectrophotometric method described in Chapter 12 is used as the cleanup step. This permits the analyst to estimate the amount of surfactant isolated before proceeding with more definitive analytical techniques. Methylene blue may be removed from the surfactant extract by passage through a cation exchange column (56). If concentration is performed by liquid-liquid extraction of the ion pair with an alkyl quaternary compound, the UV spectrum of the ion pair is identical to that of LAS alone (55).

3.   Solid Phase Extraction

(a) Ion exchange. Concentration on an anion exchange resin is the most straightforward way to separate anionic surfactants from other surfactants. This method has been used for many years, and a number of examples are given in Chapter 6. Sulfophenylcarboxylate degradation products of LAS are also concentrated by this method. In a typical application, a methanol extract of a soil or sludge sample is passed through a strong anion exchange SPE column, which is then washed with pure methanol to eliminate cationic and nonionic materials. The anionic surfactants are eluted with a methanol/water solution of HCl.

(b) Adsorption. All of the common reversed-phase HPLC packing media will quantitatively concentrate LAS from aqueous solutions. C_{18} membrane disks give quantitative recovery for isolation of LAS from water, just as do the SPE cartridges (57–59). The C_2 media adsorb less interfering compounds than do C_8 or C_{18} media. Acidification to pH 3–4 aids in retention of low MW LAS homologs. The column is typically washed with methanol/water before eluting the LAS with 100% methanol. If HPLC with fluorescence

detection is to be used, C_2 concentration should be performed prior to ion exchange treatment (60).

While concentration of LAS from water samples is quantitative with C_{18} media, recovery of sulfophenylcarboxylic acid degradation products is low under the usual operating conditions (51). Quantitative recovery from water of the lower alkyl chain length sulfophenylcarboxylates is attained only by using small aqueous sample size, pH 1, and saturated NaCl solution (61). XAD-8 resin is superior for concentration of the sulfophenylcarboxylates. Again, retention of these compounds is only effective at low pH; only LAS is retained at high pH (35). Most recent publications describe concentration of SPC on graphitized carbon black, which gives the most quantitative recoveries for degradation products (28,29,33).

C. Qualitative and Quantitative Analysis

Environmental analysis of anionics has been recently reviewed (62). Selected publications which deal specifically with analysis of environmental samples are listed in Tables 1 and 2 and discussed below. Methods for determination of aromatic sulfonates, including LAS, have been reviewed by Reemtsma (63).

1. Chromatography

Gas chromatographic and HPLC methods are used for determination of anionics in the environment. Thin layer chromatography is used only in metabolism studies of radioactive-labeled compounds. Unlike GC, HPLC does not require that the nonionic be desulfonated or derivatized. The hydrophile/lipophile nature of the surfactant molecule makes it ideal for analysis by reversed-phase LC, where the polarity of the aqueous/ organic solvent mixture as well as the pH and the ionic strength can be tailored to optimize the resolution of the anionic from other compounds. While HPLC is suitable for determination of homolog distribution, it cannot give the high resolution of individual isomers which is available from GC, but this makes quantification easier. HPLC can determine LAS metabolites, like sulfonated and unsulfonated phenylcarboxylic acids, in the same chromatogram as LAS (33,64).

Compared to GC, HPLC requires more rigorous sample pretreatment, since HPLC is more sensitive to interferences in the chromatogram than is GC. While UV detection is often used for HPLC determination of alkylarylsulfonates, fluorescence detection is superior, providing improved selectivity. Since the UV absorbance of LAS is constant on a molar concentration basis, regardless of alkyl chain length, the raw UV chromatogram can be normalized to give the molar alkyl chain length distribution of the sample without the need for calibration of each individual peak on the chromatogram. The specific response of fluorescence detection is also considered to be similar for individual LAS homologs. However, the sensitivity of fluorescence detection for LAS degradation products varies greatly. For example, the minimum detectable level of sulfophenyl acetic acid by a fluorescence HPLC procedure was found to be 0.05 ppm, while the mimimum level for p-sulfobenzoic acid was 15 ppm (65).

Liquid chromatography–mass spectrometry is still difficult to use for routine analysis. However, the technique is essential for special projects, as was demonstrated for determination of the total alkyl and ethoxy distribution of alkyl sulfate and alkylether sulfate in wastewater (66). Because of the specificity of MS detection, a simple SPE technique could be used for concentration, and an equally simple one-step HPLC separation was adequate.

Capillary electrophoresis (CE) has been demonstrated for determination of LAS in wastewater and sludge. Conditions can be optimized to give either a single peak for total LAS or a series of peaks representing the alkyl chain length distribution (67). It is too early to predict whether CE will become more useful than the established methods for environmental analysis.

Gas chromatography, most often with mass spectrometric detection, is used when information about the isomer distribution as well as the carbon number of the alkyl chain is desired. The shortcoming of GC for routine analysis is the necessity of preparing volatile derivatives of the anionics. Quantification of the dozens of peaks from the various individual homologs is also a daunting task. Field and coworkers report quantitative determination of LAS and secondary alkanesulfonates by simply injecting a solution of the tetrabutylammonium salt into the hot injection port of the GC, where the compounds are converted to the butyl derivatives (59,68).

High performance liquid chromatography is the method of choice for determination of the first biodegradation products of LAS, the sulfophenylcarboxylic acids, especially those which do not readily form the indanones and tetralones which are detectable by GC. A single HPLC analysis can give information on LAS, NPE, and the ionic degradation products sulfophenylcarboxylic acids and nonylphenoxycarboxylates (51). Exploratory work generally requires LC-MS (28,29).

Procedure: HPLC Determination of LAS in Sea Water (69)

Water samples are preserved with formalin and refrigerated until analysis. One liter of sample is taken, NaCl is added to give a total concentration of 100 g/L, 5 g $NaHCO_3$ is added, and the solution is sublated with 100 mL ethyl acetate. Three 10-min sublations are performed and the extracts are combined and evaporated to dryness. The residue is dissolved in 10 mL water and passed through an SPE C_8 column (Analytichem, 500 mg/ 2.8 mL, preconditioned with 10 mL each MeOH and water). The column is washed with 5 mL water and 3 mL 40:60 $MeOH/H_2O$. The LAS is then eluted with 5 mL MeOH, which is allowed to pass through a SPE anion exchange column (Analytichem SAX 500 mg/2.8 mL, preconditioned with 10 mL MeOH). The column is washed with 10 mL MeOH, then the anions are eluted with 2 mL 4:1 MeOH/12 M aqueous HCl. The eluate is diluted with water to about 50 mL and neutralized to pH 7 with NaOH. This solution is passed through a second preconditioned C_8 SPE cartridge, which is then washed with 5 mL water. LAS is eluted with 5 mL MeOH, the eluate is evaporated to dryness, and the residue taken up in HPLC mobile phase.

HPLC conditions: Waters C_{18} column, 3.9 × 300 mm, protected by a C_{18} guard column; 80:20 $MeOH/H_2O$, both 0.08 M in $NaClO_4$; fluorescence detection, 232 nm excitation, 290 nm emission. Individual peaks are observed for C_{10}, C_{11}, C_{12}, and C_{13} LAS.

2. Spectroscopy

The routine method for determination of low levels of anionic surfactant is the spectrophotometric methylene blue method. This is the method used by regulatory bodies to routinely monitor water pollution (70). The methylene blue method is susceptible to interference but, in many matrices, interference is not a serious problem because LAS is the dominant MBAS-positive substance. This especially includes polluted streams and sewage plant influents. For careful work, it is necessary to cross-check the results of the MBAS method with a more specific technique, such as GC or LC (71). For example, in Mississippi River water, GC-MS analysis shows that the true LAS concentration is only 1–10% of the

TABLE 1 Determination of Anionic Surfactants in the Environment

Compound determined	Sample matrix	Sample workup	Quantification	Ref.
C_{10}–C_{13} LAS: separation by alkyl chain length; partial isomer separation	Fish from bioconcentration studies	Matrix solid phase dispersion extraction with C_{18} (washed with hexane and ethyl acetate then LAS eluted with 1:1 ethyl acetate/MeOH); liquid–liquid extraction from alkaline tetrabutylammonium hydroxide/water into 3:1 CH_2Cl_2/MeOH	RP-HPLC–fluorescence ($\lambda_{ex}/\lambda_{em}$ = 225/295 nm)	33
C_9–C_{15} LAS: separation by alkyl chain length	Water, sediment, sludge	*Liquids*: SPE on C_8 (MeOH elution); *solids*: Soxhlet extraction with MeOH and SPE on anion exchange resin (methanolic HCl elution) and C_8 (MeOH elution)	RP-HPLC–UV (230 nm) or fluorescence ($\lambda_{ex}/\lambda_{em}$ = 232/290 nm) detection	21,160,161
C_{11}–C_{14} LAS: separation by alkyl chain length and point of attachment	Lake water and sediment	Extraction with MIBK; back-extraction into water after dilution with hexane	RP-HPLC–UV (222 nm)	49
C_{10}–C_{13} LAS: separation by alkyl chain length and point of attachment; APE also determined	Wastewater	SPE on C_{18}, elution with acetone	RP-HPLC–fluorescence detection ($\lambda_{ex}/\lambda_{em}$ = 230/295 nm)	23,47
C_{10}–C_{13} LAS: separation by alkyl chain length and point of attachment	Sediment and sludge	Soxhlet extraction with MeOH, centrifugation	RP-HPLC–UV (225 nm) or fluorescence ($\lambda_{ex}/\lambda_{em}$ = 230/295 nm) detection	22,23
C_{10}–C_{13} LAS: separation by alkyl chain length and point of attachment	Sludge and soil	Soxhlet extraction with MeOH (solid KOH added to sample), SPE on C_{18} (MeOH elution)	RP-HPLC–fluorescence detection ($\lambda_{ex}/\lambda_{em}$ = 225/295 nm)	162
C_{10}–C_{13} LAS: separation by alkyl chain length	Sea and estuary water	Sublation, SPE on C_8 (MeOH elution), anion exchange resin (methanolic HCl elution), and another C_8 column (MeOH elution)	RP-HPLC–fluorescence detection ($\lambda_{ex}/\lambda_{em}$ = 232/290 nm)	69

	Sample matrix	Procedure	Detection	Ref.
C_9–C_{13} LAS: separation by alkyl chain length	River water and sewage	SPE on graphitized carbon black; LAS eluted with 0.01 M tetramethylammonium hydroxide in 90:10 CH_2Cl_2/MeOH	RP-HPLC–fluorescence ($\lambda_{ex}/\lambda_{em}$ = 225/290 nm)	26,163
C_{10}–C_{14} LAS and ABS	River water	Online SPE on anion exchange resin; elution with 50:50 CH_3CN/H_2O	For LAS, RP-HPLC-UV detection (220 nm); for ABS and LAS, anion exchange HPLC–UV detection (220 nm)	164,165
C_{10}–C_{14} LAS: separation by alkyl chain length and point of attachment	River water and sewage; sediment and sludge	*Water and sewage:* SPE on Amberlite XAD-2, elution with basic and acidic MeOH and MeOH/CHCl3; *sludge:* acid hydrolysis; *sediment:* Soxhlet extraction with MeOH, acid hydrolysis; *all:* acid ethyl ether extraction, anion exchange, 2nd acid ether extraction, desulfonation	GC	18
C_{10}–C_{15} LAS: separation by alkyl chain length	Sewage sludge	Methylene blue extraction into $CHCl_3$, TLC, reaction with PCl_5 to form sulfonyl chloride derivatives	GC and GC-MS	230
C_{10}–C_{13} LAS: separation by alkyl chain length	Wastewater	SPE on poly(styrene/divinylbenzene), elution with 9:1 MeOH/triethylamine	RP-LC–electrospray ionization MS	166,167
C_{10}–C_{14} LAS: separation by alkyl chain length and point of attachment	Sediment and river water	$CHCl_3$ extraction of ion pair with methylene blue, cation exchange (EtOH elution), reaction with PCl_5 and MeOH to form methyl sulfonate derivatives (sediment is first Soxhlet extracted with 60:40 benzene/MeOH)	GC or GC-MS	168
C_{10}–C_{14} LAS: determination of alkyl chain length and point of attachment	Sediment and river water	*Sediment:* MeOH extraction; *water:* SPE on C_{18} (CH_3CN and CH_2Cl_2 elution); reaction with PCl_5 and trifluoroethanol to form trifluoroethyl sulfonate derivatives	GC-MS	72

(Continued)

TABLE 1 (Continued)

Compound determined	Sample matrix	Sample workup	Quantification	Ref.
$C_{10}-C_{14}$ ABS and LAS; distinguished from each other; alkyl distribution	Groundwater	SPE on C_2 (MeOH elution) and anion exchange (acidic MeOH elution) media	RP-HPLC–fluorescence ($\lambda_{ex}/\lambda_{em}$ = 225/290 nm) and FAB-MS	34
ABS and LAS; distinguished from each other	Groundwater	SPE on Amberlite XAD-8, MeOH elution; anion exchange	Titration; ^{13}C NMR	20
LAS; total only	Rice stems and leaves	Ultrasonic extraction with MeOH. MeOH exchanged for H_2O; cleanup on alumina (H_2O elution), SPE on C_{18} (MeOH elution)	RP-HPLC–UV detection (225 nm)	169
C_9-C_{13} LAS; total only	River water	Filtration only; filter washed with MeOH	ELISA	170
$C_{10}-C_{16}$ LAS and ABS; quantification of homologs and isomers	Sediment	Soxhlet extraction with MeOH; SPE on anion exchanger (elution with 95:5 MeOH/con HCl); derivatization; further cleanup with alumina, silica, copper	GC-MS of trifluoroethyl derivatives	42,171
$C_{10}-C_{13}$ LAS; quantification of homologs; demonstration of ion-trap MS	River water and sewage	SPE on graphitized carbon black (elution with 9:1 CH_2Cl_2/MeOH, 0.01 M in tetramethylammonium hydroxide); derivatization; further cleanup with alumina	Ion trap GC/MS of 2,2,2-trifluoroethyl esters	172
$C_{10}-C_{13}$ LAS; determination of homologs and isomers	River water and sewage	SPE on graphitized carbon black (elution with 9:1 CH_2Cl_2/MeOH, 0.02 M in tetrabutylammonium bisulfate); butyl esters formed in GC injection port	Ion trap GC-MS of butyl esters	83
$C_{10}-C_{13}$ LAS; quantification of homologs	Sewage plant material balance: wastewater, sludge, and river water	*Liquids*: SPE on C_{18} resin for the lower concentrations and C_{18} disks for the higher; MeOH extraction; extracts were cleaned up by anion exchange and additional SPE on C_{18}	RP-HPLC–fluorescence ($\lambda_{ex}/\lambda_{em}$ = 232/290 nm)	40

Analyte	Matrix	Procedure	Detection	Ref.
C_{10}–C_{13} LAS; quantification of homologs	Wastewater and sludge	Water: SPE on C_{18} (MeOH elution), anion exchange (MeOH/HCl elution); sludge: MeOH extraction and anion exchange	CE-UV (200 nm)	67
C_{10}–C_{13} LAS and C_{14}–C_{17} alkanesulfonates; quantification of homologs and isomers	Sewage sludge	SFE with tetrabutylammonium bisulfate in CO_2	GC–MS of butyl esters	43
C_{10}–C_{13} LAS and C_{14}–C_{17} secondary alkanesulfonates; quantification of homologs and isomers	Wastewater	SPE on C_{18} disks (elution with tetrabutylammonium bisulfate in $CHCl_3$)	GC–MS of butyl esters	59,68
C_9–C_{13} LAS and C_{12} alkanesulfonate; quantification of homologs and isomers	River water	SFE on XAD-4 (elution with MeOH and MeOH/ethyl acetate azeotrope) or XAD-4/C_{18} and ion exchange separation on basic alumina	RP-HPLC; detection by ion-pair extraction and fluorescence	82,173
C_{10}–C_{13} LAS, C_{13}–C_{16} alkanesulfonate, C_{12}–C_{14} alkylether sulfate	Wastewater	SPE on C_{18} (elution with hexane, ethyl ether, and mixtures thereof)	Flow injection analysis–thermospray MS	80
C_{12}–C_{18} alkyl sulfates; alkyl distribution	Wastewater and river water	SPE on C_2 (MeOH elution), anion exchange (elution with 20:80 con HCl/MeOH), cation exchange (MeOH elution), derivatization to form alkyl TMS ethers	GC-FID	174
C_{12}–C_{18} alkyl sulfates; determination of alkyl distribution	Wastewater and river water	SPE on C_2 (elution with 80:20 MeOH/2-PrOH)	RP-LC-MS	66

(Continued)

TABLE 1 (Continued)

Compound determined	Sample matrix	Sample workup	Quantification	Ref.
C_{12}–C_{15} alkylether sulfates; determination of alkyl and ethoxy (1–8) distribution	Wastewater and river water	SPE on C_2	RP-LC-MS	66
C_{12}–C_{18} alkylether sulfates; alkyl distribution	Wastewater and surface water	Anion exchange (elution with 10% HCl in MeOH), hydrolysis, $CHCl_3$ extraction, reaction with HBr to form alkyl bromides	GC-FID	175
C_{12}–C_{18} soaps; alkyl distribution	Sewage and sewage sludge	*Sludge*: preliminary Soxhlet extraction with petroleum ether to remove lipids, addition of C_{17} internal standard, conversion to soluble salts, Soxhlet extraction with MeOH. *Sewage*: no workup	RP–HPLC of *p*-bromophenacyl bromide derivative–UV detection (254 nm), or 4-bromomethyl-7-methoxy coumarin–fluorescence detection ($\lambda_{ex}/\lambda_{em}$ = 328/380 nm)	45,176,177
Perfluorinated alkyl phosphinic and phosphonic acid	Laboratory biodegradation studies	SPE on C_{18} (MeOH elution)	Thermospray MS	178
Perfluorinated C_6–C_8 carboxylic acids	Ground water	SPE on strong anion exchange disks (elution with CH_3CN and CH_3I to form methyl esters)	GC-MS	179

TABLE 2 Determination of Impurities and Degradation Products from Anionic Surfactants in the Environment

Compound determined	Sample matrix	Sample workup	Quantification	Ref.
C_4–C_{13} sulfophenylcarboxylic acids; separation by alkyl chain length; LAS also determined	Wastewater and bench study	SPE on anion exchange resin (elution with 50:50 dioxane/1 M HCl)	HPLC–UV detection; MS confirmation	85
C_1–C_7 sulfophenylcarboxylic acids, C_2–C_5 phenylcarboxylic acids, C_1–C_2 hydroxyphenylcarboxylic acids, C_2 dihydroxyphenylcarboxylic acids; separation from each other; LAS also determined	Bench study	SPE of sample (pH 1, 30% NaCl, no more than 30 mL) on C_{18} (elution with acetone or MeOH)	RP-HPLC–UV detection (215 nm)	61,64
C_6–C_{12} sulfophenylcarboxylic acids; separation by alkyl chain length; LAS also determined	Sea water	SPE on C_{18} and anion exchanger (elution with 2 M HCl in MeOH)	RP-HPLC–ion spray MS	480,181
C_4 Sulfophenylcarboxylic acids; resolution of enantiomers	Bench studies and sewage	SPE on graphitized carbon (rinse with MeOH and 70:30 CH_2Cl_2/MeOH, then elute sulfonates with 80:20 CH_2Cl_2/MeOH, 0.05 M in NH_4OAc)	CE–UV detection (223 nm)	182
C_{10}–C_{13} LAS and C_2–C_9 sulfophenylcarboxylic acids; separation by alkyl chain length; also NPE and nonylphenolcarboxylates	Sewage	Evaporation/MeOH extraction or SPE on graphitized carbon	RP-HPLC–UV, fluorescence ($\lambda_{ex}/\lambda_{em} = 225/295$ nm)	26,51
C_2–C_9 sulfophenylcarboxylic and dicarboxylic acids, sulfotetralincarboxylic acids, methyl-branched LAS and its carboxylated intermediates; separation by alkyl chain length; LAS and dialkyltetralinsulfonates also determined	Bench studies, river water, and sewage	SPE of acidified sample on graphitized carbon black (washed with MeOH and acidified 80:20 CH_2Cl_2/MeOH; back-elution with basic 80:20 CH_2Cl_2/MeOH); methyl esters of carboxylates formed	RP-LC–electrospray MS	28,29

(Continued)

TABLE 2 (Continued)

Compound determined	Sample matrix	Sample workup	Quantification	Ref.
C_{11}–C_{14} alkylbenzenes; separation by carbon number and isomer separation; correction of interference from tetrapropylene-based ABS	Sediment	Extraction with CH_2Cl_2; Cu treatment to remove S; cleanup on silica (first hexane eluate discarded; then elution of alkylbenzenes with hexane and 70:30 hexane/CH_2Cl_2)	GC-MS	183
C_{10}–C_{14} alkylbenzenes; separation by carbon number; isomer separation	Sludge, sediment	Liquid-solid extraction of wet sample with 2-PrOH, 50:50 and 10:30 MeOH/$CHCl_3$, followed by liquid-liquid extraction between 3 25-mL portions of $CHCl_3$ and 10 mL saturated NaCl + 200 mL H_2O; extract taken up in 5 mL $CHCl_3$, purified of S by chromatography on HCl-activated Cu, and cleaned up in several steps with silica	GC-MS	81
C_{10}–C_{14} alkylbenzenes; separation by carbon number; isomer separation	Sludge, sediment	Homogenization sequentially with MeOH and CH_2Cl_2; MeOH extract diluted with H_2O, extracted with hexane, and the hexane and CH_2Cl_2 extracts combined; column chromatography on alumina/silica and elution with hexane, with separation from hydrocarbons by further silica chromatography using hexane	GC: confirmation by GC-MS	184
C_{10}–C_{14} dialkyltetralinsulfonates	Groundwater	SPE on C_2 (elution with MeOH) and anion exchange (elution with acidic MeOH) media	FAB-MS	34
C_{10}–C_{14} LAS and dialkyltetralinsulfonates	Sewage, river water, and sediment	*Water*: SPE on C_8 (CH_3CN elution); *sediment*: MeOH extraction, SPE from water on C_8 with CH_3CN elution	GC-MS of trifluoroethylsulfonate derivatives	84

C_{10}–C_{13} LAS and sulfophenylcarboxylic acids; quantification of homologs; demonstration of ion-trap MS	River water and sewage	SPE on graphitized carbon black (elution with 9:1 CH_2Cl_2/MeOH, 0.01 M in tetramethylammonium hydroxide); derivatization; further cleanup with alumina	Ion-trap GC/MS of 2,2,2-trifluorethyl esters	172
C_4-3-sulfophenylcarboxylic acid; C_{12} LAS also determined	Fish from bioconcentration studies	Matrix solid phase dispersion extraction with C_{18} (washed with hexane and ethyl acetate, then LAS and SPC eluted with 1:1 ethyl acetate/MeOH, MeOH, and 1:1 MeOH/H_2O); protein precipitation; ion-pair liquid-liquid extraction of LAS; SPE of SPC on graphitized carbon black	RP-HPLC–fluorescence ($\lambda_{ex}/\lambda_{em}$ = 232/292 nm)	33
C_{10}–C_{13} LAS and dialkyltetra-linsulfonates	Sewage, river water, and groundwater	SPE on graphitized carbon (wash with H_2O, MeOH, and 90:10 CH_2Cl_2/MeOH, 0.015M in formic acid; elute LAS and DATS with 90:10 CH_2Cl_2/MeOH, 0.010M in tetramethylammonium hydroxide)	RP-HPLC–fluorescence ($\lambda_{ex}/\lambda_{em}$ = 225/295 nm)	185
C_{10}–C_{14} dialkyltetralin- and di-alkylindanesulfonates and their carboxylated intermediates, C_3–C_{10} sulfophenylcarboxylates	Groundwater	Purification by cation and anion exchange, XAD-8 resin chromatography, and liquid-liquid extraction	^{13}C NMR, GC-MS of trifluoroethylsulfonate derivatives	35
C_{12}-dialkyltetralinsulfonate and its carboxylated intermediates, C_3–C_{10} sulfophenylcarboxylates	Laboratory biodegradation media	SPE on Macherey-Nagel Chromabond HR-P reversed phase (wash with 0.01 M HCl, elute with 3 mL THF and 3 mL MeOH)	GC-MS (negative ion mode) of trifluoroethylsulfonate derivatives; total S by ICP-AES	186

MBAS value (72). The so-called "interference-limited" methylene blue test is also likely to return high values when used for unpolluted samples. The MBAS method is discussed in Chapter 12. For routine wastewater analysis, the determination is frequently automated by use of flow injection analysis. The LAS degradation product sulfophenylcarboxylic acid will contribute to the MBAS response if the alkyl chain length is greater than 6 carbons. Dialkyltetralin-and dialkylindanesulfonates are also presumed to be MBAS-active (35). There are many variations of the methylene blue method used by various regulatory agencies. These differ mainly in the amount of effort expended to eliminate interferences (73).

Because other UV-absorbing compounds are present in environmental matrices, direct UV measurement has little value for determination of LAS (74). Derivative UV spectroscopy eliminates many interferences, especially if more than one wavelength is used (75). Hellmann describes determination of LAS concentrations along the length of German rivers by second-derivative UV spectrophotometry of extracts of the water made with trichlorotrifluoroethane/chloroform and distearyldimethylammonium chloride. In most cases, additional cleanup was unnecessary (55). A similar method has been proposed for LAS determination in sewage without preliminary sample cleanup, but a rigorous comparison to results from standard methods is yet to be made (76,77).

Nowadays, it is rare for infrared techniques to be used for qualitative or quantitative analysis of environmental materials. In either case, exhaustive separation of the surfactant from other materials must first be made. It is possible for inexperienced practitioners to go far wrong when identifying materials by IR, a technique best applied to pure compounds. Most environmental extracts, even after substantial cleanup, are mixtures which give complex spectra. It requires an experienced analyst to obtain useful information from the spectrum of a mixture containing unknown materials. As a general rule, a compound cannot be said to be present unless all of its characteristic absorbance bands are exhibited by the mixture. A once-common use of IR spectroscopy was confirmation of the identity of anionic surfactants isolated by the methylene blue spectrophotometric method. By proper choice of workup procedures and bands, this approach permitted exact determination of individual types of surfactants (78).

NMR spectrometry is not used extensively in environmental analysis, because of the large sample size requirement (20). ^{13}C NMR can be used to determine whether the alkyl chain of an alkylaryl sulfonate is branched or linear. It is also applicable to measuring simple mixtures of the two. With a large sample size and careful isolation techniques, ^{13}C NMR can give fairly complete characterization of LAS, ABS, and degradation products (35).

Mass spectrometry is now the preeminent method for environmental analysis, both qualitative and quantitative (Fig. 1). LC-MS with an electrospray interface permits simultaneous determination of all important LAS homologs, isomers, and degradation products (28,29). Because of the complexity of environmental samples and the difficulties in using MS for quantitative work, this is certainly not a method for routine analysis. MS is discussed in chapter 15.

Procedure: Spectrophotometric Determination of Anionic Surfactants in
Sewage Sludge (79)

Reagents:

> *Phosphate buffer solution*: Dissolve 10 g KH_2PO_4 in water, adjust the pH to 7.5 ±
> 0.1 with NaOH, and dilute to 1 L.

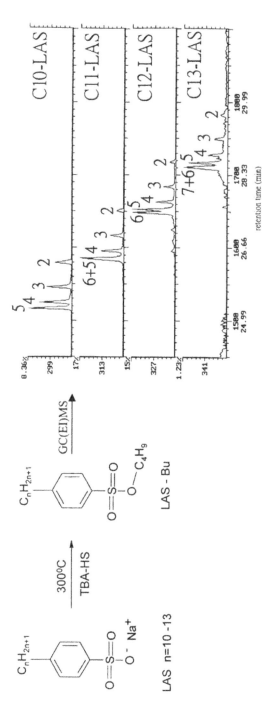

FIG. 1 GC-MS analysis of LAS isolated from river water. Shown are selected ion chromatograms, $[M - 55]^+$, for individual C_{10}–C_{13} homologs, indicating the isomer composition by point of attachment of the alkyl chain to the phenyl ring. Positive ion electron impact MS of butyl esters formed by injecting the tetrabutylammonium salts of LAS into the GC. (Reprinted with permission from Ref. 83. Copyright 1999 by Elsevier Science.)

Extraction solution: Add 4 drops 1-methylheptylamine to 100 mL *n*-hexane.

Prepare a cation exchange column from a strong cation exchange resin such as Bio-Rad AG 50W-X2, 50–100 mesh. Put into the H^+ form by washing with aqueous HCl, methanolic HCl, and then rinsing with MeOH until the eluent is free of acid. Put 10–15 mL resin in a glass column 15×200 mm. The sludge sample is isolated from the matrix by centrifugation, then freeze-dried and ground. One to two grams of sample is subjected to Soxhlet extraction with 180 mL MeOH, using a piece of glass wool to prevent the powder from being carried away with the MeOH. The extraction is continued about 2 hr, then the MeOH is evaporated to a smaller volume, transferred to a 100-mL volumetric flask, and made up to volume with additional MeOH. Transfer an aliquot of this solution (from 1 to 50 mL, expected to contain 20–150 mg of anionic surfactant if paired-ion extraction cleanup is to be used, or 200–500 mg if TLC cleanup is planned) to the cation exchange column. Rinse the solution through the column with MeOH, collecting 100 mL eluate in a 250-mL beaker. Discard the column.

Procedure 1: Purification by Paired-Ion Extraction

Evaporate the eluate to dryness. Dissolve the residue in 30 mL concentrated HCl and 100 mL water. Cover with a watch glass, and heat just to boiling, simmering for 1 hr, then remove the watch glass and evaporate to dryness on a steam bath. Dissolve the residue in 25 mL water, disregarding any insoluble portion. Make alkaline to phenolphthalein with 0.1 M NaOH and boil for about 2 min. Let cool and rinse into a 250-mL separatory funnel with water to make up to about 100 mL volume. Add 0.1 M H_2SO_4 until the pink color just disappears, then add 10 mL phosphate buffer solution. Extract four times with 25-mL aliquots of 1-methylheptylamine solution in hexane, shaking for 1 min each time. Combine the extracts in a 250-mL beaker, add 1–2 mL 1 M NaOH, and evaporate to dryness on a steam bath. It is essential that the amine be completely expelled at this stage. (There should be no residual amine odor.) Dissolve the residue with 1–2 mL methanol, then add about 50 mL water, cover with a watch glass, and boil for 15 min. Water lost by evaporation is replaced from a wash bottle, rinsing down the sides of the beaker. After cooling, transfer the solution to a 250-mL separatory funnel, neutralize to phenolphthalein with 0.1 M H_2SO_4, and add water to bring the volume to 100 mL. Determine the anionic surfactant by the methylene blue spectrophotometric procedure described in Chapter 12.

Procedure 2: Purification by Thin Layer Chromatography

The eluate from the ion exchange column is evaporated almost to dryness, then transferred to a 10-mL flask with conical bottom and evaporated to dryness. The residue is taken up in exactly 0.2 mL MeOH.

A Silica Gel F-60 plate is activated for 1 hr at 110°C and cooled in a desiccator. Apply 0.1 mL of the MeOH extract to the plate, either as a stripe or as a series of 20-mL spots. A standard containing 20 mg ABS is also spotted on the plate. The plate is developed with 6:1 $CHCl_3$/MeOH to a height of 10 cm and dried under a gentle stream of air. Development is with a spray of pinacryptol yellow, 0.1% in water. Anionic surfactants are visualized as red-orange spots with an Rf value of 0.10–0.15. The silica gel containing the surfactant is scraped from the plate and transferred to a small glass column or to a small separatory funnel plugged with glass wool. The surfactant is removed from the silica with two 5-mL MeOH rinses. The MeOH eluate is collected in a 50-mL beaker, evaporated to about 5 mL, transferred to a 250-mL separatory funnel, and made up to a volume of

100 mL with water. The anionic surfactant is determined by the methylene blue spectro-photometric procedure described in Chapter 12.

Discussion: This is the method of the analytical working group of the German Committee on Detergents. In the case of activated sludge, the blank value for sludges known to contain no anionic surfactants is in the range 0.2–0.4 parts per thousand (ppt), dry weight. Recovery of added surfactant was 80% in the 2–10 ppt range. Schröder reported only 68% recovery for secondary alkanesulfonate, but good results for LAS and an alkyl ether sulfate (80).

HPLC analysis can replace the spectrophotometric determination: The sample preparation follows Procedure 1, above, except that NaOH is not added when the hexane extract is evaporated. The residue after evaporation is dissolved in 5 mL of the HPLC mobile phase A. The separation is performed on a reversed-phase C_{18} column such as Shandon ODS or Whatman Partisil 5 ODS 3, 4.6 × 250 mm. A mobile phase gradient is used, where solution A is water, 0.1 M in $NaClO_4$, and solution B is 70:30 acetonitrile/water, both 0.1 M in $NaClO_4$. The gradient is formed from 70% B to 90% B in 25 min. Detection is by UV absorbance at 225 nm.

Procedure: Determination of Alkylbenzenes in Ocean Sediment and Sewage Sludge (81)

Wet sediment samples (about 20 g) are subjected to successive liquid-solid extraction with 40-mL portions of 2-PrOH, 50:50 $MeOH/CHCl_3$, and 10:30 $MeOH/CHCl_3$. The combined extracts are washed with 200 mL H_2O plus 20 mL saturated NaCl solution. The aqueous phase is extracted twice with 25-mL portions of $CHCl_3$ and the $CHCl_3$ phases are combined, evaporated to near dryness, and taken up in 5 mL $CHCl_3$. The extract is purified of elemental sulfur by passage through a small column packed with copper and activated with 3 M HCl. The column is rinsed with a further 5 mL $CHCl_3$ and the eluate is evaporated to near dryness. This extract is taken up in 3:1 hexane/CH_2Cl_2 and applied to a 0.9 × 35 cm column containing 7 g silica gel deactivated with 5% water. The column is eluted with 25 mL 3:1 hexane/CH_2Cl_2. This eluate is cleaned up further by transferring to a 0.5 × 18 cm column containing silica (deactivated 5 hr at 200°C). The column is eluted with 6 mL hexane. This first eluate contains aliphatic hydrocarbons and is discarded. A second fraction of 24 mL hexane contains the alkylbenzenes. This fraction is evaporated to dryness and taken up in isooctane for GC-MS analysis: J&W Durabond DB-5, 0.32 mm × 30 m, 0.25 μm film thickness; initial 3-min hold at 100°C, then ramp to 260°C at 4°C/min, then to 300°C at 20°C/min. The 1-phenyl linear alkylbenzene isomers are added to the original samples as internal standards.

If the sample is a sewage sludge, first centrifuge the sample, then analyze the precipitate in the same way as sediment. Extract hydrocarbons from the supernate with *n*-hexane and clean up with silica as described above.

3. Standards

When isolating substances from environmental matrices, it is best to use an internal standard to monitor recovery. In the case of LAS, suitable internal standards are C_8-, C_9-, and C_{15}-LAS, since these are absent from commercial products and the environment (72,82,83). Since the 1-phenyl LAS isomers of any chain length are not found in the commercial product, these are also suitable for internal standards (33). For MS analysis, deuterated analogs of anionics may be used as internal standards, but these are rarely commercially available (66).

A C_9-dialkyltetralinsulfonate may be used for DATS determination (84). For determination of alkylbenzenes, the 1-phenyl isomers (absent from the industrial product) are suitable standards (81).

Sulfophenylcarboxylic acids may be synthesized by sulfonation of the corresponding phenylcarboxylic acids (30,64). The C_3-3-SPC isomer is useful as an internal standard since it cannot be formed from commercial LAS (33). *p*-Toluenesulfonic acid has also been used (85).

Primary C_{12} and C_{18} alkanesulfonates are commercially available, absent from ordinary detergents, and therefore suitable as internal standards in alkanesulfonate analysis (43,68).

III. NONIONIC SURFACTANTS

A. Surfactants and Their Metabolites

At this writing, the predominant nonionic surfactants found in municipal wastewater are C_{12} and C_{14} alkyl ethoxylates and nonylphenol ethoxylates. The predominant degradation product is polyethylene glycol.

1. Alcohol Ethoxylates

Intermediate degradation products of an alcohol ethoxylate are PEG, long-chain carboxylates, especially mono- and dicarboxylated PEG, and short-chain acids, such as formic, acetic, glyoxylic, and oxalic acids (86). Alkyl ethoxycarboxylates are also found, as well as AE with a carbonyl group at the end of the alkyl chain. In the case of alcohol propoxylates, initial degradation converts the terminal hydroxy to an aldehyde group (80). In a thorough degradation study, these also must be determined, as well as the intact surfactant and CO_2.

Alcohol ethoxylates with branched alkyl chains are found in commercial products, although linear AEs are more common. The metabolites of branched AE are more refractory to further degradation than metabolites containing the linear chain (87).

Fluorinated nonionics of the type where the alkyl chain is perfluorinated show biodegradation of the ethoxy chain similar to AE, while the fluoroalkyl chain is apparently untouched (88).

Ozonolysis and electrochemical oxidation yield shorter ethoxylates than the starting material, fatty acid, fatty alcohols, and esters of fatty acids and ethylene glycol (89).

2. Alkylphenol Ethoxylates

Biological degradation of alkylphenol ethoxylates is generally agreed to begin with cleavage of the ethoxy chain to form shorter ethoxylates, accompanied under aerobic conditions by formation of carboxylate terminal groups. These products are referred to as alkylphenoxy carboxylic acids, with $C_9H_{19}C_6H_4OCH_2CH_2OCH_2COOH$ being the carboxylated analog most abundant in the environment. The alkyl chain is attacked later in the process, with the more common branched chains more resistant than straight chains. Degradation of the alkyl chain, again under aerobic conditions, produces terminal carboxylates, usually resulting in a dicarboxylated molecule because of prior attack on the ethoxy chain. Although ortho-substituted NPE is a small impurity in the commercial surfactant, the main para-substituted product degrades more rapidly in the environment with the result that ortho-substituted degradation products can form a high percentage of NPE metabolites in materials such as composted sewage sludge (90).

The literature is not in agreement, but it can be said that free nonylphenol is usually found in sewage sludge in the 100 ppm order of magnitude under anaerobic conditions and is not found if the sludge is digested under aerobic conditions. Chlorination of sewage effluent can sometimes lead to the formation of chlorinated and brominated analogs of APE and carboxylated APE. Complete analysis of biodegradation products should therefore include detection of intact APE, both types of carboxylated APE, phenol and alkylphenol, and determination of average degree of ethoxylation.

Better than determining the average degree of ethoxylation is the actual measurement of individual homologs, since APE with only one or two EO units is agreed to be more persistent and toxic than the original surfactant. As degradation proceeds, determination of surfactant by, for example, UV absorption, will give higher values than determination by the CTAS spectrophotometric method, since the ratio of aromatic to polyether functionality increases. Isolation techniques which discard the anionic portion of the sample will reject the carboxylated intermediates.

Photolytic degradation can eventually lead to almost complete destruction of APE. Intermediates in degradation include formates and aldehyde-terminated polyethers, as well as shorter-chain ethoxylates (91). Depending on the conditions, other products include formaldehyde and acetaldehyde, methyl and ethyl ethers of PEG, and insoluble, polymeric material (92).

Values reported for levels of APE and its metabolites vary too much to allow a statement of typical concentration. OPE is found at levels less than 10% of those of NPE, usually much less (93). Microbial degradation of the OPE metabolite octylphenoxyacetic acid results in formation of 2,4,4-trimethyl-2-pentanol (94). Ozonolysis of APE selectively destroys the aromatic ring, while electrochemical oxidation yield shorter ethoxylates than the starting material (89). A review of environmental analysis of APE appeared in 1997 (95).

3. Other Nonionics

Aerobic biodegradation of polysorbate 80, a 20-mole ethoxylate of sorbitan monooleate, results in immediate hydrolysis and degradation of the oleic acid function. Terminal hydroxyl groups on the polyethoxy chains are oxidized to the carboxylic acids. Further biodegradation is slow compared to AE (96).

B. Isolation Techniques

The analyst must remain aware that some of the metabolites of nonionics are more polar than the original surfactant, and may not be isolated by the same procedures (97).

1. Solid Phase Extraction

Solid phase extraction concentration of nonionics is usually performed with nonpolar C_{18} or XAD® adsorbants. C_{18} resins may be used to isolate carboxylated APE degradation products as well as nonionics (51,97). For PEG determination, glass cartridges are preferred, since a blank value is obtained from conventional cartridges made of polypropylene with polyethylene frits (12).

The macroreticular Rohm and Haas resins of the XAD series are excellent for the concentration of certain organic materials, including ethoxylated surfactants, from aqueous matrices. Amberlite XAD-2, XAD-4, and XAD-8 have been successfully used for determination of nonionics. Typically, the resin is first pre-extracted with organic solvents to remove interference, then water samples of as much as 50 L are allowed to flow through columns packed with small amounts (about 5 g) of the resin. Large amounts of suspended solids cannot be tolerated, since they block the column. Various solvents are used to fractionally elute surfactants and interfering compounds from the resin. Usually, hexane or petroleum ether is used to wash off oily materials prior to eluting surfactants and metabolites with more polar solvents such as ethyl ether, methanol, water, and mixtures thereof.

Nonionic surfactants are usually isolated from natural waters on XAD resin at neutral pH. This is to prevent the co-adsorption of fulvic acids which occurs under acidic conditions (98). Very hydrophilic metabolites, such as low molecular weight PEG and PEG carboxylates, may not be adequately retained on XAD resin unless the procedures are specifically optimized for these compounds (99).

Conventional ion exchange is routinely used to separate anionic and cationic material from nonionics concentrated by other means. SPE on a strong anion exchanger disk is effective in isolating carboxylated analogs of low ethoxylates of nonylphenol from wastewater (100). An unusual application of ion exchange is preparation of a polyethoxylate-specific resin by putting an anion exchange resin in the tetracyanatocobaltate(II) form with a solution of the ammonium salt. Nonionics may then be concentrated by pumping natural waters through a column of the resin, and then eluting the nonionic with ethanol. The technique gave yields of 72–85% when tested with an OPE and a sorbitan monolaurate ethoxylate (101).

Natural waters may be sampled by pumping a large quantity, two liters or more, through a column of granular activated carbon. This adsorbent is nonspecific, with most organic compounds adsorbed to some extent. However, the technique has been used in conjunction with subsequent fractionation and very specific mass spectrometric detection techniques to determine AE, APE, and APE degradation products in river water (102,103). A conventional SPE column containing octadecylsilica is suitable for concentrating APE, including nonylphenol and the mono- and diethoxylated oligomers, from sea water (104). Additional sodium chloride is added to the sample before concentration. The cartridge is rinsed free of salts with a small amount of water, then the APE and other surfactants are eluted with acetone. Nonylphenol can be recovered from water on C_{18} cartridges or disks with minimum loss if about 10% NaCl and 0.0001 M SDS is added (105).

Best recovery in concentration of APE from natural waters is attained by first passing the sample through a mixed bed ion exchange column, then concentrating APE on a C_{18} SPE column. Losses to the ion exchange column are prevented by addition of 15% methanol to the sample. Nonionic surfactants are eluted from the C_{18} column with methanol. For trace analysis, it is essential to add AE to the sample to minimize APE losses through adsorption to the glassware. Nonylphenol and low oligomers cannot be reliably concentrated by either sublation or SPE and besides are lost during solvent evaporation. They are best enriched by steam distillation (106).

Alumina can be used for cleanup of extracts during analysis of nonionic surfactants. A great many ionic and polar nonionic impurities are removed by this treatment (107). Typically, impurities are eluted from the alumina with 50:50 hexane/methylene chloride, followed by elution of the surfactants with 100:1 methylene chloride/methanol.

Solid phase microextraction, i.e., extraction of an analyte onto the surface of a small fiber inserted into an aqueous sample, was demonstrated for OPE isolation using an experimental PEG-coated fiber. The fiber was then placed in a specially modified HPLC injection port for quantitative analysis (108). Concentration of AE using a fiber coated with silicone/divinylbenzene polymer has also been demonstrated (109).

2.   Liquid-Solid Extraction and Supercritical Fluid Extraction

Marcomini and Giger report that alkylphenol and alkylphenol ethoxylates can be removed from sewage sludge by Soxhlet extraction with methanol (22). Solid NaOH is first added to the dried sludge at 20 g per 100 g sample. After concentration of the extract and dilution with an HPLC solvent, particulates must be removed by centrifugation. Field and Reed found that hot water/ethanol extraction at high, but subcritical, pressure (75°C, 150 bar) was the best method for recovering NPE carboxylates from sewage sludge. Technique is important, since without the right mixture of filter aid, proper flow through the matrix is not achieved (90). Heise and Litz used pure methanol to extract NPE. Nearly quantitative recovery is possible from soil and sediment, but only about 30% is recovered from sewage sludge. Comparable recoveries are obtained with Soxhlet extraction, ASE, and microwave-assisted extraction (15). Methanol extraction of the dried solids is suitable for removal of NPE and nonylphenoxycarboxylate from sewage (51).

Nonylphenol can be recovered from freeze-dried sediment and sewage sludge by Soxhlet extraction with 2:1 methylene chloride/methanol. p-(1,1´,3,3´-tetramethylbutyl)phenol is a suitable internal standard for the extraction, although this is sometimes already present in samples (110). Excellent recoveries are also obtained using supercritical fluid extraction with methanol-modified CO_2 (111). Microwave-assisted solvent extraction was shown to give comparable results to Soxhlet extraction in isolating added nonylphenol from environmental matrices, with great savings in time. Methylene chloride is a suitable extracting agent (105). At this stage in its development, SFE seems to offer little advantage over Soxhlet extraction for nonylphenol analysis. SFE of APE from sediment with methanol-modified CO_2 at 450 atmospheres gives recoveries in the 65–85% range (53,54).

Working with sediment, Ikonomou et al. report that pure hexane will extract low oligomers but that a more polar solvent must be added for higher oligomers. Best results were obtained with 70:30 isopropanol/hexane for Soxhlet extraction and 60:40 hexane/acetone for sonication. Sonication worked as well as Soxhlet extraction and could be performed very much faster (112).

3.   Liquid-Liquid Extraction

Liquid-liquid extraction is useful for isolation of surfactants and metabolites from laboratory degradation experiments. Because of its lack of specificity and the large volumes of solvents needed, it is less popular for analysis of the lower concentrations of surfactants in environmental matrices.

Inaba reports that AE-type nonionics may be isolated from environmental waters by toluene extraction (113). The presence of salts, as in sea water, will cause the co-extraction of anionic surfactants and other materials, so that a preliminary ion exchange step may be required prior to extraction. Liquid-liquid extraction with methylene chloride is generally applicable to isolation of ethoxylated surfactants from water (114). Liquid-liquid extraction with chloroform or methylene chloride was found to be equivalent to sublation in concentrating the NPE metabolite nonylphenoxyacetic acids from water. A pH of 2 is suitable for the extraction (35). Sublation was superior for analysis of sewage, since emulsion formation was minimized (115).

PEG is almost quantitatively removed from saturated brine by chloroform extraction and this is the principle of its determination in ethoxylates. Extraction is incomplete at one molar salt content, about 70% for the higher homologs and dropping to 12% for $n = 7$ (116).

Organics isolated from water by solid phase extraction may be separated by liquid-liquid extraction into an acidic and a basic-plus-neutral fraction following standard "priority pollutant" procedures. The acidic fraction contains the carboxylated metabolites of NPE and AE (117).

4. Steam Distillation

Continuous steam distillation combined with continuous extraction of the distillate with cyclohexane or isooctane is the most effective technique for separation of nonylphenol and its low mole EO adducts from water (106). Since isooctane has the higher boiling point, it is more suitable for recovery of the 1- and 2-mole adducts than is cyclohexane (118). The separation is nearly quantitative for homologs up to $n = 3$, with recovery dropping to only 24% for $n = 6$ (119).

C. Qualitative and Quantitative Analysis

Selected applications of analytical techniques to the determination of nonionics and their metabolites in environmental matrices appear in Tables 3–6.

1. Chromatography

Reviews have appeared recently covering chromatographic methods for determination of AE and APE in the environment (120,121).

HPLC is the method of choice for specific determination of nonionics. Complete characterization of both the alkyl chain and the ethoxylate chain is possible by using a combination of reversed-phase and normal phase HPLC. The disadvantages of conventional HPLC are (a) the need to prepare UV-absorbing or fluorescent derivatives in the case of compounds like AE or the EO/PO copolymers which do not already contain suitable chromophores and (b) the extensive sample preparation needed to eliminate interference. The 1- and 2-mole nonylphenoxyethoxylates may be determined simultaneously with NPE using reversed-phase HPLC in paired-ion mode (51).

Excellent selectivity for APE is provided by fluorescence detection. Quantification in reversed-phase mode requires only the measurement of the single peak for nonylphenol ethoxylates (octylphenol derivatives are rarely found), while normal phase chromatography permits determination of the homolog distribution (104,106). For even more selective fluorescence detection of APE, the analysis may be made with and without acetylation. Many impurities lose their fluorescence after acetylation. The change in retention time of the acetylated compounds confirms the identity of the APE peaks (93). Fluorescence detection is much more sensitive than UV detection for nonylphenol (105).

If MS detection is used, LC separation may be made in only one mode (e.g., reversed-phase separation by alkyl chain length) and without derivatization. The mass spectrometer is then capable of determining the ethoxylate distribution of each alkyl homolog. Quantification is best performed using an internal standard, since the mass spectrometer does not respond uniformly to each oligomer (112,122,123).

Gas chromatography is generally suitable for determining nonionics of low molecular weight in environmental matrices, provided that volatile derivatives are formed as discussed in Chapter 8. HBr cleavage coupled with GC determination of the dibromoethane and alkyl bromides formed can be used to monitor the concentration of ethoxylated alcohols. The surfactant must first be isolated in reasonably pure form prior to reaction. This last method cannot differentiate between different types of ethoxylated surfactants (124). More recently, GC-MS detection of alkyl bromide derivatives has been applied to determining AE in wastewater (125).

GC of the methyl esters is effective in determination of carboxylate analogs of NPE. MS detection is preferred, with 4-bromophenylacetic acid being a suitable internal standard. Single point calibration should not be used, since the calibration curve is not linear (100).

Procedure: Determination of Alkylphenol Ethoxylates in Wastewater and
River Water (106)

During all steps, the sample and work-up fractions are protected from atmospheric oxygen with a blanket of nitrogen. For determination of nonylphenol and 1- and 2-mole EO adducts, a 1-L water sample is subjected to steam distillation in a Nielsen-Kryger apparatus (Ace Glass 6555-13), using 2 mL isooctane to trap the analyte. The isooctane solution is injected directly into the HPLC. For determination of higher adducts, a 1-L sample is mixed with 150 mL MeOH and passed through a column containing 40 mL Bio-Rad 501-X8 mixed ion exchange resin and a second column containing 0.7 g Baker C_{18} adsorbant. APE is eluted from the C_{18} column with MeOH at 55°C. Five milligrams AE are added to the MeOH eluate to minimize losses and the solution is evaporated carefully to dryness at 45°C with a stream of dry nitrogen. The residue is dissolved in 25:75 CH_2Cl_2/hexane for HPLC analysis.

HPLC analysis is conducted with a Rainin Microsorb CN-bonded column, 4.6 × 250 mm, and a fluorescence detector (229 nm excitation, 310 nm emission). Solvent A is 80:20 hexane/THF, which passes through a column of activity I alumina to remove peroxides (change every 4 days) before mixing. Solvent B is 90:10 2-PrOH/H_2O. Solvents are carefully deaerated, and 100 ppm sodium sulfite is added to deaerated water to minimize free oxygen. The gradient is adjusted for good resolution, consisting approximately of 1% to 42% B in 20 min. Peak identification is checked by comparison to well-characterized commercial products, and detector response is taken as being proportional to molar concentration.

2. Spectroscopy

(a) UV spectrophotometry. Nonionics may be determined by extraction of the complex with potassium picrate and UV determination of the amount of picrate ion extracted. This procedure is described in Chapter 12. Collaborative studies indicate that this method is suitable for routine monitoring of wastewater, provided that the nonionic is first isolated by solvent sublation and purified by anion/cation exchange chromatography (126).

(b) Visible spectrophotometry. In the western hemisphere, the most widely used method for determination of nonionic surfactants in environmental matrices is the spectrophotometric procedure based on extraction of the blue potassium thiocyanatocobaltate complex in the presence of polyalkoxy groups. This procedure is discussed in detail in Chapter 12. While PEG gives a positive response to the CTAS method, carboxylated PEG does not, because the complex is not extracted into methylene chloride at neutral pH. The method may be made sensitive to carboxylated PEG by conducting the extraction at pH 2 and adding salt to decrease the aqueous solubility of the complex (99).

TABLE 3 Determination of Alcohol Ethoxylate Surfactants in the Environment

Compound determined	Sample matrix	Sample workup	Quantification	Ref.
C_{10}–C_{18} alkyl ethoxylates; alkyl distribution and distribution of 2- to 20-mole ethoxy homologs	Sewage plant streams	SPE on Amberlite XAD-2 (elution with ethyl ether, 50:50 ethyl ether/MeOH, and MeOH), liquid-liquid extraction with ethyl acetate, ion exchange, CTAS extraction with CH_2Cl_2; derivatization with phenylisocyanate	RP- and normal phase HPLC–UV detection	187
C_{12}–C_{18} alkyl ethoxylates; alkyl distribution	Sewage plant streams	Sublation, SPE on alumina (elution with 50:50 hexane/CH_2Cl_2 [discarded] and 100:1 CH_2Cl_2/MeOH); derivatization with phenylisocyanate	RP-HPLC–UV or MS detection	107
C_{12}–C_{15} alkyl ethoxylates; distribution of 3- to 20-mole ethoxy homologs	Biodegradation bench study	Sublation, ion exchange, CTAS extraction with CH_2Cl_2	Normal phase HPLC detection by evaporative light scattering or fluorescence of phenylisocyanate derivatives ($\lambda_{ex}/\lambda_{ex} = 240/310$ nm)	118
C_{12}–C_{18} alkyl ethoxylates, E = 0–20; alkyl distribution	Sewage and surface water	SPE on C_{18} disk (MeOH elution); ion exchange cleanup; derivatization with 1-naphthoylchloride	RP-HPLC–fluorescence ($\lambda_{ex}/\lambda_{em} = 300/385$ nm)	118
C_{12}–C_{16} alkyl 4- to 14-mole ethoxylates; alkyl and ethoxy distribution	Fish from bioconcentration studies	Matrix solid phase dispersion extraction with C_{18} (washed with hexane, then AE eluted with ethyl acetate and 1:1 ethyl acetate/MeOH); Al_2O_3 cleanup (washed with 1:1 hexane/CH_2Cl_2 and 100:1 CH_2Cl_2/MeOH, then eluted with additional 100:1 CH_2Cl_2/MeOH and 100:3 CH_2Cl_2/MeOH); derivatization with naphthoyl chloride; further Al_2O_3 cleanup (washed with 1:1 cyclohexane/CH_2Cl_2, then eluted with 95:5 CH_2Cl_2/MeOH)	RP-HPLC–fluorescence ($\lambda_{ex}/\lambda_{em} = 300/385$ nm)	189

C_{12}–C_{15} alkyl 9-mole ethoxylates; determination of individual homologs	River water and sewage	SPE on C_8 (elution with MeOH and 2-PrOH)	RP-LC–thermospray MS	122,134,190
C_{12}–C_{18} alkyl 1- to 18-mole ethoxylates; determination of individual homologs	River water and sewage effluent	SPE on C_{18} disk (MeOH elution); mixed bed ion exchange	RP-LC–thermospray MS	123
C_{12}–C_{18} AE; determination of individual homologs; NPE also determined	River water, drinking water, wastewater	SPE on graphitized carbon black (elution with 80:20 CH_2Cl_2/MeOH)	RP-LC–electrospray MS	191
C_{12}–C_{14} AE; determination of individual homologs; NPE also determined	Wastewater	SPE on C_{18} (elution with hexane for high ethoxylates, with 4:1 CH_2Cl_2/hexane to elute low ethoxylates; then with 9:1 MeOH/CH_2Cl_2 to elute NPE)	RP-LC–atmospheric pressure chemical ionization MS	166,167
C_{12}–C_{15} alkyl ethoxylate; resolution by alkyl chain-length	River water and sewage	SPE on C_1 (elution with 50:50 ethyl acetate/MeOH), ion exchange cleanup, HBr cleavage	GC-MS of alkyl bromide cleavage products	125

TABLE 4 Determination of Alkylphenol Ethoxylate Surfactants in the Environment

Compound determined	Sample matrix	Sample workup	Quantification	Ref.
OPE, E = 10, NPE, E = 11; ethoxy distribution and total	Sediment and sludge	Soxhlet extraction (NaOH added) with MeOH, centrifugation after addition of H_2O and acetone	RP-HPLC for total APE and normal phase-HPLC for ethoxy distribution–UV (277 nm for normal phase, 225 nm for reversed-phase) or fluorescence (225 or 230 nm excitation, 295 nm emission) detection	22
Total APE	Wastewater	SPE on C_{18} or graphitized carbon black (elution with 70:30 CH_2Cl_2/MeOH)	RP-HPLC–fluorescence detection (225 nm excitation, 290 nm emission)	26,47
Total APE and total NPE	Wastewater	SPE on graphitized carbon black (elution with 80:20 CH_2Cl_2/MeOH)	RP-HPLC–fluorescence detection (225 nm excitation, 295 nm emission), with and without derivatization with acetic anhydride; confirmation by LC–electrospray MS	93
NPE, homolog distribution for E = 1–14	Wastewater	Sublation, ion exchange	Quantification by RP-HPLC; homolog distribution by normal phase HPLC; UV (229 nm)	192
NPE, E = 10: octylphenol also determined	Fish and mussels	Matrix solid phase dispersion extraction with C_{18} (MeOH elution); Al_2O_3 cleanup of MeOH eluate	RP-HPLC–fluorescence (225 nm excitation, 301 nm emission)	193
NPE; OP, NP, 17β-estradiol and 17α-ethynylestradiol also determined	Surface water and sewage effluent	SPE on poly(styrene/divinylbenzene) disk (elution with acetone, $MeCl_2$, hexane), NP-HPLC fractionation	RP-HPLC–fluorescence (229 nm excitation, 310 nm emission)	194
APE, E = 2–26; ethoxy distribution	Sewage influent and effluent	Sublation, ion exchange, alumina cleanup (EtOH elution)	Normal phase HPLC–fluorescence detection (230 nm excitation, 302 nm emission)	195

NPE, E = 0–7; ethoxy distribution	Sewage influent and effluent	Continuous steam distillation/liquid–liquid extraction with cyclohexane	Normal phase HPLC–UV (277 nm)	119
APE, E = 1–18; ethoxy distribution	River and waste water and sediment	Ion exchange cleanup, SPE on C_{18} (MeOH elution)	Normal phase HPLC–fluorescence detection (229 nm excitation, 310 nm emission)	106,196
NPE, E = 0–17; ethoxy distribution	Biodegradation bench study	Sublation, ion exchange, CTAS extraction with CH_2Cl_2	Normal phase HPLC–fluorescence detection (230 nm excitation, 310 nm emission)	118
NPE, E = 0–19; ethoxy distribution	Industrial effluents	Sublation	Normal phase HPLC–fluorescence detection (230 nm excitation, 302 nm emission)	136,137
NPE, E = 0–17; ethoxy distribution	Sewage sludge	Supercritical fluid extraction with CO_2/H_2O	Normal phase HPLC–fluorescence detection (230 nm excitation, 300 nm emission)	197
NPE, E = 0–19; ethoxy distribution	Marine sediment	Accelerated solvent extraction of freeze-dried sample with 50:50 acetone/hexane or Soxhlet extraction (Na_2SO_4 added) with 70:30 hexane/2-PrOH or sonication with 60:40 hexane/acetone; SPE on CN phase to which was added layers of Na_2SO_4 and activated copper powder (initial deposition on cartridge with 90:10 hexane/CH_2Cl_2; rinsed with hexane; elution with acetone)	Normal phase HPLC–electrospray MS detection	112,198

(Continued)

TABLE 4 (*Continued*)

Compound determined	Sample matrix	Sample workup	Quantification	Ref.
NPE; distribution of 0- to 16-mole ethoxy homologs	Sewage influent and effluent	CH_2Cl_2 extraction of water	Electrospray MS or atmospheric pressure ionization MS	199
Total NPE; AE also determined	River water, drinking water, wastewater	SPE on graphitized carbon black (elution with 80:20 CH_2Cl_2/ MeOH)	RP-HPLC–electrospray MS detection	191
NPE; determination of individual homologs; AE also determined	Wastewater	SPE on C_{18} (elution with hexane followed by 4:1 CH_2Cl_2/hexane for AE; then with 9:1 MeOH/ CH_2Cl_2 to elute NPE)	RP-LC–atmospheric pressure chemical ionization MS	166,167

TABLE 5 Determination of Other Nonionic Surfactants in the Environment

Compound determined	Sample matrix	Sample workup	Quantification	Ref.
C_7 alkyl 7-mole propoxylate, C_6 perfluorinated alkyl 9-mole ethoxylate, C_7–C_{15} alkyldiethanolamides	Water, wastewater, and biodegradation media	SPE on C_{18}: hexane wash, then elution of alkylpropoxylates with 8:2 ethyl ether/hexane. AE with 8:2 MeOH/H_2O, and diethanolamides with MeOH (some AE eluted with MeOH)	FIA–thermospray MS	80
C_{12}–C_{16} alkylpolyglycosides	Wastewater	SPE on C_{18} cartridge or disk: MeOH elution	RP-HPLC–ELS detection	200,201
C_{12}–C_{14} alkylpolyglycosides	Bench-scale toxicity testing media	None	GC-FID	200,201
C_8–C_{16} alkylpolyglycosides	Bench-scale biodegradation media	*Water*: SPE on C_{18} cartridge (MeOH elution); *sludge*: Freeze-dry. MeOH and 98:2 MeOH/12 M HCl extraction. extracts neutralized, freeze-dry, and concentrated by SPE on C_{18} (MeOH elution)	FIA–thermospray MS and LC–thermospray MS	202
C_8–C_{12} alkylpolyglycosides	Bench-scale biodegradation media and sewage influent	*Bench studies*: direct analysis; *sewage*: SPE on poly(styrene/divinylbenzene) (MeOH elution)	RP-LC–electrospray MS	203
C_6–C_{10} perfluorinated alkyl ethoxylate; identification of individual compounds	Biodegradation media	SPE on C_{18} (elution with 60:40 hexane/ethyl ether)	FIA–thermospray MS-MS	88
Total nonionics	Activated sludge sewage plant study	Sublation, ion exchange	BiAS precipitation, derivatization HPLC for AE and APE; cleavage GC for total alkoxylates	204
Total nonionics: APE, separation of 2- to about 20-ethoxy homologs	Trickling filter sewage plant study	Sublation, ion exchange, alumina treatment (for HPLC)	BiAS precipitation, HPLC–fluorescence detection	195

Straightforward application of the analytical methods described in earlier chapters will often lead to erroneous results when examining environmental matrices. For example, analytical results based on the spectrophotometric CTAS method for nonionic surfactant determination will usually give a much higher value in sewage plant effluent than that obtained from use of more specific methods (71). It is important to check the results of the analysis of a new matrix by a second analytical method to assure that positive or negative interference is not a problem.

(c) Infrared spectroscopy.   It is sometimes possible to monitor nonionics in a well-characterized industrial wastewater stream by simply extracting with methylene chloride and measuring the IR spectrum of the extract (114).

(d) NMR spectroscopy.   Proton NMR has been demonstrated for nonspecific determination of ethoxylated material in environmental extracts. In this case, the signal from the methylene hydrogens in the polyethoxy chain is integrated (99). For greater precision, the peak may be compared to an internal standard such as pentachlorethane (127). ^{13}C NMR provides additional qualitative information, allowing identification of residues of EO/PO copolymers in environmental matrices (99).

(e) Mass spectrometry.   The most interesting developments in methodology for environmental analysis are in mass spectrometry. A number of studies have been published demonstrating GC-MS determination of AE, APE, and their degradation products. MS tends to resist efforts at precise quantitative analysis, but much progress has been made. GC-MS is more and more applied to quantitative analysis of low levels of surfactants. Direct MS, on the other hand, has so far been useful only for qualitative determination of nonionics, since the ionization process discriminates in favor of certain compounds and certain homologs. Nonylphenol is determined with great sensitivity as the pentafluorobenzyl derivative by GC-MS using chemical ionization in the negative ion mode. Under the proper GC conditions, this is an isomer-specific procedure (110,128). LC-MS with electrospray (now the most common interface for surfactants) or thermospray interface has been successfully used for determination of ethoxylates and their degradation products after preconcentration by some method (12).

3. Titration

(a) BiAS method.   The ISO method for determination of nonionic surfactants in environmental samples is based on precipitation of the nonionic with potassium tetraiodobismuthate/barium chloride (modified Dragendorff reagent), with subsequent determination of the bismuth by potentiometric titration. Alternatively, bismuth is determined by atomic absorption spectroscopy or by UV absorption of the bismuth/EDTA complex.

This procedure has official status for use in bench-scale biodegradibility testing (129) and is discussed in Chapter 17. It must be noted that this test is only effective for ethoxylated surfactants and fails altogether with alkyl polyglycosides. Waters et al. have discussed the limitations of the method for environmental analysis (130). They recommend that samples not be filtered or centrifuged before sublation, that four 10-min sublations be used, and that combined anion/cation exchange cleanup of the sublate be employed. The effect of these variables is more pronounced with raw sewage than in finished effluent analyses. For a family of effluents where the nonionic surfactant was mainly NPE, the BiAS method response was shown to be higher by factors of 2–3 for influents and 2–10 for effluents, compared to HPLC analysis (131).

(b) Potassium tetrakis(4-halophenyl)borate method. The complex formed between potassium ion and a nonionic surfactant can be extracted as an ion pair with the tetrakis(halophenyl)borate from an aqueous into an organic phase. When the nonionic surfactant is exhausted, excess titrant serves to change the color of a cationic dye in the organic phase for visualization of the end point. This method is described in Chapter 16. A collaborative study indicates that this procedure is suitable for determination of low parts per million concentrations of nonionics in wastewater, provided that sublation and ion exchange cleanup are first performed (126).

4. Electrochemistry

One research group advocates an electrochemical method, indirect tensammetry, for determination of nonionics in environmental samples. This is based on the decrease in the size of the amperometric peak of ethyl acetate in the presence of surfactants. Sensitivity is gained by preliminary extraction or sublation of the sample, or use in conjunction with the BiAS test (132,133).

5. Calibration Standards

Decylphenol monoethoxylate has been used as an internal standard to correct for incomplete recovery of nonionics (82). An elegant method of internal standard calibration for LC-MS determination of AE uses an ethoxylate of a fully deuterated fatty alcohol (134).

Pure AE homologs can be prepared by reacting the corresponding 1-bromoalkane with, e.g., triethylene glycol, hexaethylene glycol, or nonaethylene glycol (13). This approach is especially suitable for synthesis of ^{14}C-labeled materials. AE carboxylates are prepared by oxidation of AE with chromic/sulfuric acid, followed by extraction of the acids from saturated salt solution into methylene chloride (135).

Nonylphenol ethoxylates are by far the most commonly used APEs, so standardization is usually performed with NPE. Ibrahim and Wheals report calibration of APE by spiking in a single isomer, the *tert*-octylphenol 9-ethoxylate. The spike recovery is determined by measuring the apparent excess of the E = 9 peak from the HPLC chromatogram of the ethoxylate distribution of the unknown. The standard is synthesized by reacting the corresponding phenol with 1,2-*bis*(2-chloroethoxy)ethane, followed by reaction with the sodium salt of hexaethylene glycol (136,137). A synthetic internal standard, 1-(4′-methoxyphenyl)hexan-1-ol has been used for LC–fluorescence determination of APE (93).

Preparation of ^{14}C-labeled ethoxylates where the label is on the ethoxy chain is straightforward, since labeled ethylene oxide can be obtained commercially. Preparation of ring-labeled standards of APE requires the prior preparation of ^{14}C-labeled nonylphenol (138), which can then be ethoxylated (139).

The carboxylates of the 1- and 2-mole EO adducts of nonylphenol can be made by oxidizing the corresponding NPE with chromic acid in sulfuric acid. In a similar manner, carboxylated PEG can be prepared by oxidation of commercially available PEG with acidic potassium dichromate (86,99). An alternative path to nonylphenoxyacetic acid and its ethoxyacetic and diethoxyacetic acid analogs is reaction of nonylphenol and nonylphenol low mole ethoxylates with chloroacetate under alkaline conditions (51,94,100,140). Ring-brominated analogs are made by reaction of the isolated standard with bromine water (94). 2,4,4-Trimethyl-2-pentanol can be produced by hydration of 2,4,4-trimethylpentene (94).

TABLE 6 Determination of Impurities and Degradation Products from Nonionic Surfactants in the Environment

Compound determined	Sample matrix	Sample workup	Quantification	Ref.
AE photolytic degradation products: lower ethoxylates, low MW acids, PEG	Laboratory photolytic degradation study	Direct analysis or liquid-liquid extraction with CH_2Cl_2	Electrospray MS; also GC, GC-MS, NMR	205,206
Carboxylated degradation products of branched C_{12} AE	Laboratory activated sludge degradation study	SPE on graphitized carbon black (elution of neutrals with 80:20 CH_2Cl_2/CH_3OH, then elution of acids with 80:20 CH_2Cl_2/CH_3OH containing 0.05 M formic acid	RP-LC–electrospray MS	87
AE and APE degradation products: halogenated APE, AE and APE ethoxycarboxylates; detection	River and drinking water	SPE on carbon or XAD-2 (CH_2Cl_2 elution; liquid-liquid extraction with CH_2Cl_2 and ethyl ether to separate an acidic and a neutral + basic fraction; formation of methyl esters of the acidic fraction	GC-MS or FAB-MS	102,103,117
PEG ($n = 4$–35), mono-($n = 4$–13) and dicarboxylated ($n = 4$–21) PEG; semiquantitative determination	River, sea, and well water; sewage	SPE on graphitized carbon black (elution with CH_3OH and 80:20 CH_2Cl_2/CH_3OH to remove PEG, then elution of carboxylated PEG with 80:20 CH_2Cl_2/CH_3OH containing 0.02 M HCl to form the methyl esters); sea water is first acidified to control Cl^- interference	RP-LC-MS (electrospray)	12
Mono- and dicarboxylated AE and mono- and dicarboxylated PEG	Biodegradation bench test and sewage	SPE on graphitized carbon black (washed with CH_3CN and 80:20 CH_2Cl_2/CH_3CN; cartridge reversed and acids eluted with 80:20 CH_2Cl_2/CH_3CN containing 0.01 M formic acid)	RP-LC–fluorescence of 9-chloromethylanthracene derivatives	207

PEG	Wastewater	SPE on C_{18} followed by (at pH 3) poly(styrene/divinylbenzene) (MeOH elution of only the second column)	RP-LC–atmospheric pressure chemical ionization MS	166,167
PEG; approximate separation of short-chain and longer-chain compounds	River water and sewage	Sequential liquid-liquid extractions	Electrochemistry	208
OPE degradation products: lower ethoxylates, PEG, aldehydes, ethers	Laboratory photolytic degradation study	Liquid-liquid extraction with $CHCl_3$ and ethyl acetate	Radiochemical techniques; CI-MS; RP-HPLC-UV and TLC of DNB derivatives	92
OPE degradation products: lower ethoxylates, aldehydes, formates, octylphenol	Laboratory photolytic degradation study	Not described	HPLC with UV or electrospray MS detection	91
NPE degradation products: lower ethoxylates, carboxylates of 1- and 2-mole NP ethoxylate, nonylphenol	Experimental composting of wool scour sludge	Nonylphenol, NP 1 to 3-mole ethoxylates and ethoxycarboxylates: wool wax extracted from compost with CH_2Cl_2, then cleaned up on size exclusion resin (elution with 50:50 hexane/CH_2Cl_2) and derivatized with BSA silylating agent; NPE where E > 3: wool wax extracted from compost with CH_2Cl_2, then extract cleaned up by SEC, then by SiO_2 (elution of impurities with 50:25 and 50:50 hexane/ethyl acetate, then elution of NPE with 1:2 ethyl acetate/MeOH), this fraction also silylated	GC-FID, with MS confirmation	209

(Continued)

TABLE 6 (*Continued*)

Compound determined	Sample matrix	Sample workup	Quantification	Ref.
NPE degradation products: lower ethoxylates, phenol, 4-*n*-propylphenol, nonylphenol, CO_2	Laboratory photolytic degradation study	Liquid-liquid extraction with CH_2Cl_2	Normal phase HPLC-UV (277 nm) for ethoxy distribution and nonylphenol; RP-HPLC-UV (277 nm) for other phenols; headspace GC for CO_2	210
Halogenated APE, carboxylated APE analogs; homolog separation up to E = 5	Biodegradation bench studies	Liquid-liquid extraction with ethyl ether; silica gel cleanup (MeOH elution)	Derivatization with diazomethane (for determination of carboxylates) and GC-MS	211,212
2,4,4-trimethyl-2-pentanol, brominated octylphenol, and 2-aminomethoxy-3-bromo-5-(1,1,3,3,-tetramethylbutyl) phenol from degradation of the OPE metabolites octylphenoxyacetic acid and its ring-brominated analog	Bench biodegradation studies with groundwater	Liquid-liquid extraction with CH_2Cl_2	GC-MS, with or without diazomethane derivatization	94
Nonylphenol	Sludge, sediment	Homogenization sequentially with MeOH and CH_2Cl_2; MeOH extract diluted with H_2O, extracted with hexane, and the hexane and CH_2Cl_2 extracts combined; column chromatography on alumina/silica, washing with hexane and 8:1 hexane/CH_2Cl_2 before elution with CH_2Cl_2	GC of BSTFA derivative; confirmation by GC-MS	184
Nonylphenol	Sewage sludge and marine sediments	Soxhlet extraction with 2:1 CH_2Cl_2/MeOH; chromatography on alumina/Na_2SO_4; derivatization with pentafluorobenzyl bromide	GC, GC-MS	110

Nonylphenol	Sewage sludge and effluent	*Water*: acetylation and extraction with petroleum ether; *sludge*: acetylation and extraction with supercritical CO$_2$; *all*: final column chromatography on silica (elution with 50:50 CH$_2$Cl$_2$/petroleum ether	GC-MS	213
Nonylphenol	Sludge from treatment system of a terephthalic acid plant	Study of supercritical fluid extraction	GC-MS	111
Nonylphenol	Sea water	SPE on XAD-2 resin (elution by Soxhlet extraction with 90:10 CH$_3$CN/H$_2$O); further unspecified separation	GC-MS	214
Nonylphenol	Atmosphere and estuary water	*Atmosphere*: adsorption on polyurethane foam (desorption by Soxhlet extraction with petroleum ether); alumina fractionation (elution with 2:1 MeOH/CH$_2$Cl$_2$); *water*: SPE on XAD-2 resin	GC-MS	215
Octylphenol; NPE also determined	Fish and mussels	Matrix solid phase dispersion extraction with C$_{18}$ (MeOH elution): Al$_2$O$_3$ cleanup of MeOH eluate	RP-HPLC–fluorescence (225 nm excitation, 301 nm emission)	193
OP and NP; NPE, 17β-estradiol and 17α-ethynylestradiol also determined	Surface water and sewage effluent	SPE on poly(styrene/divinylbenzene) disk (elution with acetone, MeCl$_2$, hexane); NP-HPLC fractionation	RP-HPLC–fluorescence (229 nm excitation, 310 nm emission)	194
Nonylphenol and 1- and 2-mole ethoxylates	Marine water, sediment, and algae	*Water*: SPE on C$_{18}$ (acetone elution); *sediment and algae*: Soxhlet extraction with hexane; cleanup by SPE NH$_2$ media (elution with 75:25 hexane/acetone)	Normal phase HPLC–fluorescence (225 nm excitation, 304 nm emission)	216

(Continued)

TABLE 6 (*Continued*)

Compound determined	Sample matrix	Sample workup	Quantification	Ref.
Nonylphenol and 1-, 2-, and 3-mole ethoxylates	Wastewater, sludge, mussels	*Water*: liquid-liquid extraction (CH_2Cl_2); *sludge*: extraction (acetone/H_2O): cleanup with ethyl ether; *mussels*: extensive workup	GC-ECD or GC-MS of pentafluorobenzoyl chloride derivatives	128
Alkylphenol 1- and 2-mole ethoxylates	Sewage effluent	Continuous liquid-liquid extraction with hexane; silica chromatography (elution with hexane, CH_2Cl_2, and ethyl ether)	GC-MS	14
APE carboxylates; qualitative analysis	Groundwater	Purification by cation and anion exchange and liquid-liquid extraction with CH_2Cl_2	^{13}C NMR	35
Carboxylates of nonylphenol mono- and diethoxylates	River water and sewage	Liquid-liquid extraction with $CHCl_3$ or sublation, silica chromatography (MeOH elution), derivatization with MeOH/BF_3	GC-FID, GC-MS, or normal phase HPLC-UV (277 nm)	115
Carboxylates of nonylphenol mono- and diethoxylates	Sewage sludge	Supercritical fluid extraction with CO_2/H_2O, derivatization with MeOH/BF_3	GC-MS	197
Carboxylates of nonylphenol mono- and diethoxylates; low resolution separation together with LAS and sulfophenylcarboxylic acids	Sewage	MeOH extraction of dried solids or SPE on C_{18} or graphitized carbon black (elution with 70:30 CH_2Cl_2/MeOH)	HPLC-UV, fluorescence (225 nm excitation, 295 nm emission)	26,51

Compound	Sample	Extraction/Preparation	Detection	Ref.
Alkylphenol and alkylphenol 1- and 2-mole ethoxylates	Sediment and sludge	Soxhlet extraction (NaOH added) with MeOH, centrifugation after addition of acetone, H_2O, n-hexane; analysis of n-hexane layer	RP-HPLC for total APE and normal phase HPLC for ethoxy distribution; UV (277 nm for normal phase, 225 for reversed-phase) or fluorescence (225 or 230 nm excitation, 295 nm emission) detection	22
Octylphenol, nonylphenol, and nonylphenol 1-mole ethoxylate	River water, sewage, river sediment, and fish	Continuous steam distillation/solvent extraction (cyclohexane); alumina cleanup (CH_2Cl_2 elution)	GC-FID and GC-MS	217
Carboxylates of NPE 1 to 4-mole ethoxylates	Sewage and river water	SPE on anion exchange disk, treatment of disk with CH_3CN and CH_3I to form methyl esters	GC-MS	100
Carboxylates of NPE 1- to 4-mole ethoxylates	Industrial and sewage sludge	Subcritical 70:30 water/ethanol extraction (75°C, 150 bar), SPE on anion exchange disk, treatment of disk with CH_3CN and CH_3I to form methyl esters	GC-MS	90
Nonylphenol and nonylphenol 1-, 2-, and 3-mole ethoxylates; carboxylates of NPE carboxylates, E = 1 and 2, with carboxylation of both alkyl and ethoxy chains	Sewage and river water	SPE of acidified sample on graphitized carbon black (elution with 90:10 CH_2Cl_2/MeOH, 0.025 M in formic acid): formation of propyl or butyl esters in injection port	GC–ion trap MS	218,229
NPE carboxylates, E = 1–3, with carboxylation of both alkyl and ethoxy chains	Bench studies and sewage	SPE on graphitized carbon black (elution of neutrals with 80:20 CH_2Cl_2/MeOH, then elution of acids with 80:20 CH_2Cl_2/MeOH, 0.050 M in formic acid); formation of methyl esters	RP-LC with electrospray MS detection	219

(Continued)

TABLE 6 (*Continued*)

Compound determined	Sample matrix	Sample workup	Quantification	Ref.
PEG- and PPG-containing compounds, including carboxylated species	River water	SPE on XAD-8 (elution with 75:25 CH_3CN/H_2O), liquid-solid extraction with $CHCl_3$, SiO_2 chromatography of methyl derivatives, further XAD-8 SPE (elution with 25:75 CH_3CN/H_2O)	^1H and ^{13}C NMR and CTAS	99
PEG of 5–18 units; total only	River water and biodegradation media	Extraction of surfactants with ethyl acetate; extraction of PEG with $CHCl_3$	Electrochemistry	220
Alkylphenol and 1- and 2-mole EO adducts	River and wastewater and sediment	Steam distillation (isooctane extraction)	Normal phase HPLC–fluorescence detection (229 nm excitation, 310 nm emission)	106,196
Aldehydes from C_7 alkyl 1- to 13-mole propoxylates, carboxylates derived from C_6 perfluorinated alkyl 2-to 10-mole ethoxylates, and C_{12} and C_{13} alkyl 1- to 8-mole ethoxylates, PEG of 6 to 16 units	Water, wastewater, and biodegradation media	SPE on C_{18}, MeOH elution	FIA–thermospray MS	80,97
Carboxylates derived from C_6–C_{10} perfluorinated alkyl ethoxylates	Biodegradation media	SPE on C_{18}, MeOH elution	FIA–thermospray MS	88

IV. CATIONIC SURFACTANTS

A. Surfactants and Their Metabolites

The most important cationic surfactant found in the environment of most of the world is ditallowdimethylammonium chloride (also called distearyldimethylammonium chloride; alkyl chain length typically 65% C_{18} and 30% C_{16}). This compound is added to clothing as a fabric softener and is washed from the garments during subsequent laundry cycles. Typical concentrations found in countries using this material are as follows: sewage, 0.5 ppm; treated sewage, 0.05 ppm; sewage sludge, 0.3%; surface waters, 0.04 ppm or less. Monoalkyltrimethylammonium compounds are introduced to the environment mainly as impurities (2–5%) in the dialkyldimethyl surfactants, and their concentration in environmental samples is about 5% that of the dialkyl compounds. Long chain alkylnitriles, dialkylmethylamines, and trialkylamines are also thought to occur in the environment as a result of being impurities in fabric softeners (141,142).

In Europe, ditallowdimethylammonium chloride lost its preeminence in 1991 due to biodegradibility concern. Significant quantities of quaternary imidazoline derivatives are used as fabric softeners, especially in Europe. "Ester quats" are also widely used; these are ditallowdimethylammonium or imidazolium compounds which contain ester linkages to shorten their lifetime in the environment. Both these and the conventional imidazolium compounds can usually be determined by the same methods used for alkyl quaternaries. However, care must be taken to avoid analytical conditions such as extremes of pH that hydrolyze ester groups.

Laboratory studies of microbiological degradation of quats shows formation initially of a dialkylmethylamine N-oxide. This may be subsequently cleaved by two enzymatic paths, one resulting in formation of alkenes and hydroxylamines, the other yielding trialkylamine and formaldehyde (143). Fatty amines in general biodegrade with loss of the alkyl chains to give ammonia, methylamine, and dimethylamine. The main degradation intermediate of the most common ester quat, N-methyl-*bis*(tallowacyloxyethyl)-2-hydroxyethylammonium methylsulfate, is *tris*(hydroxyethyl)methylammonium methylsulfate.

B. Isolation Techniques

Cationics can be easily recovered from solution by ion exchange, extraction with an ion pairing agent, or solvent sublation. A more difficult problem is the recovery of cationics adsorbed to particulate matter in the sample. For example, they will typically be found strongly adsorbed to river sediments, rather than being dissolved in the water itself. In sewage streams, the cationics are found mainly in the suspended solids. The primary mechanism is cation exchange, although other processes also occur (144). Amphoterics show the same tendency, especially at low pH where their cationic properties are strongest (145). Gerhards and Schulz found that SPE did not give reproducible results for trace analysis of amphoterics and therefore recommend sublation (145).

A related problem is the affinity of cationics to laboratory glassware. Gerike et al. preconditioned glassware for distearyldimethylammonium chloride determination by soaking it with aqueous solutions of the surfactant overnight, then rinsing successively with water, methanol, chloroform, methanol, and water. The glassware was then reserved for this analysis (146). A few parts per million of a nonionic surfactant is usually added to liquid samples to inhibit adsorption of the cationics to the surface of the sample container. Gerhards and Schulz recommend that special glassware be used for trace analysis of amphoteric surfactants and that alkaline cleaners be avoided. They found the best results with glassware coated with an oily surfactant film from previous use (145).

Klotz determined that quantitative removal of cationics from clay minerals is quite difficult (147). He based his conclusions on experiments with a synthetic montmorillonite made with 34% distearyldimethylammonium ion rather than sodium ion. The most complete extraction was obtained with methanolic HCl, but even this resulted in only 72% recovery. Essentially complete recovery was obtained by destroying the clay matrix by evaporation with hydrofluoric acid. Similar behavior is observed with dodecylpyridinium ion (144).

Hellmann separated cationics from sewage sludge and clay minerals by extraction with 1:1:1:1 methanol/ethanol/chloroform/HCl (148,149). The sludge was first dried at 50°C, then a 10 g sample was shaken for 15 min with 100 mL mixed solvent, exercising caution to avoid excess foaming due to evolution of CO_2. The supernate was drawn off after centrifugation, and its volume measured for future reference. This extract contained many impurities; further purification was required by TLC or by precipitation with Dragendorff reagent. In later work, Hellmann found that ion exchange of sewage extracts on alumina served to separate cationics adequately for quantitative analysis (37,38). In one sludge, he found 9 ppt cationic surfactant after ion exchange, while it was not detectable before alumina treatment. This confirms that cationics are normally complexed with anionic materials, often LAS, in sludge.

Heise and Litz, using pure methanol, showed extraction efficiencies from sludge, sediment, and soil in the rather broad range of 35–100% for dihexyldimethylammonium bromide and benzyldimethylhexadecylammonium chloride. Ester quats are not stable in methanol and could not be determined (15).

Supercritical fluid extraction with CO_2 containing at least 10% methanol is effective for removing distearyldimethylammonium chloride from sediment and sewage sludge as an ion pair. Because anionic surfactants are already present in sludge, no anion need be added to insure ion pair formation. SFE is not more complete than conventional liquid-solid extraction with methanolic HCl, but is more rapid. There is some indication that less of materials other than cationic surfactants is extracted by SFE compared to conventional extraction (150,151).

When solvent sublation is used to isolate cationics, an anionic surfactant is sometimes added in excess to minimize adsorption of the cationic. This requires that the ion pair be separated later in the analysis. One method to minimize loss of cationics by adsorption to the apparatus is to pretreat the apparatus with a solution of the cationic, then rinse with water and solvent. Obviously, blank determinations must be run to confirm that this treatment is not a source of spurious high results (151). Dioctyldimethylammonium bromide has been used as an internal standard to correct for incomplete recovery of cationics (82).

C. Qualitative and Quantitative Analysis

A typical analysis of an environmental sample includes preliminary separation by liquid-liquid extraction or SPE; cleanup by ion exchange, additional extractions, or alumina adsorption; and final determination of cationic surfactant by HPLC or MS.

The general test for cationic surfactants in environmental samples is a spectrophotometric test based on formation and extraction of a colored ion pair with an anionic dye, disulfine blue. The result of this test is designated as "disulfine blue active substance," analogous to the "methylene blue active substance" and "cobalt thiocyanate active substance" used as a measure of anionic and nonionic surfactants, respectively. The disulfine blue test suffers from interference from other cationic materials, but can be made more specific by addition of cleanup steps (152). Chromatographic methods, most often HPLC, allow more specific determination of distearyldimethylammonium chloride or of other cationics. The identity of the HPLC peaks can be confirmed by IR or MS analysis.

Waters has written a well-balanced review article on environmental analysis of cationics giving a historical perspective on the development and application of the various procedures, including official methods (153). Published methods for determination of cationic surfactants in the environment are summarized in Tables 7 and 8.

1. Chromatography

Paired-ion chromatography, with detection by conductivity (146,154) or postcolumn reaction (155), is suitable for determination of quaternary amines in environmental samples, provided that the samples are first treated by extraction of the dried solids with acidified methanol and liquid-liquid partition of the ion pair with alkylarylsulfonate. For more concentrated aqueous samples, such as plant effluents, direct analysis is possible.

Procedure: Determination of Quaternary Ammonium Compounds in Sewage and River Samples by HPLC (156)

Samples are preserved with 1% formalin and stored at 4°C in polyethylene bottles. Five parts per million AE is added to minimize loss by adsorption to the container. Samples (2 L for river water, 1 L for sewage effluent, 500 mL for sewage influent) are evaporated to dryness, then extracted four times with 20 mL hot methanol containing 4% HCl. After removal of the insolubles by centrifugation, the combined extracts are evaporated to dryness, then the evaporated residue is transferred to a separatory funnel with 100 mL water and 5 mL concentrated HCl. Two milligrams LAS is added to form the ion pair with the quaternary, and the solution is extracted three times with 50 mL chloroform. The organic extract is evaporated down to about 10 mL volume, washed three times with 5 mL water, then evaporated to dryness. The extract is stored as a solid, dissolving in chloroform just prior to HPLC analysis. Dried sludge samples of 1 g are extracted with acidified methanol until the extract is colorless. Purification of the extract is as described above.

HPLC analysis is on a Whatman Partisil 5 PAC column, 4.6×250 mm, protected by a guard column also packed with a cyano/amino stationary phase. The mobile phase is an isocratic mixture of 89:10:1 chloroform/methanol/acetic acid, with the column rinsed with methanol at the end of each day.

Discussion: This method is capable of resolving mono- and dialkyl quaternaries from each other. There is also partial resolution according to alkyl chain length. Normally, four peaks are seen for the monoalkyl compounds, corresponding to chain lengths of C_{18}, C_{16}, C_{14}, and C_{12}, respectively. For the dialkyl compounds, it is usually preferable to decrease resolution so that only one sharp peak is observed. This can be accomplished by adding a small amount of acetonitrile to the mobile phase (155). The resolution of the peaks is quite sensitive to mobile phase composition, so that it may be necessary to modify the separation parameters whenever column or solvents are changed.

2. Spectroscopy

The disulfine blue spectrophotometric method is used most often for determination of cationic surfactants in the environment. Other dyes may also be used, as described in Chapter

TABLE 7 Determination of Cationic Surfactants in the Environment

Compound determined	Sample matrix	Sample workup	Quantification	Ref.
Ditallowdimethylammonium chloride, dodecyltrimethyl-ammonium chloride, hexadecylpyridininium chloride, stearyldimethylbenzyl-ammonium chloride	Waste- and river water	Extraction of ion pair from acid into CH_2Cl_2	Normal phase HPLC–conductivity detection	221,222
Distearyldimethylammonium chloride, N-dodecylpyridinium chloride, ditallowimidazolium methylsulfate, ditallow ester quat	Biodegradation test media	Paired ion extraction from acid into CH_2Cl_2	Normal phase HPLC–conductivity detection	154
Distearyldimethylammonium chloride, dodecyltrimethyl-ammonium bromide, cocoalkyl-dimethylbenzylammonium chloride, and an imidazolium compound	Sewage and activated sludge	Water: paired ion extraction from acid into $CHCl_3$, alumina cleanup ($CHCl_3$ elution); sludge: acidic MeOH extraction, paired ion extraction into $CHCl_3$, alumina cleanup ($CHCl_3$ elution)	HPLC (DIOL column)–conductivity detection	223,224
Monotallowtrimethyl-, ditallow-dimethyl-, and tritallowmethyl-ammonium chloride	River water, sewage, sludge	Extraction of dried solids with 4% HCl (v/v) in MeOH, ion pair extraction with LAS from 5% (v/v) HCl into $CHCl_3$, evaporation, dissolution in CH_2Cl_2 and washing with H_2O	Normal phase HPLC–detection by postcolumn reaction	155
C_{12}–C_{18} alkyltrimethylammonium chloride	Sewage, river water, and sludge	Evaporation, acidic MeOH extraction, paired ion extraction into $CHCl_3$	Normal phase HPLC–conductivity detection	156

Distearyldimethylammonium chloride	Sewage, soil, sludge and sediment	Evaporation of liquids and drying of solids, extraction of residues with methanolic 1M HCl, extraction into CHCl$_3$ containing LAS, SPE on anion exchanger (elution with 80:20 MeOH/CHCl$_3$ and MeOH), final washing of a CHCl$_3$ solution with water	Normal phase HPLC–conductivity detection	146
Distearyldimethylammonium chloride	Surface water	SPE on cation exchanger (elution with 50:50 acetone/H$_2$O, 0.06 M in HCl and 0.025 M in CuCl)	Normal phase HPLC–postcolumn ion pair extraction with 9,10-dimethoxyanthracene-2-sulfonate (fluorescence detection)	225
Distearyldimethylammonium chloride	Sludge and sediment	SFE (70:30 CO$_2$/MeOH, 100°C, Cu added), with or without anion exchange cleanup	Normal phase HPLC–postcolumn ion pair extraction with methyl orange (UV detection) or 9,10-dimethoxyanthracene-2-sulfonate (fluorescence detection)	151
Distearyldimethylammonium chloride	Sewage sludge	Extraction with MeOH, cleanup of CHCl$_3$ solution on alumina (elution of impurities with CHCl$_3$ and 95:5 CHCl$_3$/MeOH; elution of cationic with 5:1 CHCl$_3$/MeOH)	Spectrophotometry, IR or visible (disulfine blue)	37
Ditallowdimethylammonium chloride and dodecyltrimethyl-ammonium chloride	River water and sewage	Extraction of dried solids with 80:20, 90:10, and 100:0 MeOH/CHCl$_3$, alumina cleanup of all fractions	FAB-MS	226

(Continued)

TABLE 7 *(Continued)*

Compound determined	Sample matrix	Sample workup	Quantification	Ref.
C_{12}-, coco-, and stearyldialkyl-dimethylammonium, C_{14}–C_{18} alkylbenzyldimethylammonium, C_{12} alkylethyldimethyl-ammonium, C_{16} and stearyl alkyltrimethylammonium, C_{12}–C_{18} alkylpyridinium, benzethonium, and *N,N*-dimethyl-*N*-(2-phenoxyethyl)-1-dodecanaminium salts	River and drinking water	Carbon adsorption (desorption by CH_2Cl_2), Soxhlet extraction with CH_2Cl_2), liquid-liquid extraction into acidic, basic, and neutral fractions	FAB-MS	103
An ester quat: *N*-octadecyl-*N*-[(palmitoyloxy)-ethyl]-*N,N*-dimethylammonium chloride and its *N*-hydroxyethyl degradation product	Biodegradation test liquor and sludge	Evaporation, EtOH extraction, TLC cleanup	FAB-MS (with ^{13}C internal standard) and ^{14}C tracer studies	227
Distearyldimethylammonium chloride	Sewage sludge and river sediment	Extraction and alumina cleanup	Derivative spectrophotometry, direct (Dragendorff reagent) or indirect (LAS/methylene blue)	157
C_{12}–C_{14} alkyldimethylbenzylammonium chloride	Hospital sewage	SPE on C_{18}; elution with 50:50 MeOH/ethyl acetate, 1% in $CaCl_2$	Normal phase HPLC; postcolumn extraction/fluorescence	228
Perfluoroalkyltriethanolammonium methylsulfate	Laboratory biodegradation studies	SPE on C_{18} (MeOH elution)	Thermospray MS	178

TABLE 8 Determination of Impurities and Degradation Products from Cationic Surfactants in the Environment

Compound determined	Sample matrix	Sample workup	Quantification	Ref.
C_{13}–C_{17} alkylnitriles and C_{16}–C_{18} trialkyl- and dialkylmethyl-amines	Seawater, sediment, polychaetes, sewage, sludge	*Sediment*: extraction after freeze-drying with CH_2Cl_2 and then MeOH; *other solids*: Soxhlet extraction with CH_2Cl_2; *water and filtrates*: liquid-liquid extraction with CH_2Cl_2; *CH_2Cl_2 extracts*: chromatography on a column with Al_2O_3 at the top and SiO_2 at the bottom; sequential elution with increasing concentrations of CH_2Cl_2 in hexane, 90:10 MeOH/CH_2Cl_2, and ethyl ether	GC, GC-MS, FAB-MS	141
C_{13}–C_{17} alkylnitriles and C_{14}–C_{18} trialkyl-, dialkylmethyl-, and alkyldimethylamines	Seawater, sediment, polychaetes	*Sediment*: Soxhlet extraction of freeze-dried solids with 2:1 CH_2Cl_2/MeOH: *biota*: saponification with NaOH, ethyl ether extraction; *water*: liquid-liquid extraction; *all extracts*: chromatography on a column with Al_2O_3 at the top and SiO_2 at the bottom	GC, GC-MS	25
C_{12} trialkylamines and C_{16}–C_{18} dialkylamines: separation and quantification	Sludge, sediment	Homogenization sequentially with MeOH and CH_2Cl_2; MeOH extract diluted with H_2O, extracted with hexane, and the hexane and CH_2Cl_2 extracts combined; column chromatography on alumina/silica, washing with hexane and CH_2Cl_2 before elution with 50:50 MeOH/ethyl ether	GC; confirmation by GC-MS	184

12, or the absorbance of the ion pair with tetraiodobismuthate may be measured at several wavelengths (37,157,158). The visible spectophotometric methods generally suffer from negative interference from anionic surfactants (which are almost always present in environmental samples at higher concentrations than the cationic surfactants) because these form ion pairs which are not colored, competing with the anionic dye. When applied to environmental samples, the low concentration of cationic surfactants compared to other naturally occurring cationic materials makes positive interference with colorimetric methods a constant problem. Because of these positive and negative interferences, cleanup of the sample is almost always required prior to the final determination.

^{1}H NMR spectroscopy has seen limited use in characterizing degradation pathways in laboratory experiments. For example, it was used to show that an alkyldi(polyethoxy)benzylammonium chloride degraded such that the polyether chains were the last to disappear (159).

Mass spectrometric analysis of cationic surfactants is difficult, so MS has not been applied as much to environmental analysis of cationics as it has to anionics and nonionics.

REFERENCES

1. Federle, T. W., N. R. Itrich, Kinetics of primary and ultimate biodegradation of chemicals in activated sludge: application to LAS, *Environ. Sci. Technol.*, 1997, *31*, 1178–1184.
2. Tanaka, F. S., R. G. Wien, Specific ^{14}C-labeled surfactants: addition of homogeneous polyoxyethylene glycols to *p*-(1,1,3,3-tetramethylbutyl)phenol, *J. Labelled Compd. Radiopharm.*, 1976, *12*, 97–105.
3. Tanaka, F. S., R. G. Wien, G. E. Stolzenberg, Specific ^{14}C-labeled surfactants: addition of homogeneous polyoxyethylene glycols to 2,6,8-trimethyl-4-nonanol, *J. Labelled Compd. Radiopharm.*, 1976, *12*, 107–118.
4. Swisher, R. D., *Surfactant Biodegradation*, Marcel Dekker, Inc., New York, 1986.
5. Karsa, D. R., M. R. Porter, eds., *Biodegradability of Surfactants*, Chapman and Hall, Glasgow, 1995.
6. Schwuger, M. J., ed., *Detergents and the Environment*, Marcel Dekker, New York, 1996.
7. Tolls, J., P. Kloepper-Sams, D. T. H. M. Sijm, Surfactant bioconcentration—critical review, *Chemosphere*, 1994, *29*, 693–717.
8. Waters, J., Analysis of surfactants in laboratory test liquors and environmental samples, in M. R. Porter, ed., *Recent Developments in the Analysis of Surfactants*, Elsevier, New York, 1991.
9. Kloster, G., Surfactants and complexing agents at concentrations relevant to environmental occurrence, in Schwuger, M. J., ed., *Detergents in the Environment*, Marcel Dekker, New York, 1997.
10. Hellmann, H., Surfactants and pseudosurfactants in surface waters and in wastewaters (in German), *Tenside, Surfactants, Deterg.*, 1991, *28*, 111–117.
11. Szymanski, A., Z. Swit, Z. Lukaszewski, Preservation of water samples for the determination of nonionic surfactants, *Anal. Chim. Acta*, 1995, *311*, 31–36.
12. Crescenzi, C., A. Di Corcia, A. Marcomini, R. Samperi, PEG and related acidic forms in environmental waters by LC/electrospray/MS, *Environ. Sci. Technol.*, 1997, *31*, 2679–2685.
13. Brownawell, B. J., H. Chen, W. Zhang, J. C. Westall, Sorption of nonionic surfactants on sediment materials, *Environ. Sci. Technol.*, 1997, *31*, 1735–1741.
14. Shiraishi, H., A. Otsuki, K. Fuwa, Identification of extractable organic chemicals in sewage effluent by GC/MS, *Biomed. Mass Spectrom.*, 1985, *12(2)*, 86–94.
15. Heise, S., N. Litz, Extraction of surfactants from solid matrices (in German), *Tenside, Surfactants, Deterg.*, 1999, *36*, 185–191.

16. Ahel, M., W. Giger, APE and alkylphenol mono-and diethoxylates in environmental samples by HPLC, *Anal. Chem.*, 1985, *57*, 1577–1583.
17. Sweetman, A. J., Multi-residue method for *n*-alkanes, linear alkylbenzenes, polynuclear aromatic hydrocarbons, and 4-nonylphenol in digested sewage sludges, *Water Res.*, 1994, *28*, 343–353.
18. Osburn, Q. W., LAS in waters and wastes, *J. Am. Oil Chem. Soc.*, 1986, *63*, 257–263.
19. Taylor, P. W., G. Nickless, Paired-ion HPLC of partially biodegraded linear alkylbenzenesulfonate, *J. Chromatogr.*, 1979, *178*, 259–269.
20. Thurman, E. M., T. Willoughby, L. B. Barber, Jr., K. A. Thorn, Alkylbenzenesulfonate surfactants in groundwater using macroreticular resins and carbon-13 NMR spectrometry, *Anal. Chem.*, 1987, *59*, 1798–1802.
21. Matthijs, E., H. De Henau, Determination of LAS, *Tenside, Surfactants, Deterg.*, 1987, *24*, 193–199.
22. Marcomini, A., W. Giger, LAS, APE, and nonylphenol by HPLC, *Anal. Chem.*, 1987, *59*, 1709–1715.
23. Marcomini, A., W. Giger, Behavior of LAS in sewage treatment, *Tenside, Surfactants, Deterg.*, 1988, *25*, 226–229.
24. Fritsche, U., S. H. Hüttenhain, Fluorotensides, *Chemosphere*, 1994, *29*, 1797–1802.
25. Valls, M., P. Fernández, J. M. Bayona, Fate of cationic surfactants in the marine environment. I: Bioconcentration of long-chain alkylnitriles and trialkylamines, *Chemosphere*, 1989, *19*, 1819–1827.
26. Di Corcia, A., R. Samperi, A. Marcomini, Aromatic surfactants and biodegradation intermediates in sewage by solid-phase extraction and LC, *Environ. Sci. Technol.*, 1994, *28*, 850–858.
27. Cain, R. B., Biodegradation of anionic surfactants, *Biochem. Soc. Trans.*, 1987, *15(Suppl.)*, 7S–22S.
28. Di Corcia, A., F. Casassa, C. Crescenzi, A. Marcomini, R. Samperi, Fate of LAS and coproducts in a laboratory biodegradation test using LC/MS, *Environ. Sci. Technol.*, 1999, *33*, 4112–4118.
29. Di Corcia, A., L. Capuani, F. Casassa, A. Marcomini, R. Samperi, Fate of LAS, coproducts, and their metabolites in sewage treatment plants and in receiving river waters, *Environ. Sci. Technol.*, 1999, *33*, 4119–4125.
30. Marcomini, A., L. Cavalli, Identification and behavior of sulfophenylcarboxylic acids in LAS biodegradation screening tests, treated and untreated sewage, and surface water, *3rd CESIO Inter. Surf. Cong. Exhib. London*, 1992, *E,F,LCA*, 8–16.
31. Cavalli, L., G. Cassani, M. Lazzarin, C. Maraschin, G. Nucci, L. Valtorta, Iso branching of LAS, *Tenside, Surfactants, Deterg.*, 1996, *33*, 393–398.
32. Newsome, C. S., D. Howes, S. J. Marshall, R. A. van Egmond, Fate of anionic and alcohol ethoxylate surfactants in *Carassius auratus*, *Tenside Surfactants, Deterg.*, 1995, *32*, 498–503.
33. Tolls, J., M. Haller, D. T. H. M. Sijm, Extraction and isolation of LAS and its sulfophenylcarboxylic acid metabolites from fish, *Anal. Chem.*, 1999, *71*, 5242–5247.
34. Field, J. A., L. B. Barber, II, E. M. Thurman, B. L. Moore, D. L. Lawrence, D. A. Peake, Fate of alkylbenzenesulfonates and dialkyltetralinsulfonates in sewage-contaminated groundwater, *Environ. Sci. Technol.*, 1992, *26*, 1140–1148.
35. Field, J. A., J. A. Leenheer, K. A. Thorn, L. B. Barber, II, C. Rostad, D. L. Macalady, S. R. Daniel, Persistent anionic surfactant-derived chemicals in sewage effluent and groundwater, *J. Contam. Hydrol.*, 1992, *9*, 55–78.
36. Prats, D., M. Rodriguez, M. A. Muela, J. M. Llamas, A. Moreno, J. De Ferrer, J. L. Berna, Elimination of xenobiotics during composting, *Tenside, Surfactants, Deterg.*, 1999, *36*, 294–298.
37. Hellmann, H., Aluminum oxide as ion exchanger in the determination of cationic surfactants and LAS in sewage sludge (in German), *Z. Wasser Abwasser Forsch.*, 1989, *22*, 4–10,12.
38. Hellmann, H., Al_2O_3 and SiO_2 in the chromatographic analysis of surfactants (in German), *Fresenius' Z. Anal. Chem.*, 1989, *334*, 126–132.

39. Cavalli, L., G. Cassani, M. Lazzarin, Biodegradation of LAS and AE, *Tenside, Surfactants, Deterg.*, 1996, *33*, 158–165.

40. Schöberl, P., H. Klotz, R. Spilker, L. Nitschke, LAS monitoring (in German), *Tenside, Surfactants, Deterg.*, 1994, *31*, 243–252.

41. Feijtel, T. C. J., E. Matthijs, A. Rottiers, G. B. J. Rijs, A. Kiewiet, A. de Nijs, AIS/CESIO environmental surfactant monitoring program. Part 1: LAS monitoring study in de Meern sewage treatment plant and the receiving river Leidsche Rijn, *Chemosphere*, 1995, *30*, 1053–1066.

42. Reiser, R., H. O. Toljander, W. Giger, Alkylbenzenesulfonates in recent sediments by GC/MS, *Anal. Chem.*, 1997, *69*, 4923–4930.

43. Field, J. A., D. J. Miller, T. M. Field, S. B. Hawthorne, W. Giger, Sulfonated aliphatic and aromatic surfactants in sewage sludge by ion-pair/supercritical fluid extraction and derivatization GC/MS, *Anal. Chem.*, 1992, *64*, 3161–3167.

44. Field, J. A., T. M. Field, T. Poiger, H. Siegrist, W. Giger, Fate of secondary alkane sulfonate surfactants during municipal wastewater treatment, *Water Res.*, 1995, *29*, 1301–1307.

45. Moreno, A., J. Bravo, J. Ferrer, C. Bengoechea, Soap in different environmental compartments, *Tenside, Surfactants, Deterg.*, 1996, *33*, 479–482.

46. Matthijs, E., H. De Henau, Adsorption and desorption of LAS, *Tenside, Surfactants, Deterg.*, 1985, *22*, 299–304.

47. Marcomini, A., S. Capri, W. Giger, LAS, APE, and nonylphenol in waste water by HPLC after enrichment on octadecylsilica, *J. Chromatogr.*, 1987, *403*, 243–252.

48. Hand, V. C., G. K. Williams, Structure-activity relationships for sorption of LAS, *Environ. Sci. Technol.*, 1987, *21*, 370–373.

49. Inaba, K., K. Amano, HPLC determination of LAS in aquatic environment: seasonal changes in LAS concentration in polluted lake water and sediment, *Int. J. Environ. Anal. Chem.*, 1988, *34*, 203–213.

50. Kornecki, T. S., B. Allred, G. O. Brown, Cationic and anionic surfactant concentrations in soil, *Soil Sci.*, 1997, *162*, 439–446.

51. Marcomini, A., A. Di Corcia, R. Samperi, S. Capri, Reversed-phase HPLC determination of LAS, NPE and their carboxylic biotransformation products. *J. Chromatogr.*, 1993, *644*, 59–71.

52. Hawthorne, S. B., D. J. Miller, D. D. Walker, D. E. Whittington, B. L. Moore, Extraction of LAS using supercritical carbon dioxide and a simple device for adding modifiers, *J. Chromatogr.*, 1991, *541*, 185–194.

53. Kreisselmeier, A., H. Dürbeck, Alkylphenols and LAS in sediments applying accelerated solvent extraction, *Fresenius' Z. Anal. Chem.*, 1996, *354*, 921–924.

54. Kreisselmeier, A., H. Dürbeck, Alkylphenols, APE and LAS in sediments by accelerated solvent extraction and supercritical fluid extraction, *J. Chromatogr. A*, 1997, *775*, 187–196.

55. Hellmann, H., Anionic surfactants (LAS) in German streams using second derivative UV spectroscopy (in German), *Z. Wasser Abwasser Forsch.*, 1991, *24*, 178–187.

56. Takada, H., R. Ishiwatari, Biodegradation of linear alkylbenzenes (LABs): isomeric composition of C_{12} LABs as an indicator of the degree of LAB degradation in the aquatic environment, *Environ. Sci. Tech.*, 1990, *24*, 86–91.

57. Borgerding, A. J., R. A. Hites, LAS using continuous-flow fast atom bombardment spectrometry, *Anal. Chem.*, 1992, *64*, 1449–1454.

58. Yamini, Y., M. Ashraf-Khorassani, Extraction and determination of LAS in the aquatic environment using a membrane disk and gas chromatography, *J. High Resolut. Chromatogr.*, 1994, *17*, 634–638.

59. Krueger, C. J., J. A. Field, In-vial C_{18} Empore disk elution coupled with injection port derivatization for the quantitative determination of LAS by GC-FID, *Anal. Chem.*, 1995, *67*, 3363–3366.

60. Castles, M. A., B. L. Moore, S. R. Ward, LAS in aqueous environmental matrices by LC with fluorescence detection, *Anal. Chem.*, 1989, *61*, 2534–2540.

61. Sarrazin, L., W. Wafo, P. Rebouillon, LAS and biodegradation intermediates in sea water using SPE and RP-HPLC with UV detection, *J. Liq. Chromatogr. Relat. Technol.*, 1999, 22, 2511–2524.

62. Matthijs, E., Anionic surfactants in laboratory test liquors and environmental samples, in J. Cross, ed., *Anionic Surfactants: Analytical Chemistry*, 2nd ed., Marcel Dekker, New York, 1998.

63. Reemtsma, T., Polar aromatic sulfonates in aquatic environments, *J. Chromatogr. A*, 1996, 733, 473–489.

64. Sarrazin, L., A. Arnoux, P. Rebouillon, HPLC analysis of a LAS and its environmental biodegradation metabolites, *J. Chromatogr. A*, 1997, 760, 285–291.

65. Linder, D. E., M. C. Allen, HPLC of intact and partially biodegraded LAS, *J. Am. Oil Chem. Soc.*, 1982, 59, 152–155.

66. Popenoe, D. D., S. J. Morris, III, P. S. Horn, K. T. Norwood, Alkyl sulfates and alkyl ethoxysulfates in wastewater treatment plant influents and effluents and in river water using LC/ion spray MS, *Anal. Chem.*, 1994, 66, 1620–1629.

67. Heinig, K., C. Vogt, G. Werner, LAS in industrial and environmental samples by CE, *Analyst*, 1998, 123, 349–353.

68. Field, J. A., T. M. Field, T. Poiger, W. Giger, Secondary alkane sulfonates in sewage wastewaters by solid phase extraction and injection port derivatization GC/MS, *Environ. Sci. Technol.*, 1994, 28, 497–503.

69. Matthijs, E., M. Stalmans, Monitoring of LAS in the North Sea, *Tenside, Surfactants, Deterg.*, 1993, 30, 29–33.

70. International Organization for Standardization, Water quality—determination of surfactants—Part 1: Determination of anionic surfactants by measurement of the methylene blue index, 7875/1-1984. Available from national ISO affiliates.

71. Gledhill, W. E., R. L. Huddleston, L. Kravetz, A. M. Nielsen, R. I. Sedlak, R. D. Vashon, Treatability of surfactants at a wastewater treatment plant, *Tenside, Surfactants, Deterg.*, 1989, 26, 276–281.

72. Tabor, C. F., L. B. Barber, II, Fate of LAS in the Mississippi River, *Environ. Sci. Technol.*, 1996, 30, 161–171.

73. Zanette, M., A. Marcomini, S. Capri, A. Liberatori, Regulatory semispecific methods for the determination of surfactants in aqueous samples. Part II: Anionic Surfactants, *Ann. Chim.* (Rome), 1995, 85, 221–233.

74. Schöberl, P., Basic principles of LAS biodegradation, *Tenside, Surfactants, Deterg.*, 1989, 26, 86–94.

75. Hellmann, H., UV-spectroscopy of LAS in relevant environmental matrices—fundamentals and LAS determination under relatively problem-free conditions (in German), *Z. Wasser Abwasser Forsch.*, 1990, 23, 62–69.

76. Theraulaz, F., L. Djellal, O. Thomas, LAS in sewage using advanced UV spectrophotometry, *Tenside, Surfactants, Deterg.*, 1996, 33, 447–451.

77. Djellal, L., F. Theraulaz, O. Thomas, LAS behavior in sewage using advanced UV spectrophotometry, *Tenside, Surfactants, Deterg.*, 1997, 34, 316–320.

78. Oba, K., K. Miura, H. Sekiguchi, R. Yagi, A. Mori, Anionic surfactants in waste water by IR spectroscopy, *Water Res.*, 1976, 10, 149–155.

79. Kunkel, E., Environmental analysis of surfactants (in German), *Tenside, Surfactants, Deterg.*, 1987, 24, 280–285.

80. Schröder, H. Fr., Determination of surfactants—comparison of class-specific methods with LC/MS substance-specific analysis (in German), *Vom Wasser*, 1992, 79, 193–209.

81. Takada, H., J. W. Farrington, M. H. Bothner, C. G. Johnson, B. W. Tripp, Transport of sludge-derived organic pollutants to deep sea sediments at deep water dump site 106, *Environ. Sci. Technol.*, 1994, 28, 1062–1072.

82. Kreisselmeier, A., M. Schoester, G. Kloster, Surfactants in river water applying XAD-4/RP-18 extraction, *Fresenius' J. Anal. Chem.*, 1995, *353*, 109–111.

83. Ding, W., C. Chen, LAS in water by large-volume injection port derivatization and GC-MS, *J. Chromatogr. A*, 1999, *857*, 359–364.

84. Trehy, M. L., W. E. Gledhill, R. G. Orth, LAS and dialkyltetralinsulfonates in water and sediment by GC/MS, *Anal. Chem.*, 1990, *62*, 2581–2586.

85. Rovellini, P., N. Cortesi, E. Fideli, LAS surfactant biodegradation: monitoring of metabolic intermediates by LC-MS (in Italian), *Riv. Ital. Sostanze Grasse*, 1995, *72*, 381–389.

86. Steber, J., P. Wierich, Metabolites and biodegradation pathways of AE in microbial biocenoses of sewage treatment plants, *Appl. Environ. Microbiol.*, 1985, *49*, 530–537.

87. Di Corcia, A., C. Crescenzi, A. Marcomini, R. Samperi, LC-electrospray-MS for characterizing biodegradation intermediates of branched AE surfactants, *Environ. Sci. Technol.*, 1998, *32*, 711–718.

88. Schröder, H. Fr., Fluorine-containing surfactants—another challenge to the environment? Part 2: Nonionic surfactants (in German), *Vom Wasser*, 1992, *78*, 211–227.

89. Schümann, U., M. Gluschke, P. Gründler, L. Mikolajczyk, Reaction pathways in electrochemical oxidation of nonionic surfactants, *Tenside, Surfactants, Deterg.*, 1998, *35*, 379–386.

90. Field, J. A., R. L. Reed, Subcritical (hot) water/ethanol extraction of nonylphenol polyethoxy carboxylate industrial and municipal sludges, *Environ. Sci. Technol.*, 1999, *33*, 2782–2787.

91. Brand, N., G. Mailhot, M. Bolte, Degradation photoinduced by Fe(III): APE removal in water, *Environ. Sci. Technol.*, 1998, *32*, 2715–2720.

92. Tanaka, F. S., R. G. Wien, R. G. Zaylskie, Photolytic degradation of a homogeneous Triton X nonionic surfactant: nonaethoxylated *p*-(1,1,3,3-tetramethylbutyl)phenol, *J. Agric. Food Chem*, 1991, *39*, 2046–2052.

93. Mackay, L. G., M. Y. Croft, D. S. Selby, R. J. Wells, NPE and OPE in effluent by LC with fluorescence detection, *J. AOAC Int.*, 1997, *80*, 401–407.

94. Fujita, Y., M. Reinhard, Metabolites from the biological transformation of octylphenoxyacetic acid and its brominated analog, *Environ. Sci. Technol.*, 1997, *31*, 1518–1524.

95. Thiele, B., K. Günther, M. J. Schwuger, APE: trace analysis and environmental behavior, *Chem. Rev.* (Washington, DC), 1997, *97*, 3247–3272.

96. Baumann, U., M. Benz, E. Pletscher, K. Breuker, R. Zenobi, Biodegradation of sugar alcohol ethoxylates (in German), *Tenside, Surfactants, Deterg.*, 1999, *36*, 288–293.

97. Schröder, H. Fr., Surfactants: non-biodegradable, significant pollutants in sewage treatment plant effluents. Separation, identification and quantification by LC, flow-injection analysis-MS and tandem MS, *J. Chromatogr.*, 1993, *647*, 219–234.

98. Vojvodic, V., B. Cosovic, V. Miric, Fractionation of surface active substances on XAD-8 resin. I. Mixtures of model substances, *Anal. Chim. Acta*, 1994, *295*, 73–83.

99. Leenheer, J. A., R. L. Wershaw, P. A. Brown, T. I. Noyes, PEG residues from nonionic surfactants in surface water by ^1H and ^{13}C NMR spectrometry, *Environ. Sci. Technol.*, 1991, *25*, 161–168.

100. Field, J. A., R. L. Reed, Nonylphenol polyethoxy carboxylate metabolites of nonionic surfactants in US paper mill effluents, municipal sewage treatment plant effluents, and river waters, *Environ. Sci. Technol.*, 1996, *30*, 3544–3550.

101. Gorenc, B., D. Gorenc, A. Rosker, Preconcentration of nonionic surfactants by adsorption on ion-exchange resin in the cobaltithiocyanate form, *Vestn. Slov. Kem. Drus.*, 1986, *33*, 467–474.

102. Ventura, F., A. Figueras, I. Espadaler, J. Romero, J. Guardiola, J. Rivera, Polyethoxylated surfactants and their brominated derivatives formed at the water treatment plant of Barcelona using GC/MS and FAB-MS, *Water Res.*, 1988, *22*, 1211–1217.

103. Ventura, F., J. Caixach, A. Figueras, I. Espadaler, D. Fraisse, J. Rivera, Surfactants in water by FAB-MS, *Water Res.*, 1989, *23*, 1191–1203.

104. Marcomini, A., S. Stelluto, B. Pavoni, LAS and APE in commercial products and marine waters by reversed- and normal-phase HPLC, *Int. J. Environ. Anal. Chem.*, 1989, *35*, 207–218.

105. Chee, K. K., M. K. Wong, H. K. Lee, Optimization of sample preparation techniques for 4-nonylphenol in water and sediment, *J. Liq. Chromatogr. Relat. Technol.*, 1996, *19*, 259–275.

106. Kubeck, E., C. G. Naylor, APE, *J. Am. Oil Chem. Soc.*, 1990, *67*, 400–405.

107. Kiewiet, A. T., J. M. D. van der Steen, J. R. Parsons, Ethoxylated nonionic surfactants in influent and effluent of sewage treatment plants by HPLC, *Anal. Chem.*, 1995, *67*, 4409–4415.

108. Boyd-Boland, A. A., J. B. Pawliszyn, Solid-phase microextraction coupled with HPLC for APE in water, *Anal. Chem.*, 1996, *68*, 1521–1529.

109. Aranda, R., R. C. Burk, Nonionic surfactant by solid-phase microextraction coupled with HPLC and on-line derivatization, *J. Chromatogr. A*, 1998, *829*, 401–406.

110. Chalaux, N., J. M. Bayona, J. Albaigés, Nonylphenols as pentafluorobenzyl derivatives by capillary GC with electron capture and MS detection in environmental matrices, *J. Chromatogr. A*, 1994, *686*, 275–281.

111. Lin, J., R. Arunkumar, C. Liu, Efficiency of SFE for 4-nonylphenol in sewage sludge, *J. Chromatogr. A*, 1999, *840*, 71–79.

112. Shang, D. Y., M. G. Ikonomou, R. W. Macdonald, NPE surfactants in marine sediment using normal phase LC-ECMS, *J. Chromatogr. A*, 1999, *849*, 467–482.

113. Inaba, K., Polyoxyethylene-type nonionic surfactants in environmental waters, *Int. J. Environ. Anal. Chem.*, 1987, *31*, 63–66.

114. Andrew, B. E., Nonionic surfactants in waste water by direct extraction with FTIR detection, *Analyst*, 1993, *118*, 153–155.

115. Ahel, M., T. Conrad, W. Giger, Nonylphenoxy carboxylic acids by high-resolution GC/MS and HPLC, *Environ. Sci. Tech.*, 1987, *21*, 697–703.

116. Moldovan, Z., M. V. Delgado Luque, E. Otal Salaverri, A. Suárez, R. Andreozzi, A. Insola, J. Legrato Martínez, PEG in water by reversed-phase HPLC, *J. Chromatogr. A*, 1996, *723*, 243–249.

117. Ventura, F., D. Fraisse, J. Caixach, J. Rivera, [(Alkyloxy)polyethoxy]carboxylates in raw and drinking water by MS/MS and mass determination using fast atom bombardment and non-ionic surfactants as internal standards, *Anal. Chem.*, 1991, *63*, 2095–2099.

118. Dubey, S. T., L. Kravetz, J. P. Salanitro, Nonionic surfactants in bench-scale biotreater samples, *J. Am. Oil Chem. Soc.*, 1995, *72*, 23–30.

119. Porot, V., NPE in samples from sewage treatment plants (Toulon and Morlaix) by HPLC (in French), *Proc. 2nd World Surfactants Congress*, Paris, May, 1988, *4*, 293–302.

120. Kiewiet, A. T., P. de Voogt, Chromatographic tools for analyzing and tracking nonionic surfactants in the aquatic environment, *J. Chromatogr. A*, 1996, *733*, 185–192.

121. Marcomini, A., M. Zanette, Chromatographic determination of AE type in the environment, *J. Chromatogr. A*, 1996, *733*, 193–206.

122. Evans, K. A., S. T. Dubey, L. Kravetz, I. Dzidic, J. Gumulka, R. Mueller, J. R. Stork, Linear primary AE surfactants in environmental samples by thermospray LC/MS, *Anal. Chem.*, 1994, *66*, 699–705.

123. Cassani, G., M. Comber, A. Guarini, J. Lux, M. Hetheridge, R. Wolf, AE in environmental samples by thermospray LC/MS, *World Surfactants Congr., 4th*, 1996, *4*, 436–447.

124. Tobin, R. S., F. I. Onuska, B. G. Brownlee, D. H. J. Anthony, M. E. Comba, Ether cleavage to study the biodegradation of linear AE, *Water Res.*, 1976, *10*, 529–535.

125. Fendinger, N. J., W. M. Begley, D. C. McAvoy, W. S. Eckhoff, AE surfactants in natural waters, *Environ. Sci. Technol.*, 1995, *29*, 856–863.

126. Cozzoli, O., C. Ruffo, G. Carrer, F. Pinciroli, M. Ciambella, M. C. Grillo, N. Cogliati, M. Fiorito, A. Gellera, N. Giannelli, V. Zamboni, Ethoxylated nonionic surfactants: comparison of the precision of the iodobismuthate method and other methods (in Italian), *Riv. Ital. Sostanze Grasse*, 1993, *70*, 199–203; *Tinctoria*, 1994, *91*(5), 39–43.

127. Cavalli, L., A. Gellera, G. Cassani, M. Lazzarin, C. Maraschin, G. Nucci, Nonionics and AE: application to confirmatory biodegradation test effluents, *Riv. Ital. Sostanze Grasse*, 1993, *70*, 447–452.

128. Wahlberg, C., L. Renberg, U. Wideqvist, Nonylphenol and NPE as their pentafluorobenzoates in water, sewage sludge and biota, *Chemosphere*, 1990, *20*, 179–195.

129. International Organization for Standardization, Water quality—determination of surfactants—Part 2: Determination of nonionic surfactants using Dragendorff reagent, 7875/2-1984. Available from national ISO affiliates.

130. Waters, J., J. T. Garrigan, A. M. Paulson, Scope and limitations of the bismuth active substances procedure (Wickbold) for the determination of nonionic surfactants in environmental samples, *Water Res.*, 1986, *20*, 247–253.

131. Zanette, M., A. Marcomini, S. Capri, A. Liberatori, Regulatory semispecific methods for the determination of surfactants in aqueous samples. Part I: Nonionic surfactants, *Ann. Chim.* (Rome), 1995, *85*, 201–220.

132. Wyrwas, B., A. Szymanski, Z. Lukaszewski, Nonionic surfactants adsorbed on particles of surface water by an indirect tensammetric method combined with the BiAS separation scheme, *Anal. Chim. Acta*, 1996, *331*, 131–139.

133. Lukaszewski, Z, A. Szymanski, B. Wyrwas, Evolution and potential of the BiAS procedure for the determination of nonionic surfactants, *TrAC, Trends Anal. Chem.*, 1996, *15*, 525–531.

134. Evans, K. A., S. T. Dubey, L. Kravetz, S. W. Evetts, I. Dzidic, C. C. Dooyema, AE surfactants in environmental samples by electrospray MS, *J. Am. Oil Chem. Soc.*, 1997, *74*, 765–773.

135. Ding, W., Y. Fujita, M. Reinhard, CI mass spectra of linear alcohol polyethoxy carboxylates and polyethylene glycol dicarboxylates, *Rapid Commun. Mass Spectrom.*, 1994, *8*, 1016–1020.

136. Ibrahim, N. M. A., B. B. Wheals, APE in trade effluents by sublation and HPLC, *Analyst*, 1996, *121*, 239–242.

137. Ibrahim, N. M. A., B. B. Wheals, Oligomeric separation of APE on silica using aqueous acetonitrile eluents, *J. Chromatogr. A*, 1996, *731*, 171–177.

138. Ekelund R., Å. Bergman, Å. Granmo, M. Berggren, Bioaccumulation of 4-nonylphenol in marine animals, *Environ. Pollut.*, 1990, *64*, 107–120.

139. Ejlertsson, J., M. Nilsson, H. Kylin, Å. Bergman, L. Karlson, M. Öquist, B. H. Svensson, Anaerobic degradation of nonylphenol mono- and diethoxylates in digestor sludge, landfilled municipal solid waste, and landfilled sludge, *Environ. Sci. Technol.*, 1999, *33*, 301–306.

140. Marcomini, A., C. Tortato, S. Capri, A. Liberatori, Preparation, characterization and RP-HPLC determination of sulfophenyl and nonylphenoxy carboxylates, *Ann. Chim.* (Rome), 1993, *83*, 461–484.

141. Fernández, P., M. Valls, J. M. Bayona, J. Albalgés, Cationic surfactants and related products in urban coastal environments, *Environ. Sci. Technol.*, 1991, *25*, 547–550.

142. Maldonado, C., J. Dachs, J. M. Bayona, Trialkylamines and coprostanol as tracers of urban pollution in waters from enclosed seas: the Mediterranean and Black Sea, *Environ. Sci. Technol.*, 1999, *33*, 3290–3296.

143. Clancy, S. F., M. Thies, H. Paradies, Cope elimination observed in the biodegradation of quaternary ammonium surfactants, *Chem. Commun.* (Cambridge), 1997, *21*, 2035–2036.

144. Brownawell, B. J., H. Chen, J. M. Collier, J. C. Westall, Adsorption of organic cations to natural materials, *Environ. Sci. Technol.*, 1990, *24*, 1234–1241.

145. Gerhards, R., R. Schulz, Amphoteric surfactants in water, *Tenside, Surfactants, Deterg.*, 1999, *36*, 300–302,304–307.

146. Gerike, P., H. Klotz, J. G. A. Kooijman, E. Matthijs, J. Waters, Dihardenedtallowdimethyl ammonium compounds in environmental matrices using trace enrichment techniques and HPLC with conductometric detection, *Water Res.*, 1994, *28*, 147–154.

147. Klotz, H., Cationic surfactants in clay minerals, *Fresenius' Z. Anal. Chem.*, 1987, *326*, 155–156.

148. Hellmann, H., Extractible cationic surfactants in activated sludge (in German), *Fresenius' Z. Anal. Chem.*, 1983, *315*, 425–429.

149. Hellmann, H., Resolubilization and determination of cationic surfactants in clay minerals (in German), *Fresenius' Z. Anal. Chem.*, 1984, *319*, 267–271.

150. Breen, D. G. P. A., J. M. Horner, K. D. Bartle, A. A. Clifford, J. Waters, J. G. Lawrence, SFE and off-line HPLC analysis of cationic surfactants from dried sewage sludge, *Water Res.*, 1996, *30*, 476–480.

151. Fernández, P., A. C. Alder, M. J. F. Suter, W. Giger, Ditallowdimethylammonium in digested sludges and marine sediments by supercritical fluid extraction and LC with post-column ion-pair extraction, *Anal. Chem.*, 1996, *68*, 921–929.

152. Osburn, Q. W., Cationic fabric softener in waters and wastes, *J. Am. Oil Chem. Soc.*, 1982, *59*, 453–457.

153. Waters, J., Cationic surfactants in laboratory test liquors and environmental samples, in J. Cross and E. J. Singer, eds., *Cationic Surfactants: Analytical and Biological Evaluation*, Marcel Dekker, New York, 1994.

154. Nitschke, L., R. Müller, G. Metzner, L. Huber, Cationic surfactants in water using HPLC with conductometric detection, *Fresenius' Z. Anal. Chem.*, 1992, *342*, 711–713.

155. De Ruiter, C., J. C. H. F. Hefkens, U. A. T. Brinkman, R. W. Frei, M. Evers, E. Matthijs, J. A. Meijer, LC determination of cationic surfactants in environmental samples using a continuous post-column ion-pair extraction detector with a sandwich phase separator, *Intern. J. Environ. Anal. Chem.*, 1987, *31*, 325–339.

156. Matthijs, E., H. De Henau, Monoalkylquaternaries and assessment of their fate in domestic waste waters, river waters and sludges, *Vom Wasser*, 1987, *69*, 73–83.

157. Hellmann, H., Cationic and anionic (LAS) surfactants in sediments, suspended matter, and sludges (in German), *Z. Wasser Abwasser Forsch.*, 1989, *22*, 131–137.

158. Hellmann, H., Modified Dragendorff reagent for determination of cationic surfactants in the aquatic environment (in German), *Fresenius' Z. Anal. Chem.*, 1989, *335*, 265–271.

159. Janosz-Rajczyk, M., Biodegradation of alkyldipolyethoxybenzylammonium chloride, *Tenside, Surfactants, Deterg.*, 1992, *29*, 436–441.

160. Holt, M. S., E. Matthijs, J. Waters, Concentrations and fate of LAS in sludge amended soils, *Water Res.*, 1989, *23*, 749–759.

161. Garcia Ramon, M. T., I. Ribosa, J. Sanchez Leal, F. Comelles, LAS monitoring in the Tajo river basin, *Tenside, Surfactants, Deterg.* 1990, *27*, 118–121.

162. Comellas, L., J. L. Portillo, M. T. Vaquero, LAS degradation in sewage sludge-amended soils, *J. Chromatogr.*, 1993, *657*, 25–31.

163. Di Corcia, A., M. Marchetti, R. Samperi, A. Marcomini, LC determination of LAS in aqueous environmental samples, *Anal. Chem.*, 1991, *63*, 1179–1182.

164. Yokoyama, Y., H. Sato, Reversed-phase HPLC determination of LAS in river water at ppb levels by precolumn concentration, *J. Chromatogr.*, 1991, *555*, 155–162.

165. Yokoyama, Y., M. Kondo, H. Sato, Determination of alkylbenzenesulfonates in environmental water by anion-exchange chromatography, *J. Chromatogr.*, 1993, *643*, 169–172.

166. Castillo, M., M. C. Alonso, J. Riu, D. Barcelo, Polar, ionic, and highly water soluble organic pollutants in untreated industrial wastewaters, *Environ. Sci. Technol.*, 1999, *33*, 1300–1306.

167. Castillo, M., D. Barceló, Polar toxicants in industrial wastewaters using toxicity-based fractionation with LC/MS, *Anal. Chem.*, 1999, *71*, 3769–3776.

168. Takada, H., R. Ishiwatari, Linear alkylbenzenes in urban riverine environments in Tokyo: distribution, source, and behavior, *Environ. Sci. Tech.*, 1987, *21*, 875–883.

169. Ou, Z. Q., L. Q. Jia, H. Y. Jin, A. Yediler, T. H. Sun, A. Kettrup, Ultrasonic extraction and LC determination of LAS in plant tissues, *Chromatographia*, 1997, *44*, 417–420.

170. Fujita, M., M. Ike, Y. Goda, S. Fujimoto, Y. Toyoda, K. Miyagawa, ELISA for detection of LAS: development and field studies, *Environ. Sci. Technol.*, 1998, *32*, 1143–1146.

171. Reiser, R., H. Toljander, A. Albrecht, W. Giger, Alkylbenzenesulfonates in recent lake sediments as molecular markers for the environmental behavior of detergent-derived chemicals, *ACS Symp. Ser.*, 1997, *671*, 196–212.

172. Ding, W. H., J. H. Lo, S. H. Tzing, LAS and degradation products in water samples by GC with ion-trap MS, *J. Chromatogr. A*, 1998, *818*, 270–279.

173. Kloster, G., M. Schoester, H. Prast, Concentration of all three surfactant classes from environmental samples (in German), *Tenside, Surfactants, Deterg.*, 1994, *31*, 23–28.

174. Fendinger, N. J., W. M. Begley, D. C. McAvoy, W. S. Eckhoff, Alkyl sulfate surfactants in natural waters, *Environ. Sci. Technol.*, 1992, *26*, 2493–2498.

175. Neubecker, T. A., Alkylethoxylated sulfates in wastewaters and surface waters, *Environ. Sci. Technol.*, 1985, *19*, 1232–1236.

176. Moreno, A., J. Bravo, J. Ferrer, C. Bengoechea, Soap in various environmental matrices, *Comun. Jorn. Com. Esp. Deterg.*, 1993, *24*, 45–57.

177. Moreno, A., J. Bravo, J. Ferrer, C. Bengoechea, Soap determination in sewage sludge by HPLC, *J. Am. Oil Chem. Soc.*, 1993, *70*, 667–671.

178. Schröder, H. Fr., Fluorine-containing surfactants—another challenge to the environment? Part 1: Anionic and cationic surfactants (in German), *Vom Wasser*, 1991, *77*, 277–290.

179. Moody, C. A., J. A. Field, Perfluorocarboxylates in groundwater impacted by fire-fighting activity, *Environ. Sci. Technol.*, 1999, *33*, 2800–2806.

180. González-Mazo, E., M. Honing, D. Barceló, A. Gómez-Parra, Long-chain intermediate products from the degradation of LAS in the marine environment by SPE followed by LC/Ionspray MS, *Environ. Sci. Technol.*, 1997, *31*, 504–510.

181. Riu, J., E. Gonzalez-Mazo, A. Gomez-Parra, D. Barceló, Carboxylic degradation products of LAS in coastal water by SPE followed by LC/ionspray/MS using negative ion detection, *Chromatographia*, 1999, *50*, 275–281.

182. Kanz, C., M. Nölke, T. Fleischmann, H. P. E. Kohler, W. Giger, Chiral biodegradation intermediates of LAS by CE, *Anal. Chem.*, 1998, *70*, 913–917.

183. Zeng, E. Y., C. C. Yu, Measurements of linear alkylbenzenes by GC/MS with interference from tetrapropylene-based alkylbenzenes: calculation of quantitation errors using a two-component model, *Environ. Sci. Technol.*, 1996, *30*, 322–328.

184. Chalaux, N., J. M. Bayona, M. I. Venkatesan, J. Albaigés, Distribution of surfactant markers in sediments from the Santa Monica basin, *Mar. Pollut. Bull.*, 1992, *24*, 403–407.

185. Crescenzi, C., A. Di Corcia, E. Marchiori, R. Samperi, A. Marcomini, Simultaneous determination of LAS and dialkyltetralinsulfonates in water by HPLC, *Water Res.*, 1996, *30*, 722–730.

186. Kölbener, P., A. Ritter, F. Corradini, U. Baumann, A. M. Cook, Refractory organic carbon and sulfur in the biotransformed byproducts in commercial LAS: identifications of arylsulfonates, *Tenside, Surfactants, Deterg.*, 1996, *33*, 149–156.

187. Schmitt, T. M., M. C. Allen, D. K. Brain, K. F. Guin, D. E. Lemmel, Q. W. Osburn, HPLC determination of AE in wastewater, *J. Am. Oil Chem. Soc.*, 1990, *67*, 103–109.

188. Lux, J. A., M. Schmitt, AE in environmental matrices by HPLC/UV-fluorescence, *World Surfactants Congr., 4th*, 1996, *3*, 113–120.

189. Tolls, J., M. Haller, D. T. H. M. Sijm, Extraction and isolation of linear AE from fish, *J. Chromatogr. A*, 1999, *839*, 109–117.

190. Dubey, S. T., K. A. Evans, L. Kravetz, I. Dzidic, LC/MS analyses and sample preparation of AE in environmental samples, *World Surfactants Congr., 4th*, 1996, *4*, 395–401.

191. Crescenzi, C., A. Di Corcia, R. Samperi, A. Marcomini, Nonionic polyethoxylate surfactants in environmental waters by LC/electrospray MS, *Anal. Chem.*, 1995, *67*, 1797–1804.

192. Scarlett, M. J., J. A. Fisher, H. Zhang, M. Ronan, Dissolved NPE in waste waters by gas stripping and isocratic HPLC, *Water Res.*, 1994, *28*, 2109–2116.

193. Zhao, M., F. van der Wielen, P. de Voogt, Optimization of a matrix solid-phase dispersion method with sequential cleanup for the determination of APE in biological tissues, *J. Chromatogr. A*, 1999, *837*, 129–138.

194. Snyder, S. A., T. L. Keith, D. A. Verbrugge, E. M. Snyder, T. S. Gross, K. Kannan, J. P. Giesy, Estrogenic compounds in aqueous mixtures, *Environ. Sci. Technol.*, 1999, *33*, 2814–2820.

195. Brown, D., H. de Henau, J. T. Garrigan, P. Gerike, M. Holt, E. Kunkel, E. Matthijs, J. Waters, R. J. Watkinson, Removal of nonionics in sewage treatment plants. II: Removal of domestic

detergent nonionic surfactants in a trickling filter sewage treatment plant, *Tenside, Surfactants, Deterg.*, 1987, *24*, 14–19.

196. Naylor, C. G., J. P. Mieure, W. J. Adams, J. A. Weeks, F. J. Castaldi, L. D. Ogle, R. R. Romano, APE in the environment, *J. Am. Oil Chem. Soc.*, 1992, *69*, 695–703.

197. Lee, H., T. E. Peart, D. T. Bennie, R. J. Maguire, NPE and carboxylic acid metabolites in sewage treatment plant sludge by supercritical carbon dioxide extraction, *J. Chromatogr. A*, 1997, *785*, 385–394.

198. Shang, D. Y., R. W. Macdonald, M. G. Ikonomou, Persistence of NPE and their primary degradation products in sediments from near a municipal outfall in the Strait of Georgia, British Columbia, Canada, *Environ. Sci. Technol.*, 1999, *33*, 1366–1372.

199. Plomley, J. B., P. W. Crozier, V. Y. Taguchi, NPE in sewage treatment plants by combined precursor ion scanning and multiple reaction monitoring, *J. Chromatogr. A*, 1999, *854*, 245–257.

200. Waldhoff, H., J. Scherler, M. Schmitt, Alkyl polyglycosides—analysis of raw material: determination in products and environmental matrices, *World Surfactants Congr., 4th*, 1996, *1*, 507–518.

201. Waldhoff, H., J. Scherler, M. Schmitt, J. R. Varvil, Analysis of alkyl polyglycosides and determination in consumer products and environmental matrices, in K. Hill, W. von Rybinski, G. Stoll, eds., *Alkyl polyglycosides: technology, properties and applications*, VCH, Weinheim, 1997.

202. Schröder, H. Fr., Alkyl polyglycosides in the biological waste water treatment process—degradation behavior by LC/MS and FIA/MS, *World Surfactants Congr., 4th*, 1996, *3*, 121–135.

203. Eichhorn, P., T. P. Knepper, Metabolism of alkyl polyglucosides and their determination in waste water by LC-electrospray MS, *J. Chromatogr. A*, 1999, *854*, 221–232.

204. Brown, D., H. de Henau, J. T. Garrigan, P. Gerike, M. Holt, E. Keck, E. Kunkel, E. Matthijs, J. Waters, R. J. Watkinson, Removal of nonionics in a sewage treatment plant, *Tenside, Surfactants, Deterg.*, 1986, *23*, 190–195.

205. Sherrard, K. B., P. J. Marriott, M. J. McCormick, R. Colton, G. Smith, Electrospray MS analysis and photocatalytic degradation of polyethoxylate surfactants used in wool scouring, *Anal. Chem.*, 1994, *66*, 3394–3399.

206. Sherrard, K. B., P. J. Marriott, R. G. Amiet, R. Colton, M. J. McCormick, G. C. Smith, Photocatalytic degradation of secondary AE: spectroscopic, chromatographic, and MS studies, *Environ. Sci. Technol.*, 1995, *29*, 2235–2242.

207. Marcomini, A., G. Pojana, LC of aerobic carboxylic metabolites of aliphatic AE, *Riv. Ital. Sostanze Grasse*, 1998, *75*, 35–42.

208. Tomaszewski, K., A. Szymanski, Z. Lukaszewski, Separation of PEG into fractions by sequential liquid-liquid extraction, *Talanta*, 1999, *50*, 299–306.

209. Jones, F. W., D. J. Westmoreland, Degradation of NPE during composting of sludges from wool scour effluents, *Environ. Sci. Technol.*, 1998, *32*, 2623–2627.

210. Pelizzetti, E., C. Minero, V. Maurino, A. Sciafani, H. Hidaka, N. Serpone, Photocatalytic degradation of NPE, *Environ. Sci. Technol.*, 1989, *23*, 1380–1385.

211. Stephanou, E., M. Reinhard, H. A. Ball, Halogenated and non-halogenated octylphenol polyethoxylate residues by GC/MS using EI and CI, *Biomed. Environ. Mass Spectrom.*, 1988, *15*, 275–282.

212. Ball, H. A., M. Reinhard, P. L. McCarty, Biotransformation of halogenated and nonhalogenated OPE residues under aerobic and anaerobic conditions, *Environ. Sci. Technol.*, 1989, *23*, 951–961.

213. Lee, H., T. E. Peart, 4-Nonylphenol in effluent and sludge from sewage treatment plants, *Anal. Chem.*, 1995, *67*, 1976–1980.

214. Kannan, N., N. Yamashita, G. Petrick, J. C. Duinker, Polychlorinated biphenyls and nonylphenols in the Sea of Japan, *Environ. Sci. Technol.*, 1998, *32*, 1747–1753.

215. Dachs, J., D. A. Van Ry, S. J. Eisenreich, Estrogenic nonylphenols in the urban and coastal atmosphere of the lower Hudson River estuary, *Environ. Sci. Technol.*, 1999, *33*, 2676–2679.

216. Marcomini, A., B. Pavoni, A. Sfriso, A. A. Orio, Persistent metabolites of APE in the marine environment, *Mar. Chem.*, 1990, *29*, 307–323.

217. Lye, C. M., C. L. J. Frid, M. E. Gill, D. W. Cooper, D. M. Jones, Estrogenic alkylphenols in fish tissues, sediments, and waters from the Tyne and Tees estuaries, *Environ. Sci. Technol.*, 1999, *33*, 1009–1014.

218. Ding, W. H., S. H. Tzing, NPE and degradation products in river water and sewage effluent by GC-ion trap (tandem) MS with EI and CI, *J. Chromatogr. A*, 1998, *824*, 79–90.

219. Di Corcia, A., A. Costantino, C. Crescenzi, E. Marinoni, R. Samperi, Recalcitrant intermediates from biotransformation of the branched alkyl side chain of NPE, *Environ. Sci. Technol.*, 1998, *32*, 2401–2409.

220. Szymanski, A., Z. Lukaszewski, PEG in environmental samples by the indirect tensammetric method, *Analyst*, 1996, *121*, 1897–1901.

221. Wee, V. T., J. M. Kennedy, Quaternary ammonium compounds in river water by LC with conductometric detection, *Anal. Chem*, 1982, *54*, 1631–1633.

222. Wee, V. T., Cationic surfactants in waste- and river waters, *Water Res.*, 1984, *18*, 223–225.

223. Emmrich, M., K. Levsen, Cationic surfactants in wastewater and activated sludge (in German), *Vom Wasser*, 1990, *75*, 343–349.

224. Levsen, K., M. Emmrich, S. Behnert, Dialkyldimethylammonium compounds and other cationic surfactants in sewage water and activated sludge, *Fresenius' J. Anal. Chem.*, 1993, *346*, 732–737.

225. Gort, S. M., E. A. Hogendoorn, R. A. Baumann, P. van Zoonen, Ditallowdimethylammonium chloride at the low ppb level in surface water using SPE and normal-phase LC with on-line post-column ion-pair extraction and fluorescence detection, *Int. J. Environ. Anal. Chem.*, 1993, *53*, 289–296.

226. Simms, J. R., T. Keough, S. R. Ward, B. L. Moore, M. M. Bandurraga, Cationic surfactants in environmental matrices using FAB-MS, *Anal. Chem.*, 1988, *60*, 2613–2620.

227. Simms, J. R., D. A. Woods, D. R. Walley, T. Keough, B. S. Schwab, R. J. Larson, FAB-MS and liquid scintillation counting to determine the mechanism and kinetics of surfactant biodegradation, *Anal. Chem.*, 1992, *64*, 2951–2957.

228. Kümmerer, K., A. Eitel, U. Braun, P. Hubner, F. Daschner, G. Mascart, M. Milandri, F. Reinthaler, J. Verhoef, Benzalkonium chloride in the effluent from European hospitals by SPE and HPLC with post-column ion-pairing and fluorescence detection, *J. Chromatogr. A*, 1997, *774*, 281–286.

229. Ding, W., C. Chen, Nonylphenol polyethoxycarboxylates and their related metabolites by on-line derivatization and ion-trap GC-MS, *J. Chromatogr. A*, 1999, *862*, 113–120.

230. McEvoy, J., W. Giger, LAS in sewage sludge by high resolution GC-MS, *Environ. Sci. Technol.*, 1986, *20*, 376–383.

19
Analysis of Formulated Products

I. INTRODUCTION

Often, surfactants are determined as part of the complete analysis of a formulation such as a laundry detergent or cosmetic product, where the analyst has the goal of accounting for 100% of the sample.

Formulation analysis is a time-intensive undertaking. The larger manufacturers have laboratories dedicated to the complete analysis of their own and competitive products. For such laboratories, it is a simple matter to subject a sample to a battery of routine tests. An ordinary detergent can be characterized as to its gross composition with perhaps only a couple day's labor. On the other hand, a general analytical laboratory with little experience in detergents can expect to spend much longer, first performing qualitative analyses, then following up with quantitative determinations for each of the many components. A complete analysis, to identify the actual source of each of the surfactants and other components, can easily consume weeks, even for a laboratory experienced in the characterization. Even then, there may remain questions about trace components such as perfumes and dyes.

For this reason it is not generally possible to prepare a performance match for a competitive product on the basis of analysis alone. Also required is a skilled formulator who will use the analytical results as clues, rather than a recipe. For many purposes, such as estimating the manufacturing cost of a product or for determining whether patent infringement has occurred, an analysis accounting for considerably less than 100% is adequate.

This chapter provides a brief description of the strategies used to analyze formulations. The actual methodology for separation and analysis is described in the previous chapters. The reader is urged to also consult other works which give more detail on formulations and formulation analysis (1–5). The CTFA Compendium (6) and the publications of ASTM and the International Standards Organization are valuable sources of methods for determination of individual nonsurfactant components.

II. ANALYTICAL STRATEGIES

A. Initial Examination

A few general tests are first performed on the commercial mixture:

1. *Thermogravimetric analysis*: A commercial TGA instrument may be used, but it is sufficient for most purposes to place a gram or two of the sample in an oven and weigh after heating to 110°C and 600°C. The former value gives an estimate of the total solids in the formulation, while the latter indicates the level of inorganic material.

2. *Water content*: This is usually performed by Karl Fischer analysis, although azeotropic distillation is sometimes used. If the total of the nonvolatiles at 110°C and the water content account for 100% of the sample, then it is unnecessary to look for an organic solvent.

3. *Percent ethanol-soluble material*: The dry sample is slurried with ethanol and filtered. Liquid samples are first evaporated to dryness, then extracted. Depending upon the experience of the analyst, the ethanol may be either hot or cold, and may be either anhydrous or contain 5% water. Continuous extraction devices, such as the Soxhlet apparatus, may be used. This step separates some or all of the organic components from the matrix. Usually, all of the surfactants are present in the ethanol-soluble fraction. Continuous liquid-liquid extraction with *n*-butanol or *sec*-butanol has been suggested as an alternative to ethanol extraction of the dried solids (7). While "wet" butanol is a more universal extracting agent, it has the disadvantage of also dissolving some inorganic materials (3). Such an approach must be used with caution because hydrophilic surfactants cannot be completely removed from all matrices by partition between water and solvents; the water must first be removed (8).

4. *Spectroscopic analysis of ethanol-soluble portion*: This is usually performed by infrared spectroscopy. If the spectroscopist is experienced with these products, and if the spectral library is adequate, this step will tell the nature of the surfactant and indicate what other components are present in the ethanol-solubles. Since this fraction is usually a mixture, spectral interpretation is far from trivial. Mass spectrometry is also frequently applied to qualitative analysis of the ethanol-solubles. Experience is necessary, since ionization is quite dependent on the surfactant type (9).

5. *Emission spectrographic or X-ray fluorescence analysis of ethanol-insolubles or of "ash" (i.e., residue at 600°C)*: This indicates which inorganic compounds are present.

6. *X-ray diffraction/X-ray fluorescence analysis of original sample*: Modern X-ray diffraction techniques can be applied to mixtures as well as pure compounds. This makes them suitable for identification of specific species in powdered for-

mulations (10). X-ray fluorescence will measure elements like phosphorus, sulfur, and silicon. By using simple preliminary separation procedures, the selectivity can be increased. For example, zeolites can be filtered from an aqueous detergent solution, then quantified by XRF determination of silicon in the precipitate. Anionic surfactants can be precipitated as their calcium salts, then determined from the level of sulfur in the precipitate, correcting for interference by phosphorus (11).

7. *Demulsification*: Some formulations are emulsions of oil and water. The water may be removed from such a mixture by passage through a column of kieselguhr or diatomaceous earth (12). The water is adsorbed and the oily components, including surfactants, are eluted with petroleum ether. There is danger of ethoxylated surfactants remaining adsorbed to the silica stationary phase, so this approach must be used with caution.

B. Specific Tests

Depending on what is already known about the product, certain other preliminary tests are also performed:

1. Elemental analysis (C, H, N, S, O) of any organic fractions which IR analysis indicates are single components. This analysis is best performed by contracting with a specialized laboratory. Be sure that the analytical results are corrected for moisture content.
2. Qualitative or quantitative tests for the presence of chelating agents.
3. Liquid-solid extraction with solvents other than ethanol, or liquid-liquid extraction of an aqueous solution with chloroform or ethyl acetate, to separate organic compounds from each other.
4. Adsorption chromatography or ion exchange separation of an extract, followed by infrared analysis of the fractions (13).

The results of these preliminary tests tell the analyst which quantitative analyses should be performed next. These can include

1. GC or GC-MS analysis for solvents, if water content by Karl Fischer titration is less than weight loss at 110°C. This analysis may also be performed by fractional distillation followed by IR identification of the fractions.
2. Quantitative determination of the various organic and inorganic components detected by the qualitative tests. Most often, ion chromatography is used for determination of sulfate; phosphate and polyphosphates; chelates such as NTA, EDTA, and gluconate; and other lower molecular weight components. Capillary electrophoresis may also be used. Gas chromatography is applied to quantification of solvents and propellants. Specialized tests are required to determine polyacrylate builders: One quick semi-quantitative test is based on pyrolysis GC-MS of a water extract after addition of tetramethylammonium hydroxide. Peaks of methyl acrylate and methyl fumarate indicate the presence of acrylate and maleate polymers and/or copolymers (14).

The above protocol will normally not detect trace components, such as biostats, perfumes, and dyes. These are usually found by performing individual, specific, tests. Such tests can only be specified if the analyst has thorough knowledge of typical additives.

Nakamura and Morikawa developed a general HPLC method for separation of surfactants (15). A collection of nonionic, cationic, anionic, and amphoteric surfactants could be separated in one or two chromatographic runs on a general purpose system. If it is known that the surfactants in the unknown formulation consist of a few members of a group of limited size, an approach of this type can be very valuable in monitoring competitive products. A procedure for determining LAS and APE content of detergents consists of examining a methanol extract by reversed-phase HPLC. This yields the homolog structure for LAS and a single peak for APE. If present, APE is analyzed according to homolog structure using normal phase HPLC of 25:75 ethanol/acetone extract (16).

Precise characterization of surfactants in used formulations, such as the recycled solutions used to remove oil after metal-forming operations, normally requires some preliminary separation. For example, HPLC analysis of alkoxylated surfactants is possible if the sample is cleaned up by solid phase extraction on silica. Water is first removed by azeotropic distillation, then organic material is taken up in methylene chloride and passed through the SPE column. After elution of oil and grease with methylene chloride, the nonionic surfactants are removed with methanol (17).

III. COMPOSITION OF SOME COMMERCIAL PRODUCTS

Typical recipes for formulations can be found in the technical sections of trade publications, such as *Soap, Cosmetics and Chemical Specialties, Cosmetics and Toiletries, Household & Personal Products Industry*, and *Seifen, Öle, Fette und Wachse*. Surfactant vendors also will provide examples of formulations containing their products. These are available on the Internet; at this writing, many can be accessed through Paul Huibers's site, http://surfactants.net/. Vendor formulations are periodically compiled in summary volumes, usually without critical evaluation (18,19). While the patent literature contains many examples of formulations, this is a less valuable reference because of the large number of patents which have never been commercialized. Other volumes of the Surfactant Science Series are valuable references (20–22). It is often helpful to read information developed for formulators in order to gain perspective on what components can be expected for functional reasons (4).

While products for institutional or industrial cleaning are generally similar to those sold for household use, the former are normally more strongly alkaline or acidic, reflecting the high value of labor and the controlled conditions under which they are used. They are also more concentrated. Thus ingredients like sodium hydroxide, sodium metasilicate, phosphoric acid, and hydrochloric acid are found in industrial cleaners. Surfactants are likely to be present at lower concentrations than in household products.

A. Laundry Detergents

These normally contain anionic and nonionic surfactants, builders (such as sodium tripolyphosphate, acrylate polymers or copolymers, or sodium carbonate), ion exchangers (such as polyacrylic acid or zeolite), corrosion inhibitor (sodium silicate), processing aids, antiredeposition agent (such as sodium carboxymethylcellulose), whitener or optical brightener (stilbene derivative), dye, and perfume. They may also contain bleach (sodium perborate or percarbonate), enzymes (a protease and/or an amylase), foam suppressor, filler/processing aid (sodium sulfate), fabric softener (quaternary ammonium salt immobilized on a clay), and solubilizer. The surfactant concentration of a heavy duty powder is in the range of 5–30%. Heavy duty liquids contain 10–20% surfactant, with the unbuilt heavy duty liquids containing higher levels. Liquids will normally contain a few percent of a hydrotrope, such as sodium xylene sulfonate, to improve surfactant solubility.

Many of the inorganic ingredients of laundry detergents can be determined by X-ray diffraction analysis, sometimes without preliminary separations (10). ASTM D501 describes conventional methods for the determination of inorganic compounds in detergents (23).

Surfactants may be determined in powdered laundry detergents by HPLC after isolation of the surfactant by Soxhlet extraction with methanol or by simply stirring with methylene chloride (24). Depending on the purity of the extract, other methods may also be used for qualitative and quantitative analysis. European liquid detergents may contain an alkanesulfonate surfactant.

Builders used in low-phosphate formulations include zeolites and polyacrylates. Few methods are available for their determination. They are part of the ethanol-insoluble fraction, and the polyacrylates, at least, can be detected and approximately quantified using size exclusion chromatography (25).

B. Hair Shampoo

Shampoos normally contain a blend of cleansing agents (usually two anionic surfactants or an amphoteric and an anionic), foam booster (also a surfactant, most often an alkanolamide), conditioner, viscosity improver (inorganic salts or cellulose derivatives), opacifier, dye, perfume, biostats (often esters of p-hydroxybenzoic acid), and perhaps chelates (EDTA or citric acid). Propylene glycol or glycerin may be present to adjust solubility and viscosity. Depending on current fashion, other compounds will be present, such as vitamins and food or herb extracts (26,27).

A number of papers have appeared on the analysis of shampoos. Infrared spectrometry is used extensively to identify components. Most investigators use a combination of liquid extraction and ion exchange chromatography to separate the individual surfactants. For example, Newburger reports that alkyl sulfate may be separated with an ion exchange resin, and alkanolamide and free fatty acid determined by solvent extraction (28). Kirby and coworkers describe a more complete method, where nonionics are further separated by column chromatography on silica gel (29). Gabriel recommends first removing soap (no longer a common shampoo component) as the fatty acids by extraction of the nonvolatiles (30). The other components are then separated by ion exchange chromatography, and further characterized by gas, liquid, and thin layer chromatography.

Matissek has written an extensive treatise on the analysis of shampoos, bubble baths, and liquid soaps (31). TLC is used heavily for separation, and IR for identification. Gas chromatography and mass spectrometry are also used.

König and Waldorf discuss analysis of hair and body shampoos (32). The shampoo is first evaporated to dryness and extracted with isopropanol or 95% ethanol to separate the surfactant, which is qualitatively identified by IR. The alcohol-insoluble residue is also examined by IR in case a less-soluble surfactant remains there. The alcohol extract is then separated into anionic, cationic, nonionic, and amphoteric surfactant fractions by ion exchange. Anionics can be separated into sulfonates and carboxylates by use of strongly basic anion exchange resins in the Cl⁻ and OH⁻ form, respectively. Anionics are further characterized by TLC. Nonionics are likewise characterized by TLC. Once the components are identified, quantitative analysis is by the usual methods, described elsewhere in this volume.

C. Toothpaste

Toothpaste is composed of a base, such as sorbitol or polyethylene glycol, to which is added polishing agent (silica, phosphates, carbonates), humectant (such as glycerin), surfactant, binder, salts and buffers, flavoring agents, preservatives, and dye, as well as a fluoride compound and perhaps a tartar control additive (perhaps sodium pyrophosphate or methyl vinyl ether/maleic anhydride copolymer). Inorganic thickeners or organic gums (carboxymethyl cellulose, sodium alginate) may be present (33). In the United States, the surfactant is generally sodium lauryl sulfate or an EO/PO copolymer. Tooth powder is similar, but lacking the base and humectant (34).

König and Walldorf recommend determining surfactants after evaporating the sample to dryness and obtaining a 95% ethanol extract of the residue (35,36). Qualitative identification is made by TLC examination of the original extract, as well as by IR analysis. The extract is further partitioned prior to IR analysis: The residue is dissolved in water, and an acid ether extract is obtained. The aqueous phase is neutralized and evaporated to dryness. The residue from both phases is analyzed by IR spectroscopy. Anionic surfactants and soap are determined by two-phase titration. Nonionics are separated by ion exchange techniques from the other components and identified by IR or TLC on silica gel impregnated with oxalic acid. PEG is separated from nonionic surfactant by partition between water and butanol, and its identity confirmed by TLC.

In an extension of this work, König reports the determination of surfactants in toothpaste by HPLC (37). Here, one takes advantage of the knowledge that usually only one of a limited number of surfactants is present. The toothpaste is dissolved directly in the HPLC mobile phase, filtering out the abrasive and other insoluble matter. The characteristic retention time and peak pattern under a few standard HPLC conditions allow both identification and quantification of the surfactant.

D. Liquid Soap

The original liquid soaps are indeed solutions of soap, normally the potassium salts of coconut oil fatty acids. The more modern products usually contain no soap at all, resembling shampoos. They consist mainly of surfactant and water, with the most popular surfactant being α-olefin sulfonate. An alkanolamide is added as a foam stabilizer. Another surfactant, such as a sarcosinate, betaine, or an alkyl sulfate, may also be present. Other ingredients may be added to affect viscosity, odor, appearance, or texture.

E. Bar Soap

Most bar soap is indeed soap. A few products consist of synthetic detergent. The products most often used are isethionate esters, alcohol sulfate, and alkyl glyceryl ether sulfonate.

König has detailed the tests most often applied to soaps (38). Besides the usual analyses for water, free alkali, unsaponified matter, and the like (Chapter 1), he suggests determination of alkanolamines by TLC; determination of CO_2 and phosphate and 1-hydroxyethane-1,1-diphosphonic acid gravimetrically; determination of TiO_2 opacifier colorimetrically; chelating agents by titration with metal ions; separation of nonionic surfactants and PEG by ion exchange chromatography, with identification by IR or TLC; and detection of amino acids by the ninhydrin reaction. Carbohydrates are determined with Fehlings solution and identified by gas chromatography; perfumes by steam distillation, extraction, and gas chromatography; dyes by TLC; optical brighteners are identified by

TLC or HPLC and determined by UV absorption or fluorescence measurement, antioxidants are determined by TLC or GC; sulfite, dithionite, thiosulfate, and sulfur are determined by classical chemical reactions; and microbicides are identified by TLC.

ASTM D460 has been written for the analysis of soap and soap-based detergents (39). It is comprehensive, describing more than 30 tests, including determination of additives and builders as well as characterization of the soap itself.

F. Fabric Softeners

Although anionics and nonionics may be used in industrial applications, all products destined for domestic applications have a cationic surfactant as the active agent. Fabric softeners are sold as dispersions, to be added to the rinse water, or in dry form carried on nonwoven sheets, for use in the dryer. The most common active agent for both applications is a quaternary amine salt with high alkyl character and low water solubility, such as the chloride or methyl sulfate salt of ditallowdimethylammonium ion.

Dispersions of softeners, especially those of high concentration, often contain viscosity-reducing additives like fatty alcohols, amine ethoxylates, or acid ethoxylates. Instead of quaternary alkyl compounds, dialkyl imidazolinium compounds or ester quats are often used. Amidoamine quats like methyl-*bis*(tallowamidoethyl)-2-hydroxyethyl ammonium methyl sulfate are also popular in North America.

Fabric softener sheets normally contain quaternaries formulated as their methyl sulfate salts, since the more common chloride salts can be corrosive to the dryer. Sheets usually contain other compounds like acid ethoxylates or amine soaps to optimize properties and performance. Some fabric softeners also contain amphoterics, generally imidazolines rather than quaternary amines.

G. Other Products

Aerosol shaving cream typically contains soap, usually as the triethanolamine salt rather than the sodium salt. Other surfactants, such as AS and AES may be present, as well as a diol or polyol like propylene glycol or sorbitol. Water, fragrance, and isobutane and perhaps propane propellant are also present. In most markets, the chief ingredients are listed on the label.

Toilet tank blocks consist of two main types: One type consists mainly of paradichlorobenzene, perhaps with as much as 50% surfactant. Perfume and dye are usually added. Another type contains 50–90% surfactant, 0–20% wax, 0–40% inorganic salts, along with perfume, dye, and biocides (40).

Laundry prespotters are of two types: solvent-based and water-based. The solvent-based formulations contain petroleum distillate or chlorinated hydrocarbons and a surfactant, usually a nonionic. The water-based products may contain a small amount of ethanol or other solubilizer, as well as a surfactant, almost always nonionic.

Laundry presoaks typically contain an enzyme for breaking up protein soils; a builder such as sodium tripolyphosphate, sodium carbonate, or sodium silicate; and, usually, a surfactant. Bleach or optical brighteners may also be added.

Hand dishwashing liquid detergents contain an anionic surfactant, perhaps an alcohol ether sulfate (or, in Europe, an alkanesulfonate), as major active agent, along with a lower level of an amide or amine oxide type nonionic to improve foaming. Betaines are sometimes used as foam boosters. A hydrotrope such as an alcohol or xylene sulfonate is generally also present.

By convention, detergents designed for home dishwashing appliances are called automatic dishwash detergents, while those for institutional use are called machine dishwash detergents. The main difference is that the institutional products are more highly alkaline and are designed for use at higher temperatures. Both classes consist of alkaline (carbonates or even hydroxides) products containing phosphates or sodium citrate to control water hardness; silicates and/or aluminum compounds to control erosion of the materials washed; bleaching agent to degrade protein, remove stains, and sanitize; about 3% low-foaming nonionic surfactant; and perhaps polycarboxylates, sodium sulfate, and enzymes.

Machine dishwashing rinse aids are dilute aqueous solutions containing a surfactant. A hydrotrope and/or organic solvent is often present to maintain solubility. Nonionic surfactants are most often used and may be in admixture with cationic or anionic surfactants. Chelating agent (perhaps citric acid), fragrance, and dye may also be present. Surfactants are isolated by extraction or SPE, separated by ion exchange, identified by IR and NMR, and quantified by gravimetry.

Hard surface household cleaners are available in various types. The concentrated products contain about 10% surfactant, often nonionics. The spray-on liquids, already diluted, may only contain 1% surfactant. Powdered hard surface cleaners are more likely to contain anionic surfactants, usually only 1 or 2%. Most household cleaners are made alkaline and usually contain a sequestering agent such as sodium gluconate. A water-soluble solvent is often added. Disinfectants, perfumes, and other components are present. Surfactants are isolated by extracting the dried solids with methanol or methylene chloride (24).

Powdered cleansers, besides abrasive agent and bleach, contain about 3% surfactant, usually an anionic.

Steel wool scouring pads normally contain soap, with perhaps a small amount of another anionic or nonionic surfactant present.

Industrial metal cleaners are generally quite alkaline, containing only a few components. ASTM D800 details the analysis of these materials, giving tests for alkali, soap, silicates, phosphates, carbonates, and surfactant (41).

REFERENCES

1. Longman, G. F., *Analysis of Detergents and Detergent Products*, John Wiley & Sons, New York, 1975.
2. Milwidsky, B. M., D. M. Gabriel, *Detergent Analysis: A Handbook for Cost-Effective Quality Control*, John Wiley & Sons, New York, 1982.
3. Cullum, D. C., P. Platt, Analysis of formulated products, in M. R. Porter, ed., *Recent Developments in the Analysis of Surfactants*, Elsevier, New York, 1991.
4. Lange, K. R., ed., *Detergents and Cleaners: Handbook for Formulators*, Hanser/Gardner Publications, Cincinnati, 1994.
5. Battaglini, G. T., Analysis of detergent formulations, in L. Spitz, ed., *Soaps and Detergents: Theoretical and Practical Review*, AOAC Press, Champaign, IL, 1996.
6. Estrin, N. F., C. R. Haynes, J. M. Whelan, eds., *CTFA Compendium of Cosmetic Ingredient Composition: Methods*, Cosmetic, Toiletry & Fragrance Association, Washington, DC 20005, 1983.
7. Milwidsky, B. M., Continuous liquid/liquid extraction, *Soap Chem. Spec.*, 1969, *44(12)*, 79–80,84,86,88,117–118.
8. Marcomini, A., S. Stelluto, B. Pavoni, LAS and APE in commercial products and marine waters by reversed- and normal-phase HPLC, *Int. J. Environ. Anal. Chem.*, 1989, *35*, 207–218.

9. Ogura, I., D. L. DuVal, S. Kawakami, K. Miyajima, Surfactants in consumer products by ion-spray MS, *J. Am. Oil Chem. Soc.*, 1996, *73*, 137–142.

10. King, M. M., E. M. Sabino, XRD as a tool in the total analysis of powdered household laundry detergents, *Adv. X-Ray Anal.*, 1990, *33*, 485–492.

11. Kawauchi, A., Nonsolvent quantitation of anionic surfactants and other inorganic ingredients in laundry detergent products by X-ray fluorescence spectrometry, *J. Surfactants Deterg.*, 1999, *2*, 79–83.

12. Buschmann, N., F. Hülskötter, Titration of low ethoxylated nonionic surfactants, *Tenside, Surfactants, Deterg.*, 1997, *34*, 8–11.

13. Mandery, K., Surfactant component of an aqueous detergent (in German), *Seifen, Öle, Fette, Wachse*, 1991, *117*, 595–597.

14. Kawauchi, A., T. Uchiyama, Polycarboxylates in laundry detergents by *in situ* pyrolysis-methylation GC/MS, *J. Anal. Appl. Pyrolysis*, 1998, *48*, 35–43.

15. Nakamura, K., Y. Morikawa, Separation of surfactant mixtures and their homologs by HPLC, *J. Am. Oil Chem. Soc.*, 1982, *59*, 64–68.

16. Ballarin, B., S. Stelluto, A. Marcomini, LAS and APE in Italian household detergents, *Riv. Ital. Sostanze Grasse*, 1989, *66*, 349–353.

17. Rothbächer, H., A. Korn, G. Mayer, Nonionic surfactants in cleaning agents for automobile production (in German), *Tenside, Surfactants, Deterg.*, 1993, *30*, 165–173.

18. Bennet, H., *Chemical Formulary*, Chemical Publishing Co., New York, multi-volume, new volumes published regularly.

19. Flick, E. W., *Cosmetic and Toiletry Formulations*, 2nd ed. Vols. 1–3, Noyes Publications, Park Ridge, NJ, 1989–1995.

20. Rieger, M. M., L. D. Rhein, eds., *Surfactants in Cosmetics*, 2nd ed., Marcel Dekker, New York, 1997.

21. Lai, K., *Liquid Detergents*, Marcel Dekker, New York, 1996.

22. Showell, M. S., ed., *Powdered Detergents*, Marcel Dekker, New York, 1997

23. American Society for Testing and Materials, Sampling and chemical analysis of alkaline detergents, D501. West Conshohocken, PA 19428.

24. Marcomini, A., F. Filipuzzi, W. Giger, Aromatic surfactants in laundry detergents and hard-surface cleaners: LAS and APE, *Chemosphere*, 1988, *17*, 853–863.

25. Krusche, G., J. Illert, H. Mandery, Sodium polycarboxylates in detergents by SEC, *Chromatographia*, 1991, *31*, 17–20.

26. Fox, C., Formulation of shampoos, *Cosmet. Toiletries*, 1988, *103(3)*, 25–58.

27. Rieger, M., Surfactants in shampoos, *Cosmet. Toiletries*, 1988, *103(3)*, 59–72.

28. Newburger, S. H., Shampoos, *J. Ass. Off. Anal. Chem.*, 1958, *41*, 664–668.

29. Kirby, D. H., F. D. Barbuscio, W. Metzger, J. Hourihan, Analysis of anionic/nonionic shampoo, *Cosmetics and Perfumery*, 1975, *90(8)*, 19–23.

30. Gabriel, D. M., Cosmetics and toiletries, *J. Soc. Cosmet. Chem.*, 1974, *25*, 33–48.

31. Matissek, R., Shampoos, bubble bath preparations, and soaps with special reference to surfactants and antimicrobial substances (in German), *MvP-Ber.*, 1980, *1*, 170.

32. König, H., E. Walldorf, Skin cleaners and hair shampoos based on synthetic surfactants (in German), *Fresenius' Z. Anal. Chem.*, 1979, *299*, 1–18.

33. Prencipe, M., J. G. Masters, K. P. Thomas, J. Norfleet, Squeezing out a better toothpaste, *Chemtech*, 1995, *25(12)*, 38–42.

34. Pader, M., Surfactants in oral hygiene products, in M. M. Rieger, ed., *Surfactants in Cosmetics*, Marcel Dekker, New York, 1985.

35. König, H., E. Walldorf, Toothpastes (in German), *Fresenius' Z. Anal. Chem.*, 1978, *289*, 177–197.

36. König, H., E. Walldorf, Toothpastes (in German), *Fette, Seifen, Anstrichm.*, 1981, *83*, 281–287.

37. König, H., W. Strobel, Surfactants in toothpastes by HPLC, *Fresenius' Z. Anal. Chem.*, 1988, *331*, 435–438.

38. König, H., Soaps (in German), *Fresenius' Z. Anal. Chem.*, 1978, *293*, 295–300.
39. American Society for Testing and Materials, Sampling and chemical analysis of soaps and soap products, D460. West Conshohocken, PA 19428.
40. Block, H., Toilet bowl additives—surfactact analysis (in German), *Lebensmittelchem., Lebens-mittelqual.*, 1985, *9*, 58–70.
41. American Society for Testing and Materials, Chemical analysis of industrial metal cleaning compositions, D800. West Conshohocken, PA 19428.

Abbreviations

3A alcohol	95:5 ethanol/water denatured with 5% methanol
AAS	atomic absorption spectroscopy
ABS	branched-chain alkylaryl sulfonate
AE	alcohol ethoxylate
AES	alkyl ether sulfate
APE	alkylphenol ethoxylate
AS	alkyl sulfate
ASE	accelerated solvent extraction
ATR	attenuated total reflectance
BiAS	bismuth-active substance; i.e., material which which gives a positive response to the titration method for determination of nonionic surfactants in environmental samples
BSTFA	bis(trimethylsilyl)trifluoroacetamide
BuOH	butanol
CE	capillary electrophoresis
CI	chemical ionization
CMC	critical micelle concentration
CTAS	cobalt thiocyanate–active substance; i.e., material which gives a positive response to the spectrophotometric procedure for determination of nonionic surfactants in environmental samples
CTFA	Cosmetic, Toiletry, and Fragrance Association

DTI	ditallow imidazoline (see imidazolinium compounds)
EI	electron impact
EO	ethylene oxide
EtOH	ethanol
FAB	fast atom bombardment
FD	field desorption
FIA	flow injection analysis
FID	flame ionization detector
FT	Fourier transform
FTIR	Fourier transform infrared
GC	gas chromatography
GPC	gel permeation chromatography
HOAc	glacial acetic acid
HPLC	high performance liquid chromatography
IC	ion chromatography
ICR	ion cyclotron resonance
IR	infrared
LAS	linear-chain alkylaryl sulfonate
MBAS	methylene blue–active substance; i.e., material which gives a positive response to the spectrophotometric procedure for determination of anionic surfactants in environmental samples
MeOH	methanol
MS	mass spectrometry
MW	molecular weight
NMR	nuclear magnetic resonance
NP	nonylphenol
NPE	nonylphenol ethoxylate
OP	octylphenol
OPE	octylphenol ethoxylate
PEG	poly(ethylene gycol)
PO	propylene oxide
PPG	poly(propylene glycol)
PrOH	propanol
PVC	poly(vinyl chloride)
quat	quaternary ammonium cationic surfactant
Rf	retardation factor (in TLC analysis)
RI	refractive index
SEC	size exclusion chromatography
SFC	supercritical fluid chromatography
SFE	supercritical fluid extraction
SPE	solid phase extraction
TC	thermal conductivity
THF	tetrahydrofuran
TLC	thin layer chromatography
TMS	trimethylsilyl
UV	ultraviolet

Index